the
UNIVERSITY
of
GREENWICH

COLLECTING PLANT GENETIC DIVERSITY

Dedicato alla memoria di
mio padre e mia madre,
Sylvano e Mirella Guarino.
L.G.

COLLECTING PLANT GENETIC DIVERSITY

Technical Guidelines

Edited by

Luigi Guarino

International Plant Genetic Resources Institute, Regional Office for Sub-Saharan Africa, c/o International Laboratory for Research on Animal Diseases, PO Box 30709, Nairobi, Kenya

V. Ramanatha Rao

International Plant Genetic Resources Institute, Regional Office for Asia, the Pacific and Oceania, c/o International Development Research Centre, Tanglin, PO Box 101, Singapore 9124

Robert Reid

Department of Primary Industry, Mt Pleasant Laboratories, Pastures and Field Crops Branch, PO Box 46, Kings Meadows, Tasmania 7249, Australia

CAB INTERNATIONAL

on behalf of the

International Plant Genetic Resources Institute (IPGRI)
in association with the
Food and Agriculture Organization of the United Nations (FAO)
The World Conservation Union (IUCN) and the
United Nations Environment Programme (UNEP)

CAB INTERNATIONAL
Wallingford
Oxon OX10 8DE
UK

Tel: +44 (0)1491 832111
Telex: 847964 (COMAGG G)
E-mail: cabi@cabi.org
Fax: +44 (0)1491 833508

A catalogue entry for this book is available from the British Library.

ISBN 0 85198 964 0

Published on behalf of the

International Plant Genetic Resources Institute (IPGRI)
Via delle Sette Chiese 142
00145 Rome
Italy

in association with the

Food and Agriculture Organization of the United Nations (FAO)
The World Conservation Union (IUCN) and the
United Nations Environment Programme (UNEP)

The geographical designations employed and the presentation of material in this publication do not imply the expression of any opinion whatsoever on the part of IPGRI, FAO, IUCN or UNEP concerning the legal status of any country, territory, city or area or its authorities, or concerning the delimitation of its frontiers or boundaries. Similarly, the views expressed are those of the authors and do not necessarily reflect the views of these participating organizations.

Typeset in 10/12 pt Century by Colset Private Limited, Singapore
Printed and bound in the UK at the University Press, Cambridge

Contents

Loss of plant diversity: a call for action

H. Zedan

UNEP, PO Box 30552, Nairobi, Kenya

The world's biological diversity is a vast and often undervalued resource. Encompassing every form of life, from the tiniest microbe to the mightiest beast, biodiversity is the variety and variability of all plants, animals and microorganisms and the ecological complexes of which they are part. The earth's biodiversity – its ecosystems, species and genes – are the product of over 3000 million years of evolution. Throughout this time, small changes have accumulated in populations, resulting in a multitude of living forms closely adapted to the physical conditions they face and to each other. They supply all our food, much of our raw materials and energy and many of our medicines. Intact ecosystems also play a central role in the functioning of the biosphere. Plants are fundamental in stabilizing climate, protecting watersheds and soil and maintaining the chemical balance of the earth. When key species are lost, vital ecological services are disrupted.

Cultures from ancient times to the present day have exploited biodiversity. Ten thousand years ago neolithic farmers in the Middle East and elsewhere developed some of our most important crops and livestock from their wild relatives. They soon recognized that certain species were more suitable than others for their needs and that within populations some plants and animals had characteristics more suited to their specific requirements. They selected, bred and used these individuals in preference to others, practices that continue to this day. Through selection and breeding, human societies have developed thousands of local races of crops and livestock, each fitting a particular need in a specific physical environment and evolving in harmony with the diverse systems of land and natural resources management of which they are integral parts.

These local races are not only numerous, fulfilling a variety of needs and adapted to different conditions, but also genetically variable, which means that even in bad years at least some individuals can survive and pass on adaptability to adverse climatic conditions, water scarcity, low fertility, problem soil and aquatic systems, pests and diseases. The maintenance of this diversity has until recently been ensured by traditional systems of agriculture and land use. In some instances, it depends on introgression with wild and weedy relatives growing near by, in both natural vegetation and disturbed field margins. However, with the advent of scientific breeding some four human generations ago, new plant varieties, animal breeds and strains of microorganisms began to be developed in response to the needs of quite different, intensive production systems. Breeders assembled desired genetic traits from different varieties, and in some cases different species, gradually enabling the development of the high yielding genotypes that sustain modern societies. High performance under intensive agriculture requires genetic uniformity. This is achieved by screening existing diversity for the few characteristics that are needed at any one time, largely disregarding the remainder – until, that is, a new need arises.

The development of modern varieties, breeds, strains and production systems, while increasing productivity and thus helping to satisfy the needs of a rapidly growing world population, has also created a number of problems. Modern varieties are ill-suited to the needs of small producers, who farm with low management inputs on often marginal lands. The so-called high yielding or high performance varieties are more accurately high-response varieties, since their performance hinges on substantial external inputs (e.g. fertilizers, pesticides), which are sometimes deleterious to the environment. Without such inputs, their high potential is not realized. The new systems are also more vulnerable to the challenges of pests and diseases. Large, genetically uniform fields and herds encourage the rapid spread of pests and pathogens. Plant breeders are therefore dependent upon the availability of the pool of diverse genetic material represented by local races and wild relatives in their effort to keep one step ahead of tomorrow's unexpected calamities, since in themselves modern varieties provide too restricted a gene pool for further breeding. Without a diverse genetic reservoir to draw from, further improvement may not be possible.

Can we put a monetary value on this genetic pool? One of the major reasons why nothing is done about biodiversity loss is that national economic systems and policies fail to value the environment and its resources adequately. This is perhaps understandable. Because biodiversity is complex and information on its component parts and their interactions is incomplete, it is extremely difficult to determine accurately the economic and ecological value of the full range of goods and services that it provides. However, preliminary indications suggest that the value of biodiversity outweighs conservation costs by a significant margin. Information is gradually accumulating on the economic benefits of using genetic diversity in conventional crop breeding. Wild relatives of commercial crops have provided genetic material worth billions of dollars in higher crop yields. A few examples will illustrate the point. In Asia, by the mid-1970s genetic improvement had increased wheat production by US$2 billion (thousand million) and rice production by $1.5 billion a year by incorporating dwarfism into both crops. A 'useless' wild wheat plant from Turkey was used to give disease resistance to commercial wheat varieties worth $50 million annually to the USA alone. One gene from a single Ethiopian plant now protects some varieties of barley from yellow dwarf virus. An ancient wild relative of maize from Mexico – a perennial with resistance to seven major diseases and which can grow at high altitudes on marginal soils – can be crossed with modern varieties with potential savings to farmers estimated at $4.4 billion annually worldwide.

The contribution of genes from wild relatives has often been limited by the difficulty of making viable crosses between wild and domesticated species. The biotechnology revolution, including recent developments in recombinant DNA technology ('genetic engineering'), raises the prospect that useful traits may soon be transferred between species that could not previously be crossed by conventional means. Moreover, biotechnology permits a better understanding of how gene expression works. This knowledge can be used to facilitate the use of germplasm in the development of modern crop varieties. Advances in plant biotechnology have also led to novel and precise screening tools, for example stimulating interest in plants as sources of raw materials for new medicinal products. Biotechnology also has much to offer to the conservation of genetic resources. It is already leading to improved methods of storing germplasm of plants, animals and microbes, for example *in vitro*. However, this does not diminish the need to maintain the richest possible pool of genes. Rather, it reinforces it. The relationship between biotechnology and genetic resources is in fact one of mutual dependence. The future of biotechnology depends on conservation of a wide array of genetic diversity, its raw material. As the field of biotechnology develops, the range of future germplasm is certain to increase. The stake which the biotechnology industry holds in the conservation of biodiversity should thus not be underestimated. While breakthroughs are occurring with increasing speed, options for the future are being foreclosed by genetic erosion. The projected loss of species diversity could cripple the genetic base required for the continued improvement and maintenance of currently used species and deprive us of the potential to develop new ones.

The genetic base of the staple crops is generally very narrow. Half the Canadian prairies are planted with just one variety of wheat; the USA's entire soyabean industry stems from six plants from one place in Asia. The number of different species on which we rely for food, fibre, timber, medicines and other natural products is likewise extremely limited. Only a tiny proportion of plant and animal species have yet been tested for their usefulness. The range of species used as food, for example, is extremely limited. Of an estimated 265,000 species of plants, only about 7000 have ever been cultivated or collected for food. Of these, 20 species currently supply 90% of the world's food, and just three – wheat, maize and rice – supply more than half. Many food crops of regional or local importance have been relatively neglected by science. There is much potential to look beyond today's major crops at other species that may have value either in themselves or as sources of genes. After all, peanuts, potatoes and other crops once considered of little use are now valuable sources of food.

The case for conserving biodiversity is therefore well established on economic as well as scientific grounds. Biodiversity is essential for sustainable development, adaptation to a changing environment and the continued functioning of the biosphere – indeed, for human survival itself. However, the evidence suggests that many human activities are leading to depletion of the planet's biodiversity. The world's biological wealth is being depleted at an ever-increasing rate and this will adversely affect the well-being of people in both industrialized and developing nations.

The opportunity to exploit wild relatives as potential breeding material is increasingly limited by the degradation of ecosystems, especially in the tropics. Habitats are disappearing beneath agriculture, cities, industrial developments and dams and are being irreversibly damaged by pollution, overuse and erosion. Many species are also threatened by overexploitation, illegal trade and competition with introduced alien species. Species extinctions have increased steadily since 1600 and are now at an unprecedented high. Most of these species vanish unknown. The loss of genetic diversity within species is occurring even faster than species extinction. Breeders throughout the world are engaged in developing better and higher-yielding cultivars of crop plants to be used on an increasingly larger scale. Traditional food plants and local races of staple crops are being abandoned and lost for ever in favour of newly developed ones. Worldwide, food crop yields are increasing, but the yield is coming from ever fewer varieties. Uniformity is replacing diversity, just when the need for diversity is increasing.

Insufficient effort has been made to ensure the conservation of biodiversity in the face of the extensive destruction of plant-rich habitats, species extinction and genetic erosion. Most countries do not have a complete inventory of their plants, and most of the knowledge on their use is held by traditional societies, whose very existence is now under threat. The intense pressure on biodiversity will continue to increase unless appropriate measures for conservation and sustainable use are taken. There has been international consensus on this point for more than two decades, starting with the United Nations Conference on the Human Environment in 1972. Since then there has been the establishment of the International Board for Plant Genetic Resources (IBPGR) in 1974 (which became the International Plant Genetic Resources Institute (IPGRI) in 1994), the adoption by the Food and

Agriculture Organization (FAO) Conference in 1983 of the *International Undertaking on Plant Genetic Resources* and the establishment of the Inter-Governmental Commission on Plant Genetic Resources. Such documents as *Environmental Perspective to the Year 2000 and Beyond*, the *Report of the World Commission on Environment and Development*, *Caring for the Earth: a Strategy for Sustainable Living*, the *Global Biodiversity Strategy* and *Agenda 21* of the United Nations Conference on Environment and Development also emphasize the importance of protecting biological diversity and provide general principles for action.

But concrete action and policy reforms by governments and development agencies are still needed. The lack of sound national biodiversity strategies and action plans, adequate trained manpower and sufficient funding are the major impediments to biodiversity conservation. Two-thirds of all species occur in developing countries, particularly in the tropics, which are not able to finance investments in their conservation and sustainable use. By and large, it is the developed countries that have the technology to exploit and benefit from them. Activities will probably be proposed within the framework of the *Convention on Biological Diversity*, which will no doubt add further to the financial burden of those countries that are rich in biodiversity or that host rare or endangered species. Ways of sharing the costs and benefits of conserving genetic resources must be devised. Social, political and economic decisions by all players will be needed: those who have or need resources, and those who have or need technologies. The potential of plant genetic resources can best be exploited if they remain accessible to all users and if the information and technology on how to use them is widely disseminated. Neither having a particular, valuable genetic resource nor having the technical capability to develop new products from it should give exclusive rights of ownership or profit. No country is self-sufficient in genetic resources. While possession and custody of a potential genetic resource might be limited to one nation, the benefits from this resource can accrue to all nations. There needs to be a fair balance of benefit between custodian and consumer.

The *Convention on Biological Diversity* has begun to address these important topics, but the road ahead is difficult. Worldwide investment in the conservation of plant genetic resources has been very modest and is by no means secure. The integration of conservation and development plans is almost non-existent. National cross-sectoral cooperation is limited. The resources, power and capacity to promote conservation are unequally distributed. Local people are rarely brought into the process of planning the conservation and management of the genetic resources they shape and maintain. However, there is general agreement both in the scientific community and among the public in both industrialized and developing countries that there is a need for a global effort to conserve as much of the gene pool of crop plants as possible and to tap this reservoir for the benefit of all.

The important point is that, if conservation is to be given the priority it needs, all the institutions concerned must coordinate their efforts. Genetic resources underpin, and are threatened by, virtually every area of human activity. Only the broadest possible cooperation at both the national and the international levels can save them.

List of acronyms and abbreviations

Organizations and countries

AAB	Association of Applied Biologists
ACCT	Agence de Coopération Culturelle et Technique, France
ACP	African, Caribbean and Pacific (states)
ACSAD	Arab Centre for the Study of the Semi-Arid and Dry Areas
ACTS	African Centre for Technology Studies
AETFAT	Association pour l'Étude Taxonomique de la Flore d'Afrique Tropical
ANEN	African NGOs Environment Network
APINMAP	Asian Pacific Information Network on Medicinal and Aromatic Plants
ARCIK	African Resource Centre for Indigenous Knowledge
ARTEMIS	Agricultural Real Time Environmental Monitoring Information System
ASA	American Standards Association
ASPRS	American Society of Photogrammetry and Remote Sensing
BGCI	Botanic Gardens Conservation International
BGCS	Botanic Gardens Conservation Secretariat
BRARCIK	Brazilian Resource Centre for Indigenous Knowledge
BURCIK	Centre Burkinabè de Recherche sur le Pratiques et Savoirs Paysans
CAAS	Chinese Academy of Agricultural Sciences
CABI	CAB INTERNATIONAL
CARDI	Caribbean Agricultural Research and Development Institute
CATIE	Centro Agronómico Tropical de Investigación y Enseñanza
CGIAR	Consultative Group on International Agricultural Research
CIAT	Centro Internacional de Agricultura Tropical
CICA	Centro de Investigaciones de Cultivos Andinos, Peru
CIKARD	Center for Indigenous Knowledge for Agriculture and Rural Development

CIKO	Cameroon Indigenous Knowledge Organization
CIMMYT	Centro Internacional de Mejoramiento de Maiz y Trigo
CIP	Centro Internacional de la Papa
CIRAD	Centre de Coopération Internationale en Recherche Agronomique pour le Développement, France
CIRAN	Center for International Research and Advisory Networks
CIS	Commonwealth of Independent States
CLADES	Consorcio Latinoamericano Sobre Agroecologia y Desarollo
CMI	Commonwealth Mycological Institute
CNR	Consiglio Nazionale per la Ricerca, Italy
CPC	Centre for Plant Conservation
CPCS	Commission de Pédologie et de Cartographie des Sols, France
CRES	Centre for Resource and Environmental Studies, Australia
CSIRO	Commonwealth Scientific and Industrial Research Organization, Australia
CTA	Technical Centre for Agriculture and Rural Cooperation
DANIDA	Danish International Development Agency
DIANA	Data and Information Available Now in Africa
DSE	Deutsche Siftung für Internationale Entwicklung, Germany
ECP/GR	European Cooperative Programme for Crop Genetic Resources Networks
ELC	Environmental Law Centre (IUCN)
ELCI	Environment Liaison Centre International
EMBRAPA	Empresa Brasilera de Pesquisas Agropecuarias
ENDA	Environment and Development Action in the Third World
EPTA	Expanded Program of Technical Assistance
ETI	Expert-Centre for Taxonomic Identification
EUCARPIA	European Association for Research on Plant Breeding
FAO	Food and Agriculture Organization of the United Nations
GATT	General Agreement on Tariffs and Trade
GDR	German Democratic Republic
GEF	Global Environment Facility
GEMS	Global Environment Monitoring System (UNEP)
GHARCIK	Ghana Resource Centre for Indigenous Knowledge
GIEWS	Global Information and Early Warning System on Food and Agriculture (FAO)
GPS	Global Positioning System
GRAIN	Genetic Resources Action International
GRID	Global Resources Information Database (UNEP)
GTZ	Deutsche Gesellschaft für Technische Zusammenarbeit, Germany
HEM	Harmonization of Environmental Measurement Programme
HRAF	Human Relations Area Files
IABG	International Association of Botanic Gardens
IARC	International Agricultural Research Centre
IBP	International Biological Programme
IBPGR	International Board for Plant Genetic Resources
IBTA	Instituto Nacional de Tecnología Agropecuaria, Bolivia
ICARDA	International Centre for Agricultural Research in the Dry Areas
ICITV	Institut de la Carte International du Tapis Végétal
ICRAF	International Centre for Research in Agroforestry

ICRISAT	International Crops Research Institute for the Semi-Arid Tropics
IDRC	International Development Research Centre, Canada
IDS	Institute of Development Studies, UK
IEEE	Institute of Electrical and Electronics Engineers
IGER	Institute of Grassland and Environmental Research, UK
IGU	International Geographical Union
IIE	International Institute of Entomology
IIED	International Institute for Environment and Development, UK
IIP	International Institute of Parasitology
IITA	International Institute for Tropical Agriculture
ILCA	International Livestock Centre for Africa
ILDIS	International Legume Database and Information Service
ILEIA	Information Centre for Low-External-Input and Sustainable Agriculture
ILO	International Labour Organization
ILRAD	International Laboratory for Research on Animal Diseases
IMI	International Mycological Institute
INBio	Instituto Nacional de Biodiversidad, Costa Rica
INIAA	Instituto Nacional de Investigaciones Agrarias y Agroindustriales, Peru
INIAP	Instituto Nacional de Investigaciones Agropecuarias, Ecuador
INIBAP	International Network for the Improvement of Banana and Plantain
INRIK	Indonesian Resource Center for Indigenous Knowledge
INTAGRES	International Agricultural Research – European Service
INTERAISE	International Environmental and Natural Resource Assessment Information Service
IOPI	International Organization for Plant Information
IPGRI	International Plant Genetic Resources Institute
IPK	Institut für Pflanzengenetik und Kulturpflanzenforschung (Institute of Plant Genetics and Crop Plant Research), Germany
IPPC	International Plant Protection Convention
IRAT	Institut de la Recherche Agronomique Tropical et des Cultures Vivrieres, France
IRED	Innovations et Reseaux pour le Développement
IRRI	International Rice Research Institute
ISO	International Standards Organization
ISRIC	International Soil Reference and Information Centre
ITC	International Institute for Aerospace Survey and Earth Sciences
IUCN	The World Conservation Union
IUFRO	International Union of Forestry Research Organizations
KENGO	Kenya Energy Non-Governmental Organization
KENRIK	Kenya Resource Centre for Indigenous Knowledge
KIT	Royal Tropical Institute, The Netherlands
LEAD	Leiden Ethnosystems and Development Program
LRDC	Land Resources Development Centre
MAB	Man and the Biosphere Programme
MARCIK	Madagascar Resource Centre for Indigenous Knowledge
MIRCEN	Microbiological Resource Centre

MSDN	Microbial Strains Data Network
MUCIA	Midwest Universities Consortium for International Activities
NARS	National Agricultural Research Systems
NASA	National Aeronautics and Space Administration, USA
NBPGR	National Bureau of Plant Genetic Resources, India
NGO	non-governmental organization
NifTAL	Nitrogen Fixation by Tropical Agricultural Legumes Project
NIRCIK	Nigerian Centre for Indigenous Knowledge
NISER	Nigerian Institute of Social and Economic Research
NOAA	National Oceanic and Atmospheric Administration, USA
NORAGRIC	Norwegian Centre for International Agricultural Development
NRI	Natural Resources Institute, UK
ODI	Overseas Development Institute, UK
ORSTOM	Institut Français de Recherche Scientifique pour le Developpement en Cooperation
PAGE-PACA	Groupe de Recherche et de Developpement sur le Patrimonie Genetique et Vegetal de la Region Provence-Alpes-Côte d'Azure, France
PCA	Philippines Coconut Authority
PHIRCIKSD	Philippines Resource Center for Indigenous Knowledge and Sustainable Development
PGRC/E	Plant Genetic Resources Centre, Ethiopia
PROSEA	Plant Resources of South-East Asia
PUDOC	Centre for Agricultural Publishing and Documentation
QDPI	Queensland Department· of Primary Industries, Australia
RAFI	Rural Advancement Foundation International
REPPIKA	Regional Program for the Promotion of Indigenous Knowledge in Asia
RIDSCA	Mexican Research, Teaching and Service Network on Indigenous Knowledge
SADC	Southern Africa Development Community
SADCC	Southern Africa Development Coordination Conference
SAPRIS	South-East Asian Plant Resources Information System
SARCIK	South African Resource Centre for Indigenous Knowledge
SATCRIS	Semi-Arid Tropical Crops Information Service (ICRISAT)
SEARICE	South-East Asian Regional Institute for Community Education
SEPASAL	Survey of Economic Plants of the Arid and Semi-arid Lands (Royal Botanic Gardens, Kew)
SSC	Species Survival Commmission (IUCN)
SLARCIK	Sri Lanka Resource Centre for Indigeneous Knowledge
SPOT	Systeme Probatoire pour l'Observation de la Terre
TDWG	International Working Group on Taxonomic Databases for Plant Sciences (formerly Taxonomic Databases Working Group)
TRIPS	Trade Related Intellectual Property Rights
UK	United Kingdom
UNCED	United Nations Conference on Environment and Development
UNDP	United Nations Development Programme
UNEP	United Nations Environment Programme
Unesco	United Nations Educational, Scientific and Cultural Organization
UNITAR	United Nations Training and Research Institute

UNSO	United Nations Sudano-Sahelian Office
UPOV	International Union for the Protection of New Varieties of Plants
UPWARD	User's Perspective With Agricultural Research and Development (CIP)
URURCIK	Uruguay Resources Centre for Indigenous Knowledge
USA	United States of America
USAID	United States Agency for International Development
USDA	United States Department of Agriculture
USSR	Union of Soviet Socialist Republics
VERCIK	Venezuelan Resource Secretariat for Indigenous Knowledge
WARDA	West Africa Rice Development Association
WCMC	World Conservation Monitoring Centre
WIPO	World Intellectual Property Organization
WMO	World Meteorological Organization
WRI	World Resources Institute
WWF	World Wide Fund for Nature

Other abbreviations

AEZ	agroecological zone
ART	Andean root and tuber (crops)
ASCII	American Standard Code for Information Interchange
AVHRR	Advanced Very High Resolution Radiometer
BBTV	banana bunchy top virus
BRAHMS	Botanical Research and Herbarium Management System
CBD	Convention on Biological Diversity
CD-ROM	compact disk, read-only memory
CITES	Convention on International Trade in Endangered Species of Wild Fauna and Flora
CMV	cucumber mosaic virus
CTAB	hexadecyltrimethylammonium bromide
CVA	canonical variates analysis
DBH	diameter at breast height
DBMS	database management system
DCW	Digital Chart of the World
DELTA	Descriptive Language for Taxonomy
DNA	deoxyribonucleic acid
DW	dry weight
FDA	fluorescein diacetate
GIS	geographic information system
ha	hectare
HRV	High Resolution Visible
ICBN	International Code of Botanical Nomenclature
ICNCP	International Code of Nomenclature of Cultivated Plants
IK	indigenous knowledge
IPPC	International Plant Protection Convention
IPR	intellectual property rights
ISBN	International Standard Book Number

ITF	International Transfer Format for Botanic Garden Records
MC	moisture content
MSS	Multi-Spectral Scanner
MW	molecular weight
NDVI	Normalized Difference Vegetation Index
PBR	plant breeders' rights
PCA	principal components analysis
PCO	principal coordinates analysis
PCR	polymerase chain reaction
PRA	participatory rural appraisal
RAPD	randomly amplified polymorphic DNA
RFLP	restriction fragment length polymorphisms
RH	relative humidity
RMS	root mean square
RRA	rapid rural appraisal
SAMMDATA	South American Monthly Meteorological Database
SLR	single lens reflex
TM	Thematic Mapper
TRADIS	Tropical Agricultural Development Information System
VAM	vesicular-arbuscular mycorrhiza
VDU	visual display unit
WW	wet weight

A brief history of plant germplasm collecting

1

Co-sponsors

Plant genetic diversity is the key component of any agricultural production system – indeed, of any ecosystem. Without it, no natural, evolutionary adjustment of the system (agricultural or natural) to changing environmental and biotic conditions would be possible. Farmers would not be able to spread the risk of crop failure or experiment with and refine crop varieties to suit their tastes and changing needs. Agricultural science and forestry would not have the basic raw materials for their introduction, domestication and improvement programmes. For development to be sustainable, conservation and use of genetic diversity must be at its core. Because the world is dynamic, this need for diversity is continuous. It is also increasing, because the number of people that must be fed, kept warm, housed and cured is increasing.

For thousands of years, wild habitats and farmers with their fields, orchards and home gardens have been sufficient to ensure the conservation, within the framework of change dictated by natural and artificial selection, of the vital natural resource that is the genetic diversity of plants. However, these systems have recently come under increasing pressure from demographic, socioeconomic and technological change. In some parts of the world, they have been under such pressure for hundreds of years. The results have been habitat fragmentation and even destruction, the abandonment of traditional agricultural and natural resources management practices and the replacement of farmers' landraces by modern cultivars. Species have always become extinct and landraces gone out of fashion, of course, but the current pace is unprecedented, and variety is being replaced by uniformity, rather than more, different variety. The result is a loss of genetic material, an irreversible erosion of genetic diversity. Active measures, both institutional and grass-roots, to ensure the conservation of plant genetic resources have thus taken on an increasing urgency of late.

Two strategies have been followed, their complementarity now recognized by the *Convention on Biological Diversity* (Articles 8 and 9), *Agenda 21* (Chapters 14 and 15) and the *Global Biodiversity Strategy* (WRI *et al.*, 1992). One approach involves the fostering and protection of the systems in which wild species have evolved, and crops have been developed. Biosphere Reserves, national parks and other kinds of protected areas, sustainably managed forests and the efforts of (in particular, though not exclusively) non-governmental organizations (NGOs) in on-farm and community-level crop conservation are examples of the application of such an *in situ* strategy. However, it is also possible for the conservation process to take another course. Propagules and other plant parts may be collected from the field and transferred to a suitable form of *ex situ* storage facility. Orthodox seeds and pollen may be maintained in cold storage of different types, whole plants and perennating organs in a field collection, arboretum or botanic garden and *in vitro* samples in the tissue culture laboratory under slow growth conditions or cryopreservation. In the gene bank and botanic garden, germplasm is both conserved and made available for study and use. Different species and situations will require different mixes of *in situ* and *ex situ* approaches.

Collecting germplasm – the first step in *ex situ* conservation and clearly a prerequisite for use of the material – is not as easy as it may sound. It is not simply a matter of being at the right place at the right time (though even this can on occasion be difficult enough) and putting a few seeds in a bag. Target species must be found and correctly identified. A decision must be taken as to what plant part or parts to collect. An attempt must be made to capture maximum diversity for the amount of material collected and resources expended. The material must be kept viable under often difficult field conditions. The germplasm must be carefully documented if it is to be useful to the eventual user.

Even deciding what plants to collect can be difficult. The genetic diversity of a given species extends beyond its taxonomic boundaries. The gene pool of a crop includes not only traditional local forms (landraces) but also wild and weedy relatives. The entire gene pool forms the basic unit of conservation and must be the ultimate target of genetic resources collecting. Crop relatives are not the end of the story, however. There are well over a quarter of a million plant species on this planet, and only a small proportion falls in the gene pools of current crops. Thousands of species are exploited by local communities and their livestock but are not domesticated. Then there are forest trees, the thousands of species used in traditional medicine and the vast ranks of ornamentals. Many plants are important in land management and habitat restoration or rehabilitation. Which other species have the potential to prove significant in the future as sources of food, medicines, energy and industrial products? What are the keystone species of ecosystems? Which are the most threatened?

Farmers have always recognized the value of exotic plants and novel crops and varieties. In the Andes, they have gathered for centuries at regional fairs to exchange planting material and learn from each other. In eastern Sierra Leone, women expect to take seeds and plants from each other's farms to meet immediate consumption needs. They also do so to obtain desired planting material, which may also be given as gifts during visits to family members in distant villages and towns (Leach, 1991). In some parts of India, it is customary for a bride to bring a gift of rice seeds grown by her family to her new husband's home. The state has also contributed to the process of germplasm collecting and transfer. Almost a thousand years ago, Emperor Chen-Tsung introduced Champa varieties of rice from Vietnam to China's Yangtze Delta, perhaps the first large-scale germplasm introduction. Plant collecting has in fact been one of the most powerful stimuli to exploration down the ages. Some 3500 years ago, Queen Hatshetput of Egypt dispatched a collecting expedition to Punt: the bas reliefs at the Deir al-Bahar temple show live frankincense trees, precious for their perfumed gum, being carried along in a procession. Though germplasm was not the kind of riches they primarily had in mind, the Iberian colonists engaged in plant collecting 500 years ago in the New World, unwittingly starting a major programme of germplasm exchange with the Old World which is still continuing and which is helping to feed both. Three centuries later, Captain Bligh was leading a germplasm collecting expedition which included two horticulturalists from the Royal Botanic Gardens, Kew, when the famous mutiny occurred. The *Bounty* was transporting breadfruit seedlings from Tahiti to the West Indies. The idea had been Sir Joseph Banks', famous for botanical collecting in Canada, New Zealand and points between.

In modern times, the institutional roots of the kind of germplasm collecting exemplified by the *Bounty*'s mission can be traced back to the hospital gardens of the Muslim world and the herb and medicinal gardens of the monasteries of medieval Europe, the precursors of the 'physic' gardens established in Renaissance Italy in connection with the first universities. The development of botanic gardens from the largely medicinal gardens of the 16th and 17th centuries was largely determined by a desire to increase the variety of collections. To this end, plant collectors were sent out to various parts of the world to bring back species new to cultivation, initially to the Near East, later to the Americas, southern Africa, Australia and the Far East. Orangeries and the first glasshouses greatly expanded the range of plants that could be cultivated. As the exploration of the tropics by the emerging colonial powers continued, the first botanic gardens outside Europe were created, such as those at the Cape of Good Hope (South Africa), Pamplemousses (Mauritius), Buitenzorg (now Bogor, Indonesia), Calcutta (India) and Bath (Jamaica). These served as introduction and acclimatization centres for a wide range of crops, fruits, spices and ornamentals (Smith, 1986; Heywood, 1990).

The largest network of botanic gardens was that created in the former British colonies under the aegis of the Royal Botanic Gardens at Kew. These were responsible for the movement of vast amounts of germplasm around the world. The flow was not only North–South, but also South–South, as material introduced through a botanic garden in one country was sent for trial in other parts of the world with a suitable climate. European botanic gardens also served as staging posts. The Royal Botanic Garden at Edinburgh, for example, received seedlings of coffee from Java via the Hortus Botanicus in Amsterdam and sent some on to Nyasaland (today Malawi). In the opposite direction, Leiden Botanic Garden served as an intermediary for the transfer of the vanilla orchid from South America to Java. Kew received 70,000 rubber seeds from the Brazilian Amazon in 1876, of which 2397 germinated; almost 2000 seedlings were then sent on to the botanic garden at Heneratgoda (Sri Lanka), which had been specially created to receive them, and smaller numbers to Calcutta and Singapore. The scale of operations was often vast. Between 1873 and 1876, almost 3.5 million seedlings of chinchona, a species newly introduced through Kew, were made available to local planters by the botanic garden at Hagkala (Sri Lanka).

Another important mechanism for collecting and distributing germplasm was the seed exchange system that was formalized in the 18th century as the *Index Seminum*. Through this mechanism, seeds and other propagules are offered for free exchange between botanic gardens throughout the world. Today, some 800 botanic gardens regularly issue such seed lists, which are analogous to the accession catalogues of gene banks. Prior to this, the holdings of botanic gardens were frequently listed in specially published catalogues, which often contain invaluable information about the first introduction of plants into cultivation.

A major contribution of botanic gardens to plant introduction has been in the field of ornamental plants. An important figure was George Forrest, who was sent by the Royal Botanic Garden, Edinburgh, to western China and Bhutan at the beginning of this century. Among his introductions were many *Rhododendron* and *Primula* species, for which Edinburgh is now famous, and *Meconopsis* and *Lilium* species. Plant collecting expeditions from Edinburgh to China continue to the present day, in collaboration with local botanists and botanic gardens.

The number of species introduced into cultivation by botanic gardens probably runs to 80,000 or more. Many of these are represented by only small samples, sometimes single individuals, of limited value as genetic resources. However, in the last decade many botanic gardens have adopted more specialized sampling procedures, and over 120 maintain seed banks. These range from fully equipped facilities like that of the Royal Botanic Gardens, Kew at Wakehurst Place to simple collections in vials in deep freezes. Many more gardens maintain significant conservation collections of growing plants.

Botanic gardens have not, however, been the only modern institutions instigating scientific germplasm collecting, or the only recipients and guardians of the material. The private sector has been active, in the form of commercial nurseries, seed companies, gardening societies and the like. Nowadays, the field increasingly includes NGOs and other local grass-roots organizations. Government bodies dealing with agriculture and forestry took their cue from the early botanic gardens. In the USA, for example, the Office of Foreign Seed and Plant Introduction of the Department of Agriculture sponsored many expeditions in the early 1900s. The Act which established the Department under President Lincoln set out its tasks as 'to collect . . . new and valuable seeds and plants; to test, by cultivation, the value of such of them as may require such tests; to propagate such as may be worthy of propagation, and to distribute them among agriculturists' (Berg *et al.*, 1991). Among its outstanding collectors were David Fairchild, Frank Meyers, Joseph F. Rock and Wilson Popenoe, but even diplomats and military personnel serving abroad contributed. They continued a tradition of plant introduction into the USA that goes back to the first settlers and to the 'botanist-kings' of the early years of the republic, people like Benjamin Franklin, George Washington and Thomas Jefferson, who knew all about the activities of botanic gardens like Kew.

Initially, the material obtained by the Office of Foreign Seed and Plant Introduction was maintained in living collections or as seeds stored at ambient temperature and grown out every year. In the late 1940s, as it became clear that only a very small percentage of the tens of thousands of accessions collected in the previous half-century could be accounted for in living collections, four regional centres for storing seeds under medium-term conditions (i.e. at about $0°C$) were established in the USA. These were joined in 1958 by the National Seed Storage Laboratory, which was later upgraded for long-term storage (i.e. at about $-15°C$).

Perhaps the most significant of the national germplasm institutions of the early part of this century was the All-Union Institute of Plant Introduction in St Petersburg, Russia. Established in 1890, in the 1920s it began to house the extensive worldwide collections being amassed by Nikolai I. Vavilov and his colleagues. It was later renamed the N.I. Vavilov All-Union Scientific Research Institute of Plant Industry (VIR) in honour of his achievement. Vavilov, in many ways the pioneer of scientific, systematic germplasm collecting, gathered some 50,000 samples of crop plants in over 50 countries in the 1920s and 1930s. His work set a pattern, as can be seen by what happened in potato collecting in Central and South America. In the wake of Soviet missions in 1925–27 came American, German, Swedish and British collectors in the 1930s, including Jack Hawkes, later one of the founders of the worldwide movement to conserve plant genetic resources. This was followed by collecting by the Latin American countries themselves, culminating in the life-work of such legendary plant explorers as Carlos Ochoa. The Centro Internacional de la Papa (CIP), one of the centres of the Consultative Group on International Agricultural Research (CGIAR), is the heir to this international collecting programme.

The Vavilov Institute acquired long-term seed storage capability in the 1970s. Smaller national gene banks were also established around that time in various European countries, Australia, Canada and Japan, in some cases as the culmination of national collecting efforts going back almost a century. The year 1990, for example, marked the twentieth anniversary of the establishment of the gene bank at Braunschweig in Germany but also the hundredth anniversary of the International Agricultural and Forestry Congress, held in Vienna, at which Emanuel Ritter von Proskowetz and Franz Schindler reported on the importance of landraces in agriculture. The former's collection of Moravian barley landraces is still justly famous.

In the 1960s, the realization began to grow that developing countries (and, indeed, local communities) must be integrated to a much greater extent into a global plant genetic resources system. That was where crop genetic diversity was not only most abundant, but also most at risk, as the Green Revolution got into its stride. The Food and Agriculture Organization of the United Nations (FAO) had actually initiated international discussions as far back as 1947 and introduced the first international newsletter on crop genetic resources in 1957, the precursor of today's *Plant Genetic Resources Newsletter*. Following a Technical Meeting which its Plant Production and Protection Division convened in 1961, a Panel of Experts on Crop Germplasm Exploration and Introduction was set up in 1965, with Sir Otto Frankel as chair. A similar panel on forestry was established in 1968, followed by an Expert Consultation on Forest Genetic Resources. In several meetings held over nine years, the crop experts set priorities for exploration, drafted proposals for an international network of genetic resources centres and developed guidelines for international cooperation in seed conservation. The International Biological Programme (IBP) was closely involved in this process. One of the more important outcomes was Frankel's *FAO/IBP Survey of Crop Genetic Resources in their Centres of Origin*, published in 1973. FAO's Crop Ecology and Genetic Resources Unit started work in 1968, including wide-ranging germplasm collecting, for example by Erna Bennett, herself the author of pioneering papers. As awareness increased of the importance of plant genetic diversity in sustainable development, and of the threat it was facing, systematic worldwide collecting for long-term conservation began in earnest.

FAO Technical Conferences on plant genetic resources were held in 1967, 1973 and 1981. In addition to Sir Otto Frankel and Erna Bennett, R.O. Whyte, Jack Harlan, T.T. Chang, Jack Hawkes and others played an important part in organizing these early meetings. In 1972, the UN Conference on the Human Environment, held in Stockholm, gave FAO responsibility for the establishment of an International Genetic Resources Programme, in addition to leading to the setting up of the United Nations Environmental Programme (UNEP). FAO then submitted to the CGIAR a proposal which eventually led, in 1974, to the establishment of the International Board for Plant Genetic Resources

(IBPGR) by the CGIAR. Placed within the institutional framework of FAO, the mandate of IBPGR was to help coordinate plant genetic resources activities worldwide.

Germplasm collecting was one of IBPGR's main activities in the early years, always in collaboration with the national plant genetic resources programmes that it was at the same time helping to strengthen and in some cases start. Collaboration with international agricultural research centres (IARCs) both inside and outside the CGIAR was also strong. Many IARCs have germplasm collecting and conservation programmes of long standing: for example, Carlos Ochoa worked at CIP from 1971 and T.T. Chang at the International Rice Research Institute (IRRI). The initial stress at IBPGR, as at the commodity IARCs, was on the broad-scale, worldwide collecting of germplasm of the major food crops, which were thought to be most threatened by genetic erosion. More recently, work on such staples has concentrated on gap-filling and collecting for special purposes. Since the mid-1980s, increasing emphasis has been placed by the plant genetic resources community, including IBPGR, on forages, multipurpose trees, wild crop relatives and less well-known crops. Forestry species have recently been added to IBPGR's (and the CGIAR's) mandate. By the beginning of the 1990s, IBPGR had sponsored some 650 missions in about 130 countries, for a total of almost 200,000 samples collected. In 1974 there were ten long-term seed storage facilities in the world, nine in developed countries. By 1990 there were 89 in developed countries and 39 in developing countries, storing some three and a half million accessions collected by national, regional and international organizations, mostly during the previous 20 years or so.

In late 1991 an international agreement was signed by five governments establishing the International Plant Genetic Resources Institute (IPGRI) as the independent successor to IBPGR. IPGRI, in close partnership with FAO and other international organizations, continues to be actively involved in germplasm collecting and seeks above all to work with and strengthen the collecting and other conservation activities of national plant genetic resources systems in developing countries. These systems can be anything from a single institute with the facilities for *ex situ* conservation to a committee representing a national network of crop-specific agricultural research institutes, university departments and botanic gardens. The private sector may also be involved. The extent of coordination in such a system varies widely from country to country. The priorities of a national programme will depend on needs, on national research and development capacity and on the diversity of germplasm within the country, but collecting will usually feature to at least some extent. National plant genetic resources systems are the building-blocks of the global conservation effort. Increasingly, they include NGOs and grass-roots organizations, an 'informal' sector to set besides the 'formal', institutional sector represented by gene banks, botanic gardens and arboreta.

National programmes and their collecting activities have been supported by both bilateral and multilateral development agencies. They also collaborate among themselves in collecting. An example is the long-standing collaboration between the gene bank of the Institute of Plant Genetics and Crop Plant Research, at Gatersleben in Germany, and the national programmes of Cuba and other countries. In some cases, collaboration among the national programmes of developing countries is supported, technically and/or financially, by regional organizations. Examples are the Regional Plant Genetic Resources Centre of the Southern Africa Development Community (SADC) in Lusaka, Zambia, and the Centro Agronómico Tropical de Investigaçión y Enseñanza (CATIE) at Turrialba, Costa Rica.

An international framework for these disparate activities is provided by the Resolution of the 1983 FAO Conference, which established the Commission on Plant Genetic Resources as a global forum for plant genetic resources debate. The Commission, whose Secretariat is provided by the Plant Production and Protection Division of FAO, has met five times since 1983 and now numbers 123 member countries. Its role in monitoring the implementation of the principles of the International Undertaking on Plant Genetic Resources and in drawing up an International Code of Conduct for Plant Germplasm Collecting and Transfer is discussed in Chapter 2.

There is thus nothing new about germplasm collecting. It has been done, and continues to be done, by all kinds of different people – conservationists, ecologists, botanists, foresters, breeders, geneticists, extension officers, development workers, local communities themselves – who have different immediate interests and concerns. They have also had to deal with a huge range of plants – wild and cultivated, annual and perennial, woody and herbaceous, outbreeding and inbreeding, seed-producing and vegetatively propagated. Until relatively recently, they have not had much in the way of guidelines to follow. A turning-point was the publication in 1970 of *Genetic Resources in Plants – Their Exploration and Conservation*, edited by Sir Otto Frankel and Erna Bennett. This volume grew out of the 1967 FAO Technical Conference and included several papers giving valuable practical advice to the germplasm collector. Another, equally important work emerged after the 1973 meeting: *Crop Genetic Resources for Today and Tomorrow*, edited by Sir Otto Frankel and Jack Hawkes (1975). In 1972 a *Manual for Field Collectors of Rice* (Chang *et al.*, 1972) was published, the fruit of a decade of experience in rice collecting by IRRI, but as the title implies its scope is limited. The same is also true of the Centro International de Agricultura Tropical (CIAT)'s *Handbook for the Collection, Preservation and Characterization of Tropical Forage Germplasm Resources* (Mott and Jiménez, 1979).

From the mid-1970s, there was a rapid increase in the number of publications on the theoretical and practical aspects of germplasm collecting, particularly as regards crops. Much of this information was synthesized by Prof. Jack Hawkes in his *Crop Genetic Resources Field Collection Manual* (Hawkes, 1980), co-sponsored by IBPGR and the European Association for Research on Plant Breeding (EUCARPIA). More specialized guides followed. For example, FAO produced publications on collecting forestry species (e.g. Ffolliott and Thames, 1983; FAO, 1985). Marchenay's (1986; 1987) 'Guide pratique' and 'Guide méthodologique' apply the methods refined over the previous 15 years or so to the specific problem of collecting traditional crop varieties in France. Kenya Energy Non-Governmental Organization's (KENGO) *How to Collect, Handle and Store Seeds* (Mboye and Kiambi, n.d.) is an example of a developing country NGO publication in the field, with the emphasis on 'easily digestible and synthesized technical information'.

Following a series of Conservation Conferences held at Kew, beginning in 1975 with *The Function of Living Plant Collections in Conservation and Conservation-Oriented Research and Public Education* (Simmons *et al.*, 1976), the botanic garden sector began to play an increasingly important role in the collecting and conservation of germplasm, especially of wild species, including medicinal plants and ornamentals. The *Botanic Gardens Conservation Strategy*, published in 1989 by the World Wide Fund for Nature (WWF), the World Conservation Union (IUCN) and the Botanic Gardens Conservation Secretariat (today Botanic Gardens Conservation International, BGCI), focused attention on the priorities for germplasm conservation by botanic gardens. BGCI has also produced a set of *Guidelines for the* Ex Situ *Conservation of Germplasm by Botanic Gardens* (BGCI, 1993). The Centre for Plant Conservation (CPC) at St Louis has published 'Genetic sampling guidelines for conservation collections of endangered plants' (CPC, 1991). Various workshops and symposia on genetic resources conservation, plant introduction by botanic gardens and related topics have been published (e.g., He *et al.* 1990; Hernández-Bermejo *et al.*, 1991; Hamann, 1992). The *Guidelines on the Conservation of Medicinal Plants* (WHO *et al.*, 1993) include sections on collecting and conservation.

As the 1990s began, it became evident, however, that there was no single publication available which could be consulted by the prospective collector of plant germplasm for generic as well as specific, and theoretical as well as practical, information. Some topics had not previously been dealt with fully, other fields had advanced very rapidly. It was to fill this gap that IBPGR, FAO, UNEP and IUCN agreed to cooperate in the publication of the present volume of *Technical Guidelines* for plant germplasm collectors, aimed both at the newcomer to the field and the experienced worker faced with new technical challenges.

A major impetus to this effort was the UN Conference on Environment and Development (UNCED), held in Rio de Janeiro in June 1992 and the spiritual successor of the 1972 Stockholm Conference, which in different ways led to the establishment of both UNEP and IBPGR. The UNCED Conference adopted *Agenda 21*, a global environment and development programme which recognizes the crucial importance of plant genetic resources conservation – both *in situ* and *ex situ* – to future development prospects. The *Convention on Biological Diversity* was also opened for signature at the Conference, and came into effect in late 1993. This will have a profound impact on plant germplasm collecting policies and programmes. Article 9 of the *Convention* calls upon countries to 'adopt measures for the *ex situ* conservation of components of biological diversity, preferably in the country of origin of such components'. One of these measures must of course be a technically sound germplasm collecting programme, and it is hoped that this book will provide the basis on which such a programme can be constructed and run. Much remains to be done. This was underlined as recently as 1991, when the Final Consensus Report of the Keystone International Dialogue Series on Plant Genetic Resources noted that: 'The inadequacy of most current collections is widely recognized. Even in major crops, there are important areas of diversity that remain to be sampled, and some areas where past sampling was inadequate or faulty may need to be revisited.'

References

Berg, T., Å. Bjornstad, C. Fowler and T. Skroppa (1991) *Technology Options and the Gene Struggle*. NORAGRIC, Aas.

BGCI (1993) *Guidelines for the* Ex Situ *Conservation of Germplasm by Botanic Gardens*. Botanic Gardens Conservation International and Cabildo Insular de Gran Canaria, Las Palmas de Gran Canaria.

Chang, T.T., S.D. Sharma, C. Roy Adair and A.T. Perez (1972) *Manual for Field Collectors of Rice*. IRRI, Los Baños.

CPC (1991) Genetic sampling guidelines for conservation collections of endangered plants. In: Falk, D.A. and K.E. Holsinger (eds) *Genetics and Conservation of Rare Plants*. pp. 209–238. Oxford University Press, New York.

FAO (1985) *A Guide to Forest Seed Handling (With Special Reference to the Tropics)*. FAO Forestry Paper 20/2, compiled by R.L. Willan. FAO, Rome.

Ffolliott, P.F. and J.L. Thames (1983) *Collection, Handling, Storage and Pretreatment of Prosopis Seeds in Latin America*. FAO, Rome.

Frankel, O.H. and E. Bennett (eds) (1970) *Genetic Resources in Plants – their Exploration and Conservation*. Blackwell Scientific Publications, Oxford.

Frankel, O.H. and J.G. Hawkes (eds) (1975) *Crop Genetic Resources for Today and Tomorrow*. Cambridge University Press, Cambridge.

Hamann, O. (ed.) (1992) *Ex situ* conservation in botanical gardens. *Opera Botanica* 113.

Hawkes, J.G. (1980) *Crop Genetic Resources Field Collection Manual*. IBPGR and EUCARPIA, Rome.

He, Shan-an, V.H. Heywood and P.S. Ashton (eds) (1990) *Proceedings of the International Symposium on Botanic Gardens*. Jiangsu Science and Technology Publishing House, Nanjing.

Hernández-Bermejo, E., M. Clemente and V.H. Heywood (eds) (1991) *Conservation Techniques in Botanic Gardens*. Koeltz Scientific Books, Koenigstein.

Heywood, V.H. (1990) Botanic gardens and the conservation of plant resources. *Impact of Science on Society* 158:121–132

Leach, M. (1991) Women's vegetable production in eastern Sierra Leone. In: Camman, L. (ed.) *Peasant Household Systems*. Proceedings of an International Workshop. 3–5 April 1990. Feldafing. pp. 186–208. Deutsche Siftung für Internationale Entwicklung (DSE), Feldafing.

Marchenay, P. (1986) *Prospection et Collecte des Variétés Locales de Plantes Cultivées*. Groupe de Recherche et de Développement sur le Patrimonie Génétique et Végétal de la Région Provence–Alpes–Côte d'Azure (PAGE-PACA), Hyères.

Marchenay, P. (1987) *A la Recherche des Variétés Locales de Plantes Cultivées*. Groupe de Recherche et de Développement sur le Patrimonie Génétique et Végétal de la Région Provence–Alpes–Côte d'Azure (PAGE-PACA), Hyères.

Mboye, A. and K. Kiambi (n.d.) *How to Collect, Handle and Store Seeds*. KENGO, Nairobi.

Mott, G.O. and A. Jiménez (eds) (1979) *Handbook for the Collection, Preservation and Characterization of Tropical Forage Germplasm Resources*. CIAT, Cali.

Simmons, J.B., R.I. Beyer, P.E. Brandham, G.Ll. Lucas and V.T.H. Parry (eds) (1976) *Conservation of Threatened Plants*. Plenum Press, New York.

Smith, N.J.H. (1986) *Botanic Gardens and Germplasm Conservation*. Harold L. Lyon Arboretum, University of Hawaii Press, Honolulu.

WHO, IUCN and WWF (1993) *Guidelines on the Conservation of Medicinal Plants*. IUCN, Gland.

WRI, IUCN and UNEP (1992) *Global Biodiversity Strategy. Guidelines for Action to Save, Study and Use Earth's Biotic Wealth Sustainably*. WRI, IUCN and UNEP, Washington DC.

Legal issues in plant germplasm collecting

2

Co-sponsors

No country is self-sufficient in plant genetic resources, and until recently this was thought to be sufficient to ensure the free flow of germplasm among nations. However, the concepts of ownership, sovereignty and intellectual property rights (IPR) have increasingly been invoked of late in discussions of how best to conserve and use plant genetic resources. This has resulted from the growing realization that germplasm has real economic value. At the same time, awareness has grown that the repercussions of overexploitation of natural resources, of the extinction of species and the disappearance of crop landraces, of environmental damage and of habitat destruction can be global, transcending national boundaries and short-term financial considerations. The legal context of germplasm collecting has as a result become more complex. This chapter will briefly describe this evolving debate and how it affects germplasm collecting.

Towards an international system for *ex situ* germplasm conservation

A strategy evolved in the 1970s to deal with the problem of the conservation of plant genetic resources based on the concept of a worldwide system of gene banks providing a combination of long-term seed storage in base collections and short- or medium-term storage in active collections (Box 2.1). First the Food and Agriculture Organization (FAO) and later the International Board for Plant Genetic Research (IBPGR), with the support of a panel of experts, asked adequately equipped gene banks around the world to accept the responsibility for the long-term conservation of the global or regional base collections of given crops. The

13

BOX 2.1
Some definitions

Base collections of germplasm are stored at low temperature for long-term, secure conservation. Accessions are only removed:

- for regeneration, when their viability has declined below an acceptable standard;
- to provide material for an active collection for regeneration, if the stocks held by the active collection are more than two or three regeneration cycles removed from the original material;
- when stocks of an accession are no longer available from an active collection.

Currently, base collections are only maintained for orthodox seeds. In contrast to recalcitrant seeds, which die if so treated, orthodox seeds may be dried to low moisture content, sealed in airtight containers and stored at low temperature (usually 0°C to −20°C) for considerable periods. *In vitro* base gene banks are currently at the research stage.

In contrast, material in **active collections** is continuously being removed, whether for regeneration, multiplication, characterization, evaluation or distribution. Currently, active collections are maintained for orthodox seed, which are dried and stored at temperatures above 0°C but below 15°C. *Ex situ* collections of living plants under field or nursery conditions are often called **field gene banks**. In general, they fall within the category of active gene banks and are used for material which would be difficult to maintain as seeds (e.g. species with recalcitrant seeds) or when it is desired to maintain particular genotypes. *In vitro* active gene banks are currently at the pilot stage.

Working collections are also called breeders' collections or research collections. Storage of material in these collections is usually under ambient conditions or in air-conditioned rooms. They may include special genetic stocks such as breeders' lines and mutants.

---

expectation was that close cooperation between the global base collection of a given crop and the many active collections of the crop gene pool around the world would develop. Active collections would deposit a duplicate set of their holdings of a particular species in an appropriate base collection. There would also be safety storage of duplicates of base collection accessions in at least one other gene bank, preferably in another country.

Such a system of base and active collections is not yet fully operational. One of the main reasons is that it requires extensive movement of germplasm among countries, and this has presented problems. Countries have an understandable wish (indeed, duty) to protect their agriculture and natural habitats from outside pests, including weeds and diseases. Germplasm consignments may be contaminated with pests, or the plant itself may be a potential pest. Under the provisions of the *International Plant Protection Convention* (IPPC) of 1951, this may result in legal restrictions on germplasm movement (Chapter 17). Constraints also arise, however, because germplasm has value. If something is perceived as having value, sooner or later ownership will be claimed over it and restrictions put on its availability. On the other hand, the relatively unimpeded flow of germplasm and information is necessary for the efficient conservation and use of plant genetic resources.

As is discussed below, a resolution to this debate has been sought within the FAO Global System for the Conservation and Utilization of Plant Genetic Resources, which has tried to provide

- a forum for discussion, the Commission;
- a flexible international framework, the Undertaking;
- the beginnings of a financial mechanism, the Fund.

The ownership of germplasm

Plant breeders' rights

Genetic resources have sometimes been thought of as a common good, the common heritage of humanity. In commercial plant breeding, however, genotypes have economic value. With the commercialization of agriculture and the increasing importance and development costs of modern, scientifically bred varieties, plant diversity is becoming an ever more valuable resource. The costly efforts that have had to be undertaken to safeguard the plant breeder's raw material have added further value to the germplasm kept in gene banks. To protect this investment, plant breeders' rights (PBR) are granted by some countries to plant breeders to exclude others from producing and selling propagating material of a protected variety for a period of 15–25 years. A protected improved variety is legally 'owned' by the breeder.

This principle is enshrined in the *Convention of the International Union for the Protection of New Varieties of Plants* (UPOV Convention). In order to be eligible for PBR protection, a variety must be distinct, uniform and stable in its essential characteristics and not yet commercialized. The maintenance of free availability of genetic resources was an important objective when the UPOV Convention was drawn up. The so-called breeders' exemption allows plant breeders to use without restriction protected varieties in the production of new varieties. At the same time, farmers are allowed the reusable part of the harvested material of protected varieties as seeds for the next year's planting (farmers' exemption).

In a recent revision of the UPOV Convention (UPOV, 1991), however, an optional restriction of the farmers' exemption has been introduced giving individual countries the choice of whether or not to grant farmers the right to save seeds for future sowing. There are currently 21 member countries which have adopted this revised UPOV Convention, mainly European countries, the USA, Canada and Japan. For most developing countries, the benefits of membership do not as yet outweigh the drawbacks, in the form of administrative costs and loss of access to protected varieties (Belcher and Hawtin, 1991). However, the recently concluded agreement on trade-related intellectual property rights (TRIPs) negotiated during the Uruguay Round of talks under the General Agreement on Tariffs and Trade (GATT) requires signatory nations to introduce legislation for intellectual property protection of plant varieties. Argentina and Paraguay were the first developing countries to begin the process of becoming signatories to the UPOV Convention. Other developing countries (e.g. Chile and Cuba) already have analogous legislation.

Patent rights

PBR is not the only IPR system which has been brought into play. The emergence of modern biotechnology has diffused the definition of plant genetic resources to include not only whole plants but their individual constituent elements, down to tissues, genes or even fragments of deoxyribonucleic acid (DNA). Biotechnological research, an important output of which is crop varieties, is increasingly being undertaken by private institutions or results from the growing relationship between universities and public institutions on the one hand and private industry on the other. The industrial complex generally considers the traditional protection of varieties through PBR inadequate in a biotechnological age. A number of industrialized countries have responded to this argument by considering the expansion of the definition of patentable material to include plants or parts thereof.

Patent rights can be granted to inventors to exclude others from imitating, manufacturing, using or selling a patented process or product for commercial use for a period of usually 17–20 years. In return for the patent, the inventor discloses how the invention works, so that knowledge is available to the public. In order to obtain a patent, the process or product has to be novel, useful and non-obvious. Furthermore, the patent must relate to a technology for which patents are permitted. Many countries do not grant patents on pharmaceuticals and some prohibit patents on agricultural innovations.

Innovations on living organisms are in many countries not patentable, but this is changing (Belcher and Hawtin, 1991). In the USA patents have been granted for specific plant and animal varieties. In contrast to PBR, patent protection could give the patent holder the authority to restrict use of the patented variety for breeding purposes. Patent protection is also available in a number of countries for plants that contain a novel gene. To qualify, the gene must not be found in nature; it must be novel in the sense that it was created by the inventor or transferred to a species in which it is not found naturally (Barton and Siebeck, 1991). Such patents on genes seem to imply that the holder of the patent could prohibit others from engaging in unauthorized commercial activity involving any plant material of the protected species. This protection might even be extended to closely related species to which the protected gene could be transferred through conventional breeding techniques (Barton, 1991). An even more controversial development is the granting by the US Patent and Trademark Office of a patent on a plant characteristic, irrespective of the process by which the characteristic was arrived at.

To some extent in reaction to these trends, the concept of national sovereignty began to be invoked by countries to assert ownership of germplasm of certain species within their borders which were deemed to be important to the national interest. Examples are the restricted availability of germplasm of coffee from Ethiopia, rubber from Brazil, spices from Indonesia, black pepper from India and pyrethrum from Kenya. Some countries established a practice of releasing germplasm only in exchange for training, technology or other kinds of support of the national programme. This policy has found an echo in FAO's Code of Conduct for Plant Germplasm Collecting and Transfer and the *Convention on Biological Diversity* (CBD) (see below).

The FAO global system for the conservation and utilization of plant genetic resources

As part of the developing debate, discussions took place during the 1983 FAO Conference which led to a resolution establishing the Commission on Plant Genetic Resources as a global forum where the donors and users of germplasm, of funds and of technology could meet on an equal footing to try to resolve the kinds of issues outlined above. The Commission has since met in 1985, 1987, 1989, 1991 and 1993. Its Secretariat is provided by the Plant Production and Protection Division of FAO.

One of the main tasks of the Commission is monitoring the implementation of the principles of the International Undertaking on Plant Genetic Resources, a non-binding agreement also drawn up in 1983. Its aim is to ensure that these resources – especially species of present or future economic and social importance – are identified, collected, conserved, evaluated and made available without restriction. Since 1983, 110 countries have adhered to the Undertaking (140 countries are either members of the Commission or have adhered to the Undertaking). In accordance with one of its articles and a memorandum of understanding between the two organizations, gene banks featuring in the IBPGR (now the International Plant Genetic Resources Institute (IPGRI)) register of base collections are beginning to be incorporated into a network of base collections under the auspices and/or jurisdiction of FAO.

In the Undertaking, plant genetic resources are taken to include: cultivated varieties in current use and newly developed; obsolete cultivars; primitive cultivars (landraces); wild and weedy species, near relatives of cultivated varieties; and special genetic stocks (including élite and current breeders' lines and mutants). Many industrialized countries opposed the view that special genetic stocks should be the object of the agreement. They argued that these cannot be freely exchanged as national legislation in these countries provides for private property rights on crop varieties in the form of PBR and patents. Many developing countries, on the other hand, argued that special genetic stocks are largely derived from germplasm originating from within their boundaries, and that these genetic resources should be made available for free, just as the relatively unimproved germplasm originally was.

These problems were discussed by the second and third sessions of the Commission and there was full consensus on how to overcome them. Two resolutions of the FAO Conference of 1989 were added as Annexes to the Undertaking, providing an 'agreed interpretation'. They recognize not only plant breeders' rights but also farmers' rights, as the legitimate demands of, respectively, the donors of technology and those of germplasm, both of whom are to be compensated for their contributions. Originally based on the principle that plant genetic resources are part of the common heritage of humanity, with its Complementary Resolutions and Annexes, the Undertaking is now based on the principle of the sovereign rights of nations over the plant genetic resources within their borders.

Following the recommendations of the Undertaking, an International Fund for Plant Genetic Resources was officially established by FAO in 1988. It is meant to provide a channel for governmental and non-governmental organizations and individuals to support the conservation and use of plant genetic resources. As such, it is expected to become a critical element in ensuring the equitability of the global system. The developing countries in particular recognize the Fund as an appropriate mechanism for the realization of rewards for breeding and maintaining landraces. The argument is that; just as scientists are rewarded for their labour in creating breeding lines and commercial varieties, farmers have a right to receive material compensation for their efforts over the centuries in conserving, improving and making available plant genetic resources.

The Fund is currently voluntary. A different approach was suggested by the results of the Keystone International Dialogue Series. The Keystone Centre, a US organization dedicated to the arbitration of conflicts, brought together interested parties from all sides of the debate in a series of discussions starting in 1988. Early sessions refined the concept of farmers' rights and recognized the role of informal innovation systems in generating and conserving plant genetic resources. The consensus report arising from the final plenary session in 1991 argued that any fund 'designed to act as an analog to breeders' rights and patents with mandatory royalty payments' should itself be mandatory (Belcher and Hawtin, 1991).

The 1991 session of the FAO Conference, in a resolution which forms a further Annex to the Undertaking, endorsed the principles of nations having sovereign rights over their plant genetic resources and of the availability of breeders' lines and farmers' breeding material (i.e. landraces) being at the discretion of their developers during the period of development. The resolution also endorsed the view that farmers' rights should be implemented through an international fund that would also be used to support conservation and sustainable development of plant genetic resources. In order to determine the funding needed, FAO, IBPGR (now IPGRI) and other relevant organizations were asked to prepare a periodical report on the state of the world's plant genetic resources and a global plan of action on plant genetic resources. The Conference agreed that the first state of the world report and the plan of action will be prepared through the Fourth International Technical Conference on Plant Genetic Resources, to take place in 1996.

The contribution of the United Nations Conference on Environment and Development (UNCED)

The debate has not stopped at the admittedly vague dividing line between crop genetic resources and the wild flora. It has extended to biodiversity as a whole, playing a central role in the process leading up to (and following) the United Nations Conference on Environment and Development (UNCED), held at Rio de Janeiro, Brazil in June 1992.

The CBD was opened for signature at UNCED. It entered into force on 29 December 1993, when it was ratified by the thirtieth country. The CBD recognizes biodiversity as a 'common concern', rather than a common heritage, of humanity. Article 15 states that: 'Recognizing the sovereign rights of States over their natural resources, the authority to determine access to genetic resources rests with the national governments and is subject to national legislation.' It adds that access to genetic resources, where granted, shall be subject to the 'prior informed consent' of the donor of the genetic resources. National legislation should promote the 'fair and equitable' sharing of benefits from the commercial use of resources on 'mutually agreed terms'. Duesing (1992) (quoted by Gollin, 1993) summarizes Articles 15 and 16 as suggesting the following ways whereby a country can so benefit: (i) participation in research using the resource; (ii) receiving technology which embodies or uses the resource; and (iii) sharing the financial benefits realized from commercial exploitation of the resource. The World Conservation Union (IUCN) Environmental Law Centre (ELC) has prepared an explanatory guide to the CBD.

The CBD distinguishes between germplasm already collected and germplasm to be collected in accordance with its provisions. Germplasm collecting after the CBD's coming into force, in a country party to the CBD, is subject to the provisions of the CBD regarding national sovereignty. Several countries have started to develop the necessary national policies and legislation.

The Code of Conduct for Plant Germplasm Collecting and Transfer

In 1989, the FAO Commission requested its secretariat to prepare an International Code of Conduct for Plant Germplasm Collecting and Transfer. The Code is intended to form an important tool in regulating the collecting and transfer of plant genetic resources and their associated information (including indigenous knowledge), with the aim of facilitating access to these resources and promoting their use and development on an equitable basis. Along with other international and regional organizations, national programmes and experts, IBPGR (as it then was) had an input into the Code of Conduct. A draft Code was presented to the FAO Conference in 1991, and while the Conference agreed, in general, with its contents, it noted that further elaboration by the Commission was necessary. A new version of the Code was adopted by the Commission in 1993. A copy is provided in this chapter's Appendix 2.1. Its guiding principle is that though 'the conservation of plant genetic resources is a common concern of humankind', 'nations have sovereign rights over their plant genetic resources in their territories'.

The basic provision of the Code is that countries should regulate germplasm collecting through the issuing of collectors' permits. These are quite separate from the export and import permits and phyto-sanitary certificates that countries currently require for the movement of plant material across borders, and which, as already mentioned, are discussed fully in Chapter 17. There is a parallel here with the CBD, whose Article 9 (taken together with Article 15) 'provides a basis for domestic regulation of access to biodiversity, through, for example, collectors' agreements and access restrictions' (Gollin, 1993). Article 14 of the Code enumerates the ways in which the benefits of germplasm collecting could be shared with local communities, farmers and host countries. In 1993, the FAO Commission on Plant Genetic Resources adopted a Revision of the International Undertaking on Plant Genetic Resources, subsequently endorsed by the FAO Council, recommending full harmonization of the Undertaking, including the Code, with the CBD.

The issue of permits for biodiversity research in general, including plant collecting, is discussed by Janzen *et al.* (1993). Laird (1993) and Downes *et al.* (1993) discuss what contracts for access to biodiversity might look like, in the light of both national and international law, including the CBD. Barton and Siebeck (1994) discuss how 'material transfer agreements' could be used by a country to regulate access to germplasm collected within its borders but stored in gene banks abroad, for example in a base collection.

Other relevant national and international legislation

One of the responsibilities of collectors (together with donors, curators and users of germplasm) that the Code of Conduct for Plant Germplasm Collecting and Transfer emphasizes is that of minimizing the adverse effect of collecting on the environment and on biodiversity. Article 9 of the CBD makes the same point. Collecting germplasm should not contribute to genetic erosion or damage the ecosystem. In fact, of course, many countries already have national laws restricting the collecting of plant species, especially threatened species, protecting their habitats and regulating designated protected areas such as national parks. An analysis of national legislation protecting wild plants and their habitats is provided by Klemm (1993). Clearly, national private property laws will also affect access to particular sites and taking of specimens, but land may also be protected by traditional rules and customary law. These are incorporated into, or at least recognized by, national law in some countries. There are also two international agreements that are relevant to plant germplasm collectors in that they provide for the protection of sites, the *World Heritage Convention* (1972) and the *Convention on Wetlands of International Importance Especially as Waterfowl Habitat* (1971), also known as the *Ramsar Convention*.

The collecting and movement of threatened species and their products is subject to the provisions of another international treaty, the *Convention on International Trade in Endangered Species of Wild Fauna and Flora* (CITES). CITES, which entered into force in 1975 and now has more than 115 member countries, bans commercial international trade in an agreed list of endangered species and regulates and monitors trade in others that might become endangered. CITES has established a worldwide system of controls on international trade in threatened species and their products by stipulating that government permits are required for such trade. Enforcement of CITES is the responsibility of member countries, usually via the customs service. Countries are required to establish management and scientific authorities for the purpose of enforcing CITES regulations and to submit reports, including trade records, to the CITES secretariat in Switzerland. To ensure effective enforcement, the secretariat acts as a clearing house for the exchange of information and liaison between the member countries and with other authorities and organizations. The World Conservation Monitoring Centre (WCMC) maintains a database on the international trade in CITES-listed species (Chapter 10).

There are also relevant international regulations in the area of documenting indigenous knowledge. The World Intellectual Property Organization (WIPO)/United Nations Educational, Scientific and Cultural Organization (Unesco) Model Law on Folklore may restrict the use of material such as photographs and recordings. Downes *et al.* (1993) discuss the relevance of such agreements as the *International Covenant on Economic, Social and Cultural Rights*, the *Draft Declaration of the Rights of Indigenous People*, the International Labour Organization's (ILO) Convention No. 169 and Unesco's *Convention on the Means of Prohibiting and Preventing the Illicit Import, Export and Transfer of Ownership of Cultural Property*.

IUCN ELC maintains a database of national and international environmental conservation instruments, including wild plant species protection laws and regulations, which is available for consultation on request.

References

Barton, J.H. (1991) Patenting life. *Scientific American* 264:40–46.

Barton, J.H. and W.E. Siebeck (1991) *Intellectual Property Issues for the International Agricultural Research Centres.* CGIAR, Washington DC.

Barton, J.H. and W.E. Siebeck (1994) *Material Transfer Agreements in Genetic Resources Exchange – the Case of the International Agricultural Research Centres.* Issues in Genetic Resources No. 1. IPGRI, Rome.

Belcher, B. and G. Hawtin (1991) *A Patent on Life. Ownership of Plant and Animal Research.* IDRC, Ottawa.

Downes, D., S.A. Laird, C. Klein and B. Kramer Carney (1993) Biodiversity prospecting contracts In: Reid, W.V., S.A. Laird, C.A. Meyer, R. Gámez, A. Sittenfeld, D.H. Janzen, M.A. Gollin and C. Juma (eds) *Biodiversity Prospecting: Using Genetic Resources for Sustainable Development.* pp. 255–287. WRI Publications, Baltimore.

Duesing, J.H. (1992) The Convention on Biological Diversity: its implications on biotechnology research. *Agro Food Industry Hi-tech* 3:19

Gollin, M.A. (1993) An intellectual property rights framework for biodiversity prospecting. In: Reid, W.V., S.A. Laird, C.A. Meyer, R. Gámez, A. Sittenfeld, D.H. Janzen, M.A. Gollin and C. Juma (eds) *Biodiversity Prospecting: Using Genetic Resources for Sustainable Development.* pp. 159–197. WRI Publications, Baltimore.

Janzen, D.H., W. Hallwachs, R. Gámez, A. Sittenfeld and J. Jimenez (1993) Research management policies: permits for collecting and research in the topics. In: Reid, W.V., S.A. Laird, C.A. Meyer, R. Gámez, A. Sittenfeld, D.H. Janzen, M.A. Gollin and C. Juma (eds) *Biodiversity Prospecting: Using Genetic Resources for Sustainable Development.* pp. 130–157. WRI Publications, Baltimore.

Klemm, C. de (1993) *Biological Diversity, Conservation and the Law.* IUCN, Gland.

Laird, S.A. (1993) Contracts for biodiversity prospecting. In: Reid, W.V., S.A. Laird, C.A. Meyer, R. Gámez, A. Sittenfeld, D.H. Janzen, M.A. Gollin and C. Juma (eds) *Biodiversity Prospecting: Using Genetic Resources for Sustainable Development.* pp. 99–130. WRI Publications, Baltimore.

UPOV (1991) *International Convention for the Protection of New Varieties of Plants.* UPOV, Geneva.

Useful address

IUCN Environmental
Law Centre (ELC)
Adenauerallee 214
D-53113 Bonn
Germany
Tel: +49 228 2692231
Fax: +49 228 2692250

APPENDIX 2.1

International Code of Conduct for Plant Germplasm Collecting and Transfer

Contents

Chapter I
Objectives and Definitions

Article 1: Objectives

This Code has the following objectives:

1.1 to promote the conservation, collection and use of plant genetic resources from their natural habitats or surroundings, in ways that respect the environment and local traditions and cultures;

1.2 to foster the direct participation of farmers, scientists and organizations in countries where germplasm is collected, in programmes and actions aimed at the conservation and use of plant genetic resources;

1.3 to avoid genetic erosion and permanent loss of resources caused by excessive or uncontrolled collection of germplasm;

1.4 to promote the safe exchange of plant genetic resources, as well as the exchange of related information and technologies;

1.5 to help ensure that any collecting of germplasm is undertaken in full respect of national laws, local customs, rules and regulations;

1.6 to provide appropriate standards of conduct and to define obligations of collectors;

1.7 to promote the sharing of benefits derived from plant genetic resources between the donors and users of germplasm, related information and technologies by suggesting ways in which the users may pass on a share of the benefits to the donors, taking into account the costs of conserving and developing germplasm;

1.8 to bring recognition to the rights and needs of local communities and farmers, and those who manage wild and cultivated plant genetic resources and in particular to promote mechanisms:

(a) to facilitate compensation of local communities and farmers for their contribution to the conservation and development of plant genetic resources; and

(b) to avoid situations whereby benefits currently derived from plant genetic resources by these local communities and farmers are undermined by the transfer or use by others of the resources.

Article 2: Definitions

2.1 'Collector' means a legal or natural person that collects plant genetic resources and related information.

2.2 'Curator' means a legal or natural person that conserves and manages plant genetic resources and related information.

2.3 'Donors' means a country or legal or natural person that makes available plant genetic resources for collection.

2.4 'Farmers' rights' means the rights arising from the past, present and future contributions of farmers in conserving, improving, and making available plant genetic resources, particularly those in the centres of origin/diversity. These rights are vested in the international community, as trustee for present and future generations of farmers, for the purpose of ensuring full benefits to farmers, and supporting the continuation of their contributions, as well as the attainment of the overall purposes of the International Undertaking.[1]

2.5 '*Ex situ* conservation' means the conservation of plant genetic resources outside their natural habitat.

2.6 'Genetic erosion' means loss of genetic diversity.

2.7 '*In situ* conservation' means the conservation of plant genetic resources in the areas where they have naturally evolved, and, in the case of cultivated species or varieties, in the surroundings where they have developed their distinctive properties.

2.8 'Plant genetic resources' means germplasm or genetic material of actual or potential value.

2.9 'Plant germplasm' or 'genetic material' means the reproductive or vegetative propagating material of plants.

2.10 'Sponsor' means a legal or natural person that sponsors, financially or otherwise, a plant collecting mission.

2.11 'User' means a legal or natural person that utilizes and benefits from plant genetic resources and related information.

Chapter II
Nature and Scope of the Code

Article 3: Nature of the Code

3.1 The Code is voluntary.

3.2 The code recognizes that nations have sovereign rights over their plant genetic resources in their territories and it is based on the principle according to which the conservation and continued availability of plant genetic resources is a common concern of humankind. In executing these rights, access to plant genetic resources should not be unduly restricted.

3.3 The Code is addressed primarily to governments. All relevant legal and natural persons are also invited to observe its provisions, in particular those dealing with plant exploration and plant collection, agricultural and botanical activities and research on endangered species or habitat conservation, research institutes, botanical gardens, harvesting of wild plant resources, agroindustry including pharmaceutical plants and the seed trade.

3.4 The provisions of the Code should be implemented through collaborative action by governments, appropriate organizations and professional societies, field collectors and their sponsors, and curators and users of plant germplasm.

3.5 FAO and other competent organizations are invited to promote full observance of the Code.

3.6 The Code provides a set of general principles which governments may wish to use in developing their national regulations or formulating bilateral agreements on the collection of germplasm.

---

[1] This definition is extracted from the FAO Conference Resolution 5/89.

Article 4: Scope

4.1 The Code describes the shared responsibilities of collectors, donors, sponsors, curators and users of germplasm so as to ensure that the collection, transfer and use of plant germplasm is carried out with the maximum benefit to the international community, and with minimal adverse effects on the evolution of crop plant diversity and the environment. While initial responsibility rests with field collectors and their sponsors, obligations should extend to parties who fund or authorize collecting activities, or donate, conserve or use germplasm. The Code emphasizes the need for cooperation and a sense of reciprocity among donors, curators and users of plant genetic resources. Governments should consider taking appropriate action to facilitate and promote observance of this Code by sponsors, collectors, curators and users of germplasm operating under their jurisdiction.

4.2 The Code should enable national authorities to permit collecting activities within its territories expeditiously. It recognizes that national authorities are entitled to set specific requirements and conditions for collectors and sponsors and that sponsors and collectors are obliged to respect all relevant national laws as well as adhering to the principles of this Code.

4.3 The Code is to be implemented within the context of the FAO Global System on Plant Genetic Resources, including the International Undertaking and its annexes. In order to promote the continued availability of germplasm for plant improvement programmes on an equitable basis governments and users of germplasm should endeavour to give practical expression to the principles of farmers' rights.

Article 5: Relationship with the other legal instruments

5.1 The Code is to be implemented in harmony with:
(a) the Convention on Biological Diversity and other legal instruments protecting biological diversity or parts of it;
(b) the International Plant Protection Convention (IPPC) and other agreements restricting the spread of pests and diseases;
(c) the national laws of the host country; and
(d) any agreements between the collector, host country, sponsors and the gene bank storing the germplasm.

Chapter III
Collectors' permits

Article 6: Authority for issuing permits

6.1 States have the sovereign right, and accept the responsibility, to establish and implement national policies for the conservation and use of their plant genetic resources and, within this framework, should set up a system for the issuance of permits to collectors.

6.2 Governments should designate the authority competent for issuing permits. This authority should inform proposed collectors, sponsors and the other agencies of the government's rules and regulations in this matter, and of the approval process to be followed, and of follow-up action to be taken.

Article 7: Requesting of permits

To enable the permit issuing authority to arrive at a decision to grant or to refuse a permit, prospective collectors and sponsors should address an application to the issuing authority to which they:
(a) undertake to respect the relevant national laws;

(b) demonstrate knowledge of, and familiarity with, the species to be collected, their distribution and methods of collection;

(c) provide indicative plans for the field mission – including provisional route, estimated timing
of expedition, the types of material to be collected, species and quantities – and their plans
for evaluation, storage and use of the material collected; where possible, the sort of benefits
the host country may expect to derive from the collection of the germplasm should be
indicated;

(d) notify the host country of the kind of assistance, that may be required to facilitate the
success of the mission;

(e) indicate, if the host country so desires, plans for cooperation with national scholars,
scientists, students, non-governmental organizations and others who may assist or benefit
from participation in the field mission or its follow-up activities;

(f) list, so far as it is known, the national and foreign curators. to whom the germplasm and
information is intended to be distributed on the completion of the mission; and

(g) supply such personal information as the host country may require.

Article 8: Granting of permits

The permit issuing authority of the country in which a field mission proposes collecting plant
genetic resources should expeditiously:

(a) acknowledge the application, indicating the estimated time needed to examine it;

(b) communicate to the collectors and sponsors of the proposed collecting mission its decision.
In case of a positive decision, conditions of collaboration be established as soon as possible
before the mission arrives in the country, or begins fieldwork. If the decision is to prohibit
or restrict the mission, whenever possible, the reasons should be given and, where
appropriate, an opportunity should be given to modify the application;

(c) indicate, when applicable, what categories and quantities of germplasm may or may not
be collected or exported, and those which are required for deposit within the country;
indicate areas and species which are governed by special regulation;

(d) inform the applicant of any restrictions on travel or any modification of plans desired by
the host country;

(e) state any special arrangement or restriction placed on the distribution or use of the
germplasm, or improved materials derived from it;

(f) if it so desired, designate a national counterpart for the field mission, and/or for subsequent
collaboration;

(g) define any financial obligation to be met by the applicant including possible national
participation in the collecting team, and other services to be provided; and

(h) provide the applicant with the relevant information regarding the country, its genetic
resources policy, germplasm management system, quarantine procedures, and all relevant
laws and regulations. Particular attention should be drawn to the culture and the society
of the areas through which the collectors will be travelling.

Chapter IV
Responsibilities of Collectors

Article 9: Pre-collection

9.1 Upon arrival in the host country, collectors should acquaint themselves with all research
results, or work in progress in the country, that might have a bearing on the mission.

9.2 Before fieldwork begins, collectors and their national collaborators should discuss, and to
the extent possible, decide on practical arrangements including: (i) collecting priorities,
methodologies and strategies, (ii) information to be gathered during collection, (iii) processing

and conservation arrangements for germplasm samples, associated soil/symbiont samples, and voucher specimens, and (iv) financial arrangements for the mission.

Article 10: During collection

10.1 Collectors should respect local customs, traditions, and values, and property rights and should demonstrate a sense of gratitude towards local communities, especially if use is made of local knowledge on the characteristics and value of germplasm. Collectors should respond to their requests for information, germplasm or assistance, to the extent feasible.

10.2 In order not to increase the risk of genetic erosion, the acquisition of germplasm should not deplete the populations of the farmers' planting stocks or wild species, or remove significant genetic variation from the local gene pool.

10.3 When collecting cultivated or wild genetic resources, it is desirable that the local communities and farmers concerned be informed about the purpose of the mission, and about how and where they could request and obtain samples of the collected germplasm. If requested, duplicate samples should be also left with them.

10.4 Whenever germplasm is collected, the collector should systematically record the passport data, and describe in detail the plant population, its diversity, habitat and ecology, so as to provide curators and users of germplasm with an understanding of its original context. For this purpose, as much local knowledge as possible about the resources (including observations on environmental adaptation and local methods and technologies of preparing and using the plant) should be also documented; photographs may be of special value.

Article 11: Post-collection

11.1 Upon the completion of the field mission, collectors and their sponsors should:
 (a) process, in a timely fashion, the plant samples, and any associated microbial symbionts, pests and pathogens that may have been collected for conservation; the relevant passport data should be prepared at the same time;
 (b) deposit duplicate sets of all collections and associated materials, and records of any pertinent information, with the host country and other agreed curators;
 (c) make arrangements with quarantine officials, seed storage managers and curators to ensure that the samples are transferred as quickly as possible to conditions which optimize their viability;
 (d) obtain, in accordance with the importing countries' requirements, the phytosanitary certificate(s) and other documentation needed for transferring the material collected;
 (e) alert the host country and the FAO Commission on Plant Genetic Resources about any impending threat to plant populations, or evidence of accelerated genetic erosion, and make recommendations for remedial action; and
 (f) prepare a consolidated report on the collecting mission, including the localities visited, the confirmed identifications and passport data of plant samples collected, and the intended site(s) of conservation. Copies of the report should be submitted to the host country's permit issuing authority, to national counterparts and curators, and to the FAO for the information of its Commission on Plant Genetic Resources and for inclusion in its World Information and Early Warning System on PGR.

11.2 Collectors should take steps to promote observance of the Code by the curators and users to whom they have passed the germplasm which they have collected. Where appropriate, this might be by means of agreements with curators and users consistent with Articles 13 and 14.

Chapter V
Responsibilities of Sponsors, Curators and Users

Article 12: Responsibilities of sponsors

12.1   Sponsors should take steps to ensure, as far as is possible and appropriate, that collectors of collecting missions which they sponsor abide by the Code, particularly Articles 9, 10 and 11.

12.2   Sponsors should, as far as is possible and appropriate, establish agreements with curators of the germplasm collected under missions that they sponsor to ensure that curators abide by the Code particularly Article 13. Such agreements should, as far as is possible and appropriate, ensure that subsequent curators and users of the collected germplasm also abide by the Code.

Article 13: Responsibilities of curators

13.1   In order to be able to identify in the future the origin of the samples, curators should ensure that the collectors' original identification numbers, or codes, continue to be associated with the samples to which they refer.

13.2   Curators of the collected germplasm should take practical steps to ensure, as far as is possible and appropriate, that future enquiries from the local communities and farmers who have provided the original material, and the host country, are responded to, and the samples of the plant germplasm collected are supplied upon request.

13.3   Curators should take practical steps, *inter alia* by the use of material transfer agreements, to promote the objectives of this Code including the sharing of benefits derived from collected germplasm by the users with the local communities, farmers and host countries as indicated in Article 14.

Article 14: Responsibilities of users

Without prejudice to the concept of farmers' rights, and taking into account Articles 1.7 and 1.8, users of the germplasm, should, to benefit the local communities, farmers and the host countries, consider providing some form of compensation for the benefits derived from the use of germplasm such as:

(a) facilitating access to new, improved varieties and other products, on mutually agreed terms;

(b) support for research of relevance to conservation and utilization of plant genetic resources, including community-based, conventional and new technologies, as well as conservation strategies, for both *ex situ* and *in situ* conservation;

(c) training, at both the institutional and farmer levels, to enhance local skills in genetic resources conservation, evaluation, development, propagation and use;

(d) facilitate the transfer of appropriate technology for the conservation and use of plant genetic resources;

(e) support for programmes to evaluate and enhance local land races and other indigenous germplasm, so as to encourage the optimal use of plant genetic resources at national, subnational, and farmers and community level and to encourage conservation;

(f) any other appropriate support for farmers and communities for conservation of indigenous germplasm of the type collected by the mission; and

(g) scientific and technical information obtained from the germplasm.

Chapter VI
Reporting, Monitoring and Evaluating the Observance of the Code

Article 15: Reporting by governments

15.1 Governments should periodically inform the FAO Commission on Plant Genetic Resources of actions taken with regard to the application of this Code. When appropriate, this may be effected in the context of the yearly reports provided under Article 11 of the International Undertaking on Plant Genetic Resources.

15.2 Governments should inform the FAO Commission on Plant Genetic Resources of any decision to prohibit or restrict proposed collecting missions.

15.3 In cases of non-observance by a collector or sponsor of the rules and regulations of a host country regarding the collecting and transfer of plant genetic resources, or the principles of this Code, the government may wish to inform the FAO Commission on Plant Genetic Resources. The collector and sponsor should receive copies of this communication, and have the right to reply to the host country with copy to the FAO Commission. At the request of collectors or their sponsors, FAO may provide a certificate stating that no unresolved complaints are outstanding about them under this Code.

Article 16: Monitoring and evaluating

16.1 Appropriate national authorities and the FAO Commission on Plant Genetic Resources should periodically review the relevance and effectiveness of the Code. The Code should be considered a dynamic text that may be brought up to date as required, to take into account technical, economic, social, ethical and legal developments and constraints.

16.2 Relevant professional associations and other similar bodies accepting the principles embodied in this Code may wish to establish peer review ethics committees to consider their members' compliance with the Code.

16.3 At a suitable time, it may be desirable to develop procedures for monitoring and evaluating the observance of the principles embodied in this Code under the auspices of the FAO Commission on Plant Genetic Resources which, where invited to do so by the parties concerned, may settle differences that may arise.

An introduction to plant germplasm exploration and collecting: planning, methods and procedures, follow-up

J.M.M. Engels[1], R.K. Arora[2] and L. Guarino[3]

[1]*IPGRI, Via delle Sette Chiese 142, 00145 Rome, Italy:* [2]*IPGRI, c/o National Bureau of Plant Genetic Resources, Pusa Campus, New Delhi 110012, India:* [3]*IPGRI, c/o International Laboratory for Research on Animal Diseases, PO Box 30709, Nairobi, Kenya.*

This chapter deals mainly with the planning that plant germplasm collecting trips require, but also with some general considerations regarding their conduct and with some important follow-up activities. It acts as an introduction to the rest of this volume of *Technical Guidelines*, pointing to its various other chapters along the way, which provide more information on specific topics. This has been planned as a reference work. It cannot hope to provide all the information that will be required by collectors, but will at least point to the best sources of information. It seeks to set a common standard that can be followed by all those involved in germplasm collecting to ensure that the material they collect is viable, representative and adequately documented when it comes to be used. The volume is divided into four main parts. The first, Before Setting Out, deals in detail with the scientific and logistical planning of germplasm collecting missions and programmes. The second, In the Field, is concerned with specific collecting techniques and procedures. This is followed by Back at Base. The final part contains Case-Studies chosen to represent a range of species and types of collecting missions.

The importance of preparatory studies has been well summarized by Bunting and Kuckuck (1970):

> The ecological appreciation needed in plant explorations must be based on broad studies of the region as a whole, completed as far as possible before the practical work of exploration is undertaken. The plant explorer should not regard himself as simply a collector of potentially useful novelties or curiosities: he is a research worker consciously seeking to understand and record the bases of the adaptation to their general and specific environment of the plant forms and agricultural systems and methods which he encounters. He can do this best only if he carries out his practical work against the background of knowledge in depth of the ecological, human and agricultural characteristics of the

area. To assemble such knowledge must take labour and time spent in mastering the relevant literature and the reports of earlier investigators. Such preparatory work is essential in all research, and plant exploration is no exception.

However, planning does not stop at such technical preparation. Logistical aspects also need to be considered, for example the size and composition of the collecting team, transport needs and the various permits that will be necessary. Both technical and logistical planning are discussed in this chapter after a brief account of some of the different reasons why plant germplasm collecting may be necessary and of the various kinds of missions that may be undertaken, given different motivations, targets and constraints. Germplasm collecting tactics, logistics, preparations and procedures are also discussed by, for example, Bennett (1970), Chang et al. (1972), Harlan (1975), Hawkes (1980), Arora (1981), Chang (1985) and Astley (1991).

Reasons for collecting germplasm

There are over a quarter of a million plant species on earth. There is a case to be made, based on the precautionary principle, that all are equally legitimate targets for collecting. The demand for germplasm (ranging from individual genes to co-adapted gene complexes to entire genotypes or even populations) is unpredictable and dynamic. There is no way of telling what tomorrow's needs may be, and what plants may be able to fulfil them. The more diversity is conserved and made available for future use, the better the chances of fulfilling future demand. In practice, however, some prioritization is necessary, as to both species and geographic regions. Germplasm collecting can be expensive, and funds are usually limited. It is therefore important to have clear justification for any collecting expedition. If outside support is being sought for collecting, some kind of proposal will probably have to be prepared (see below, 'The collecting proposal'). Among other things, this will need to spell out in detail why the work is thought to be necessary.

The main reasons that can be put forward for collecting germplasm of a given gene pool in a given area are that:

- it is in danger of genetic erosion or even extinction;
- a clear need for it has been expressed by the users, at national level or internationally;
- the diversity it represents is missing from, or insufficiently represented in, existing *ex situ* germplasm collections;
- more needs to be known about it.

These are not mutually exclusive. Germplasm may be both threatened and useful, and there may be gaps both in collections of a gene pool and in what is known about it.

Important as germplasm collecting may be, it is essential to remember that it is not the end of the story. It needs to be seen as simply one facet of a conservation strategy that may also include an *in situ* component, for example. A successful collecting programme does not mean that one can stop worrying about conservation of the target gene pool.

Rescue collecting

If genetic diversity is imminently threatened in an area and *in situ* conservation methods are not feasible or insufficient, germplasm collecting may be warranted. Examples are the emergency programme that has been undertaken in the areas which are being affected by the Ataturk Dam in southeast Turkey and forage collecting on Mt Kilimanjaro at the high altitudes possibly in danger from global warming. Clearly, other things being equal, small populations (in numbers or spatial extent) and taxa or genotypes with restricted distributions will be most at risk in such circumstances, and will tend to be high priorities for rescue collecting. Chapter 4 deals with some of the methods that can be used to estimate the threat of genetic erosion and with recent initiatives in the worldwide monitoring of this threat.

Genetic erosion – the loss of genetic diversity – may come about for a number of often interrelated reasons. Chapter 4 also discusses some of the different ways these may be classified. Important agents include socioeconomic and agricultural change. The replacement of numerous, diverse traditional crop types by fewer, uniform modern varieties has become a global threat to traditional landraces of many of the major crops, especially since the 1960s. In many areas, locally important traditional food crops are being replaced by exotic staples or by cash crops. In some cases, agriculture is abandoned completely, as rural populations leave the land and drift to the cities. It is increasingly being recognized that the loss of cultural diversity that often results from socioeconomic change can be an important factor in the loss of biodiversity (e.g. McNeely, 1992).

Irreversible loss of genetic diversity can also be caused by the overexploitation of a species in the wild, for example by overgrazing in the case of forages or by uncontrolled harvesting from their natural habitat in the case of medicinal plants, firewood species, timber trees, etc. The exploitation of a particular wild species (e.g. a timber tree in tropical rain forest) can lead to the incidental removal of other, non-target species. This can extend to complete habitat destruction. Dam and road construction, the spread of cities and the cutting down of tropical rain forest for ranching are other examples of processes leading to habitat fragmentation and loss.

The occurrence of natural calamities such as disease epidemics, extended drought periods and floods can also result in the loss of genetic diversity. Often, these are exacerbated, if not actually caused, by human interference. Wholly artificial calamities that can threaten biodiversity include pollution (including the sort that many believe will result in global warming) and the introduction of alien organisms (e.g. competitors, predators, pests). Wars and civil disturbances have also led to genetic erosion, as the communities which developed and maintained landraces and managed forests and rangelands are dislocated and dispersed.

Collecting for immediate use

Local communities are continuously collecting germplasm for immediate use. This ranges from farmers exchanging planting material, as they have done for millennia, to local people collecting tree seeds from the wild for community forestry projects. However, in the formal sector, 'using germplasm' in general means incorporating it into a crop breeding or plant introduction and selection programme. Plant breeders usually maintain their own active collections consisting of carefully selected genotypes, but there is a continuous need for new, specific traits and combinations of traits in introduction, selection, domestication and improvement programmes, allowing new problems to be solved and new demands to be met. A close linkage between germplasm collecting and germplasm use has often proved to be the most effective way of guaranteeing its conservation.

Modern biotechnological tools have dramatically increased the possibility of transferring genes from one species to another. One of the immediate consequences is a widening of the pool from which a specific gene can be drawn. This means that users will increasingly be interested in species which are taxonomically and genetically less closely related to the species with which they are primarily concerned, e.g. material in the tertiary gene pool of crops as well as the primary and secondary (Chapter 6). However, there are useful species beyond the wild relatives of crops (IUCN, 1992). Such germplasm often does not need much enhancement to make an impact. Many species are used by communities and their livestock but do not reach the status of crops. Forages, multipurpose trees and many wild fruits are examples. These are increasingly becoming the target of systematic collecting. Some 25,000 plant species have been used in herbal medicines. Other species are important in land management and habitat restoration or rehabilitation, or as the 'keystones' of ecosystems. 'The reintroduction of plants will become an increasingly utilized strategy in plant conservation and protected area management' (Maunder, 1992). For a germplasm sample to be useful in a crop breeding programme it may be sufficient for it to include at least one copy of all common alleles. For other uses, in particular if germplasm samples are to be used directly as adapted populations, for example in a reintroduction programme, this may not be sufficient, and it may be required for samples to be accurately representative of allelic frequencies. As Chapter 5 discusses, this may require somewhat different sampling strategies.

One possible source of useful germplasm is existing *ex situ* collections, either within the country or abroad, which have been characterized (and possibly evaluated) or which include material from geographic areas or habitats likely to harbour desired traits. If the required material is not available in gene banks, it may be necessary to collect it. For example, if cold resistance is required in a particular crop, and landrace material or wild relatives from high-altitude or high-latitude areas is not available in existing collections, such regions would be high priorities for collecting. Clearly, if material of the target species from particular areas has already been collected, an especially strong case will need to be made for new collecting there. It is possible to make such a case if, for example, the material in gene banks is unavailable to the prospective user or available only in inadequate quantities, for example because of low viability. Many older samples were not collected with a view to reflecting variation within the original population (e.g. perhaps only a few individuals were collected at a site, or selective collections of particular phenotypes were made) and, in addition, may have undergone genetic drift and erosion during storage. Documentation may have been inadequate. Also, since sometimes germplasm is only collected from more easily accessible sites within a region (e.g. along roadsides or in markets), careful analysis will be required to judge the true ecogeographic coverage of past collecting. Repeat collecting may also be necessary if it is wanted to estimate genetic erosion in an area (see below).

Gap-filling for future use

Immediate user need is an important reason for collecting, but material not considered particularly useful today may turn out to be vital tomorrow. Agronomic problems and priorities change, ecosystems now believed to be safe may need rehabilitation in the future. Potential future use is also a legitimate justification for collecting germplasm. Within gene pools, because much genetic variation is associated with variation in environmental factors, ecological conditions which are not represented in existing collections will be accorded high collecting priorities, as well as missing genotypes and taxa.

There are, however, cases of a 'random' distribution of genes over the geographic range of a crop. Examples include cold tolerance in rice germplasm from tropical regions and the occurrence of disease resistance in germplasm from outside the geographic range of the pathogen. Collecting targeted on areas characterized by under-represented ecological conditions or of high environmental, taxonomic or genetic diversity is therefore not enough, and there should also be adequate overall coverage of the geographic distribution of gene pools. Such gap-filling (like re-collecting in inadequately covered areas) is increasingly important in the major food staples that were the targets of worldwide collecting during the early days of the plant genetic resources conservation movement.

Research

Developing a comprehensive knowledge base of the target gene pool is an important motivation for collecting. Germplasm is often needed so that particular research problems may be resolved. Examples of such problems are the mating systems of species, their taxonomic boundaries, the evolutionary relationships among taxa, and where and how cultivated forms were domesticated. Taxonomically unique or isolated, rare and narrowly endemic species often deserve a high priority in research collecting. The development of an appropriate sampling strategy will be considerably facilitated if there is adequate knowledge of the distribution of genetic diversity among and within populations, such as may emerge from preliminary research collecting (see below). Collecting for research purposes will require a somewhat different approach from conservation collecting, a topic that is explored further in Chapter 6.

Opportunistic collecting

Germplasm is sometimes collected on an opportunistic basis during a mission originally targeted on quite different species, characters or ecological conditions. Striking phenotypic features, occurrence in unusual situations and novel or interesting local uses are reasons that can lead to such collecting. Germplasm collecting may also be an incidental part of other activities, for example ethnographic studies, or botanical studies mainly concentrating on the collecting of herbarium specimens.

Types of collecting missions

Because of resource constraints, many collecting missions organized by national plant genetic resources programmes have very long and varied lists of target species. Typically, all the field crops grown in an area will be sampled, or all the forage species. This stands in marked contrast to the type of collecting generally organized by plant breeders. Such missions generally have a sharp focus on particular gene pools, species or even genotypes. Of course, these are the ends of a continuum, rather than rigid categories, and an expedition or programme will often contain elements of both strategies. Whether the target species are wild or cultivated will also fundamentally affect the character of the collecting trip, and thus the planning that will be needed. Also important is whether the trip is an intial survey/exploration or a return visit. Finally, the modalities of collecting will to some extent depend on the character of the organization(s) involved.

Multi-species vs. species-specific collecting

In a multi-species collecting mission, a region is targeted and an attempt is made to sample as much as possible of the diversity of as many species as possible. Such missions may be described as being area-driven. Usually, they are planned when no systematic collecting in the area has been conducted before and/or when the area is difficult to reach and future visits are therefore unlikely. The motives for multi-species missions have frequently more to do with conservation than immediate use, an extreme example being emergency rescue collecting in areas imminently threatened by genetic erosion.

Some problems with this kind of collecting should be noted. The most important is the relatively restricted knowledge the collectors are bound to have of many of the species they will be dealing with. This means that it will not be possible to follow an optimal sampling strategy for all the species, interesting and perhaps unique material which an expert would have recognized will be missed and the information on each sample will not be as complete as it might have been. It will therefore be all the more important to tap the large store of indigenous knowledge about plants and the environment maintained by local communities. Furthermore, it may not be possible to collect many potentially interesting species or landraces within the course of the mission because of differences in maturation time. Finally, different kinds of species may require radically different collecting techniques and even equipment. For all these reasons, a multi-species collecting mission will sometimes need to be focused at least to some extent, usually on a 'plant category'. Examples might include collecting Andean root and tuber crops in Ecuador or forages in the semiarid regions of Kenya.

Species-specific (or gene-pool-specific) missions, in contrast, tend to be driven by the eventual users of the germplasm, typically breeders and plant introduction people. The targets of such missions may be relatively broad (e.g. wild *Phaseolus* spp. in southern Mexico) or very specific indeed (e.g. the high-lysine gene in barley in the highlands of Ethiopia), in which case specialists like pathologists, entomologists and microbiologists may also be involved. Generally, species-specific missions are less complicated to plan than multi-species missions. The ecogeographic distribution of the target material will be known in more detail, including the specific habitats (or even precise localities) in the case of wild species, and the maturation period will be better documented and more restricted. In addition, the team members will have detailed knowledge of the gene pool and will probably be familiar with material collected during previous missions.

Wild species vs. crop collecting

There are great differences between collecting wild species and collecting crops. For wild species, the collecting window will be much narrower, because ripe seeds are generally quickly shed and are usually not available to the collector once this happens (though in some cases some collecting from the ground may be possible). In contrast, crop seeds usually stay on the plant. In any case, collecting can be done at times other than when a crop ready for harvest is in the field, by visiting harvested stacks, farmers' stores and markets. Timing will thus have to be more precise for wild species than for related crops, but there will also be more variation in fruiting time both among and within wild populations, so that repeat visits and longer stays in the field may be necessary. The sources of information on which to base decisions on timing will be different. Herbaria, for example, will be one of the main sources of data on wild species, but tend not to have much cultivated material.

Populations of wild species are also usually more difficult to find. More time will be spent looking for collecting sites in wild species collecting than in crop collecting. Populations may be widely scattered and in comparatively inaccessible places. Wild species collectors therefore tend to need camping equipment more than crop collectors. It may be possible to use herbarium and other records to locate potential collecting sites with precision during the planning stage, but the collector also needs to carefully research the habitat preferences of target taxa in order to start developing a 'search image' for likely collecting sites, as well as for the species itself (N. Maxted, pers. comm.). This will be perfected through actual field experience of populations of the target species. A search image is simply a set of indicators that appear to be correlated with the presence of the target species. The indicators could be physical (e.g. outcrops of a particular rock type or a particular physiographic feature) or biotic (e.g. a conspicuous associated species or the spoor of animals known to feed on the plant). For tree species in open communities, the search image also includes the shape and colour of the crown (Chapter 23); in closed communities, the remains of leaves and fruit on the ground. Though the concept of the search image is most useful in the context of wild species collecting, it also has a place in crop collecting. For example, areas where landraces of the target crop are still likely to be found may be identified by the presence of so-called 'indicator crops' of traditional indigenous agriculture (Hammer *et al.*, 1991). As the collecting mission progresses and the search image is refined, time spent searching unsuccessfully for collecting sites will probably decrease.

Within wild populations, individuals may be widely scattered, few, small and/or inconspicuous. Again, with time in the field the collector will develop a search image for the target species, but it will still generally be more time-consuming collecting a given number of seeds of a wild species than of a related crop. However, wild species are generally more outbreeding than related crops, so, other things being equal, it may be acceptable to collect fewer individuals at each site and at fewer sites. Taxonomic identification will be more difficult than with crops. Some of the general problems and some possible aids to identification are discussed in Chapter 11. On the other hand, intraspecific variation is usually much more extensive and complicated in crops, leading to a variety of often contradictory approaches to its classification (Chapter 7). Collecting herbarium specimens for confirmation of field identifications is essential in wild species collecting but rarer in crop collecting. At the ideal time for seed collecting, however, it may be almost impossible to accurately identify plants, because this often requires the examination of floral structures. This may necessitate two-stage collecting, as described below.

Different kinds of passport data descriptors are needed for crops and wild species, which usually means that at least partly different collecting forms are used (Chapter 19). Wild species collectors tend to gather more ecological information at collecting sites. Though indigenous knowledge will be important in all collecting, it is essential in crop collecting (Chapter 12). As described in Chapter 18, the participatory methodology that this requires can be time-consuming. Compared with wild species collectors, crop collectors will tend to spend more time at the collecting site, in discussion with local people, than looking for and travelling between sites.

Crop collectors have a wide range of possible sources of germplasm on which to draw. The ideal is collecting from farmers' fields, gardens and orchards at harvest time, but farm stores, markets, shops and even family seed companies (Crisp and Forde-Lloyd, 1992) are legitimate and useful sources. Though crop collectors are more likely to visit markets, these can also be useful to wild species collectors. Fruits harvested from the wild, for example, can often be found in local markets, as well as some non-germplasm products of wild species (e.g. leaves or bark of medicinal plants, wood from timber trees, etc.). In interviews, these can be used to locate populations. Markets are often good sources of information on the diversity (both interspecific and intraspecific) available in the area they serve. They are sometimes the only sources of germplasm. They can serve a large hinterland, making them ideal first collecting stops in a target area. The areas covered by different markets are also often a good basis on which to divide up a target region for stratified sampling.

There are also problems associated with collecting from markets. They often contain a biased sample of the material available at large, as farmers may grow some crops and landraces exclusively for home consumption. The material on market stalls often consists of a number of populations mixed together and some postharvest selection by the farmer may have occurred. There can certainly be genetic differences between material sampled at random from farm stores and the material that is selected by farmers from their stores for planting (Voss, 1992). Also, the material in markets may come from an area quite distant from the collecting site. It will be impossible to record many morphological details and such sampling information as the number of plants collected. Material collected in markets (and farm stores) may have low viability compared with freshly harvested material, and larger samples than normal should be collected. Ambient conditions can be used to estimate how much loss of viability is likely to have occurred in market samples since harvesting (Chapter 20).

The differences between collecting wild species and crops, in particular as regards timing, mean that it is usually difficult to collect both categories during a collecting mission. Perhaps the most common case of this is collecting the crop together with its weedy relatives and any hybridization products.

Single-visit vs. multiple-visit collecting

Most collecting projects consist of a single, fairly short visit to the target area. Shortage of resources usually militates against repeat visits. However, there are several reasons why it may be necessary for collectors to plan to visit a target area more than once.

There may be considerable variation within a region in the timing of fruiting, due to latitudinal, altitudinal and climatic differences. For example, the fruit tree *Uapaca kirkiana* fruits in June in southern Malawi but in January in the north of the country, a distance of about 600 km. The peak harvesting time of wheat and barley may differ by a month or more within relatively restricted areas in mountainous regions such as the Atlas Mountains or the Himalayas, depending on altitude. A single, relatively short, visit may result in missing later- and earlier-maturing material, within a population as well as among populations. In species with indeterminate flowering, fruits produced at different times may be the result of pollination by genetically quite different sources. If there are two or more growing seasons in the region, material specifically adapted to one may be missed, though in the case of crops it may be available from farm stores. If the visit is too early or too late for seed collecting, vegetative or *in vitro* collecting may be possible, or local people or organizations may be commissioned to collect seeds at the next opportunity. Otherwise, a return visit by the collecting team will be necessary.

Year-to-year variation will also be overlooked by single visits. Collecting the same wild population of an annual forage in a below-average rainfall year and in an above-average year may well recover significantly different variation. In some years, it may not be possible to collect at all, for example if rainfall has been particularly low or pest populations particularly high. Some species have alternate flower type in successive years. If populations are small and net reproductive output poor, it may not be possible to collect the target number of individuals in any one year without adversely affecting the demography of the population. Collecting would then have to be spread over a number of years (CPC, 1991).

As mentioned in the previous section, it is often impossible to correctly identify wild plants at the ideal time for germplasm collecting. One possible solution is to defer taxonomic determination to the stage when the collected material is grown out for multiplication or characterization. However, this may considerably delay use of the material. Anyway, in the case of the wild relatives of some root and tuber crops, it may be difficult to even find the plants when vegetative propagules are ready for collecting, as the tops will have died back. A better solution, if resources are available, is to organize the collecting in two stages. At the preliminary, exploration or reconnaissance stage, which is carried out at flowering time, target populations are located, the material identified (or, at any rate, herbarium specimens collected for later identification) and the site (or even individual plants) marked. This may also be the best time to collect microsymbionts (Chapter 26). The sites are then revisited at a suitable time for germplasm collecting. An additional advantage of a reconnaissance stage is that it will make it easier to accurately estimate the optimal time for germplasm collecting.

A preliminary survey visit could also involve collecting material for genetic diversity analysis, the results of which could then be used to formulate a more efficient germplasm sampling strategy. The material collected could be leaf fragments to be used in isozyme or deoxyribonucleic acid (DNA) variation studies or germplasm to be characterized for morphological characters (Chapter 6). Such survey work will tend to be of the sort described as coarse-grid by Bennett (1970) and Hawkes (1980). This involves collecting systematically (typically every certain number of kilometres) on a broad scale, covering all major environments. It is followed up with fine-grid collecting in specific areas of interest as revealed by analysis of the material collected.

The exploration stage could also be dedicated to the acquisition of socioeconomic, ethnographic and ethnobotanical data, in particular if adequate background information in these fields is not available from the literature. 'Gatekeepers' and key local collaborators may be identified, market days noted, harvesting time pin-pointed and a checklist of landrace names begun. All this would make a later germplasm collecting visit more efficient.

Repeat visits will also be necessary to monitor genetic erosion at a site or in a region (Chapter 39). This could be done by re-collecting and comparing levels of genetic diversity in the two samples, as measured by morphological, biochemical or molecular markers. Indigenous knowledge could also be tapped, for example by comparing the number of different landrace names recorded during the two visits (Guarino *et al.*, 1992) or by actually asking local people if they used to grow landraces in the area which have now disappeared.

Institutional (formal sector) vs. community (informal sector) collecting

Not all germplasm collecting is organized centrally and carried out by formal sector institutions such as national gene banks, agricultural research centres (national and international) and the like. There are also informal, local systems of germplasm exchange and improvement in which farmers participate on a continuous basis, as they have done since the beginning of agriculture. The formal sector has not always fully recognized the crucial role such systems have played in the development and conservation of crop genetic resources. This is changing, however, as the work of local non-governmental organizations (NGOs) (Chapter 37), community groups and other grass-roots organizations in the field becomes more widely known, and as the activities of subsistence farmers themselves are better and more fully documented.

The different approaches of the formal and informal sectors in safeguarding diversity, from collecting to use, are summarized by Mooney (1992). The key difference is that the institutional strategy has largely been one of *ex situ* germplasm storage in gene banks, while the community strategy is generally one of *in situ* conservation. Mooney (1992) describes the latter as follows:

> Cultivars are kept as part of the farming system or, where possible, in small plots for endangered cultivars and/or seed samples are cleaned, dried, stored under cool/dry conditions within the community and monitored by local people knowledgeable about the species.

This difference in emphasis is reflected in how germplasm exploration and collecting is done. Thus, the institutional strategy is described as being based on large-scale ecogeographic surveys concentrating on globally important species. Collecting missions are said to be of relatively short duration and to cover relatively large areas, so that any one community will generally be visited only once. Collectors are mainly crop-specific specialists, often from outside the country. Evaluation is carried out in laboratories and research stations. In contrast, the community strategy is described as being based on consultations with local plant-users, a long-term exercise of survey and monitoring covering locally important species within a socioecological unit. Collecting is community-based, with little if any outside involvement. It is continuous, taking place over the whole growing season, for example, rather than at a single point during it. Evaluation is based on indigenous knowledge gathered in discussion with local users, and germplasm is documented using local classification systems and languages.

These are somewhat extreme characterizations of the two approaches. Thus, it is generally recognized that collectors should take into account not just ecogeographic but also socioeconomic and cultural factors in planning their work. The two sectors are beginning to borrow methods and techniques from one another, so that the differences in their approaches are becoming less and less marked and their complementarity more obvious. People from local communities are being trained in conventional crop genetic improvement and formal sector collectors in the use of novel social science methodologies such as participatory rural appraisal. The importance of indigenous knowledge in plant genetic resources documentation is increasingly being recognized by the formal sector, and that collecting must be an active collaboration with local communities for pragmatic as much as ethical and legal reasons.

The formal sector is thus slowly realizing that collecting does not necessarily have to take place in discrete missions setting off from a central place and returning after a few weeks in the field. A national gene bank could work through locally based extension or forestry services, traditional institutions, farmers' organizations, grass-roots NGOs or schools, for example. It would need to provide training in sampling strategy, seed handling and documentation and possibly some material (collecting bags, maps, collecting forms etc.) and funds, but could then rely on a network of trained people within such organizations to do the collecting in particular target areas and forward samples for long-term storage. A national programme could even employ the media (newspapers, radio, posters, etc.) to elicit germplasm directly from growers (Chapter 39).

Before setting out: technical planning

Once the decision has been made that collecting is necessary, the technical and logistical planning can begin. The aims of technical planning are:

- to develop a sampling strategy appropriate to the target region, species and plant parts (seed, vegetative propagules, pollen, etc.);
- to ascertain the optimum timing for collecting;
- to decide what collecting equipment and techniques will be used;
- to assemble the documentation it will be necessary or useful to take to the field.

Chapters 5 and 6 discuss the elements of a basic sampling strategy and some of the reasons for its modification. Refinement of such a strategy to suit individual cases requires research. The starting-point for this research is previous collecting work. Chapter 8 deals with obtaining information on existing germplasm collections. This will include: (i) passport data, which should say exactly where and when conserved material was collected, and can be used to estimate its adaptation and phenology; and (ii) characterization and evaluation data, which in conjunction with passport data can be used to pin-point the occurrence of specific traits and to describe the pattern of distribution of variation.

Information on the environment will also be necessary. It can show where climatic or ecological conditions exist which are likely to be associated with morphological or physiological traits of interest (e.g. cold tolerance in high-altitude areas) or which have not previously been explored. Environmental information (including topography, geology, soil, climate, vegetation and floristics) will continue to be necessary once the initial decision to collect has been made. In the absence of data on the distribution of genetic variation from previous field studies, a collecting strategy for conservation purposes should be based on sampling the widest possible range of ecogeographical and agricultural conditions. Environmental data will also be needed to determine the timing of collecting, as variation in latitude and altitude within the region to be sampled will lead to variation in flowering and fruiting time. Chapter 9 deals with obtaining and using published environmental data. Such data are increasingly available in digital form, a subject that is introduced in Chapter 16.

The environment is an interaction of natural and human factors, of course, and the human dimension is no less important in this context than climate and soil. Crops and farming systems evolve and migrate in the hands of people. Local communities use and protect wild plants. Human diversity and biodiversity are thus inextricably linked. A knowledge of the socioeconomic setting (Chapter 9) of the target region and of its history and ethnography (Chapter 12) will help collectors to understand more fully what they find and collect it more efficiently. It will also help plant breeders to use it better.

Information on the environment (natural and human) of the target area must be synthesized with information on the target taxa. This will include data on distribution, phenology, genetic diversity, reproductive biology and ethnobotany. For crops, information on pests and cultural practices will also be needed. The taxonomy and autecology of wild species will need to be researched. Collectors must have as clear an idea as possible of the seed storage behaviour of their target species. Some information on these topics will be available from the documentation of existing germplasm collections, as discussed above. Other sources are:

- literature: Chapter 13 deals with bibliographic databases in agriculture. The literature on specific topics is discussed in Chapters 7 (infraspecific classification of crops), 10 (taxonomy and ecology of wild species), 12 (indigenous knowledge systems) and 17 (pests);
- factual databases: Chapters 10, 12 and 17 (subjects as above);
- herbaria: Chapter 14.

Chapter 14 also describes, together with Chapter 15, how to bring together, organize and analyse ecogeographic data. Some of these analyses will be greatly facilitated by geographic information system (GIS) technology, as discussed in Chapter 16.

The result of technical planning should be a list of specific target areas within the overall target region, each with a list of the target material (landraces, crops, wild species) likely to be found there. There should also be a master checklist of target taxa, annotated with such information as distribution, local name, etc. (Hammer, 1991). In some cases, for example rare wild species, actual named localities will have been identified. A priority should be attached to each target area, and a tentative estimate of the optimum time for germplasm collecting according to the preferred method (seed, vegetative material, pollen, etc.). It is generally preferable to err on the side of earliness, especially for wild species, but some species can only really be efficiently collected in bulk at full maturity (e.g. many dry-zone *Acacia* species).

Not all of the information gathered in the course of technical planning will need to be taken into the field. Chapters 9, 10, 11, 12 and 17 also make suggestions as to what documentation collectors should have with them when they set out. This will include maps of various kinds, botanical keys and other taxonomic identification aids, annotated checklists of target taxa, descriptions of pests and disease symptoms and such ethnographic information as annotated lists of important terms in local languages.

If the background information required for efficient germplasm collecting is not available in the sources outlined above, and resources permit, it may be necessary to obtain it first-hand in the field in a preliminary exploratory survey or reconnaissance of the target area, as already discussed.

Specialized collecting equipment may or may not be needed in the field. Most germplasm collecting is seed collecting. This is generally done by hand, either by stripping or cutting inflorescences or shaking them over some kind of container (e.g. a plastic tray or tarpaulin). The seeds are then stored in cotton or paper bags. Little equipment is required for this, beyond gardeners' gloves and secateurs, except in the case of tall trees (Chapter 23). Mechanical means have sometimes been employed, allowing large samples of seeds to be gathered in a short time, but they require that the target species be growing in pure stands in gentle topography (Young and Young, 1986). In any case, the damage that such methods can inflict on seeds means that they are not recommended for plant genetic resources conservation purposes, though they may be suitable if the material is exclusively for immediate use, for example revegetation work. The handling and storage of seeds (both orthodox and recalcitrant) during collecting trips are discussed in Chapter 20. Incidentally, environmental data will be needed in this connection too, to decide whether active drying of orthodox seeds in the field will be necessary.

However, environmental and species-specific factors may combine to dictate the collecting of other plant parts – i.e. vegetative cuttings (possibly *in vitro*), vegetative propagules, whole individuals and pollen – either in addition to, or instead of, seeds. Examples include vegetatively propagated crops and forages in overgrazed or drought-stricken areas. The relevant specialized techniques and equipment, and the reasons why they must sometimes be used, are discussed in detail as follows:

- vegetative material: Chapters 21 (roots and tubers), 22 (forages) and 23 (woody perennials);
- *in vitro* material: Chapter 24;
- pollen: Chapter 25.

The germplasm collected must be adequately documented if it is to be efficiently conserved and used. This will require the gathering of data in the field (including indigenous knowledge) and, often, ancillary reference specimens. This must be planned for, as again it may require specialized techniques and equipment. These are discussed as follows:

- specimens for isozyme and DNA analysis: Chapter 6;
- indigenous knowledge: Chapter 18;
- passport data: Chapter 19;
- pest specimens: Chapter 17;
- microsymbionts: Chapter 26;
- herbarium specimens: Chapter 27.

A number of national, regional and international gene banks have accepted responsibility for global or regional base collections of particular crops. This information is available from the Food and Agriculture

Organization (FAO) and the International Plant Genetic Resources Institute (IPGRI). Arrangements should be made for the deposit in at least two such base collections of duplicates of all the material collected. (Restrictions may be placed by the authorities of the country of origin of the collection on the further use of this material, for example through material transfer agreements (Chapter 2).) This is a crucial part of the technical planning process. In the same way, arrangements must be made for the processing and storage of ancillary specimens.

Before setting out: logistical planning

The technical planning of a collecting mission will in practice be undertaken in parallel with logistical planning, the two affecting each other. By logistical planning is meant the practical arrangements that have to be made so that the technical planning can be implemented efficiently and successfully.

These arrangements will be particularly important (and time-consuming) for collectors planning to work abroad. No germplasm collecting can take place without the knowledge, agreement and participation of the national authorities. Collecting permits (Chapter 2) and export and import permits (Chapter 17) should be obtained and plans made about the use of vehicles and the participation of local scientists and support staff. Separate documents may be necessary sanctioning internal travel, ranging from letters of introduction to local government officials to a detailed 'ordre de mission'. All this should be finalized well ahead of time (at least six months may be required), but communication will need to be continuous on both logistical and technical issues right up to the start of the trip, for example so that a provisional timing for the visit can be altered on the basis of actual weather conditions. In some cases, it may be necessary to organize missions at relatively short notice, in response to news of mast fruiting or of good rains. IPGRI maintains databases on national plant genetic resources programmes both at its headquarters and at its regional offices, and works to facilitate contacts among national programmes, for example via crop networks.

Collectors working within their own country also need to be aware that they are part of a wider national system. For example, the collecting programme of the national gene bank may make use of the experience of forestry workers, extension staff or a local NGO in the target area, as already pointed out. Local contacts are as essential in such cases as for outside collectors. Breeders and other potential users of the material collected will need to be consulted at the planning stages and possibly included in the collecting team. Similarly, the collecting activities of a breeder, university researcher or NGO will benefit from the expertise of the national gene bank, and it will certainly be a good idea for the resulting collections to be duplicated at the national gene bank.

Size and composition of the collecting team

Institutional collecting is normally carried out by small teams of three to five people in total, sometimes multidisciplinary. One person will normally be the driver, who in the interest of safety should have no other duties. Larger teams should certainly be avoided. They tend to cause undue disruption of local life. They are also more difficult to transport, manage and coordinate, and more expensive.

It is always an advantage if the prospective end-users of the target material take part in the collecting mission, and not just when the collecting is for immediate use. A rescue or gap-filling mission organized by a national gene bank will benefit from the participation of specialists in the main target species, who will usually come from a different institute. At the very least, such people should be closely involved in the planning stages. For example, Ethiopia's Plant Genetic Resources Centre (PGRC/E) and the National Bureau of Plant Genetic Resources (NBPGR) in India, which have national responsibility for germplasm collecting, regularly carry out joint missions with breeders from other national agricultural research institutes.

Chapters 18 and 38 discuss the importance of participation by social scientists experienced in the target region along with biological scientists who are experts on the target crops, and also explore the gender issue in team composition. It is essential that at least one member of the team speak each of the languages likely to be encountered, or a lingua franca. Chapter 17 argues for the inclusion of a plant pathologist in collecting teams in some circumstances. Teams collecting wild species often include a herbarium taxonomist. A microbiologist may be helpful if collecting root symbionts is an important aspect of the work (Chapter 26). A specialist may be necessary if *in vitro* collecting is to take place (Chapter 24). A collecting team should have a leader, or coordinator, in overall charge of administrative and logistical arrangements, though clearly day-to-day decisions will be made on a consensus basis. A certain amount of division of tasks (both technical and mundane, for example cooking) and responsibilities will be necessary, and this should be worked out at an early stage. Rotas should be set up for routine, especially end-of-day, tasks (see below).

The driver should be familiar with the vehicle and its accessories, and be able to carry out basic repairs if needed. Ideally, he/she should also have some first-hand knowledge of the target region, but in any case should have experience of off-road driving in the conditions likely to be encountered (e.g. sand-dunes, rocky mountain roads, etc.).

It is essential for the efficiency and eventual impact of the collecting, as much as for ethical reasons, to see the local people and communities who agree to share germplasm and information as active participants in the collecting process, in effect *ad hoc* members of the team. This will be especially important in crop collecting. Local people should in fact be involved right from the planning stage, if possible, and the inclusion of a local representative in the collecting team as guide, go-between, 'gatekeeper' and key informant may also be considered (Chapter 18).

Transport

It is increasingly only the more remote areas that still hold valuable genetic resources. Landraces, for example, may already have disappeared from the more easily accessible villages and districts. In many regions, some coarse-grid collecting along the major roads may already have been done to rescue such material. Collectors should be ready to visit areas which are accessible only by pack-animal or on foot if necessary. A recent mission to the Aïr Mountains of central Niger, for example, was done by camel caravan and one in Lesotho on horseback. Local porters and guides will be essential in such cases. In wetland areas, it will probably be necessary to employ canoes and small boats. Advice should be sought from local people, past collectors and other recent travellers through the region on the best mode of transport in particular areas at different times of the year.

However, collecting teams for the most part move around the collecting region by four-wheel-drive vehicle, occasionally pairs of vehicles if the team is large or the terrain so rough and remote that a backup is deemed necessary. The occasionally restricted availability of vehicles is often the major obstacle in organizing a collecting mission. This is an aspect of logistical planning that should be addressed and resolved very early on. Travel to the collecting region may be in the same vehicle(s), but on occasion it may be necessary to travel by public transport to a central locality in the target region, where vehicles for travel within the region could then be arranged. Collecting vehicles usually belong to one or more of the institutes collaborating in the mission, but private and hired cars, and even taxis and public transport, have also been successfully used as a last resort.

Preferably, the collecting vehicle should have a winch and roof-rack (with a waterproof cover). It should be completely covered and lockable. Tyres suitable for off-road driving will be essential. Other necessary accessories are listed in Box 3.1.

Box 3.1
Vehicle accessories

- Basic set of spare parts.
- Toolkit.
- Two spare tyres, pump and pressure gauge.
- Puncture repair kit with plentiful supply of patches.
- Heavy-duty jack and tyre levers.
- Spare petrol cans, large funnel and plastic tubing.
- Chain or nylon rope.
- Shovel and pick.

Harlan (1975) suggests that motorcycles may be suitable collecting vehicles in some terrains. They could be transported by four-wheel-drive car as far as practical, say the head of an isolated valley, and used to collect along the valley itself, and then the procedure repeated at the next valley. Pack-animals and walking would be the alternatives. Helicopters have been used to reach very isolated or inaccessible places for botanical and germplasm collecting, but, understandably, not very often. Collectors may be able to make use of regular helicopter flights. For example, a germplasm collector hitched a ride on one occasion on the weekly helicopter flight organized by the air force in Oman to supply a series of isolated mountain villages.

The itinerary

A provisional itinerary should be drawn up at an early stage of the planning process, showing the main target areas (or even precise localities) to be visited within the overall target region, as derived from the technical planning, the roads, tracks, paths or rivers to be followed in reaching each of these and the proposed timings of each visit. The mode of transport for each leg should also be specified. Quite apart from the obvious technical reasons for preparing such an itinerary, it may be necessary for purely logistical reasons. Some countries require permits for internal travel (whether by national or foreign researchers), which may have to specify dates and places. Collecting permits may also require specific areas to be mentioned. Letters of introduction to local government officials are often useful, and their preparation again will require some rough idea of the itinerary to be followed.

Maps will clearly be needed in planning the itinerary, but local contacts are essential for advice on the feasibility of following particular routes at different times of the year (Hawkes, 1980). Such information needs to be updated whenever possible and reconfirmed before finally setting out. Local contacts will also be able to comment on the likely availability of petrol, lodging, food and water along the proposed route, which may have to be altered on the basis of such information. For example, if petrol is not available along a particular stretch of the proposed itinerary beyond the capacity of the vehicle tank, then clearly either spare petrol cans would have to be procured for the mission or an alternative route or mode of transport found. If lodging is unlikely to be found, the team will have to camp out, and equipment and supplies will be necessary. If these are not available, again the route will have to be changed. The final proposed itinerary will thus be a pragmatic compromise between the ideal based on the technical planning and what will actually be feasible on the ground given the available resources. It will no doubt have to be adapted further in the field, as unforeseen problems and opportunities arise.

Sampling intensity along the proposed itinerary will vary according to the strategy being followed. Where the target species and the environment are relatively uniform, or if only the genetic variation associated with broad geographic variation in environmental factors is of interest, one can plan to collect relatively infrequently, perhaps at regular, predetermined intervals (e.g. every 10 or 20 km), on a coarse-grid basis. Where the target species and the environment are variable – and that includes the human (cultural) environment – one can plan to sample relatively more frequently. For wild species, this might be whenever a new, distinct combination of ecological conditions occurs; for crops, it might be in each market area. If research at the planning stage (or, indeed, previous collecting or a preliminary survey) has highlighted specific genotypes or specific areas of particular interest, or if the focus of the collecting is variation within a given agroecological zone, the sampling could be of a more intensive character in some areas, i.e. fine-grid. Sampling intensity can be used to estimate the likely number of samples that will be collected, a figure that can usefully be included in the collecting proposal.

In practice, the most common procedure will be a combination of coarse-grid and fine-grid sampling, with the team staying for a certain period in one locality, using this as a temporary base (with a fair degree of intensive sampling within a readily accessible distance), and then moving on to another target locality. There would be further collecting, on a systematic basis (with the interval depending on environmental heterogeneity), when travelling between the localities. Advantage could be taken of stops at such temporary bases to dry material (seed samples and herbarium specimens) and arrange for the dispatch of samples to the gene bank if necessary.

In addition to temporary bases, most missions will have some kind of home base, a place where the team can assemble and equipment and supplies be brought together. Various necessary end-of-mission activities can also be carried out here, such as packing seeds for dispatch and completing passport data forms (see below). In regions where the climate necessitates active drying of seed, a base should be found where this can be carried out. The base may be the national gene bank or its substations. If these are too far from the collecting region, a suitably equipped local institute could act as base. As a last resort, hotel rooms and private houses have successfully been pressed into service as collecting bases. Arrangements for a particular place acting as base should be finalized at the planning stage.

Duration of the collecting mission

Based on the proposed itinerary, it will be possible to estimate how long the team will be in the field. In addition, it will be necessary to schedule time at base both before going into the field and after coming back. It might be that the itinerary turns out to be too long for the time and funds likely to be available, and may have to be altered. Most institutional collecting lasts less than a month. Apart from anything else, it is recommended that this be the maximum time that elapses between collecting seeds and their receipt at the gene bank (Chapter 20).

A month is usually more than sufficient for single-species (especially single-crop) missions, in particular if they are well timed. However, if the target region is very diverse in climate and topography, the resulting variation in harvesting time may mean that a short visit will miss genetic diversity in some areas. Different crops and landraces may be harvested at widely different times even in a single village. If the mission is to be multi-crop, more than one visit to a given village may be necessary during the course of the mission, which will add to the overall duration of the trip. Longer missions are more expensive and may not be feasible if participants have other commitments and the vehicle is needed for other purposes. The team may have to plan to go back to base after a certain period in the field so that some members may drop out, others join, and seed samples may be dispatched to the gene bank. An important advantage of community-based collecting is that, as a continuous process, it can expect to cover the growing season(s) more thoroughly.

Equipment

The fact that specialized collecting equipment may be required has already been alluded to, and reference made to the relevant individual chapters which describe it. However, general travelling equipment may also be necessary during a collecting mission in addition to basic vehicle accessories. This will be particularly true if the mission involves camping out in remote areas, which is perhaps more common in wild species collecting, though not unknown in crop collecting. Whether camping will be necessary should emerge from careful consideration of the proposed itinerary in consultation with people knowledgeable about the target region.

Rural communities are usually extremely hospitable, but lodging and feeding a collecting team, even for a single night, may constitute a significant drain on local resources, and collectors should try to be as little of a burden, and as little disruptive of local life, as possible. On the other hand, refusing hospitality may be considered offensive in some societies. Offering to pay may be equally out of the question, though a contribution in kind to a meal, or a parting gift, may be acceptable. Collectors will need to be extremely sensitive to local customs in this as in other aspects of their relationship with rural communities. In any case, they should never count on hospitality, but be prepared to cope on their own at all times as regards food, water and a place to sleep.

An important part of logistical planning is making sure any necessary equipment, and any food supplies that might also be needed, will be available at the right place (probably the home base) in good time. General travelling equipment and supplies may be divided into camping and medical. Bennett (1970) and Hawkes (1980) give comprehensive lists of both categories. Hatt (1982) is a good general guide for travellers and also gives useful checklists of medical and personal supplies and equipment. Box 3.2 gives a brief summary of camping needs. Exactly how

---

Box 3.2
Camping equipment

- Tents and accessories.
- Tarpaulin and ropes.
- Waterproof sleeping-bags.
- Mosquito netting.
- Camp-beds or air mattresses.
- Small folding table and chairs.
- Battery-operated hand torches and spare batteries.
- Cooking stove (butane, solid-fuel, etc.) and spare fuel.
- Matches.
- Cooking pots and utensils, plates, mugs, cutlery, etc.
- Candles and/or lamp (battery or gas).
- Water containers (both large cans and individual bottles).

---

much of this equipment will actually be taken into the field will depend very much on the mode of transport and on the climate. If space is at a premium in the collecting vehicle, as might be the case if vegetative material is being collected, or much of the travelling is to be on foot, then clearly a folding table and chairs are likely to be a low priority. In some climates, a tent may well be unnecessary. On the other hand, torches and water-bottles, for example, are likely to be present among the equipment of all collecting missions, whether camping is planned or not.

Box 3.3 lists some medical supplies likely to be useful on collecting expeditions. This should not be treated as anything more than a general guide. Ready-made first-aid kits (and more comprehensive medical kits) are available on the market in many countries. These come with detailed instruction for the use of individual components, and are thus a better option for collectors than simply putting together a kit themselves *ad hoc*. Many institutes provide basic medical kits in their field vehicles. WHO (1993) provides detailed health advice for international travellers.

The collecting proposal

To put the planning into practice, funding will be needed. As has already been mentioned, whether this comes from within the collecting institute(s) or from some outside source, some kind of written proposal will probably be necessary. The collecting proposal should address the issues of what, where, why, when, how and who by including the following results of the technical and logistical planning process:

- a justification of the collecting by area and species, including evidence of genetic erosion, user need, gaps in existing collections and/or gaps in knowledge;

Box 3.3
Medical supplies

- Water-purifying tablets.
- Insect-repellent cream.
- Antihistamine cream.
- Antiseptic cream or wipes.
- Antibiotic tablets and cream.
- Fungal infection remedies.
- Antacid tablets.
- Antidiarrhoeal tablets.
- Sachets of oral rehydration solution.
- Eyewash.
- Oil of cloves for toothache.
- Lipsalve.
- Aspirin, paracetamol or other pain-killer.
- Antimalarial tablets for both prophylaxis and treatment.
- Snakebite sera.
- Disposable hypodermic syringes.
- Cotton wool.
- Splints.
- Bandages and plasters.
- Scissors.

- a prioritized list of target species or plant categories, each with a note of what plant part(s) will be collected, of any reference and ancillary specimens that will be collected, of any specialized techniques that will be used and of the proposed protocol for the division and distribution (including to base collections) of germplasm and ancillary specimens;
- a sketch map showing the location of the proposed target region and of specific target areas (and actual localities) within the region, with priorities based on an explicitly stated sampling strategy;
- an itinerary, including tentative dates, timings and proposed mode of travel;
- a description of proposed follow-up activities, e.g. characterization, evaluation and/or use of the material;
- a list of the people and organizations involved, and their respective roles;
- a budget, which will depend on such considerations as the size of the collecting team, the length of time to be spent in the field, the distance to be travelled, the mode of travel and the equipment that will be necessary.

To what outside agencies can a collecting proposal be submitted? The collecting activities of the national plant genetic resources systems of developing countries, whether in the formal or informal sector, are occasionally supported by bilateral and multilateral development agencies. National programmes the world over also collaborate with each other. Such collaboration may be on a bilateral basis, in crop networks or under the aegis of a regional structure. An example of bilateral collaboration is the Gatersleben gene bank's collecting in Cuba and elsewhere (Chapter 39). An example of a regional organization which supports collecting is the Southern Africa Development Community (SADC) Regional Plant Genetic Resources Centre in Lusaka. International agricultural research centres (IARCs), both inside and outside the Consultative Group on International Agricultural Research (CGIAR), are also possibilities. IPGRI can advise prospective germplasm collectors on the possibilities for funding and collaboration.

In the field

It is important, once in the field, to establish a daily routine to find collecting sites and to collect at the site. The concept of the search image and its importance in locating potential collecting sites, especially of wild species, has already been discussed. It has also already been stressed that an integral part of the collecting routine, both as regards locating potential collecting sites and at the site itself, will be consultations with local people, in particular when collecting crops. However, local people will also be vital in the more practical aspects of collecting: it is good practice to take every opportunity to ask about the situation ahead (e.g. where the road is going, whether it is practicable, whether petrol, water or accommodation is available in the next village, etc.), in order to confirm and supplement previously obtained information.

Once a collecting site has been identified, it must be described on a collecting form and the population sampled and also described. Some definitions are in order here. The formal definition of the *population* is those conspecific individuals among which gene flow normally takes place, i.e. an interbreeding group of plants. The spatial extent of the population will depend on the pollination and dispersal systems of the target species and requires experimental study for its estimation (Chapter 6). Therefore, for practical purposes the population is usually defined by collectors not in strict genetic terms but ecologically, as those conspecific individuals found within a restricted area under relatively homogeneous ecological conditions. How restricted? Chapter 5 presents guidelines for the minimum number of individuals that should be collected from each population. Pragmatically, this, together with guidelines on the minimum distance that should be kept between sampling points to avoid excessive sampling of clones and closely related individuals, will largely determine the minimum size of the area to be sampled. This ecologically uniform and distinct area is the *collecting site*, and the collecting form aims to document the conditions which characterize and distinguish it. In crop collecting, the site is usually the farmer's field, garden plot or orchard, or perhaps the whole village in the case of some fruit trees. In wild species collecting, it will not be so easy to demarcate. In general, however, if two adjacent areas differ in some feature of the physical, chemical or biotic environment, they should be taken as being separate collecting sites.

The *sample* is material chosen from the population (in whatever manner) for *ex situ* conservation. There may be more than one sample from a given population (separate random and selective samples, for example) and more than one species sampled at a site. Strictly speaking, for any given species, by definition there should be only one population sampled per site. Thus, distinct samples of a particular species from different microenvironments within a location, sampled according to a stratified procedure, should be treated as coming from separate populations inhabiting distinct sites, so that the nature of the microenvironmental differences can be recorded on the collecting form. In practice, to speed things up in the field, such samples are often said to come from the same 'site', but distinct *subsites* or *microsites*. This concept may also be useful in crop collecting, for example when landraces are collected from different farmers' or households' fields or plots in the same village.

Chapter 5 discusses the issues of how to select individual plants for inclusion in the sample, of the number of such plants and of the number of seeds to be collected per individual. A baseline criterion for conservation collecting is to gather a bulk sample of equal numbers of seeds from some 50 individuals, sampling at intervals along a number of transects across the collecting site. Modifications of this basic strategy to suit particular species and purposes are discussed in Chapters 5 and 6.

Reference has already been made in the context of technical planning to individual chapters dealing with specialized collecting techniques and data documentation. There may be people on the team specifically included to deal with such specialized tasks, for example a microbiologist or a herbarium taxonomist. In any case, each team member's duties at the collecting site should be clearly set out. In particular, the jobs of documenting the site and population and of actually gathering the germplasm can to some extent be separated. Chapter 22 gives an example of how this might work in practice. Though the context is one of collecting vegetative material, the principles would be the same when collecting seed. First, the limits of the collecting site must be decided on, and how exactly sampling should take place. Then, while one or two team members collect germplasm, perhaps transecting the site in different directions, another can be starting to fill in site information on the collecting form, before joining the others in gathering germplasm or, perhaps, associated samples such as herbarium material, soil samples, etc. Population information is filled in later by all collectors together, when collecting at the site is finished, along with sample information. The amount of time spent documenting site and population will vary, but certain passport data descriptors must always be recorded. A minimum set is proposed in Chapter 19. Herbarium specimens should always be prepared when collecting wild species. Samples are then labelled and packaged as necessary. Chapter 18 shows how participation by the local community in the day-to-day conduct of a germplasm collecting expedition might be organized.

The state of germplasm and other samples should be checked throughout the course of the mission as part of the daily routine. Time should be set aside for such activities at the end of each day. The most important tasks will be the following (there are details in the appropriate chapters):

- Seed and vegetative (including *in vitro*) samples should be checked on a regular basis for insect and fungal attack, and infected samples either treated or discarded.
- If seed samples (or other material) are being actively dried, the silica gel may have to be changed and dry samples removed.
- Samples of fleshy fruits in plastic bags will need to be aerated regularly and checked for rotting.
- Drying papers in herbarium presses must be changed every couple of days and dry specimens removed and packaged together in bundles.
- Samples may need to be sent back to the gene bank at various stages during the course of the mission, rather than waiting until the end.

There will also be various documentation activities to be completed, including the daily diary entry. A few descriptors on collecting forms can also be left until the end of the day (or the end of the mission), for example latitude and longitude and other data which can be read off a map on which the location of the site was previously marked. The end of the day is also a suitable time to check that all of the samples collected that day are accounted for and that the collecting numbers accompanying them tally with those noted on collecting forms and field notebook.

It is good practice to prepare a checklist of end-of-day activities and a roster assigning such tasks to team members on a cyclical basis. It will often be a good idea to hold team meetings at the end of each day to summarize and discuss findings and plan the next day's work.

Not everything can be planned for, however. Collectors can decide on a broad strategy beforehand, but will need to be flexible in their tactics in the field both to overcome unexpected difficulties and to take advantage of unexpected opportunities. Many of the difficulties will be logistical: unpassable roads, rivers in flood, vehicle breakdown and the like. Some will be technical. The visit may be taking place too early or too late in the season, for example. In crop collecting, this may require visiting more markets and collecting from farmers' stores. In wild species collecting, the decision may be made to turn the germplasm collecting mission into a herbarium survey and come back later for germplasm, or to change the itinerary to cover areas where the timing might be more suitable. If target populations are rarer than originally thought, more individuals than planned might be collected at each site. If populations turn out to be small, and it is not possible to collect the required number of individuals in each, it might be decided to collect more populations than planned to compensate. If the target species turns out to be heavily grazed or not in seed, it might be necessary to improvise a vegetative collecting method.

As for the opportunities, collectors should always be on the look-out for interesting and unusual germplasm: a population of a species (even one that is not a priority target) in an atypical ecological setting, a crop being grown or used in a peculiar way, or the odd healthy spike in an infested field. The basic strategy is to collect from the largest possible number of different and distinct local environments. These can be defined to some extent at the planning stage, but one must watch out for them in the field too. If a track presents itself that is not marked on the map, collectors must be prepared to temporarily abandon the original itinerary if this is thought likely to result in obtaining more, and more interesting, diversity of the target taxa.

Back at base

Once back at home base, on completion of the fieldwork, collectors will still need to carry out some important tasks before the mission can be deemed successfully completed. These tasks, described in detail in Chapter 28, are:

- sorting and preparing germplasm samples and any reference and ancillary specimens;
- collating, completing and editing the collecting data;
- distributing the germplasm samples, reference and ancillary specimens and collecting data.

These back-at-base tasks will require a certain amount of planning. In the first place, sufficient time and resources should be allocated to them. Sorting samples requires space. So does ambient drying of seeds and herbarium specimens. Active drying of seeds before dispatch requires either specialized equipment or chemical supplies. A computer will be necessary for the documentation of passport data. All these should be available at the base on completion of the mission. Some equipment and supplies may have to be brought in from elsewhere by the collectors themselves.

Arrangements should also be made for the distribution of germplasm and other samples. The agreement of institutes to accept germplasm for storage and follow-up work, recalcitrant seeds for immediate planting, *in vitro* material for processing, *Rhizobium* nodules for isolation, herbarium vouchers and pest specimens for identification, soil samples for analysis, etc. must be obtained before the mission actually starts. Recipients must then be warned of the imminent arrival of material once it has been sent. Dispatch of biological specimens will require phytosanitary certificates, which will need to be obtained once back at base.

The final task in collecting is reporting on the mission. This is crucial if potential users are to be kept up to date on the availability of germplasm in *ex situ* storage. What a mission report should contain and where it can be published are discussed in Chapter 29.

Careless collecting: two points to watch

Collecting may cause more problems than it solves. In particular, there are two dangers associated with careless collecting: the damage of fragile populations and the introduction of pests and competitors.

Excessive sampling from a small population may endanger its chances of survival. Given the likely investment of time, effort and resources, collectors will understandably be keen not to return empty-handed or with insufficient material, but a conservation ethic must be kept in mind at all times. The collector must be careful not to collect so much material from a population as to adversely affect its demography. A potential collecting site should be passed over if it is deemed impossible to collect germplasm from it without damaging the target population or the habitat. This point is clearly made in FAO's Code of Conduct for Plant Germplasm Collecting and Transfer (Chapter 2).

Also, germplasm samples may be diseased, infested with insects or contaminated with weeds. Care must be taken to prevent the movement of such pests around the world (Chapter 17). The species being collected may itself be actually or potentially noxious. Introduced species are a major cause of biodiversity loss and will probably be the biggest cause of species loss in the future (W.A. Strahm, pers. comm.). Precautions need to be taken when collecting and introducing species that could become naturalized and compete with the native flora or introgress with local species.

The case-studies

What has been presented in this chapter is perhaps an ideal model. However, rarely will it be possible to organize and run a collecting project or programme exactly as one would have wished, or indeed as was planned. Actual examples are therefore in many ways more informative than any ideal model. For this reason, the final section of this volume presents a number of case-studies chosen to reflect a wide range of species (and therefore collecting problems), geographic areas and types of collecting programmes. This is not to say that the range is exhaustively covered. However, the ten examples included here are illustrative of the variety of ways that collecting may be organized and carried out and describe how highly specific problems were overcome, which it was not possible to address in great detail in the more general chapters of this volume. The types of collecting described in the case studies are as follows:

Chapter 30   Worldwide collecting of a wide range of wild species for evaluation and selection by national and international programmes.

Chapter 31   Ecogeographic study of a tree genus with recalcitrant seeds in its centre of diversity in a collaborative project involving various national institutes and international conservation organizations.

Chapter 32   Collecting vegetatively propagated root and tuber crops of local importance by a national programme.

Chapter 33   Collecting vegetative material of a crop and its wild relatives in its centre of diversity by an IARC in collaboration with a national programme, with later *in vitro* culturing in a neighbouring country for disease screening and safe transfer.

Chapter 34   Worldwide collecting of a major cereal staple and its wild relatives over many years by an IARC in collaboration with national programmes.

Chapter 35   Collaborative collecting of the wild species within the genus of a crop, all geocarpic, by a number of national programmes within the geographic region of the distribution of the genus, with support from other interested national programmes and IARCs.

Chapter 36   Collecting rare and endangered local wild species by a botanic garden.

Chapter 37   Collecting socioeconomically important forestry species by an NGO.

Chapter 38   Collecting a staple root crop and the indigenous knowledge (IK) associated with it by an interdisciplinary, international team.

Chapter 39   A national institute's collaboration with several other national programmes on a wide range of crop gene pools.

Conclusion: the ingredients of success

In summary, the success of germplasm collecting programmes may be seen to depend on keeping in mind the following basic points.

Plan well ahead

The importance of thorough research and planning in the success of field collecting cannot be overemphasized. Information on the distribution of the target species (and of genetic variation within the species), on breeding system, on fruiting time (and how this varies geographically), on seed storage characteristics and on appropriate collecting techniques will all be crucial to the success of collecting. Also important will be background data on the physical, biotic and human environment of the target area. As a general rule, at least as much time should be allocated to planning and preparation as to field collecting. Usually, planning will take considerably longer.

Involve local people

Active local participation is an ethical (and, increasingly, a legal) imperative, but the efficiency of collecting will also depend on it, in particular crop collecting. Deciding on the timing of collecting, locating target populations, developing a suitable sampling strategy and documenting the collection will all benefit from local participation.

Be prepared to be flexible

Good timing of collecting is essential for wild species, especially those that shed their seeds over a short time span. A profile of seed maturation and how it is affected by location (latitude and elevation) and climate should be developed for each target species. This may not be enough, however. Local observers can be engaged to keep a watch on weather conditions and the developing fruiting season. The collecting team should be ready to go into the field at short notice if local sources report that the fruit is shedding early (e.g. due to a hot, windy spell) or is being destroyed by seed- or fruit-eating animals. It should be flexible enough to make last-minute changes in the itinerary to take account of new information while in the field.

For woody species, collecting in mast years is an efficient use of resources. Considerably more seeds can be collected for a given input of time and resources. The likelihood is also increased that the seeds collected are of the best genetic quality (due to a reduced proportion of selfed seed) and of high viability (due to a higher proportion of fertilized ovules and lower incidence of insect attack). Again, there should be enough flexibility in a collecting programme to allow missions to be undertaken at relatively short notice in a year of mast seed production or postponed in poor seed years.

Develop a search image

The speedy location of populations of target species is essential to the success of collecting expeditions. A considerable amount of time can be wasted searching for populations if research has been inadequate and a rudimentary search image has not been developed. It is always an advantage if collecting teams include an experienced person capable of rapidly and accurately locating target species and ascertaining whether they are carrying a useful seed crop (i.e. one in which the desired quantity of viable seeds can be efficiently collected within a reasonable period).

Choose collecting and processing techniques with care

In any given situation, the collector must be able to quickly determine the most efficient harvesting technique for each species and processing technique for each sample. With woody species, it may be necessary to employ two or three different harvesting techniques for a population. Seed extraction procedures may vary for a species depending on the maturity of the fruit. When dealing with species that are not well known, the ability to develop and improvise new collecting and processing techniques may be crucial to the success of the mission.

Document the collection scrupulously

Germplasm without passport data (both indigenous knowledge and the results of scientific measurement) is less useful than it could be. Collecting forms should be filled in conscientiously, at the collecting site. In all crop collecting and much wild species collecting, this will require consultations with local people. All the descriptors in the recommended minimum set should be recorded, and any others as time allows and circumstances require. Samples should be numbered legibly and unambiguously.

Take trouble with samples

The material collected should be of the highest possible quality (i.e. viability), and it is crucial to keep it that way. The key to preventing deterioration of samples is frequent inspection. At the first sign of deterioration, appropriate remedial action should be taken. If fruit samples begin sweating or show signs of mould, they should quickly be opened up and dried out and affected material discarded if necessary. It is imperative that samples of clean, dry seeds be stowed in a manner that prevents wetting. In all cases, it is best to get samples back to the gene bank as rapidly as possible, and in any case within a month.

Think about safety

Due attention to safety is vital for continuing, successful collecting programmes. This may require that vehicles travel in pairs on certain routes, and that some areas be avoided altogether. When necessary, collectors must be provided with, and make use of, safety equipment and clothing. Particular care needs to be taken in the operation of vehicles, secateurs and rifles, and in tree climbing.

Follow up

A successful collecting programme does not end once the team is back from the field. The germplasm (and associated data) must be properly stored in duplicate base collections, voucher specimens must be deposited in herbaria and identified as necessary, and so on. Collectors themselves often follow up their fieldwork with characterization, evaluation and other experimental work on the germplasm they have collected. A report must also be written, and circulated widely, so that interested scientists worldwide may be informed of the availability of potentially useful material.

Follow the Code of Conduct for Plant Germplasm Collecting and Transfer

Acknowledgements

The section on the ingredients of success was for the most part written by Lex Thomson.

References

Arora, R.K. (1981) Plant genetic resources explorations and collection, planning and logistics. In: Mehra, K.L., R.K. Arora and S.R. Wadhi (eds) *Plant Exploration and Collection.* pp. 46–54. National Bureau of Plant Genetic Resources Science Monograph 3. NBPGR, New Delhi.

Astley, D. (1991) Exploration: methods and problems of exploration and field collecting. *Biological Journal of the Linnean Society* 43:11–22.

Bennett, E. (1970) Tactics of plant exploration. In: Frankel, O.H. and E. Bennett (eds) *Genetic Resources in Plants – Their Exploration and Conservation.* pp. 157–179. Blackwell Scientific Publications, Oxford.

Bunting, A.G. and H. Kuckuck (1970) Ecological and economic studies related to plant exploration. In: Frankel, O.H. and E. Bennett (eds) *Genetic Resources in Plants – Their Exploration and Conservation.* pp. 181–188. Blackwell Scientific Publications, Oxford.

Chang, T.T. (1985) Collection of crop germplasm. *Iowa Sate Journal of Research* 54:349–364.

Chang, T.T., S.D. Sharma, C. Roy Adair and A.T. Perez (1972) *Manual for Field Collectors of Rice.* IRRI, Los Baños.

CPC (1991) Genetic sampling guidelines for conservation collections of endangered plants. In: Falk, D.A. and K.E. Holsinger (eds) *Genetics and Conservation of Rare Plants.* pp. 209–238. Oxford University Press, New York.

Crisp, P. and B. Forde-Lloyd (1992) A different approach to vegetable germplasm collection. *FAO/IBPGR Plant Genetic Resources Newsletter* 48:11–12.

Guarino, L., H. Chadja and A. Mokkadem (1992) Wheat collecting in southern Algeria. *Rachis* 10:23–25.

Hammer, K. (1991) Checklists and germplasm collecting. *FAO/IBPGR Plant Genetic Resources Newsletter* 85:15–17.

Hammer, K., G. Laghetti, S. Cifarelli and P. Perrino (1991) Collecting in northeastern Italy using the indicator-crop method. *FAO/IBPGR Plant Genetic Resources Newsletter* 86:39–40.

Harlan, J.R. (1975) Practical problems in exploration. Seed crops. In: Frankel, O.H. and J.G. Hawkes (eds) *Crop Genetic Resources for Today and Tomorrow.* pp. 111–115. Cambridge University Press, Cambridge.

Hatt, J. (1982) *The Tropical Traveller.* Pan Books, London.

Hawkes, J.G. (1980) *Crop Genetic Resources Field Collection Manual.* IBPGR and EUCARPIA. Rome.

IUCN (1992) *Species and Global Change. A Report of a Workshop to Review Critical*

Changes Affecting the Redistribution of Plant Species and Populations and Their Establishment in New Environments as a Consequence of Global Change. IUCN, Gland.

McNeely, J. (1992) Nature and culture: conservation needs both. *Nature and Resources* 28:37–43.

Maunder, M. (1992) Plant reintroduction: an overview. *Biodiversity and Conservation* 1:51–61.

Mooney, P.R. (1992) Towards a folk revolution. In: Cooper, D., R. Vellvé and H. Hobbelink (eds) *Growing Diversity.* pp. 125–138. Intermediate Technology Publications, London.

Voss, J. (1992) Conserving and increasing on-farm genetic diversity: farmer management of varietal bean mixtures in central Africa. In: Mook, J.L. and R.E. Rhoades (eds) *Diversity, Farmer Knowledge and Sustainability.* Cornell University Press, Ithaca.

WHO (1993) *International Travel and Health.* World Health Organization, Geneva.

Young, J.A. and C.G. Young (1986) *Collecting, Processing and Germinating Seeds of Wildland Plants.* Timber Press. Portland.

I

BEFORE SETTING OUT

Assessing the threat of genetic erosion

4

L. Guarino

*IPGRI, c/o International Laboratory for Research on Animal
Diseases, PO Box 30709, Nairobi, Kenya.*

The causes of genetic erosion

Agenda 21 states that 'the current decline in biodiversity is largely the result of human activity and represents a serious threat to human development'. But exactly what kinds of human activity are to blame, and what are the other factors involved? Collectors need to answer such questions so that they can target for priority collecting regions and species that are particularly at risk.

Many attempts have been made to list the threats faced by plant diversity, both wild and cultivated. WRI *et al.* (1992a) provide a detailed analysis of the indirect or underlying causes of biodiversity loss. WCMC (1992) quotes the following factors as currently endangering biodiversity:

- habitat loss or modification, often associated with habitat fragmentation
- over-exploitation for commercial or subsistence reasons
- introduction of exotic species which may compete with, prey on or hybridize with native species
- disturbance and uprooting
- incidental take
- disease
- limited distribution.

Muchiru (1985) lists essentially the same agents – habitat loss, overexploitation, introduced species and indirect effects – but includes agricultural development as a separate factor. Of course, habitat loss or disturbance may in turn be due to a variety of causes, and agricultural development can take many forms. Such lists can therefore be made quite detailed. Gomez-Campo *et al.* (1992), for example, present a very

67

comprehensive checklist of specific factors potentially affecting the persistence and genetic diversity of individual populations of wild plants (Box 4.1). Dahl and Nabhan (1992) discuss the threats endangering the genetic diversity of cultivated plants, from global environmental change and international economic pressures to crop-specific problems. They provide a list of the threats perceived by grass-roots organizations, arranged in decreasing order of importance, and suggest that this 'can be used as an evaluation tool for any local community wishing to impede genetic erosion' (Box 4.2). Clearly, it can also be used to assess the danger of such erosion taking place.

Looking at the problem from the other side, Brush (1993) lists four factors which are important in preserving crop diversity, i.e. in limiting the rate of genetic erosion: (i) fragmentation of farm holdings, allowing farmers to maintain landraces in at least one field; (ii) increasing cultivation of marginal land, where landraces tend to have an advantage over modern varieties; (iii) economic isolation, creating market distortions which give landraces a competitive advantage; and (iv) cultural values and preferences for diversity. The contention is that in many cases adoption of modern varieties does not result in the complete replacement of landraces, but reaches an asymptote.

Box 4.1

- Drainage works that destroy humid habitats and lower the water-table in adjacent areas.
- Dam building and the resultant flooding.
- Clearing of land for agriculture.
- Change in agricultural techniques, particularly increasing use of chemicals and heavy machinery.
- Forestry plantation.
- Decrease in pollinator populations due to increasing insecticide use.
- Overgrazing by domestic livestock or wild herbivores.
- Scrub regeneration as a result of lessened grazing pressure on pasture.
- Increased or decreased frequency of forest fires.
- Water pollution.
- Air pollution.
- Contamination of the soil.
- Industrialization and urbanization.
- Tourism and touristic development.
- Road construction.
- Mining and quarrying.
- Intensification of traditional exploitation.
- Horticultural collecting.
- Competition with introduced plants.
- Genetic contamination by hybridization with other species.
- Introduced pests and diseases.
- Small population size.

Box 4.2

- Introduction of modern varieties and exotic crops.
- Loss of seed-saving and vegetative propagation skills.
- Acculturation of traditional caretakers (or their death).
- Change in economic base.
- Land conversion to industrial agriculture.
- Destruction (urbanization) of habitat and farmland.
- Herbicide and pesticide impact.
- Environmental contamination.
- Introduction of exotic pests.
- Loss of seeds to pests.
- Net reduction in the number of farmers.
- Inadvertent crossing of varieties.

One of the most comprehensive attempts to catalogue the threats to biodiversity is presented by UNEP (1993). Article 7 of the *Convention on Biological Diversity* enjoins countries to identify and monitor 'the components of biodiversity important for its conservation and sustainable use' and the processes and activities which threaten them. Country studies on biodiversity are being prepared by many countries to meet this objective. International coordination for their preparation is being provided by the United Nations Environment Programme (UNEP). UNEP (1993) presents a set of guidelines for country studies. It includes a section on defining the threats to biodiversity. Threats are classified into four main generic categories, as follows:

- external socioeconomic factors
- direct human threats: local impact
- direct human threats: regional/global impact
- natural hazards

Under each rubric are listed a number of specific factors, as well as the kinds of information that are required in order to be able to determine an appropriate response.

Measuring the risk of genetic erosion

Each factor on the kind of checklist discussed in the previous section could be scored as present or absent for any given area, wild population or local community, giving an assessment of overall risk of genetic erosion. At a more sophisticated level, each risk could be scored as to temporal and spatial remoteness from the site, area or population under consideration, duration, severity, reversibility, to what extent action has already been taken on the species in other areas, quality of information available, etc. (IBPGR, 1986; UNEP, 1993).

Goodrich (1987) takes such a quantitative approach in developing a model that can be used to estimate the threat of genetic erosion that a particular taxon (wild or cultivated) faces in a defined area. The model is based on scoring a variety of factors – biological, environmental and socioeconomic – and summing the factor scores to give a total which increases in magnitude with increasing threat of genetic erosion. It can be used to compare the threat of genetic erosion that a given taxon is facing in different equivalent areas, or the relative threat to different taxa in an area. Comparisons should be made using only those parameters for which data are available for all areas or taxa being compared. A somewhat modified version of the model is presented at the end of this chapter in Appendix 4.1.

For this and similar models to be used, a substantial amount of information will need to be gathered. Sources will include agriculture, forestry and environment departments and ministries, the local representatives of bilateral and multilateral development agencies, national and international conservation bodies, seed companies, etc. Some information will be available in formal published form. For example, there is a catalogue of seed production projects in the African, Caribbean and Pacific (ACP) countries (Delhove, 1992). Many countries have carried out environmental assessments of various kinds, and the International Environmental and Natural Resource Assessment Information Service (INTERAISE), established in 1991, has published a useful annotated bibliography of such environmental country profiles in the *1993 Directory of Country Environmental Studies* (WRI *et al.*, 1992b). Some sources of relevant information on environmental change are discussed in Chapter 9. Sources of information on conservation activities are discussed in Chapter 10. The international agricultural research centres (IARCs) will also have relevant data, for example on the release and spread of modern crop varieties (e.g. Dalrymple, 1986a, b).

Data from formal sources such as national agricultural surveys and impact assessment studies are only part of the picture, however, and the role of grass-roots organizations such as local non-governmental organizations (NGOs) is particularly important in this context (e.g. Dahl and Nabhan, 1992). As Muchiru (1985) points out, 'NGOs are composed of a wide network of people ... [and] ... are therefore well placed to monitor development projects that may have negative impact on the environment.' Some data will have to be collected first-hand in the field, through interviews with farmers and direct observation, either during the course of collecting or in preliminary surveys. Repeat collecting visits to given areas some years apart are invaluable sources of information on genetic erosion (Chapter 39).

The erosion of biodiversity: global monitoring systems

The data on specific areas, populations and species coming out of the kinds of studies outlined above can be integrated at the national level to estimate the danger to a country's biodiversity as a whole. UNEP (1993) presents a very comprehensive list of key parameters for monitoring biodiversity at the country level. It also suggests, however, that a useful alternative is to monitor a much more restricted number of parameters, a so-called minimum set of indicators of change. It goes on to support the recommendation of the *Global Biodiversity Strategy* (WRI *et al.*, 1992a) that an early-warning network of national centres be set up to monitor potential threats to biodiversity, including crop and livestock diversity, listing the parameters that such a network would need to monitor. The Food and Agriculture Organization (FAO) Commission on Plant Genetic Resources has similarly suggested that an early-warning system be set up for plant genetic resources, to identify gaps and emergency situations.

References

Brush, S.B. (1993) *In situ* conservation of landraces in centres of crops diversity. Paper delivered at the 'Symposium on Global Implications of Germplasm Conservation and Utilization'. 85th Annual Meeting of the American Society of Agronomy. 8 November 1993. Cincinnati, Ohio.

Dahl, K. and G.P. Nabhan (1992) *Conservation of Plant Genetic Resources. Grassroots Efforts in North America.* ACTS Press, Nairobi.

Dalrymple, D.G. (1986a) *Development and Spread of High-Yielding Wheat Varieties in Developing Countries.* USAID, Washington DC.

Dalrymple, D.G. (1986b) *Development and Spread of High-Yielding Rice Varieties in Developing Countries.* USAID, Washington DC.

Delhove, G.E. (1992) *Seed Programmes and Projects in ACP Countries.* CTA, Wageningen.

Gomez-Campo, C. and collaborators (1992) *Libro Rojo de Especies Vegetales Amenazadas de España Peninsular e Islas Balneares.* Ministerio de Agricultura y Alimentacion, Madrid.

Goodrich, W.J. (1987) Monitoring genetic erosion: detection and assessment. Unpublished consultancy report. IBPGR, Rome.

IBPGR (1986) *Genetic Erosion: Monitoring and Assessment.* AGPG:IBPGR/86/99. IBPGR, Rome.

Muchiru, S. (1985) *Conservation of Species and Genetic Resources. An NGO Action Guide.* Environment Liaison Centre, Nairobi.

UNEP (1993) *Guidelines for Country Studies on Biological Diversity.* UNEP, Nairobi.

WCMC (1992) *Global Diversity: Status of the Earth's Living Resources.* Chapman and Hall, London.

WRI, IUCN and UNEP (1992a) *Global Biodiversity Strategy.* WRI, Washington DC.

WRI, IIED and IUCN (1992b) *1993 Directory of Country Environmental Studies.* WRI, Washington DC.

APPENDIX 4.1
A model for quantifying the threat of genetic erosion

FACTOR   SCORE

1. General

1.1   Taxon distribution
        • Rare   10
        • Locally common   5
        • Widespread or abundant   0

1.2   Drought
        • Known to have occurred in two or more consecutive years   10
        • Occurring on average one or more times every ten years, but not in
          consecutive years   5
        • Occurring less than once every ten years on average   0

1.3   Flooding
        • Area known to be very flood prone   10
        • Area not known to be flood prone   0

1.4   Accidental fires
        • Area known to be very prone to fires   10
        • Area not known to be prone to fires   0

1.5   Potential risk from global warming
        • Summit areas or low-lying coastal areas   0

2. Crop species

2.1   Area under the crop
        • Declining rapidly   10
        • Increasing or static   0

2.2   Modern cultivars of the crop
        • Available and used by >70% of farmers   15
        • Available and used by 50–70% of farmers   10
        • Available and used by <50% of farmers   5
        • Not yet available, but introduction planned   2
        • Not available   0

2.3   Performance of agricultural services
        • Very strong, and biased towards modern varieties   10
        • No agricultural services   0

2.4   Mechanization
        • Tractors used by >30% of farmers   10
        • Animal traction used by >50% of farmers   5
        • Manual labour used by >50% of farmers   0

2.5 Herbicide and fertilizer use
- \> 50% of farmers 10
- 25% of farmers 5
- None 0

2.6 Farming population
- Declining rapidly 10
- Increasing or static 0

3. Wild species

3.1 Extent of wild habitat of target species within study area
- Very restricted (< 5%) 15
- Restricted (5–15%) 10
- 15–50% 5
- Extensive (> 50%) 0

3.2 Conservation status of target species
- Species not known to occur in any protected area 10
- Species known to occur within a protected area, but protection status poor or unknown 5
- Species known to occur within a protected area, and protection status good 0

3.3 Extent of use of wild habitat of target species
- Industrial exploitation 15
- Exploitation by surrounding populations (e.g. fuelwood gathering from nearby towns) 10
- Hunting and gathering by small local communities 2
- Completely protected 0

3.4 Extent of use of target species
- Industrial exploitation 15
- Exploitation by surrounding populations 10
- Local exploitation 5
- Protected or not used 0

3.5 Agricultural pressure on wild habitat
- Large-scale cultivation within habitat margins 15
- Subsistence cultivation areas within habitat margins 12
- Land suitable for cultivation, cultivated areas within 3 km of habitat margins 10
- Land suitable for cultivation, cultivated areas within 3–10 km of habitat margins 5
- Land unsuitable for cultivation 0

3.6 Human population growth rate per year
- \> 3% 10
- 1–3% 5
- < 1% 0

3.7 Availability of agricultural land
- \> 70 ha km^{-2} cultivated 10
- 30–70 ha km^{-2} cultivated 5
- < 30 ha km^{-2} cultivated 0

3.8   Species palatability
      • High   10
      • Medium   5
      • Low   0

3.9   Ratio of present livestock density to estimated carrying capacity
      • >1.0   10
      • 0.5–1.0   5
      • <0.5   0

3.10 Average proximity to borehole or other all-year round water supply
      • <10 km   10
      • 10–20 km   5
      • >20 km   0

3.11 Distance to major population centre
      • <20 km   10
      • 20–50 km   5
      • >50 km   0

3.12 Distance to major road
      • <10 km   10
      • 10–30 km   5
      • >30 km   0

3.13 Distance to development projects (irrigation scheme, tourism complex, mining site, hydroelectric power scheme, land reclamation scheme)
      • <20 km   10
      • 20–50 km   5
      • >50 km   0

A basic sampling strategy: theory and practice

5

A.H.D. Brown[1] and D.R. Marshall[2]

[1]CSIRO Division of Plant Industry, GPO Box 1600, Canberra, ACT 2601 Australia: [2]Plant Breeding Institute, University of Sydney, NSW 2006 Australia.

Introduction

One of the most important and difficult tasks facing plant germplasm collectors is defining the most appropriate sampling strategy for a particular species and region. It is important because the plant collector acts at the crucial interface between the genetic diversity that history has left as our endowment, and what will be conserved in collections for immediate and future use (Namkoong, 1988). In the case of rapidly eroding genetic resources, the collector finally controls what will survive for the future. Defining the strategy is often difficult because species differ in crucial ways, many plant populations have complex genetic structures and samples may be used in a variety of different ways.

Marshall and Brown (1975) considered the issue of optimal sampling strategies for use in the genetic conservation of crop plants. The focus of that paper was relatively narrow, however. It defined a strategy for populations under imminent threat of extinction, in particular traditional landraces of some major crops (notably wheat, rice and barley) in danger of replacement on a very broad scale because of the spread of modern cultivars. The strategy sought to optimize the use of the resources available for emergency action programmes to save samples of such material.

Over the last 15 years, the emphasis in the collecting of crop genetic resources, at least with respect to the major world food crops, has changed radically. In particular, 'crisis-driven', broad-scale, crop-specific programmes are no longer appropriate. Much of the material that was targeted in the mid-1970s has either been collected or lost. Rather, the collecting of crop genetic resources has entered a new phase. This involves three elements. The first is a greater emphasis on the collecting and conservation of germplasm of the wild and weedy relatives of the major crops. The second is an increased focus on the many nationally or locally important crops used for food, fibre, medicine or fuel. Examples include the tropical and subtropical fruits of southeast Asia and Latin America, the unique high-elevation crops of the Andean region, traditional medicinal plants and the leafy vegetables used worldwide in subsistence agriculture. Few of these plants are included in scientific improvement programmes. The emphasis is on collecting for direct use. The third element is the sampling and resampling of major crop germplasm to fill specific gaps, to replace lost or poorly representative samples or to meet current specific needs of breeders, for which the material in hand is deficient. The important point is that the sampling of crop germplasm needs to be more discerning, and collectors now generally need to meet specific objectives defined in advance.

Information on the kinds and amounts of genetic variation in target species populations and its distribution in the target region is critically important in developing efficient sampling strategies. Over the last 15 years, knowledge of the genetic structure of plant populations has increased markedly (e.g. Doebley, 1989; Hamrick and Godt, 1989). This provides a more secure base on which to develop robust and relevant sampling technologies. In addition, further research has been undertaken by a number of authors, in a variety of contexts, on the theory of crop germplasm sampling (e.g. Oka, 1975; Marshall and Brown, 1981; 1983; Yonezawa, 1985; Namkoong, 1988; Chapman, 1989; Brown, 1992).

The aim of this chapter is to provide an up-to-date overview of the theory and practice of crop germplasm collecting taking account of these changes. Using simple theoretical models, the impact of varying sample size and distribution on the effectiveness of germplasm sampling will be explored. This will provide a basis for advice on how to respond, in terms of sampling strategy, to the particular practical problems that collectors face in the field.

Theoretical background

Allard (1970) first identified clearly the critical problem facing plant explorers. He stressed that most plant species contain remarkable stores of genetic variation and consist of millions of different genotypes. Indeed, in many species each plant is genetically unique. There are several exceptions to this, even in naturally occurring populations. In particular, populations of self-pollinated or clonally propagated colonizers or crops can be depauperate in genetic variation and consist of only one or a few genotypes. Yet these species store considerable variation among populations. As a result, a plant collector can hope to sample only a fraction of the variation that occurs in nature. It is important that this fraction be as large as possible and contain the maximum amount of useful (now and in the future) variation.

Allard (1970) also recognized that collectors as well as end-users of germplasm have limited time and resources at their disposal. Thus the problem is to define a sampling procedure that yields the maximum amount of useful genetic variation, within a specified and limited number of samples (Marshall and Brown, 1983).

Measurement of variation

How should plant collectors conceive of the genetic variation that they are aiming to sample? The genetic variation present in a set of populations can be described by a large array of parameters (e.g. allele and genotype frequencies, gene diversities, heterozygosity levels, disequilibrium coefficients (see Weir, 1990)). However, from the standpoint of sampling genetic resources, the basic parameter for each population is the allelic richness or the number of distinct alleles at a single locus. In practice, when an estimate of this parameter is made, it is usually the average number of alleles for a large number of marker loci after the sample is taken. This parameter is the basic one for our purposes because later users of the genetic resources can adjust the frequencies of specific desired alleles at will. Breeders might use the single copy of an allele for disease resistance irrespective of its frequency in the original population or in the sample. Thus, the allelic richness of a sample is a direct measure of its value.

Several populations

Sampling strategies for more than one population depend on two crucial parameters. The first is the extent of genetic divergence among populations. Marshall and Brown (1975) recognized four types of alleles: (i) common, widely distributed; (ii) common, locally distributed; (iii) rare, widely distributed; and (iv) rare, locally distributed. They argued that collecting and conserving the first class of alleles present no problem. They will almost certainly be included even in small samples from a few populations. In contrast, adequate *ex situ* conservation of the last two classes will ultimately be limited by the population sizes which we are prepared to collect and conserve, and thus by the resources available. Critical to any collecting strategy are therefore traits or alleles that are locally common. Marshall and Brown (1975) thus argue that the key indicator for optimal sampling is the number of such alleles that each population possesses, i.e. the number of alleles that attain appreciable frequencies in only one population or in a few adjacent populations. Their conservation will largely depend on the identification of populations closely adapted to specific environments and agricultural practices. This definition is similar to that of Slatkin's 'private alleles' (e.g. Slatkin and Takahata, 1985), except that only private alleles with frequencies exceeding some value (e.g. 0.05) are considered. Other things being equal, populations with a higher number of locally common alleles deserve priority.

The second parameter for optimal sampling from many populations is the variation among them in the level of genetic variation (Marshall and Brown, 1975; Schoen and Brown, 1991). This aspect is reflected in the distribution (the range and pattern) of numbers of alleles per locus. As a general rule, populations that have higher values of diversity are genetically richer and merit larger samples.

Number of alleles in the sample

As mentioned above, the values of diversity parameters cannot be known ahead of sampling. Indeed, the actual number of alleles in a population is a difficult parameter to estimate because its value in a sample steadily increases with increasing sample size. However, theoretical computations based on hypothetical distributions of allele frequencies can be made. The results from such theory can be checked with those from numerical simulations of sampling from actual or conjectural examples to test the effect of strategy on allele recovery (Marshall and Brown, 1975; Yonezawa and Ichihashi, 1989).

The neutral allele theory of Kimura and Crow (1964) is the most useful for this purpose as its sampling theory has been well developed (Brown and Briggs, 1991). Thus, for example, the number of selectively neutral alleles (k) in a sample of S random gametes, from an equilibrium population of size N, at a locus with mutation rate u, is approximately:

$$k \approx \theta \log_e[(S + \theta)/\theta] + 0.6 \tag{1}$$

where $\theta = 4Nu > 0.1$ and $S > 10$. This formula shows that the expected number of alleles in a sample increases in proportion to the logarithm of sample size. In contrast, the resources required to collect the individuals at a site (once the collector is at the site) increase in direct proportion to the sample size. Thus, there is a diminishing return (in terms of collecting new alleles) per unit of cost for increasing sample size, which becomes progressively more wasteful of resources.

A benchmark criterion

Given the diminishing rewards for effort expended on any one population, is there a reasonable objective that would guide the collector as to when a sufficient sample was in hand? It has previously been suggested that the objective should be to include in the sample at least one copy of 95% of the alleles that occurred in the target population at frequencies greater than 0.05 (Marshall and Brown, 1975). While the biological basis for this criterion is debatable (see Marshall (1989) for further discussion and references and Krusche and Geburek (1991) for an argument for more conservative values when dealing with forest trees), the point is that either increasing the certainty level higher than 95% or dropping the critical allele frequency below 0.05 drastically increases sample size, with only marginal gains. A sample of 59 random unrelated gametes from the population is sufficient to attain this objective. This would be assured by collecting and bulking seeds or vegetative material from 30 randomly chosen individuals in a fully outbreeding sexual species, or from 30 random genotypes in an apomictic species, or from 59 random individuals in a self-fertilizing species. A sample of 50 individuals from each population will be considered as a benchmark. Factors that lead the collector to increase or decrease this sample size are discussed later.

Basic sampling strategy

A full statement of a basic sampling strategy can be set out as follows.

Number and location of sampling sites

Before the mission begins, the collector should assemble the available information on the kinds of environments occurring in the target region and on the pattern of distribution of the target species. Based on these data, the region is then roughly divided into a limited number of areas, clearly distinct because of ecological (i.e. physiographic, edaphic, climatic), botanical, agricultural or cultural differences. The total time available can then be divided among the areas according to travelling convenience, prevalence of the targets and any perceived or known differences in genetic diversity among areas. More sites should be sampled in areas where the target species is more common, or where it is evidently more variable for conspicuous polymorphisms.

Delimiting the population, and thus the sampling site, can be problematic in wild species collecting (Chapters 3 and 6), but in crop collecting the site is usually taken as being the farmer's field or orchard, the farmer's store or the market stall. Is there a benchmark figure for the total number of such sampling sites for each species in a region? Analogous statistical and cost/benefit arguments conceivably apply at the mission level to those made above at the population level. This would suggest that a set of about 50 sites comprises a justifiable sample. However, this can only be a weak guide. Unlike the members of a single population, the potential sites in a given area could conceivably be completely different from each other and yield no redundancy upon sampling. Moving on and collecting an individual at a new site is usually preferable to collecting an additional individual at the current site. Nevertheless, it is helpful to start with a specific target of 50 sites per species per region, and vary this target up or down when clear reasons exist for doing so.

Number of individual plants sampled at a site

As already stated, the sample size at a site should be about 50 individuals. This figure should be increased to take account of the following factors: (i) any splitting and duplication of the sample that will take place (clearly, if only two seeds are collected from an individual, a three-way splitting of the sample will leave one subsample short); (ii) any suspicion that seeds from some individuals in the sample are not fully viable; and (iii) possible loss of some individuals in the sample in transport and quarantine. The aim is that each sample at the time of entry into each of the gene banks conserving the material should trace back to at least 50 original individuals. If it is not possible to collect from 50 individuals at each site because populations are small or shattering has already started, more sites should be sampled.

Choice of individuals

A much discussed question in sampling technique for crop genetic resources is whether individuals should be taken strictly at random or biased as to phenotype or microsite (e.g. Marshall and Brown, 1975, 1981; Porceddu and Damania, 1992). Random sampling is generally the most reliable and desirable method, particularly for crop populations or market samples because subpopulation structure is unlikely to be present. In contrast, natural populations of wild species often evolve local subpopulation structure and hence stratified random sampling is appropriate for them (e.g. separate random sampling of individuals in different microsites). In the case of a crop field which consists of a mechanical mixture of species (e.g. durum and bread wheat), clearly a separate random sample should be made from each of the components (see below, 'Mixed populations'). Biased sampling of rare phenotypic variants in a population is to be avoided, except when such plants clearly merit separate and distinct recognition (e.g. a rare disease-free individual in a heavily diseased field). Such samples may be taken in addition to, but not instead of, population samples, and should receive separate collecting numbers.

Strict randomness of sampling requires that every plant at the site have an independent and equally likely chance of inclusion in the sample. Ward (1974), for example, has described a simple way in which two people may collect such a sample. This is often impossible or impractical to achieve, however (Marshall and Brown, 1983). In practice, collectors usually sample at systematic or random intervals along a number of transects. Systematic sampling is easiest and spreads the sample over the population but can be biased if variation is periodic in the population (Brown and Briggs, 1991). The starting-points and directions of the transects and the position of sampling points along them can be chosen according to a randomization procedure. It will often be advantageous to keep a minimum distance between sampling points to avoid excessive sampling of closely related individuals and repeated sampling of clones (Chapters 21, 22 and 23).

Number and type of propagules per plant

The final decision for the plant collector concerns the kind(s) and amount(s) of material to be collected from each plant chosen for sampling. The material ranges from pollen, seeds and vegetative cuttings or propagules (such as tubers, bulbs or corms) to whole individuals (Brown and Briggs, 1991). Collecting seeds differs from collecting vegetative material in a number of ways. In particular, it is more restrictive for the timing of the mission. Otherwise, seed, being the organ of dispersal and storage, enjoys a number of advantages, for example less bulk and easier handling and storage. Pollen is even less bulky, but its storage is difficult and it is at the moment not possible to recover plants from pollen. Cuttings are often the most convenient vegetative samples to make, but they may require the use of *in vitro* techniques. In perennials, the removal of whole individuals should be avoided, especially when to do so would destroy the source population.

The genotypes in a sample of vegetative material are exactly those of the sampled parent plants, whereas those in a seed sample depend on the breeding and pollination system. With uniparental reproduction (self-fertilization or agamospermy), seeds will closely resemble the parent plant. In contrast, the seeds of an outbreeder will differ from the parent and show within-family diversity (Yonezawa and Ichihashi, 1989). In outbreeding species, seeds from several fruits should be gathered from each individual sampled, if possible, rather than from a single fruit, to increase the diversity of genes in the sample. Also, fruits should be collected from all parts of the crown of trees, because these may have been pollinated by different pollen sources. Similar numbers of seeds should be collected from each individual sampled.

The need for sufficient material, rather than genetic principles, largely determines the number of propagules per plant to sample. If available, enough seeds or cuttings should be sampled to provide for the division of the samples among collaborators and other recipients and avoid immediate multiplications. Some gene banks have lower limits on the number of seeds per sample that will be accepted for storage, for example if multiplication of the material is not likely to be prohibitively expensive. On the other hand, quarantine facilities or the space available in the collecting vehicle or in the field genebanks (for root and tuber crops and species with recalcitrant seeds, for example) may limit the total volume of material that can be collected. When this is the case, it would be better to maximize the number of plants sampled and reduce the number of propagules per plant.

Modifications to basic strategy for different species

The basic sampling strategy developed above thus consists of the following four elements:

- sample about 50 populations in an ecogeographic area or mission;
- sample about 50 individual plants in each population;
- in general, sample individuals at random at each site, but sample separately within distinct local microenvironments if the habitat at the site is heterogeneous;
- sample sufficient seeds or vegetative material per plant to assure representation of each original plant in all duplicates.

Species – both cultivated and wild – differ in a number of life-history, ecological and genetic attributes that will require amendment of this basic strategy. We now consider the more important of these attributes individually and their effect upon sampling in practice. Each point will assume a comparison of species differing only for the attribute in question, while other features are held in common.

Distribution

The spatial occurrence of a species will often play a major role in limiting or expanding the options available to collectors.

Geographic range

A species found only in a narrow geographic range would merit sampling from fewer sites, but with an increased number of individuals at each site, and an increased number of propagules per individual.

Local abundance

Likewise, for species (particularly wild species) that are locally rare, it may be very time-consuming or impossible to meet a target of 50 individuals per population. Compensation for limited numbers locally can be made by increasing the number of sites and by increasing the number of propagules per plant. Clearly, the problem of choosing which individuals to sample from the populations of species that are locally rare may not arise. Brown and Briggs (1991) give guidelines for the collecting of endangered species of ten individuals at each of up to five sites.

Interpopulation migration

Migration rates are likely to differ among species, whether the agents of migration are natural (wild species), inadvertently human (weeds) or deliberately human (crops). When migration rates appear to be high, populations are more likely to share most of their genes and less likely to diverge. Hence the sampling of fewer but more widely spread populations is appropriate.

Habitat diversity

Species that grow in a wide range of ecological situations are more likely to have diverged genetically among their different habitats (ecotypification). In such species, an increased number of populations or distinct subpopulations should be sampled at the expense of the number of individuals per population.

Life history

Among plant species, life-history traits are in part correlated with features of their distribution (see above) and genetic system (see below). For example, perennial fruit trees are commonly outbreeding whereas selfing is common in many annual crops. Differences in life-history traits lead to little modification of the first two elements of the basic strategy because they do not appear to have a clear-cut effect on the disposition of genetic variation within and among populations. For instance, out-breeding annuals and outbreeding perennials tend to have similar levels of genetic divergence among populations (Hamrick and Godt, 1989). However, when considered on their own, several life-history traits affect sampling procedures by determining the kind and amount of collectable material.

Duration of life cycle

The collecting of perennial species may be less dependent on seasonal timing of the mission as vegetative material is available for collecting throughout the year. It may be possible to arrange return to a site should the collected material prove not viable or insufficient. Therefore, the number of propagules from a site need not be as high as for an annual species.

Population age structure

Populations of perennial species can either consist of individuals of the same age (e.g. orchards, plantations) or possess an age structure (as in most natural populations). In the case of age-structured populations, individuals should be sampled at random irrespective of size or age to maximize genetic diversity, because different cohorts may be divergent genetically. If distinct age cohorts are obvious, a stratified random sample can be made.

Vegetative reproduction

Species that produce organs of vegetative reproduction, or cuttings of which are viable, add further collecting options, as discussed above. Where both seeds and vegetative material are available, it is advisable to include samples of both kinds, particularly when the species is poorly known or when quarantine procedures may have uncertain outcomes. How to label such samples with appropriate collecting numbers is discussed in Chapter 19.

Fecundity

Clearly, fecundity has direct effects on the fourth element of the basic strategy, namely the number of propagules per plant. The sampling pattern may need adjustment in the case of species that produce few seeds per individual. Sampling from more individuals at a site could compensate for scarcity of seeds. Biasing the sample towards the most fecund individuals should be avoided.

Determinacy of flowering and seed maturation

Species with highly synchronized flowering (and hence maturity of fruit) require well-timed missions. On the other hand, indeterminate flowering may mean that only a portion of the population is ready for sampling at the time of the visit. As such variation may be due to genetic variation in response to photoperiod, it is important to sample from plants with different maturities. Thus, variation in maturity would affect the choice of individuals for the sample and whether other types of propagules than seeds should be included. Some wild species (e.g. perennial *Glycine* species) combine a determinate chasmogamous flowering habit with a relatively indeterminate fruit production from cleistogamous flowers. This adds to the flexibility of sampling.

Genetic system

Various aspects of the genetic system determine the apportionment of genetic variation among and within populations and therefore substantially affect sampling strategy.

Mating system

Differences among species in mating system and variation within the species profoundly influence all four elements of the basic strategy. In sampling outbreeding species, the number of populations in an ecogeographic region can be reduced and the number of individuals per site increased without great loss of efficiency. The spatial scale of local differentiation in outbreeders is likely to be larger than for autogamous species, and the collecting of seeds will in itself lead to the sampling of dispersed sources of pollen. Hence there is less reason for locally stratified sampling. Open-pollinated progeny arrays are likely to be genetically variable and include some inbred seeds. Seed viability and seedling vigour could vary and the sample size per plant should allow for this.

Under self-fertilization or apomixis, populations diverge for both the alleles they contain and the amount of genetic polymorphism. Hence it is important to sample a large number of populations even at the expense of the average number of individuals at a site. In addition, with selfing species, it is important to be on the lookout for populations that have an exceptionally large amount of polymorphism and to increase the sample size of these. Natural populations of autogamous species can possess local subpopulation structure that justifies stratified random sampling at collecting sites. It is important to realize that the seeds sampled from a plant of a species with a predominantly uniparental mating system are very similar to one another genetically. Hence the total sample size should not be inflated with large numbers of seeds from few original plants.

Pollination mode

The mode of pollination, in particular whether pollination is by animals or by wind, affects the genetic structure of progeny arrays from a single fruit. Under wind pollination, such arrays tend to be half-sibs, as they are the products of many sources of pollen. Under animal pollination, the array of seeds from a single fruit in many species largely has the same male parent. This implies that the sample from each individual in animal-pollinated species should include seeds from several randomly chosen fruits. In addition, animal-pollinated outbreeding species tend to show more population divergence than do wind-pollinated outbreeders but less than is shown by selfers (Hamrick and Godt, 1989). As already noted, population divergence justifies an emphasis on the sampling of more populations.

Conspicuous polymorphism

Populations of some species can show a range of levels of morphological polymorphisms. As already noted, when some populations of a species appear to be much more polymorphic than others, it is sensible to increase the sample size in the richer populations. This is especially important in inbreeding species, where the range in the level of polymorphism is likely to be wider than in outbreeders and the link between genetic variation in marker genes and genetic variation throughout the whole genome is stronger.

Mixed populations

Fields of landraces may sometimes consist of mechanical mixtures, deliberately composed by the farmer. Examples range from a field with very different species, like wheat and barley or durum and bread wheat, or with differently maturing strains of a vegetable crop, to mixtures of clones of noble sugarcane each diagnostically marked by its own stalk colour pattern. If possible, the collector should seek from the farmer samples from the original seed lots of each component of such mixtures. Otherwise, the collector must decide between two options. Option 1 is to make a separate sample of each component in the field. Option 2 is to regard the field as a single population and therefore make a single random sample. Option 1 is clearly the appropriate one when two crop species are mixed, whereas option 2 is more appropriate if the components may have interbred to produce the seeds to be sampled, as for mixed races of maize or of an outbreeding vegetable. For cases intermediate between these two extremes, collectors should take a single random sample, because they can rely on the general statistical robustness and simplicity of such samples and avoid giving an extreme bias to any rare type. More guidance on this issue is given in Chapter 18, where the importance of the farmers' knowledge in deciding how to collect is stressed.

Modifications to the basic strategy when sampling for specific goals

Sometimes, a collecting mission is undertaken to meet specific goals, such as the collecting of further genes for resistance to a particular disease from a region where the disease is known to occur. When collecting objectives are specific, how does the sampling strategy differ from that appropriate for generalized collecting?

The rare variant

The first point of departure is the treatment of rare alleles, as they may be more important than in the case of the sampling theory for generalized collecting. A rare allele might be the basis of the desired phenotype. Alternatively, a population that would meet the explicit objective of the mission might occur at only one kind of site (e.g. a waterlogged or saline habitat), which could be rare in the target region. In such cases, more individuals per site, or more sites, should be sampled.

Consider the case when the nominated target is coded by a specific allele. Let us assume the frequency of this allele in the populations is p. The size of sample in terms of the number of random gametes (S) required to be 95% certain of including at least one copy of the target allele or genotype is:

$$S \approx -3/\log_e[1 - p] \tag{2}$$

From this formula, the sample sizes required for an allele or character that is increasingly rare ($p = 0.05, 0.03, 0.01, 0.001$) are 59, 99, 299 and 2999. The size increases at an exponential rate as the allele is progressively rarer. The same computations apply for the number of sites if p represents the frequency of the desired site and S the required number of sites to sample to be 95% certain of a successful mission.

Thus, the collector may well decide to increase substantially the sample size at a site that evidently meets a specific objective of the mission, especially if the desired variants are likely to be rare and restricted to that site.

More than one copy

The second major point of departure in sampling for specific goals compared with generalized collecting is the adequacy of a strategy that assures just one copy of the desired variant. If the mission is seeking certain genetic variants, it may need more than just one copy of a desired allele. The sampling should provide the allele in sufficient numbers to guard against its later loss, and provide it in a variety of genetic backgrounds. The size required to be assured of a specified number of copies of an allele can be calculated using the relevant sampling theory as follows.

Sedcole (1977) has computed the sample size (S) required to be 95% certain of recovering a minimum number (r) of plants with a trait that occurs in a population with frequency p. Table 5.1 lists a selection of these values of S. An approximate formula for S for relatively infrequent traits ($p < 0.20$) is:

$$S \approx \{r + 1.645\sqrt{r} + 0.5\}/p \tag{3}$$

The values in Table 5.1 show that the required sample size increases with increasing required number of copies, as would be expected. However, the increase is less than proportionate. Thus, increasing the sample size

Table 5.1. The sample size (S) required to be 95% certain to recover a minimum number (r) of plants with a trait that occurs in a population with frequency p. (From Sedcole, 1977.)

p	Number of copies be recovered (r)								
	1	2	3	4	5	6	8	10	15
0.25	11	18	23	29	34	40	50	60	84
0.125	23	37	49	60	71	82	103	123	172
0.0625	47	75	99	122	144	166	208	248	347
0.03125	95	150	200	246	291	334	418	500	697
0.015625	191	302	401	494	584	671	839	1002	1397
0.05 (from formula (3))	63	97	127	156	184	211	264	314	438

fivefold (for $p = 0.05$, from 63 to 314) increases the assured number of copies tenfold.

A relatively large sample is thus justified in missions that are aimed at finding several examples of specific rare and valuable variants. The final figure will depend on the practical limits on handling larger samples weighed against the number of copies that are actually needed.

Representativeness

Rather than simply being 95% certain of the presence of a given number of copies of common alleles in the sample, the collector may want the sample to accurately reflect the frequencies of alleles in the source population. This may be the case when the material is for direct use as an adapted population, rather than for indirect use as the source for alleles to be incorporated in a breeding programme. Marshall and Brown (1983) suggest that under these conditions sample size should be considerably larger, about 200 individuals. Whereas for inbreeders the number of seeds collected per individual has little effect on the fidelity of the sample, for outbreeders increasing the number of seeds per plant reduces the variance of sample allele frequencies, though with diminishing returns.

Modifications to the basic strategy when resampling a region

The 'basic sampling strategy' (see above) was devised for the situation where little or no detailed information is available on the genetic structure of the target populations. This may be because no previous samples have been taken from the region, or no reports of the nature of its populations have been published. However, in many cases, partial information is at hand about the ecological distribution of the target species and its genetic variation from previous sampling. The sources of information, which are discussed in other chapters, include gene-bank databases, Floras[1] and monographs, the route maps of previous collectors and their herbarium specimens with associated field records.

A specific example of when resampling of a region is particularly justified is the replacement of samples in a gene bank that have become degraded. Resampling is also appropriate in populations that are undergoing temporal coevolutionary changes in response to changes in cropping systems or in the prevalence or genetic structure of other interacting organisms such as predators, pathogens and weeds. Monitoring genetic erosion will also require repeated sampling.

Several workers have noted that sampling in two or more episodes can be a very efficient strategy, though it is clearly not always feasible. For example, Jain (1975) advocated a two-step sequential method with the first collecting season on a coarse grid and a second round on a finer scale in areas of particular interest. However, just how the collector is to modify the sampling strategy in the light of previous sampling is open to discussion.

In the case of resampling, the general strategic questions facing the collector (modified from Nabhan, 1990) are as follows:

- Should the collector give particular emphasis to uncollected or largely unstudied areas within the region? The risk of doing this is that the species may be rare or absent from such areas. The benefit is that any new samples could add new ecogeographic representation to the collection.
- Should the collector emphasize areas where the previous data and samples indicate high genetic diversity? The risk here is that the new samples may contain much that is redundant with the accessions already in gene banks. The benefit is that the collector will be assured of getting a very rich collection.
- Should the collector aim for specific ecotypes or landraces reported to exist but no longer available from gene banks? The disadvantage of this strategy is that such specific localities are often very scattered and the material may have subsequently disappeared from the site. For crop populations in particular, the field situation is likely to be relatively labile. The advantage is that the specific-target strategy has a good chance of locating at least some of the targets.

Nabhan (1990) has given a detailed case-study of how to use the data from previous explorations in devising a strategy for the sampling of wild *Phaseolus* species in northern Mexico. In this example, the primary data are herbarium records of species occurrence in various geographic subregions. Graphs of species richness against the number of samples from each subregion indicate that the subregions that may be under-

---

[1] See footnote p. 96.

collected and would repay further visits. This amounts to making relatively more samples in the areas of higher diversity. Chapter 15 discusses areographic methods such as the ones used by Nabhan (1990) in more detail. But what should be the degree of bias or weighting towards more diverse areas in sampling?

From the standpoint of genetic conservation, the objective of a resampling project is to obtain a maximum amount of new genetic diversity additional to that already available in gene banks. Specific guidance as to the appropriate weighting of effort among the various subregions can be sought from the sampling theory of selectively neutral alleles in finite populations in terms of its basic parameter $\theta (=4 \times$ effective population size $\times$ mutation rate). The question is: What is the optimum allocation of a fixed total sampling effort among several independent populations with varying levels of genetic diversity? It can be shown that the number of alleles expected in the total sample is a maximum when the total sampling effort is divided among the populations in direct proportion to the value of θ in each population (D.J. Schoen and A.H.D. Brown, in prep.). The value of θ should be the average over many loci, the same loci being tested in all the populations.

How is it possible to estimate the values of θ in practice? If no genetic data are available and identity by descent does not vary greatly among populations, then an approximate relative estimate of θ is the population size. Simply put, this implies taking samples from each subregion in proportion to the commonness of the species in that subregion. If data on genetic polymorphism are available on a comparable set of loci, the estimates of θ for each locus are obtained by the formula $\theta = h/(1 - h)$, where h is the gene diversity or probability that two gametes sampled at random from the population (subregion) differ at the locus (Weir, 1990). The number of samples from a subregion is then taken in proportion to the average value of θ in that subregion. Finally, if comparative estimates of genetic variance are available for quantitative characters, the number of samples from each subregion should be in proportion to its genetic variance. This follows from the theory that links such variance with the parameter θ.

Conclusions

Sampling is a critical step in the conservation of plant genetic resources. Therefore, the onus is on the collector to obtain the richest collection for a given expenditure of effort. This is ensured not by collecting large quantities of material, but by a judicious division of the target region into ecogeographical areas and collecting a limited number (about 50) of random individuals from many populations in each area, up to a total of about 50 populations for the mission. These guidelines can readily be adapted to take account of biological differences among species. If the mission has specific goals, it will be appropriate to increase the sample size when the material at a locality appears to meet (or is likely to meet) one of these goals. In the case of resampling, evidence as to genetic divergence among populations can be used in the division of the region into relatively homogeneous areas, and evidence of genetic richness can be used to adjust sample size in terms of the number of samples for each area. In this way, the samples will· recover a high percentage of the alleles on offer.

References

Allard, R.W. (1970) Population structure and sampling methods. In: Frankel, O.H. and E. Bennett (eds) *Genetic Resources in Plants – Their Exploration and Conservation.* pp. 97–107. Blackwell Scientific Publications, Oxford.

Brown, A.H.D. (1992) Human impact on plant gene pools and sampling for their conservation. *Oikos* 63:109–118.

Brown, A.H.D. and J.D. Briggs (1991) Sampling strategies for genetic variation in *ex situ* collections of endangered plant species. In: Falk, D.A. and K.E. Holsinger (eds) *Genetics and Conservation of Rare Plants.* pp. 99–119. Oxford University Press, New York.

Chapman, G.C. (1989) Collection strategies for the wild relatives of field crops. In: Brown, A.H.D., O.H. Frankel, D.R. Marshall and J.T. Williams (eds) *The Use of Plant Genetic Resources.* pp. 263–279. Cambridge University Press, Cambridge.

Doebley, J.W. (1989) Isozymic evidence and the evolution of crop plants. In: Soltis, D.E. and P.S. Soltis (eds) *Isozymes in Plant Biology.* pp. 165–191. Dioscorides Press, Portland.

Hamrick, J.L. and M.J. Godt (1989) Allozyme diversity in plant species. In: Brown, A.H.D., M.T. Clegg, A.L. Kahler and B.S. Weir (eds) *Plant Population Genetics, Breeding and Genetic Resources.* pp. 43–63. Sinauer Associates, Sunderland.

Jain, S.K. (1975) Population structure and the effects of breeding system. In: Frankel, O.H. and J.G. Hawkes (eds) *Crop Genetic Resources for Today and Tomorrow.* pp. 15–36. Cambridge University Press, Cambridge.

Kimura, M. and J.F. Crow (1964) The number of alleles that can be maintained in a finite population. *Genetics* 49:725–738.

Krusche, D. and Th. Geburek (1991) Conservation of forest gene resources as related to sample size. *Forest Ecology and Management* 40:145–150.

Marshall, D.R. (1989) Crop genetic resources – current and emerging issues. In: Brown, A.H.D., M.T. Clegg, A.L. Kahler and B.S. Weir (eds) *Plant Population Genetics, Breeding and Genetic Resources.* pp. 370–391. Sinauer, Sunderland.

Marshall, D.R. and A.H.D. Brown (1975) Optimum sampling strategies in genetic conservation. In: Frankel, O.H. and J.G. Hawkes (eds) *Crop Genetic Resources for Today and Tomorrow.* pp. 53–80. Cambridge University Press, Cambridge.

Marshall, D.R. and A.H.D. Brown (1981) Wheat genetic resources. In: Evans, L.T. and W.J. Peacock (eds) *Wheat Science – Today and Tomorrow.* pp. 21–40. Cambridge University Press, Cambridge.

Marshall, D.R. and A.H.D. Brown (1983) Theory of forage plant collection. In: McIvor, J.G. and R.A. Bray (eds) *Genetic Resources of Forage Plants.* pp. 135–148. CSIRO, Melbourne.

Nabhan, G.P. (1990) *Wild* Phaseolus *Ecogeography in the Sierra Madre Occidental,*

Mexico. Systematic and Ecogeographic Studies on Crop Genepools 5. IBPGR, Rome.

Namkoong, G. (1988) Sampling for germplasm collections. *HortScience* 23:79–81.

Oka, H.I.(1975) Consideration on the population size necessary for conservation of crop germplasm. In: Matsuo, T. (ed.) *Gene Conservation - Exploration, Collection, Preservation and Utilization of Genetic Resources.* pp. 57–84. JIBP Synthesis Volume 5. University of Tokyo Press, Tokyo.

Porceddu, E. and A.B. Damania (1992) *Sampling Strategies for Conserving Variability of Genetic Resources in Seed Crops.* Technical Manual No. 17. ICARDA, Aleppo.

Schoen, D.J. and A.H.D. Brown (1991) Intraspecific variation in population gene diversity and effective population size correlates with the mating system in plants. *Proceedings of the National Academy of Sciences* 88:4494–4497.

Sedcole, J.R. (1977) Number of plants necessary to recover a trait. *Crop Science* 17:667–668.

Slatkin, M. and N. Takahata (1985) The average frequency of private alleles in a partially isolated population. *Theoretical Population Biology* 28:314–331.

Ward, D.B. (1974) The 'ignorant man' technique of sampling plant populations. *Taxon* 23:325–330.

Weir, B.S. (1990) *Genetic Data Analysis.* Sinauer Associates, Sunderland.

Yonezawa, K. (1985) A definition of the optimal allocation of effort in conservation of plant genetic resources – with application to sample size determination for field collection. *Euphytica* 34:345–354.

Yonezawa, K. and H. Ichihashi (1989) Sample size for collecting germplasm from natural plant populations in view of the genotypic multiplicity of seed embryos borne on a single plant. *Euphytica* 41:91–97.

Strategies for the collecting of wild species

6

R. von Bothmer[1] and O. Seberg[2]

[1]*Department of Plant Breeding Research, Swedish University of Agricultural Sciences, S-268 31 Svalöv, Sweden: [2]Botanical Laboratory, Copenhagen University, Gothersgade 140, DK-1123 Copenhagen K, Denmark.*

Introduction

The human race relies on a relatively small number of crops for its survival and, as world population increases, ever more pressure is placed on these limited resources. Plant breeding is a continuous search for new sources of diversity (i.e. genes) which can be incorporated into advanced material in an effort to ease this pressure. Lately, demand for new sources of genetic variation has been increasing rapidly. There is keen interest both in developed and in developing countries in novel variation, even completely new species, to be used in breeding or introduced into cultivation.

Due to rapid technical developments in gene transfer in recent years, the diversity of wild species has become more accessible for use in breeding. However, our knowledge of the wild and weedy relatives of most crops is still fragmentary. Basic information on species delimitation, chromosome numbers, distribution and genetic variation is in many cases still imperfect or completely lacking. At the same time, genetic erosion has also increased alarmingly in many parts of the world, and with it national and international concern. There are thus several reasons why systematic collecting of germplasm of wild and weedy species may be done:

- collecting for taxonomic, phylogenetic and biosystematics research;
- collecting for genetic diversity study and conservation;
- collecting for immediate use in a breeding programme.

Collecting and conservation of wild species is generally most effective when there is a clear and specific need for the material, whether for research or exploitation. These different objectives can certainly not be entirely separated from each other but the plant collector needs to carefully consider the aims of each collecting effort, because, as is described in this chapter, the choice of an appropriate strategy will to some extent depend on it.

The main area of research of the authors is the tribe Triticeae of the family Poaceae, and that is where the majority of the examples used in this chapter will come from. The principles and recommendations are, however, generally applicable to wild crop relatives, and indeed potentially useful wild species in general, for instance forages, of which in fact some species in the Triticeae are examples. The tribe Triticeae comprises about 350 species (Dewey, 1984; Löve, 1984). Generic limits are matters of some controversy (Baum *et al.*, 1987; Gupta and Baum, 1989; Seberg, 1989). Phylogenetic relationships within the tribe are rather complicated (Kellog, 1989; Frederiksen and Seberg, 1992), polyploidy and extensive hybridization being the main factors responsible for the intricate patterns of variation that are observed (e.g. Löve 1982, 1984; Dewey, 1984).

The gene pool concept

Central to the study of genetic diversity in wild crop relatives is the concept of the gene pool (Harlan and de Wet, 1971). This may be considered a broad summary of cytogenetic and biosystematic data and genome relationships in a particular plant group. The value of the gene pool concept has been its direct application in plant breeding. Though higher priority may be given to those species within the gene pool that are most easily used and those that are endangered, it is the gene pool as a whole that is the unit of study and conservation and must therefore be the target of collecting.

The primary gene pool consists of the taxa, including cultivated, weedy and wild forms of a crop, among which there are no sterility barriers and gene transfer is therefore straightforward. The secondary gene pool consists of all taxa that will cross with the crop, but from which gene transfer is difficult. The tertiary gene pool consists of taxa from which gene transfer is very difficult due to strong sterility barriers.

To take the case of cultivated barley, *Hordeum vulgare*, as an example, the primary gene pool comprises ssp. *vulgare* (i.e. breeding lines, commercial varieties and landraces) and ssp. *spontaneum*, which is the progenitor and closest wild relative of cultivated barley (Fig. 6.1; Brown, 1992). There are no problems in gene transfer from subsp. *spontaneum* to cultivated barley. However, immediate use in breeding programmes is not possible, since several unwanted traits of the wild form (e.g. shattering and shrunken seeds) are transferred simultaneously with the desired characters. There must therefore be a prebreeding programme, including some generations of backcrosses to barley and repeated selection of desired genotypes (Lehmann and Bothmer, 1988).

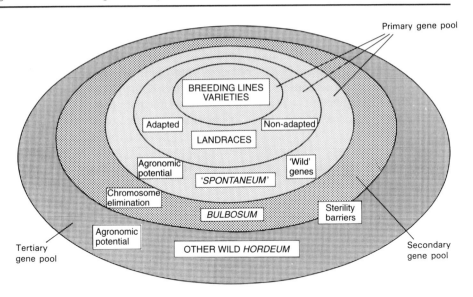

Fig. 6.1. The gene pool of cultivated barley (*Hordeum vulgare*).

The secondary gene pool of barley includes only one species, *H. bulbosum*, a perennial Mediterranean species from which gene transfer to barley through sexual hybridization is putatively possible (Pickering, 1988; Bothmer *et al.*, 1992; Xue and Kasha, 1992). This species has also been of great importance for the production of barley haploids through selective elimination of the *H. bulbosum* chromosomes in the interspecific hybrid (Kasha and Kao, 1970; Lange, 1971; Thörn, 1992).

The tertiary gene pool of barley comprises the remaining species of the genus *Hordeum* (Bothmer *et al.*, 1992). In the near future, however, improvements in the techniques for gene transfer will probably result in the inclusion of the whole of the Triticeae (Bothmer, 1992). As pointed out by Marshall (1990), molecular technologies may in the near future make the gene pool concept obsolete.

Collecting for taxonomic, phylogenetic and biosystematic research

The first step in planning any collecting programme for wild species will be a background study of the taxonomy and distribution of the target group, often called an ecogeographic survey (Chapter 14). Such surveys should be planned for the gene pool as a whole, even if it may only be possible to collect relatively small parts of the distribution area at any one time. Floras and monographs are the primary sources of such information, though floristic databases will be increasingly important in this respect (Chapter 10). A Flora[1] is an account of the plants found in a specified region, usually providing means of identifying the different taxa. A monograph, in contrast, is a comprehensive treatment of the available taxonomic data pertaining to all the taxa in a specified group (less comprehensive treatments are sometimes referred to as revisions, conspectuses or synopses, in decreasing order of thoroughness (Stace, 1984)). A monograph will usually be more taxonomically rigorous, for example as to nomenclature, than a Flora, but some Floras (often called 'critical') are designed as series of monographs, each treatment being produced by an expert on the group (Stace, 1984; Funk, 1993).

The information in Floras, monographs and other taxonomic literature is only as good as the plant material available to the authors. Distribution maps, for example, sometimes reflect the collecting routes of botanists more than the actual distribution of the plants. Thus, the distribution map of the South American species *Elymus cordilleranus* (Fig. 6.2) shows an aggregation of populations around La Paz in Bolivia, Lima in Peru and along the Panamerican Highway in Ecuador. This is known not to reflect the true distribution of the species or its habitat preferences very faithfully. Similarly, taxonomic treatments go out of date as more material and thus data become available and species concepts change. In many groups, numerous taxonomic names of various

Fig. 6.2. The distribution of *Elymus cordilleranus* in South America from herbarium records.

---

[1] Editors' note. Notice the upper-case initial letter, and compare with the lower case initial letter of 'flora' referring to the ensemble of plant species found in an area. Following Stace (1984), this convention will be used throughout this volume.

ranks have been proposed, often referring to biologically insignificant morphological variants extracted from a virtually continuous range of variation. Even in an intensively studied group such as the Triticeae many entities remain ambiguous as to correct name and delimitation. Until such taxonomic issues are settled it will be difficult to interpret data on habitat preferences, distribution, etc. An important reason for collecting germplasm of wild species, particularly crop relatives, is thus to clarify their taxonomy and genetic relationships, for example by hybridization, mating systems and cytogenetic studies. This is a prerequisite for the optimal use of genetic resources in breeding and introduction programmes and for making informed decisions on conservation strategy.

In the planning process, published information must be supplemented with visits to a number of relevant herbaria, both international and local (Chapter 14). *H. muticum* is a diploid species of northern Argentina, Bolivia and Peru (Fig. 6.3; Bothmer *et al.*, 1991). It has only been possible to establish the northern part of this range through extensive studies of older herbarium material. The southern part of the distribution is rather well represented in germplasm collections, but there is no germplasm conserved from the north. *H. stenostachys* and the hexaploid *Elymus breviaristatus* ssp. *scabrifolius* are both perennial species native to northern Argentina, Uruguay and southern Brazil (Bothmer *et al.*, 1980). These species are far more evenly collected over most of their distribution than *H. muticum*. However, a few restricted areas (mainly in Brazil and Uruguay) still need work. Consideration of the passport data associated with past germplasm collecting in conjunction with other information on distribution often reveals such geographical gaps in existing collections. These will be high-priority areas for conservation collecting. However, though some effort must certainly be invested in finding out about existing collections, experience shows that quite often even areas from which some material is already in gene banks must be re-collected. Reasons range from the unavailability of conserved material to the poor ecological coverage of collections (Chapter 3).

Regions where large numbers of taxa in the target gene pool are concentrated (centres of diversity) will be priorities when collecting to elucidate taxonomy and relationships, though they may not necessarily also show high levels of genetic diversity within taxa. In the Triticeae, examples of regions with relatively large numbers of species are southern South America for the genus *Hordeum*, Central Asia for *Elymus* and southwest Asia for *Triticum–Aegilops*. However, such regions are often the most intensively covered by previous workers. Geographical regions and habitats which have not already been much visited by botanists, and from which information is therefore scarce, will also be important targets.

Special consideration should also be paid to isolated populations, and those on the edges of the ecogeographical distribution of the gene

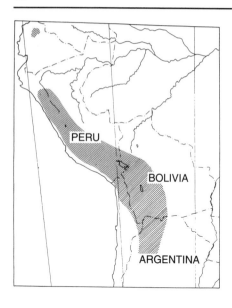

Fig. 6.3. The distribution of *Hordeum muticum* in South America from herbarium records.

pool. Populations in marginal areas or in unusual, distinctive or isolated habitats have a higher likelihood of representing distinct taxa, or at least of harbouring unique traits, though they tend to be genetically more uniform than more central populations. Apparently homogenous regions may reveal species exclusively found in scattered microhabitats, showing an island-like distribution (Hamrick, 1983). Only a very preliminary indication of this may usually be gained from studies of the taxonomic literature and of herbarium material. For example, it will rarely be possible to determine that a taxon is confined to a particular soil type from a Flora or monograph. This may sometimes be ascertained from herbarium labels or the ecological literature, but often it has to be documented by collectors themselves in the field.

'Critical' species groups showing complicated patterns of variation (e.g. introgression) need particular attention during collecting. They may be undergoing active differentiation and often there are no sterility barriers or extrinsic isolation mechanisms among the entities, which are sometimes not easily distinguishable even after close examination or crossing experiments, let alone in the field. An example is the *H. patagonicum* group from southern South America, which grows primarily in the steppes of Patagonia. It is a diploid perennial complex, probably of recent origin, which is in a phase of active differentiation in response to ecological and geographic heterogeneity (Bothmer *et al.*, 1986, 1988). Five subspecies are recognized, but the variation pattern is much more complex than is suggested by the taxonomy. Often, 'typical' representatives of two or even three of the subspecies grow together and intermediate forms are frequent. The *H. brevisubulatum* complex is an older group which occurs in Asia with a wide distribution from western

Turkey to eastern China (Bothmer, 1979; Dewey, 1979; Landström *et al.*, 1984; Baum and Bailey, 1991). It is a perennial, polyploid complex with a self-incompatibility system and thus with very heterozygous populations. Five major subgroups, usually treated as subspecies, have been recognized, and these are mainly allopatric. However, in the contact zones between the subspecies, much morphological variation is found. This is obviously the result of hybridization, segregation and introgression. Obviously, the different subspecies have not developed any intrinsic sterility barriers.

Zones of primary and secondary intergradation are of particular importance for collecting living material for studies of speciation. There is also evidence for higher frequencies of rare alleles in hybrid zones (Barton and Hewitt, 1985). During both the fieldwork and the subsequent investigations, the different morphotypes should be kept separate as far as possible, though collectors should be careful not to spend valuable time collecting what may be nothing more than malformed individuals. Bulk seed samples of each variant should be collected separately, each accompanied with a voucher herbarium specimen, but samples need not usually be large (ten individuals). Vegetative samples of each morphotype would provide a useful comparison with the seed samples. Variation may be cryptic, i.e. confined to anatomical or cytological features not readily observable in the field. For example, *H. bulbosum* occurs in diploid and tetraploid cytotypes which cannot be easily separated morphologically. If such cryptic forms as chromosome races are known to be sympatric and likely to coexist at a site, collecting can be on an individual plant basis. Collecting should be concentrated in areas and sites where the taxa meet and where transitional types occur, but it is also important to make a series of reference collections from areas and populations where the different entities are not in contact.

It is very important in all wild species collecting to supplement each germplasm sample with a herbarium specimen. The most frequent source of confusion in follow-up work with material raised from seeds is incorrect field determinations. This confusion can often be avoided, or at any rate diminished, if a herbarium specimen of the original population is made at the time of collecting. Often, good herbarium material cannot be collected at the optimal time for seed collecting. Two-stage visits may be necessary in such cases. Painted wooden stakes, coloured plastic tags and the like can be used to mark target populations. Particularly careful note should also made of the exact location of sampled populations.

Documenting the environment at the collecting site is important in all germplasm collecting, but particularly in wild species collecting. Related taxa are sometimes separated more by differences in habitat preferences than in morphology, and the ecology of the collecting site may thus hold clues as to the identity of the material. Such passport data as latitude, altitude, soil type and pH, drainage and associated species will also help future users screening collections for particular adaptive traits. Population information should also be documented in the field. The different parameters of importance (e.g. population size, pattern) are discussed in Chapter 19. Such information may be used in the assessment of the conservation status of the target species. The conservation status (number of populations, size of population, risks faced) of most species in the secondary and tertiary gene pools of crops is at best incompletely known. When we do have such information, it may be based on old records. The status of many species and populations may have changed considerably, for example due to urbanization, changes in agricultural systems or land use, deforestation or pollution.

It is an important obligation of collectors to gather information about any changes in the frequency and geographical extent of taxa, and the reasons for such changes. Thus, there is little doubt that *H. secalinum* has declined during the last 80–100 years due to changes in farming practices, in particular the increased use of fertilizers. In Sweden and parts of Denmark it has become very rare and threatened (Bothmer and Jacobsen, 1980). *H. intercedens*, which has been recorded from southern California in the USA to adjacent Baja California in Mexico, was formerly rather abundant in the vernal pool habitats which were very characteristic of this area. Due mainly to urbanization the species has decreased markedly in California in the last 50 years, but it is still fairly abundant in Baja California (Bothmer and Jacobsen, 1982; Bothmer *et al.*, 1982; Baum and Bailey, 1988a). A similar example is *H. arizonicum*, native to a very small area in the southern USA and northern Mexico. The species has now disappeared from most of its former localities. When canals with concrete walls were built for irrigation the species lost its natural habitat, the banks of small ditches or creeks, and it is now very close to extinction (Bothmer and Jacobsen, 1982; Craig and Fedak, 1985; Baum and Bailey, 1988b). This kind of information is essential in evaluating the different possible actions that could be taken to ensure the adequate conservation of the gene pool, and the active collector is often the person best placed to obtain it.

However, care must be taken as a species may change in frequency and in extent of distribution for entirely natural reasons. In comparatively well-known areas like northern Europe it may be quite easy to judge whether a species is declining for natural reasons or not, as in the case of *H. secalinum*. In other cases it may be more difficult. In species like *H. erectifolium* and *H. guatemalense*, only recently described and known from only a single population or a very restricted area, it may not be possible to judge whether the populations are declining or increasing (Bothmer *et al.*, 1985; Bothmer and Jacobsen, 1989). Populations of such a species should be carefully monitored, and possibly *in situ* conservation efforts instituted to complement *ex situ* activities.

Collecting for genetic diversity study and conservation

The material chosen for long-term storage in a gene bank should be representative of the genetic variation within each species within the target gene pool. Populations within a species may diverge genetically to different extents, possibly resulting in ecotypes and/or clines. Geographical areas and individual populations themselves may be genetically homogeneous or heterogeneous. If a population or species is in immediate danger of severe genetic erosion or even extinction, it will not be possible to await detailed scientific investigations indicating what should be preserved. The only thing to do is to sample as quickly as possible following the basic strategy set out in Chapter 5. Ideally, however, a detailed analysis of the amplitude, partitioning and ecogeographic pattern of distribution of genetic diversity should precede germplasm collecting. This will allow the sampling effort to be directed as efficiently as possible, for example at the within-population level at the expense of the between-population level (or vice versa), at particular geographical areas of high overall genetic diversity, and perhaps at some species more than others.

Genetic diversity studies may be carried out using a variety of morphometric, biochemical and molecular methods. Isozyme electrophoresis has been the technique most widely used in studying the organization of genetic diversity in plant species (Schaal *et al.*, 1991; May, 1992). For example, Crawford's (1991) admittedly not exhaustive list records isozyme work on some 38 genera between 1983 and 1987 alone. Variation in seed storage proteins has also been used to study genetic diversity, the process of domestication and homologies among genomes (Gepts, 1990). However, advances in molecular techniques, including the perfecting of the polymerase chain reaction (PCR), have meant that DNA sequencing, restriction fragment length polymorphism (RFLP) analysis and randomly amplified polymorphic DNA (RAPD) analysis are becoming increasingly easily accessible and popular (Schaal *et al.*, 1991; Aquadro *et al.*, 1992; Hoelzel and Green, 1992). With the advent of these molecular methods in recent years, the range of species being covered has been increasing rapidly. Clegg (1990) and Schaal *et al.* (1991) compare and contrast isozyme and molecular methods for the detection and study of genetic variation.

It will clearly be worth searching the literature before embarking on a new genetic diversity study (Chapter 13). It may be that some genetic diversity data on the target species have already been published, though the study perhaps covered only a portion of the range and/or a limited number of traits or markers. If published data are lacking or inadequate, the collector may decide to carry out an exploratory, fairly coarse-grid genetic diversity survey of the target species prior to detailed germplasm collecting. Two types of approaches are possible. In one, germplasm is collected (either seeds or vegetative organs), grown out near the laboratory and material then harvested for analysis. The other approach involves the sampling of leaves directly from the field, and the transport of these samples back to the laboratory for analysis. The methodologies may be summarized as follows:

1.   (a) Seeds are collected from a number of individuals in each population, keeping the seeds from each individual separate. Seeds can then be tested directly (e.g. for seed storage proteins) back at the laboratory and/or germinated and the seedlings tested for isozymes and/or molecular markers. (Morphological traits may also be measured on the adult plants.) At least ten seeds should be collected per plant, if possible, to allow the determination of the maternal genotype with reasonable statistical confidence (e.g. Brown and Allard, 1970).
(b) Whole plants or appropriate plant parts (i.e. organs of vegetative spread or perennation) are collected and transported to the laboratory alive. There, they are transplanted, grown out and maintained for later use, or sampled and tested immediately, or samples are taken and freeze-dried or stored in liquid nitrogen. Appropriate temporary storage methods for vegetative material are discussed in Chapters 21, 22 and 23.
2.   Samples of vegetative tissue (usually leaves or leaf fragments) are collected in the field and transported to the laboratory for analysis. Like method 1(b), this recovers the actual genotypes present in the population, rather than the genetic 'potential' represented by seeds. How the tissue samples are stored during transport will to some extent depend on whether proteins or DNA are to be analysed. Proteins are more temperature-labile, but contamination is more of a problem with DNA (May, 1992). Tissue samples should be stored moist and cool while in the field. If the laboratory is nearby, the samples can be taken there in an ice chest and then frozen in liquid nitrogen, pulverized and stored at $-70°C$ until they need to be used. If the laboratory is at some distance, a better option may be to freeze-dry samples at some kind of base nearer the collecting site (Hamrick and Loveless, 1986b). It may be possible to carry liquid nitrogen containers into the field. For DNA, if refrigeration is not possible the best method of tissue preservation is rapid drying in individual vials containing anhydrous $CaSO_4$ or silica gel (Pyle and Adams, 1989; Chase and Hills, 1991; Milligan, 1992). Rogstad (1992) suggests preserving leaves in a saturated solution of NaCl and hexa-decyltrimethylammonium bromide (CTAB). Excessive temperatures and ultraviolet (UV) light (sunlight) should still be avoided. It is best to harvest the freshest material available.

Hamrick and Loveless (1986b) compare collecting leaf material and collecting seeds for genetic diversity surveys. An important advantage of leaf material is that collecting may be done over a longer period. With leaves, it is also easier to sample the entire population in an unbiased way, including all age classes and, in dioecious species, both sexes. Finally, fewer analyses will be necessary, as maternal genotypes can be determined directly, rather than having to be estimated (at least ten pro-genies are usually required to determine the genotype of a fruiting individual). Collecting seeds of course has the advantage that some will remain as a germplasm sample which it will be possible to store and use, whether in biosystematic studies or a breeding programme. A certain number of seeds from each individual could be bulked to constitute a population sample for conservation, though it is unlikely that such samples will be sufficiently large for immediate long-term storage, or represent enough individuals. Alternatively, a complementary popula-tion sample of seeds for conservation can be collected at the same time as the individual plant samples are gathered for genetic diversity studies.

Nei (1987) discusses how many loci and how many individuals per locus should be studied to accurately estimate genetic parameters such as average heterozygosity and gene diversity in a population (see also Hoelzel and Dover, 1991). His recommendation is that as large a number of loci be examined as possible. As few as 20 individuals per locus should be sufficient to keep the bias of the heterozygosity estimate reasonably low, as long as large numbers of loci (30–70) are investigated. If the number of loci that can be investigated is limiting, or when hetero-zygosity is high, sampling more individuals will improve the precision of estimates of genetic parameters such as diversity and distance. If the number of loci is about 25, the number of individuals should certainly not be less than 20. This compares with about ten (bulked) for taxonomic studies (as already suggested) and 50 or more (again bulked) for conser-vation collections (Chapter 5). To accurately reflect the frequencies of alleles in the source population, rather than simply being 95% certain of the presence of a copy of all common alleles in the sample (the basis for the figure of 50), Marshall and Brown (1983) suggest that sample size should be about 200 individuals. More detailed, special-purpose genetic studies may require even larger samples. For example, in order to test whether or not two diploid populations with estimated allele frequencies of two alleles of 0.50/0.50 and 0.55/0.45 differ at the 90% confidence level, one would have to sample 2081 individuals.

This discussion has somewhat begged the question of what con-stitutes the population. Delimiting the population is much more difficult for wild species than for crops, which are usually grown in fairly well-defined fields. In most wild species germplasm collecting, the population is usually pragmatically taken to be all the individuals of a particular taxon found in a particular, fairly ecologically homogeneous place at a particular time. In genetic diversity studies, however, a stricter defini-tion may be necessary, i.e. a local group of individuals of a particular taxon within which free exchange of genes is occurring in nature. This is sometimes referred to as the 'gamodeme' (Briggs and Walters, 1984). The spatial extent of gamodemes can be estimated from studies of pollen movement and seed dispersal (Levin and Kerster, 1974). The general impression that has emerged from such studies is that 'geneflow is quite restricted in plants and gamodemes are therefore small' (Briggs and Walters, 1984). Thus, species occurring in more or less continuous vegetation types over vast areas, such as the Russian steppe or the Argentinian pampas, may nevertheless be organized in relatively small gamodemes isolated by distance. Sampling of such species can be done in a systematic way, say every so many kilometres.

Sampling of individuals should be at random within the spatial extent of the gamodeme, or stratified random in the case of obvious environmental heterogeneity (Chapter 5). A minimum distance should be kept between sampling points, ideally dictated by consideration of the average size of clones in vegetatively spreading species and the leptokurtic pattern of dispersal of seeds and fruits.

If it is not possible to carry out preliminary genetic diversity surveys prior to germplasm collecting, and no previous work has been done on the target species, generalizations based on reviews of isozyme studies can provide rough predictions of the genetic structure of unstudied plant species (Hamrick et al., 1991). Many such reviews are available, assessing the influence of different parameters of life history, reproductive biology, and distribution on overall genetic variation within species, and how this is partitioned within and among populations (e.g. Hamrick et al., 1979, 1981, 1991, 1992; Hamrick, 1983, 1989; Loveless and Hamrick, 1984; Hamrick and Loveless, 1986a; Govindaraju, 1988; Hamrick and Godt, 1990; Schoen and Brown, 1991; Loveless, 1992). On average, inbreeding species show more variation among populations than outbreeders, and annuals more than perennials. On the other hand, individual populations of outbreeders are usually more variable than populations of inbreeders, and those of widespread species more than those of narrow endemics. Schoen and Brown (1991) suggest that preliminary surveys of genetic diversity will be more important for inbreeders than for outbreeding species, where there is comparatively less variation among populations in gene diversity.

There are some problems with using such generalizations to develop germplasm sampling strategies (Hamrick et al., 1991). First, the kinds of life history and other parameters considered typically only explain less than 50% of species-to-species variation in genetic diversity parameters. Species-specific factors are thus equally important. However, it seems clear that in order to effectively study the genetic variation of a species its breeding and pollination systems must be known. These should certainly be investigated before large-scale collecting is started (e.g. Brown, 1990). Some work may already have been carried out, of course. Fryxell (1957), for example, catalogues the mating systems of some 1200 plant species. Richards (1986) is a more up-to-date review. Crane and Walker (1984) provide information on the mode of pollination of crops. If the literature is no help and a special study is not possible, a qualified guess may in some cases be possible, for example based on the size and morphology of the reproductive organs. Thus, in the

Triticeae and other grasses, long-anthered species with large styles are usually outbreeders whereas short-anthered species with small stigmas are inbreeders.

It may not be possible to use the results of isozyme and molecular studies to predict variation in other traits, in particular quantitative morphological traits. Though results often correspond, this is by no means always the case (Schaal *et al.*, 1991). Large-scale clinal and ecotypic variation is sometimes not revealed by isozyme studies (Falkenhagen, 1985). Morphologically heterogeneous species are sometimes quite uniform at the biochemical or molecular levels and vice versa. As Marshall (1990) points out, there is nothing unexpected or unusual about this. There are good reasons, for example, why the kinds of morphometric differences used to distinguish infraspecific taxa, in particular domesticated forms from related wild taxa, often develop prior to divergence at isozyme loci (Crawford, 1991).

Hamilton (1994) argues that quantitative genetic variation and the structure of genetic correlations should be given at least as much importance, if not more, in formulating a sampling and conservation strategy as biochemical and molecular markers. While agreeing that quantitative traits are important, Marshall (1990) points out that the argument will be irrelevant to most collectors, who usually have almost no information at all on the population genetic structure of their target species.

> In such cases it would appear reasonable to suggest that the most effective strategy would be to spend only sufficient time at each site to ensure the collection of common genes, and to visit as many sites as possible, so as to maximise the opportunity of sampling those populations that are variable for each class of character or gene.

This is the basis of the collecting strategy set out in Chapter 5. Of course, genetic variation often follows environmental variation, though the proportion of genetic variation regarded as explicable by ecogeographic factors rarely exceeds 50% (Chapman, 1989). In the absence of genetic diversity information, climatic, physiographic and edaphic factors can be used to delimit areas which are different from each other but relatively homogeneous internally. In first-stage exploration for conservation, as many as possible of these areas should be visited and sampled. If environmental variation is clinal, sampling should be done systematically along the gradient. In the case of two-stage collecting, Marshall (1990) advises the prudent approach of ensuring that the most variable populations for each class of characters are sampled.

Collecting for use in a breeding programme

Breeders and plant introduction workers often have very specific requirements for particular traits. For example, there may be an urgent need for a source of resistance to an abiotic stress (e.g. salinity, drought, frost, etc.) or to a pest. One possible way of obtaining desired traits is to systematically screen available gene bank holdings of the target species and related taxa in the primary and secondary gene pools. If the traits are not available in existing collections, collecting missions deliberately aimed at searching for them in suitable areas will have to be considered. Even if target material is already present in collections, however, further collecting in the areas where the desired trait(s) or gene(s) is (are) most likely to occur may be justified. For example, a wider range of genetic backgrounds may be sought. Gaining first-hand experience of the variation within and among populations in the field is also an important consideration.

Characterization and evaluation data may be available for particular collections (or parts of collections), but the distribution of desired traits in the gene pool is often not known in detail. However, their presence can often be predicted with some confidence from consideration of the ecogeographical passport data associated with the accession. If material adapted to particular climatic conditions is being sought, only those ecogeographic regions satisfying these requirements will be targeted (homoclime strategy). Searching for germplasm with tolerance to abiotic stress should start in areas where the species has been exposed to the stress factor for a considerable period. Thus, possible sources of salt tolerance in *Hordeum* would be the *H. patagonicum* complex in South America and *H. bogdanii* in Central Asia.

Similarly, in the case of pest resistance, collecting will be focused on areas where there is a long documented history of coexistence of the host and the pest. Anikster *et al.* (1976) and Moseman *et al.* (1990) found a correlation between the occurrence of resistance to *Puccinia hordei* in wild barley (*H. vulgare* subsp. *spontaneum*) and the distribution of *Ornithogalum* species, which are the alternate hosts for the pathogen. However, it should be remembered that resistance may occur anywhere (Harlan, 1978).

Some very detailed investigations have been carried out of the distribution of disease resistance with wild crop relatives. For example, Moseman *et al.* (1985) screened 687 accessions of wild emmer (*Triticum dicoccoides*) from Israel for resistance to leaf rust (*Puccinia recondita* f.sp. *tritici*) and found a clear spatial pattern in the distribution of resistance over the country. The geographic distribution of resistance to stripe rust (*P. striiformis*) in wild emmer is described by Nevo *et al.* (1986). He explained the pattern with reference to rainfall, evaporation and temperature. Several genes were found to be involved in the resistance reactions. Resistance to stem rust (*P. graminis* f.sp. *tritici*) in wild emmer was found to be negatively correlated with two ecological factors, namely altitude and number of hot dry days (Nevo *et al.*, 1991). Accessions collected in marginal habitats where wild emmer was growing poorly and had lower grain weight were more susceptible to powdery mildew than were accessions collected at more optimal *T. dicoccoides* sites (Moseman *et al.*, 1984). The authors recommended collecting wild barley (*H. vulgare* ssp. *spontaneum*) as well as wild emmer from optimal habitats in areas where the powdery mildew pathogen is prevalent.

Spatial variation may occur even within populations, as shown by the distribution of resistance genes to powdery mildew in *H. vulgare* ssp. *spontaneum* (Segal *et al.*, 1987). There was a correlation between the occurrence of different hordeins (storage proteins) and resistance genes. Such correlation may facilitate the rapid screening of large collections. Epperson (1990) gives guidelines for the sampling of populations with a view to investigating the spatial pattern of genetic variation within them.

In summary, when collecting for specific traits, the target area should be chosen after studying the available ecogeographic and pest information (including information on the distribution of alternate hosts), analysing existing characterization and evaluation data and/or screening gene-bank accessions for the desired (or correlated) trait. Relatively small areas should be selected for careful, intensive sampling. Comparatively large random samples should be collected from populations likely to harbour the desired trait (Chapter 5). In addition to a population sample, a selective sample (e.g. of healthy individuals in an infested field) can be collected if the expression of the target trait is obvious and stable.

References

Anikster, Y., J.G. Moseman and I. Wahl (1976) Paradise specialization of *hordei* Otth. and sources of resistance in *Hordeum spontaneum* C. Koch. In: Gaul, H. (ed.) *Barley Genetics 3*. pp. 468–469. Verlag Karl Themig, Munich.

Aquadro, C.F., W.A. Noon and D.J. Begun (1992) RFLP analysis using heterologous probes. In: Hoelzel, A.R. (ed.) *Molecular Genetic Analysis of Populations*. pp. 115–157. IRL Press, Oxford.

Barton, N.H. and G.M. Hewitt (1985) Analysis of hybrid zones. *Annual Review of Ecology and Systematics* 15:113–148.

Baum, B.R. and L.G. Bailey (1988a) A taxonomic study of the annual *Hordeum depressum* and related species. *American Journal of Botany* 66:401–408.

Baum, B.R. and L.G. Bailey (1988b) A taxonomic investigation of *Hordeum arizonicum* (Poaceae: Triticeae) with reference to related species. *Canadian Journal of Botany* 66:1848–1855.

Baum, B.R. and L.G. Bailey (1991) A numerical taxonomic investigation of the *Hordeum brevisubulatum* aggregate. *Canadian Journal of Botany* 69:2011–2019.

Baum, B.R., J.R. Estes and P.K. Gupta (1987) Assessment of the genomic system of classification in the Triticeae. *American Journal of Botany* 74:1388–1395.

Bothmer, R. von (1979) Revision of Asiatic taxa of *Hordeum* sect. *Stenostachys*. *Botanisk Tidsskrift* 74:117–147.

Bothmer, R. von (1992) The wild species of *Hordeum*: relationships and potential use for improvement of cultivated barley. In: Shewry, P.R. (ed.) *Barley: Genetics, Biochemistry, Molecular Biology and Biotechnology*. pp. 3–18. CAB International, Wallingford.

Bothmer, R. von and N. Jacobsen (1980) A taxonomic revision of *Hordeum secalinum* and *H. capense. Botanisk Tidsskrift* 75:223–235.

Bothmer, R. von and N. Jacobsen (1982) The wild species of *Hordeum* in North America with special reference to the two endangered species *H. intercedens* and *H. arizonicum.* Internal IBPGR report, Rome.

Bothmer, R. von and N. Jacobsen (1989) Interspecific hybridization with *Hordeum guatemalense. Genetica* 79:147–151.

Bothmer, R. von, N. Jacobsen and E. Nicora (1980) Revision of *Hordeum* sect. *Anisolepis* Nevski. *Botaniska Notiser* 133:539–544.

Bothmer, R. von, N. Jacobsen, R.B. Jorgensen and E. Nicora (1982) Revision of the *Hordeum pusillum* complex. *Nordic Journal of Botany* 2:307–321.

Bothmer, R. von, N. Jacobsen and R.B. Jorgensen (1985) Two new American species of *Hordeum* (Poaceae). *Willdenowia* 15:85–90.

Bothmer, R. von, B.E. Giles and N. Jacobsen (1986) Crosses and genome relationships in the *Hordeum patagonicum* group. *Genetica* 71:75–80.

Bothmer, R. von, B.E. Giles and N. Jacobsen (1988) Taxonomy and variation in the *Hordeum patagonicum* group (Poaceae). *Botanische Jahrbuecher für Systematik Pflanzengeschichte und Pflanzengeographie* 109:373–384.

Bothmer, R. von, N. Jacobsen, C. Baden, R.B. Jorgensen and I. Linde-Laursen (1991) *An ecogeographical study of the genus* Hordeum. Systematic and Ecogeographic Studies of Crop Genepool 7. IBPGR, Rome.

Bothmer, R. von, O. Seberg and N. Jacobsen (1992) Genetic resources in the Triticeae. *Hereditas* 116:141–150.

Briggs, D. and S.M. Walters (1984) *Plant Variation and Evolution.* 2nd edition. Cambridge University Press, Cambridge.

Brown, A.H.D. (1990) Genetic characterization of plant mating systems. In: Brown, A.H.D., M.T. Clegg, A.H. Kahler and B.S. Weir (eds) *Plant Population Genetics, Breeding, and Genetic Resources.* pp. 145–162. Sinauer Associates Inc., Sunderland.

Brown, A.H.D. (1992) Genetic variation and resources in cultivated barley. In: Munch, L. (ed.) *Barley Genetics VI.* Volume II. pp. 669–682. Munksgaard International Publishers, Copenhagen.

Brown, A.H.D. and R.W. Allard (1970) Estimation of the mating system in open-pollinated maize populations using isozyme polymorphisms. *Genetics* 66:133–145.

Chapman, C.G.D. (1989) Collection strategies for the wild relatives of field crops. In: Brown, A.H.D., O.H. Frankel, D.R. Marshall and J.T. Williams (eds) *The Use of Plant Genetic Resources.* Cambridge University Press, Cambridge.

Chase, M.W. and H.H. Hills (1991) Silica gel: an ideal material for field preservation of leaf samples for DNA studies. *Taxon* 40:215–220.

Clegg, M.T. (1990) Molecular diversity in plant populations. In: Brown, A.H.D., M.T. Clegg, A.H. Kahler and B.S. Weir (eds) *Plant Population Genetics, Breeding, and Genetic Resources.* pp. 98–115. Sinauer Associates Inc., Sunderland.

Craig, I.L. and G. Fedak (1985) *Hordeum arizonicum* – threatened with extinction. *Cereal Research Communications* 13:269–271.

Crane, E. and P. Walker (1984) *Pollination Directory of World Crops.* International Bee Research Association, London.

Crawford, D.J. (1991) *Plant Molecular Systematics.* John Wiley & Sons, New York.

Dewey, D.R. (1979) The *Hordeum violaceum* complex of Iran. *American Journal of Botany* 66:166–172.

Dewey, D.R. (1984) The genomic system of classification as a guide to intergenetic hybridization with perennial Triticeae. In: Gustafson, J.P. (ed.) *Gene Manipulation in Plant Improvement*. pp. 209–279. Plenum, New York.

Epperson, B.K. (1990) Spatial patterns of genetic variation within plant populations. In: Brown, A.D.H., M.T. Clegg, A. Kahler and B.S. Weir (eds) *Plant Population Genetics, Breeding and Genetic Resources*. pp. 229–253. Sinauer Associates, Sunderland.

Falkenhagen, E.R. (1985) Isozyme studies in provenance research in forest trees. *Theoretical and Applied Genetics* 69:335–347.

Frederiksen, S. and O. Seberg (1992) Phylogenetic analysis of the Triticeae (Poaceae). *Hereditas* 116:15–19.

Fryxell, P.A. (1957) Mode of reproduction in higher plants. *Botanical Review* 23:135–233.

Funk, V.A. (1993) Uses and misuses of floras. *Taxon* 42:761–772.

Gepts, P. (1990) Genetic diversity in seed storage proteins in plants. In: Brown, A.H.D., M.T. Clegg, A.H. Kahler and B.S. Weir (eds) *Plant Population Genetics, Breeding, and Genetic Resources*. pp. 64–82. Sinauer Associates Inc., Sunderland.

Govindaraju, D.R. (1988) Relationship between dispersal ability and levels of gene flow in plants. *Oikos* 52:31–35.

Gupta, P.K. and B.R. Baum (1989) Stable classification and nomenclature in the Triticeae: desirability, limitations and prospects. *Euphytica* 41:191–197.

Hamilton, M.B. (1994) *Ex situ* conservation of wild plant species – time to reassess the genetic assumptions and implications of seed banks. *Conservation Biology* 8:39–49.

Hamrick, J.L. (1983) The distribution of genetic variation within and among natural plant populations. In: Schonewald-Cox, C.M., S.M. Chambers, B. MacBryde and L. Thomas (eds) *Genetics and Conservation. A Reference for Managing Wild Animal and Plant Populations*. pp. 335–348. Benjamin Cummings, Menlo Park.

Hamrick, J.L. (1989) Isozymes and the analysis of genetic structure in plant populations. In: Soltis, D.E. and P.S. Soltis (eds) *Isozymes in Plant Biology*. pp. 73–105. Advances in Plant Sciences Series Volume 4. Dioscorides Press, Portland.

Hamrick, J.L. and M.J.W. Godt (1990) Allozyme diversity in plant species. In: Brown, A.H.D., M.T. Clegg, A.H. Kahler and B.S. Weir (eds) *Plant Population Genetics, Breeding, and Genetic Resources*. pp. 43–63. Sinauer Associates Inc., Sunderland.

Hamrick, J.L. and M.D. Loveless (1986a) The influence of seed dispersal mechanisms on the genetic structure of plant populations. In: Estrada, A. and T.H. Fleming (eds) *Frugivores and Seed Dispersal*. pp. 211–233. Junk, The Hague.

Hamrick, J.L. and M.D. Loveless (1986b) Isozyme variation in tropical trees: procedures and preliminary results. *Biotropica* 18:201–207.

Hamrick, J.L., Y.B. Linhart and J.B. Mitton (1979) Relationship between life history characteristics and electrophoretically-detectable genetic variation in plants. *Annual Review of Ecology and Systematics* 10:173–200.

Hamrick, J.L., J.B. Mitton and Y.B. Linhart (1981) Levels of genetic variation in trees: influence of life history characteristics. In: Conkle, M.D. (ed.) *Isozymes in North-American Forest Trees and Forest Insects*. pp. 35–41. Technical Report 48. Pacific SW Forage and Range Experimental Station, Berkeley.

Hamrick, J.L., M.J.W. Godt, D.A. Murawski and M.D. Loveless (1991) Correlations between species traits and allozyme diversity: implications for conservation biology. In: Falk, D.A. and K.E. Holsinger (eds) *Genetics and Conservation of Rare Plants*. pp. 75–86. Oxford University Press, New York.

Hamrick, J.L., M.J.W. Godt and S.L. Sherman-Broyles (1992) Factors affecting levels of genetic diversity in woody plant species. *New Forests* 6:95–124.

Harlan, J.R. (1978) Sources of genetic defence. *Annals of the New York Academy of Sciences* 287:345–356.

Harlan, J.R. and J.H.J. De Wet (1971) Towards a rational classification of cultivated plants. *Taxon* 20:509–517.

Hoelzel, A.R. and G.A. Dover (1991) *Molecular Genetic Ecology.* Oxford University Press, Oxford.

Hoelzel, A.R. and A. Green (1992) Analysis of population-level variation by sequencing PCR-amplified DNA. In: Hoelzel, A.R. (ed.) *Molecular Genetic Analysis of Populations.* pp. 159–187. IRL Press, Oxford.

Kasha, K.J. and K.N. Kao (1970) High frequency of haploid production in barley (*Hordeum vulgare* L.). *Nature* 225:874–876.

Kellog, E.A. (1989) Comments on genomic general in the Triticeae (Poaceae). *American Journal of Botany* 76:796–805.

Landström, T., R. von Bothmer and D.R. Dewey (1984) Genomic relationships in the *Hordeum brevisubulatum* complex. *Canadian Journal of Genetics and Cytology* 26:569–577.

Lange, W. (1971) Crosses between *Hordeum vulgare* and *H. bulbosum.* 2. Elimination of chromosomes in hybrid tissues. *Euphytica* 20:181–194.

Lehmann, L. and R. von Bothmer (1988) *Hordeum spontaneum* and land races as gene resources for barley breeding. In: Jorna, M.L. and L.A.J. Slootmaker (eds) *Cereal Breeding Related to Integrated Cereal Production.* pp. 190–194. PUDOC, Wageningen.

Levin, D.A. and H.W. Kerster (1974) Gene flow in seed plants. *Evolutionary Biology* 7:139–220.

Löve, A. (1982) Generic evolution of the wheatgrasses. *Biological Abstracts* 101:199–212.

Löve, A. (1984) Conspectus of the Triticeae. *Feddes Repertorium* 95:425–521.

Loveless, M.D. (1992) Isozyme variation in tropical trees: patterns of genetic organization. *New Forests* 6:67–94.

Loveless, M.D. and J.L. Hamrick (1984) Ecological determinants of genetic structure in plant populations. *Annual Review of Ecology and Systematics* 15:65–95.

Marshall, D.R. (1990) Crop genetic resources: current and emerging issues. In: Brown, A.D.H., M.T. Clegg, A. Kahler and B.S. Weir (eds) *Plant Population Genetics, Breeding and Genetic Resources.* pp. 367–388. Sinauer Associates, Sunderland.

Marshall, D.R. and A.H.D. Brown (1983) Theory of forage plant collection. In: McIvor, J.G. and R.A. Bray (eds) *Genetic Resources of Forage Plants.* pp. 135–148. CSIRO, Melbourne.

May, B. (1992) Starch gel electrophoresis of allozymes. In: Hoelzel, A.R. (ed.) *Molecular Genetic Analysis of Populations.* pp. 5–27. IRL Press, Oxford.

Milligan, B.G. (1992) Plant DNA isolation. In: Hoelzel, A.R. (ed.) *Molecular Genetic Analysis of Populations.* pp. 59–88. IRL Press, Oxford.

Moseman, J.G., E. Nevo, M.A. El-Morshidy and D. Zohary (1984) Resistance of *Triticum diococcoides* to infection with *Erysiphe graminis tritici. Euphytica* 33:41–47.

Moseman, J.G., E. Nevo, Z.K. Gerechter-Amitai, M.A. El-Morshidy and D. Zohary (1985) Resistance of *Triticum dicoccoides* collected in Israel to infection with *Puccinia recondita tritici. Crop Science* 25:262–265.

Moseman, J.G., E. Nevo and M.A. El-Morshidy (1990) Reactions of *Hordeum spontaneum* to infection with two cultures of *Puccinia hordei* from Israel and United States. *Euphytica* 49:169–175.

Nei, M. (1987) *Molecular Evolutionary Genetics.* Columbia University Press, New York.

Nevo, E., Z. Gerechter-Amitai, A. Beiles and E.M. Golenberg (1986) Resistance of wild wheat to stripe rust: predictive method by ecology and allozyme genotypes. *Plant Systematics and Evolution* 153:13–30.

Nevo, E., A. Gerechter-Amitai and A. Beiles (1991) Resistance of wild emmer wheat to stem rust: ecological, pathological and allozyme associations. *Euphytica* 53:121–130.

Pickering, R.A. (1988) The attempted transfer of disease from *Hordeum bulbosum* L. to *H. vulgare* L. *Barley Genetics Newsletter* 18:5–8.

Pyle, M.M. and R.P. Adams (1989) *In situ* preservation of DNA in plant specimens. *Taxon* 38:576–581.

Richards, A.J. (1986) *Plant Breeding Systems*. Unwin Hyman, London.

Rogstad, S.H. (1992) Saturated NaCl–CTAB solution as a means of field preservation of leaves for DNA analyses. *Taxon* 41:701–708.

Schaal, B.A., W.J. Leverich and S.H. Rogstad (1991) Comparison of methods for assessing genetic variation in plant conservation biology. In: Falk, D.A. and K.E. Holsinger (eds) *Genetics and Conservation of Rare Plants*. pp. 123–134. Oxford University Press, New York.

Schoen, D.J. and A.H.D. Brown (1991) Intraspecific variation in gene diversity and effective population size correlated with the mating system in plants. *Proceedings of the National Academy of Sciences of the USA* 10:4494–4497.

Seberg, O. (1989) Genome analysis, phylogeny and classification. *Plant Systematics and Evolution* 166:159–171.

Segal, A., K.H. Dorr, G. Fischbeck, D. Zohary and I. Wahl (1987) Genotype composition and mildew resistance in a natural population of wild barley, *Hordeum spontaneum*. *Plant Breeding* 99:118–127.

Stace, C.A. (1984) *Plant Taxonomy and Biosystematics*. Edward Arnold, London.

Thörn, E. (1992) Selective chromosome elimination in barley: the *bulbosum* system. Possibilities and limitations in plant breeding. PhD Thesis. Swedish University of Agricultural Sciences, Svalov.

Xue, J. and K.J. Kasha (1992) Transfer of a dominant gene for powdery mildew resistance and DNA from *Hordeum bulbosum* into cultivated barley (*H. vulgare*). *Theoretical and Applied Genetics* 84:771–777.

Classifications of intraspecific variation in crop plants

7

P. Hanelt and K. Hammer

IPK, Corrensstr. 3, D-06466 Gatersleben, Germany.

Introduction

Germplasm collectors must be thoroughly familiar with what is known of the variation present within their target taxon if they are to sample it efficiently. In crops this can be many times greater than in wild plants, especially for species which were domesticated early and have been widely spread around the world. Such variation is the result of both natural and artificial selection pressures. The latter may be conscious or unconscious, and result from the application of diverse agricultural practices and from the disparate and changing demands of growers for specific agronomic and other properties. Variation may be in morphological, anatomical, karyological, physiological, biochemical and molecular characters. Exploratory genetic diversity surveys using biochemical or molecular markers may be useful preliminaries to germplasm collecting (Chapter 6). Most relevant for the collector in the field, however, will be variation in morphological traits and in ecological adaptation. Making use of a scheme for the classification of the morphological variation within a crop can help collectors keep track of what they are finding and compare the diversity of different areas (Chapter 19).

The term 'intraspecific' is used here to refer to variation within a cultivated taxon, but it should be pointed out that the crop in a wild–weedy-crop complex is often given subspecific rank (e.g. Harlan and de Wet, 1971). Morphological intraspecific variation has been studied in many crops, though often for only a limited part of their geographic range or for a restricted set of characters. From 1978 to 1993 the literature on crop taxonomy (and evolution) has been reviewed in a series of publications by staff at the Institute of Plant Genetics and Crop Plant Research (IPK), Gatersleben. A full list of these reviews is given by Hanelt *et al.* (1993b). This series has now been discontinued, its task taken over by the new *Plant Genetic Resources Abstracts* (Chapter 13). Zeven and de Wet (1982) and Schultze-Motel (1987) summarize information on the taxonomy and evolution of different crops. Floras sometimes consider variation within crop plants, though never in much detail. However, there are also specialized Floras dealing only with cultivated plants (see below).

There is general agreement about the necessity and importance of such studies (Mansfeld, 1953; Baum, 1981) in both applied and theoretical applications, ranging from the investigation of the history of the domestication of plants and their subsequent evolution to the characterization of germplasm. However, the procedures used to develop the classifications and the resulting schemes themselves are extremely diverse and a generally agreed approach has not yet emerged (Hanelt, 1986). Two extreme types of schemes may be recognized:

- complex hierarchical taxonomic subdivisions of a cultivated plant taxon, with many infraspecific taxa at several taxonomic ranks between the species and cultivar level (e.g. Dorofeev and Korovina, 1979; Nechanský and Jirásek, 1967);
- relatively simple, non-structured, special-purpose schemes with a few main groups (e.g. de Wet, 1978).

Because selfing results in the splitting up of variation within a crop into distinct homozygous lines, autogamous species tend to be relatively easier to classify in detail into many groups than allogamous species. In the past, this has led to over splitting, a trend that has been somewhat reversed by genetic studies.

The different methods of approaching the infraspecific taxonomy of crops are discussed in this chapter in so far as they may be relevant to the needs of collectors. For further details, see Hanelt (1986) and Hanelt *et al.* (1993a). The focus is on the literature in languages other than English, which tends to be somewhat overlooked.

Classifications

A classification scheme for classifications of the intraspecific variation of crop plants has been proposed by Hanelt (1986):

1. Formal taxonomic classifications:
    (a) diagnostic–morphological;
    (b) phenetic–numerical;
    (c) ecogeographic.
2. Informal taxonomic classifications:
    (a) diagnostic–morphological;
    (b) phenetic–numerical;
    (c) genetic.
3. Mixed classifications.

Two principal types of approaches are distinguished, formal taxonomic and informal classifications. Whereas in the former formally recognized categories are used (more or less) according to the rules of the International Code of Botanical Nomenclature (ICBN) and the International Code of Nomenclature of Cultivated Plants (ICNCP), informal classifications use non-standard categories. In informal classifications, therefore, nomenclatural problems resulting from the use of the ICBN and the ICNCP (and the fact that the two codes are not always compatible (e.g. Brandenburg and Schneider, 1988)) are avoided. However, a broadly accepted designation of a group is not guaranteed and, therefore, communication of information on the material under study is more difficult.

Formal taxonomic classifications

Diagnostic–morphological classifications

Usually these classifications are based on a few, easily recognizable morphological characters and allow a rapid overview of variation within a crop. Several major publication projects have been based on this type of infraspecific classification, e.g. the *Flora of Cultivated Plants of the Soviet Union* and *The Cultivated Plants of Hungary* (Máthé and Priszter, 1982). R. Mansfeld, the founder of the Gatersleben school of taxonomy, which has studied the infraspecific classification of several important cereal, legume and vegetable crops, provides a typical example with his morphological system of *Triticum aestivum* (Mansfeld, 1951). He considers 12 characters and organizes intraspecific variation into more than 400 varieties, each differing from related ones in only one character. Mansfeld's (1950) scheme for *Hordeum vulgare* is another example. Below the species level he applied the category of convariety (defined by Alefeld, 1866; cited by Helm, 1964) and accepted five, defined by major spike characters: convar. *vulgare* (convar. *hexastichon*), convar. *intermedium*, convar. *distichon*, convar. *deficiens* and convar. *labile*. Formerly, some of these convarieties had even been described as separate species (not least, by Linnaeus). There are some differences in geographic distribution and even some barriers among them which may indicate that this category has some biological significance. Varieties, of which 191 are described, are purely artificial entities, however. In fact, such classifications are as a rule rather artificial, especially at lower taxonomic levels.

The same principles have been applied to *Papaver somniferum*. Based on the classification of Danert (1958), Hammer (1981) developed a system containing three subspecies. Ssp. *setigerum* is the wild ancestor; ssp. *somniferum* and ssp. *songaricum* are both cultivated. The cultivated subspecies differ in having sulcate lobes of the stigmatic disc with dentate margins vs. flat lobes with entire margins. These characters have been considered as very important by *Papaver* taxonomists and also show clear geographic differentiation. The con-variety level is defined by indehiscent vs. dehiscent capsules, another important character indicating different stages of domestication (Hanelt and Hammer, 1987). The variety level is based on seed colour, resulting from selection pressure under domestication, and other characters.

Such classifications can be very useful to the plant collector. Since the morphological entities they define can be recognized relatively easily, they can be used as the basis of field checklists. Rapid comparison of different areas with regard to the variation found there is possible and gaps in collections can be identified. Assessments of variation at different times based on such classifications have been used to estimate genetic erosion, for example in Sicily (Prestianni, 1926; Perrino and Hammer, 1983).

In some cases, the classifications are of restricted applicability because they deal with the cultivated flora of a rather restricted area (e.g. Máthé and Priszter, 1982). However, even country Floras of cultivated plants may employ a comprehensive concept of taxa, allowing them to be used for even a worldwide survey. The *Flora of Cultivated Plants of the Soviet Union* is perhaps the best example. Important recent contributions, in addition to the already mentioned *Triticum* volume are: Kazakova (1978), Makaseva (1979), Fursa and Filov (1982), Smaraev and Korovina (1982), Girenko and Korovina (1988), Kobyljan-skij (1989), Kobyljanskij and Lukjanova (1990). The morphological classifications from the Gatersleben school are listed by Hammer (1981). In addition to the already mentioned *Hordeum vulgare*, *Papaver som-niferum* and *Triticum aestivum* studies, there are works on *Beta vulgaris*, *Brassica oleracea*, *Glycine max*, *Linum usitatissimum*, *Lycopersicon esculentum*, *Nicotiana rustica*, *N. tabacum* and *Pisum sativum*.

A potential problem with such schemes is that the availability of the publications describing them may be limited. Many are not available in English and may be difficult to obtain. As a result, some older classifica-tions, such as that of Percival (1921) on *Triticum*, are sometimes used even today in the English-language literature.

Phenetic–numerical classifications

These classifications consider a large number of characters. Various multivariate mathematical methods are used to calculate similarities among types and identify groupings. There are several examples (reviewed by Schultze-Motel, 1987) but none is particularly convincing in the context of formal taxonomy.

Ecogeographic classifications

Such classifications have been developed by the Vavilov school based on the hypothesis that, in an area where selection pressures by environmental factors, cultivation practices, progagation methods, etc. are relatively homogeneous, a crop will tend to have a certain genetic integrity (Vavilov, 1940). An example of ecogeographic classification is that proposed by Flaksberger (1935) for *Triticum aestivum*, which includes two subspecies, 15 proles and six subproles.

New taxonomic categories have often been introduced. Groups are largely defined by their geographic origin and by characters which reflect the agricultural and ecological conditions to which they are adapted (e.g. reproductive phenology, pest and disease resistance, growth characters, etc.). In general, field experiments are necessary to verify the results and to incorporate new accessions into such a classification. Therefore, they may not be directly applicable during the collector's fieldwork. However, they may be very useful for the characterization of collections, facilitating use of the material by breeders. There is still no bridge between formal ecogeographic classifications and the use of an ecogeographic approach in fieldwork (Chapter 14).

Informal taxonomic classifications

Diagnostic–morphological classifications

There is no example of this kind of approach describing the full extent of variation within a crop. There are, however, some regional studies. The classification of French bush bean cultivars is one. They have been arranged into three categories, i.e. groups, sections and classes (Anon., 1983). Pod characters (11 character states) are used for the differentiation of groups and sections, and leaf colour, pod length, colour of unripe pods and length of bracts (12 character states) for the differentiation of classes. The resulting system comprises five groups, 14 sections and many classes. Another example, also from *Phaseolus vulgaris*, shows that the input of biochemical methods (in this case, phaseoline types) can lead to phylogenetically more relevant groupings within an informal classification of the diagnostic–morphological type (Hammer, 1992).

Phenetic–numerical classifications

One of the best examples of this type of classification is the study of the South American cultivars of cassava (*Manihot esculenta*) by Rogers and Fleming (1973). They used 55 character states and defined 19 groups of cultivars. Within these groups there is a high degree of phenotypic similarity, and evidently also considerable genetic similarity. New material can be easily incorporated into the proposed classification scheme. However, the evaluation of the basic data for this type of study is very time-consuming.

Genetic classifications

This type of classification is only possible in crops whose genetics have been well studied, such as *Pisum sativum* (Blixt, 1979), where the genes responsible for the expression of many different characters are known. In peas there has also been an attempt to combine a formal diagnostic and a genetic classification (Lehmann and Blixt, 1984). It is difficult to incorporate new material into such classifications. Test crosses are necessary and multifactorial characters cannot be included at all. Somewhat different are classifications based on genomic composition. An example is that of Simmonds for the edible fruit-bearing bananas (Simmonds, 1966; Simmonds and Weatherup, 1990). These are classified by reference to ploidy ($2\times$, $3\times$, $4\times$) and the genomic contribution made by two diploid wild species (AA *Musa acuminata* and BB *Musa balbisiana*). Some 15 characters are used to distinguish among cultivar groups.

Mixed classifications

There is no single classification approach suitable for all possible demands. Different aims can be achieved with different types of classifications. A combination of classifications has been proposed by Hanelt (1972) for *Vicia faba*. A formal diagnostic classification into two subspecies, three varieties and six subvarieties, based mainly on seed size, form and structure of pods, was combined with an informal classification into 14 races, based mainly on ecogeographic data. A similar approach has been used for *Citrullus lanatus* (Fursa, 1981).

Conclusions

It is well known that most of the more important and widespread crop species are characterized by an enormous amount of intraspecific variation. Familiarity with this is essential for the effective collecting of plant genetic resources. There are many publications on the infraspecific taxonomy of crop plants, but many have been written in languages other than English, in particular the papers of the Vavilov school. A variety of methods have been proposed for the classification of crop plants. Most appropriate for collectors seem to be ones based on easily recognizable characters of the gross morphology. Variation in such characters can be used to establish taxonomically formal or informal diagnostic classifications. These will be no less useful for the later management of collections than for the collector in the field.

References

Anon. (1983) *Description et essai de classificaton de variétiés de Haricot Nain*. INRA, Versailles.

Baum, B.R. (1981) Taxonomy of the infraspecific variability of cultivated plants. *Kultur-pflanze* 29:109–239.

Blixt, S. (1979) Some genes of importance for the evolution of the pea in cultivation (and a short presentation of the Weibullsholm-P.G.A. collection). In: Zeven, A.C. and A.M. van Harten (eds) *Broadening the Genetic Basis of Crops.* pp. 195–202. PUDOC, Wageningen.

Brandenburg, W.A. and F. Schneider (1988) Cultivar grouping in relation to the International Code of Nomenclature for Cultivated Plants. *Taxon* 37:141–147.

Danert, S. (1958) Zur Systematik von *Papaver somniferum* L. *Kulturpflanze* 6:61–88.

de Wet, J.M.J. (1978) Systematics and evolution of *Sorghum* sect. *Sorghum* (Gramineae). *American Journal of Botany* 65:477–488.

Dorofeev, V.F. and O.N. Korovina (eds) (1979) Psenica [Wheat]. In: Breznev, D.D. (ed.) *Kul'turnaja Flora SSSR.* Vol. 1. Kolos, Leningrad.

Flaksberger, C.A. (1935) Wheat. In: Wulff, E.V. (ed.) *Flora of Cultivated Plants.* Vol. 1. State Agricultural Publishing Co., Moscow and Leningrad.

Fursa, T.B. (1981) Intraspecific classification of water-melon under cultivation. *Kultur-pflanze* 29:297–300.

Fursa, T.B. and A.I. Filov (1982) Tykvennye [Cucurbitaceae – *Citrullus, Cucurbita*]. In: Breznev, D.D. (ed.) *Kul'turnaja Flora SSSR.* Vol. 21. Kolos, Moscow.

Girenko, M.M. and O.N. Korovina (eds) (1988) Listovye ovoscnye rastenija [Leafy vegetables – asparagus, rhubarb, sorrel, spinach, purslane, garden cress, dill, chicory, lettuce]. In: Dorofeev, V.F. (ed.) *Kul'turnaja Flora SSSR.* Vol. 12. Agropromizdat, Leningrad.

Hammer, K. (1981) Problems of *Papaver somniferum* classification and some remarks on recently collected European poppy land-races. *Kulturpflanze* 29:287–296.

Hammer, K. (1992) Neu- und altweltliche Bohnen. *Vorträge für Planzenzüchtung* 22:162–166.

Hanelt, P. (1972) Die infraspezifische Variabilität von *Vicia faba* L. und ihre Gliederung. *Kulturpflanze* 20:75–128.

Hanelt, P. (1986) Formal and informal classifications of the infraspecific variability of cultivated plants – advantages and limitations. In: Styles, B.T. (ed.) *Infraspecific Classification of Wild and Cultivated Plants.* pp. 139–156. Clarendon Press, Oxford.

Hanelt, P. and K. Hammer (1987) Einige infragenerische Umkombinationen und Neubeschreibungen bei *Brassica* L. und *Papaver* L. *Feddes Repertorium* 98:553–555.

Hanelt, P., G. Linne von Berg and M. Klaas (1993a) Taxonomische Untersuchungen zur infraspezifishchen Variabilität bei Kulturpflanzen. *Vorträge für Pflanzen-züchtung* 25:212–227.

Hanelt, P., R. Fritsch, K. Hammer, J. Kruse, K. Pistrick and J. Schultze-Motel (1993b) Taxonomy and evolution of cultivated plants: literature reviews. *FAO/IBPGR Plant Genetic Resources Newsletter* 91/92:24.

Harlan, J.R. and de Wet, J.M.J. (1971) Towards a rational taxonomy of cultivated plants. *Taxon* 20:509–517.

Helm, J. (1964) 100 Jahre Kulturpflanzentaxonomie. Dr Friedrich Alefeld zum Gedächtnis. *Kulturpflanze* 12:75–92.

Kazakova, A.A. (1978) Luk [Allium]. In: Breznev, D.D. (ed.) *Kul'turnaja Flora SSSR.* Vol. 10. Kolos, Leningrad.

Kobyljanskij, V.D. (ed) (1989) Roz [Rye]. In: Dorofeev, V.F. (ed.) *Kul'turnaja Flora SSSR.* Vol. 2, Part 2. Agropromizdat, Leningrad.

Kobyljanskij, V.D. and M.V. Lukjanova (eds) (1990) Jacmen' [Barley]. In: Kriveenko, V.I.

(ed.) *Kul'turnaja Flora SSSR*. Vol. 2, Part 1. Agropromizdat, Leningrad.

Lehmann, Ch.O. and S. Blixt (1984) Artificial infraspecific classification in relation to phenotypic manifestation of certain genes in *Pisum*. *Agri Hortique Genetica* 42:49–74.

Makaseva, R.Ch. (ed.) (1979) Zernovye bobosye kul'tury [Grain legumes]. In: Breznev, D.D. (ed.) *Kul'turnaja Flora SSSR*. Vol. 4, Part 1. Kolos, Leningrad.

Mansfeld, R. (1950) Das morphologische System der Saatgerste, *Hordeum vulgare* L. s.l. *Züchter* 20:8–24.

Mansfeld, R. (1951) Das morphologische System des Saatweizens, *Triticum aestivum* L. s.l. *Züchter* 21:41–60.

Mansfeld, R. (1953) Zur allgemeinen Systematik de Kulturpflanzen I. *Kulturpflanze* 1:138–155.

Máthé, I. and S. Prisater (1982) *The Cultivated Plants of Hungary*. Akadémiai Kiadó, Budapest.

Nechanský, F. and V. Jirásek (1967) Systematische Studie über kultivierte Sommerastern (Gartenastern) – *Callistephus chinensis* (L.) Nees (Asteraceae). *Preslia* 39: 122–150.

Percival, J. (1921) *The Wheat Plant. A Monograph*. Duckworth & Co., London.

Perrino, P. and K. Hammer (1983) Sicilian wheat varieties. *Kulturpflanze* 31:229–279.

Prestianni, P. (1926) *I Frumenti Coltivati in Sicilia*. Commissione Provinciale per la Propaganda Granaria di Caltanissetta, Caltanissetta.

Rogers, D.J. and H.S. Fleming (1973) A monograph of *Manihot esculenta* with an explanation of the taximetric methods used. *Economic Botany* 27:1–113.

Schultze-Motel, J. (1987) Numerisch-taxonomische Studien an *Triticum* L. und *Aegilops* L. zur Theorie der Klassifizierung von Kulturpflanzen. *Kulturpflanze* 35:53–97.

Simmonds, N.W. (1966) *Bananas*. 2nd edition Longman, London.

Simmonds, N.W. and S.T.C. Weatherup (1990) Numerical taxonomy of cultivated bananas. *Tropical Agriculture* 67:90–92.

Smaraev, G.E. and O.N. Korovina (eds) (1982) Kukuruza [Corn]. In: Breznev, D.D. (ed.) *Kul'turnaja Flora SSSR*. Vol. 6. Kolos, Moscow.

Vavilov, N.I. (1940) The new systematics of cultivated plants. In: Huxley, J. (ed.) *The New Systematics*. pp. 549–566. Oxford University Press, Oxford.

Zeven, A.C. and J.M.J. de Wet (1982) *Dictionary of Cultivated Plants and Their Regions of Diversity*. PUDOC, Wageningen.

Sources of information on existing germplasm collections

8

M.C. Perry and E. Bettencourt

IPGRI, Via delle Sette Chiese 142, 00145 Rome, Italy.

Introduction

The planning that takes place before a germplasm collecting mission will ultimately determine its success. Since germplasm collecting has been going on for a considerable period of time and collections of many different species are now maintained by many organizations, part of such planning must be to learn as much as possible about any relevant previous work. Doing this may provide:

- firm justification for collecting particular species in a particular area at a particular time, or reasons for the modification of the original intentions and timing of the collecting mission;
- a better understanding of particular geographic areas that have already been explored for germplasm and of the genetic diversity found there;
- an alternative method of acquiring required germplasm through exchange.

This chapter will examine the various sources of information that may be available on existing germplasm collections. It will be assumed that the collector begins the planning of the collecting well in advance of the fieldwork. Depending on the information source, it may be necessary to request information up to a year or more ahead of time and the collector needs to schedule accordingly.

Sources of information

There are two broad categories of information on past collecting work, both of which will need to be consulted by the collector before setting out:

- general information on any relevant past collecting mission;
- specific information on individual samples collected during a mission and later included as accessions in a germplasm collection.

The key points of entry into the subject are existing germplasm collections. To help collectors locate pertinent collections, directories of germplasm collections are published by the International Plant Genetic Resources Institute (IPGRI) for most of the major groups of crops. These publications, a list of which is provided at the end of this chapter (Appendix 8.1) are available from IPGRI and many libraries. They list, on a species basis, all collections of which IPGRI is aware and give the following information on each:

- name and address of the collection/curator;
- number of accessions;
- status of samples (wild, cultivated, etc.);
- geographic coverage (country where samples were originally collected);
- documentation status (manual, computerized, etc.);
- availability and exchange regulations;
- level of characterization/evaluation;
- information on other institutes where duplicates of the collection are kept.

This information is a direct restructuring of IPGRI's in-house database of the world's germplasm collections. The database is continuously updated. If needed, up-to-date listings from the database for given species can be obtained from IPGRI headquarters in Rome as well as from its regional offices.

The information in the database and the directories of germplasm collections can be used to identify the best sources of the detailed information that will be required in planning a germplasm collecting mission or programme. A number of institutions have accepted responsibility for maintaining global or regional base collections of particular species, forming the basis of a proposed Food and Agriculture Organization (FAO) network of base collections. This is noted in their entry in the appropriate directory of germplasm collections and a list of these gene banks is available in IBPGR (1990). When the major holdings of germplasm of the target and any other relevant species from the target region, adjacent regions or other areas of interest have been identified, the curator(s) may be contacted for details both of the collecting missions and of the collections themselves.

A separate database is being developed by IPGRI for forestry species (TRESOURCE). Another major source of information on such species is the Forest Resources Division of FAO. It maintains information, available upon request, on forest genetic resources collections and on *in situ* conservation of forest genetic resources. It publishes an annual newsletter, *Forest Genetic Resources Information* (see below). Together with the International Union of Forestry Research Organizations (IUFRO), FAO's Forest Resources Division has also published a *Directory of Forestry Research Organizations* (FAO, 1993). This records whether each organization listed has a genetic resources programme. Carlowitz (1991) lists sources of seeds and microsymbiont inoculants for multipurpose trees and shrubs. National Research Council (1991) has an appendix listing sources of forestry species germplasm for research.

General collecting mission information

General information on a past collecting mission can be obtained from mission reports, which may be published or unpublished, and collectors' notebooks.

Collecting mission reports may be available fairly soon after the mission is completed and provide an important supplement to the directories of germplasm collections, which lag some years behind the collecting. Many collectors publish accounts of their missions in the *Plant Genetic Resources Newsletter* (the continuation of the *FAO Plant Introduction Newsletter* and the *FAO/IBPGR Plant Genetic Resources Newsletter*), which is issued four times a year. FAO's *Forest Genetic Resources Information* publishes mission reports as well as other kinds of information, including a list of recent literature of interest. Other relevant internationally available publications include various crop-specific newsletters (e.g. those of crop networks and those of the international agricultural research centres (IARCs), such as the International Center for Agricultural Research in the Dry Areas (ICARDA)'s *Rachis* for wheat and barley) and specialized national agricultural research journals in countries with large national plant genetic resources programmes (e.g. Germany's *Kulturpflanze*, continued internationally as *Genetic Resources and Crop Evolution* in 1992, the *Australian Plant Introduction Review* and the *Indian Journal of Plant Genetic Resources*).

Crop-specific *CAB Abstracts* may be consulted to track down published mission reports. A more general publication, *Plant Genetic Resources Abstracts*, produced in collaboration with IPGRI, has been available from 1992 (this is also available on CD-ROM). Its subject index lists collecting missions by species. Hawkes *et al.* (1983) is a useful bibliography derived from various *CAB Abstracts* for the period 1976–1983. See Chapter 13 for information on bibliographic databases in agriculture, including CD-ROMs.

Accounts of collecting missions published in these sources usually include general data on the natural and human environment of the collecting region (climate, topography, land use, agricultural systems and practices, etc.), the number of samples of each species collected, the collecting route taken, the degree of genetic erosion encountered and any important general features of the germplasm collected. They occasionally have limited information on individual accessions. The gene bank(s) where the material is stored is also normally specified. If it is not, the information can be obtained from the collector or the institute involved. The curator may then be contacted for fuller specific details on the samples collected.

Not all mission reports are published. To locate unpublished reports, the institute that sponsored or actually carried out the collecting mission should be contacted. The directories of germplasm collections will assist in locating institutes. The institute listed as maintaining the germplasm should be contacted, with a request that the query be forwarded accordingly if another institute within the country was in fact responsible for collecting the germplasm. In addition to national programmes, other sources of unpublished mission reports are IPGRI, the commodity IARCs and other crop-specific and regional research centres. A digest of collecting activities is usually available in the annual reports of these organizations. IPGRI also publishes regional bulletins which include information on recent collecting activities.

The institute to which germplasm collected through missions sponsored by the International Board for Plant Genetic Resources (IBPGR) (and now IPGRI) has been sent is recorded in another of IPGRI's databases. IPGRI does not maintain germplasm, but it does hold reports and passport data (either on paper collecting forms or in dBASE or ASCII files) on the germplasm collected on the missions it has sponsored. Each sample collected is divided into at least two subsamples. One is left in the country in which it was collected and the other is sent to an existing germplasm collection that has agreed to maintain and make it available for distribution, perhaps after regeneration and multiplication. Often, this institute will further duplicate the material. IPGRI's database contains the names and addresses of each of the institutes in which duplicates were originally deposited by the collector. This information is available on request.

The data in the database from which the directories of germplasm collections are derived to some extent duplicates that found in the database on IBPGR/IPGRI collecting missions. Generally, the directories and the associated database provide a more reliable indication of what is maintained in collections since germplasm may have died or not been successfully regenerated upon receipt. They should always be consulted as a primary source of information.

The notes and diaries collectors keep in the field may also provide valuable information, particularly those of collectors working for botanical gardens. Collectors often deposit their notebooks in their institute's library. The older botanical gardens have collectors' notebooks dating back many decades, comprising an invaluable source of historical information. For more recent collections, copies of notebooks may sometimes be obtained directly from the collectors themselves. Limited accession-specific data will be found in these documents. Information which may be found in notebooks but not always in reports includes the accessibility (and exact method of access) of specific locations, problems encountered, ethnobotanical observations, and casual observations of interesting species which were not collected.

Accession-specific data

Accession-specific data are often essential for proper planning, but very time-consuming to acquire and work with. For this reason, and because it is difficult to anticipate needs at the beginning of the planning process, as many data as possible should be acquired, consistent with one's capacity to process them. Accession-specific data include passport, characterization and evaluation data.

1. Passport data are information about a germplasm sample and the collecting site, recorded at the time of collecting. Examples are collector's name and institute, collecting date, collecting number, botanical name of species and any vernacular names, location of collecting site (name of locality and longitude and latitude), sample type (seed, vegetative, etc.), sample status (wild, cultivated, etc.), and site environmental characteristics such as altitude, topography and various soil features.
2. Characterization data are observations collected on traits that are highly heritable, can be seen with the naked eye and are fairly consistently expressed in all environments. Examples of such characters, which for some 60 crops are included in descriptor lists published by IPGRI, are flower colour, pod size, leaf shape, time to flowering, etc. A list of IPGRI descriptor lists is provided at the end of this chapter (Appendix 8.2).
3. Evaluation descriptors represent characters whose measurement is strongly influenced by the environment in which measurement is made, so that replication over years and/or location is needed for an accurate portrayal of accession-specific variation. Examples include number of fruits per plant and yield. Descriptor lists also have some evaluation characters.

The IARCs maintain germplasm collections of their mandate crops and their wild relatives. They also maintain passport, characterization and possibly some evaluation data. Besides the information on the material they hold, some IARCs may maintain databases with global or regional scope (also referred to as central crop databases). For example, the Centro Internacional de Mejoramiento de Maiz y Trigo (CIMMYT) has a regional database on maize and ICARDA maintains a global database on wild *Triticum* and *Aegilops* species as well as a regional (Mediterranean) forages database. The European Cooperative Programme for Crop Genetic Resources Networks (ECP/GR) has sponsored the setting up of 22 central databases as of 1991 (Perret, 1990). All of these include passport data (9–30 fields) but some also include a minimum set of characterization descriptors. A total of 140,750 accessions are represented, of *Helianthus*, forage grasses and legumes, *Hordeum*, *Avena*, *Prunus*, *Allium*, *Brassica*, *Beta*, *Pisum*. IPGRI can be contacted for more details on these databases.

Many gene banks publish printed catalogues and distribute them more or less widely. These usually contain data for passport and characterization descriptors. A printed catalogue may be very difficult to use, as the order of the entries is not always what is required for the particular task in hand, and much searching may have to be done. The data may, however, be available in computer readable format from the institute. Data on the CIMMYT maize holdings are available on CD-ROM, for example. Any request for such data should include precise details on which accessions are of interest and which data fields. It is not always reasonable to request all the data for a particular crop or species, or for all the crops from a given country, as this may represent several thousands of accessions and this size of data set may prove unmanageable. Journals such as *Economic Botany* and *Euphytica* occasionally publish the results of germplasm characterization and evaluation (Chapter 13 has more on literature sources).

Accessions of wild species are often accompanied by herbarium voucher specimens collected either in the field or during any subsequent growing-out. Such a voucher should be easy to locate as its identification number and place of deposit should be part of the passport data of the associated germplasm accession. Voucher specimens are normally collected at least in duplicate and often in triplicate, to be stored in the herbarium of the collecting institute, the national herbarium and a major international herbarium.

Botanic gardens are also involved in germplasm collecting and maintain a wide range of species, particularly if endangered or rare, taxonomically important or of ornamental value. Details of the live specimens they maintain are usually available on request. Collecting mission reports are not generally published, but summaries of collecting activities may be available in the annual report and similar publications. The periodical *Botanic Gardens Conservation News* often carries news of new additions and established collections. National data systems of botanical garden holdings are planned for Australia, China, the former USSR and the USA. Information on botanic gardens is summarized in the *International Directory of Botanic Gardens* (Heywood *et al.*, 1991). The typical entry includes the following information:

- name, address and status of the institution and names of director and curator(s);
- area, latitude and longitude, altitude and rainfall;
- taxa in cultivation;
- special collections;
- conservation collections;
- special gardens;
- associated nature reserves and natural vegetation in the garden;
- herbarium;
- seed list and seed bank;

- catalogue and records system;
- research and other facilities (e.g. library).

Many botanic gardens are now cooperating in the standard recording of their specimen data. This is called the International Transfer Format for Botanic Garden Records (ITF) and is available in a number of languages, including for example Chinese. Botanic Gardens Conservation International (BGCI, formerly the Botanic Gardens Conservation Secretariat) maintains a database, increasingly incorporating ITF records, containing information on about 1500 of the world's botanical gardens and their resources, *ex situ* conservation collections and their origins.

References

Carlowitz, P.G. von (1991) *Multipurpose Trees and Shrubs – Sources of Seeds and Inoculants*. ICRAF, Nairobi.

FAO (1993) *Directory of Forestry Research Organizations*. FAO Forestry Paper 109. FAO, Rome.

Hawkes, J.G., J.T. Williams and R.P. Croston (1983) *A Bibliography of Crop Genetic Resources*. IBPGR, Rome.

Heywood, C.A., V.H. Heywood and P. Wyse Jackson (1991) *International Directory of Botanic Gardens*. 5th edition. Koeltz Scientific Books, Koenigstein.

IBPGR (1990) *1989 Annual Report*. IBPGR, Rome.

National Research Council (1991) *Managing Global Genetic Resources. Forest Trees*. National Academy Press, Washington DC.

Perret, P.M. (1990) European germplasm programme marks progress with new focus on crop networks. *Diversity* 6:15–18.

APPENDIX 8.1
IPGRI directories of germplasm collections

Cereals (1990)	*Avena, Hordeum*, millets, *Oryza, Secale, Sorghum, Triticum, Zea* and pseudocereals
Food legumes (1985)	*Glycine*
Food legumes (1989)	*Arachis, Cajanus, Cicer, Lens, Lupinus, Phaseolus, Pisum, Psophocarpus, Vicia* and *Vigna*
Forages (1992)	Legumes, grasses, browse plants and others
Industrial crops (1981)	*Camellia, Cocos, Piper, Saccharum* and *Theobroma*
Industrial crops (1989)	*Beta, Coffea, Gossypium, Elaeis* and *Hevea*
Root and tuber crops (1986)	Aroids, *Dioscorea, Ipomoea, Manihot, Solanum* and others
Temperate fruits and nuts (1989)	*Actinidia, Amelanchier, Carya, Castanea, Corylus, Cydonia, Diospyros, Fragaria, Juglans, Malus, Mespilus, Morus, Olea, Pistacia, Prunus, Pyrus, Ribes, Rosa, Rubus, Sambucus, Sorbus, Vaccinium* and others
Tropical and subtropical fruits and tree nuts (1993)	*Anacardium, Ananas, Annona, Artocarpus, Carica, Citrus, Ficus, Mangifera, Musa, Passiflora, Persea, Phoenix, Psidium* and others

Vegetables (1990)   *Abelmoschus, Allium, Amaranthus,* Brassicaceae, *Capsicum,*
                                     Cucurbitaceae, *Lycopersicon, Solanum* and others

Directory of European Institutions Holding Crop Genetic Resources Collections (3rd edition, 1986)

In preparation: Oil crops
                                     Root and tuber crops

APPENDIX 8.2
Descriptor lists published by IPGRI

Anacardium occidentale (1986)
Ananas comosus (1991)
Arachis hypogea (1992)
Avena sativa (1985)
Beta (1991)
Brassica and *Raphanus* (1990)
Brassica campestris (1987)
Cajanus cajan (1993)
Carica papaya (1988)
Carthamus tinctorius (1983)
Chenopodium quinoa (1981)
Cicer arietinum (1993)
Citrus (1988)
Colocasia (1980)
Dioscorea (1980)
Echinochloa millet (1983)
Elaeis guineensis (1989)
Eleusine coracana (1985)
Forage grasses (1985)
Forage legumes (1984)
Fragaria vesca (1986)
Glycine max (1984)
Gossypium (revised, 1985)
Helianthus (cultivated and wild) (1985)
Hordeum vulgare (1982)
Ipomoea batatas (1991)
Lens culinaris (1985)
Lupinus (1981)
Malus (apple) (1982)
Mango mangifera (1989)
Medicago (annual) (1991)
Musa (1984)
Oryza (1980)
Oxalis tuberosa (1982)
Panicum miliaceum and *P. sumatrense* (1985)
Paspalum scrobiculatum (Kodo millet) (1983)
Pennisetum glaucum (1981)
Phaseolus acutifolius (1985)
Phaseolus coccineus (1983)
Phaseolus lunatus (1982)

Phaseolus vulgaris (1982)
Prunus (cherry) (1985)
Prunus armeniaca (apricot) (1984)
Prunus domestica (plum) (1985)
Prunus dulcis (almond) (1985)
Prunus persica (peach) (1985)
Psophocarpus tetragonolobus (revised, 1982)
Pyrus communis (pear) (1983)
Secale cereale and *Triticale* (1985)
Sesamum indicum (1981)
Setaria italica and *S. pumila* (1985)
Solanum melongena, S. aethiopicum, S. macrocarpon (and others) (1990)
Solanum tuberosum (cultivated) (1977)
Sorghum bicolor (1993)
Triticum and *Aegilops* (1989)
Tropical fruits (1980)
Vicia faba (1985)
Vigna aconitifolia and *V. trilobata* (1985)
Vigna mungo and *V. radiata* (revised, 1985)
Vigna radiata (mung bean) (1980)
Vigna subterranea (Bambara groundnut) (1987)
Vigna unguiculata (1983)
Vitis vinifera (1983)
Xanthosoma (1989)
Zea mays (1991)

In preparation: *Arracacia xanthorhiza, Capsicum, Dioscorea, Elettaria cardamomum, Fagopyrum esculentum, Hordeum, Juglans, Persea americana, Piper nigrum, Psidium*

Published information on the natural and human environment

9

G.C. Auricht[1], R. Reid[2] and L. Guarino[3]

[1]South Australian Research and Development Institute, Northfield Research Laboratories, GPO Box 1671, Adelaide, South Australia, 5001 Australia: [2]Department of Primary Industry, Pastures and Field Crops Branch, Mt Pleasant Laboratories, PO Box 46, Kings Meadow, Tasmania 7249, Australia: [3]IPGRI, c/o International Laboratory for Research on Animal Diseases, PO Box 30709, Nairobi, Kenya.

Introduction

Plant genetic resources collectors need information on different aspects of the environment of their target region. They need it to help them in a number of important tasks, which are listed in Box 9.1. In collecting plant germplasm for conservation, if direct information on genetic diversity is not available, the recommended strategy is the sampling of the widest possible range of different ecogeographic conditions (Chapter 5). It will thus be necessary to use data on features of the physical, biotic and human environment (such as climate, soil, vegetation and land use) in conjunction with data on the distribution of the target taxa derived from the literature, herbaria and existing germplasm collections, if collectors are to capture the greatest possible diversity of germplasm. Regions with a wider range of ecogeographic conditions will tend to be higher collecting priorities. Distribution data for target species often consist of nothing more than the name of a place: maps, atlases and gazetteers will be necessary to locate such sites. Knowing the range of conditions occurring in a given target area will also allow the collector to design a more user-friendly collecting form, actually listing the range of vegetation, soil types, etc. likely to be encountered (Chapter 19).

Certain types of information, for example on human demography, land use, deforestation and desertification hazard, can be useful guides in assessing the likelihood or severity of genetic erosion in an area, and hence also in determining the priority and urgency of collecting. In this category will also come data on climate change.

If the aim is to procure germplasm of particular taxa or genotypes with known ecological requirements or adapted to particular ecological conditions, again published sources of information on the environment will be essential planning tools. Where do acid soils occur in the target

---

Box 9.1

Uses of environmental information

- To define ecogeographically and agroecologically homogeneous areas, which can be used as the basis for stratified sampling.
- To estimate the extent of biodiversity in an area on the basis of the extent of heterogeneity in the physical, biotic or human setting.
- To develop better collecting forms.
- To estimate the threat and extent of genetic erosion.
- To predict the presence/absence of a species or genotype in a given area, on the basis of its ecological preferences.
- To predict the best timing for collecting.
- To find their way around in the field.
- To estimate the rate of loss of seed viability after collecting.
- To fully document the collection.

---

region? Where is rainfall below 500 mm? Are there any areas where these conditions coincide? These are the kinds of questions that collectors will often need answers to.

Information on the environment will also be necessary in deciding when to visit an area: the approximate timing of the flowering and fruiting season can be estimated from monthly temperature and rainfall data, and the estimate refined on the basis of actual weather reports. Climatic data can also be used to identify areas where seed viability loss after collecting is likely to be unacceptable, so that the collector's stay in the area can be minimized (Chapter 20).

Once in the field, maps of various kinds will be necessary to adapt the collecting route and to actually find localities. The environmental information gathered together before the mission will also be important in documenting collecting sites. The passport data associated with each germplasm sample will include not only data collected in the field at the collecting site, but often also data gleaned from maps and other reference sources (Chapter 19). These will be useful in interpreting the data that are collected when material is characterized and evaluated, for example. In turn, such analyses will inform not only the use of conserved germplasm, but also future exploration, allowing collecting strategies to be refined and improved.

A broad overview of the environment, complete with maps, explanatory text and bibliography can be found in the *Atlas of the Environment* (Lean *et al.*, 1990). Tables of national-level statistics on the state of the environment are also published annually as *World Resources* (WRI *et al.*, 1994). Data from such sources will be useful in developing global strategies and priorities, but the collector will also need sources of more detailed information. These are the subject of this chapter. Most detailed data on the environment will be in the form of published maps, though they may also be presented as tables and diagrams. However, geographically referenced environmental data are also available in digital form. Increasingly, countries are digitizing their maps of the physical environment and their biodiversity information. Though some digital data sets will be briefly mentioned here by way of introduction, a fuller discussion can be found in Chapter 16.

Types of environmental data

Geographic, climate and socioeconomic data sets are discussed below in turn, with examples and some information on sources.

Geography

Geography is used broadly here to include administrative boundaries, topography (forming a base data layer), geology, soils and land use. Various kinds of maps can provide information on different aspects of these topics. Digital base data layers are discussed in more detail in Chapter 16.

Cadastral maps

These show administrative boundaries. They can be useful when documenting the location of a collecting site, for which a primary administrative unit, and if possible secondary and tertiary units, should be quoted. In large and federal countries (e.g. India), maps of individual states or provinces are common.

Road maps

These are vital in planning routes both before setting out and from day to day in the field. They display the road network, the locations of cities and towns and sometimes administrative boundaries and the more obvious physical features, usually at a relatively small scale, typically about 1 : 1,750,000.

Topographical or relief maps

These show altitude contours and spot heights, in addition to roads, towns, lakes, rivers, etc. Sometimes, they include basic data on vegetation and land use. They will be useful in identifying areas of high environmental diversity and may be necessary for taking altitude readings in the field. Topographical maps at 1 : 250,000 or better are commonly taken to the field and annotated to show the location of collecting sites. Digital relief surfaces (or terrain models) for various regions are either already available or under development (Chapter 16).

Hydrological maps

These give information on groundwater and surface drainage as well as showing more obvious hydrological features, such as lakes, rivers, streams, swamps and dams, which are often of interest to the collector. Some 70 countries are contracting parties of the *Convention on Wetlands of International Importance Especially as Waterfowl Habitat* (also known as the *Ramsar Convention*). Many have carried out wetlands surveys in consultation with the Ramsar Convention Bureau and other interested organizations, such as the World Conservation Union (IUCN). One of the more important outputs of such surveys are maps of wetlands distributions within the country, using the standard classification system adopted as part of the Convention.

Geological maps

These show the distribution of different kinds of solid rocks and unconsolidated material. Most geological maps use as mapping units stratigraphic (i.e. age) categories rather than categories of more immediate interest to the plant collector (e.g. ones based on mineralogy). They may or may not include structural features such as faults and folds. The Commission for the Geological Map of the World publishes 11 *Listes des Cartes Géologiques Nationales et Internationales* (Parry and Perkins, 1987). There are also guides to available geological maps available on-line and on Compact Disk Read-Only Memory (CD-ROM) (see below).

Soil maps

Many different classification systems for soils are in use, ranging from local or national schemes to the international classifications of the Food and Agriculture Organization–United Nations Educational, Scientific and Cultural Organization (FAO–Unesco), the US Department of Agriculture (USDA) and the Commission de Pédologie et de Cartographie des Sols (CPCS). The FAO system (FAO–Unesco–ISRIC, 1988) was initially intended as a legend to the *Soil Map of the World* at 1:5 million, but it is in fact a soil classification and has been widely treated as such. The map is available in digital form. The US soil taxonomy (USDA, 1975) is widely used throughout the Americas, as is the CPCS (1977) system in francophone countries. Many countries use their own detailed national classification systems, but increasingly such systems are congruent with international classifications. The International Soil Reference and Information Centre (ISRIC) is preparing a worldwide digital database for soils and terrain on a scale of 1:1 million, converting existing soil maps to a common format (Chapter 16). Details of soil profile, texture, depth and colour may be shown on the more detailed soil maps. Specialized soil degradation maps may show the level of salinity, the degree of soil erosion, the presence of toxic substances, fertility and structure decline, etc. White (1983) reviews published soil mapping for Africa. FAO keeps an index of soil maps.

Land use maps

These show how the human population interacts with and exploits the land and the vegetation cover. Again, a local classification system may be used, or a more standard framework. Some examples of land use and farming systems classifications, and references, are given in Chapter 19. Land use maps may be very detailed, showing the distribution of individual crops. The *World Atlas of Agriculture* includes land use and relief maps and country monographs (Committee for the World Atlas of Agriculture, 1976).

Vegetation maps

These may employ a local classification, or one of the more widely adopted schemes, for example White's (1983), Unesco's (1973) and that used in the series of maps by the Institut de la Carte Internationale du Tapis Végétal (ICITV) (Chapter 10). In more specialized maps, the condition of the vegetation may also be shown, for example giving details of forest clearing, overgrazing, etc. This kind of information often comes from remote sensing. Chapter 10 deals with the published sources of information on the vegetation and ecology of an area, including maps.

Terrain maps

Various different types of terrain maps are in use. In one type, geomorphological physiographic units are derived from aerial photography and satellite imagery. More complex maps use a combination of physiography, soils, geology, climate and vegetation to define the different mapping units or land facets within land systems. Examples are the Land Resource Studies published by the Land Resources Development Centre (LRDC), UK, covering southern, central and western Africa. FAO, the Commonwealth Scientific and Industrial Research Organization (CSIRO) and the International Institute for Aerospace Survey and Earth Sciences (ITC), The Netherlands, have also used land system mapping extensively. Hilwig (1987) gives a list of terrain, land use, land evaluation, vegetation and other surveys employing remote sensing methods.

Protected areas maps

Many countries produce detailed maps of individual protected areas, often for tourist use, as well as more general maps of the whole country showing the distribution of national parks and the like. A good example of the former is the *Road Map of the Etosha National Park*, published by the Ministry of Wildlife, Nature Conservation and Tourism of Namibia, which in addition to roads and important sights also shows and describes the different vegetation types of the park. Basic protected area information is also often included in other types of maps, for example road maps and land use maps. The World Conservation Monitoring Centre (WCMC) has a digital data set showing protected areas worldwide. Chapter 10 considers sources of information on protected areas.

Climate

Climate may be described in terms of individual parameters, such as temperature and rainfall, or by means of synoptic combinations of such parameters, which are believed to have increased predictive power. Such data may be presented as text descriptions, tables, diagrams or maps.

To the plant germplasm collector, the most relevant climate data will be annual, seasonal or monthly means, extremes and ranges for temperature and rainfall, and perhaps such secondary, derived statistics as the difference between average temperature in the hottest and coldest months and the relative lengths of wet and dry seasons. Climate data should be long-term norms. A 30-year period is the rule, but 25-year and ten-year periods produce satisfactory means for precipitation and temperature respectively, although they are not quite as satisfactory for extremes; in any case, means should be at least over five years (Reid, 1980). A very basic list of parameters would include:

- mean annual temperature (°C);
- mean minimum temperature of the coldest month (°C);
- mean maximum temperature of the warmest month (°C);
- mean annual rainfall (mm);
- seasonality of rainfall (summer, winter, bimodal or uniform);
- number of dry months (i.e. months with mean monthly rainfall <60 mm in the tropics or <30 mm in the subtropics, latitude >23.5°);
- incidence of frost.

A fuller list would include monthly figures for rainfall and for mean, minimum and maximum temperature. The absolute minimum and maximum temperature are sometimes recorded. The incidence of drought and snow can also be important. Measures of the reliability and predictability of rainfall (e.g. mean deviation as % of annual mean) are often more relevant than average values in semiarid areas.

Martyn (1992) gives a 'general presentation of the climates of the continents and oceans, plus more detailed discussions of the climates of large states and groups of smaller states'. The bibliography includes references to climatic atlases and other data sources. The World Meteorological Organization (WMO) also publishes a *Bibliography of Climatic Atlases and Maps*. Good examples of major climatic atlases are those published by WMO itself (e.g. WMO *et al.*, 1979). Another extensive general descriptive work is the multivolume *World Survey of Climatology* (Landsberg, 1969–76), which includes not only tables of monthly means but also detailed descriptions of the climates of the continents. The *Ecosystems of the World* series is a useful source of general data on climate and other features of the physical environment (see Chapter 10 for references).

Such broad-scale, synoptic studies are of course ultimately based on raw time-series data from individual meteorological stations. There are about 200,000 stations collecting meteorological data worldwide, about a quarter keeping records in digital form, but they are by no means evenly spread around the world, nor do they all gather the same information to comparable precision. It is often difficult to find out exactly what data are available from the national meteorological service, unless it is formally published (e.g. Servicio Meteorologico Nacional Mexico, 1976), even for national plant genetic resources programmes attached to the same government ministry. However, WMO publishes the *INFOCLIMA Catalogue of Climate System Data Sets*, which brings together information on the existence and availability of climate data sets (usually time series, rather than long-term means), as provided to WMO by countries and specialized data centres.

Long-term means for meteorological stations are available in published form in a number of international accounts in addition to national publications. Specht (1988), for example, brings together information from Mediterranean-type environments worldwide. Another useful collection, this one presenting climate data for the whole world, has been published by the British Meteorological Office (Anon., 1965). The compilation of Smithsonian Institution (1944) is similar. A series of publications for Asia, Africa and Latin America presents the average values of the main agroclimatic parameters in the form of monthly and yearly tables (FAO, 1984, 1985, 1987). These data pertain to the main observing stations and are grouped by country. In addition to these average values, rainfall has been given particular attention and synthetic monthly rainfall probability tables are presented in the second part of each publication, giving a fairly uniform coverage of the main agroecological zones. FAO maintains a database of climate information (FAOCLIM). The CD-ROM *World Weather Disc* (Weather Disc Association) contains climatological records dating back to the 18th century from thousands of weather stations worldwide.

The international agricultural research centres (IARCs) also hold climatic data in digital and other forms, and publish them in various ways. For example, the Centro Internacional de Agricultura Tropical (CIAT) has a climate database including information from several thousand stations in Latin America (the South American Monthly Meteorological Database (SAMMDATA)) and Africa (Jones, 1987). Examples of IARC publications in this field include general works such as Huke (1982) and Sivakumar *et al.* (1984) and crop-specific studies such as Virmani *et al.* (1991) and Carter *et al.* (1992).

Though long-term climate means for meteorological stations are usually presented in various types of tables, a standard climate diagram is sometimes used. This has months on the horizontal axis, and shows the annual course of mean monthly temperature and rainfall superimposed on each other against vertical axes along which 1°C is equivalent to 2 mm of rainfall. This relationship approximates evaporation, so that drought is denoted when the precipitation curve falls below the temperature curve. Walter and Lieth (1960–67) and Walter *et al.* (1975) display in atlas form about 10,000 such climate diagrams from all over the world, summarizing up to 11 temperature parameters as well as showing the mean monthly temperature and rainfall curves.

Numerous attempts have been made to classify climate on a global or regional basis by combining together a range of climatic parameters in some kind of predictive model. A list of the major classification schemes is given by Young (1987). The most famous are:

- the Köppen (1936) classification, based on mean annual and mean monthly rainfall and temperature;
- the Holdridge (1967) life zones system, based on potential evapotranspiration ratio;
- the FAO (1978–81) agroecological zones (AEZ) system, based on temperature during the growing period, length of growing period and seasonality of rainfall;
- on a more restricted scale, Emberger's (1955) classification of Mediterranean environments, based on mean annual rainfall, mean minimum temperature of the coldest month and mean maximum temperature of the warmest month.

Young (1987) recommends that the Köppen classification be employed to characterize sites broadly and the FAO AEZ system for more specific characterization. There are also various national systems in use. For example, Kenya's system of agroclimatic zones divides the country into seven regions based on the ratio of mean annual rainfall to mean annual potential evapotranspiration and nine regions based on mean annual temperature (Kenya Soil Survey, 1982). The IARCs have been very active in agroclimatic and, by incorporating soils and other data, agroecological characterization, both of regions and of the growing environments of specific crops (e.g. Carter, 1987). Young (1987) lists some major agroclimatic surveys of countries and regions. Of course, there is no reason why collectors should adopt one of the standard agroclimatic classification systems. Given the raw data from the meteorological stations in a target region, multivariate analyses may be used to define climatically homogeneous areas (Chapter 15).

An important use of climate data is in the timing of collecting (see Box 9.2). Accounts of the phenology of wild species in the literature, herbarium label data, germplasm passport data, published information on sowing and harvesting times and so on will be important for this, but may need to be interpreted on the basis of climate data and fine-tuned on the basis of the actual weather at the time of collecting. Thus, if it is known from the literature and other records what is the most appropriate time for collecting seeds in a given area, a comparison of the climate of that area with that of a nearby target area from which there are no phenological data should make it possible to decide at least the order in which the two areas should be visited. A 5°C fall in mean annual temperature, for example, as will generally occur with a 1000 m gain in altitude, may delay crop maturity by three to four weeks. It is sometimes useful to plot species phenological data, e.g. peak flowering

---

Box 9.2
The timing of seed collecting: a case-study

Kiambi (1992) discusses the collecting of seeds of *Tamarindus indica* in Kenya. Much of Kenya is characterized by a bimodal rainfall pattern. Some tree species flower during one or the other, some during both, though there may be genetic differences in populations flowering at different times of the year. *Tamarindus*, however, has only one reproductive cycle per year, normally flowering towards the end of the long rains or just after. This normally means April–May, depending on the year. The time from flowering to seed maturation is four to six months, depending largely on temperature. The peak *Tamarindus* seed collecting season in Kenya is thus between August (lower altitudes) and mid-November (higher altitudes).

---

time or time of harvest, against such parameters as latitude (which controls photoperiod), altitude or mean annual temperature for a range of sites or years. On the issue of time of harvest, a note of caution is necessary. The harvest time of safflower in southeast China was communicated to a prospective collector as being May. On the spot, it turned out that this was the time when the petals, which are used for medicinal purposes, were harvested, and not the seeds. Similar misunderstandings may occur when collecting tuber crops.

A major determinant of plant growth and flowering is rainfall (except for species of permanently wet areas). Knowledge of when the rainy season(s) normally starts and ends is crucial in the preliminary planning of collecting, but long-term climate averages may not be a particularly useful indication of exactly the best time to visit an area in any given year. Source(s) of up-to-the-minute information will be necessary, which may include both official weather reports and the direct testimony of people who are in the target area or have only recently left it. FAO runs a Global Information and Early Warning System on Food and Agriculture (GIEWS), which assesses weather conditions and crop prospects on a regular basis. Remote sensing of the target area can also be a useful source of real-time phenology information, particularly in semiarid areas, where rainfall is erratic and unpredictable in both space and time (Chapter 15).

It is not just the timing of reproductive events but also their success that is affected by climatic factors. Houle and Filion (1993), for example, present models for predicting cone and viable seed production and seed mass in *Pinus banksiana* on the basis of a number of climatic variables. Such information will be valuable in deciding whether a given year is likely to be a good seed year, and therefore whether collecting should go ahead or be postponed.

Socioeconomic data

Knowing about the human environment in an area is just as important to the germplasm collector (in particular the crop collector) as ecogeographic or climatic information. Human diversity and biodiversity are inextricably linked. Chapter 12 considers this in more detail, and also discusses sources of information on human cultures in general and their relationship with plants in particular. Not all human diversity may be classified as cultural, however. There are also socioeconomic differences both between and within communities. Socioeconomic data sets that may be of relevance to germplasm collectors will include:

- human population data (age and size of settlements, demography, ethnic or linguistic composition, population movements);
- agricultural survey data (farming systems, production of different crops in different areas, size of holdings, type of land tenure, etc.);
- economic indicators (e.g. household income, commodity prices, employment);
- infrastructure (roads, railways, towns, markets, dams, development projects, etc.);
- level of services (education, extension, research, health, financial, etc.).

Such socioeconomic data will be important in predicting the extent and danger of genetic erosion (Chapters 4 and 12). They may also be important in developing an appropriate sampling strategy. For example, collecting crop landraces and/or the indigenous knowledge associated with them might be stratified within different climatic or ecological zones on the basis of socioeconomic group as defined by size of holding, household income or age of settlement. Production data will identify areas that are optimal and areas that are marginal for a crop, which may be due to cultural and socioeconomic as much as agroecological reasons. Again, these could be targeted as distinct strata.

What to look for: scale, accuracy and age

In planning germplasm collecting, maps are generally the most convenient form of display of environmental information. The importance of scale at this stage will depend on the aims of the mission. Relatively small-scale maps (e.g. 1 : 250,000) will be sufficient for most plant collecting programmes aiming to cover a whole country or a relatively large area of a country. Such maps were used in an International Board for Plant Genetic Resources (IBPGR) project to collect forages throughout the Italian island of Sicily, for example. A programme of rice collecting in southern and western Madagascar used the 1 : 500,000 maps issued by the Institut National de Géodesie et Cartographie: four sheets covered the target region. Larger-scale (e.g. 1 : 50,000), more detailed maps will, however, be preferable for missions that are more narrowly targeted, either ecologically or geographically. An example is the collecting of forage material adapted to low rainfall and sandy acidic soils in southern Portugal carried out by the South Australian Department of Agriculture in collaboration with the Portuguese national programme in 1992.

Ideally, only maps at scales of about 1 : 250,000 or larger should be used to determine latitude and longitude in the field (see Chapter 19 for an alternative to reading latitude and longitude off maps). Some road and general touring maps show latitude and longitude lines, which may be an important detail if topographic maps are not available. Also important in such circumstances is that road maps taken into the field should show more than just the main paved roads. Road maps for a given country often differ substantially in the degree of detail they show, and collectors will generally need to journey to relatively out-of-the-way areas.

Maps of rainfall and temperature will generally portray information as contour lines of equal values (called isopleths in general, isohyets and isotherms for rainfall and temperature respectively in particular). This allows the collector to determine a range for any collecting site with respect to each environmental variable (e.g. 250–500 mm mean annual rainfall). A single figure can be arrived at by taking into account the perpendicular distance from the site to the two nearest isopleths, but this may not give a very accurate result. Tables, on the other hand, give accurate information, but only for a specific range of points, usually the larger towns in a region, where meteorological stations are located. The information for the station closest to a given collecting site is often quoted as part of the documentation of the site, along with the distance between the site and the station.

Alternatively, interpolation can be carried out using a number of nearby stations. Interpolation of data from a grid of point sources is a difficult problem and may give inaccurate results in areas of great topographic heterogeneity, where environmental factors will vary widely over short distances. Chapter 15 describes how to derive equal-value contours from point data by hand. Most geographic information system (GIS) packages can carry out such analyses (Chapter 16). A digital terrain model (a quantitative representation of land-form in digital form) can be used to produce climate surfaces predicting temperature and rainfall parameters with good accuracy, and there will be such models for the whole world before too long (Chapter 16). The International Centre for Research in Agroforestry (ICRAF)'s Multipurpose Tree and Shrub Database programme also has a facility for deriving three temperature parameters from latitude, longitude and altitude data, based on a multiple regression model using FAO's FAOCLIM agroclimatic database (Carlowitz *et al.*, 1991).

The age of sources of environmental data can vary greatly. Whether this will be important will depend on the kind of data presented. For example, topographical data will not generally change (unless, for example, large dams are constructed), whereas land use or vegetation maps may relatively quickly become outdated. Road maps may also date very quickly in rapidly developing areas. Older geological and soil maps may use outdated classification systems. A gene-bank sample is a kind of 'time capsule' – a freezing of the otherwise continual process of plant evolution. Given the time frame over which collections are stored and the rate at which the environment can change, it is important that the site data associated with an accession be as relevant as possible to the time at which it was collected. (See Box 9.3 for the effects of possible global climate change.)

Socioeconomic data also need to be as up to date as possible (though historical data can also be of interest) and at a district (tertiary administrative division) level or below. The results of surveys and censuses need to be treated with caution. For example, conventional surveys typically do not capture women's socioeconomic and agricultural roles adequately unless a special effort is made to address their interests, interviewers are present at times when women are free to be interviewed, and the location of the interview encourages women's participation. Hill (1984) is an interesting commentary on the problems associated with official socioeconomic statistics.

Box 9.3
Global climate change and germplasm collecting

The possibility of global climate change presents an important challenge to the plant genetic resources conservation community. Not only may temperatures rise as a result of the 'greenhouse effect', but rainfall may also be affected, and in different ways in different regions. Some climates may be created which have no precedent in the recent history of the earth.

Regions and species that are particularly vulnerable to the complex alterations of climatic patters which many global warming models predict will need to be accorded a high priority by germplasm collecting programmes. Some of these priority areas are (IUCN, 1992; Markham et al., 1993):

1. The high altitudes of mountains at low and middle latitudes: as well as being particularly threatened by the rise in temperatures, these areas are often very high in endemism.
2. Low-lying coastal areas: these may be subject to rising sea levels.
3. Arid and semiarid continental tropical and subtropical areas: as well as temperatures increasing, rainfall in these areas may not only decline overall but also become more unpredictable, with increased frequency of droughts.
4. Wetlands: temporary and shallow freshwater wetlands in particular face the threats of desiccation and increased salinity.
5. Temperate and boreal forests: as tree species will migrate at their own, different paces, in general slower than the northward shift in climatic conditions, some isolated and relic populations are likely to become extinct.

For a review of climate change and plant genetic resources conservation, see also Jackson et al. (1990).

Where to look

How much environmental information may be available varies enormously from country to country and sometimes also within countries. Some areas will be covered by little more than a road map, others by a range of large-scale thematic maps. Some may have no meteorological stations, others several. Germplasm collectors thus really have two separate problems. One is to find out whether the maps and other forms of information they require are available. The second is to obtain them, if indeed they exist. These problems may be as acute for a national plant genetic resources programme as for a collector planning to work in a foreign country.

Some sources of information on published maps and other forms of environmental data have already been alluded to. Thus, though there is no single international bibliography on mapping, there are specialized published bibliographies on vegetation, geological and climate maps (see above and Chapter 10). There may also be national bibliographies and catalogues. The crucial reference is Parry and Perkins' (1987) *World Mapping Today*, which provides listings of current topographic and resource mapping on a country-by-country basis and describes the organization and structure of national mapping activities. There are also sections on each of the continents and the world as a whole. Hopkins and Jones (1983) list atlases, gazetteers and bibliographies (in the fields of physical geography, geology, soil, climatology, botany, agriculture, agricultural economics, anthropology, human geography, etc.) on a country and regional basis.

Maps are sometimes published in scientific journals. Wise (1975) lists the more relevant publications. Chapter 13 describes modern methods of searching the literature. There are several bibliographic databases specifically in the field of the environment. For example, the Tropical Agricultural Development Information System (TRADIS) is a bibliographic database developed by LRDC and available at CAB International (CABI), covering non-conventional literature on land resources in tropical countries, from the 1960s onwards. *Geobase*, available on-line via Dialog, covers the worldwide literature on geography, geology, ecology and related disciplines. France's Bureau de Recherches Géologiques et Minières offers both a regular hard-copy *Géocarte Information* service, and an on-line database of earth science maps. The *1993 Directory of Country Environmental Studies* (WRI *et al.*, 1992), 'an annotated bibliography of environmental and natural resources profiles and assessments' mainly covering the period 1987–92, is also available as a database on diskette from the publishers. The bibliographic database *Focus On: Global Change* (Institute for Scientific Information, ISI) may also be of use to collectors. Available on diskette only, it brings together published information from different disciplines on human interaction with the environment, including deforestation, climate change and environmental legislation. Table 9.1. gives relevant bibliographic databases available on CD-ROM.

Table 9.1. CD-ROM bibliographic databases on human interaction with the environment.

Name	Contents	Coverage
Earth Sciences Disc (US Geological Survey (USGS))	(i) Directory of earth science databases; (ii) guide to published geological maps; (iii) citations to materials held in the USGS library	From 1975
On-line Computer Library Center (OCLC) Environment Library	Worldwide coverage of environment-related sources	From 1960s
Environmental Periodical Bibliography	Worldwide collection of bibliographic records on environment issues and research	From 1973

The United Nations Environment Programme (UNEP) Global Environment Information Exchange Network (INFOTERRA) should also be mentioned in the context of information sources. It links users requiring information on, for example, pollution, desertification or environmental law to 155 national focal points, ten regional service centres and over 6000 sources, including bibliographic information sources like CABI. The main tool for accessing the most appropriate source(s) for a given enquiry is the *International Directory of Sources*, which is both a digital database and a hard-copy publication.

Turning now to actually obtaining maps, perhaps the easiest way is in atlas form. A standard global atlas is virtually a necessity for any collecting programme, national, regional or international. Many are available, in many languages, but two have become standards (in English), namely the *Times Atlas of the World* (Comprehensive Edition) and the *National Geographic Atlas of the World* (6th edition). Many national and regional atlases are also available, giving not only topographic but also thematic information of various kinds, for example on agriculture. Federal countries often have atlases on individual states. Details of national atlases and gazetteers are given by Parry and Perkins (1987).

Loose large-scale maps are generally more difficult to obtain than atlases. Both are, however, usually available in national libraries and the libraries of universities and relevant government departments and institutes (see case-study in Box 9.4). Global Resources Information Database (GRID)'s Meta-Database (Chapter 16) includes details of the environmental information holdings (electronic and otherwise) not only of GRID itself, but also of national and other institutions (as of mid-1993, covering Central and South America and Africa). It will clearly not be possible to annotate maps held by libraries or other institutions or take them into the field. With older maps that are out of copyright, it is sometimes possible to make copies of various kinds. Major libraries such as the British Library, for example, run high-quality map-copying services for material in the public domain. Permission from the publisher is needed for material still covered by copyright. However, for the most part collectors will need to purchase maps.

Leaving aside the kind of basic road maps that are available in many general bookshops in large cities the world over, and thus usually present little problem to acquire, maps may be purchased either directly from their publishers (or their agents, the distributors: a list is provided at the end of the chapter) or from map jobbers, companies who are prepared to procure maps from a variety of publishers and pass them on to the user, either in specialized shops or by mail. Publishers of the kind of large-scale topographic, geological and soil maps necessary to germplasm collectors tend to be national government agencies, which will generally have an outlet for the purchase of maps in the capital city. Development agencies also assist countries in publishing mapping. One example among many is the Institut Français de Recherche Scientifique pour le Développement en Coopération (formerly Office de la Recherche Scientifique et Technique Outre-Mer (ORSTOM)) and their *Atlas de la Nouvelle Calédonie et Dépendances* (ORSTOM, 1981). Again, details are given in Parry and Perkins (1987). The catalogues of publishers, agents and jobbers are a very important source of information on available maps, quite apart from also usually including an order form and instructions for purchasing. Parry and Perkins (1987) list the more important map jobbers, mainly from Europe and the USA. Some addresses are given at the end of the chapter. The leading map jobber in the world is said to be GeoCenter ILH. Geoscience Resources specialize in geological maps.

Relevant socioeconomic data may be available in published form. For example, there may be a national agricultural bulletin or journal, and a national atlas may have sections on demography and agriculture. Whether this is the case or not, the obvious immediate source is the statistical department of the ministry responsible for agriculture. The records of local extension offices may have more detailed statistics on their district. There may also have been household surveys primarily aimed at sectors other than the agricultural (e.g. health) and conducted by other ministries, which may nevertheless contain relevant socioeconomic data. CABI's *Rural Development Abstracts* and *World Agricultural Economics and Rural Sociology Abstracts* may be useful to the plant germplasm collector looking for published socioeconomic data.

FAO, the IARCs and regional agricultural research organizations can be useful secondary sources of socioeconomic information. Several IARC publications combine socioeconomic and agroecological data for a specific crop. An example is Carter *et al.* (1992), and consideration of the methods of data acquisition used in that study may be helpful. Two main repositories of contemporary information on cassava production in Africa were tapped, the FAO statistics library at FAO headquarters, Rome, and the Economics Section of CIAT's Cassava Program. The actual sources used ranged from the *World Atlas of Agriculture* to more detailed national studies. For example, for Chad, total production for the country in 1982 from the 1984 Statistik des Aulandes, a publication of the German

---

BOX 9.4
Map availability: a case-study

The situation in Kenya can be taken as a fairly typical example. There is a *Catalogue of Maps* available for the country, describing the various maps produced by or available through the Survey of Kenya, the government department responsible for mapping. Parry and Perkins (1987) reproduce much of this information.

    The maps listed can be obtained from the Public Map Office in Nairobi, or through the Survey Offices in seven other major cities. Some may also be obtained through a number of agents, both in Kenya and abroad (including GeoCentre and Edward Stanford), and all these are listed with their addresses in the *Catalogue*. The maps listed include relatively small-scale road maps of the whole country, maps of national parks, various thematic maps, series of topographic maps at 1 : 250,000 and 1 : 50,000 and aerial photographs.

    Topographic maps and aerial photographs are available for sale only with the permission of the Director of Surveys, for other government departments as for the general public. The *Catalogue* includes an order form. The Survey maintains an Aerial Photographic Library in Nairobi where photographs may be inspected and orders placed.

---

statistical office, was converted into hectares using typical yields from a 1985 agricultural census in Mali, and for distribution among prefectures older data from a 1972/73 agricultural census were used. For historical data, extensive searches of pre-independence literature were carried out at the Agricultural University of Wageningen, the African Library of Brussels, the African Studies Centre in Leiden, the British Museum and the Food Policy Research Institute at Stanford University.

Conclusions and recommendations

High-quality environmental data are essential for the proper planning and execution of a plant germplasm collecting mission or programme and for the interpretation of data from the characterization and evaluation of the material collected. The quality and type of information available vary greatly, but the more information and the greater the detail which can be obtained, the better. (See Box 9.5 for a case-study.)

Collectors should make an effort to obtain data, preferably in map form, on roads, soils, relief, mean annual rainfall and vegetation in the target area. In addition, if available, data on temperature, geology, land use, protected areas and socioeconomic factors will assist in ensuring the best possible targeting of effort. Land system studies based on remote sensing are a useful further layer of information.

As a bare minimum, collectors will therefore need atlases – one of the standard international atlases plus a national one if available. The problem with these is that they will tend to be at relatively small scales and awkward to take to the field. So, ideally, collecting programmes should also have a full set of 1 : 250,000 or larger-scale maps, and small-scale road maps of the region they cover. For information on the availability of national atlases and of loose large-scale maps, the key reference is Parry and Perkins (1987). There are also more specialized information sources, and some of the more important of these are mentioned in this chapter and in Chapter 10 (e.g. White (1983) on the vegetation and general ecology of Africa). Collecting programmes should have access to the sections relevant to their target regions in these and the other main reference sources, so that an agenda for the acquisition of maps and other environmental data can be planned. The catalogue of the government department responsible for mapping is also a must, if available.

For climate data, if a national climate or agroecological classification study is available, perhaps in a national climate or more general atlas or in a paper published in a scientific journal, the collector will need to have access to it, or at least to one of the international climatic atlases, along with the relevant FAO AEZ study and a general synoptic work on the target region, for example the relevant section in the *World Survey of Climatology* series. For detailed point data on climate, the publications of the national meteorological bureau are the obvious starting-point. Information on these should be obtained from the *INFOCLIMA Catalogue of Climate System Data Sets*. If the data are not easily available from the mainly national sources listed in the *Catalogue*, one or more of the international compilations or databases described in this chapter will need to be obtained.

As for socioeconomic data, it is highly desirable for crop collecting programmes to have access to the latest agricultural census or survey data.

In the first instance, data will need to be procured as hard copy. Some digital data sets have been mentioned, and, if collecting programmes do not have their own GIS technology, there is always the possibility of cooperation with organizations that do. The days are perhaps not far off when collectors will be able to go into the field with a laptop computer containing a full range of digitized topographic, soil, climate, vegetation, etc. maps of the target area in a GIS, perhaps including satellite imagery, with a Global Positioning System (GPS) receiver feeding directly into it. However, until then, collectors will plan their work mainly using conventional maps, and then take some of these into the field to annotate.

What maps and other data on the environment will need to be taken into the field? It is essential to take topographic maps of the specific target area(s) at a scale of about 1 : 250,000 or larger (in multiple copies, ideally, as described in Chapter 19), on which collecting site localities will be marked, plus smaller-scale road maps of the general region to be travelled through. Especially when collecting wild species, it will usually be advantageous, and sometimes essential, to also take along large-scale soil, vegetation, geological and land use maps, in particular the first two. However, these will mostly be used at the stage of mission planning, to identify specific areas to be searched, for example, which could then be

---

Box 9.5

The mapping requirements of a collecting programme: a case-study

Again taking Kenya as a representative example, what follows is a list of the mapping and other information that a national collecting programme might consider assembling. The Gene-bank of Kenya at Muguga just outside Nairobi actually uses such a collection of sources in planning its collecting:

- 1 : 1,750,000 road maps of the entire country;
- full set of 1 : 250,000 topographic maps;
- selected 1 : 50,000 maps;
- special maps of national parks;
- *Exploratory Soil Map and Agro-climatic Zones Map of Kenya*, at 1 : 1,000,000 (Kenya Soil Survey, 1982);
- 1 : 250,000 map of *Kenya Vegetation* (Trapnell *et al.*, 1966–69), supplemented by the relevant sections of White (1983);
- *Simplified Soil Maps of Kenya* (Siderius, 1979);
- *Farm Management Handbook of Kenya* (Jaetzold and Schmidt, 1983) and *Range Management Handbook of Kenya* (Schwartz *et al.*, 1991), giving district-level climate tables, rainfall maps, agroecological zone maps, soil maps, agricultural (including livestock) statistics and results of Small Farms Survey of 1977.

---

marked on the topographic maps to be used in the field, and also to develop lists of possible categories for the collecting form descriptors on these subjects. It will rarely be necessary to take raw climatic or socioeconomic data into the field. However, it may be useful to annotate topographic maps with such data, for example the annual rainfall in different areas, or the name(s) of the ethnic group(s) found there.

References

Anon. (1965) *Tables of Temperature, Relative Humidity and Precipitation for the World.* Vols 1–6. HMSO, London.

Carlowitz, P.G. von, G.V. Wolf and R.E.M. Kemperman (1991) *Multipurpose Tree and Shrub Database – An Introduction and Decision-Support System. User's Manual, Version 1.0.* ICRAF, Nairobi.

Carter, S.E. (1987) Collecting and organizing data on the agro-socio-economic environment of the cassava crop: case study of a method. In: Bunting, A.H. (ed.) *Agricultural Environments.* pp. 11–29. CAB International, Wallingford.

Carter, S.E., L.O. Fresco and P.G. Jones, with J.N. Fairbairn (1992) *An Atlas of Cassava in Africa: Historical, Agroecological and Demographic Aspects of Crop Distribution.* CIAT, Cali.

Committee for the World Atlas of Agriculture (1976) *World Atlas of Agriculture.* Istituto Geografico de Agostini, Novara.

CPCS (1977) *Classification de sols.* Documents des Laboratoires de Géologie et Pédologie. École Nationale Supérieur Agronomique, Grignon.

Emberger, L. (1955) Une classification biogéographique des climats. *Travaux du Laboratoires de Botanique et de Géologie Université de Montpellier, Série Botanique* 7:3–43.

FAO (1978–81) *Report on the Agroecological Zones Project.* Vols I–IV. World Soil Resources Report 48/1–4. FAO, Rome.

FAO (1984) *Agroclimatological Data. Africa.* Plant Production and Protection Series No. 22. FAO, Rome.

FAO (1985) *Agroclimatological Data. Latin America and the Caribbean.* Plant Production and Protection Series No. 24. FAO, Rome.

FAO (1987) *Agroclimatological Data. Asia.* Plant Production and Protection Series No. 25. FAO, Rome.

FAO–Unesco–ISRIC (1988) *Revised Legend of the FAO–Unesco Soil Map of the World.* World Soil Resources Reports No. 60. FAO, Rome.

Hill, P. (1984) The poor quality of official socioeconomic statistics relating to the rural tropical world: with special reference to South India. *Modern Asia Studies* 8:491–514.

Hilwig, F.W. (1987) Methods for the use of remote sensing in agro-ecological characterization and environmental monitoring. In: Bunting, A.H. (ed.) *Agricultural Environments.* pp. 221–245. CAB International, Wallingford.

Holdridge, L.R. (1967) *Life Zone Ecology.* Tropical Science Centre, San José.

Hopkins, S.T. and D.E. Jones (1983) *Research Guide to the Arid Lands of the World.* Oryx Press, Phoenix.

Houle, G. and L. Filion (1993) Interannual variations in the seed production of *Pinus banksiana* at the limit of the species distribution in northern Quebec, Canada. *American Journal of Botany* 80:1242–1250.

Huke (1982) *Agroclimatic and Dry Season Maps of South, Southeast and East Asia.* IRRI, Los Baños.

IUCN (1992) *Species and Global change. A Report of a Workshop to Review Critical Changes Affecting the Redistribution of Plant Species and Populations and Their Establishment in New Environments as a Consequence of Global Change.* IUCN, Gland.

Jackson, M.T., B.V. Ford-Lloyd and M.L. Perry (eds) (1990) *Climatic Change and Plant Genetic Resources.* Belhaven Press, London.

Jaetzold, R. and H. Schmidt (1983) *Farm Management Handbook of Kenya. Vol. II. Natural Conditions and Farm Management Information.* Ministry of Agriculture, Kenya and GTZ, Nairobi.

Jones, P.G. (1987) Current availability and deficiencies in data relevant to agro-ecological studies in the geographic area covered by the IARCs. In: Bunting, A.H. (ed.) *Agricultural Environments.* pp. 69–83. CAB International, Wallingford.

Kenya Soil Survey (1982) *Exploratory Soil Map and Agro-climatic Zones Map of Kenya, 1980.* Kenya Soil Survey, Nairobi.

Kiambi, K. (1992) *Tamarind. Tamarindus indica.* Indigenous Trees Training Series. KENGO, Nairobi.

Köppen, W. (1936) *Handbuch der Klimatologie.* Borntrager, Berlin.

Landsberg, H.E. (ed.) (1969–76) *World Survey of Climatology.* Vols 1–15. Elsevier Scientific Publishing, Amsterdam.

Lean, G., D. Hinrichsen and A. Markham (1990) *Atlas of the Environment.* WWF, London.

Markham, A., N. Dudley and S. Stolton (1993) *Some Like It Hot. Climate Change, Biodiversity and the Survival of Species.* WWF, Gland.

Martyn, D. (1992) *Climates of the World.* Elsevier Scientific Publishing, Amsterdam.

ORSTOM (1981) *Atlas de la Nouvelle Calédonie et Dépendances.* ORSTOM, Paris.

Parry, R.B. and C.R. Perkins (1987) *World Mapping Today*. Butterworths, Borough Green.

Reid, R. (1980) The collection and use of climatic data in pasture plant introduction. In: Clements, R.J. and D.G. Cameron (eds) *Collecting and Testing Tropical Forage Plants*. CSIRO, Canberra.

Schwartz, H.J., S. Shaabani and D. Walther (1991) *Range Management Handbook of Kenya*. Ministry of Livestock Development, Nairobi.

Servicio Meteorologico Nacional Mexico (1976) *Normales Climatologicas Periodo 1941–1970*. Direccion General de Geografia y Meteorologia, Mexico City.

Siderius, W. (1979) *Simplified Soil Maps of Kenya*. Kenya Soils Survey, Nairobi.

Sivakumar, M.V.K., M. Konate and S.M. Virmani (1984) *Agroclimatology of West Africa: Mali*. Information Bulletin No. 19. ICRISAT, Patancheru.

Smithsonian Institution (1944) *World Weather Records*. Smithsonian Miscellaneous Collections. Smithsonian Institution, Washington DC.

Specht, R.L. (1988) *Mediterranean-type Ecosystems. A Data Source Book*. Kluwer Academic Publishers, Dordrecht.

Trapnell, C.G. *et al.* (1966–69) *Kenya Vegetation*. Map 1 : 250,000, in colour. Directorate of Overseas Surveys, Surbiton.

Unesco (1973) *International Classification and Mapping of Vegetation*. Ecology and Conservation 6. Unesco, Paris.

USDA (1975) *Soil Taxonomy*. US Agricultural Handbook 436. USDA, Washington.

Virmani, S.M., D.G. Faris and C. Johansen (1991) *Agroclimatology of Asian Grain Legumes (Chickpea, Pigeonpea and Groundnut)*. ICRISAT, Patancheru.

Walter, H. and H. Lieth (1960–67) *Klimadiagramm-Weltatlas*. Fischer, Jena.

Walter, H., E. Harnickell and D. Mueller-Dombois (1975) *Climate-diagram Maps of the Individual Continents and the Ecological Climatic Regions of the Earth*. Springer-Verlag, New York.

White, F. (1983) *The Vegetation of Africa. A Descriptive Memoir to Accompany the Unesco/AETFAT/UNSO Vegetation Map of Africa*. Unesco, Paris.

Wise, D. (1975) Selected geographical and cartographical serials containing lists and/or reviews of current maps and atlases. *SLA Geography and Map Division Bulletin* 102:42–45.

WMO, Unesco and Cartographia (1979) *Climatic Atlas of North and Central America*. WMO, Paris.

WRI, IIED and IUCN (1992) *1993 Directory of Country Environmental Studies*. WRI, Washington DC.

WRI, UNEP and UNDP (1994) *World Resources 1994-95. A Guide to the Global Environment*. Oxford University Press, Oxford.

Young, A. (1987) Methods developed outside the international agricultural research system. In: Bunting, A.H. (ed.) *Agricultural Environments*. pp. 43–63. CAB International, Wallingford.

Useful addresses

Map jobbers

GeoCenter ILH	Edward Stanford Ltd.	Geoscience Resources
Postfach 80 08 30	12–14 Long Acre	2990 Anthony Road
D-7000	London WC2E 9LP	Burlington, NC 27215
Stuttgart 80	UK	USA
Germany		

Distributors

The Marketing
Department
John Bartholomew &
Son Ltd
Duncan Street
Edinburgh EH9 ITA
UK
Road maps (countries,
regions and continents)
and detailed tourist
maps in English

Roger Lascelles
Cartographic and Travel
Publisher
47 York Road
Brentford
Middlesex TW8 OQP
UK
Road maps of Europe,
Morocco and Turkey in
English

Nelles Verlag GmbH
Schleissheimer Str 371b
D8000, Munich 45
Germany
Road maps of South
and Southeast Asia
in German

Freytag and Berndt
M. Artaria
1071 Vienna
Austria
European and East
African road maps in
German

Kummerly & Frey
CH-3001 Bern
Switzerland
Wide range of road maps
in German and French

Institut Geographique
National
136, rue de Grenelle
75700 Paris
France
Road and travel maps of
Europe and francophone
Africa in French

International Travel
Map Productions
PO Box 2290
Vancouver, BC V6B
3W5
Canada
Travel maps of North
and South America

Geo Projects
PO Box 113 5294
Beirut
Lebanon
Political/road maps of
the Middle East in
English

Map Studio
PO Box 4482
Cape Town 8000
South Africa
Political/road maps of
southern Africa in
English

Runaway Publications
176 South Creek Road
Dee Why, NSW 2099
Australia
Political/road maps of
Australia and the Pacific

Others

World Meteorological
Organization (WMO)
CP 2300
41 av. Giuseppe Motta
CH-1211 Genève 2
Switzerland
Tel: +41 22 4308111
Fax: +41 22 7342326
Telex: 23260

INFOTERRA
Programme Activity
Centre
UNEP
PO Box 30552
Nairobi
Kenya
Tel: +254 2 230800
Fax: +254 2 226949
Telex: 22018 UNEP KE
E-mail: UNE 002

Land Resources Develop-
ment Centre (LRDC)
Ministry of Overseas
Development
Tolworth Tower,
Surbiton
Surrey KT6 7DY
UK

Institut de la Carte
Internationale du Tapis
Végétal (ICITV)
Université Paul Sabatier
Allée Jules Guesde 39
F-31062 Toulouse Cedex
France
Tel: +33 61 5322352

Published sources of information on wild plant species

10

H.D.V. Prendergast

Economic Botany Section, Royal Botanic Gardens, Kew, Richmond, Surrey TW9 3AB, UK.

Introduction

Germplasm collectors need to be able to identify plants, to determine what their accepted names are and to understand something about where they grow, if they are to locate and recognize their target material and communicate their findings to others. This chapter's aim is to assist collectors by providing a guide to the world's taxonomic and ecological literature on vascular (especially flowering) plants. Also included are works on the identification of seeds and fruits, since this is how most germplasm is collected. A conservation theme is introduced as well. If genetic resources continue to dwindle as they are doing now, collectors of the 21st century may well find little left outside protected areas. Though this chapter deals with the printed literature, there are a box and an appendix on some relevant databases. There is more on information sources on other, in particular electronic, media in Chapters 13 and 14.

Instead of listing all the literature needed by all collectors of wild species – clearly a formidable task beyond the scope of just a single chapter of a book – a methodology is presented that should lead anyone to the relevant sources. This methodology begins with and is based on a number of 'Key works' and is supported by sections listing large-scale floristic, ecological and bibliographic works. The problem has been to reduce the literature available to a manageable but effective minimum. Thus, in general, anything pertaining to just one country or to just one group of plants has been omitted. Taxonomic works are thereby confined mainly to Floras with an international coverage (e.g. *Flora of Tropical East Africa*) although Floras of some very large countries (e.g. China) are also included. Similarly, ecological works comprise mainly those with a large national, international or even continental approach. Monographs, revisions, other national Floras and vegetation descriptions on a smaller scale should be easy to locate via the key works or the works listed in the other sections. Throughout this chapter there is a leaning, reinforced by the unapologetic breaking of the selection criteria, towards works about botanically less well-known parts of the world, particularly the tropics.

Most of what is cited is fairly recent – 1980 or later. Time in the field is valuable and the collector needs to rely on the most up-to-date works available (Box 10.1). A new taxonomic revision, for example, may not only simplify the identification of species in a difficult group but also provide new insights into such matters as distribution, breeding system and ecology. Sometimes, however, an older work may still be the best or the only one available.

There are two unusual features of this chapter. One is the inclusion wherever possible or appropriate of the International Standard Book Number (ISBN) of the works cited. The other is a list of some international booksellers of botanical literature. Both should simplify the collector's task in acquiring literature.

Inevitably, this chapter reflects some personal bias. Just as there are some regions of the world which are reasonably well known to the author and others with which he is not so familiar, there must also be works which have been omitted but which would have been the choice of others. Nevertheless, whatever the subject of the collector's search, the hope is that this backbone of a bare minimum of publications will be as useful and relevant to the gingers (Zingiberaceae) as to the grasses (Poaceae), and to Thailand as to Togo.

Key works

The methodology begins with key works, those works considered essential to every collector and which act as 'signposts' to the rest of the relevant botanical literature.

First, however, it should not be overlooked that quicker than any search for relevant literature is the right answer from someone already well acquainted with the flora of the target collecting area. As well as pressed plant specimens, national and international herbaria also house taxonomists and other botanists. The eighth edition of *Index Herbariorum* (Holmgren *et al.*, 1990) lists the 7627 recorded staff members (and their botanical interests) of 2639 herbaria in 147 countries. There is an index to such specialists in a separate companion volume (Holmgren and Holmgren, 1992). In many cases, especially where there are major regional Floras in preparation, it is worth contacting specialists in the larger international herbaria as well as those working more locally (Chapter 14). The inclusion of the periodicals and serial works (such as Floras) published by each herbarium means that the collector can obtain updates of much of the information presented below. Full postal addresses, along (where appropriate) with telephone, fax and

Box 10.1
What to take to the field?

Clearly, collectors cannot carry an entire library into the field. Depending on target species and areas, they should have at least a local Flora or copies of the relevant pages from a more regional one, however. Where there is a high chance of finding unrecorded species, it is also worthwhile having keys from Floras of adjacent areas. Collectors concentrating on specific taxonomic groups should take the relevant excerpts from Floras (keys, descriptions, illustrations, etc.), or the latest monograph. If one has not already been published, it is useful to develop annotated checklists of the taxa likely to be found in the target area, or simply of the target taxa.

The value of illustrated popular field guides should not be underestimated. The best of them are not only fine examples of how to condense the maximum amount of information into the minimum of space, but also confirm the dictum that 'a picture is worth a thousand words'. The pictures in even the most basic field guide can often stimulate helpful discussion on plant names, uses and localities with local people.

An ecological work, especially one identifying vegetation zones, is often more helpful in the field than simply a vegetation map.

Beyond this, it is perhaps preferable to be underburdened with literature than the opposite: burying one's head into the (not always reliable) printed word can take valuable field time away from the collector's main task – finding and gathering germplasm. The time for reading is at the planning stage.

Two examples may be instructive. During a recent seed-collecting trip in southern Madagascar aimed at 'useful' wild species in general, the only floristic literature taken to the field was the one-volume *Flore et Végétation de Madagascar* (Koechlin *et al.*, 1974), as opposed to the multivolume *Flore de Madagascar et de Comores*, and relevant extracts from White's (1983) *The Vegetation of Africa*. The sections of *Madagascar: Revue de la Conservation et des Aires Protégées* (Nicoll and Lagrand, 1989) relevant to the south of the country were also taken along. The book provides useful administrative as well as scientific information on nature reserves and other protected areas in Madagascar.

On a collecting mission for *Sesbania* spp. in Zambia, copies of the latest revision relevant to the area were taken to the field (Lewis, 1988), along with a number of different keys (e.g. to vegetative characters or pod characters only) produced by the KEY function of the Descriptor Language for Taxonomy (DELTA) computer package on the basis of descriptive data provided in the revision (Chapter 11). A small-scale vegetation map and extracts from White (1983) were also taken.

telex numbers, facilitate the collector's task in making the first steps towards acquiring the relevant botanical literature. Databases of botanical specialists are discussed in Chapter 14.

Although collectors need not be full-time taxonomists, they do need to be aware of recent taxonomic thinking, such as the describing of new species (about 2500 per year at present) and the changing names of others. Since 1893 the standard international reference has been *Index Kewensis* (Royal Botanic Gardens, Kew, 1893–). It lists the place of publication of new and changed names of seed-bearing plants from family level downwards (before 1972, only genera and species). It appears in five-yearly supplements (the last, published in 1991, covering 1986–1990) and since 1986 in yearly volumes as *Kew Index*. It is available on microfiche (to 1975) and on CD-ROM. Since 1935, references to botanical illustrations have been incorporated in *Index Kewensis*, taking over the task of *Index Londinensis*.

The most comprehensive publication listing worldwide botanical literature is *The Kew Record of Taxonomic Literature*, an annual from 1971 to 1987 and, as the *Kew Record*, a quarterly since (Royal Botanic Gardens, Kew, 1971–). It now

> lists references to all publications relating to the taxonomy of flowering plants, gymnosperms and ferns. In addition to sections on taxonomic groups, there are references on phytogeography, nomenclature, chromosome surveys, chemotaxonomy, Floras and botanical institutions; also papers of taxonomic interest in the fields of anatomy and morphology, palynology, embryology, and reproductive biology, and relevant bibliographies and biographies.

Of greatest potential interest to collectors is the floristics section. In the 1990 volume alone there are more than 800 references to books and journal articles. Their scope is broad, ranging from Volume 3 of the *Flora of Inner Mongolia*, to the vegetation of swamps in Côte d'Ivoire, the caatinga plants of Brazil and collections from islands of the South Pacific. Plant atlases and vegetation maps are also mentioned. References are listed under various geographic subdivisions of the world and within these by country and/or administrative unit. It is worth noting that the number of references cited for a country depends largely on its size (which may be anything from a continental mass to an oceanic atoll of a few square kilometres) and the extent to which it has been, and continues to be, studied botanically. Whereas some countries may have almost as many active taxonomists as species, others may have vast areas never visited by a single botanist. Whatever the relative impact and increase in knowledge made by a new publication (and this will be more for a poorly known area), the *Kew Record* is an invaluable publication in which individual countries can easily be scanned for all sorts of botanical works published since 1971. *Kew Record* is computerized from 1982 onwards, but for internal (Kew) use only.

If there is any work which expands to book length what is presented here, it is that of Davis *et al.* (1986). Produced primarily for conservationists, it addresses the questions: 'Where can I find out about the flora of any country, which species in that flora are threatened, and who may be trying to save them?' Data sources are presented for each country and island group in the world, with the sections on floristics, vegetation, checklists and Floras, and field guides being of greatest value here. Other sections contain lists of voluntary organizations and useful addresses and there is an appendix with a comprehensive bibliography of general and regional references.

For a discursive, in-depth and more retrospective compilation of Floras alone, the indispensable source is the handbook of Frodin (1984). With a cut-off date of 1980, it covers

> all of the most generally useful and/or comprehensive Floras, enumerations, lists, and related works for different parts of the world in an approximately uniform fashion within a single, not too bulky volume. References to more specialized or extensive bibliographies, guides, and indices are provided throughout the work for the use of persons with interests in a particular geographic unit or units, thus enabling it also to act as a bibliography of bibliographies.

The latter is also a good statement of this chapter's aim. References are easily located either by author or by geography in two separate indexes and an appendix lists more than 500 (largely) biological journals (this list is itself an invaluable pointer to taxonomic journals worldwide). A revised edition of Frodin's *Guide* is in preparation (D. Frodin, pers. comm.). The thoroughness of the present edition is what has made it possible to omit from the list presented here virtually all Floras completed before 1980. Increasingly, computerized Floras and checklists are being produced. There is no overall guide to these yet, but the subject is introduced in Appendix 10.1.

It is fortunate that the tropics, where unquestionably most botanical and genetic resources work still needs to be done, have been specifically addressed by the compilation of Campbell and Hammond (1989). Forty-two of the book's chapters each covers a tropical country or region (e.g. Thailand, Equatorial Africa) for which there is given a brief vegetation description, a list of vegetation maps and an assessment of the extent of past, and the need for future, collecting. This brings up to date works such as Hedberg and Hedberg's (1968) compilation. While all chapters have useful references, some (e.g. those on Fiji by J. Ash and S. Vodonaivalu and on eastern, extra-Amazonian Brazil by S.A. Mori) have very extensive bibliographies whose coverage of ecological as well as taxonomic works complements and updates that provided by Frodin (1984). Collecting methods (e.g. of tropical germplasm and palm specimens) are also reviewed.

Ecological literature has no formal equivalent to what Frodin (1984) has provided for Floras but there are two outstanding works which none the less fulfil such a role, one globally and one for Africa alone. Takhtajan (1986) is the most modern synthesis and classification of world plant geography. The vegetation and endemic families of the areas covered by the different levels of classification are described, starting with kingdom (e.g. the Neotropical) and descending through subkingdom and region to province (e.g. the West Indian). At every step there are numerous references, providing an entry to the world's ecological literature. White (1983), on the other hand, deals only, but indispensably, with Africa and its offshore islands. Each of the continent's floristic regions is described in terms of geographic position and area, geology and physiography, climate and flora (e.g. number of endemic families and species), but the bulk of the work is devoted to descriptions of the vegetation types within the regions. The amount of information in this work is prodigious, including a list of about 2000 references. Mainly ecological, they can be accessed either by author or by country or floristic region. That few of them are dated later than 1980 does not detract from the immense value of this work to the collector in Africa. There is also a species index.

Highly diverse areas are likely to be special priorities for germplasm collectors. Different approaches to the definition of areas of high biodiversity have been developed, e.g. by National Academy of Science (1980), Myers (1988) and Mittermeier and Werner (1990), but the initiative by the World Wide Fund for Nature and the World Conservation Union (IUCN) to produce a guide to such regions is particularly to be welcomed. A publication called *Centres of Plant Diversity: A Guide and Strategy for their Conservation* is being produced (WWF and IUCN, in prep.). This directory will contain data sheets on each of about 250 critical sites. There will also be regional overviews. For each site there will be data on: geography, vegetation, flora, useful plants, social and environmental values, economic assessment, threats and conservation. The critical requirement for inclusion of a site is that it should have a high total number of species and/or high endemism. Consideration was also given to whether candidate sites are threatened by destruction or contain important plant gene pools, a diverse range of habitats or unique edaphic conditions. A volume on Europe, the Middle East and southwest Asia was published in 1994 and the final two volumes are planned for 1995.

By the very nature of the work, every germplasm collector is, and should be, a conservationist too. For literature on conservation the main access is the bibliography made by the Royal Botanic Gardens, Kew and the Threatened Plants Unit, World Conservation Monitoring Centre (1990). Its 10,500 references, mostly dating from 1970 onwards, span a broad range of scale, from papers on an individual threatened species, to floristic and vegetation accounts of centres of plant diversity, and to directories on protected areas in various global regions. As well as information on how quickly habitats, vegetation and species are disappearing, the collector can also get a good idea of the conservation 'climate' of a country: for example, whether a Plant Red Data Book (based on the format of Lucas and Synge's (1978) world overview) or threatened species list has been produced. For more on the World Conservation Monitoring Centre (WCMC), see Box 10.2. For forestry species, the Food and Agriculture Organization (FAO) is also a useful source of information on conservation: National Research Council (1991) provides a list of forestry species reported as threatened, mainly in the *IUCN Plant Red Data Book* and various FAO publications.

These key works refer directly or indirectly to much of the literature listed later. It is appropriate, however, to mention four other works of value in a different sense, works of a more general, reference nature. A

Box 10.2
The World Conservation Monitoring Centre (WCMC) databases

Information on the conservation status of species and ecosystems is crucial to the germ-plasm collector: the threat of genetic erosion is one of the prime motivators of collecting, and will continue to be as long as the threats of desertification, deforestation and global warming persist. A primary source of information on the world's biological diversity and the disparate threats it is facing is WCMC, a joint undertaking of IUCN, the United Nations Environment Programme (UNEP) and the World Wide Fund for Nature based in the UK (for address, see Appendix 10.1). Information from 'published literature, unpublished reports and government reports, conservation organizations, and a wide network of contacts and cor-respondents throughout the developing world' (McNeely *et al.*, 1990) is maintained by WCMC in a set of databases on:

- the distribution, ecology and status (e.g., population size, threats, conservation category and occurrence in captivity or in cultivation) of species whose conservation situation is causing concern (about 52,000 plants);
- key sites of high biodiversity (in particular tropical forest sites but also wetlands and coral reefs) which, given adequate protection, would ensure the survival of a significant percentage of the world's plants;
- the location, history, conservation importance and management of the 16,000 or so protected areas of the world;
- the international trade in the *Convention on International Trade in Endangered Species of Wild Fauna and Flora* (CITES)-listed species.

There is also a Global Biodiversity Database of mapped digital data. The user interface, the Biodiversity Map Library, allows users to browse the database, which holds data on moist tropical forests, protected areas, mangroves, wetlands, coral reefs, Antarctica and biogeography.

Excerpts, digests and analyses of the information in these databases are also available in printed form. For example, in early 1995 WCMC will publish the first comprehensive global list of threatened plants, based on data in the first of the databases listed above. This will attempt to incorporate all national Red Data Books and Red Data Lists, and will supersede the *IUCN Plant Red Data Book*. Data from the protected areas databases con-tribute to the *Conservation Atlases of Tropical Forests* and data from the critical sites database to the *Centres of Plant Diversity* project. The information in the protected areas database is published periodically in the *United Nations List of Protected Areas* and the *IUCN Register of Threatened Protected Areas of the World* as well as in more detailed regional IUCN directories (a list of these and other relevant IUCN publications is given in this chapter).

WCMC works closely with the Species Survival Commission (SSC), one of six volunteer commissions of IUCN. *Biodiversity in Sub-Saharan Africa and its Islands* is the first of a series of regional studies by the SSC (a study of South America is planned for publication). WCMC, IUCN and national institutions are also collaborating on a project to develop a database cataloguing all sources of biodiversity information on East Africa, both within and outside the region. If successful, this pilot project, which started in 1993, will be extended to other geographic regions.

A major new publication by WCMC is *Global Biodiversity: Status of the Earth's Living Resources* (1991). In the wake of the UN Conference on Environment and Development (UNCED), WCMC is also playing a catalytic role in the production of more detailed, national-level biodiversity country studies. Article 6 of the Convention on Biological Diversity calls on contracting parties to develop national biodiversity strategies and action plans, and the UNEP Country Study Programme is the mechanism for this.

theoretical background to flowering plant classification, such as that proposed by Cronquist (1988), would provide the collector with a useful framework for understanding and identifying species in an unfamiliar flora. On the other hand, when confronted by a strange name rather than plant, the collector can do no better than turn to Mabberley (1989). Based on Cronquist's system of classification, it lists all generic and family names of vascular plants, giving data on features such as floral and fruit structure, economic uses and geographic distribution. Its predecessor, in scope, style and inspiration, was Willis (1973), which, although outdated and largely based on the Englerian system, is still a useful complement to Mabberley, not least because of its inclusion of numerous synonyms. The latest in this line of publications is Brummitt (1992), which is 'a listing of the genera of vascular plants of the world according to their families, as recognized in the Kew Herbarium, with an analysis of relationships of the flowering plant families according to eight systems of classification'. In addition, there is a section in which all accepted generic names are listed family by family – a unique feature in modern botanical literature.

Full citations of the chosen key works are given below.

Brummitt, R.K. (comp.) (1992) *Vascular Plant Families and Genera*. Royal Botanic Gardens, Kew. 0947643435.

Campbell, D.G. and H.D. Hammond (eds) (1989) *Floristic Inventory of Tropical Countries*. New York Botanical Garden, Bronx. 0893273333.

Cronquist, A. (1988) *The Evolution and Classification of Flowering Plants*. New York Botanical Garden, Bronx. 0893273325.

Davis, S.D., S.J.M. Droop, P. Gregerson, L. Henson, C.J. Leon, J.L. Villa-Lobos, H. Synge and J. Zantovska (1986) *Plants in Danger – What Do We Know?* IUCN, Gland. 2880327075.

Frodin, D.G. (1984) *Guide to Standard Floras of the World*. Cambridge University Press, Cambridge. 0521236886.

Holmgren, P.K. and N.H. Holmgren (1992) Plant specialists index. *Regnum Vegetabile* 124:1–394.

Holmgren, P.K., N.H. Holmgren and L.C. Barnett (1990) *Index Herbariorum. Part 1: The Herbaria of the World*. International Association for Plant Taxonomy by New York Botanical Garden, Bronx. 0893273589.

Mabberley, D.J. (1989) *The Plant-Book. A Portable Dictionary of the Higher Plants*. Cambridge University Press, Cambridge. 0521340608.

Royal Botanic Gardens, Kew (1971–) *The Kew Record of Taxonomic Literature*. Her Majesty's Stationery Office, London. Annual to 1986, quarterly since 1987.

Royal Botanic Gardens, Kew (1893–) *Index Kewensis. Names of Seed-bearing Plants at the Rank of Family and Below*. Oxford University Press, Oxford. (Davies, R.A. and R.M. Lloyd (eds) (1991) Supplement 19: 1986–1990. 0198546769). Available on CD-ROM.

Royal Botanic Gardens, Kew, and Threatened Plants Unit, World Conservation Monitoring Centre (1990) *World Plant Conservation Bibliography*. Threatened Plants Unit, World Conservation Monitoring Centre and Royal Botanic Gardens, Kew. 0947643249.

Takhtajan, A. (1986) *Floristic Regions of the World*. Translated by Theodore J. Crovello with the assistance and collaboration of the author and under the editorship of Arthur Cronquist. California University Press, Berkeley. 0520040279.

White, F. (1983) *The Vegetation of Africa. A Descriptive Memoir to Accompany the Unesco/AETFAT/UNSO Vegetation Map of Africa*. Unesco, Paris. 9231019554.

Willis, J.H. (1973) *A Dictionary of the Flowering Plants and Ferns*. 8th edition, revised by H.K. Airy Shaw. Cambridge University Press, Cambridge.

WWF and IUCN (in prep.) *Centres of Plant Diversity: A Guide and Strategy for Their Conservation*. Edited by S.D. Davis, V.H. Heywood and O. Herrera-MacBryde. WWF and IUCN.

The following publications have also been referred to in this section.

Hedberg, I. and O. Hedberg (eds) (1968) *Conservation of Vegetation in Africa South of the Sahara*. Acta Phytogeographica Suecica 54, Uppsala.

Koechlin, J., J.L. Guillamet and P. Morat (1974) *Flore et Végétation de Madagascar*. Cramer, Vaduz.

Lewis, G.P. (1988) *Sesbania* Adans. in the Flora Zambesiaca region. *Kirkia* 13:11–51.

Lucas, G. and H. Synge (comps) (1978) *The IUCN Plant Red Data Book*. IUCN, Morges (Threatened Plants Committee, c/o Royal Botanic Gardens, Kew). 2880322022.

McNeely, J.A., K.R. Miller, W. Reid, R.A. Mittermeier and T. Werner (1990) *Conserving the World's Biological Diversity*. WRI, IUCN, Conservation International and World Bank, Washington DC and Gland.

Mittermeier, R.A. and T.B. Werner (1990) Wealth of plants and animals unites 'megadiversity' countries. *Tropicus* 4:1–15.

Myers, N. (1988) Threatened biotas: 'hotspots' in tropical forests. *Environmentalist* 8:1–20.

National Academy of Science (1980) *Research Priorities in Tropical Biology*. Committee on Research Priorities in Tropical Biology, National Academy of Science, Washington DC.

National Research Council (1991) *Managing Global Genetic Resources. Forest Trees*. National Academy Press, Washington DC. 0309040345.

Nicoll, M.E. and O. Lagrand (1989) *Madagascar: Revue de la Conservation et des Aires Protégées*. WWF, Gland. 2880850274.

WCMC (1992) *Global Biodiversity: Status of the Earth's Living Resources*. Chapman & Hall, London. 0412472406.

International, regional and large national Floras and floristic works

Figure 10.1 outlines the regions covered by the following works (listed by title) and by older works which have yet to be replaced (e.g. *Flora of West Tropical Africa*). Editors' or authors' names have only been included when they are consistent throughout the appearance of the relevant publication.

Circumpolar Arctic Flora. Polunin, N. (1959) Oxford University Press, Oxford.

Flora Europaea. Tutin, T.G., V.H. Heywood, N.A. Burges, D.H. Valentine, S.M. Walters and D.A. Webb (eds) (1964–1980) Cambridge University Press, Cambridge. 2nd edition of Vol. 1 in preparation.

Flora Iranica. Rechinger, K.H. (ed.) (1963–) Akademischer Druk und Verlagsanstalt, Graz. 168 of *c.* 175 parts published by 1991. Completion due before 2000.

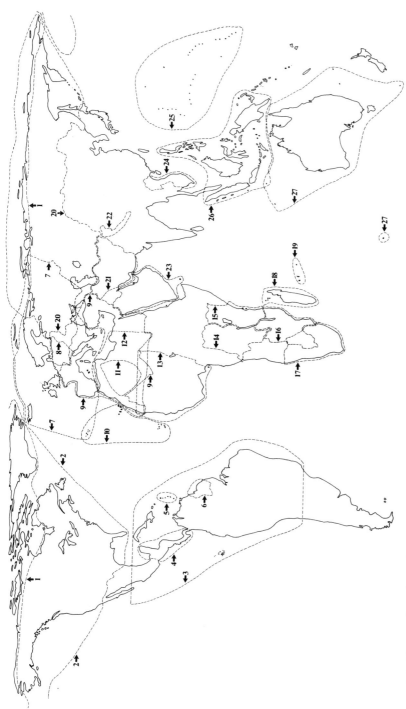

Fig. 10.1. World map (based on the Peters Equal Area Projection) showing boundaries of regions covered by international Floras and floristic works. 1, Circumpolar Arctic Flora; 2, Flora of North America; 3, Flora Neotropica; 4, Flora Mesoamericana; 5, Flora of the Lesser Antilles. Leeward and Windward Islands; 6, Flora of the Guianas; 7, Flora Europaea; 8, Illustrierte Flora von Mitteleuropa; 9, Med-checklist; 10, Flora of Macronesia; 11, Flore du Sahara; 12, Flore de l'Afrique du Nord; 13, Flora of West Tropical Africa[1]; 14, Flore d'Afrique Centrale; 15, Flora of Tropical East Africa; 16, Flora Zambesiaca; 17, Flora of Southern Africa; 18, Flore de Madagascar et des Comores; 19, Flore des Mascareignes: La Réunion, Maurice, Rodrigues; 20, Flora of the USSR[2]; 21, Flora Iranica; 22, Flowers of the Himalaya; 23, Flora of the Arabian Peninsula; 24, Flore du Cambodge, du Laos et du Vietnam; 25, Flora of Micronesia; 26, Flora Malesiana; 27, Flora of Australia.

[1]Hutchinson, J. and J.M. Dalziel (1954–1972) Revised edition edited by F.N. Hepper. Crown Agents for Overseas Governments and Administrations, London.
[2]For discussion of boundary changes during production of this Flora see Bobrov, E.G. (1965) Nature 205:1046–1049. Changes occurred to the western boundary in Vols 11 and 12 and to both the western and eastern (Pacific islands) boundaries in Vols 14 and 15. Boundaries on this map are those for

Flora Malesiana. (1948–) Nijhoff, The Hague; then Kluwer Academic Publishers, Dordrecht; from July 1991 Rijksherbarium, Leiden. Eight volumes on flowering plants (Series I) and two volumes on pteridophytes (Series II) wholly or partially published by 1991. Completion date uncertain (but distant).

Flora Mesoamericana. (1992–) Missouri Botanical Garden; Instituto de Biología, Universidad Nacional Autónoma de México; British Museum (Natural History). Two of seven volumes in press. Completion due *c.* 2000.

Flora Neotropica. (1968–) Hafner Publishing Co., New York (monographs 1–14); then New York Botanical Garden, Bronx, for Organization for Flora Neotropica. 57 volumes published by early 1992.

Flora of the Arabian Peninsula. Miller, A.N. *et al.* (1994–) Edinburgh University Press, Edinburgh. Five volumes to be published, the first in 1994. Completion due *c.* 2000.

Flora of Australia. (1981–) Australian Government Publishing Service, Canberra. 12 of 50 volumes (two cover oceanic islands) on the vascular flora published by 1992. Completion due *c* 2001–2005.

Flora of the Guianas. (1985–) Koeltz Scientific, Konigstein. Ten fascicles on phanerogams (Series A) and one on ferns and fern allies (Series B) published by 1991. Completion not due before 2000.

Flora of the Lesser Antilles. Leeward and Windward Islands. Howard, R.A. (ed.) (1974–) Arnold Arboretum, Harvard University, Jamaica Plain. Six volumes published by 1989.

Flora of North America. Morin, N. (convening ed.) (1993) Oxford University Press, Oxford. First two volumes appeared in 1993.

Flora of Macronesia. Checklist of vascular plants. 3rd edition Hansen, A. and P. Sunding (1985) *Sommerfeltia* 1:1–167. 8299052768.

Flora of Micronesia. Fosberg, F.R. and M.-H. Sachet (1975–) Smithsonian, Washington DC. Four fascicles published by 1985 and a fifth one in press.

Flora of Southern Africa. (1963–) Government Printer for Botanical Research Institute, Pretoria. All or parts of 15 volumes on flowering plants out of a total of 33 planned volumes published by 1991. A few more should appear in the current format but beyond them revisions are likely to be in other journals or series.

Flora of Tropical East Africa. (1949–) Balkema, Rotterdam. *Circa* 7500 of 11,000 species covered by 1991. Completion due 2000.

Flora of the USSR. (1963–) Translation of *Flora URSS* (Komarov, V.L. and B.K. Shishkin, 1933–1964). Israel Program for Scientific Translations, Jerusalem. 22 (1–21, 24) of 30 volumes published by 1979. Also: Shishkin, B.K. (ed.) (1990) Vol. 25. Bishen Singh Mahendra Pal Singh, Dehra Dun. 8121100488.

Flora Zambesiaca. (1960–) Flora Zambesiaca Managing Committee, London. 3867 of *c* 9889 species published by March 1991 (and 2100 written but unpublished). Completion due by 2000.

Flore d'Afrique Centrale (Zaïre–Rwanda–Burundi). (1948–) Jardin Botanique National de Belgique, Meise. *Circa* 158 of 203 families of spermatophytes and 12 of 39 pteridophyte families wholly or partially published by 1989. Three more spermatophyte families in press.

Flore de l'Afrique du Nord. Maire, R. (1952–) Éditions Lechevalier, Paris. 16 volumes published by 1987. (For index of volumes 1–16 see Carazo-Montijano, M. and C. Fernandez-Lopez (1990). Murillo, Jaen. 8460075451.)

Flore de Madagascar et des Comores. (1936–) Muséum National d'Histoire Naturelle, Paris. *Circa* 148 of 189 families wholly or partially published by 1991 (but some treatments very old).

Flore des Mascareignes: La Réunion, Maurice, Rodrigues. (1976–) Sugar Industry
  Research Institute, Mauritius; l'Office de la Recherche Scientifique et Technique
  Outre-mer, Paris; Royal Botanic Gardens, Kew. *Circa* 2100 of 3000 species
  written up (and 93 of 203 families published) by 1991.

Flore du Cambodge, du Laos et du Vietnam. (1960–) Muséum National d'Histoire
  Naturelle, Paris. 25 parts published by 1990.

Flore et Végétation du Sahara. Ozenda, P. (1983) 2nd edition. Centre National de la
  Recherche Scientifique, Paris. 2222046149.

Flowers of the Himalaya. Polunin, O. and A. Stainton (1984) Oxford University Press,
  Oxford.

Flowers of the Himalaya. A Supplement. Stainton, A. (1988) Oxford University Press,
  Delhi. 0192177567.

Illustrierte Flora von Mitteleuropa. Hegi, G. (1966–) Parey, Berlin. Many fasciscles of 3rd
  edition published.

*Med-checklist: A Critical Inventory of Vascular Plants of the Circum-Mediterranean
  Countries.* (1984) 2nd edition. Conservatoire et Jardin botaniques de la Ville de
  Genève, Genève. Four of six volumes published by 1989.

Other relevant works, not, however, included in Fig. 10.1, are listed
below, this time by author. Again, editors' or authors' names have only
been included in full when they are consistent (otherwise, they are listed
as 'Various'). Translated titles are in brackets.

Adam, J.-G. (1971–1983) *Flore Descriptive des Monts Nimba (Côte d'Ivoire, Guinée,
  Libéria).* Centre National de Recherche Scientifique, Paris. 222202837X. Six
  volumes.

Bhandari, M.M. (1990) *Flora of the Indian Desert.* MPS Repros, Jodhpur. 8185304130.
  Covers the eastern end of the Saharo-Sindian botanical region, which extends to
  the Atlantic coast of the Sahara.

Cook, C.D.K. (1990) *Aquatic Plant Book.* SPB Academic Publishing, The Hague.
  9051030436. World guide to 407 aquatic genera of spermatophytes and
  pteridophytes.

Fosberg, F.R., M.-H. Sachet and R. Oliver (1979) A geographical checklist of the Microne-
  sian Dicotyledoneae. *Micronesica* 15:41–295.

Fosberg, F.R, M.-H. Sachet and R. Oliver (1982) Geographical checklist of the Microne-
  sian Pteridophyta and Gymnospermae. *Micronesica* 18:23–82.

Gentry, A.H. (1993) *A Field Guide to the Families and Genera of Woody Plants of NW
  South America (Colombia, Ecuador and Peru).* Conservation International,
  Washington DC. 1881173011.

Greene, S.W. and D.W.H. Walton (1975) An annotated checklist of the sub-Antarctic and
  Antarctic vascular flora. *Polar Research* 17:473–484.

Grey-Wilson, C. and M. Blamey (1993) *Mediterranean Wild Flowers.* Harper Collins,
  London. 0002199017.

Hedberg, I. and S. Edwards (1990–) *Flora of Ethiopia and Eritrea.* Addis Ababa University, Addis Ababa; Asmara University, Asmara; Swedish Agency for Research
  Cooperation with Developing Countries, Uppsala. First volume (Vol. 3: Pitto-
  sporaceae to Araliaceae) published in 1990. Vols 2 (Magnoliaceae to Euphor-
  biacea) and 7 (Poaceae) to be published in 1994. Eight volumes planned, the last
  in 1998. 9197128503.

Kartesz, J.T. (1994) *A Synonymized Checklist of the Vascular Flora of the United States,
  Canada and Greenland.* Timber Press, Portland. 0881922048, 694 and 820.

Lebrun, J.-P. and A.L. Stork (1991, 1992) *Enumération des Plantes à Fleurs d'Afrique Tropicale*. Vol. I. *Généralités et Annonaceae à Pandanaceae*. Vol. II. *Chrysobalanaceae à Apiaceae*. Conservatoire et Jardin Botanique de la Ville de Genève.

Ohba, H. and S.B. Malla (eds) (1988–) *The Himalayan Plants*. University Museum, University of Tokyo. Two of five volumes published by 1991. Mainly Nepal.

Various (1959–) *[Flora Reipublicae Popularis Sinicae]* (1959–) Science Press, Beijing. 65 of 125 books published by 1991; completion due 2000. See *Flora of China*.

Various (1963–) *North American Flora*. New York Botanical Garden, Bronx, Series II. Five parts on vascular plants published by 1990.

Various (1980) *Conspectus Florae Orientalis. An Annotated Catalogue of the Flora of the Middle East*. Israel Academy of Sciences and Humanities, Jerusalem. Four fascicles published by 1987.

Various (1992–) *Flora of China*. This is a 'project based on the revision, updating, and condensation of the *Flora Reipublicae Popularis Sinicae*' (*Flora of China Newsletter* 1 : 1), undertaken by Missouri Botanical Garden, Arnold Arboretum (Harvard University), California Academy of Sciences, Smithsonian Institution and all major botanical institutions at national, provincial and local levels in China. 25 volumes to be completed in 15 years.

Development agencies occasionally publish floristic works. These are often less widely known than those produced by herbaria. Some examples are listed below, by author.

Boudet, G. and J.-P. Lebrun (1986) *Catalogue des Plantes Vasculaires du Mali*. Institut d'Élevage et de Médecine Vétérinaire des Pays Tropicaux, Maisons-Alfort. 2859851186. There are similar recent studies of Burkina Faso and Djibouti by the same publisher.

Hawthorne, W. (1990) *Field Guide to the Forest Trees of Ghana*. Natural Resources Institute, Chatham, for the Overseas Development Administration. 0902500341.

Maydell, H.-J. von (1990) *Arbres et Arbustes du Sahel*. Deutsche Gesellschaft für Technische Zusammenarbeit (GTZ), Eschborn. 3823611976. Also available in English. GTZ also publishes a Flora of Togo.

Tailfer, Y. (1989) *La Forêt Dense d'Afrique Centrale: Identification Pratique des Principaux Arbres*. Technical Centre for Agriculture and Rural Cooperation (CTA), Wageningen.

Vivien, J. and J.J. Faure (1985) *Arbres des Forêts Denses d'Afrique Centrale*. Ministère des Relations Extérieurs, Coopération et Développement, Paris; Agence de Coopération Culturelle et Technique, Paris. 2110847964; 9290280654.

World, continental, international and large national ecological works

Single works

Acocks, J.P.H. (1988) Veld types of South Africa. *Memoirs of the Botanical Survey of South Africa* 57:1–146. 0621113948.

Aleksandrova, V.D. (1980) *The Arctic and Antarctic: Their Division into Geobotanical Areas*. Cambridge University Press, Cambridge. 0521231191.

Aleksandrova, V.D. (1988) *Vegetation of the Soviet Polar Deserts*. Cambridge University Press, Cambridge. 0521329981.

Baas, P., K. Kalkman and R. Geesink (eds) (1989) *The Plant Diversity of Malesia*. Proceedings of the *Flora Malesiana* Symposium. August 1989. Leiden, Netherlands. Kluwer Academic Publishers, Dordrecht. 0792308832.

Backer, A.P., D.J.B. Killick and D. Edwards (1986) A plant ecological bibliography and thesaurus for southern Africa up to 1975. *Memoirs of the Botanical Survey of South Africa* 52:1–216. 0621088714.

Balgooy, M.M.J. van (ed.) (1984) *Pacific Plant Areas*. Rijksherbarium, Leiden. A projected fifth volume has apparently not appeared.

Balsler, H. and J.L. Luteyn (eds) (1992) *Paramo. An Andean Ecosystem Under Human Influence*. Academic Press, London. 0124604420.

Barbour, M.G. and W.D. Billings (eds) (1988) *North American Terrestrial Vegetation*. Cambridge University Press, Cambridge. 0521261988.

Beadle, N.C.W. (1981) *The Vegetation of Australia*. Cambridge University Press, Cambridge. 0521241952.

Browicz, K. (and J. Zielinski) (1978–) *Chorology of Trees and Shrubs in South-West Asia*. Polish Academy of Sciences Institute of Dendrology, Kornik, Poznan (Vol. 1) then Polish Scientific Publishers, Warsaw. Eight volumes published by 1991. 8101105283 (Vol. 8).

China Vegetation Commission (1980) [*Vegetation of China*]. Science Publishing House, Beijing.

Davis, P.H., P.C. Harper and I.C. Hedge (1971) *Plant Life of Southwest Asia*. First Plant Life of Southwest Asia Symposium. Botanical Society of Edinburgh, Edinburgh.

Dong-Sheng, L. (ed.) (1981) *Geological and Ecological Studies of Qinghai-Xizang Plateau*. Vol. 2. *Environment and Ecology of Qinghai-Xizang Plateau*. Science Press, Beijing and Gordon & Breach, New York. 067760220. Tibet.

Eiten, G. (1983) *Classifiçao da Vegetaçao do Brasil*. Conselho Nacional de Desenvolvimento Científico e Tecnológico, Brasilia.

Ellenberg, H.H. (1988) *Vegetation Ecology of Central Europe*. Cambridge University Press, Cambridge. 0521236428.

Engel, T., W. Frey and H. Kürschner (eds) (1991) *Contributiones Selectae ad Floram et Vegetationem Orientis*. Third Plant Life of Southwest Asia Symposium. Flora et vegetatio mundi IX. Cramer, Berlin. 3443660010.

FAO and UNEP (1981) *Tropical Forest Resources Assessment Project (in the Framework of GEMS). Forest Resources of Tropical Africa. Part I. Regional Synthesis. Part II. Country Briefs*. FAO, Rome.

FAO and UNEP (1981) *Tropical Forest Resources Assessment Project (in the Framework of GEMS). Forest Resources of Tropical Asia. Part I: Regional Synthesis. Part II. Country Briefs*. FAO, Rome.

FAO and UNEP (1981) *Tropical Forest Resources Assessment Project (in the Framework of GEMS). Los Recursos Forestales de la América Tropical. Part I. Regional Synthesis. Part II. Country Briefs*. FAO, Rome.

Frankenberg, P. and K. Dieter (1980) *Atlas der Pflanzenwelt des Nordafrikanischen Trockenraumes. Computerkarten Wesentlicher Pflanzenarten und Pflanzenfamilien*. Arbeiten aus der Geographischen Instituten der Universität Bonn. Schwarzbold, Witterschlick b. Bonn.

Friis, I.B. (1992) *Forests and Forest Trees of Northeast Tropical Africa*. Kew Bulletin Additional Series XV. Her Majesty's Stationery Office, London. 0112500056.

Goodall, D.W. and R.A. Perry (eds) (1979) *Arid-land Ecosystems: Structure, Functioning and Management.* Vol. 1. International Biological Programme. Cambridge University Press, Cambridge. 052121842X.

Graham, A. (ed.) (1973) *Vegetation and Vegetational History of Northern Latin America.* Papers presented as part of the Symposium 'Vegetation and Vegetational History in Northern Latin America' at the American Institute of Biological Sciences Meeting. Bloomington, Indiana. 1970. Elsevier Scientific Publishing, Amsterdam. 0444410562.

Haslam, S.M. (1987) *River Plants of Western Europe: The Macrophytic Vegetation of Watercourses of the European Economic Community.* Cambridge University Press, Cambridge. 0521264278.

Hedge, I.C. (ed.) (1986) *Plant Life of Southwest Asia.* Second Plant Life of Southwest Asia Symposium. *Proceedings of the Royal Society of Edinburgh, B* 89:1–319.

Horvat, I., V. Glavac and H. Ellenberg (1974) *Vegetation Südosteuropas. Vegetation of Southeast Europe.* Fischer, Stuttgart. 3437301683.

Hou, H.-Y. (1984) Vegetation of China with reference to its geographical distribution. *Annals of the Missouri Botanical Garden* 70:509–549.

Jain, S.K. (1983) *Flora and Vegetation of India – An Outline.* Botanical Survey of India, Howrah.

Kingdom, J. (1990) *Island Africa.* Collins, London.

Larsen, K. and L. Holm-Nielsen (1979) *Tropical Botany.* Academic Press, London. 012437350X.

Lawson, G.W. (ed.) (1986) *Plant Ecology in West Africa – Systems and Processes.* John Wiley & Sons, Chichester. 0471903647.

Le Houérou, H.N. (1989) *The Grazing Land Ecosystems of the African Sahel.* Springer-Verlag, Berlin. 3540507914.

Léonard, J. (1981–) *Contribution à l'Etude de la Flore et de la Végétation des Déserts d'Iran.* Jardin National de Belgique, Meise. Nine fascicles published by 1989. Concerns deserts from Atlantic to India.

Lovett, J.C. and S.K. Wasser (1993) *Biogeography and Ecology of the Rain Forests of Eastern Africa.* Cambridge University Press, Cambridge. 0521430836.

Merrill, E.D. (1981) *Plant Life of the Pacific World.* Tuttle, Rutland. 0804813701. Reprint of 1945 edition.

Mueller-Dombois, D., F.R. Fosberg and R. McQueen (in prep.) *Vegetation of the Pacific Islands.*

Puig, H., F. Blasco and M.F. Bellan (1981) *Vegetation Map of South America: Explanatory Notes.* Unesco, Paris. 923001933X.

Quézel, P. (1965) *La Végétation du Sahara. Du Tchad à la Mauritanie.* Fischer Verlag, Stuttgart.

Radovsky, F.J., P.H. Raven and S.H. Sohmer (eds) (1984) *Biogeography of the Tropical Pacific.* B.P. Bishop Museum, University of Kansas, Lawrence. 0942924088.

Ramamoorthy, T.P., R.A. Bye, A. Lot and J.E. Fa (eds) (1992) *Biological Diversity of Mexico. Origins and Distributions.* Oxford University Press, Oxford.

Rzedowski, J. (1978) *Vegetación de México.* Limusa, México. 9681800028.

Schnell, R. (1987) *Flore et Végétation de l'Amérique Tropicale.* Vol. 1. *Généralités, les Flores, les Formations Forestières Denses et les Formations Mesophiles.* Masson, Paris. 2225803846.

Schnell, R. (1987) *Flore et Végétation de l'Amérique Tropicale.* Vol. 2. *Les Formations Xériques, les Peuplements des Montagnes, la Végétation Aquatique et Littorale, les Plantes Utiles.* Masson, Paris. 2225803854.

Seeliger, U. (ed.) (1991) *Coastal Plant Communities of Latin America.* Academic Press, London. 0126345503.

Steenis, C.G.G.J. van (1979) Plant-geography of east Malesia. *Botanical Journal of the Linnean Society* 79:97–180.

Steentoft, M. (1988) *Flowering Plants in West Africa*. Cambridge University Press, Cambridge. 0521261929.

Stoddart, D.R. (1983) Floristics and ecology of western Indian Ocean islands. 1. Introduction. *Atoll Research Bulletin* 275.

Unesco (1973) *International Classification and Mapping of Vegetation. Classification Internationale et Cartographie de la Végétation. Clasificación Internacional y Cartografía de la Vegetation*. Unesco, Paris. 9230010464.

Walter, H. (1974) *Die Vegetation Osteuropas, Nord- und Zentralasiens*. Fischer, Stuttgart. 3437201336.

Walter, H. (1979) *Vegetation of the Earth and Ecological Systems of the Geo-biosphere*. 2nd edition. Springer-Verlag, Berlin. 3540904042.

Whitmore, T.C. (1984) A new vegetation map of Malesia at scale 1 : 5 million. *Journal of Biogeography* 11:461–471.

Whitmore, T.C. (1984) *Tropical Rain Forests of the Far East*. Clarendon Press, Oxford. 0198541368.

Series

Beihefte zum Tübinger Atlas des Vorderen Orients. Ludwig Reichert, Wiesbaden

Baierle, H.U., A.M. El-Sheikh and W. Frey (1985) *Vegetation und Flora im Mittleren Saudi-Arabien (at-Ta'if-ar-Riyad)*. 3882262664.

Frey, W. and W. Probst (1978) *Vegetation und Flora des Zentralen Hindukus (Afghanistan)*. 3882260211.

Kürschner, H. (ed.) (1986) *Contribution to the Vegetation of Southwest Asia*. 3882262974.

Zohary, M. (1982) *Vegetation of Israel and Adjacent Areas*. 3882261250.

Dissertationes Botanicae. Cramer, Vaduz, then Berlin

Cleef, A.M. (1981) *The Vegetation of the Paramos of the Columbian Cordillera Oriental*.

Correll, D.S. and Correll, H.B. (1982) *Flora of the Bahama Archipelago (including the Turks and Caicos Islands)*.

Deil, U. (1984) *Zur Vegetation im Zentralen Rif (Nordmarokko)*. 3768213862.

Konig, P. (1987) *Vegetation und Flora im Sudwestlichen Saudi-Arabien (Asia, Tihama)*. 3443640133.

Miehe, G. (1990) *Langtang Himal. Flora und Vegetation als Klimazeiger und -zeugen im Himalaya. A Prodromus of the Vegetation Ecology of the Himalayas*. 3443640702.

Ecology/biogeography works. Junk, The Hague

Gressitt, J.L. (ed.) (1982) *Biogeography and Ecology of New Guinea*. Vol. 1 9061930944.

Mani, M.S. (ed.) (1974) *Ecology and Biogeography in India*. 9061930758.

Stoddart, D.R. (ed.) (1984) *Biogeography and Ecology of the Seychelles Islands*. 906193107X.

Ecosystems of the World. Elsevier Scientific Publishing, Amsterdam

Thirty ecosystems are to be covered. Some relevant examples are given.

Bourlière, F. (ed.) (1983) *Tropical Savannas.* 0444420355.
Breymeyer, A. (ed.) (1991) *Managed Grasslands. Regional Studies.* 0444429980.
Castri, F. di, D.W. Goodall and R.L. Specht (1981) *Mediterranean-type Shrublands.* 044441858X.
Chapman, V.J. (ed.) (1977) *Wet Coastal Ecosystems.* 0444415602.
Evenari, M., I. Noy-Meir and D.W. Goodall (eds) (1985) *Hot Deserts and Arid Shrublands.* Vol. 12A. 044442282X.
Evenari, M., I. Noy-Meir and D.W. Goodall (eds) (1986) *Hot Deserts and Arid Shrublands.* Vol. 12B. 044442296X.
Gore, A.J.P. (ed.) (1983) *Mires: Swamp, Bog, Fen and Moor. Regional Studies.* 0444420045.
Lieth, H. and M.J.A. Werger (eds) (1989) *Tropical Rain Forest Ecosystems. Bio-geographical and Ecological Studies.* 0444427554.
Lugo A., S. Brown and M. Brinson (eds) (1990) *Forested Wetlands.* 0444428127.
Ovington, J.D. (ed.) (1983) *Temperate Broad-leaved Evergreen Forests.* 0444420916.
Röhrig, E. and B. Ulrich (eds) (1991) *Temperate Deciduous Forests.* 0444885994.
Specht, R.L. (ed.) (1979) *Heathlands and Related Shrublands. Descriptive Studies.* 044441701X.
West, N.E. (ed.) (1983) *Temperate Deserts and Semi-deserts.* 0444419314.

Environmental Management Development in Indonesia

Future works in this series will include coverage of Bali, Irian Jaya, Java and Kalimantan and will be published by Periplus Editions, Singapore.

Whitten, A.J., S.J. Damanik, J. Anwar and N. Hissyam (1987) *The Ecology of Sumatra.* Gadjah Mada University Press, Yogyakarta. 9794200352.
Whitten, A.J., M. Mustafa and G.S. Henderson (1987) *The Ecology of Sulawesi.* Gadjah Mada University Press, Yogyakarta. 9794200484.

IUCN (mainly) directories and similar publications. IUCN, Gland and Cambridge
Carp, E. (comp.) (1980) *Directory of Western Palearctic Wetlands.* IUCN and UNEP. 2880323002.
Collins, N.M., J.A. Sayer and T.C. Whitmore (eds) (1991) *The Conservation Atlas of Tropical Forests: Asia and the Pacific.* 0333539923.
IUCN (1982) *IUCN Directory of Neotropical Protected Areas.* Tycooly International, Dublin.
IUCN (1987) *Directory of Wetlands of International Importance: Sites Designated Under the Convention on Wetlands of International Importance Especially as Water-fowl Habitat.* 2831700140.
IUCN (1990) *IUCN Directory of South Asian Protected Areas.* 2831700302.
IUCN (1992) *Protected Areas of the World. A Review of National Systems.* 1. *Indo-malaya, Oceania, Australia and Antarctic.* 283170906. 2. *Palearctic.* 2831700914. 3. *Afrotropical.* 2831700922. 4. *Nearctic and Neotropical.* 2831700930.

IUCN and UNEP (1987) *IUCN Directory of Afrotropical Protected Areas*. 2880328047.
IUCN and Unesco (1987) *Directory of Biosphere Reserves*. Unesco, Paris.
Mepham, R. and S. Mepham (1990) *Directory of African Wetlands*. 2880329493.
Paine, J.R. (1991) *IUCN Directory of Protected Areas in Oceania*. 2831700698.
Sayer, J.A., C. Harcourt and N.M. Collins (1992) *The Conservation Atlas of Tropical Forests: Africa*. 0333577574.
Scott, D.A. (comp.) (1989) *A Directory of Asian Wetlands*. 2880329841.
Scott, D.A. and M. Carbonell (comp.) (1986) *A Directory of Neotropical Wetlands*. International Waterfowl and Wetlands Research Bureau, Slimbridge. 2880325048.
Stuart, S.N., R.J. Adams and M.D. Jenkins (1990) *Biodiversity in Sub-Saharan Africa and its Islands*. Occasional Papers of the IUCN Species Survival Commission No. 6. 2831700213.

Key Environments. Pergamon, Oxford in collaboration with IUCN

Some relevant examples are given.

Cloudsley-Thompson, J.L. (ed.) (1985) *Sahara Desert*. 0080288693.
Cranbrook, Earl of (ed.) (1988) *Malaysia*. 0080288669.
Jolly, A., P. Oberle and R. Albignac (eds) (1984) *Madagascar*. 0080280021.
Perry, R. (ed.) (1984) *Galapagos*. 0080279961.
Prance, G.T. and T.E. Lovejoy (eds) (1985) *Amazonia*. 0080307760.

Plant Resources of South-East Asia (PROSEA) Handbooks. PUDOC and PROSEA Publications Office, Wageningen

These are given in chronological order of publication (Vols 5(2) and 7, on minor timbers and bamboos respectively, are planned for 1995).

Maesen, L.J.G van der and S. Somaatmadja (eds) (1989) *PROSEA 1: Pulses*. PUDOC, Wageningen. 902200984X.
Verheji, E.W.M. and R.E. Coronel (eds) (1991) *PROSEA 2: Edible Fruits and Nuts*. PUDOC, Wageningen. 9022009866.
Lemmens, R.H.M.J. and N. Wulijarni-Soetjipto (eds) (1991) *PROSEA 3: Dye and Tannin-producing Plants*. PUDOC, Wageningen. 9022009874.
Mannetje, L.'t and R.M. Jones (eds) (1992) *PROSEA 4: Forages*. PUDOC, Wageningen. 9022010325.
Soerianegara, Ì. and R.H.M.J. Lemmens (eds) (1993) *PROSEA 5(1): Timber Trees. Major Commercial Timbers*. PROSEA Publications Office, Wageningen. 902201033.
Dransfield, J. and N. Manokaran (eds) (1993) *PROSEA 6: Rattans*. PROSEA Publications Office, Wageningen. 9022010570.
Siemonsma, J.S. and Kasem Piluek (eds) (1993) *PROSEA 8: Vegetables*. PROSEA Publications Office, Wageningen. 9022010589.

Botanical bibliographies

Albrecht, J. (1993) *Tropical Forest Conservation and Development. A Bibliography*. Expanded Program of Technical Assistance (EPTA) and Midwest Universities Consortium for International Activities (MUCIA), Madison.

Balgooy, M.M.J. van (ed.) (1984) Bibliography of Pacific and Malesian plant maps of phanerogams. 3rd supplement. In: Balgooy, M.M.J. van (ed.) *Pacific Plant Areas.* Vol. 4. Rijksherbarium, Leiden.

Bullock, A.A. (1978) *Bibliography of South African Botany (up to 1951).* National Botanical Institute, Pretoria. 0621047139.

Byrne, J.E. (ed.) (1979) *Literature Review and Synthesis of Information on Pacific Island Ecosystems.* Office of Biological Services, US Fish and Wildlife Service, Washington DC.

Chen Sing-chi, Li Jiao-lan, Zhu Xiang-yun and Zhang Zhi-yun (1993) *Bibliography of Chinese Systematic Botany.* Science Press, Beijing.

DeFilipps, R.A. (1987) A bibliography of plant conservation in the Pacific Islands: endangered species, habitat conversion, introduced biota. *Atoll Research Bulletin* 311.

Demiriz, H. (1993) *An Annotated Bibliography of Turkish Flora and Vegetation.* Scientific and Technical Research Council of Turkey, Ankara.

Downey, J.A. and C. Morton (1990) Index to American botanical literature. *Bulletin of the Torrey Botanical Club* 117:476–497. A regularly updated bibliography in this journal.

Fourie, D.M.C. (1990) *Guide to Publications on the Southern African Flora.* National Botanical Institute, Pretoria. 0620154993.

Giess, W. (1989) *Bibliography of South West African Botany.* South West African Scientific Society, Windhoek. 0949995460.

Haddad, S.Y. (1988) Bibliography on the flora of Iraq. *Bulletin of the Iraq Natural History Museum* 8:149–167.

Hampshire, R.J. and D.A. Sutton (1988) *Flora Mesoamericana. A Preliminary Bibliography of the Mesoamerican Flora.* Missouri Botanical Garden, St Louis; Instituto de Biología, Universidad Nacional Autónoma de México, Mexico; British Museum (Natural History), London. 0565010689.

Kerkham, A.S. (comp.) (1988) *Southern African Botanical Literature* 1600–1988 (SABLIT). South African Library, Cape Town. 0869680838.

Küchler, A.W. (ed.) (1965–80) *International Bibliography of Vegetation Maps.* University of Kansas Libraries, Lawrence. Five volumes, the last a 2nd edition.

Kunming Institute of Botany (1984) *A List of the Works and Monographs of Kunming Institute of Botany, Academia Sinica (1958–1983).* Scientific Archives of Kunming Institute of Botany, Academia Sinica, Kunming.

Lebrun, J.-P. and A.L. Stork (1977) *Index 1935–1976 des Cartes de Répartition des Plantes Vasculaires d'Afrique.* Conservatoire et Jardin Botanique de la Ville de Genève.

Lejoly, J. and K. van Essche (1983) Additions to the AETFAT Library from June 1992 to May 1993. Classification per country, family and area through the TAXAT software. *AETFAT Bulletin* 41:42–100. A regularly updated listing in this journal of new additions to the AETFAT Library at the Université Libre de Bruxelles.

Lorence, D.H. and R.E. Vaughn (1992) *An Annotated Bibliography of Mascarene Plant Life.* National Tropical Botanical Garden of the USA, Hawaii.

Marticorena, C. (1993) *Bibliografía Botánica Taxonómica de Chile.* Monographs in Systematic Botany of the Missouri Botanical Garden No. 41. Missouri Botanical Garden, St Louis.

Mill, S.W., D.P. Gowing, D.R. Herbst and W.L. Wagner (1988) *Indexed Bibliography on the Flowering Plants of Hawaii.* Miscellaneous Publication 34. Bishop Museum, Honolulu.

Miller, A.G., I.C. Hedge and R.A. King (1982) Studies in the flora of Arabia: 1. A botanical
    bibliography of the Arabian peninsula. *Notes of the Royal Botanic Garden, Edin-
    burgh* 40:43–61.

Nayar, M.P. (1984–86) *Key Works to the Taxonomy of Flowering Plants of India.* Five
    volumes. Botanical Survey of India, Howrah.

Pope, G.V. and R.K. Brummitt (1991) *A Bibliography of Flora Zambesiaca.* Royal Botanic
    Gardens, Kew. 094764332X.

Rechinger, K.H. (1989) Fifty years of research in the Flora Iranica area (1937–1987). In:
    Tan, K. (ed.) *The Davis and Hedge Festschrift: Plant Taxonomy, Phyto-
    geography and Related Subjects.* pp. 301–349. Edinburgh University Press,
    Edinburgh. 0852246382.

Sims, R.W., P. Freeman and D.L. Hawksworth (eds) (1988) *Key Works to the Fauna and
    Flora of the British Isles and Northwestern Europe.* 5th edition. For the
    Systematics Association by Clarendon, Oxford. 0198577060.

Steenis, C.G.G.J. van (1978) Bibliography. *Flora Malesiana Bulletin* 31:3089–3171.

Stork, A.L. and J.-P. Lebrun (1981) *Index des Cartes de Répartition des Plantes
    Vasculaires de l'Afrique.* Complément 1935–1976, supplément 1977–1981 avec
    Addendum A–Z. Étude Botanique de Géneve 8. Institut d'levage et de Médecine
    Vétérinaire des Pays Tropicaux, Maisons-Alfort.

Stork, A.L. and J.-P. Lebrun (1985) *Index des Cartes de Répartition des Plantes
    Vasculaires de l'Afrique.* Complément 1935–1981, supplément 1982–1985. Étude
    Botanique 14. Institut d'Élevage et de Médecine Vétérinaire des Pays Tropicaux,
    Maisons-Alfort.

Taylor, M.S. (comp.) (1991) *Flora and Ethnobotany of Madagascar: Contributions
    Towards a Taxonomic Bibliography.* 2nd draft. Missouri Botanical Garden,
    St Louis.

Thothathri, K. and A.R. Das (1984) *Bibliography on the Botany of the Eastern
    Himalayas.* Botanical Survey of India, Howrah.

Tikhomirov, V.N., A.V. Shcherbakov and L.V. Denisova (1987) [Main sources of informa-
    tion on the flora of USSR reservations.] *Byulleten Moskovskogo Obshchestva
    Ispytatelei Prirody Otdel Biologicheskii* 92:119–136.

Tralau, H. (and successors) (1969–) *Index Holmiensis. A World Index of Plant Distribu-
    tion Maps.* Scientific Publishers, Zurich (Vols 1–4). and Swedish Museum of
    Natural History, Stockholm (Vols 5–6). Pteridophytes, gymnosperms,
    monocotyledons and dicotyledons to F completed by 1988.

Vidal, J.E., Y. Vidal and P. Ho (1988) *Bibliographie Botanique Indochinoise de 1970 à
    1985: Documents pour la Flore du Cambodge, du Laos et du Vietnam.* Muséum
    National d'Histoire Naturelle, Paris. 2856541836.

Wang, Z. (1983) [Bibliography of Chinese Botany.] Three volumes. Chinese Botanical
    Society, Beijing.

Zanoni, T.A. (1986) Bibliografia de la flora y de la vegetación de la isla Española: 2. Adi-
    ciones. *Moscosoa* 4:39–48. Haiti and Dominican Republic.

Zanoni, T.A. (1989) Bibliografia botánica del Caribe: 2. *Moscosoa* 5:349–374.

Zanoni, T.A., C.R. Long and G. McKiernan (1984) Bibliografia de la flora y de la vegeta-
    ción de la isla Española. *Moscosoa* 3:1–61. Haiti and Dominican Republic.

Fruit and seed identification

Berggren, G. (1969–1981) *Atlas of Seeds and Small Fruits of Northwest European Plant Species*. Swedish Natural Science Research Council, Stockholm. Three volumes published. 9172604980 (Vol. 3).

Bergstrom, D.M. (1986) An atlas of seeds and fruits from Macquarie Island. *Proceeding of the Linnean Society of New South Wales* 109:69–90.

Delorit, R.J. and C.R. Gunn (1986) *Seeds of Continental United States Legumes (Fabaceae)*. Agronomy Publications, Wisconsin. 0961684704.

Dickison, W.C. (1984) Fruits and seeds of the Cunoniaceae. *Journal of the Arnold Arboretum* 65:149–190.

Durani, S. and A.M. Kak (1988) Aquatic and wetland vegetation of the north western Himalaya: 28. Seed identification of some economic aquatic and wetland plants of north western Himalaya. *Journal of Economic and Taxonomic Botany* 11:443–457.

Flood, R.J. and G.C. Gates (1986) *Seed Identification Handbook*. National Institute of Agricultural Botany, Cambridge. 0948851074.

Gunn, C.R. (1984) *Fruits and Seeds of Genera in the Subfamily Mimosoideae (Fabaceae)*. Agricultural Research Service Technical Bulletin No. 1681. USDA, Washington DC.

Gunn, C.R. (1991) *Fruits and Seeds of Genera in the Subfamily Caesalpinoideae (Fabaceae)*. Agricultural Research Service Technical Bulletin No. 1755. USDA, Washington DC.

Gunn, C.R. and J.V. Dennis (1991) *World Guide to Tropical Drift Seeds and Fruits*. Quadrangle, New York. 0812906160.

Gunn, C.R. and C.A. Ritchie (1988) *Identification of Disseminules Listed in the Federal Noxious Weed Act*. Agricultural Research Service Technical Bulletin No. 1719. USDA, Washington DC.

Hanf, M. (1983) *The Arable Weeds of Europe with Their Seedlings and Seeds*. BASF United Kingdom Ltd., Hadleigh.

Harden, G.J. and J.B. Williams (1979) *Fruits: A Guide to Some Common and Unusual Fruits Found in Rainforests*. University of New England, Armidale.

Kumar Das, D. (1987) *Edible Fruits of Bangladesh Forests*. Forest Research Institute, Chittagong.

Monod, T. (1977) Fruits et graines de Mauritanie (suite). *Bulletin du Muséum National d'Histoire Naturelle (Paris), 3e Série No. 461 Bot.* 32:73–128.

Monod, T. (1979) Fruits et graines de Mauritanie (suite). *Bulletin du Muséum National d'Histoire Naturelle (Paris), 4e Série* 1:3–51.

Montgomery, F.H. (1977) *Seeds and Fruits of Plants of Eastern Canada and Northeastern United States*. Toronto University Press, Toronto.

Morton, J.F. (1987) *Fruits of Warm Climates*. J.F. Morton, 20534 SW 92 Ct, Miami, FL 33189. 0961018410.

Nesbitt, M. and J. Grieg (1989) A bibliography for the archaeobotanical identification of seeds from Europe and the Near East. *Circaea* 7:11–30.

Ng, F.S.P. (1991 and 1992) *Manual of Forest Fruits, Seeds and Seedlings*. Malayan Forest Record No. 34. Forest Research Institute of Malaysia, Kepong. 983959205 (Vol. 1) and 9679991547 (Vol. 2).

Ogden, E. and R.S. Mitchell (1990) *Identification of Plants with Fleshy Fruits*. New York State Museum, Albany. 15555411883 (with diskette). Covers northeastern USA and southeastern Canada.

Prévost, M.-F. (1983) Les fruits et les graines des espèces végétales pionnières de Guyane
    Française. *Terre et Vie* 38:121–145.
Puerto, O. del and E. Sequeira (1981) Identificación de semillas de malezas (4). (Amaran-
    thaceae, Chenopodiaceae, Geraniaceae, Plantaginaceae.) In: 4ta Reunión, técnica,
    Montevideo, 2–3 diciembre 1981. Cátedra de Botánica, Facultad de Agronomía,
    Universidad de la República 33.
Roosmalen, M.G.M. van (1985) *Fruits of the Guianan Flora*. Institute of Systematic
    Botany, Rijksuniversiteit, Utrecht. 9090009876.
Sa'ad, F.M. (1980) Identification of weed seeds encountering the crops cultivated in
    Egypt. *Notes from the Agricultural Research Centre Herbarium (Egypt)* 5:25–41.
Sparke, C.J. and G.H. Williams (1983) *The Identification of Seeds in Bracken-infested
    Hill-soils*. West of Scotland Agricultural College, Auchincruive.
Teel, W. (1984) *A Pocket Directory of Trees and Seeds in Kenya*. KENGO, Nairobi.
Veevers-Carter, W. (1991) *Riches of the Rain Forest. An Introduction to the Trees and
    Fruits of the Indonesian and Malaysian Rain Forests*. Oxford University Press,
    Oxford. 0195889894.
Wickens, G.E. (1979) The propagules of the terrestrial flora of the Aldabra Archipelago,
    Western Indian Ocean. *Atoll Research Bulletin* 229.
Young, J.A. and C.G. Young (1992) *Seeds of Woody Plants in North America*. Dioscorides
    Press, Portland. 0931146216

International booksellers

Many collectors will have access to a library containing at least some
of the works cited above; many others, however, will not. It is clearly
of limited value to know about a work but not be able to see it, so a short
list (provided by the Library of the Royal Botanic Gardens, Kew) is
appended here of some international booksellers of both in-print and out-
of-print works. The regular bulletins or catalogues published by some of
these booksellers (e.g. the Natural History Book Service) are well worth
receiving since they can keep collectors in touch with new and recent
titles, in much the same way as does the *Kew Record*.

Clarke's Bookshop
211 Long Street
Cape Town 8001
South Africa
Tel: +27 21 235739
Fax: +27 21 236441
Both in-print and out-of-
    print titles

F. Fluck-Wirth (Krypto)
CH-9053 Teufen AR
Switzerland
Tel: 71 331687
Fax: 71 331664
Both in-print and
    out-of-print titles

Koeltz Scientific Books
PO Box 1360
D-6240 Konigstein
Germany
Tel: +49 6174
    4492/3189
Fax: +49 6174 1634
Both in-print and out-of-
    print titles

Lubrecht & Cramer
RD No. 1, Box 244
Forestburgh
New York 12777
USA
Tel: +1 914 7948539
Fax: +1 914 7917575
Both in-print and
    out-of-print titles

Natural History Book
    Service Ltd
2 Wills Road
Totnes
Devon TQ9 5XN
UK
Tel: +44 1803 865913
Fax: +44 1803 865280
E-mail: nhbs@gn.apc.org
In-print titles only

The Rural Store
Lowdens Road
Kilmore West
Victoria 3764
Australia
Tel: +61 57 821118/2283
Fax: +61 57 810183
In-print titles only

Wheldon & Wesley Ltd
Lytton Lodge
Codicote
Hitchin
Hertfordshire SG4 8TE
UK
Tel: +44 1438 820370
Fax: +44 1438 821478
Both in-print and out-of-
  print titles

Acknowledgements

Special thanks are due to many colleagues in the Herbarium, Jodrell Laboratory and Library at Kew; C. Sperling (United States Department of Agriculture); M. Beasley and R. Hampshire (The Natural History Museum, London); R. DeFilipps and F. Fosberg (Smithsonian Institution, Washington); R. Flood (National Institute of Agricultural Botany, Cambridge); S. Mori (The New York Botanical Garden); and S. Smaridge (Natural History Book Service, Totnes).

APPENDIX **10.1**
Plant diversity databases
International Organization for Plant Information (IOPI)

In the past few years, botanical and other biodiversity databases have proliferated. In developing countries, institutions involved in such work have often received support from outside agencies and institutions. For example, the Nature Conservancy in the US has helped to establish a network of 85 national and subnational computer-based biodiversity information centres in the Americas. Intergraph Corporation and Costa Rica's National Biodiversity Institute (INBio) have jointly developed a computerized Biodiversity Information Management System. The Missouri Botanical Garden, the Muséum National d'Histoire Naturelle in Paris and two herbaria in Madagascar are collaborating on producing an electronic Conspectus of the Vascular Plants of Madagascar. Conservation International, World Wide Fund for Nature (WWF) and WCMC have also been active in this field.

Numerous plant databases (listing locally accepted names and synonyms, some distributional data, perhaps ethnobotanical information such as local names and uses) now exist or are in preparation, usually at herbaria, universities, conservation bodies, biodiversity centres or similar institutions. They use a variety of management software and also vary widely in structure and content. They may cover anything from a single protected area to the whole country, and the whole flora or selected taxa. Often based on actual herbarium specimens, they may or may not be taxonomically reliable on a wider scale.

The databases listed below are the principal large plant databases in usable form at present and are nomenclaturally reliable across regions. The International Organization for Plant Information (IOPI) has established a register of plant databases at the Royal Botanic Garden, Edinburgh which it updates as information becomes available. Contact may be made through R.J. Pankhurst (see below) for more information. IUCN and WCMC are collaborating on the development of regional listings of biodiversity databases and other data sets, starting in East Africa. The International Union of Biological Sciences Commission on Taxonomic Databases (TDWG) is developing guidelines and standards for botanical databases (Chapters 14 and 19).

IOPI is itself preparing a world nomenclatural database of all vascular plants, the *Global Plant Checklist*, which will be taxonomically edited to ensure comparability throughout the world. This will act as a core module to which other information will be added. The first version is planned for 1997. Further information can be obtained from the Secretary:

Alex George
18 Barclay Road
Kardinya
WA 6163
Australia
Fax: +61 9 3371655
E-mail: iopi@cjb.unige.ch

For more information on IOPI and to register for its electronic news service, contact David Green (david.gree@anu.edu.anu).

1. Name: BONAP
   Address: Department of Biology
            University of North Carolina
            Chapel Hill
            NC 27599-3280
            USA
            Tel: +1 919 9620578
            Fax: +1 910 9621625
   Contact: John T. Karteaz
   Content: Nomenclature and distribution of plants of the USA and neighbouring areas

2. Name: ERIN (Environmental Resources Information Network)
   Address: GPO Box 636
            Canberra
            ACT 2601
            Australia
   Contact: Arthur Chapman
            E-mail: arthur@erin.gov.au
   Content: Biodiversity of Australia, including plant taxonomic data

3. Name: ESFEDS (European Science Foundation European Documentation System)
   Address: Royal Botanic Garden
            Edinburgh EH3 5LR
            UK
            Tel: +44 131 5520382
            Fax: +44 131 5527171
   Contact: R.J. Pankhurst
            E-mail: rjp@castle.ed.ac.uk
   Content: Nomenclature, distribution and bibliography of the Flora Europaea

4. Name: Grass Genera of the World
   Address: Taxonomy Unit
            Research School of Biological Sciences
            Australian National University
            GPO Box 475
            Canberra, ACT 2601
            Australia

Contact: L. Watson
Content: Nomenclature, morphology, anatomy, distribution etc. of 778 grass genera worldwide

5. Name: (a) IK (Index Kewensis)
   (b) Cactaceae checklist
   Address: Royal Botanic Gardens, Kew
   Richmond
   Surrey TW9 3AB
   UK
   Tel: +44 181 9401171
   Fax: +44 181 9481197
   Contact: Bill Loader
   E-mail: vxxloadr@cms.am.rdg.ac.uk
   Content: (a) Nomenclature and place of publication of world angiosperms
   (b) World checklist of Cactaceae (David Hunt)

6. Name: ILDIS (International Legume Database and Information Service)
   Address: Department of Biology
   University of Southampton
   Bassett Crescent East
   Southampton SO9 3TU
   UK
   Tel: +44 1703 59444
   Fax: +44 1703 594269
   Contact: Frank Bisby
   Content: Nomenclature, bibliography, distribution and biodiversity data on Leguminoseae worldwide

7. Name: ING (Index Nominum Genericorum)
   Address: Department of Botany
   Smithsonian Institution
   Washington DC 20560
   USA
   Tel: +1 202 3574362
   Fax: +1 202 7862563
   Contact: Ellen Farr
   E-mail: mnhbo001@sivm.si.edu
   Content: All plant genera names with authorities and publication data

8. Name: MED-CHECKLIST
   Address: Conservatoire et Jardin Botaniques
   CP 60
   CH-1292 Chambesy
   Geneva
   Switzerland
   Tel: +41 22 7326969
   Fax: +41 22 7384597
   Contact: Catherine Zellweger
   E-mail: zellweger@cjb.unige.ch
   Content: Checklist of vascular plants for coutries around the Mediterranean

9. Name: Flora Mesoamericana
Address: Missouri Botanical Garden
          PO Box 299
          St Louis
          Missouri 63166–299
          USA
          Tel: +1 202 3574362
          Fax: +1 202 7862563
Contact: Doug Stevens
Content: Nomenclature, distribution and bibliography for American countries; also
          specimen data for Missouri Botanical Garden

10. Name: Conspectus Florae Orientalis
Address: Herbarium
          Botany Department
          Hebrew University
          Jerusalem 91904
          Israel
Contact: Clara Heyn
Content: Nomenclature, distribution and bibliography for Near East

11. Name: PRECIS (Pretoria Computerized Information System)
Address: Botanical Research Institute
          Private Bag X1O1
          Pretoria
          South Africa
          Tel: +27 12 861164
          Fax: +27 12 861194
Contact: T.H. Arnold
Content: Herbarium catalogue and biodiversity of plants of southern Africa; a by-product,
          the Plants of Southern Africa Database, is available on diskette

12. Name: USDA (US Department of Agriculture) Families and Genera
Address: Agricultural Research Service
          Beltsville Agriculture Research Center
          Beltsville
          MD 20705–2350
          USA
Contact: Joseph H. Kirkbride, Jr.
          E-mail: jkirkbride@asrr.arsusda.gov
Content: World list for flowering plants

13. Name: Czerepanov List from Flora USSR
Address: Komarov Botanical Institution
          Russian Academy of Sciences
          Prof. Popov St. 2
          St Petersburg 197376
          Russia
Contact: Dimitri Geltman
Content: Nomenclature of plants from ex-USSR

14. Name: WCMC

Address: World Conservation Monitoring Centre
219c Huntingdon Road
Cambridge CB3 0DL
UK
Tel: +44 1223 277314
Fax: +44 1223 277136
E-mail: info@wcmc.org

Contact: Harriet Gillett

Content: Nomenclature, distribution and bibliography of threatened and useful plants worldwide

Aids to taxonomic identification

11

N. Maxted[1] and R. Crust[2]

[1]*School of Biological Sciences, University of Birmingham, Edgbaston, Birmingham B15 2TT, UK:* [2]*Department of Biological Sciences, University of Zimbabwe, PO Box MP 167, Mount Pleasant, Harare, Zimbabwe.*

Introduction

The vast majority of germplasm collected for *ex situ* conservation in the 1970s and early 1980s was crop germplasm. However, since then, increasing attention has been focused on wild species, including crop relatives, forages, multipurpose trees and shrubs and forestry species. This has brought with it a special set of problems for collectors, not the least of which is the taxonomic identification of the plants. It is essential to obtain its correct scientific name if a specimen is to be connected to the wealth of information that may be known about the taxon to which it belongs. Germplasm that is not accurately identified in the field may remain uncollected, because its value is not recognized when encountered. Misidentification of any material that is collected will lead to spurious results when the germplasm is studied and used. The growing out and reidentification of several large germplasm collections (Maxted and Bisby, 1986, 1987; Maxted, 1989, 1992) have established that a significant proportion of the material was either wrongly identified or not identified at all. Material incorporated into gene banks as 'unknown legume species' or '*Eleusine* sp.' is of limited conservation value and is difficult to use.

Identification, or 'determination', of a plant specimen involves two steps. First, there is the choice of a taxon name to which the specimen may be referred, i.e. the act of recognizing or establishing the taxon to which a specimen belongs. Second, there is the decision as to what the correct name is, if in fact more than one name has been applied to the taxon. The correct name for a taxon is a legitimate name which is accepted according to the rules of the International Code of Botanical Nomenclature (ICBN) regarding the choice of names. Legitimate names are those which are in accordance with ICBN rules of publication and are thus available for consideration as the acceptable name (Stace, 1989).

The second step requires a knowledge of the history of the taxon in the botanical literature, a subject which is introduced in Chapter 10. This chapter focuses on the mechanics of the first step in identification, i.e. attaching a name to the specimen. The difficulties encountered in doing this may be due to a number of factors (Abbott *et al.*, 1985). Some taxa are genuinely difficult to distinguish or show different features at different stages of their life history. However, it is undeniable that the available field identification aids are frequently poorly designed, and therefore difficult to use, particularly for those who lack formal taxonomic training and are unfamiliar with the terminology they employ. Some are also inaccurate. Collectors must learn to make the best use possible of what identification aids may already be available, but must also be occasionally ready to develop their own.

Identification aids

Specimens are commonly identified to species, but if lower taxonomic entities have been described they could be named to subspecies, variety or form. The correct identification of the specimen is achieved by comparing its characteristics to the set of 'key' characteristics possessed by each species. If the specimen's characteristics fall within the range of key characteristics of a particular species, then the specimen is identified as a representative of that species. The range of the key characteristics for each species will have been previously determined by a detailed study of a broad range of specimens representing that species.

There are two approaches to identification, matching and elimination. Matching identification involves the comparison of the specimen to taxon descriptive data or some form of exemplar, such as a named herbarium sheet. Clearly, trying to match a specimen to one of a large number of possible taxa could be impossibly time-consuming. Some method is needed to narrow down the possibilities. Identification by elimination involves comparing a specimen to a set of mutually exclusive short descriptions, as in a printed key, and making a decision as to which fits the specimen best, repeating the process for another set of descriptions until only one taxon remains. Often, identification will begin by elimination, and proceed by matching when the range of possible taxa has been narrowed down to manageable proportions.

Four types of aids to identification are described below:

- single-access keys;
- multi-access keys;
- illustrations;
- computer-based aids.

Single-access keys

A taxonomic key is a logical arrangement of the distinguishing features of all the members of a taxonomic group, designed to assist in the naming of unidentified members of that group. Keys are commonly based on gross morphological features, characteristics that are readily observable in the field, laboratory or herbarium. Traditional keys are commonly referred to as being:

- dichotomous, meaning that at each step the user is presented with two brief, diagnostic descriptions, only one of which should match the specimen being identified;
- single-access, meaning there is only one point of entry into the key, via the choice between the first pair of descriptions;
- sequential, meaning that once the first choice is made the user is faced with a second choice of alternative diagnostic information and then a third and so on, until by making the final choice the name of the specimen is obtained;
- diagnostic, meaning that the few characters used in the key distinguish the taxa included, but are not intended to provide a full description of the taxon.

The nature of the key design means that the process of identification thus follows a specific, structured sequence until the identification is finally made.

The two alternative descriptions provided in a dichotomous key are referred to together as a 'couplet'. Each individually is a 'lead'. The characters used for identification should be constant and stable for the particular species and the two leads in a couplet should be comparable and yet mutually exclusive without any overlap. Not all single-access keys are dichotomous: they can be polychotomous, presenting the user with more than two leads at any one point. For practical reasons, keys that use triplets or quadruplets are not commonly used, however. In a good key, the couplets will refer to easily observable (at most with the aid of a hand lens) characteristics that divide the taxa into two approximately equal groups. If the characters divide the initial set of taxa repeatedly into two in this way, the average number of questions one needs to answer to identify a specimen will be fewer, and thus the chances of making an error in identification are reduced and the speed of identification increased.

Single-access keys may be monothetic, if only one character is used in each lead, or polythetic, if several characters are used together. An example of a polythetic key to a group of vetches (the legume genus *Vicia*) is shown in Key 11.1.

There are two basic styles of single-access key, the parallel (or bracketed) and the yoked (or indented). The polythetic key provided in Key 11.1 illustrates the parallel style. The same key is shown in the yoked style in Key 11.2. The essential difference between the two styles of key is that in the parallel style all the components of a set of related

Key 11.1

Parallel key to a group of vetches.

1. Lower stipules subentire, 2–4 mm long; flowers 6–14 mm; seeds sculptured; tendrils mostly simple **2**
   – Lower stipules distinctly toothed, > 4 mm long; flowers (10–)14–36 mm; seeds smooth; tendrils simple or branched **3**
2. Flowers 5–12 mm; legume falcate, distal end beaked; seed surface ruminate-reticulate .... *V. cuspidata*
   – Flowers 9–15 mm; legume not falcate, distal end unbeaked; seed surface tuberculate .... *V. lathyroides*
3. Wings purplish; legume rounded in cross-section; hilum < third seed circumference **4**
   – Wings cream or yellow; legume distinctly laterally flattened in cross-section; hilum > third seed circumference **6**
4. Annual, stolons absent; standard cream, yellow or purple **5**
   – Perennial, stoloniferous; standard purple .... *V. pyrenaica*
5. Standard face purple; wings reddish purple .... *V. sativa*
   – Standard face yellow; wings bluish purple .... *V. barbazitae*
6. Aril present; flower (19–)24–36 mm .... *V. grandiflora*
   – Aril absent; flower 20–21 mm .... *V. qatmensis*

descriptions are printed together, whereas in the yoked style all the possibilities which depend on part of a description are set down and exhausted first, before other parts of a description are taken up. Neither style of key has a clear botanical advantage over the other, but parallel keys tend to take up less space because there is no need for indentation.

The single-access key is the traditional tool used for biological identification, but any biologist who has had to use one professionally will agree that it has many undesirable features. The user may not be able to proceed past the first couplet because neither of the leads provided fits the specimen or it may not be possible to score the characters. For example, the first lead may involve seed characters, and the specimen may still be only at the flowering stage. Also, users may not understand the technical terminology used. This may halt progress through the key or may result in errors when interpreting the specimen's character states, resulting in misidentification. A glossary should be used to check terms (see below) or the key should be annotated before use with definitions of all unfamiliar terms. When measurements are used in the key, several structures should ideally be looked at, but the specimen might have only one example, again causing uncertainty. One may also find that neither of the two alternative descriptions in a couplet appears to describe the specimen accurately. In this case one has to act pragmatically and follow the option that supplies the better fit, or try both choices and then compare full descriptions of the two answers with the specimen.

In fact, once an identification is achieved, the specimen should

Key 11.2
Yoked key to a group of vetches.
1. Lower stipules subentire, 2–4 mm long; flowers 6–14 mm; seeds sculptured; tendrils mostly simple **2**
   2. Flowers 5–12 mm; legume falcate, distal end beaked; seed surface ruminate–reticulate .... *V. cuspidata*
   2. Flowers 9–15 mm; legume not falcate, distal end unbeaked; seed surface tuberculate .... *V. lathyroides*
1. Lower stipules distinctly toothed, > 4 mm long; flowers (10–)14–36 mm; seeds smooth; tendrils simple or branched **3**
   3. Wings purplish; legume rounded in cross-section; hilum < third seed circumference **4**
      4. Annual, stolons absent; standard cream, yellow or purple **5**
         5. Standard face purple; wings reddish purple .... *V. sativa*
         5. Standard face yellow; wings bluish purple .... *V. barbazitae*
      4. Perennial, stoloniferous; standard purple .... *V. pyrenaica*
   3. Wings cream or yellow; legume distinctly laterally flattened in cross-section; hilum > third seed circumference **6**
      6. Aril present; flower (19–)24–36 mm .... *V. grandiflora*
      6. Aril absent; flower 20–21 mm .... *V. qatmensis*

always be compared with a detailed botanical description of the taxon in order to check that no mistake has been made during the keying-out process. A botanical description is a logically laid-out statement, technically worded, of the characteristics of a taxon. If this checking is not done, then keying-out errors will remain undetected. If the description does not match the specimen or if, when using the key, neither of the leads seems appropriate, then it is possible that a mistake has been made in using the key. To attempt to locate the error the user must return to the first couplet and repeat the entire identification process, hoping that the error can be spotted and corrected.

Floras, monographs, revisions and other taxonomic works include keys and descriptions, sometimes supplemented by illustrations (see below). Collectors must know how to select the most appropriate aid to help identify their specimen. In attempting to identify a tree species from Zimbabwe, the collector might start out by using a key to angiosperm families. This key should enable identification of the specimen to the Ebenaceae, for example. A key to genera might then be used to identify the specimen as a *Diospyros* species. A key including all 500 *Diospyros* species will not be required for the final step if it is known that the specimen to be identified is native to Zimbabwe. Then the specimen is likely to be one of the 13 native Zimbabwean *Diospyros* species and so the key provided in *Flora Zambesiaca* (Brenan *et al.*, 1983) would be appropriate. A difficulty may arise when a Flora is not available for the region being visited or the species of interest is not included in the Flora. The collector could then use identification aids for an adjoining area or those included in broader taxonomic treatments, if these are available. However, keys included in a Flora of a specific region are designed to work with the species found in that region and if used for an adjacent region may not provide an accurate identification, because the two regions are likely to have a similar but not identical range of taxa. Chapter 10 deals with finding the most appropriate taxonomic treatment for a given area or species. When using a key for the first time, it is important to read carefully any introductory comments on format details, abbreviations, etc. This will also reveal whether any special tools (in particular, a hand lens) will be necessary.

Multi-access keys

Multi-access keys were developed to overcome some of the problems associated with single-access keys. A multi-access key does not force the user to go through the character set in a specific, preordained sequence. It allows the user to ignore particular characters and still obtain an identification. For example, if the specimen lacks fruit and seed, the identification can be based on vegetative and flower characters alone.

The early forms of multi-access or polyclave keys were produced using edge-punched cards. Each card, representing a taxon, has a series of labelled holes around its edges, each representing a particular character state. If the character state is absent, then the edge of the card is punched out at the appropriate point, making an indentation out of the hole. Depending on the character state of the specimen, the user selects a character and inserts a needle through the pack of taxon cards. When the needle is lifted, the cards (taxa) which do not match the specimen fall out of the pack, leaving those that possess the selected character state hanging from the needle. This process of selecting characters, projecting a needle through appropriate character state holes and excluding cards, is continued until only one card remains, the card for the taxon to which the specimen belongs (Stace, 1989).

One reason for the limited use to which punched-card keys have been put has been the practical difficulties in designing and producing the cards. Tabular keys are a more commonly used form of multi-access key and can take a variety of different forms. An example of a tabular key for the *Vicia* taxa of Keys 11.1 and 11.2 is provided in Table 11.1, with species in the columns and characters in the rows. The table is filled with coded character states for each taxon. For example, *V. pyrenaica* has the code P for the character 'life form', while the other taxa have the code A, indicating that *V. pyrenaica* is perennial and the other taxa are annuals. To identify a new specimen, the user records its characteristics using the same codes and row structure as the tabular key and then compares the specimen data with each column of the table. A match between the user's score card and a particular taxon provides the identification.

A refinement of the tabular format was proposed by Sinker (1975), who devised the lateral key. An example of a lateral key for the *Vicia*

Table 11.1. Tabular key to a group of vetches. For abbreviations see p. 189.

	V. barbazitae	V. cuspidata	V. grandiflora	V. lathyroides	V. pyrenaica	V. qatmensis	V. sativa
1. Life form	A	A	A	A	P	A	A
2. Stipule length (mm)	4–6.5	2.5–4	4–6.5	2–4	4–6	4–5.5	4–12
3. Number stipule teeth	H/D	H	H/D	H	H/D	D	H/D
4. Tendril form	S/B	S	S/B	S	S/B	B	S/B
5. Flower length (mm)	15–26	7–14	19–33	5–14	15–25	20–21	15–28
6. Standard colour	Y	L	Y	G	G	Y	G
7. Wing colour	G	L	Y	G	G	Y	G
8. Legume cross-section shape	R	R/C	C	R	R	C	R
9. Legume curvature	N	F	N	N	N	N	N
10. Legume shape	U	B	U	U	U	U	U
11. Seed surface	S	T	S	T	S	S	S
12. Hilum shape	O	R	E	R	O	E	O
13. Aril presence	–	–	+	–	–	–	–

taxa is provided in Table 11.2. The style is very similar to the tabular key, but in this variant there is a column for each character state. Characters are separated by a vertical line and each cell in the table is filled by a symbol which indicates if the character state is present or absent in that taxon. To identify a specimen the user constructs a one-row structure on paper similar to the structure of the lateral key. A description for the specimen is then prepared, recording the presence or absence of each character state in the table. Having completed the scoring, the score sheet is then slid down the table, keeping the columns of the lateral key and score sheet aligned, until a match is found, providing the identification.

The various printed forms of multi-access keys are now being superseded, at least in part, by computerized keys. It seems likely that in the future the use of printed multi-access keys will be restricted to identification guides designed for use in the field, where it may remain less convenient to use a computer.

Illustrations

One of the problems non-specialists have in learning to use traditional keys is the amount of technical botanical terminology involved. Where is the calyx? What is a capitate style? Ideally, identification aids should either avoid such technical terms altogether or explain them. If illustrations (line drawings, photographs, paintings, etc.) of the key features of the species are provided, recourse to complex terminology may be avoided. Many botanical field guides now include illustrations as well as detailed text including scientific and vernacular names, diagnostic descriptions, distribution maps, phenology, ecological preferences, geographic distribution and conservation status. Information on whether illustrations are available for any particular plant group can be obtained from the relevant taxonomic revisions or monographs, where they should be cited, or from a survey of the appropriate catalogues in a good botanical library. *Index Londinensis* is a key to the botanical illustrations of flowering plants and ferns of the world up to 1935. After that date, *Index Kewensis* took up the task (Chapter 10).

The use of illustrations, when combined with technical terminology, can, in addition to facilitating identification, also help the user to learn and understand the meaning of botanical terms. Useful illustrated morphological glossaries are sometimes provided by Floras and field guides (e.g. Blundell, 1987) and may also be found in various botanical textbooks (e.g. Radford *et al.*, 1974; Radford, 1986). Bell (1991) is a specialized illustrated guide to morphological terminology as applied to flowering plants. Botanical dictionaries (e.g. Jackson, 1928; Stearn, 1983; Tootill, 1984) are also helpful in this context.

The relative efficacy of using line drawings, paintings or photographs to aid identification is a matter of subjective assessment and individual preference. However, one problem associated with using photographs for identification is that they can only show what is

Table 11.2. Lateral key to a group of vetches.

Character codes	A	P	<4	≥4	H	D	S	B	<14	≥14	Y	L	G	Y	L	G	R	C	N	F	U	B	S	T	R	O	E	+	–
State numbers	1	2	3	4	5	6	7	8	9	10	11	12	13	14	15	16	17	18	19	20	21	22	23	24	25	26	27	28	29
V. barbazitae	■	○	○	■	■	○	*	■	○	■	■	○	○	○	○	■	■	○	■	○	■	○	■	○	○	■	○	○	■
V. cuspidata	■	○	■	○	■	*	■	○	■	○	○	■	○	○	■	○	*	■	○	■	○	■	○	■	■	○	○	○	■
V. grandiflora	■	○	○	■	■	○	*	■	○	■	■	○	○	■	○	○	■	○	■	○	■	○	■	○	○	○	■	■	○
V. lathyroides	■	○	■	○	■	*	■	■	■	○	○	○	■	○	○	■	■	○	■	○	■	○	○	○	○	○	○	○	■
V. pyrenaica	○	■	○	■	*	■	*	■	○	■	○	○	■	■	○	○	■	○	■	○	■	○	■	○	○	■	■	○	■
V. qatmensis	■	○	○	■	○	■	○	■	○	■	■	○	○	■	○	○	■	○	■	○	■	○	○	○	○	○	■	○	■
V. sativa	■	○	○	■	*	■	*	■	○	■	○	○	■	○	○	■	○	○	■	○	■	○	■	○	○	■	○	○	■

Abbreviations used in Tables 11.1 and 11.2. A annual, P perennial; H hastate, D dentate; S simple, B branched; Y yellow, L lilac, G purple; C laterally compressed, R round; N not falcate, F falcate; U unbeaked, B beaked; S smooth, T tuberculate; O oval, E elongated; + present, – absent; / – or +; ■ state present, * state rarely present, ○ state absent.

observed at a particular time in a two-dimensional image, whereas with a drawing or painting the illustrator can enhance the observed two-dimensional image to include features that may be less obvious on an individual specimen at that time or in that particular plane of view. For this reason, it is less likely that a specimen could be accurately identified by comparison with photographs. As a general rule drawn illustrations are preferable, to be used in conjunction with other aids.

Computer-assisted identification

Computers have been used by taxonomists in two main ways:

- to produce printed keys;
- for interactive identification.

Production of printed keys

Several key-generating programs have been devised (Pankhurst, 1991). Perhaps the most user-friendly and widely used is KEY (Dallwitz, 1974). KEY is an integral part of the DELTA (Descriptive Language for Taxonomy) package (Dallwitz, 1980), a simple and versatile taxonomic coding system. The International Union of Biological Sciences Commission on Taxonomic Databases (TDWG) has endorsed the DELTA format for recording and exchanging descriptive data. The DELTA package (which is relatively inexpensive) comes with a set of programs which can be used to organize and manage taxonomic data. KEY generates bracketed and tabular keys from data coded into DELTA format (see Box 11.1 for case-study). There are a number of options which can be used to modify the basic key-generating algorithm, making it possible to experimentally produce numerous keys in a few minutes, compared with the several days which it might take to produce even just one key by hand. If the characters have been well structured and have been carefully chosen on the basis of the requirements of identification (rather than classification), good workable keys can be produced. Invariably, once a draft key has been generated by KEY, further editing with a word processor is likely to be required, but this extra work is minimal compared with the time saved in key construction.

Interactive identification programs

There have been many attempts to produce computerized keys since the 1960s, one of the earliest being that of Boughey *et al.* (1968). Most of these efforts have not found an application outside their laboratory of origin, either because they have been experimental, have been tied to specific taxa or have used an inappropriate identification paradigm. Examples of contrasting approaches to interactive key generation and use are found in Wilson and Partridge (1986), Atkinson and Gammerman (1987), O'Shea (1988) and Colosimo *et al.* (1991).

The most successful systems seem to be those that employ a computerized multi-access key, or polyclave. In its simplest form, a polyclave

---

Box 11.1
Case-study: use of identification software in a collecting project

The International Centre for Research in Agroforestry (ICRAF), the International Board for Plant Genetic Resources (IBPGR), the International Livestock Center for Africa (ILCA), and national programmes in five Southern Africa Development Community (SADC) countries mounted a regional collecting programme for woody *Sesbania* spp. in 1993. To aid identification, software from two sources was used – the DELTA system (see text) and EXPLORE (an experimental program written by Richard Crust of the University of Zimbabwe). DELTA was used to put together a workable taxa/characters matrix from published sources, using CONFOR to help ensure consistency and accuracy. The DELTA program KEY was then used to generate a series of dichotomous keys. By using DOS batch files, it was possible to produce dozens of such keys in a very short time. The most plausible keys were then tested in the field, discovering in the process that the literature was occasionally somewhat inaccurate in the documentation of variation among taxa. The DELTA files were updated to take this into account, and another series of dichotomous keys produced. These included country-specific keys, keys based only on vegetative and/or fruit characters and keys biased towards common or important species (e.g. *S. sesban*) or particularly important characters. After further testing during the collecting missions, the original data files were again revised and corrected, and then used as input to EXPLORE, an interactive polyclave identification program intended for training plant collectors.

---

is a taxonomic data matrix stored in the computer's memory. For example, each row in the matrix might represent a taxon and each column a character. The program manipulates the matrix in response to input from the user. The choice of characters is decided by the user, but the program may assist by ranking the characters and presenting them in different ways, perhaps in a new sequence. A common method is to order the unused characters by separation coefficient (Gyllenburg, 1963), information function (Shannon and Weaver, 1949) or some similar parameter. These are dynamic values that indicate how evenly each remaining, unused character divides the residual set of candidate taxa. As each character is used, at least one column of the matrix is discarded. A variable number of rows may also be discarded as taxa are eliminated from the search. Discarded rows and columns may be restored, however, at the user's request. Thus, the active part of the matrix diminishes or expands until identification is complete, i.e. when only one taxon row remains, giving the identification of the unknown specimen.

The versatility of the computerized polyclave makes it easy to provide identification by simultaneous matching as well as sequential elimination. One of the most valuable benefits is that it is possible to allow the user to increase the accuracy of the identification. When a dichotomous key is used, many of the eliminated taxa may differ from the unknown specimen by just a single character but, because of the structure of the key, these taxa are excluded. The computer polyclave user may set an arbitrary threshold for the minimum number of differences accepted before any taxon can be eliminated from the search.

A good example of a generalized computer polyclave is INTKEY, one of the programs in the DELTA package. INTKEY is a command-line driven program that can be used with any descriptive data set coded in DELTA. It demonstrates most of the functions described above, and can also be used for 'information retrieval', e.g. obtaining diagnostic descriptions. Watson *et al.* (1989) describe, with examples, how INTKEY has been used to disseminate information about the grass genera of southern Africa. Another example of a package that can be used with different descriptive data sets is CABIKEY (I. White, pers. comm.). This is a polyclave program which is perhaps most suitable for educational purposes. It is menu-driven, with displays appearing in windows. Information concerning the context in which identification takes place can be integrated into the data. At any time, the CABIKEY user can call up text descriptions, diagnoses and illustrations of both taxa and characters. CABIKEY data sets are entered by arrangement with the program author, and so, unlike INTKEY, it is not suitable for the casual or experimental user. The suite of programs described by Pankhurst (1986) allows the user to compare and match a specimen against data sets held in memory.

Current trends in database technology are moving towards what is generally known as 'multimedia', the integration of text with illustrations, video animation and sound. The Expert-centre for Taxonomic Identification (ETI) at the University of Amsterdam is developing sophisticated taxonomic software that uses multimedia concepts. Its identification software (IdentifyIt) is a computer polyclave that uses expert-system methods. It points to ways in which advanced technology may be used for more efficient taxonomic identification. The computer hardware required to use ETI's software and databases is likely to become more widely available in the near future.

Future prospects

Current concerns about the rapid loss of biodiversity are imposing demands on the taxonomic community that can only be met by overhauling traditional practices and applying appropriate up-to-date techniques. Single-access keys and the taxonomic terminology required to use them cannot adequately meet the growing requirement for accurate specimen identification. There is thus immense potential for the application of computerized identification. There have, however, been a number of practical difficulties that have prevented this potential from being fully realized. For example, the development of reliable software is very expensive and getting taxonomic data into a suitable format is tedious, time-consuming and error-prone. It is also difficult to design a user interface which is intuitively simple to use, especially for someone more familiar with traditional printed keys, and yet which takes account of all the possibilities held out by the computerized polyclave. Nevertheless, it is reasonable to predict that these problems will be resolved before long.

Multi-access keys are a step forward, but further developments are required. These may include more use of expert systems, deploying not just raw taxonomic data but taxonomic knowledge and expertise (e.g. Diederich and Milton, 1988), and large-scale systems that directly link taxonomic expertise with the applied disciplines that make use of it, such as plant breeding and phytochemistry. There is a pressing need for methods that assist taxonomists in understanding their data, rather than merely processing it.

Acknowledgements

We are particularly grateful to L. Guarino and R.N. Lester for their comments on the manuscript.

References

Abbott, L.A., F.A. Bisby and D.J. Rogers (1985) *Taxonomic Analysis: The Use of Computers, Models and Data Bases in Biological Taxonomy*. Columbia University Press, New York.

Atkinson, W.D. and A. Gammerman (1987) An application of expert systems technology to biological identification. *Taxon* 36:705–714.

Bell, A.D. (1991) *Plant Form: An Illustrated Guide to Flowering Plant Morphology*. Oxford University Press, Oxford.

Blundell, M. (1987) *Collins Guide to the Wild Flowers of East Africa*. William Collins Sons & Co., London.

Boughey, A.S., K.W. Bridges and A.G. Ikeda (1968) *An Automated Biological Identification Key*. Research Series 2. Museum of Systematic Biology, University of California, Irvine.

Brenan, J.P.M., E. Launert, E.J. Mendes and H. Wild (1983) *Flora Zambesiaca*. Vol. 7, Part 1. Royal Botanic Gardens, Kew.

Colosimo, A., E. Rota and P. Omodeo (1991) A Hypercard program for the identification of biological specimens. *Computer Applications in the Biosciences* 7:63–69.

Dallwitz, M.J. (1974) A flexible computer program for generating identification keys. *Systematic Zoology* 23:50–57.

Dallwitz, M.J. (1980) A general system for coding taxonomic descriptions. *Taxon* 29:41–46.

Diederich. J. and J. Milton (1988) NEMISYS: an expert system for nematode identification. In: Fortuner, R. (ed.) *Nematode Identification and Expert Systems Technology*. pp. 9–18. NATO ASI Series A, Vol. 162. Plenum, New York.

Gyllenburg, H.G. (1963) A general method for deriving determination schemes for random collections of microbial isolates. *Annals of the Finnish Academy of Sciences. Series A, IV. Biology* 69:1–23.

Jackson, B.D. (1928) *A Glossary of Botanical Terms*. Duckworth, London.

Maxted, N. (1989) *Bulking, Characterisation and Duplication of Vicieae Germplasm at ICARDA, Syria. Final Report.* IBPGR, Rome.

Maxted, N. (1992) Identification of Vicieae and Cicereae germplasm held by the USDA. Unpublished report to USDA.

Maxted, N. and F.A. Bisby (1986) *Wild Forage Legume Collection in Syria. Final Report.* IBPGR, Rome.

Maxted, N. and F.A. Bisby (1987) *Wild Forage Legume Collection in Turkey. Phase I: South-west Turkey. Final Report.* IBPGR, Rome.

O'Shea, B.J. (1988). An 'expert-system' approach to bryological identification. In: Glime, J.M. (ed.) *Methods in Bryology.* Proceedings of the Bryological Methods Workshop. July 1987. Mainz, Germany. pp. 321–326. Hattori Botanical Laboratory, Japan.

Pankhurst, R.J. (1986) A package of computer programs for handling taxonomic databases. *Computer Applications in the Biosciences* 2:33–39.

Pankhurst, R.J. (1991) *Practical Taxonomic Computing.* Cambridge University Press, Cambridge.

Radford, A.E. (1986) *Fundamentals of Plant Systematics.* Harper & Row, New York.

Radford, A.E., W.C. Dickinson, J.R. Massey and C.R. Bell (1974) *Vascular Plant Systematics.* Harper & Row, New York.

Shannon, C.E. and W. Weaver (1949) *The Mathematical Theory of Communication.* University of Illinois Press, Urbana.

Sinker, C.D. (1975) A lateral key to common grasses. *Bulletin of the Shropshire Wildlife Trust* 31:11–18.

Stace, C.A. (1989) *Plant Taxonomy and Biosystematics.* 2nd edition. Arnold, London.

Stearn, W.T. (1983) *Botanical Latin.* David & Charles, Newton Abbot.

Tootill, E. (ed.) (1984) *The Penguin Dictionary of Botany.* Penguin Books, London.

Watson, L., G.E. Gibbs Russell and M.J. Dallwitz (1989) Grass genera of southern Africa: interactive identification and information retrieval from an automated data bank. *South African Journal of Botany* 55:452–63.

Wilson, J.B. and T.R. Partridge (1986) Interactive plant identification. *Taxon* 35:1–12.

Secondary sources on cultures and indigenous knowledge systems

12

L. Guarino

IPGRI, c/o International Laboratory for Research on Animal Diseases, PO Box 30709, Nairobi, Kenya.

Introduction

Data from scientific observations and experiments are used routinely by plant germplasm collectors to plan their work and document their collections. Measurements of environmental features of collecting sites taken in the field (e.g. soil pH, altitude) or estimated from their values at nearby sites (e.g. temperature, rainfall) help define the adaptation of the material collected and can thus guide multiplication, evaluation and use. Observations of plant characters made during characterization and evaluation trials will be of interest to breeders and other users and can also be helpful in directing future collecting by identifying areas of high diversity and pin-pointing where desired traits occur.

There is, however, another sort of data that can tell us about plants and the places where they are found: that derived from local or indigenous knowledge (IK). Farmers give names to the different landraces they cultivate and know their properties and requirements; nomadic pastoralists know where and when the plants grow which their livestock like to eat; tropical rain-forest dwellers know which nut can be eaten and which bark can be pounded to make a poison. Customs, cults, rites, taboos, legends, myths and folklore all speak of the relationship between people and plants, a relationship that is based on long-term, intimate experience and is often crucial to survival. The term 'traditional knowledge' is also sometimes used for this, but has been generally rejected as implying that the knowledge of local people is somehow static (G.D. Prain, pers. comm.). Richards (1985) states that local farming practice is 'not a matter of "traditions" refined by a long process of trial and error and handed down from generation to generation, but of active innovation and invention ... in the recent past'. In fact, both processes will be important.

No less than scientific data, the knowledge of plants, of animals (including pests) and of the environment that local men and women acquire, refine, maintain and exchange (usually orally, but also in writing and by observation) can help in making decisions about what and how to sample and about the use of germplasm. The two domains, the scientific and the indigenous, are complementary. Most importantly, in contrast to reductionist scientific knowledge, IK is interdisciplinary, holistic and diachronic, an approach that farming systems research and related techniques seek to emulate. It is, however, important to recognize that IK has limitations. They include its uneven (and often limited) distribution within communities, the fact that its transfer by oral means is error-prone, and its occasional fragility in the face of disturbance (IDS Workshop, 1989). This means that specialized methodologies are needed for its study and use. The documentation of indigenous botanical knowledge – the study of indigenous peoples and their relationship with, and use of, plants – is ethnobotany, a branch of ethnography (Plotkin, 1989; Given and Harris, 1992). Martin (1994) is a general guide to ethnobotanical fieldwork. Chapter 18 presents a methodology for documenting IK specifically of plant genetic resources. Chapter 19 discusses in more detail the specific topics on which information derived from both scientific observation and indigenous knowledge is needed in documenting a germplasm collection. This chapter discusses the role of IK in plant germplasm collecting and describes the various secondary sources of ethnographic data, in particular ethnobotanical data.

The importance of IK – of, that is, the knowledge of farmer, homemaker, herder, traditional healer – in the conservation of biodiversity has not perhaps been fully recognized in the past. However, this is changing, in particular as a result of the relatively recent involvement of social scientists in so-called formal sector plant genetic resources work and of the efforts of non-governmental organizations (NGOs) in the informal sector (e.g. the various papers in Cooper *et al.*, 1992). Indeed, Article 8 of the *Convention on Biological Diversity* now enjoins each Contracting Party to

> Subject to its national legislation, take action to respect, preserve and maintain knowledge, innovations, and practices of indigenous and local communities embodying traditional lifestyles relevant for the conservation and sustainable use of biological diversity and promote their wider application with the approval and involvement of the holders of such knowledge, innovations and practices and encourage the equitable sharing of the benefits arising from the utilisation of such knowledge, innovations and practices.

The role of women in this context is particularly emphasized in the preamble of the *Convention*. In Africa, for example, women make up perhaps 70% of the agricultural labour force and 80% of food producers, they undertake 60–90% of the marketing and they do all the processing of basic foodstuffs (Anon., 1993). They are thus the main custodians of crop-related knowledge, though there is often a discrepancy between female workloads and experience on the one side and their social status (and power) on the other. Even where women's direct production tasks are more limited, they will probably be responsible for pre-sowing, processing, storage or cooking activities, and thus hold key information on germplasm performance and quality. Further, their general lack of decision-making power on land use means women often cultivate the main field crops at marginal sites, and thus have specialist knowledge of landrace performance with respect to problem soils or environments. In addition, women's trading networks and kin relationships are often the main channels for acquisition and exchange of germplasm (Jiggins, 1990).

Of course, communities have always changed, and continue to do so. For example, worldwide over 30% of rural households are now headed by women owing to male out-migration, and in some areas the proportion can be as high as 70%. The accompanying shifts in tasks and responsibilities and in labour inputs is bringing about far-reaching changes in management practices and in selection criteria (Jiggins, 1986). IK is thus not static: like the germplasm itself, it evolves. Unfortunately, however, it is also sometimes as threatened with erosion and extinction as the germplasm. In many areas it is fading even faster, as the cultural assimilation of rural populations outstrips even deforestation. D.A. Posey (quoted by Khalil *et al.*, 1992) has estimated that one Amerind group has vanished every year in one way or another since the beginning of the century: with each has gone 'an accumulated wealth of millennia of human experience and adaptation'. IK disappears when native people are stripped of their land or when war dislocates societies, but also when young people in contact with the outside world start to embrace the view that traditional ways are illegitimate and irrelevant.

Preserving and documenting the dwindling resource represented by IK are not an optional adjunct to the conservation of germplasm, but an integral and necessary part of the process. This is perhaps most obvious in the case of crops. A landrace may be defined as a set of populations (or clones) of a crop species developed and maintained by farmers and recognized by them as all belonging to the same entity. Landraces have also been called 'primitive varieties' or 'traditional varieties', but terms such as 'farmer's varieties' and 'folkseeds' are perhaps more appropriate than either (Mooney, 1992; but see Cromwell (1990) for a different use of the term 'farmer's variety'). Landraces are defined and delimited by what farmers and other everyday users know about them just as much as modern varieties, the products of scientific plant breeding, are defined by their pedigree and performance in trials. As Berg *et al.* (1991) point out, 'behind any named folk variety there is knowledge'. Collecting landraces (but also medicinal plants, forages and other species used by local communities) while ignoring the dimension of local knowledge cannot but be wasteful at best, hopelessly flawed at worst.

IK recorded on collecting forms can affect the later use of conserved germplasm no less than the results of formal evaluation and screening trials. But it is not just that breeding and introduction programmes will have had part of their work already done for them if conserved germplasm is accompanied by IK. The very aims and procedures of such programmes should be informed, even dictated, by what farmers and other local users of germplasm require and need, and collecting is an excellent opportunity to document such information. Farmers have been fulfilling their own requirements and needs for centuries. As Vellvé (1992) puts it: 'farmers have been breeders ever since agriculture began, ... breeders have been scientists only for the past two hundred years or so'. The scientific community needs to learn how farmers have been working, if their weaknesses are to be overcome and their strengths built on. At a time of shortages of staff and resources, the formal sector cannot afford to ignore the accumulated experience of local people, and in particular local women, in solving their own conservation and development problems.

IK is thus increasingly being recognized as crucial in agricultural research, extension and development in general (e.g. Brokensha *et al.*, 1980; McCorkle, 1989; Warren *et al.*, 1990, 1994; Warren, 1991). Some examples from the Consultative Group on International Agricultural Research (CGIAR) may be instructive. The Centro Internacional de Agricultura Tropical (CIAT) and Centro Internacional de la Papa (CIP) have been leaders in on-farm experimentation, the Internatioal Rice Research Institute (IRRI) rice breeders have used farmers' evaluations of germplasm to guide their work and International Crops Research Institute for the Semi-Arid Tropics (ICRISAT) and International Institute of Tropical Agriculture (IITA) have shown that understanding indigenous soil classification and management systems can assist farming systems research (Warren, 1993). On the role of women in particular, CIAT has pioneered the involvement of women bean farmers in the breeding process, IRRI has expanded understanding of postharvest rice characterization through the Women in Asian Rice Farming Systems Network and ICRISAT has demonstrated the value of collaborative work between entomologists and poor women sorghum growers in breeding for insect resistance.

Relevance of IK in germplasm collecting

Everyday users of germplasm have information that can play an important, and sometimes a decisive, role in several aspects of the collecting process and beyond. In particular, they have knowledge of:

- the vernacular names of landraces, wild plants and their pests;
- the local criteria for distinguishing among them, and their relationship to each other in any folk taxonomy;

- their appearance, properties, environmental preferences and uses;
- the places and habitats where they may be found, and the rules of access to them;
- the agricultural and management practices with which they are associated;
- the origin (history) of planting material, including any selection practices that may have been applied;
- the character of any changes in farming practice, land management and natural habitats.

The specific tasks in which such information will be important to the collector are listed in Box 12.1, and are discussed in more detail below.

Box 12.1
Uses of IK in germplasm collecting

- Locating target areas and material.
- Deciding what to collect, and how.
- Documenting the collection.
- Assessing the 'completeness' of collections.
- Understanding the origin and distribution of diversity, and the rules of access to it.
- Assessing the extent and threat of genetic erosion.

Locating and accessing target areas and material

Knowing the name in the local vernacular of a wild species, crop or landrace of interest can significantly facilitate the process of finding it, especially if it is rare. For example, it was possible to find the tree *Punica protopunica* at a number of 'new' sites on the island of Socotra (Republic of Yemen) through the help of local people because the correct local name was known. Finding another species, *Dirachma socotrana*, proved more difficult due to a confusion over its Socotri name (Ba'azara *et al.*, 1991). Not all wild species have distinct vernacular names, however, especially if they have no particular local uses and/or are inconspicuous. In contrast, it is unusual for this to be the case with landraces, though problems may occur with synonymy or when the same name (e.g. a 'category name' referring to the circumstances of acquisition) is used to refer to more than one morphological entity (Richards, 1991). In ethnically or linguistically diverse areas the names and uses of wild plants and landraces may change drastically within short distances.

Local people have knowledge not only of the distribution of particular wild plants, but also of the existence and location of 'sanctuaries' of high diversity (G.D. Prain, pers. comm.). Indeed, as A. Gupta (quoted by Khalil *et al.*, 1992) points out, these are often actively protected by communities, who may regard niches of plant and animal abundance as sacred groves, the residences of ancestral spirits or deities. They may also fulfil a more practical function, serving as fall-backs, reserves to meet contingencies and for lean seasons and bad years (Chambers, 1990). Use of these areas, as of communal grazing areas and the like, may be under strict control, and collectors will need to be aware of this, so that permission for access can be asked of the appropriate community authorities.

Chambers (1990) makes the point that much subsistence agriculture works by creating or altering 'distinct, small-scale environments which differ from their surroundings, presenting sharp gradients or contrasts in physical conditions internally and/or externally'. Examples are silt-trap fields, pockets of fertile soil (e.g. termitaria), flood recession zones, patches of high groundwater, etc. Such specialized, in many cases marginal, environments are generally missed by conventional soil surveys and land systems studies because of their small size and dispersion. They are also often neglected by agricultural professionals, who tend to be male, have shorter time horizons than farmers and concentrate on staple and cash crops. These environments are often tended by women, may take years to develop, often feature 'unimportant' crops and do not fit in easily with conventional station-based research. Finding them requires collectors to go out of their way to talk to local people.

Locating target material requires not only being in the right place, but also being there at the right time. Local knowledge is often the best guide not only to where a particular wild species, crop, landrace or area of high diversity may be found, but also to the optimal timing of collecting. Farmers will know where harvesting is late in their area and which landraces mature early, for example, and pastoralists where the grass is in seed.

Deciding what to collect, and how

Farmers will be able to help the collector avoid modern cultivars, recently introduced exotic material and duplicates. When material with particular characteristics that are not easily observable in the field is being sought, indigenous knowledge (even just the vernacular name of a plant or landrace) can provide crucial clues. The classic case of this is medicinal plants, where ethnobotany has proved a valuable short cut to the identification of those plants that are likely to be of interest to medicine (Schultes, 1986; Waterman, 1989). Forages provide another example. The decision to collect a particular little-known species as a potential forage will usually be determined by its appearance (e.g. habit, leafiness), actual field observations of grazing and/or its taxonomic proximity to better-known species. Local knowledge of its acceptance by livestock can be much more precise than any of these, for example as regards differences in acceptability to different kinds of livestock and at different stages of growth. Thus, among Fulbe pastoralists in semiarid West Africa, the names of grasses will change with their quality as feed as they mature after rain, the quality of pasture being linked mainly to the taste of the milk produced by the herd feeding on it (Bonfiglioli, 1992). Barrau (1989) discusses how ethnobotany can aid the search for new food and industrial crops.

The strategy and tactics used in collecting will also be informed by IK. Farmers will be able to define the extent of local market areas and seed exchange networks, which could form the basis of a stratification for sampling. The participation of farmers will be crucial when collecting from mixed fields. For example, the collector will need to know if the individual phenotypic entities in a heterogeneous field have been maintained separately and only mixed at sowing, before a decision can be made about how to collect. This issue is explored further in Chapter 18.

Documenting the collection

IK can help collectors decide if a particular crop field or wild population should be collected and, if so, how. It should, however, also form an important part of the documentation of any germplasm samples that are collected. Documenting IK on the properties and adaptations of germplasm – the kind of cultural practices and management that a landrace is subjected to, its requirements and susceptibilities, why it is favoured or disliked and how its products are processed, how a certain wild plant is used, etc. – can be seen as part of the characterization and evaluation process. Indeed, it is that process carried out from the everyday user's point of view. Local people's knowledge of landraces develops over generations of first-hand observation of crucial features of their appearance and performance in a variety of environments, through good years and bad years, 'observation which has to be keenly executed since farmers' lives literally depend on it' (V.D. Nazarea-Sandoval, pers. comm.).

There are numerous cases of the names of landraces reflecting not just appearance but properties such as days to heading and cooking quality (Boster, 1985; Hamon and Hamon, 1991; Richards, 1991). Sorghum landraces identified by Ethiopian farmers as of superior food quality (being referred to by such names as 'milk in my mouth' and 'squirts out honey', for example) were found by breeders to contain high levels of lysine and protein (Brhane and Yilma, 1979). The indigenous vegetable *Gynandropsis gynandra* is taken by pregnant women in western Kenya because it is said to relieve dizziness and generally make them stronger. Chemical analysis has revealed particularly high levels of iron (dizziness is a common symptom of iron deficiency syndrome in pregnant women), calcium and vitamin C (Opole, 1991). There is also evidence that farmers are aware of differences among landraces in their resistance to pests; they certainly have considerable knowledge of the biology of pests, and of pest control methods (Altieri, 1993).

Such knowledge about species and landraces is often systematized in folk classifications (e.g. Conklin, 1972; Brown, 1985; Brush, 1986; Berlin, 1992). These can be remarkably congruent with the results of scientific approaches, at both the interspecific (Alcorn, 1984) and the intraspecific (Asfaw, 1990; Quiros *et al.*, 1990) levels. They can also provide new and useful insights. Folk taxonomies are not, however, invariable. Ellen (1982) points out that they are 'extremely flexible, vary considerably within a culture, contain different and contradictory organizational structures and appear generally pretty messy'. Though, as he goes on to say, 'this is not to suggest that they are somehow without structure', it does mean that a certain amount of methodological sophistication is necessary in studying them.

In addition to the properties of crop germplasm, its history also needs to be documented. Is the origin of the material to be traced back to just a few seeds from a single mother plant? Was the seed lot from which the material is descended only recently introduced to the collecting area from a place that is very different agroecologically? Is the sample taken from seeds already selected for planting? An understanding of the workings of local seed production and exchange systems will help to characterize the origin, genetic base and degree of adaptation of germplasm (Cromwell, 1990). The local names of crops and landraces can sometimes be used to deduce origin. For example, Esquivel and Hammer (1988) used linguistic as well as historical evidence to trace the geographical source of crops grown in Cuba today.

There is IK of the environment as much as of plants and pests. Niamir (1990) gives examples of descriptive IK of climate, soils, geomorphology (including groundwater) and vegetation types from Africa. Folk taxonomies and descriptions of land types, farming systems, soils and vegetation can all help in characterizing the collecting site, complementing scientific descriptions by highlighting those features of the environment most relevant to everyday users of the land (Johnson, 1974; Rhoades, 1990; Tabor *et al.*, 1990; Nazarea-Sandoval, 1991). For example, Tabor and Hutchinson (1994) note that the three riparian landscapes recognized in the valley of the Senegal River by local farmers are mapped as essentially identical in conventional soil surveys but differ in how often, for how long and at what time of year they are flooded and are therefore managed and used quite differently.

Assessing the completeness of collecting

Local men and women will know which crops, and which varieties of each crop, are grown in their village or district or are being sold in the local markets. People will also know which trees in their area are good firewood sources, which produce palatable fodder, which are suitable for building, and so on. Bearing in mind the problems of synonymy and category names, a checklist can be compiled based on such information which can act as a guide to collecting in a given area. In his work in Peru, for example, Berlin (1985) asked his local collaborators 'to produce comprehensive written inventories of all recognized, named plant taxa in the local flora and then to monitor carefully that list as collections were made'. Hammer (1991) discusses the use of annotated botanical checklists, incorporating local names, in crop germplasm collecting.

Understanding the origin and distribution of diversity, and the rules of access to it

The preamble to the action plan proposed by the International Society of Ethnobiology at its first congress in 1988 states that 'there is an inextricable link between cultural and biological diversity'. Understanding the diversity within a crop in an area (which is crucial in developing a sampling strategy) means understanding the people who grow it just as much as understanding the climate and soils of the region and the distribution of wild relatives and pests. This is because the pattern of diversity in crops is the result of an interaction between the genetic make-up of the plants and not just environmental (e.g. climate, soil) and biotic (e.g. relatives, pests) but also human factors.

Landraces are at least partly shaped by what may be referred to as the informal plant breeding and seed production and supply systems. There are often well-defined patterns of germplasm exchange within and between communities, sometimes stretching over large areas, based on kin groups and the extended family. Farmers also obtain landraces from local markets and during occasional journeys outside their home areas. Evidence that farmer selection is widespread in the tropics is given by Clawson (1985) and Brush (1986). Xolocotzi (1987) points out that differences between Cuban and Mexican maize are due to the fact that it is prepared and eaten in different ways in the two countries, which has led to selection for different properties. The work of Boster (1984, 1985) shows that among the Aguaruna of the Peruvian Amazon diversity in cassava is sought for its own sake. He recognizes selection for mere perceptual distinctiveness as well as for locally valued characters (not necessarily conventional 'agronomic' characters) such as taste and cooking quality. Such idiosyncratically selective maintenance, in conjunction with farmer-to-farmer exchange, occasional introduction and random loss, has meant that there is 'a core of common widely shared and widely known [cassava] cultivars and a much larger number of rarer cultivars known only by small numbers of women' (Boster, 1984). This is a common situation. Even knowledge of the fact that multiplication of open-pollinated varieties requires reproductive isolation is documented from traditional farming systems (Berg *et al.*, 1991). All this can only really be investigated by observing local people and asking them questions.

In wild species also, human action is often responsible for the maintenance of diversity within particular ecosystems through such management practices as grazing, burning and cutting in particular ways or at particular times. Examples range from Aboriginal Australians burning the bush to sheep grazing on the English Downs. Individual species are often protected against overexploitation by traditional systems of ownership or rights of use over the land or the plants themselves, often underpinned by a concept of the spiritual value of the land. The collector must be aware of these rules and conform to them. Osemeobo (1992) points out that plants of economic or of social and medicinal value such as *Garcinia cola* and *Piper quineense* are protected in Nigeria under communal land tenure systems. Barrow (1992) discusses tree rights among the Turkana, and how they have worked to preserve a crucial resource in the semiarid regions of Kenya. An early European visitor to Australia, Sir George Gray, noted in 1841 that 'the natives have ... a law that no plant bearing seeds is to be dug up after it has flowered' (quoted by Harlan, 1989). Niamir (1990) reviews the range and plant tenure, management, monitoring and improvement practices of African pastoralists.

Assessing the reasons for and extent and danger of genetic erosion

To what extent farmers adopt modern varieties to replace their multiplicity of local landraces ultimately depends on the extent to which the varieties offered by scientific plant breeding and the formal seed industry better satisfy their household livelihood strategy. This in turn will be shaped not just by what is usually, and rather narrowly, defined as 'culture' (belief, art, moral law, custom, religion, etc.), but also by such socioeconomic factors as access to land, labour and capital, government macroeconomic initiatives and the influence of extension workers and other 'modernizers'.

For millions of resource-poor farmers in marginal areas, it is still local cultivars that serve them best, though there may be considerable turnover of landraces within a community as novel types arise by hybridization and volunteer seeding or are introduced and are then either maintained or eventually rejected. This preference for landraces (as for one landrace over another) may be difficult to account for in conventional on-station agronomic evaluation trials (e.g. Carney, 1980; Jackson *et al.*, 1980). In the Wadi Hadramaut region of southern Yemen, for example, farmers are resisting the introduction of higher-yielding, modern varieties of wheat partly because of the greater tolerance of the local landraces to increasingly saline irrigation water, but also because these are taller and much of the value of the crop in the area lies in the straw, which is essential in making mud bricks (pers. obs.). Jackson *et al.* (1980) also discuss the different reasons, human and natural, why apparently 'inferior' potato genotypes survive in Andean fields. Brush (1993) describes three cases of continued maintenance of landraces by farmers who have also adopted modern varieties, and discusses the factors which promote this (Chapter 4).

The socioeconomic and cultural context is no less important a factor in understanding genetic erosion in some wild species. It is common for Turkana women in northern Kenya to say that 'eusugu' (*Zanthoxylum chalybeum*), an alternative to tea leaves, 'is moving further and further up the hills' (Anon., 1992). Tyler *et al.* (1992) describe how changes in the management of old permanent pastures in the UK, some in use since medieval times, are threatening diversity.

It is often the abandonment of traditional management practices and tenure systems, sometimes as a result of misguided development efforts, that is threatening range vegetation in many arid and semiarid areas of the world (Gilles, 1988). Documenting how indigenous management of individual species and of vegetation as a whole operates, as discussed in the previous section, will help in predicting what will happen if it should stop. Conant (1989), for example, describes how the movement of several hundred pastoralist Pokot families from the Masol Plains of Kenya resulted in massive changes in the vegetation, as revealed by Landsat data. The Sahel Oral History Project of SOS Sahel has used hundreds of interviews with elderly people to document how development has affected land use practice, land tenure and farming and pastoral systems in the region (Cross and Barker, 1991). Though satellite imagery and written historical sources can make a contribution, such oral testimony is often the only source of information on change – whether in the vegetation of an area, in the extent of cultivation of a crop or landrace, in the cultural practices being used or in the range or abundance of a wild species.

Secondary sources of information on human cultures

Clearly, if previous attempts have been made to document how a community uses wild plants or grows crops, the germplasm collector will need to become familiar with them. This will be as critical as consulting and collating Floras, taxonomic monographs, soil maps, climatic data and the results of agricultural and socioeconomic surveys, and not just to the plant genetic resources worker planning to collect in a foreign country. As McArthur Crissman (1989) points out in the case of Kenya, for example, 'ethnic affiliation is correlated with choice of crops and even occupations'. There is considerable ethnic, cultural and linguistic diversity within countries, and national plant genetic resources programmes need to be aware of it and its link to the diversity not just of crops, but of wild species too.

It is not just strictly ethnobotanical information that will be relevant to the germplasm collector, however. Indigenous ecological knowledge will be important, for example folk soil taxonomies. More general ethnographic information will also be pertinent. Knowing the local system of measuring area, distance, weight and time will be essential in making sense of discussions with local people, as will knowing about the system of land tenure, when market days fall and who the decision-makers are in the community. Secondary sources on these subjects may well be very numerous, generating a considerable amount of complex and contradictory textual information, which will need somehow to be abstracted and organized. A record will need to be kept of conflicting statements, of information that is probably imprecise or no longer valid and of any gaps in the documentation.

One useful approach to the ordering of such data is to develop 'profile memos', bringing together and summarizing information (with references to sources, including the date of the observations) on particular individual topics of interest to the collector. Thus, for example, for each area or ethnic (or language) group to be visited during the collecting there would be memos on:

- the phonetic system commonly used to render local words into the collector's language, and the local script, if any;
- the local systems for the measurement of time, distance, area, weight and volume;
- different crops and wild species, and the relationships among them in any folk taxonomy;
- any folk nomenclatures and taxonomies of land types, soils, vegetation, etc.;
- different farming systems and agricultural and pastoral practices;
- significant places and geographical features in the target area;
- important people, social groupings and indigenous institutions;
- relevant laws, customs, taboos and restrictions, in particular as regards tenure of, and access to, land and natural resources;
- significant occasions (e.g. holidays and festivities, market days, village temple days, etc.);
- how gender, age, class, ethnicity and other socioeconomic and cultural factors affect access to and control of resources, including plant genetic resources.

The information in some of these memos could then be worked up into annotated glossaries of local terms, suitable for taking into the field. (Clearly, published dictionaries, if available, can also be extremely useful in the field.) In the case of the terms for soils, vegetation types, farming systems, etc., these could also be incorporated into the collecting forms, suitably defined (Chapter 19). A special case of such glossaries would be the annotated checklists of wild species, crops and landraces already mentioned, against which the material collected can be ticked off and which can be updated during the course of the collecting (Hammer, 1991). Some of the information, in coded form, will also find its way into the databases constructed as part of ecogeographic surveys (Chapter 14).

Literature

Where can the prospective collector obtain this kind of information? The main source will be the literature – historical and current, formal and grey. Travellers, explorers, conquerors and colonists have often written about the plants that are grown or gathered by the peoples among whom they have found themselves. They include ibn Battuta writing of his African and Asian travels, Marco Polo and his *Il Milione*, the story of his travels to China, composed in a Genoese prison, and the Franciscan friar Bernardino de Sahagon recording the customs of the Aztecs in the *Historia General de las Cosas de la Nueva España*. Arrogance, bias, ethnocentricity and insensitivity are all too obvious in the lesser exponents of this tradition, which, however, finds specialized, professional expression in modern scientific anthropology and ethnography. The work on the Nupe of west-central Nigeria described by Blench (1989) is a good example of how the diverse (in accuracy and attitude as much as style) writings of explorers, traders, colonial administrators and missionaries, as well as ethnologists, linguists, anthropologists and historians, in this case spanning almost two centuries, can be used to build up a picture of changes in crop repertoire and farming systems in an area.

At the other extreme from the anthropological and ethnographic literature, both amateur and professional, limited information on the vernacular names and uses of wild plants can occasionally be gleaned from purely taxonomic works such as Floras and botanical checklists, especially local ones. Floras are of course usually compiled in herbaria, and specimen labels (as well as the notebooks of botanical collectors) can therefore be another useful source of ethnobotanical information. Examples of the kind of ethnobotanical information available on herbarium labels may be found in Altschul (1968, 1970, 1973) and in such unpublished sources as the East African Herbarium Card Catalogue on Plant Uses (cited by Peters *et al.*, 1992). This herbarium also has a card catalogue of local names; both are being upgraded to computer databases. In agriculture, Richards (1985) quotes examples of colonial departments of agriculture in West Africa recording different aspects of local practice, including the vernacular names of species and landraces. Often, this was a preliminary to trying to replace them (usually with disastrous results), but such was by no means always the case. The sophistication, adaptability and appropriateness of local farming practices were occasionally recognized even in the colonial context, not usually otherwise particularly conducive to the development of such views. Official agricultural censuses, specialized surveys, the reports of extension workers and other grey literature can still occasionally provide information on local practices, making the archives of municipal libraries (Marchenay, 1987), district departments of agriculture and local extension offices potentially important sources.

Crop germplasm collectors have gathered ethnobotanical data in the field no less than botanists collecting for herbaria, but until relatively recently this has unfortunately often been similarly haphazard and unsystematic. As a result, collecting mission reports and the notebooks of germplasm collectors are often better sources of ethnobotanical insights than the admittedly more easily accessible passport data associated with collections (the equivalent of herbarium labels).

Inevitably, ethnographic sources suffer on occasion from a lack of botanical and agricultural expertise in their recording of IK relating to wild plants and crops. The usefulness of botanical, agricultural and genetic resources works in this context is often similarly limited by a lack of ethnographic or linguistic expertise. Travel and other amateur literature often suffers from a lack of both. Care should therefore be exercised in interpreting ethnobotanical data in non-specialist sources. Bisset (1990) gives examples of problems that have arisen in using the ethnographic literature as a source of information on medicinal and toxic plants. Data in older sources may no longer be valid. A local plant name or use quoted in a travel book, or even a Flora, and remarks such as 'said to be liked by camels' and 'used as a diuretic' on a herbarium label, are to be treated with due caution. In many cases, the information will be no more than anecdotal. There will also be misunderstandings. Kuchar (1989) tells the story of learning that the local name recorded on a herbarium label of *Leucas urticifolia* from Somalia translated from the local language as 'It's just a plant'. Familiarity with the local language is an absolute requirement for the collecting of names in particular and ethnobotanical information and indigenous knowledge in general. Burkill *et al.* (1985) discuss further the problems involved in documenting vernacular names.

Botanical and ethnographic expertise come together, however, in the growing ranks of specialized ethnobotanical and economic botany studies, perhaps the most typical example of which is the 'useful-plants work'. Useful-plants works range from those with a regional (e.g. Burkill *et al.*, 1985) or country-wide scope (e.g. Abbiw, 1990) to those dealing with single ethnic groups (e.g. Riley and Brokensha, 1988). They may deal with all useful plants in a flora, as in these studies, or with a single category of:

- use: e.g. edible plants (e.g. Peters *et al.*, 1992) or medicinal plants;
- taxonomy: e.g. Balick and Beck's (1990) *Useful Palms of the World* or Stevels (1990) on the legumes traditionally grown in Cameroon;
- plant type: e.g. Maydell (1990) on Sahelian trees and shrubs.

One of the most thorough surveys of useful plants is that of the Plant Resources of South-East Asia (PROSEA) Programme. Based in Bogor, and a foundation under Indonesian law with an international charter, PROSEA collects, evaluates and summarizes knowledge on useful plants in southeast Asia. It produces handbooks by commodity (eight by 1994, see Chapter 10) and runs the South-East Asian Plant Resources Information System (SAPRIS), a documentation system which includes six linked databases (e.g. Jansen and Siemonsma, 1992). Though not basically ethnobotanical in character, the handbooks do include information on local names and uses, for example, as well as taxonomic, ecological and agronomic information.

One category of use – medicinal plants – has given rise to a particularly extensive specialist literature, evidence of the enormous importance of such plants in not just traditional but also modern medicine. This literature includes many very large-scale works, for example those in the *Medicinal Plants of the World* series and such monumental efforts as Schultes and Raffauf (1990). Smaller-scale research is published in specialized periodicals such as *Economic and Medicinal Plant Research, Journal of Ethnopharmacology, Fitoterapia* and *Planta Medica*. There is a 'Bibliography of Herbal Medicine' in Lewis and Elvin-Lewis (1977).

To gain entry to the older specialized literature on useful plants on the basis of scientific name or plant product, a possible first step is Uphof's (1968) *Dictionary of Economic Plants*. Kunkel (1984) is an updating of Uphof (1968) and similar works, but the information it supplies is very limited (and restricted to plants that are consumed) and it quotes only secondary sources. Schultze-Motel (1986) also documents the uses of cultivated plants. It has a species index, a list of species according to uses and a bibliography which includes ethnobotanical works. None of these works concerns itself with the landrace level of variation, however.

Published ethnobotanical data at the landrace level within crops is in fact fairly limited, and is mainly to be found in the reports of investigations of single ethnic groups or small geographic areas. There are exceptions, however, for example Yen's (1974) wide-ranging ethnobotanical study of the sweet potato in Oceania, which draws on such disparate sources as ethnographies, dictionaries and other linguistic works, histories, archaeological works and *The Journals of Captain James Cook*. Small-scale studies of landraces, as well as of useful (including medicinal) plants, are published in periodicals such as *Advances in Economic Botany, Economic Botany, Human Ecology, Journal d'Agriculture Tropicale et de Botanique Appliquée* (continued as *Journal d'Agriculture Traditionelle et de Botanique Appliquée*), *Journal of Economic and Taxonomic Botany, Journal of Ethnobiology* and *Journal of Natural Products*. Another relevant journal is *Agriculture and Human Values*.

To keep track of the current literature, *Plant Genetic Resources Abstracts*, produced by CAB International (formerly the Commonwealth Agricultural Bureaux), and the International Plant Genetic Resources Institute (IPGRI), includes an 'Ethnobotany and socioeconomics' section. Nowadays, bibliographic databases can greatly simplify literature searches. Chapter 13 deals with agricultural databases, both on-line and on CD-ROM. The Royal Botanic Gardens, Kew, maintain an Economic Botany Bibliographic Database. Among relevant published bibliographies, Hawkes *et al.* (1983), also mentioned in Chapter 13, includes sections on 'Archaeology, palaeoethnobotany and ethnobotany' and 'Crop ecology, agroecology and agricultural systems' as well as on individual crops. Lawani *et al.* (1979), Graham (1986) and McCall (1988) are useful sources on indigenous farming systems. Mathias-Mundy *et al.* (1992) is a bibliography on indigenous tree-based farming systems, including home gardens. Niamir (1990) and Shepherd (1992) are bibliographies of traditional vegetation management and protection methods.

There is, unfortunately, no geographical guide to the worldwide literature on ethnobotany along the lines, for example, of what Frodin (1984) has done for the floristic literature, though Uphof (1968) includes a bibliography arranged by geographical area. There is, however, a useful geographical guide to the world's cultures, Price's (1990) *Atlas of*

World Cultures. This consists of a set of maps which physically locate some 3500 human cultures and an alphabetical index which points the researcher to the appropriate map(s) and to the literature. Using this atlas, collectors should be able to determine which cultural groups are found within an unfamiliar target area, and thus gain entry to the general ethnographic literature, which could in turn lead to more specialized ethnobotanical works. Each culture in Price's (1990) index is cross-referenced to its listing in Murdock's (1967) classic *Ethnographic Atlas* and his Human Relations Area Files (HRAF) (Murdock *et al.*, 1983). Rhoades (1988) notes that, in building up the HRAF, Murdock 'collected detailed data on the incidence and distribution of cultivated plants from over 2,000 ethnographic sources' for Africa alone. There is also information on technology, social patterns, economics, language, and so on. Originally called the Cross-Cultural Survey, Murdock's project was to compile a database of descriptive information on human cultures worldwide. HRAF has grown to a large-scale research organization devoted to the compilation of information that facilitates cross-cultural comparative study. Some of the material in HRAF is available on a series of CD-ROMs (*Cross-Cultural CD*).

There are also national-level ethnographic atlases and bibliographies. An example of the former is Merwe (1983), which, among other things, gives information on the distribution of different population groups in Namibia. Parry and Perkins (1987) review mapping (including such thematic mapping as ethnographic atlases) on a country-by-country basis (Chapter 9). Ellen's (1984) manual of *Ethnographic Research* has a section on 'Getting into the literature', which covers the *International Bibliographies of Social Sciences* series and ethnographic archives. Hopkins and Jones (1983) list national and regional bibliographies in anthropology and human geography. The Center for Indigenous Knowledge for Agriculture and Rural Development (CIKARD) at Iowa State University has a documentation unit and library that will be of help in searching the ethnographic literature (McKiernan, 1989). It produces a series of Bibliographies in Technology and Social Change, of which Mathias-Mundy *et al.* (1992) is one.

An important limitation of the literature, especially older sources, is the general neglect of women's knowledge, though exceptions do exist, such as William Lawson's *The Countrie Housewife's Garden*, published in 1617. Travellers' notes and ethnographies of the 19th and 20th centuries generally make no or only passing reference to women's knowledge of plant species and uses. More recent literature, such as *Rural Women in Pakistan Farming Systems Research* (PARC, 1988), is available for most regions of the world for the main crops, but collectors may need to access the probably unfamiliar territory of women's studies libraries. *Rural Women* (Kubisz, 1992) is an annotated bibliography of this literature. Similarly, agricultural and extension departments are often poorly informed concerning the role of women in farming, are staffed mainly by men and in general contact only few women farmers. Staff of home economics departments might be of help, depending on their resources to run field-based activities.

Though wild plant lore and agricultural knowledge are for the most part transmitted orally through such varied channels as performing arts, deliberate instruction, debate and conversation, their writing down by a community itself is just as widespread and ancient a practice as that of travellers, anthropologists and plant collectors recording the names and uses of plants among the different communities they encounter. New Kingdom papyrus manuscripts from ancient Egypt, for example, list hundreds of medicinal herbs and preparations. In India, the medical system of the Ayurveda, known through a vast scholarly literature in Sanskrit and other languages, refers to over 3000 plant species, many still used in the same way today. It was not very long after the fall of the Aztec empire that two of the survivors wrote, in Latin, the book of medical botany now known as the Badianus manuscript. There are local texts describing farming practices and crop varieties from many ages and cultures, from Yemeni agricultural calendars to such Chinese treatises as *Skilful Hands Create the World*, a 17th century AD heir to a 3000-year-old tradition. Historical written sources such as these may be available in modern translations, but in many cases will need to be deciphered by experts. Nowadays, the informal sector (local NGOs and other grass-roots organizations) is very active in this field (e.g. Cooper *et al.*, 1992). An instructive example is provided by the agroecological project of the Agroecologia Universidad Cochabamba in Bolivia (Rist, 1991). Project staff have been collaborating with local people in documenting their local knowledge in the form of 'fichas', simple printed information notes. These 'have provided a method of horizontal, farmer to farmer and community to community, communication, thus increasing the communities' ability to support each other in dealing with common problems'.

Databases

Bibliographic databases have already been alluded to. There are also, however, factual databases bringing together ethnobotanical information from the literature, herbarium labels and expert opinion. One can keep track of developments in this field through various specialist newsletters (see next section).

At the farming system level, the International Centre for Research in Agroforestry (ICRAF) has developed a database on agroforestry practices (Oduol *et al.*, 1988). The Survey of Economic Plants of the Arid and Semi-arid Lands (SEPASAL), based at the Economic Botany Section of the Royal Botanic Gardens, Kew, compiles information on indigenous plant names and uses as well as taxonomy, ecology and distribution. Its publications include *Forage and Browse Plants for Arid and Semi-arid Africa* (IBPGR and Royal Botanic Gardens, Kew, 1984). A coding system for economic uses has been developed by SEPASAL (Chapter 19). The International Legume Database and Information Service

(ILDIS) has the aim of developing a database containing basic nomenclatural, distributional and descriptive information on the legumes of the world, from the literature and data at the Herbarium at Kew. Uses are also recorded, linked to bibliographic records. The checklist of Lock (1989) has been compiled from the data on African species.

Other examples of databases on a global or regional scale which include information on local uses and other ethnobotanical data are: the US Department of Agriculture (USDA)'s Minor Economic Plant Species Database (Duke, 1983); ICRAF's Multipurpose Tree Database (Carlowitz, 1984; Carlowitz *et al.*, 1991); ACSAD's Arab Data Bank for Arid Zone Plants; and PROSEA's SAPRIS, already alluded to. Some examples of databases specifically on medicinal plants are listed in Box 12.2. The database maintained by the gene bank of the Institut für Pflanzengenetik und Kulturpflanzenforschung at Gatersleben in Germany contains the local names and uses of the cultivated plants of Libya, North Korea, Cuba and southern Italy (Knüpffer, 1992; K. Hammer, pers. comm.). National and smaller-scale databases on indigenous plants and their uses are proliferating rapidly, as are taxonomic and biodiversity databases in general, to which they are often linked (Chapter 10). An early example among many is the database of Maya indigenous plant knowledge and useful plants of Mexico, which was developed in conjunction with the *Flora of Veracruz* Project (Gomez-Pompa and Nevling, 1988). Others are the databases being developed at the East African Herbarium in Nairobi, Kenya, from long-standing card catalogues of uses and local names, which have already been mentioned.

At the landrace level within crops, local name and some indication of uses and local management practices (for example, sowing and harvesting time) are commonly included as part of the passport information in the germplasm databases of national programmes and regional and international institutions involved in crop genetic resources conservation. The systematization in databases (electronic or otherwise) of the full range of IK associated with the landraces maintained by rural societies is only just beginning, however. Recent work on sweet potatoes in the Philippines and Irian Jaya by CIP User's Perspective With Agricultural Research and Development (UPWARD) is a notable example, but is still on a relatively small scale (e.g. Nazarea-Sandoval, 1990; Prain, 1993; Chapter 38).

Expert sources: a global community

Jain *et al.* (1986) have produced *A World Directory of Ethnobotanists*, which may help to identify experts on particular topics. Also useful in this will be professional societies such as the International Society of Ethnobiology and the Society for Economic Botany. The latter publishes the journal *Economic Botany* and a newsletter, *Plants and People*. There are also a number of important international programmes in the field of IK in general and botanical IK in particular. Unesco, the World Wide

Box 12.2
Some databases on medicinal plants

- The World Health Organization-funded Natural Products Information System database (NAPRALERT) has bibliographic references, numerical data and textual information on biochemistry, pharmacology and indigenous uses (Loub *et al.*, 1985).
- The regional bibliographic database and referral database of information sources, research institutions and experts of the United Nations Educational, Scientific and Cultural Organization (Unesco)-supported Asian Pacific Information Network on Medicinal and Aromatic Plants (APINMAP) bring together information from 11 national nodes.
- The database of the Istituto Mexicano para el Estudio de las Plantas Medicinales has data on the uses of Mexican plants in traditional medicine extracted from the literature (Loub and Farnsworth, 1984).
- The database of the Chinese University of Hong Kong contains information on traditional Chinese medicine (Loub and Farnsworth, 1984).
- PHARMEL is a database of information on medicinal plants collected on ethnobotanical expeditions organized by France's Agence de Coopération Culturelle et Technique (ACCT) in 11 countries, mostly in West Africa; a standard methodology for data gathering has been developed (Waechter and Lejoly, 1990).
- NEMOBASE holds fieldwork and literature data on traditional uses of plants in France (Dos Santos, 1990).
- The AYURBASE project aims to compile data from the Ayurveda system of Indian medicine (Mazars, 1990).

Fund for Nature (WWF)-International and the Royal Botanic Gardens, Kew, have recently launched the People and Plants Programme, for example. This supports ethnobotanists studying and recording plant uses with local communities in tropical countries. Unesco Canada/MAB (Man and the Biosphere Programme) is developing an international programme on traditional ecological knowledge, and a quarterly newsletter is being published (*TEK Talk*).

Perhaps the most important development, however, is the growth of a network of national, regional and international IK resource centres. CIKARD is collaborating with the Center for International Research and Advisory Networks (CIRAN) and the Leiden Ethnosystems and Development Programme (LEAD), both in the Netherlands, in publishing the *Indigenous Knowledge and Development Monitor*, a quarterly newsletter for this global network. First published in early 1993, this absorbed the CIKARD newsletter, *CIKARD News*. It gives information on current research projects, databases under development and being planned, recent publications, etc. A list of IK resource centres, taken from the latest issue, is provided in Appendix 12.1 at the end of this chapter. CIRAN is planning an inventory of existing databases containing information relevant to the global IK network – bibliographic, factual and relating to expert sources.

In addition to IK resource centres, universities, herbaria and museums are important sources of relevant expertise and publications. As for NGOs, periodicals such as *Ecoforum* (published by Environment Liaison Centre International, which acts as the NGOs' link to the United Nations Environment Programme (UNEP)), *EcoAfrica* (African NGOs Environment Network), *IRED Forum* (Innovations et Réseaux pour le Développement) and *Seedling* (Genetic Resources Action International) report on their activities and initiatives worldwide. *Development Education and Exchange Papers* is a periodic review of Food and Agriculture Organization (FAO) and NGO programmes and publications in agricultural and rural development. The September 1993 issue was entirely dedicated to plant genetic resources issues. An NGO networking system on indigenous technology and innovation is being established through the efforts of the Indian Institute of Management at Ahmedabad. Its publications include Gupta *et al.* (1990) and the quarterly newsletter *Honey Bee*. In collaboration with a committee of NGOs, IPGRI is developing a directory of African NGOs involved in plant genetic resources work, expected to be ready in 1994.

Another pertinent organization is the Information Centre for Low-External-Input and Sustainable Agriculture, which publishes *ILEIA Newsletter* quarterly (e.g. issue 4/89 is entirely devoted to IK), in addition to bibliographies and a register of organizations. *International Ag-Sieve* is a useful 'sifting of news about regenerative agriculture' published by the Rodale Institute.

Expert sources with a specific understanding of gender-related issues include: the Association of Women in Development, the International Federation of Women in Agriculture, the Associated Country Women of the World, the Women in Rice Farming Systems Network and the Association of Farming Systems Research-Extension. The Association of Farming Systems Research–Extension, an international society organized to promote the development and dissemination of methods and results of participatory on-farm research and extension, publishes the *Journal for Farming Systems Research–Extension*. The Rural Sociological Society has a Rural Women in Economic Production Research Group. The Rural Sociological Society publishes the journal *Rural Sociology*, the newsletter *The Rural Sociologist* and a directory listing members alphabetically, by geographical region, world regions of interest and area of competence.

Since 1993, CIKARD, CIRAN, the Honey Bee Network and several other organizations have been facilitating the electronic mailing list INDKNOW. This provides an open forum for discussion of IK and related issues. More information can be obtained from Preston Hardison at cied@u.washington.edu.

Conclusion: the need for participatory collecting

With increasing recognition of the fundamental role that farmers continue to play in generating and maintaining the diversity of landraces, and indeed of the role of traditional societies in general in developing the many uses of plants, wild and cultivated, has come the acknowledgement that they must be involved much more profoundly than has perhaps been the case in the past in the process of systematic germplasm conversation. On-farm conservation (e.g. Altieri and Merrick, 1987; Merrick, 1990; Brush, 1991; Worede, 1991; various papers in Cooper *et al.*, 1992) and 'memory-banking' IK of landraces within rural communities are clearly part of this (Nazarea-Sandoval, 1990), but *ex situ* conservation is no less important than *in situ* and IK is equally central to both.

Having collected whatever background ethnographic information on their target species and target region may be available, germplasm collectors – nationals as much as foreigners – can perhaps approach the task of documenting IK for themselves in the field with more confidence, and certainly with more sensitivity. Empathy and familiarity with (and respect for) the local culture are of course necessary for such work, but not sufficient. The active participation of the community is essential. After all, who is better placed to understand a culture than someone born into it? As pointed out earlier, there is nothing new about the documentation by a community itself, by the men and women who make it up, of its agricultural and botanical knowledge. Farmer participation (e.g. Farrington and Martin, 1988; Amanor, 1989) is increasingly recognized as a vital way not only of doing better, more relevant research, but of empowering communities at the same time. Chapter 18 discusses this more fully, and describes how participatory germplasm/IK collecting might work in practice.

Acknowledgements

I am grateful for the comments and suggestions of Janice Jiggins, Virginia Nazarea-Sandoval, Gordon Prain, Toby Hodgkin and Esbern Friis-Hansen. I would also like to thank all the participants of the Wageningen Agricultural University/ENDA-Zimbabwe/CGN/GRAIN Seminar on 'Local Knowledge and Agricultural Research' (Brodensbury Park Hotel, Nyanga, Zimbabwe, 28 September to 2 October 1992) and of the Intermediate Technology Development Group/Plan International Workshop on 'Collecting and Feeding Back Existing Local Knowledge' (Embu, Kenya, 27–30 September 1993).

References

Abbiw, D.K. (1990) *Useful Plants of Ghana.* Intermediate Technology Publications and Royal Botanic Gardens, Kew, London.

Alcorn, J.B. (1984) *Huastec Mayan Ethnobotany.* University of Texas Press, Austin.

Altieri, M.A. (1993) Ethnoscience and biodiversity: key elements in the design of sustainable pest management systems for small farmers in developing countries. *Agriculture, Ecosystems and Environment* 46:257-272.

Altieri, M.A. and L.C. Merrick (1987) *In situ* conservation of crop genetic resources through maintenance of traditional farming systems. *Economic Botany* 41:86-96.

Altschul, S.V.R. (1968) Useful food plants in herbarium records. *Economic Botany* 22:293-296.

Altschul, S.V.R. (1970) Ethnopediatric notes in the Harvard University Herbaria. *Lloydia* 33:195-198.

Altschul, S.V.R. (1973) *Drugs and Foods from Little-known Plants: Notes in Harvard University Herbaria.* Harvard University Press, Cambridge.

Amanor, K. (1989) *340 Abstracts on Farmer Participatory Research.* ODI Agricultural Research and Extension Network Paper No. 5. Overseas Development Institute, London.

Anon. (1992) Reaching the woman. *Indigenous Food Plant Programme Newsletter* 10:1-2.

Anon. (1993) A woman's rightful place? *Spore* 44:1-3.

Asfaw, Z. (1990) An ethnobotanical study of barley in the Central Highlands of Ethiopia. *Biologisches Zentralblatt* 108:51-62.

Ba'azara, M., L. Guarino, A. Miller and N. Obadi (1991) *Dirachma socotrana* - back from the brink? *Oryx* 25:229-232.

Balick, M.J. and H.T. Beck (eds) (1990) *Useful Palms of the World. A Synoptic Bibliography.* Columbia University Press, New York.

Barrau, J.F. (1989) The possible contribution of ethnobotany to the search for new crops for food and industry. In: Wickens, G.E., N. Haq and P. Day. (eds) *New Crops for Food and Industry.* pp. 402-410. Chapman and Hall, London.

Barrow, E.G.C. (1992) *Tree Rights in Kenya. The Case of Turkana.* ACTS Press, Nairobi.

Berg, T., A. Bjornstad, C. Fowler and T. Skroppa (1991) *Technology Options and the Gene Struggle.* NORAGRIC Occasional Papers Series C. Development and Environment No. 8. NORAGRIC, Aas.

Berlin, B. (1985) Contributions of native American collectors to the ethnobotany of the Neotropics. In: Prance, G.F. and J.A. Kallunki (eds) *Ethnobotany in the Neotropics. Advances in Economic Botany* 1:24-33. The New York Botanical Garden, New York.

Berlin, B. (1992) *Ethnobiological Classification: Principles of Categorization of Plants and Animals in Traditional Societies.* Princeton University Press, Princeton.

Bisset, N.G. (1990) The ethnographic approach to ethnopharmacology: a critique. In: Fleurentin, J., P. Cabalion, G. Mazars, J. Dos Santos and C. Younos (eds) *Ethnopharmacology: Sources, Methods, Objectives.* ORSTOM and Société Française d'Ethnopharmacologie, Paris.

Blench, R. (1989) The evolution of the cultigen repertoire of the Nupe of West-Central Nigeria. *Azania* 24:51-63.

Bonfiglioli, A.M. (1992) *Pastoralists at a Crossroads. Survival and Development Issues in African Pastoralism.* UNICEF/UNSO Project for Nomadic Pastoralists in Africa, Nairobi. Final Version, November 1992.

Boster, J.S. (1984) Classification, cultivation and selection of Aguaruna cultivars of

Manihot esculenta (Euphorbiaceae). In: Prance, G.F. and J.A. Kallunki (eds) *Ethnobotany in the Neotropics. Advances in Economic Botany* 1:34–47. New York Botanical Garden, New York.

Boster, J.S. (1985) Selection for perceptual distinctiveness: evidence from Aguaruna cultivars of *Manihot esculenta. Economic Botany* 39:310–325.

Brhane, G. and K. Yilma (1979) The traditional culture and yield potential of the Ethiopian high lysine sorghums. *Ethiopian Journal of Agricultural Science* 1:29–40.

Brokensha, D.W., D.M. Warren and O. Warner (1980) (eds) *Indigenous Knowledge Systems and Development.* University of America Press, Boston.

Brown, C.H. (1985) Mode of subsistence and folk biological taxonomy. *Current Anthropology* 26:43–53.

Brush, S.B. (1986) Genetic diversity and conservation in traditional farming systems. *Journal of Ethnobiology* 35:70–88.

Brush, S.B. (1991) A farmer-based approach to conserving crop germplasm. *Economic Botany* 45:153–165.

Brush, S.B. (1993) *In situ* conservation of landraces in centres of crops diversity. Paper delivered at the 'Symposium on Global Implications of Germplasm Conservation and Utilization'. 85th Annual Meeting of the American Society of Agronomy. 8 November 1993. Cincinnati, Ohio.

Burkill, H.M., J.M. Dalziel and J. Hutchinson (1985) *The Useful Plants of West Tropical Africa.* Royal Botanic Gardens, Kew.

Carlowitz, P.G. von (1984) *Multipurpose Trees and Shrubs: Opportunities and Limitations – the Establishment of a Multipurpose Tree Database.* Working Paper 17. ICRAF, Nairobi.

Carlowitz, P.G. von, G.V. Wolf and R.E.M. Kemperman (1991) *Multipurpose Tree and Shrub Database – An Introduction and Decision-Support System. User's Manual, Version 1.0.* ICRAF, Nairobi.

Carney, H.J. (1980) *Diversity, Distribution and Peasant Selection of Indigenous Potato Varieties in the Mantaro Valley, Peru: a Biocultural Evolutionary Process.* Social Science Department, Working Paper Series 1980-3. CIP, Lima.

Chambers, R. (1990) *Microenvironments Unobserved.* Gatekeepers Series No. 22. IIED, London.

Clawson, D.L. (1985) Harvest security and intraspecific diversity in traditional tropical agriculture. *Economic Botany* 35:70–88.

Conant, F.P. (1989) The Pokot way with thorny shrubs: a case example. In: McKell, C.M. (ed.) *The Biology and Utilization of Shrubs.* pp. 593–602. Academic Press, London.

Conklin, H.C. (ed.) (1972) *Folk Classification: A Topically Arranged Bibliography.* Yale University, New Haven.

Cooper, D., R. Vellvé and H. Hobbelink (eds) (1992) *Growing Diversity.* Intermediate Technology Publication, London.

Cromwell, E.A. (ed.) (1990) *Small Farmer Seed Diffusion Mechanisms: Lessons from Africa, Asia and Latin America.* ODI Agricultural Research and Extension Network Paper No. 21. Overseas Development Institute, London.

Cross, N. and R. Barker (eds) (1991) *At the Desert's Edge. Oral Histories from the Sahel.* Panos/SOS Sahel, London.

Dos Santos, J.R. (1990) NEMOBASE: Système d'informatique sur les usages populaires de la flore. In: Fleurentin, J., P. Cabalion, G. Mazars, J. Dos Santos and C. Younos (eds) *Ethnopharmacology: Sources, Methods, Objectives.* ORSTOM and Société Française d'Ethnopharmacologie, Paris.

Duke, J.A. (1983) The USDA Economic Botany Laboratory's database on minor economic plant species. In: *Plants: the Potentials for Extracting Protein, Medicines and Other Useful Chemicals*. Workshop Proceedings. pp. 196–214. US Congress Office of Technology Assessment, Washington DC.

Ellen, R. (1982) *Environment, Subsistence and System*. Cambridge University Press, Cambridge.

Ellen, R.F. (ed.) (1984) *Ethnographic Research*. Academic Press, London.

Esquivel, M. and K. Hammer (1988) The 'conuco' – an important refuge of Cuban plant genetic resources. *Kulturpflanze* 36:451–463.

Farrington, J. and A. Martin (1988) *Farmers' Participation in Agricultural Research*. ODI Occasional Paper No. 9. Overseas Development Institute, London.

Frodin, D.G. (1984) *Guide to Standard Floras of the World*. Cambridge University Press, Cambridge.

Gilles, J.L. (1988) Slippery grazing rights: using indigenous knowledge for pastoral development. In: Whitehead, E.E. and C.F. Hutchinson (eds) *Arid Lands: Today and Tomorrow*. Proceedings of an International Research and Development Conference. 20–25 October 1985. Tucson, Arizona. Belhaven Press, London.

Given, D.R. and W. Harris (1992) *Techniques and Methods of Ethnobotany As an Aid to the Study, Use and Conservation of Biodiversity. A Training Manual*. Commonwealth Science Council, London.

Gomez-Pompa, A. and L.I. Nevling (1988) Some reflections on floristic databases. *Taxon* 37:744–775.

Graham, G.A. (ed.) (1986) *A Farming Systems Research Bibliography of Kansas State University's Vertical File Materials*, revised edition. Farming Systems Research Paper Series No. 4. Kansas State University, Manhattan.

Gupta, A., J. Capoor and R. Shah (1990) *Inventory of Peasant Innovations for Sustainable Development. An Annotated Bibliography*. Center for Management in Agriculture, Indian Institute of Management, Ahmedabad.

Hammer, K. (1991) Checklists and germplasm collecting. *Plant Genetic Resources Newsletter* 85:15–17.

Hamon, S. and P. Hamon (1991) Future prospects of the genetic integrity of two species of okra (*Abelmoschus esculentus and A. caillei*) cultivated in West Africa. *Euphytica* 58:101–111.

Harlan, J.R. (1989) Self perception and the origins of agriculture. In: Swaminatham, M.S. and S.L. Kochhar (eds) *Plants and Society*. pp. 5–23. Macmillan, London.

Hawkes, J.G., J.T. Williams and R.P. Croston (1983) *A Bibliography of Crop Genetic Resources*. IBPGR, Rome.

Hopkins, S.T. and D.E. Jones (1983) *Research Guide to the Arid Lands of the World*. Oryx Press, Phoenix.

IBPGR and Royal Botanic Gardens, Kew (1984) *Forage and Browse Plants for Arid and Semi-arid Africa*. IBPGR, Rome.

IDS Workshop (1989) Farmers' knowledge, innovations, and relation to science. In: Chambers, R., A. Pacey and L.A. Thrupp (eds) *Farmer First. Farmer Innovation and Agricultural Research*. pp. 31–38. Intermediate Technology Publications, London.

Jackson, M.T., J.G. Hawkes and P.R. Rowe (1980) An ethnobotanical field study of primitive potato varieties in Peru. *Euphytica* 29:107–113.

Jain, S.K., P. Minnis and N.C. Shah (1986) *A World Directory of Ethnobotanists*. Society of Ethnobotanists, Lucknow.

Jansen, P.C.M. and J.S. Siemonsma (1992) PROSEA. Data gathered on plants in Southeast Asia. *Prophyta* 46:52–55.

Jiggins, J. (1986) *Gender-related Impacts and the Work of the International Agricultural*

Research Centres. CGIAR Study Papper No. 17. World Bank, Washington DC.

Jiggins, J. (1990) Gender issues and agricultural technology development. In: Altieri, M.A. and S.B. Hecht (eds) *Agroecology and Small Farm Development.* CRC Press, Boca Raton.

Johnson, A. (1974) Ethnoecology and planting practices in swidden agricultural system. *American Ethnologist* 1:87–101.

Khalil, M.H., W.V. Reid and C. Juma (1992) *Property Rights, Biotechnology and Genetic Resources.* ACTS Press, Nairobi.

Knüpffer, H. (1992) The database of cultivated plants of Cuba. In: Hammer, K., M. Esquivel and H. Knüpffer (eds) '... *y tienen faxones y fabas muy diversos de los nuestros* ...'- *Origin, Evolution and Diversity of Cuban Plant Genetic Resources.* Vol. 1. pp. 202–212. Institut für Pflanzengenetik und Kulturpflanzenforschung, Gatersleben.

Kubisz, K.M. (ed.) (1992) *Rural Women.* CAB International, Wallingford.

Kuchar, P. (1989) *The Plants of Somalia: an Overview and Checklist.* Central Rangelands Development Project Technical Report No. 16. CRDP, National Range Agency, Mogadishu.

Kunkel, G. (1984) *Plants for Human Consumption.* Koeltz Scientific Books, Koenigstein.

Lawani, S.M., F.M. Alluri and E.N. Adimirah (1979) *Farming Systems in Africa. A Working Bibliography 1930-1978.* G.K. Hall & Co., Boston.

Lewis, W.H. and M.P.F. Elvin-Lewis (1977) *Medical Botany: Plants Affecting Man's Health.* John Wiley & Sons, New York.

Lock, J.M. (1989) *Legumes of Africa. A Checklist.* Royal Botanic Gardens, Kew.

Loub, W.D. and N.R. Farnsworth (1984) Utilisation de l'informatique pour la mise au point de produits naturels. *Impact* 136:371–381.

Loub, W.D., N.R. Farnsworth, D.D. Soejarto and M.L. Quinn (1985) NAPRALERT: Computer handling of natural product research data. *Journal of Chemical Information and Computer Sciences* 25:99–103.

McArthur Crissman, L. (1989) *Evaluation, Choice and Use of Potato Varieties in Kenya.* Social Sciences Department Working Paper 1989-1. CIP, Lima.

McCall, M.K. (1988) *Indigenous Technical Knowledge in Farming Systems and Rural Technology: a Bibliography on Eastern Africa.* Working Paper. Technology and Development Group. Twent University of Technology, Twent.

McCorkle (1989) Toward a knowledge of local knowledge and its importance for agricultural RD&E. *Agriculture and Human Values* 6:4–12.

McKiernan, G. (1989) The CIKARD international documentation unit and library of indigenous knowledge for agriculture and rural development. In: Warren, D.M., L.J. Slikkerveer and S. Oguntunji Titilola (eds) *Indigenous Knowledge Systems: Implications for Agriculture and International Development.* pp. 86–90. Iowa State University, Ames.

Marchenay, P. (1987) *A la Recherche des Variétés Locales de Plantes Cultivées.* PAGE-PACA, Hyères.

Martin, G. (1994) *Ethnobotany and Plant Conservation.* Chapman and Hall, London.

Mathias-Mundy, E., O. Muchena, G. McKierkan and P. Mundy (1992) *Indigenous Technical Knowledge of Private Tree Management: A Bibliographic Report.* Technology and Social Change Program, Iowa State University, Ames.

Maydell, H.-J. von (1990) *Arbres et arbustes du Sahel: leurs caractéristiques et leurs utilisations.* Deutsche Gesellschaft für Technische Zusammenarbeit (GTZ), Eschborn.

Mazars, G. (1990) Le projet AYURBASE. In: Fleurentin, J., P. Cabalion, G. Mazars,

J. Dos Santos and C. Younos (eds) *Ethnopharmacology: Sources, Methods, Objectives*. ORSTOM and Société Française d'Ethnopharmacologie, Paris.

Merrick, L. (1990). Crop genetic diversity and its conservation in traditional agroecosystems. In: Altieri, M.A. and S.B. Hecht (eds) *Agroecology and Small Farm Development*. pp. 3–13. CRC Press, Boca Raton.

Merwe, J.H. van der (ed.) (1983) *National Atlas of South West Africa (Namibia)*. Directorate Development Co-ordination, Windhoek.

Mooney, P.R. (1992) Towards a folk revolution. In: Cooper, D., R. Vellvé and H. Hobbelink (eds) (1992) *Growing Diversity*. pp. 125–138. Intermediate Technology Publication, London.

Murdock, G.P. (1967) *Ethnographic Atlas*. University of Pittsburgh Press, Pittsburgh.

Murdock, G.P. *et al.* (1983) *Outline of World Cultures*. 6th edition. Human Relations Area Files, New Haven.

Nazarea-Sandoval, V.D. (1990) Memory banking of indigenous technology of local farmers associated with traditional crop varieties: focus on sweet potato. In: *Proceedings of the Inaugural Planning Workshop on the User's Perspective With Agricultural Research and Development*. pp. 180–195. CIP, Los Baños.

Nazarea-Sandoval, V.D. (1991) Ethnoagronomy and ethnogastronomy: on indigenous typology and use of biological resources. *Agriculture and Human Values* 8:121–131.

Niamir, M. (1990) *Herders' Decision-making in Natural Resources Management in Arid and Semi-arid Africa*. Community Forestry Note 4. FAO, Rome.

Oduol, P.A., P. Muraya, E.C.M. Fernandes and P.K.R. Nair (1988) The agroforestry systems database at ICRAF. *Agroforestry Systems* 6:253–270.

Opole, M. (1991) Women's indigenous knowledge base in the translation of nutritional and medicinal values of edible local plants in western Kenya. In: Prah, K.K. (ed.) *Culture, Gender, Science and Technology in Africa*. pp. 81–96. Harp Publications, Windhoek.

Osemeobo, G.J. (1992) Land use issues on wild plant conservation in Nigeria. *Journal of Environmental Management* 36:17–26.

PARC (1988) *Rural Women in Pakistan Farming Systems Research*. PARC, Islamabad.

Parry, R.B. and C.R. Perkins (1987) *World Mapping Today*. Butterworths, Borough Green.

Peters, C.R., E.M. O'Brien and R.B. Drummond (1992) *Edible Wild Plants of Sub-Saharan Africa*. Royal Botanic Gardens, Kew.

Plotkin, M.J. (1989) Traditional knowledge of medicinal plants – the search for new jungle medicines. In: Akerele, O., V. Heywood and H. Synge (eds) *The Conservation of Medicinal Plants*. pp. 53–63. Cambridge University Press, Cambridge.

Prain, G.D. (1993) Mobilizing local expertise in plant genetic resources research. In: de Boef, W.S., K. Amanor, K. Wellard with T. Bebbington (eds) *Cultivating Knowledge: Genetic Diversity, Farmer Experimentation and Crop Research*. Intermediate Technology Publications, London.

Price, D.H. (1990) *Atlas of World Cultures*. Sage Publications, London.

Quiros, C.F., S.B. Brush, D.S. Douches, K.S. Zimmerer and G. Huestis (1990) Biochemical and folk assessment of variability of Andean cultivated potatoes. *Economic Botany* 44:254–266.

Rhoades, R.E. (1988) The reference file method: an eclectic approach for improving agroecological and crop data of developing countries. In: *The Social Sciences at CIP*. Report of the Third Social Science Planning Conference. pp. 118–128. CIP, Lima.

Rhoades, R.E. (1990) The coming revolution in methods for rural development research.

In: *Proceedings of the Inaugural Planning Workshop on the User's Perspective With Agricultural Research and Development.* pp. 196–210. CIP, Los Baños.

Richards, P. (1985) *Indigenous Agricultural Revolution.* Unwyn Hyman, London.

Richards, P. (1991) Mende names for rice: cultural analysis of an agricultural knowledge system. In: Tillmann, H. (ed.) *Proceedings of the Workshop on Agricultural Knowledge Systems and the Role of Extension.* Institut für Agrarsoziologie, Universitaet Hohenheim, Stuttgart.

Riley, B.W. and D. Brokensha (1988) *The Mbeere in Kenya. 2. Botanical Identities and Uses.* University Press of America, Lenham.

Rist, S. (1991) Participation, indigenous knowledge and trees. *Forests, Trees and People Newsletter* 13:30–36.

Schultes, R.E. (1986) Ethnopharmacological conservation: a key to progress in medicine. *Opera Botanica* 92:217–224.

Schultes, R.E. and R.F. Raffauf (1990) *The Healing Forest: Medicinal and Toxic Plants of the Northwest Amazonia.* Dioscorides Press, Portland.

Schultze-Motel, J. (ed.) (1986) *Rudolf Mansfeld. Verzeichnis Landwirtschaftlicher und Gärtnerischer Kulturpflanzen.* Springer-Verlag, Berlin.

Shepherd, G. (1992) *Managing Africa's Tropical Dry Forests. A Review of Indigenous Methods.* Overseas Development Institute, London.

Stevels, J.M.C. (1990) *Légumes Traditionels du Cameroun.* Wageningen Agricultural University Papers, Wageningen.

Tabor, J.A. and C.F. Hutchinson (1994) Using indigenous knowledge, remote sensing and GIS for sustainable development. *Indigenous Knowledge and Development Monitor* 2:2–6.

Tabor, J.A., D.W. Kilambya and J.M. Kibe (1990) *Reconnaissance Survey of the Ethnopedology in the Embu, Meru, Machakos and Kitui Districts of Kenya's Eastern Province.* University of Missouri and USAID, Nairobi.

Tyler, B.F., K.H. Chorlton and I.D. Thomas (1992) Activities in forage grass genetic resources at the Welsh Plant Breeding Station, Aberystwyth. *FAO/IBPGR Plant Genetic Resources Newsletter* 88/89:37–42.

Uphof, J.C.T. (1968) *Dictionary of Economic Plants.* Verlag von J. Cramer, Lehre, Germany.

Vellvé, R. (1992) *Saving the Seed.* Earthscan Publications Ltd., London.

Waechter, P. and J. Lejoly (1990) PHARMEL: banque de données de médecine traditionelle et de pharmacopée. In: Fleurentin, J., P. Cabalion, G. Mazars, J. Dos Santos and C. Younos (eds) *Ethnopharmacology: Sources, Methods, Objectives.* ORSTOM and Société Française d'Ethnopharmacologie, Paris.

Warren, D.M. (1991) *Using Indigenous Knowledge in Agricultural Development.* Discussion Paper No. 127. World Bank, Washington DC.

Warren, D.M. (1993) Using IK for agriculture and rural development: current issues and studies. *Indigenous Knowledge and Development Monitor* 1:7–10.

Warren, D.M., L.J. Slikkerveer and S. Oguntunji Titilola (1990) (eds) *Indigenous Knowledge Systems: Implications for Agriculture and International Development.* pp. 86–90. Iowa State University, Ames.

Warren, D.M., D. Brokensha and L.J. Slikkerveer (1994) (eds) *Indigenous Knowledge Systems: the Cultural Dimensions of Development.* Kegan Paul International, London.

Waterman, P.G. (1989) Bioactive phytochemicals – the search for new sources. In: Wickens, G.E., N. Haq and P. Day. (eds) *New Crops for Food and Industry.* pp. 378–390. Chapman and Hall, London.

Worede, M. (1991) Crop genetic resource conservation and utilization: an Ethiopian perspective. In: *Science in Africa: Achievements and Prospects.* pp. 103–123. AAAS, Washington DC.

Xolocotzi, E.H. (1987) Experiences leading to a greater emphasis on man in ethnobotanical studies. *Economic Botany* 41:6–11.

Yen, D.E. (1974) *The Sweet Potato in Oceania.* Bishop Museum Press, Honolulu.

Additional reading

Agar, M. (1980) *The Professional Stranger: An Informal Introduction to Ethnography.* Academic Press, New York.

Carroll, C.R., J.H Vandermeer and P.M. Rosset (eds) (1990) *Agroecology.* McGraw-Hill, New York.

Conklin, H.C. (1956) An ethnoecological approach to shifting agriculture. *Transactions of the New York Academy of Science, Series II* 17:133–142.

Crick, M. (1982) Anthropology of knowledge. *Annual Review of Anthropology* 11:287–313.

Hunn, E. (1985) The utilitarian in folk biological classification. In: Dougherty, J. (ed.) *Directions in Cognitive Anthropology.* University of Illinois Press, Chicago.

Inglis, J.T. (ed.) (1993) *Traditional Ecological Knowledge. Concepts and Cases.* International Program on Traditional Ecological Knowledge and IDRC, Ottawa.

Johannes, R.E. (1989) *Traditional Ecological Knowledge: A Collection of Essays.* IUCN, Gland.

Knight, C.G. (1980) Ethnoscience and the African farmer: rationale and strategy. In: Brokensha, D.W., D.M. Warren and O. Warner (eds) *Indigenous Knowledge Systems and Development.* University of America Press, Boston.

Metzer, D. and G. William (1966) Some procedures and results in the study of native categories: Tzeltal firewood. *American Anthropologist* 68:389–407.

Sturtevant, W.C. (1964) Studies in ethnoscience. In: Romney, A.K. and R.G. D'Andrade (eds) *Transcultural Studies in Cognition. American Anthropologist* 66:99–331.

Warren, D.M., L.J. Slikkerveer and S. Oguntunji Titilola (eds) (1990) *Indigenous Knowledge Systems: Implications for Agriculture and International Development.* Iowa State University, Ames.

Useful addresses

Some international and networking NGOs, northern and southern

African NGOs Environment Network (ANEN)
PO Box 53844
Nairobi
Kenya
Tel: +254 2 28138
Telex: 25331 ANEN KE

Consorcio Latinoamericano Sobre Agroecologia y Desarollo (CLADES)
Casilla 97
Correo 9
Santiago
Chile
Tel: +56 2 2341141
Fax: +56 2 2338918

Environment and Development Action in the Third World (ENDA)
BP 3370
Dakar
Senegal
Tel: +221 225565
Fax: +221 222695
Telex: 51456 SG

Environment Liaison
  Centre International
  (ELCI)
PO Box 72461
Nairobi
Kenya
Tel: +254 2 562015
Fax: +254 2 562175
Telex: 23240 ELC KE

Genetic Resources
  Action International
  (GRAIN)
Jonqueres 16
6° D
08003 Barcelona
Spain
Tel: +34 3 3105909
Fax: +34 3 3105952

Honey Bee Centre for
  Management in
  Agriculture
Indian Institute of
  Management
Ahmedabad-380015
India
Fax: +91-272-427896
E-mail:
anilg@iimahd.ernet.in

Information Centre for
  Low-External-Input
  and Sustainable
  Agriculture (ILEIA)
ETC Foundation
PO Box 64
3830 AB Leusden
The Netherlands
Tel: +31 33 943086
Fax: +31 33 940791
Telex: 79380 ETC NL

Innovations et Réseaux
  pour le
  Développement
  (IRED)
3, rue de Varembe, case
  116
1211 Geneva 20
Switzerland
Tel: +41 22 341716
Telex: 289450

Overseas Development
  Institute (ODI)
Regent's College
Regent's Park
London NW1 4NS
UK

The Panos Institute
  (publishers of the
  sustainable
  development periodical
  Panoscope)
9 White Lion Street
London N1 9PD
UK
Tel: +44 171 2781111
Fax: +44 171 2780345
Telex: 9419293

1717 Massachussetts
  Ave.
Suite 301
Washington DC 20036
USA
Tel: +1 202 4830044
Fax: +1 202 4833059

31 rue de Reuilly
75012 Paris
France
Tel: +33 1 43792935
Fax: +33 1 43799135

Rodale Institute
222 Main St.
Emmanus, PA 18098
USA

Rural Advancement
  Foundation
  International (RAFI)
130 Slater Suite 750
Ottawa
Ontario K1P 6E2
Canada
Tel: +1 613 5650900
Fax: +1 613 5948705

South-east Asian
  Regional Institute for
  Community Education
  (SEARICE)
PO Box EA31
Ermita, Manila
Philippines
Fax: +254 2 742352

Some other relevant organizations

Associated Country
  Women of the World
50 Warwick Square
London SW1V 2AJ
UK

Association of Farming
  Systems
  Research–Extension
Dr T. Finan, Secretary
Bureau of Applied
  Research in
  Anthropology
University of Arizona
Tucson
AZ 85721
USA

c/o Dr C. Lightfoot
International Center for
  Living Aquatic
  Resources
  Management
  (ICLARM)
MC PO Box 1501
Makati
Metro Manila 1299
Philippines

Human Relations Area
  Files (HRAF)
755 Prospect Street
PO Box 2054
New Haven, CT 06520
USA

International Federation
  of Women in
  Agriculture
Dr C. Prasad, Secretary
  General
Krishni Anusandham
  Bhavan
Pusa
New Delhi 110012
India

Rural Sociological
  Society
P.C. Jobes, Treasurer
Department of
  Sociology, Wilson Hall
Montana State
  University
Bozeman, MT 59717
USA

Society of Economic
  Botany
New York Botanical
  Gardens
Bronx, NY 10458-5126
USA

International Program
  on Traditional
  Ecological Knowledge
Canadian Museum of
  Nature
PO Box 3443
Station D
Ottawa, Ontario
Canada K1P6PN

International Society for
  Ethnobiology
Ms Katy Moran
3521 S. St., NW
Georgetown,
  Washington DC.
USA

Women in Rice Farming
  Systems Network
IRRI
PO Box 933
1009 Manila
Philippines

APPENDIX 12.1
Indigenous knowledge resource centres

Established Centres

International
Center for International Research and Advisory Networks (CIRAN)
Dr G.W. von Liebenstein, Director
PO Box 29777
2509 LS The Hague
The Netherlands
Tel: +31 70 4260321
Fax: +31 70 4260329

Center for Indigenous Knowledge for Agriculture and Rural Development (CIKARD)
Dr D.M. Warren, Director
318 Curtiss Hall
Iowa State University
Ames
Iowa 50011
USA
Tel: +1 515 2940938
Fax: +1 515 2941708

Leiden Ethnosystems and Development Programme (LEAD)
Dr L.J. Slikkerveer, Director
Institute of Cultural and Social Studies
University of Leiden
PO Box 9555
2300 RB Leiden
The Netherlands
Tel: +31 71 273469 or 273472
Fax: +31 71 273619

Regional
African Resource Centre for Indigenous Knowledge (ARCIK)
Prof. A. Phillips, Director
Dr T. Titilola, Research Coordinator
Nigerian Institute of Social and Economic Research
(NISER)
PMB 5 – UI Post Office
Ibadan
Nigeria
Fax: +234 22 416129 or +234 1 614397

Regional Program for the Promotion of Indigenous Knowledge in Asia (REPPIKA)
Dr Evelyn Mathias-Mundy, Coordinator
International Institute of Rural Reconstruction (IIRR)
Silang
Cavite 4118
Philippines
Tel: +63 2 9699451 or 582659
Fax: +63 2 5222494
E-mail: iirr@phil.gn.apc.org

National
Brazilian Resource Centre for Indigenous Knowledge (BRARCIK)
Prof. D.A.J. Cancian, Director
UNESP, Dept. Biologica
14870.000 Jaboticabal SP
Brazil
Tel: +55 163 232500
Fax: +55 163 224275
E-mail: uejab@brfapesp.bitnet

Centre Burkinabè de Recherche sur les Pratiques et Savoirs Paysans (BURCIK)
Dr B.E. Dialla, Director
BP 7047
Ouagadougou
Burkina Faso
Tel: +226 362835
Fax: +226 336517

Cameroon Indigenous Knowledge Organization (CIKO)
Prof. C.N. Ngwasiri, Director
Private Sector Research Institution
PO Box 170
Buea
Southwest Province
Cameroon
Tel: +237 322685
Fax: +237 322106

Ghana Resource Centre for Indigenous Knowledge (GHARCIK)
Dr M. Bonsu, Interim Director
School of Agriculture
University of Cape Coast
Cape Coast
Ghana
Tel: +233 42 22409 or 24809
Telex: 2552 UCC GH

Indonesian Resource Center for Indigenous Knowledge (INRIK)
Prof. K. Adimihardja, Director
Department of Anthropology
University of Padjadjaran
Bandung 40132
Indonesia
Tel: +62 22 81594 or 832728
Fax: +62 22 431938

Kenya Resource Centre for Indigenous Knowledge (KENRIK)
Dr Mohamed Isahakia, Acting Director
The National Museums of Kenya
PO Box 40658
Nairobi
Kenya
Tel: +254 2 742131
Fax: +254 2 741424

Madagascar Resource Centre for Indigenous Knowledge (MARCIK)
Ms Juliette Ratsimandrava
c/o Centre d'Information et de Documentation Scientifique et Technique
BP 6224
Antananarivo 101
Madagascar
Fax: +261 2 32123/20422

Mexican Research, Teaching and Service Network on Indigenous Knowledge (RIDSCA)
Dr A. Macia-Lopez, Director
Colegio de Postgraduatos (CEICADAR)
Apartado Postal 1–12
CP 72130
Col. La Libertad
Puebla, Pue.
Mexico
Tel: +52 22 48088 or 480978 or 480542

Nigerian Resource Centre for Indigenous Knowledge (NIRCIK)
Dr J.O. Olukosi, Coordinator
Institute for Agricultural Research
Ahmadu Bello University
PMB 10044, Zaria
Nigeria
Tel: +234 69 50571
Fax: +234 69 50891
Telex: 75248 NITEZ NG

Philippines Resource Center for Indigenous Knowledge and Sustainable Development (PHIRCIKSD)
Dr R.C. Serrano, National Coordinator
Philippine Council for Agriculture, Forestry and Natural Resources Research and Development
  (PCACRD)
Los Baños
Laguna
Philippines
Tel: +632 94 50015 to 50020
Fax: +63 94 50016
Telex: 40860 PARRS PM

South African Resource Centre for Indigenous Knowledge (SARCIK)
Prof. M.H. Cohen, Co-Director
The Institute for Indigenous Theory and Practice
110 Long Street
8001 Cape Town
South Africa
Tel: +27 21 242012
Fax: +27 21 262466

Sri Lanka Resource Centre for Indigenous Knowledge (SLARCIK)
Dr R. Ulluwishewa
University of Sri Jayewardenapura
Department of Geography
Gangodawila, Nugegoda
Sri Lanka
Tel: +94 1 552028
Fax: +94 1 500544

Uruguay Resources Centre for Indigenous Knowledge (URURCIK)
Pedro de Hegedus, Coordinator
CEDESUR
Casilla Correo 20.201
Codigo Postal 12.900
Sayago, Montevideo
Uruguay
E-mail: pdh@agrocs.edu.ey

Venezuelan Resource Secreteriat for Indigenous Knowledge (VERSIK)
Dr C. Quiroz, National Coordinator
Centre for Tropical Alternative Agriculture and Sustainable Development (CATADI)
University of the Andes, Nùcleo 'Rafael Rangel'
Apartado Postal #22
Trujillo 3102
Estado Trujillo
Venezuela
Tel/Fax: +58 73 33667

Centres being established
Regional/subregional centres: European Resource Center for Indigenous Knowledge, Trans-Andean
   Resource Center for Indigenous Knowledge

National centres: Australia, Benin, Bolivia, Colombia, Costa Rica, India, Namibia, Nepal, Peru,
   Tanzania, Zimbabwe

Bibliographic databases for plant germplasm collectors

13

J.A. Dearing[1] and L. Guarino[2]

[1]IPGRI, Via delle Sette Chiese 142, 00145 Rome, Italy: [2]IPGRI, c/o International Laboratory of Research on Animal Diseases, PO Box 30709, Nairobi, Kenya.

At various points in this volume, the collector is referred to the literature. For example, Chapter 9 deals with the specialist taxonomic, floristic, ecological and conservation literature and Chapter 10 with information sources on the physical and socioeconomic environment. Other chapters deal with the literature on such specific topics as the taxonomy of cultivated species (Chapter 7), the cultural setting of target areas (Chapter 12) and crop pests (Chapter 17). Chapter 20 recommends that the literature be searched prior to collecting to ascertain the seed storage type of target species. Chapter 24 urges workers developing *in vitro* collecting procedures to search the literature for references to successful methods of regenerating plants from *in vitro* explants. These are all extensive topics and all that it has been possible to do in many cases is refer to published bibliographies and some of the major review works. More efficient ways of gaining access to the literature are needed.

This is a general problem in science. As more and more scientific information is generated around the world, it becomes increasingly difficult to keep track of developments in a given field. This growth in the scientific literature has led to the appearance of abstracts journals and more recently to computerized databases. Three of the major agricultural databases are AGRIS (managed by the Food and Agriculture Organization (FAO)), AGRICOLA (US National Agricultural Library) and CAB Abstracts (CAB International, or CABI). The printed hardcopy versions of these databases are, respectively: *Agrindex*, *Bibliography of Agriculture* and 50 specialized CABI abstracting journals. Some of the CABI publications are mentioned in other chapters, for example *Seed Abstracts* in Chapter 20 in connection with acquiring information on the seed storage type of particular species. These large bibliographic databases are available through 'hosts' such as DIALOG, DIMDI and ESA-IRS, which provide the necessary computing facilities to store, update and search them. The prospective user must register with and obtain a password from a host service before access to the database is permitted. In order to search a database the user must have a computer, communications software, a modem and a telephone line. Such on-line searching has an obvious advantage over manual searching of printed abstract journals in that it is much more rapid and flexible. Also, the databases are updated much more frequently than their printed versions are produced. However, some knowledge of the host command language will be necessary. Also, costs are high: they include telecommunications and host connection charges and a fee for each record extracted from the database.

In the developing world on-line searching is often not a viable option due to the occasional unreliability of telecommunications and the problem of recurring expenditure. However, in the past few years compact disk read-only memory (CD-ROM) technology for information storage has become readily and relatively cheaply available. As a result, rapid access to large bibliographic and other databases is now possible anywhere in the world without the problems associated with on-line searching. Over 600 megabytes of data may be stored on a standard 12 cm plastic optical compact disk, the equivalent of 300,000 written pages of information. A special CD-ROM reader is needed to read the disk. Readers can be purchased as separates but very often are now included as an integral part of a personal computer. The software supplied with the CD-ROM databases usually allows complex searches on multiple keywords and downloading into word processing or personal database files. A CD-ROM database may not be as up to date as its on-line counterpart: hosts update the databases monthly, weekly or even daily but CD-ROM updates are sent to users usually only quarterly and in some cases annually. Also, though a CD-ROM stores a vast amount of information, some databases are spread on to two or three disks, and unless one has a multiple CD-ROM drive it is not possible to search the whole database in one go.

Information units and libraries in various international organizations involved in agricultural research and development – for example FAO and the international agricultural research centres (IARCs) – provide collaborators with access to bibliographic databases. The programmes of the IARCs depend on strong information support and therefore make extensive use of both in-house and externally available bibliographic databases in printed, CD-ROM and on-line formats. The International Crops Research Institute for the Semi-Arid Tropics (ICRISAT), for example, runs the Semi-Arid Tropical Crops Information Service (SATCRIS) using its in-house SATCRIS bibliographic database, which includes data downloaded from CAB Abstracts and AGRIS as well as books and materials acquired by its library. Users of the ICRISAT library also have access to AGRIS, AGRICOLA and CAB Abstracts on CD-ROM. Many IARCs regularly produce printed bibliographies and abstracts lists specific to their mandate crops and regions. For example, the Centro Internacional de Mejoramiento de Maiz y Trigo (CIMMYT) publishes the *Cassava Bibliographic Bulletin*, a quarterly alert service on cassava. The International Board for Plant Genetic Resources (IBPGR) produced a one-off listing of crop genetic resources literature covering the period 1975–83 extracted from various CABI abstracting journals and other sources. However, CABI now collaborates with The International Plant Genetic Resources Institute (IPGRI) to produce the quarterly abstracts journal *Plant Genetic Resources Abstracts* and the CD-ROM PlantGeneCD. Details of the information services of FAO and the IARCs are given in Appendix 13.1.

Some regional agricultural research organizations also provide on-line access to commercial databases, as well as developing their own. Many developing countries now have national agricultural information services equipped with CD-ROM readers. For example, AGRIS on CD-ROM is distributed to AGRIS Participating Centres. The Technical Centre for Agricultural and Rural Cooperation (CTA) has supplied personal computers, CD-ROM readers, laser printers, bibliographic databases on disk and training to many agricultural libraries in African, Caribbean and Pacific (ACP) countries. CTA also runs a Question-and-Answer Service and has published specialized bibliographies. Fisher *et al.* (1990), sponsored by CTA, describe agricultural information resource centres worldwide. Keenan and Wortley (1994) is also published by CTA. Lilley (1994) is another general guide to sources of information in agriculture and horticulture while Davis (1987) provides a survey of information sources for botanists.

Table 13.1 gives a list of agricultural and plant sciences bibliographic databases that may be useful to germplasm collectors in gathering background information on their target taxa and region during the planning stages of their work. Coverage of the on-line versions of databases usually goes further back in time than that of CD-ROM versions (databases that are only available on CD-ROM are marked with an asterisk). In addition to PlantGeneCD, there are further specialized CD-ROMs by CABI on agricultural economics, horticulture, soil science and crop protection.

Bibliographic databases in the field of the environment are referred to in Chapter 9. A *Directory of Online Databases* is published twice a year by Gale Research in the USA and the UK, giving details of databases and the on-line host services providing access to them. It is also available on-line through Data-Star, on magnetic tape and on diskette. The contact address is given below. For a full list of CD-ROMs available, consult *CD-ROMs in Print 1991*, which is published annually by Meckler, or Mitchell (1991). To keep track of developments in the field, including new issues, a useful journal is *CD-ROM End User.*

Table 13.1. Agricultural and plant sciences bibliographic databases useful to germplasm collectors.

Name	Contents	Coverage
AGRICOLA (US Department of Agriculture (USDA))	Records of material received by US National Agricultural Library and associated institutions; mostly US publications	From 1970
AGRIS (FAO)	Worldwide coverage of food, agriculture and fisheries, including input from 160 national centres	From 1975
Biological Abstracts (BioScience Retrieval Services) On-line version: Biosis Previews	Current research in the biological sciences; covers journals from 100 countries	On-line from 1969
CAB Abstracts (CABI)	Worldwide coverage of journals, books, conferences and reports covering agriculture, forestry and allied disciplines, including social sciences and aspects of human medicine	On-line from 1972. CD-ROM from 1984
Coffeeline (International Coffee Organization)	Periodical articles, reports, thesis and audiovisual material on coffee	From 1973. Not available on CD-ROM
Compact International Agricultural Research Library Basic Retrospective Set (Consultative Group on International Agricultural Research (CGIAR))*	Full text of CGIAR international research centre publications	From 1962 to 1986
Musadisk (International Network for the Improvement of Banana and Plantain (INIBAP))	Worldwide literature on cultivation, harvest, transport and trade of bananas and plantains (MUSALIT). Information on researchers and institutions (BRIS). Factual data (e.g. statistics). List of accessions held in collections. Expert system for cultivar identification	CD-ROM forthcoming. MUSALIT and BRIS on-line at INIBAP
PlantGeneCD (CABI)*	Worldwide coverage of crop genetics, plant breeding and plant genetic resources	From 1973

Table 13.1. *continued*

SCI CDE (Institute for Scientific Information) On-line version: SCISEARCH	CD-ROM version of *Science Citation Index*, containing bibliographic information from 3100 international science journals	On-line from 1974. CD-ROM from 1980
SESAME (Centre de Coopération Internationale en Recherche Agronomique Pour le Développement (CIRAD))*	French literature on agricultural research and rural development in the tropics	From 1989
TREECD (CABI)*	Forestry information from CAB printed journals *Forestry Abstracts, Forest Products Abstracts* and *Agroforestry Abstract*	From 1939
TROPAG & RURAL (Royal Tropical Institute (KIT); since 1993, the *Abstracts* in collaboration with CIRAD and Natural Resources Institute (NRI)).	Agriculture and rural development in tropics and subtropics, including grey literature from developing countries; corresponds to *Abstracts on Tropical Agriculture* and *Abstracts on Rural Development in the Tropics*	From 1975

* Only available on CD-ROM.

References

Davis, E.B. (1987) *Guide to Information Sources in the Botanical Sciences*. Libraries Unlimited, Littleton.

Fisher, R.C., J.C. Peterson, J.W. Beecher, J.S. Johnson and C. Boast (1990) *Agricultural Information Resource Centres: A World Directory*. IAALD, Urbana.

Keenan, S. and P. Worteley (1994) *Agricultural Bibliographic Information Sources*. CTA, Wageningen.

Lilley, G.P. (ed.) (1994) *Information Sources in Agriculture and Horticulture*. Butterworths, London.

Mitchell, J. (ed.) (1991) *The CD-ROM Directory 1991*. 5th edition. TFPL Publishing, London.

Useful addresses

Major bibliographic databases

*Directory of on-line
  databases*
Gale Research Inc.
835 Penobscot Bldg.
Detroit, MI 48226–4094
USA
Tel: + 1 313 9612242
Fax: + 1 313 9616815
Telex: 810 221 7087

AGRICOLA
US Department of
  Agriculture
Science and Education
  Administration
Technical Information
  Systems
Room 300
National Agricultural
  Library Building
Beltsville, MD 20705
USA
E-mail: jmangin@
  asrr.arsudsa.gov

AGRIS
AGRIS/CARIS
  Coordinating Centre
FAO
Via delle Terme di
  Caracalla
00100 Rome
Italy
Tel: + 39 6 57974993
Fax: + 39 6 57973152,
  57975155 or 5782610
Telex: 625852 FAO I

BIOSIS
BioScience Retrieval
  Services
2100 Arch Street
Philadelphia
PA 19103–1399
USA
Tel: + 1 800 5234806

SESAME
Centre de Coopération
  Internationale en
  Recherche
  Agronomique Pour le
  Développement
  (CIRAD)
BP 5035
34052 Montpellier
  Cedex 1
France
Tel: + 33 67 615800
Fax: + 33 67 615820
Telex: 670871
  INFRANCA

CAB Abstracts
CAB International
Wallingford
Oxfordshire OX10 8DE
United Kingdom
Tel: + 44 1491 832111
Fax: + 44 1491 833508
Telex: 847964
  COMAGG G
E-mail: cabi@cabi.org

TROPAG & RURAL
Royal Tropical Institute
  (KIT)
Department of Informa-
  tion and
  Documentation
Mauritskade 63
1092 AD Amsterdam
The Netherlands
Tel: + 31 20 5688711
Fax: + 31 20 5688444
Telex: 15080 KIT

Plant Genetic Resources Abstracts can be ordered from CABI UK or any of the following
addresses:

North America
845 North Park Avenue
Tucson
Arizona 85719
USA
Tel: + 1 800 5284841
  or + 1 602 6217897
Fax: + 1 602 6213816

Asia
PO Box 11872
50760 Kuala Lumpur
Malaysia
Tel: + 60 3 2552922
Fax: + 60 3 2551888
Telex: 28031 CABI MA

*Caribbean and Latin
  America*
Gordon Street
Curepe
Trinidad and Tobago
Tel: + 809 6624173
Fax: + 809 6632859
Telex: 24438 CARIRI

Hosts

Dialog Information
  Services Inc.
3460 Hillview Avenue
Palo Alto
CA 94304
USA
Tel: + 1 415 8582700
Fax: + 1 415 8587069
Telex: 334499 DIALOG

DIMDI
Weisshausstrasse 27
PO Box 420580
5000 Cologne 41
Germany
Tel: + 49 221 47241
Fax + 49 221 411429
Telex: 8881364 DIM

Data-Star
Plaza Suite
114 Jermyn Street
London SW1Y 6HJ
UK
Tel: + 44 171 9305503
Fax: + 44 171 9302581

ESA-IRS
Via Galileo Galilei
PO Box 64
00044 Frascati
Italy
Tel: + 39 6 94011
Fax: + 39 6 9401361

STN International
c/o Japan Information
  Center of Science and
  Technology
CPO Box 1478
Tokyo 100
Japan
Tel: + 81 3 5816411
Fax: + 81 3 5816446

CD-ROMs

SilverPlatter
One Executive Park
Newton Lower Falls
MA 02162 1449
USA
Tel: + 1 617 9691332
Fax: + 1 617 9695554

SilverPlatter publishes many CD-ROM databases, including AGRICOLA, AGRIS, BIOSIS, CAB-CD and TROPAG & RURAL.

Other sources of information on information sources

Technical Centre for Agriculture and Rural Cooperation (CTA)
Postbus 380
NL-6700 AJ Wageningen
The Netherlands
Tel: + 31 8380 60400
Fax: + 31 8380 31052
Telex: 30169 CTA

CTA is mandated under the Lomé Convention to work to provide ACP states with better access to information, research, training and innovations in agricultural development and extension. It has co-published *Agricultural Information Resource Centres: A World Directory 1990* (with the International Association of Agricultural Information Specialists (IAALD); see References), a two-volume directory of *Tropical Agriculture Information Sources* (Vol. 1: European Community; Vol. 2: ACP countries) and a booklet on *Online Information Sources for Tropical Agriculture*. CTA also publishes a bimonthly bulletin, *Spore*.

CGNet Services
1024 Hamilton Court
Menlo Park, California
USA 94025
Tel: + 1 415 325 3061
Fax: + 1 415 325 2313
Telex: 490 000 5788 CGN
E-mail: 157:CGI100; cgnet@cgnet.com

CGNet Services is a not-for-profit organization which provides advice on, and assistance with, information technology to researchers in developing countries, including on-line access to information sources, electronic mail, internetworking and CD-ROM subscriptions.

Appendix 13.1
Information resources of FAO and the commodity IARCs

The basis of the information recorded here is the Information Resources of the International Agricultural Research Centres (INTAGRES) brochure. INTAGRES has kindly agreed to allow its reproduction. The information in the brochure was updated by means of a questionnaire sent out to all the relevant organizations in late 1993.

Note: Under E-mail, the Dialcom address is given, followed by a semicolon and the Internet address.

Food and Agriculture Organization of the United Nations (FAO)

Subject coverage

Agriculture, forestry, fisheries, nutrition.

In-house databases (selected)

• ESHD Data Base: agrarian reform, agricultural education, demography, human resources, rural development, rural women.
• FAO Documentation Data Base (FAODOC).
• FAO Library Monographs Data Base (FAOLIB).
• FAO Library Serials Data Base. Access software: CDS/ISIS.
• Population Documentation Centre Data Base (PDC): demographic statistics, population policy, rural development, rural women.

External databases

• AGRIS (CD-ROM).
• BEASTCD (CD-ROM).
• CAB Abstracts (CD-ROM).
• DIALOG.
• ESA-IRS.
• TREECD (CD-ROM).

- TROPAG & RURAL (CD-ROM).
- VETCD (CD-ROM).

Networking and cooperative systems

- AGLINET (International Agricultural Libraries Network).
- AGRIS Input Centre.

Information Unit publications

- *FAO Documentation – Current Bibliography.*

Contact information

David Lubin Memorial Library
FAO
Via delle Terme di Caracalla
00100 Rome
Italy
Tel: +39 6 57973703
Fax: +39 6 57973152, 57975155 or 5782610
Telex: 625852 FAO I

International Network for the Improvement of Banana and Plantain (INIBAP)

Subject Coverage

Musa (bananas and plantains).

In-house databases

- BRIS: trilingual database on *Musa* researchers, research projects and institutions. Access software: Micro-CDS/ISIS.
- MUSALIT: trilingual bibliographic database including abstracts, conventional and non-conventional literature on *Musa* compiled by INIBAP, IRFA/CIRAD (France), UPED (Panama) and (soon) PCARRD (Philippines). Access software: Texto, Micro-CDS/ISIS.

External databases

- Agritrop (on-line).
- SESAME (CD-ROM).

Networking and cooperative systems

- Regional networks for Asia/Pacific, Latin America/Caribbean, West and Central Africa, East Africa.

Information Unit publications

- *Directory of Researchers Working on Banana and Plantains.*
- *Directory of Musa Research Projects and Institutions.*
- *Infomusa,* an international magazine on bananas and plantains.
- *Musarama,* the International Bibliographic Abstract Journal on Banana and Plantain.

Contact information

Information/Documentation Unit
INIBAP
Parc Scientifique Agropolis
34397 Montpellier Cedex 5
France
Tel: +33 67 611302
Fax: +33 67 610334
Telex: 490376 F
E-mail: INIBAP; inibap@cgnet.com

International Crops Research Institute for the Semi-Arid Tropics (ICRISAT)

Subject coverage

Sorghum, pearl millet, pigeonpea, chickpea, groundnut, farming systems for the semiarid tropics, soils, agroclimatology and the socioeconomics of the semiarid tropics related to agriculture and agricultural development.

In-house databases

- Aflatoxin Contamination of Groundnuts. Access software: Micro CDS/ISIS.
- *Busseola fusca* Database. Access software: Micro CDS/ISIS.
- SATCRIS: Includes books acquired by the library, apart from data downloaded from CABI and AGRIS. Access software: BASIS and Micro CDS/ISIS.

External databases

- AGRICOLA (CD-ROM).
- AGRIS (CD-ROM).
- Books in Print (CD-ROM).
- CAB Abstracts (CD-ROM).
- DIALOG.
- IARC Union Catalog of Serials.
- Pesticide Disk (CD-ROM).
- SESAME (CD-ROM).

Networking and cooperative systems

- AGRIS Input Center.
- AGLINET (International Agricultural Libraries Network).

- CLAN (Cereals and Legumes Asia Network).
- West African Farming Systems Research Network.

Information Unit publications

- *Chickpea and Pigeon Pea Prompts* (CABI–ICRISAT) – to end 1993.
- *Forthcoming Conferences in Agriculture and Related Sciences: A Selected List.*
- *Groundnut Prompts* (CABI–ICRISAT) – to end 1992.
- *International Chickpea Newsletter.*
- *International Pigeonpea Newsletter.*
- *International Arachis Newsletter.*
- *Sorghum and Millets Abstracts* (CABI–ICRISAT) – to end 1992.

Specialized information services

Semi-Arid Tropical Crops Information Service (SATCRIS).

Contact information

Library and Documentation Services
ICRISAT
Patancheru
AP 502 324
India
Tel: +91 842 224016
Fax: +91 842 241239
Telex: 422203 or 4256366 ICRI IN
E-mail: ICRISAT; icrisat@cgnet.com

International Center for Integrated Mountain Development (ICIMOD)

Subject coverage

Agriculture, population, employment, roads, watershed management, tourism, environment, energy, women in mountain development and natural resources within the Hindu Kush Himalayas.

In-house databases

- Bibliographic. Access software: CDS/ISIS.
- Serials. Access software: CDS/ISIS.

External databases

- AGRIS.
- CAB Abstracts.
- Food, Agriculture and Science (CD-ROM).

Networking and cooperative systems

- Mountain Information Exchange Network.

Information Unit publications

- *Bibliographies.*
- *New Documents in Library* (bimonthly).
- *New Serials in Library* (fortnightly).

Contact information

Mountain Documentation and Information Exchange Division
ICIMOD
PO Box 3226
Kathmandu
Nepal
Tel: +977 1 525313
Fax: +977 1 524509
Telex: 2439 ICIMOD NP
Cable: ICIMOD Kathmandu

Asian Vegetable Research and Development Center (AVRDC)

Subject coverage

Chinese cabbage, mungbean, onion, garlic, peppers, shallots, sweet potato, tomato, soyabeans.

In-house databases

- AVLIB: Bibliographic records on Chinese cabbage, garlic, onion, mungbean, pepper, shallot, soyabean, sweet potato, tomato and other vegetables. Access software: MINISIS.
- Chinese: Bibliographic records in Chinese. Access software: MINISIS.
- HOLD: Titles and holdings of journal collections. Access software: MINISIS.
- INSM: Institution directory. Access software: MINISIS.
- THESMAST: Controlled keywords with hierarchical structure. Access software: MINISIS.

External databases

- AGRICOLA (CD-ROM).
- AGRIS (CD-ROM).
- CAB Abstracts (CD-ROM).

Networking and cooperative systems

- IAALD (International Association of Agricultural Information Specialists).
- Science and Technology Library Network of Republic of China.

Information Unit publications

- *Annotated Bibliography of Diamondback Moth.*
- *Bibliography of Soybean Rust, 1895–1986.*
- *Bibliography of Tropical and Subtropical Soybeans, 1970–1982.*
- *Diseases and Insect Pests of Mungbean and Blackgram: A Bibliography.*
- *Hot Peppers and Sweet Peppers* (943 citations).
- *Radishes* (566 citations).

Contact information

Information and Documentation
AVRDC
PO Box 42
Shanhua, Tainan 74199
Taiwan, Republic of China
Tel: +886 5837801
Fax: +886 5830009
Telex: 73560
Cable: Asveg Shanhua
E-mail: AVRDC; avrdc@cgnet.com

Centro Internacional de Mejoramiento de Maiz y Trigo (CIMMYT)

Subject coverage

Wheat, maize, triticale, breeding, crop management, plant physiology, plant pathology, entomology, biotechnology, food sciences, soils, agricultural economics.

In-house databases

- AGRB: Imported references from AGRIS dealing with cereal production. Access software: BASIS.
- CABI: Imported references and abstracts on maize and wheat from CABI. Access software: BASIS.
- INTD: CIMMYT internal documents. Access software: BASIS.
- LIBR: All books, reprints, reports acquired by the library since 1986. Access software: BASIS.
- SERI: Titles and holdings of CIMMYT subscriptions to journals, newsletters and other serial publications. Access software: BASIS.

External databases

- AGRICOLA (CD-ROM).
- AGRIS (CD-ROM).
- AGRISEARCH (CD-ROM).
- ARIES (CD-ROM).
- Bancos Bibliograficos Mexicanos I, II and III (CD-ROM).
- Books in Print (CD-ROM).

- CD-ROM Directory 1992 (CD-ROM).
- DIALOG.
- Food, Agriculture and Science (CD-ROM).
- Food and Human Nutrition (CD-ROM).
- IARC Union Catalog of Serials (CD-ROM).
- KIT Abstracts (CD-ROM).
- LIBRUNAM (CD-ROM) (Book Titles from the·National Autonomous University of Mexico).
- LIS (CD-ROM).
- MICROSOFT Bookshelf (CD-ROM).
- POPLINE (CD-ROM).
- SESAME (CD-ROM).
- TROPAG & RURAL (CD-ROM).
- ULRICH'S Serials Directory (CD-ROM).
- World Weather Disk (CD-ROM).

Networking and cooperative systems

- AGRIS Input Center.
- AGLINET (International Agricultural Libraries Network).
- AIBDA (Interamerican Association of Agricultural Librarians and Documentalists).
- ANBAGRO (Mexican Agricultural Librarian Association).
- IARC Information Network in Latin America.
- IFLA.

Information Unit publications

- *Maize Abstracts* (CABI–CIMMYT).
- *Scientific Information Bulletin*.
- Wheat, Barley and Triticale Abstracts (CABI–CIMMYT).

Contact information

Scientific Information Unit
CIMMYT
Apartado Postal 6–641
06600 Mexico DF
Mexico
Tel: +52 5 7269091 or 9542100
Fax: +52 5 9541069
Telex: 1772023-CIMTME
Cable: CENCIMMYT
E-mail: CIMMYT; cimmyt@cgnet.com

International Livestock Center for Africa (ILCA)

Subject Coverage

Sub-Saharan Africa livestock production, cattle meat and milk, small ruminant meat and milk, animal traction, animal feed resources, trypanotolerance, livestock policy, resource use.

In-house databases

- ILCA: Bibliographic description (including keywords) of publications produced by ILCA staff. Access software: MINISIS.
- ILCABIB: Includes subsets of AGRIS and CABI databases and bibliographic description (including keywords and abstracts) of all documents acquired by the center library. Access software: MINISIS.
- Mail: Addresses of recipients of ILCA publications. Access software: MINISIS.
- Serial: Titles and holdings of ILCA subscriptions to journals and serials publications. Access software: MINISIS.
- TRACTION: Bibliographic description (including keywords and abstracts) of documents on animal traction. Access software: MINISIS and CDS/ISIS.

External databases

- AGRICOLA (CD-ROM).
- AGRIS (CD-ROM).
- Books in Print (CD-ROM).
- CAB Abstracts (CD-ROM).
- Computer Select (CD-ROM).
- Development Activity Information (CD-ROM).
- IDRIS (Inter-Agency Development Research Information). Access software: MINISIS.
- PADISDEV (Pan-African Documentation and Information System). Access software: MINISIS.
- SESAME (CD-ROM).

Networking and cooperative systems

- AGRIS Input Center.
- IAALD (International Association of Agricultural Information Specialists).
- MUG (MINISIS Users' Group).

Information Unit publications

- *Accessions Bulletin.*
- *Country Catalogs.*
- *ILCA In-Print.*
- *Serials Holdings.*
- *Special Bibliographies.*

Contact information

Information Services
ILCA
PO Box 5689
Addis Ababa
Ethiopia
Tel: +251 1 613215
Fax: +251 1 611892
Telex: 21207 ILCA ET
Cable ILCAF ADDIS ABABA
E-mail: ILCA; ilca@cgnet.com

International Institute of Tropical Agriculture (IITA)

Subject coverage

Maize, cowpea, cassava, soyabeans, yams, plantain, bambara groundnuts, sustainable agricultural systems in the humid and subhumid tropics.

In-house databases

- ACQ: All books, journals, reprints and audiovisual materials acquired since 1984. Access software: BASIS.
- CAT: Contains references to bibliographic information on journals, books, pamphlets, analyticals, articles, microfiche and microfilm, slides, audio tapes, maps and video cassettes. Access software: BASIS.
- CIR: Records referring to scientists and National Agricultural Research System (NARS) staff.
- STAC: Serial Inventory Control.
- VEN: Records of vendors to IITA library.

External databases

- AGRICOLA (CD-ROM).
- Agricultural Library (CD-ROM).
- AGRIS (CD-ROM).
- AGRISEARCH (CD-ROM).
- Books in Print (CD-ROM).
- CAB Abstracts (CD-ROM).
- Dissertation Abstracts (CD-ROM).
- Food, Agriculture and Science (CD-ROM).
- FSTA (CD-ROM).
- IARC Union Catalog of Serials (CD-ROM).
- Science Citation Index (CD-ROM).
- SESAME (CD-ROM).
- Social Science Citation Index (CD-ROM).
- TROPAG & RURAL (CD-ROM).
- ULRICH's Serials Directory (CD-ROM).
- World Weather Disk (CD-ROM).

Networking and cooperative systems

- AGRIS Input Center.
- AGLINET (International Agricultural Libraries Network).
- IAALD (International Association of Agricultural Information Specialists).

Information Unit publications

- *A Guide to the Library and Documentation Center.*
- *A Record of IITA Publications, 1967–1991.*
- *Banbara Groundnut (Voandzeia subterranea Thours): Abstracts of World Literature.*
- *Bibliography of Yams and the Genus* Dioscorea.

- *Bibliography of Plantains and Other Cooking Bananas.*
- *Cassava Bacterial Blight: Abstracts of World Literature.*
- *Catalogues of Serials.*
- *Cowpeas (*Vigna unguiculata *L. Walp): Abstracts of World Literature.*
- *Farming Systems in Africa: A Working Bibliography.*
- *Guide to the IITA Library Database.*
- *Guide to Selected Publications on Maize Available in the Library.*
- *Guide to Selected Publications on Rice Available in the Library.*
- *Library and Documentation Staff Manual.*
- *The Winged Bean (*Psophocarpus tetragonolobus*) and other* Psophocarpus *Species: Abstracts of World Literature.*
- *Tropical Grain Legume Bulletin.*

Specialized information centres

International Grain Legume Information Center (IGLIC).

Contact information

Library and Documentation Center
IITA
Oyo Road
PMB 5320
Ibadan
Nigeria
Tel: +234 22 400300 to 400319
Fax: +234 1 610650 or +229 301466 (via IITA Benin)
Fax: (INMARSAT) 874–1772276 (Ibadan)
Telex: 31417 or 31159 TROPIB NG
Cable: TROPFOUND IKEJA
E-mail: IITA; iita@cgnet.com

Mailing address:
c/o L.W Lambourn & Co.
Carolyn House
26 Dingwall Road
Croydon CR9 3EE
UK
Tel: +441 81 6869031
Fax: +441 81 6818583
Telex: 851 946979 LWL G

International Rice Research Institute (IRRI)

Subject coverage

Rice and rice-based cropping systems, genetic evaluation and use, control and management of rice pests and diseases, irrigation water management, soil microbiology, climate and rice, constraints on rice yields, consequences of new technology, cropping systems, machinery development and testing.

In-house databases

- *Azolla* Database.
- Rice Literature Database. Access software: CARDBOX-PLUS.

External databases

- AGRICOLA (CD-ROM).
- Biological Abstracts (CD-ROM).
- CAB Abstracts (CD-ROM).
- Food, Agriculture and Science (CD-ROM).
- IARC Union Catalog of Serials (CD-ROM).

Networking and cooperative systems

- AGLINET (International Agricultural Libraries Network).

Information Unit publications

- *25 Years of IRRI Theses and Dissertations: An Abstract Bibliography.*
- *A Bibliography of Rice Literature Translations Available in the IRRI Library and Documentation Center.*
- *International Directory of Rice Workers.*
- *International Bibliography of Rice Research* (with annual supplements and 5 year cumulative indexes).
- *International Bibliography on Cropping Systems.*
- *International Bibliography on* Azolla.
- *List of Japanese Rice Literature* (quarterly) (English and Japanese editions).
- *Publications of the International Agricultural Research and Development Centers* (CGIAR–IRRI).
- *Rice Literature Update* (quarterly).
- *Theses and Dissertations on Rice Available in Library of IRRI.*
- *Upland Rice Research: An International Bibliography.*

Contact information

Library and Documentation Center
IRRI
PO Box 933
1099 Manila
Philippines
Tel: +63 2 8181926
Fax: +63 2 8182087 or 8188470
Telex: 54365 RICE PM via ITT; 22456 IRI PH via RCA; 63786 RICE PN via EASTERN
Cable: RICEFOUND, MANILA
E-mail: IRRI; irri@cgnet.com

West Africa Rice Development Association (WARDA)

Subject coverage

West African rice development.

In-house databases

- Bibliographic (WARBI): Conventional and non-conventional literature in WARDA library network, focused on West Africa rice literature. Access software: CDS/ISIS.
- Serials (PERIO): Library serials holdings and subscription management. Access software: CDS/ISIS.
- NARS: Rice research institutions and rice scientists. Access software: CDS/ISIS.
- Institutional.
- Statistical.

External databases

- AGRICOLA (CD-ROM).
- AGRIS (CD-ROM).
- CAB Abstracts (CD-ROM).
- CIDARC (CD-ROM and on-line).
- TROPAG & RURAL (CD-ROM).

Networking and cooperative systems

- WARDA Library Network.
- WARIS (West Africa Rice Information Systems).

Information Unit publications

- *NARS Directories.*
- *Recent Accessions Lists.*
- *Specialized Bibliographies.*
- *Tables of Contents.*
- *World Rice References for West Africa.*

Contact information

Library and Documentation Centre
WARDA
01 BP 2551
Bouake
Côte d'Ivoire
Tel: +225 633242, 632396 or 634514
Fax: +225 634714
Telex: 69138 ADRAO CI, BOUAKE
Cable: ADRAO BOUAKE CI
E-mail: WARDA; warda@cgnet.com

International Centre for Research in Agroforestry (ICRAF)

Subject coverage

Agroforestry systems and practices, multipurpose trees and shrubs and methodologies for agroforestry research.

In-house databases

- AFBIB: Bibliographic database, including books, reprints and reports in the ICRAF library covering world literature on agroforestry and its related subjects. Access software: Micro CDS/ISIS.
- EPPER: Bibliographic database of journals and newsletters held in the ICRAF library. Access software: Micro-CDS/ISIS.

External databases

- AGRICOLA (CD-ROM).
- AGRIS (CD-ROM).
- CAB Abstracts (CD-ROM).
- SESAME (CD-ROM).
- TREECD (CD-ROM).

Networking and cooperative systems

- AGRIS Input Center.
- CARIS.
- IAALD (International Association of Agricultural Information Specialists).
- INFOTERRA (UNEP's Global Environment Exchange Network).
- KENISIS (Kenya Micro CDS/ISIS Users Group).
- Kenya Agricultural Research Library.
- NIG (Nairobi Information Group).

Information Unit publications

- *Agroforestry Abstracts* (CABI–CRAF)
- *Agroforestry for Development in Kenya: An Annotated Bibliography.*
- *Agroforestry Literature: A Selected Bibliography.*
- *Agroforestry Literature: A Selected Bibliography on Sub-Saharan Africa.*
- *Economic Analysis of Agroforestry Projects: An Annotated Bibliography.*
- *Grevillea robusta: An Annotated Bibliography.*
- *ICRAF Bi-monthly Accessions List.*
- *Technology Monitoring and Evaluation in Agroforestry Projects: An Annotated Bibliography.*

Contact information

Information Programme
ICRAF
PO Box 30677
Nairobi
Kenya
Tel: +254 2 521450
Fax: +254 2 521001
Telex: 22048
Cable: ICRAF
E-mail: ICRAF; icraf@cgnet.com

Centro Internacional de la Papa (CIP)

Subject coverage

Potato, sweet potato and Andean root and tuber crops (biological and socioeconomic aspects), plant pathology, plant breeding, plant physiology, biotechnology, seed technology, food sciences, germplasm, postharvest and tropical agriculture.

In-house databases

- AGRIS: Potato and sweet potato. Access software: BASIS.
- CABI: Potato and sweet potato. Access software: BASIS.
- CIP Database: Bibliographic references from CIP library.
- Potato and Sweet Potato Network Database: Individuals and institutions involved in potato and sweet potato research. Access software: Software 1032.
- Publishing Procedures of Agricultural Journals. Access software: BASIS.

External databases

- AGRIS (CD-ROM).
- BIOSIS.
- DIALOG.
- Food, Agriculture and Science (CD-ROM).
- SESAME (CD-ROM).

Networking and cooperative systems

- AGRIS Input Center.
- AIBDA (Interamerican Association of Agricultural Librarians and Documentalists).
- IAALD (International Association of Agricultural Information Specialists).
- IARC Information Network in Latin America.

Information Unit publications

- *Accession Lists.*
- *CIP Bibliographies.*
- *Serials Publications Catalog.*

Contact information

Information Unit
CIP
Apartado 5969
Lima
Peru
Tel: +51 14 354354, 354283
Fax: +51 14 351570
Telex: 25672 PE
Cable: CIPAPA, Lima
E-mail: CIP; cip@cgnet.com

Centro Internacional de Agricultura Tropical (CIAT)

Subject coverage

Germplasm development for common beans, cassava, rice and forages in tropical ecosystems; natural resources management; land use and sustainable development for agricultural production, particularly in hillsides, forest margins and savannah ecosystems.

In-house databases

- CINFOS: References and abstracts covering world literature on common beans, cassava and tropical pastures. Access software: CDS/ISIS.
- CATAL: References to books, journals, maps, audiovisuals and audiotutorials in the general collection as well as the collection of Latin American literature on rice. Access software: Micro CDS/ISIS.
- SERIAD: Bibliographic information and holdings for journals, newsletters and serial publications at CIAT. Access software: Micro CDS/ISIS.

External databases

- AGRICOLA (CD-ROM).
- AGRIS (CD-ROM).
- Biological Abstracts (CD-ROM).
- CAB Abstracts (CD-ROM).
- Compact International Agricultural Library Basic Retrospective Set (CD-ROM).
- DIALOG.
- Food Science and Technology Abstracts (CD-ROM).
- SESAME (CD-ROM).
- TROPAG & RURAL (CD-ROM).

Networking and cooperative systems

- AGLINET (International Agricultural Libraries Network).
- AGRIS Input Center.
- AIBDA (Interamerican Association of Agricultural Librarians and Documentalists).
- IAALD (International Association of Agricultural Information Specialists).

- IARC Information Network in Latin America.
- SNICA (National Sub-System of Information in the Agricultural Sciences).

Information Unit publications

- *Boletin bibliografico.*
- Brochures and user guides on information services.
- Quick Bibliographies series.
- Specialized bibliographies on many aspects of cassava, common bean and tropical pasture research as well as others on the small farmer and rural development.

Contact information

Information and Documentation Unit
CIAT
Apartado Aereo 6713
Cali
Colombia
Tel: +57 23 675050
Fax: +57 23 647243
Telex: 05769 CIAT CO
Cable: CINATROP
E-mail: CIAT; ciat-library@cgnet.com

International Center for Agricultural Research in the Dry Areas (ICARDA)

Subject coverage

Agriculture in West Asia and North Africa, faba bean, kabuli chickpea, lentil, barley, durum wheat, bread wheat, forage legumes, forage grasses, stress tolerance, disease resistance, farming systems, supplementary irrigation and agro-ecological zoning.

In-house databases

- Directory of Faba Bean and Lentil Research Workers. Access software: Micro-CDS/ISIS.
- ILDOC: ICARDA Library and Documentation Database. Access software: Micro-CDS/ISIS.
- LENS: Comprehensive bibliographic database on lentil. Access software: Micro-CDS/ISIS.
- Networking: Individuals and institutions.

External databases

- AGRICOLA (CD-ROM).
- AGRIS (CD-ROM).
- CAB Abstracts (CD-ROM).
- KSU.
- SESAME (CD-ROM).

Networking and cooperative systems

• AGRIS Input Center.

Information Unit publications

• *Faba Bean in AGRIS.*
• *FABIS Newsletter.*
• *Lentil in AGRIS.*
• *LENS Newsletter.*
• *RACHIS Newsletter.*

Specialized information services

FABIS: Faba Bean Information Services.
LENS: Lentil Experimental News Service.
RACHIS: Barley and Wheat for WANA Researchers

Contact information

Library
ICARDA
PO Box 5466
Aleppo
Syria
Tel: +963 21 550465, 551280
Telex: 331206, 331208, 331263
Cable: ICARDA-Aleppo

International Plant Genetic Resources Institute (IPGRI)

Subject coverage

Plant genetic resources.

In-house databases

• IPGRI library holdings.
• IBPGR project reports.

External databases

• AGRICOLA (CD-ROM).
• AGRIS (CD-ROM).
• CAB Abstracts (CD-ROM).
• Food, Agriculture and Science (CD-ROM).
• SESAME (CD-ROM).
• PlantGeneCD (CD-ROM).

- TREECD (CD-ROM).
- TROPAG & RURAL (CD-ROM).

Networking and cooperative systems

- AGRIS Input Center.

Information Unit publications

- *Plant Genetic Resources Abstracts* (IPGRI–CABI).

Contact information

Library
IPGRI
Via delle Sette Chiese 142
00145 Rome
Italy
Tel: +39 6 51892214
Fax: +39 6 5750309
Telex: 4900005332 (IBRU UI)
E-mail: IPGRI; ipgri@cgnet.com

International Center for Living Aquatic Resources Management (ICLARM)

Subject coverage

Aquaculture, coastal resource management and fisheries.

In-house databases

- LIBRI: Database containing library holdings such as books and monographs.
- NAGA: Selected journal articles on tropical aquatic resources.

External databases

- DIALOG.
- ASFA (CD-ROM).

Networking and cooperative systems

- PASFIS.
- SEAFIS.
- IAMSLIC.

Contact information

Library and Documentation Center
ICLARM
MC, PO Box 1501
Makati, Metro Manila 1299
Philippines
Tel: +63 2 8180466, 8189283
Fax: +63 2 8163183
Telex: ETPI 64794 ICLARM PN; 4900010376 ICL UI (USA)
Cable: ICLARM Manila
E-mail: ICLARM; iclarm@cgnet.com

Ecogeographic surveys

14

N. Maxted[1], M.W. van Slageren[2] and J.R. Rihan[3]

[1]*School of Biological Sciences, University of Birmingham, Edgbaston, Birmingham B15 2TT, UK:* [2]*Genetic Resources Unit, ICARDA, PO Box 5466, Aleppo, Syria:* [3]*Department of Biology, Biomedical Sciences Building, University of Southampton, Southampton SO9 3TU, UK.*

Introduction

To make the most efficient use of limited resources, plant germplasm collectors must have a clearly defined set of target taxa, and must know as much as possible about where (and when) to find these plants within their general target region. Much time and effort can be wasted if collectors do not know enough about the geographic distribution, ecology, phenology and diversity of the plants they are looking for before setting out into the field.

Conditions at the localities inhabited by a species will be characterized by more or less specific environmental constraints. The passport data associated with herbarium specimens, germplasm accessions and other plant records can be used to identify these constraints. For example, if passport data for a particular species or genotype indicate that in the past it has only been found on limestone scree slopes above 2000 metres in southwest Asia, then localities occurring within these parameters are clearly where one should initially look, if further material is being sought. A combination of ecological and geographic passport data from existing collections and the literature can be used to predict where plants may be found and when they are likely to be ready to collect.

Plant collectors are thus like detectives: they gather and analyse the clues in passport data in order to trace the plants that interest them. This is the essence of ecogeographic investigations, the subject of this chapter. IBPGR (1985) summarizes the three major components of ecogeography as the study of:

- distributions of particular species in particular regions and ecosystems;
- patterns of intraspecific diversity;

- relationships between ecological conditions and the survival or frequency of variants.

It concludes that: 'Field data provide a basis for determining how to
maximize the sampling of genetic diversity. Ecogeographic information
can be used to locate significant genetic material and representative
populations can then be monitored.'

An ecogeographic investigation can focus on various levels of the
taxonomic hierarchy. It can be concerned with informal categories such
as landraces, cultivar groups or races within a crop (Chapter 7); or formal
categories such as subspecies, varieties, and forms within a species or
species within a series, genus or tribe. It can also deal with wild or
cultivated taxa. The kinds of data will be largely the same, though in
practice there will be a difference in the sources of the data. For example,
among herbaria, only the few specialist institutes that concentrate on
cultivated material (e.g. the N.I. Vavilov Institute for Plant Industry in
St Petersburg, Russia) will have collections which do not under-
represent the variation found within crops. Thus, gene banks will usually
be better sources of ecogeographic data on cultivated material than
herbaria.

An ecogeographic model

The steps involved in undertaking an ecogeographic study will be
discussed below. First, a definition: an ecogeographic study is a process
of gathering and synthesizing taxonomic, geographic and ecological
data. The results are predictive and can be used to assist in the formulation of collecting and conservation priorities.

The difference between a 'study' and a 'survey' is one of degree.
Ehrman and Cocks (1990) provide a good example of an ecogeographic
study of the annual legumes of Syria. They present a detailed analysis
of the climatic and soil characteristics that influence the distribution of
the annual legume species they sampled throughout the country. They
suggest that species diversity and seed production are related to annual
rainfall and that populations in the drier areas face greater threat of
genetic erosion. Based on their analysis of the ecogeographic data, they
propose a detailed list of conservation priorities. However, this was only
possible because of the very detailed ecogeographic data gathered over
several years by the authors. Clearly, considerable time and resources
are required to undertake such a study. If ecogeographic data are to be
used as a routine part of collecting and conservation, then the quicker,
less expensive, option of undertaking a survey is likely to be favoured.
A survey will focus on collating data recorded by other plant collectors,
rather than obtaining new data. It may be restricted to a literature
search and gathering passport data from herbarium specimens and gene
bank accessions.

In practice, all conservation activities, collecting not least among them, are necessarily preceded by some form of ecogeographic data collation and analysis. Though this may not follow exactly the methodology proposed here, most ecogeographic surveys or studies will be articulated in three phases, as follows:

- Phase I – Project design:

1. Project commission.
2. Identification of taxon expertise.
3. Selection of target taxon taxonomy.
4. Delimitation of the target region.
5. Identification of taxon collections.
6. Designing and building the ecogeographic database structure.

- Phase II – Data collection and analysis:

1. Listing of germplasm conserved.
2. Survey of taxonomic, ecological and geographic data sources.
3. Collection of ecogeographic data.
4. Data verification.
5. Analysis of taxonomic, ecological and geographic data.

- Phase III – Product generation:

1. Data synthesis.
2. Ecogeographic database, conspectus and report.
3. Identification of conservation priorities.

Phase I – Project design

Project commission

Ecogeographic projects may start in a variety of different ways. An individual collector may simply decide it is necessary to gather some background data prior to setting off for the field, or an international organization may commission a full ecogeographic study of a particular target gene pool as a preliminary to developing a comprehensive conservation strategy. Whatever the case, a taxon or taxa from a defined geographic region must be considered to be of sufficient interest to warrant time-consuming and possibly expensive background research to support subsequent collecting, conservation and use. The range of the study may vary from one species in a restricted area to a whole genus throughout the world, e.g. *Arachis* species in Brazil, *Aegilops* sect. *Sitopsis* in the Near East or *Hordeum* species worldwide. An example of a project commission is provided by Edmonds (1990) in the report of her herbarium survey of the genus *Corchorus* in Africa. She states that:

> A general survey of *Corchorus* L. species was commissioned by the International Jute Organization to provide the necessary background

data on which future germplasm collecting expeditions could be based
... the survey was required to identify those wild species for potential
use in the future genetic improvement of jute, in addition to identifying
the countries and locations where collecting expeditions would be most
profitable.

Care should be taken in identifying a specialist to undertake or
supervise ecogeographic studies. The taxonomy of wild species is sometimes difficult, identification aids often lacking or of poor quality, and
retrieving data from older herbarium specimens presents special problems. The ecogeographer need not be an expert in the target taxon, but
should have some background knowledge of the group and be experienced in the use of identification aids. Misidentification of material will
diminish the predictive value of the data collected. Employing a specialist to gather ecogeographic data might at first sight be considered
extravagant, but many taxon specialists may be willing to undertake or
supervise such a study, if they consequently had the opportunity to see
material from the taxon's centre of diversity to which they might not
otherwise have access. The specialist should also, if possible, have a good
understanding of the geography of the region to be studied, especially
in the case of local studies. This is illustrated by Sánchez and Ordaz
(1987), who found local geographic and ecological expertise invaluable in
their study of *Zea mexicana* (teosinte) in Mexico. Local expertise may
also prove vital in trying to decipher locality details from specimen
labels written by hand several decades ago.

Identification of taxon expertise

The acquisition of ecogeographic data will prove much easier if advice
is sought from taxon experts at an early stage. They will be able to
advise on the accepted taxonomy of the group, recommend (and perhaps
provide) possibly obscure Floras, monographs and other literature,
advise on any relevant databases, suggest which herbaria and/or gene
banks should be visited and provide the ecogeographer with useful
local connections. The authors of relevant scientific papers will probably be the first contact points. *Index Herbariorum* (Holmgren *et al.*,
1990) and its companion volume (Holmgren and Holmgren, 1992) list
the researchers associated with different herbaria and what their
specialities are. Increasingly, herbaria are acquiring electronic mail
facilities, and there is a list of Plant Taxonomists Online (contact Jane
Mygatt, jmygatt@bootes.unm.edu). A database of experts in botany
and mycology worldwide is maintained at the University of Oulu,
Finland (contact Anne Jäkäläniemi at the Department of Botany,
anne.jakalaniemi@oulu.fi).

Selection of target taxon taxonomy

It is clearly important to have a good taxonomic understanding of the target group prior to undertaking an ecogeographic study. This can be obtained from various sources, in particular target taxon specialists, monographs, recent revisions of the group and, increasingly, taxonomic databases (see section on 'Survey of taxonomic, ecological and geographic data sources' below and Chapter 10). These will help the ecogeographer determine the generally accepted classification of the group, which will list the taxa currently considered members of the target group, their accepted names and the more common synonyms. The taxonomic limits to the study will thus be set.

There may be various alternative classifications of the target group. The ecogeographer must consider these and make a decision as to which one to adopt. Wild species are usually described using a combination of morphological characteristics. A classification using the biological species concept, where genetic data are given greater importance, may be more appropriate if the aim is to conserve maximum genetic variation in the target taxon. However, there are few such biologically based classifications available. They tend to be restricted to well-known crop plants and their allies, where the genetic relationships among the taxa have been extensively studied and the make-up of the gene pool is relatively well understood. Increasingly, however, genetic diversity studies are also being carried out on wild plants using biochemical and molecular markers.

Knowing the accepted classification of a group will provide leads to other literature: iconography, distribution maps, identification aids, autoecological studies, ethnobotanical investigations (including checklists of vernacular names), bibliographies. More obscure groups may lack a recent revision or monograph, but the researcher must still collate whatever published taxonomic data are available to provide the backbone to the study.

Delimitation of the target region

The target region under study may be restricted by the terms of reference of the project commission, but if it is unspecified the taxon should be studied throughout its range. The commissioning agent may restrict the survey to a specific area (e.g. the Sahel, Vietnam, South America) if the area is fairly clearly defined and/or a complete study would be too costly in time and resources. However, restriction of the target region to save resources in the short term may ultimately prove to be a false economy. Multiple studies of the same taxon, possibly by different authors, are likely to form a less coherent whole. Rihan (1988), who undertook an ecogeographic survey of the forage *Medicago* species of the Mediterranean and adjacent arid/semiarid areas, found that her target region was floristically ill-defined. The natural distribution of the species she was studying did not coincide with the target region she was commissioned to cover. Both of these considerations may unnecessarily limit the predictive value of the ecogeographic study (see also Funk, 1993; Stressey, 1993).

Having established the limits of the target region, additional information on the target taxon can be obtained from local Floras and field guides. These can provide more detailed information on local geographic distribution and ecological preferences. Guides to which Floras cover which parts of the world are provided by Frodin (1984) and Davis *et al.* (1986). Information on sources of information on wild species, including more on these works, is provided in Chapter 10.

Identification of taxon collections

The researcher undertaking an ecogeographic study will need to visit the major herbarium collections of target taxon specimens from the target region. Travelling to herbaria may be expensive and so the selection of which ones are to be visited is crucial. Target taxon and region specialists will be able to suggest which herbaria and libraries the ecogeographer should concentrate on. Part 1 of *Index Herbariorum* (Holmgren *et al.*, 1990) records for each herbarium the historical plant collections conserved there. Part 2 (1–7), published by various authors between 1954 and 1988 in volumes 2 (A–D), 9 (E–H), 86 (I-L), 93 (M), 109 (N–R), 114 (S) and 117 (T–Z), of *Regnum Vegetabile*, an occasional series of the International Association for Plant Taxonomy, is an alphabetical index of the most important plant collectors, giving the present location of their specimens.

The important collections to be seen during the study fall into two categories: major international herbaria and local herbaria in the target region. The relative advantages and disadvantages of two categories of herbarium for the ecogeographer are given in Table 14.1. Because each kind of herbarium has its own strengths and weaknesses, it is important that both should be visited in the course of an ecogeographic survey. Another factor that should be considered when selecting which herbaria to visit is their age. This is especially important for the smaller, regional herbaria. Recently established herbaria are likely to contain a higher proportion of recently collected specimens. These specimens commonly have more comprehensive and more legible passport data than older collections, so newer herbaria are more likely to yield better-quality ecogeographic data.

Davis and Heywood (1973) stress in a similar context that it is important to sample material from as many herbaria as possible, so that a true estimate of within-taxon variation can be made. Likewise, the broader the sampling of ecogeographic data associated with herbarium specimens or germplasm, the more likely the data will prove ecologically and geographically predictive.

Some of the ecogeographic data included in the database compiled

Table 14.1. Relative advantages and disadvantages of major international herbaria and regional herbaria.

	Advantages	Disadvantages
Major international herbaria	1. Broad taxonomic coverage, possibly material used in the production of revisions and monographs 2. Broad international geographic coverage, possibly material used in the production of local Floras 3. Skilled researchers available to provide general advice 4. Appropriate taxonomic and geographic specialists 5. Type material of target taxa 6. Good botanical library	1. Predominance of old collections, making extraction of passport data more difficult and likely predictive value lower 2. Geographic names associated with older collections sites may have changed more recently.
Regional herbaria	1. Good local regional coverage of target region 2. Better-documented material, as the herbarium is likely to have been more recently established 3. Regional specialists present, who can assist in deciphering local geographic names	1. Limited resources for herbarium maintenance 2. Lack of target taxon specialists 3. Limited botanical library

by Maxted (1990) were taken direct from the author's own germplasm collection database. This illustrates the point that ecogeographic data can equally well be obtained from the passport data recorded by previous germplasm collectors. The importance generally given by plant genetic resources workers to such data means that the passport data associated with germplasm accessions will often be of a higher standard than that associated with herbarium specimens. However, systematic germplasm acquisition programmes have only been established relatively recently. As a result, for many species herbarium specimens may provide the only source of detailed ecogeographic data. The sources of information on existing germplasm collections are discussed further below and in Chapter 9.

Designing and building the ecogeographic database structure

The ecogeographic data for the target taxon can be recorded on paper, but, for those who have access to (and experience of) computers, it is much more efficient to collate data directly into a database. Hardware and software requirements need to be considered early on. For example, a portable computer will be necessary if most data collation is to be done away from the base institution. If data are likely to be numerous, faster (and probably more expensive) machines will be preferred.

It is not uncommon to use a word-processing package for data management, but this has serious limitations. For serious users, a database management system (DBMS) is recommended. The main advantages of using a DBMS are (Date, 1981; Painting *et al.*, 1993):

- data capture and editing are easier;
- there is less chance of introducing errors while copying data between formats;
- searching and retrieving data for reports are easier;
- complicated sorting and indexing on multiple fields are possible;
- identification of duplicate records is possible;
- the structure of a file can be altered in response to changing information needs.

Many DBMSs are available, differing in their flexibility, ease of use and capabilities. Examples include dBASE IV, FoxPro (or FoxBASE), Paradox, etc. (Tatian, 1993). ORACLE and INFORMIX[1] are more complicated packages. Users should determine the type of DBMS (and, indeed, computer hardware) that is being used by any collaborators they may have. This may constrain their own choice, though many DBMSs can export a dBASE-compatible database or a flat-file format ASCII file, allowing data exchange.

Specialized packages are available for the management of herbarium label and related information. Examples include TROPICOS (Pankhurst, 1991) and the Botanical Research and Herbarium Management System (BRAHMS) developed by D. Filer (Oxford Forestry Institute). However, these may not be entirely appropriate for the ecogeographer's purposes, who will probably therefore need to develop his or her own database structure before data collation can begin. Maxted (1991) and Pankhurst (1991) discuss the design and construction of databases in taxonomy and related fields. More general practical advice on database design is given by Painting *et al.* (1993).

In general, database structure should be kept as simple as possible. A database is composed of records, equivalent to the horizontal lines in

---

[1] dBASE IV is a trademark of Ashton-Tate Corp., FoxPro and FoxBASE are trademarks of Fox Software Inc; ORACLE is a trademark of Oracle Corp.; and INFORMIX is a trademark of Informix Software Inc.

a report, and fields, each equivalent to a single column in a report. Each record could describe a separate herbarium sheet, accession, country or species, for example. Each field would then provide specific details about a different aspect of the specimen, country or species.

Fields should be defined for all information that can be expressed in a limited number of words or numbers. It is not reasonable to enter complete narrative information for all records if the narrative is more than a couple of lines (80 characters) long. Data in this form will be difficult to search and retrieve. In general, if a field contains more than ten words or word-and-number combinations, it should be divided into separate fields. Indeed, if a field can be split, it should be. Combining different data into a single field should certainly be avoided. For example, latitude and longitude can be entered in a number of ways, but it is best to have separate numeric fields for degrees and minutes and a character field for hemisphere (E, W, N, S). Similarly, if measurements are given in the original data in a variety of units (e.g. distances in kilometres and miles), there should be a numeric field for the data and a separate character field in which the unit used is specified; this will allow later transformation of the data into a common unit. Data in numeric fields can be manipulated mathematically, which is not the case for character fields, but many software packages will turn missing data or blanks in numeric fields into zeros. Room for extra characters should be allowed when setting field widths. Field names should be unique, descriptive and simple (Painting *et al.*, 1993).

When it is possible to enter more than one piece of data for a given field and record, these should be accommodated in separate records or in separate fields. For example, if variation within an accession in a continuous numerical descriptor like plant height exists and needs to be shown, create separate fields for maximum and minimum value or for mean and standard error. In the case of discrete descriptors, if a variety of flower colours occur in a species, for example, and a particular population appears mixed from a herbarium or germplasm sample, the options are as follows: (i) open separate records for each state if the descriptor is particularly significant and the number of different states large; (ii) note the presence/absence or frequency of each state (colour form, in this case) in a separate field; or (iii) note the most common state in one descriptor, the second most common in another, and so on, to the extent desired. Instead of separate fields, a single field could be used, with the data separated by standard delimiters, but this may complicate retrieval and analysis (but see Hintum, 1993). When a particular system has been chosen for recording a descriptor, this should be rigidly adhered to in building up the database. The 'rules' to be followed for each descriptor should be written down before starting to enter data.

Codes and/or a standard wording should be adopted whenever possible in entering data. This promotes data consistency. It can also speed up data entry and verification and facilitate exchange. A listing must be maintained of what the codes mean. In many cases, it is possible to use accepted standard codes, but one system should be selected and used consistently. Examples include standard codes for herbaria (Holmgren *et al.*, 1990), for authors of plant names (Brummitt and Powell, 1992) and for political units (International Standards Organization, 1981). The International Union of Biological Sciences Commission on Taxonomic Databases (TDWG) was established to facilitate data standardization and data exchange between botanical databases. It is producing sets of standard codes for botanical data, for example for botanical recording units (Hollis and Brummitt, 1992). Standards in preparation include ones for: economic use; habitat, soil and landscape; life-form; and plant occurrence and status. Information can be obtained from the TDWG Secretariat, based at the Missouri Botanical Garden. Since 1963 the World Conservation Union (IUCN) has been working on a system for describing the conservation status of species (for history, see Fitter and Fitter, 1987). A quantitative system for determining categories of threat is being developed (Mace *et al.*, 1992; IUCN, 1994). The United Nations Environment Programme (UNEP)/GEM's Harmonization of Environmental Measurement (HEM) Programme aims to 'enhance the compatibility and quality of information on the state of the environment worldwide'. It collates and disseminates information on environmental data, including models and classification schemes. It is working with the World Conservation Monitoring Centre (WCMC) on a classification of vegetation. Published germplasm descriptor lists (which include passport, characterization and evaluation data) should be examined before beginning to develop a database (Chapter 8).

A problem that commonly occurs in database construction is that of data repetition. For example, in a database of information on the provenance of herbarium specimens, an entry in a PROVINCE field will always be associated with the same COUNTRY field: Chiapas is always in Mexico, Uttar Pradesh always in India, etc. One could clearly have a single file with both PROVINCE and COUNTRY fields, but this would be uneconomic in terms of both data-inputting time and computer storage and more prone to inputting errors, as there may be dozens of records from Uttar Pradesh for which the word 'India' would have to be entered, for example. A better alternative is to create a second database file containing only PROVINCE and COUNTRY fields, each province being listed only once. One could then have the main file of locality information containing the PROVINCE field but no COUNTRY field and link this to the second file through the common field PROVINCE to access the country information. A DBMS that allows the processing of data in linked files in this way is sometimes called relational, though the technical definition of the term is somewhat stricter (Pankhurst, 1991).

A simple linked file structure, in this case dBASE files, is illustrated in Fig. 14.1 (Maxted, 1990). Only linking fields are shown. The example illustrates the inclusion of the three basic kinds of ecogeographic data in the one database: ecological, geographic and taxonomic. Files GEN-NOME, SECTNOME and TAXONOMY contain the taxonomic data,

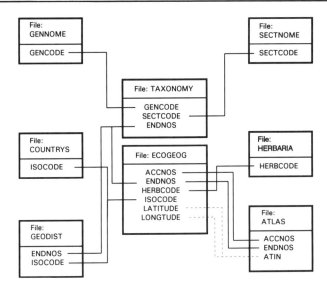

Fig. 14.1. Ecogeographic database file relations (Maxted 1990). Only fields within files that are linked to fields within separate files are shown. Solid lines connecting fields indicate direct links and hatched lines indicate partial links.

file ECOGEOG contains the bulk of the ecological data and files COUNTRYS, GEODIST and ATLAS contain the geographic data. One file, HERBARIA, contains curatorial data, the addresses of the herbaria visited. ATLAS contains the input data for a mapping program, which plotted dot-distribution maps using the latitude and longitude data held in the database. However, the program used by Maxted (1990) required the locality data in a specific input format, so the field ATIN is derived from the fields LATITUDE and LONGTUDE, but not identical to them. The field ENDNOS contains the taxon identification code and illustrates the role of the linking field. When building up the main ECOGEOG file, time was saved by entering the species name as a code. However, the full Latin name and generic section corresponding to each taxon code are found in the file TAXON.

As stressed by Painting *et al.* (1993), a register should be kept of field specifications (full descriptor name, field name, field width, field type, data entry rules, data validation rules, indexing, whether a linking field, file name).

Consistency is important when entering data into the database. It assists retrieval and report generation as well as sorting routines. Consistency will be better if data are entered all at once, rather than in fits and starts, as data entry conventions will be more easily remembered. Entries should be made as they will be required to look in a final report: the software should not be relied on to correct errors in data entry. For example, genus name should have the first character entered as upper-case. A hand-held scanner can be used to acquire optical images of herbarium labels and the like for later printing, examination and entry into the database.

Herbaria, especially young and rapidly developing institutions, are increasingly entering their specimen information in in-house databases, for example using the TROPICOS·or BRAHMS software already mentioned. For her ecogeographic survey of Mediterranean *Medicago*, Rihan (1988) obtained specimen passport data on diskette from the databases of the herbarium of the Madrid Botanical Garden, while Maxted (1990) used geographic data from the Vicieae Project Database (Bisby, 1984) in his ecogeographic study of *Vicia* subgenus *Vicia*. If a new data set is received on diskette, it needs only to be appended to the existing database file. Before doing so, however, it should be verified that the two files have compatible structures (fields of the same name, length and type), or data may be lost or appended incorrectly. Diskettes should also always be checked for computer 'viruses'.

Phase II – Data collection and analysis

Listing of germplasm conserved

Before embarking on the detailed data collation phase of the project, current conservation activities should be reviewed. If sufficient genetic variation of the target taxon from the target region is already safely conserved either *ex situ* or *in situ*, then there may be little justification for further collecting. Details of what material is currently being conserved can be obtained from the catalogues and databases of botanical gardens, gene banks and *in situ* conservation areas. Identifying these sources may prove time-consuming, but taxon experts may help guide the ecogeographer (Chapter 10). The International Plant Genetic Resources Institute (IPGRI) produces international directories of germplasm collections on a crop basis and also maintains a parallel database, which may be queried on demand (Chapter 9).

Care must be taken when interpreting information on current genebank or botanical garden holdings. The material held may be incorrectly identified, though it may be possible to check the identification by consulting voucher material or identifying living material. The actual quantities of germplasm available could also be misleading: gene banks and botanical gardens are encouraged to duplicate their holdings in other collections, so the total number of accessions held around the world can give a false impression of the genetic diversity conserved. The ecogeographer should also consider that, although accessions may be held in a gene bank, the material may for various reasons be unavailable to potential users and so create a false impression of a taxon's conservation status.

Survey of taxonomic, ecological and geographic data sources

Chapters 9 and 10 deal with sources of published information on the environment and on wild species, respectively, and Chapter 12 with sources of ethnographic, especially ethnobotanical, information. Increasingly, however, information is becoming available in media other than the conventional printed literature. Abstracts of publications (and, in some cases, the full text) may be available on microfiche, compact disk read-only memory (CD-ROM) or on-line bibliographic databases, for example (Chapter 13). There may have been other attempts to survey herbarium label information, the results of which may or may not be formally published. For example, some herbaria and other organizations are developing floristic and indigenous knowledge databases (Chapters 10 and 12). Herbaria may hold some label data in card catalogues. The example of the East Africa Herbarium in Nairobi may be instructive. It maintains card catalogues on plant uses (cited by Peters *et al.*, 1992) and on local names gleaned from herbarium labels and other sources. Databases are being developed from both catalogues.

Increasingly, computer networks, particularly the academic network known as the Internet, are being used as sources of information. The Internet links together some one million computers worldwide, which means that there are probably tens of millions of users. Many scientific interest groups have been set up on the Internet. Software, such as listserv and Usenet, supports electronic discussion groups and distributes electronic newsletters and scientific papers. In addition, many important information resources, such as university libraries and public domain software and databases, are being made available on the Internet. Compilations of listservs, Usenet news groups and information archives of relevance to biologists are provided by Dr Una Smith's 'A Biologist's Guide to Internet Resources' (smith-una@yale.edu). TAX-ACOM (taxacom@harvarda.harvard.edu) is perhaps the best-known mailing list on taxonomy and related subjects. How to gain access to Internet resources is described by Krol (1992).

The main categories of data on taxa that may be obtained from the literature and other information sources are listed in Box 14.1 (*indicates data that could be coded). The collation of much of this information may be undertaken while visiting major herbaria, which often have good botanical libraries attached and some of which have access to the Internet.

Collection of ecogeographic data

The kinds of information that the ecogeographer may obtain from the passport data associated with herbarium specimens and germplasm accessions are given in Box 14.2 (* indicates data that could be coded). Characterization and evaluation data could be added for germplasm accessions. This is an extensive (though not exhaustive) list, and it is

Box 14.1

- Accepted taxon name.*
- Locally used taxon name.*
- Botanical description.
- Taxonomic affinities.
- Distribution within the target region.*
- Timing and periodicity of local flowering and fruiting.*
- Habitat preference.*
- Topographic preference.*
- Soil preference.*
- Geological preference.*
- Climatic and microclimate preference.*
- Pollination and breeding system.*
- Germination requirements.
- Seed storage type.*
- Dispersal system.
- Genotypic and phenotypic variation, including karyotype.
- Biotic interactions, including seed predation (pests, pathogens, herbivores).
- Archaeological information (e.g. palynology).
- Ethnobotanical information (e.g. vernacular name, local uses).
- Conservation status* (e.g. IUCN Red List status).

unlikely that all items will be recorded during a given study. There are certain data items, however, that must be recorded for the study to yield predictive results, and these are shown in bold.

The herbaria of the world contain millions of specimens and the number of specimens of any one target taxon can be vast, but the scope of ecogeographic investigations will be limited by the availability of time and resources. During the course of an ecogeographic project several thousand specimens of the target taxon may be seen. Each of these specimens will require identification: the scientific names written on herbarium sheets should always be checked. However, only a proportion of the specimens seen and identified will be selected to have their passport data recorded in the database. The researcher must be discriminating. Specimens are more likely to be selected if they have detailed ecogeographic passport data or if they show features of particular taxonomic, ecological or geographic interest, i.e. they are odd or rare forms, come from unusual environments or are found on the edges of their natural range. Maxted (1990) found that data from about a third of the specimens seen during the study were finally included in the ecogeographic database.

It is important that the ecogeographer place particular emphasis on obtaining reliable specimen locality data for those specimens which are to be included in the database. Ideally, only specimens for which latitude

Box 14.2

- Herbarium, gene bank or botanical garden where specimen is deposited.*
- Name(s) of collector(s) and collecting number.
- **Collecting date* (to derive flower and fruiting time)**.
- **Sample identification.***
- Phenological data* (presence of flowers and/or fruits).
- **Locality,* latitude and longitude or even greater detail if possible**.
- **Altitude.***
- Soil type.*
- **Habitat.***
- Vegetation type.*
- Slope and aspect.*
- Land use and/or farming system.*
- Phenotypic variation.
- Evidence of pests and pathogens.*
- Competitive ability.*
- Palatability.*
- Vernacular names.
- Local uses.*

and longitude data are recorded or for which these data can be established should be selected for inclusion in the database. In practice, it may be advisable to include specimens with two levels of detail, those for which full latitude and longitude details can be obtained and those with major country subunit detail (i.e. province or state) (Rhoades and Thompson, 1992). Specimens that lack even this lower level of geographic data should not be included unless they are particularly noteworthy.

Early plant collectors could not have predicted the detailed analysis that would subsequently be based on the information they recorded. This is therefore often very limited: locality data may be ambiguous and ecological details missing. Older specimen labels are almost invariably handwritten, which adds the problem of having to read the script, which may be in a foreign language. Herbarium staff may be able to help decipher semi-illegible labels. During the ecogeographic study of Vicieae and Cicereae from the southern republics of the ex-Soviet Union (N. Maxted, in prep.) assistance from workers in local herbaria proved invaluable not only in herbarium label interpretation, but also in the precise identification of specimen localities.

A gazetteer of local geographic names and localities will also help in this. There is no comprehensive world gazetteer yet available, but the *Atlas of the World* (Times Books, 1988) contains an extensive gazetteer and the *Official Standard Names Gazetteer* is being constructed and is available in country volumes from the US Board of Geographical Names.

Herbaria sometimes develop their own unpublished gazetteers, perhaps in a card catalogue or computer database. Details of the localities mentioned in standard Floras are sometimes published in separate volumes or in appendices. As pointed out by Forman and Bridson (1992), some problems may be encountered in finding localities in gazetteers: (i) the word on the specimen label may not be a locality; (ii) the name of the locality may have changed; (iii) political and administrative boundaries may have moved; (iv) a variant or incorrect spelling of the locality name may be being used; and (v) the name may be common and therefore have more than one entry in the gazetteer. Older maps, atlases and gazetteers, and even travel books, can be a useful source of localities if names or boundaries have changed (see also Room, 1979). If names have been transliterated from other scripts (e.g. Arabic, Chinese or Cyrillic), the various possible alternative renderings should be checked. In the case of common names, some effort will have to be made to at least partly reconstruct the collector's route.

This will also be necessary if specimen locality data are of poor quality. Fortunately, many collectors keep detailed collecting notebooks and/or diaries, which are usually to be found at the herbarium or gene bank conserving their collections. Almost invariably handwritten in field conditions, they may prove time-consuming to decipher. If notebooks are not available, it may be possible to tentatively estimate where a given specimen may have been collected by comparing its collecting date and/or collecting number with those of specimens whose localities can be recognized. Herbarium determination lists, arranged in collecting number order, are useful in reconstructing itineraries. There is an extensive secondary literature on early collectors, which can help in tracing itineraries. *Taxonomic Literature* (generally known as *TL-2*), published as volumes 94 (A–G), 98 (H–Le), 105 (Lh–O), 110 (P–Sak), 112 (Sal–Ste), 115 (Sti–Vuy) and 116 (W–Z)) of *Regnum Vegetabile* under the editorship of F.A. Stafleu and R.S. Cowan (1976–1988), has a 'Bibliography and biography' section for each author listing thousands of secondary literature references on itineraries. Volumes 125 (A–Ba) and 130 (Be–Bo) are supplements. The reports of germplasm collecting missions usually include a map showing the itinerary followed.

Forman and Bridson (1992) provide a list of biographies and gazetteers that may assist in the pin-pointing of important herbarium specimen collecting localities. It is arranged by regions (plus a general section), but does not include Europe and North America.

Specimens should be positively selected to represent the breadth of geographic and ecological conditions under which the target taxon is found. It is desirable to collect detailed passport data from a broad range of representative specimens. Duplicating data should be avoided.

Recently collected specimens often have higher-quality passport data which is easier to read, being often typewritten. These data are also more likely to have remained current. Extensive use of specimens collected in past decades may provide important details about changes in distributions, but is likely to yield less useful information about contemporary populations. This might present a special problem when trying to locate populations of threatened, rare and restricted taxa. For example, recent collecting activities (Slageren, 1990) have indicated that populations of *Aegilops uniaristata* seem unlikely to have survived in Turkey beyond 1900, so the widely available herbarium specimens of this taxon from that country would provide a false indication of its distribution and frequency.

Though fruit characters are important in some groups, flowers are required to identify most species, which means that herbarium specimens are usually of flowering material. Such material will help give an indication of the most appropriate time for germplasm collecting, but will not be as useful as material that is in fruit, especially mature fruit. The ecogeographer should be particularly on the lookout for fruiting material of the target species. Samples of fleshy fruits may be preserved in alcohol, rather than dried and pressed, and may be stored separately within the herbarium institute.

The database will inevitably contain many gaps. In general, it is much easier to extract curatorial or geographic than ecological data from herbarium specimens. This is illustrated by the percentages of fields containing data in the central file of the ecogeographic database compiled by Maxted (1990). Few of the specimens included in the database lacked collector's name, collecting number, collecting date and locality details, though the degree of detail recorded for the latter varied considerably. Ecological details were found to be much less commonly recorded on herbarium specimen labels: soil type was recorded in about 25% of cases, altitude in about 55% and habitat 65%.

It should be noted that the amount of ecological information gathered in a study such as Maxted's (1990) may be unusual. This is because a relatively high proportion of the specimen data included was taken directly from the author's own germplasm collection database and not from the herbarium specimens of other collectors. In the latter case, the data are likely to be much less complete. This point is illustrated by Rihan's (1988) ecogeographic study of *Medicago* species. The data she recorded were taken entirely from herbarium specimens. The percentage of entries containing soil type, altitude and habitat data declined to 15, 26 and 52% respectively, while the amount of locality information recorded was approximately equal in both projects.

The ecogeographer may be able to infer various features of collecting sites (e.g. latitude and longitude, geology, soil, altitude) from locality data by reference to appropriate maps or databases (Chapters 9 and 16). Whether this will be possible will depend on the precision of the locality data available, the topography of the collecting area and the precision of the environmental data required. For example, if the collecting site is situated on a gently undulating plain, then a crude estimate of altitude may be gained from the locality data, as the altitude is unlikely to vary significantly in the vicinity of the collecting site. However, if the site is situated in a mountainous area, then the altitude is likely to vary markedly within relatively short distances and so estimates of site altitude based on locality might be very wide of the mark unless this is extremely precisely specified. This kind of secondarily derived data should be flagged in the database to distinguish it from data derived directly from herbarium labels. Specimens should also be flagged, if possible, to denote the accuracy of locality data, whether available in the original record or derived from other locality information; in practice, this may mean giving a code for the scale of map used to pin-point a locality and read off coordinates (Rhoades and Thompson, 1992).

Ecogeographers will be faced with the question of how many specimens should be entered into the database before the amount of extra information gained from each additional specimen fails to significantly increase the predictive value of the data set. There is no specific answer. However, the compiler should be on the look out for the point when novel ecogeographic combinations no longer occur in the specimens being examined and the latitudinal/longitudinal extent of the distribution of the species has ceased to expand: the full range of geographic and ecological niches that the taxon inhabits will then probably have been recorded in the database.

After the database is set up and all the available data have been entered, the compiler will probably find that there are still a number of gaps in many of the fields for many records. A record should be marked (flagged) when all possible information has been collected. This will help to determine the completeness of the data set. If many of the records have significant amounts of missing data, certain kinds of mathematical analysis will not be possible or will give misleading results.

Data verification

Before the database can be deemed complete, errors must be spotted and corrected. A lot of errors can be avoided by appropriate time-of-entry features of the software, such as allowing only a limited number of valid responses (perhaps listed in a menu) or storing a default value. Once all the data have been entered, verification needs to be carried out. Tatian (1993) discusses the subject and suggests the following checks:

1. Range and rule checks: are some values outside the allowed range?
2. Inter-record checks: have some records been entered twice?
3. Visual comparison with original forms (or double entry of all data): have some records not been entered, or entered in incomplete fashion?
4. Interfile checks: are data consistent among files?

A useful way of carrying out range and rule checks involves indexing the database (i.e. rearranging the records in alphabetical or numerical order) on each field in turn. Records with typing errors, invalid and out-of-range entries, etc. in the indexed field can then be easily picked out by browsing through the file. Mapping latitude and longitude data may reveal errors if particular localities are shown up as obvious outliers in impossible places (e.g in the middle of a lake). Collectors often send duplicate sets of herbarium specimens to different international herbaria. Germplasm accessions are also commonly duplicated. The compiler should search the database for these duplicates and be aware of their possible effect on data analysis. The database software should be able to pick out records that match for a set of fields.

Analysis of taxonomic, ecological and geographic data

The raw ecological and geographic data included in the database can be analysed to help identify the habitats favoured by the target taxa and the geographic limits of its distribution. Useful aids to the interpretation of ecogeographic data are tables and bar charts indicating the number (or percentage) of specimens seen from different geographic or ecological units (e.g. climate type, soil type, aspect, shading characteristics, habitat). It should be possible to get the data for such charts directly from the database: a DBMS will usually be able to count the number of records which have particular entries in a specified field or fields. Data arranged in this fashion will help to characterize the ecological niche of the target taxon. For continuous ecological factors, such as altitude, latitude and soil pH, correlation with the frequency of occurrence of specimens along the gradient can be calculated. Correlation of morphological characters with environmental gradients will help to indicate clinal adaptation, in both wild and cultivated material.

One of the most thoroughly statistically tested ecogeographic data sets is that reported by Cocks and Ehrman (1987), Ehrman and Cocks (1990) and Ehrman and Maxted (1990) for the annual legumes in Syria. These authors undertook comprehensive fieldwork over several years, during which they gathered extensive ecogeographic data and were able to use these data to predict potential areas of conservation. They divided Syria into climatic regions and then recorded the percentage of sites for each region where each annual legume species was found. The authors also studied the percentage of sites of each soil type in which various taxa were found. The influence of both climatic factors and soil alkalinity on the distribution of various species was clearly demonstrated.

Such methods deal with one environmental factor at a time, or a single morphological variable. Ecogeographic data, however, are multivariate, in that two or more items of data are available for each record (e.g. each collecting site, germplasm accession or herbarium specimen). Ehrman and Cocks (1990) used various methods of cluster analysis on their environmental data to classify the collecting sites into groups or classes (clusters) the members of which had climates which were more similar overall (rather than as regards any one single variable) to one another than they were to members of any other class. A more detailed discussion of multivariate data analysis is provided in Chapter 15.

Another approach to the study of ecogeographic data involves

mapping collecting sites. These distribution maps can be used in conjunction with topographical, climate, geological or soil maps. Stace
(1989) stresses the importance of the means of visually displaying plant
distribution. This can take two forms: (i) shading or enclosing an area
with a single line, or (ii) using various kinds of dot-distribution maps.

The use of an enclosing line is ambiguous, as it provides no indication of the frequency of the taxon within the region. A single outlying
specimen might erroneously suggest that the taxon is continuously present throughout an entire region. The occurrence of a species is often
sparse at the periphery of its range and there is rarely a distinct cut-off
line. Indicating presence in this manner also means that any variation
due to local ecological and geomorphological factors within the individual provinces or countries cannot be shown. The problems associated
with enclosed line maps can be illustrated with an example taken from
Edmonds' (1990) ecogeographic survey of African *Corchorus*. Figure 14.2
shows the distribution of *C. aestuans*. A crude enclosed line map would
shade the entire area from northwestern Uganda to southern Tanzania,
which would not bring out the fact that in Uganda the species is represented by a single record and the majority of records are from southeastern Kenya and central and eastern Tanzania. Enclosing line maps
do have advantages, however. Westman (1991), for example, used
such maps in conjunction with climatic isoline maps to calculate the

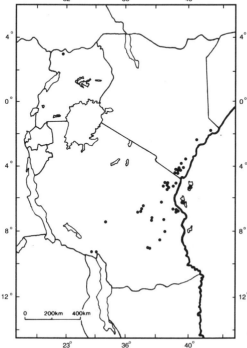

Fig. 14.2. Dot-distribution map for *Corchorus aestuans* in east Africa (taken from Edmonds, 1990).

percentage of the total areal distribution of various species in each climatic category in California.

Enclosing line maps can be used to indicate concentration of species. Such isoflor maps do not show actual species distributions: each line is a contour delimiting a greater or lesser concentration of species. Species distributions are superimposed on to a single map then contour lines are drawn around areas of the map with the same number of species. An example of an isoflor map for *Vicia* sect. *Narbonensis* is shown in Fig. 14.3 (Maxted, 1990). Of course, isoflor maps can be produced for infraspecific taxa within species (e.g. subspecies, crop landraces) as much as for species within sections or genera.

To represent distribution patterns in detail there is a general trend towards the use of dot-distribution maps (Stace, 1989). These may be of two types: (i) presence is indicated in subregions (e.g. grid squares) within the study region, or (ii) spots represent actual localities. The former option places the onus on the researcher to determine the presence or absence of the taxon in each square. This ensures evenness of coverage, but will be impractical if the target region is large and time allowed for the study limited. Species mapping is discussed in detail by Miller (1994).

Morphological or ecological information can also be superimposed

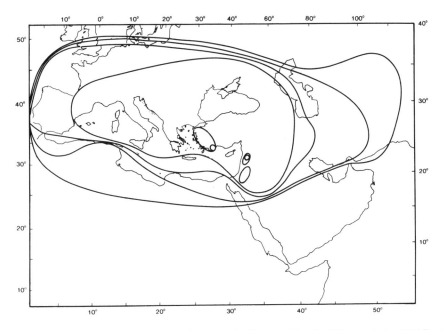

Fig. 14.3. Isoflor map for *Vicia* sect. *Narbonensis* in Europe, North Africa and the Middle East (taken from Maxted, 1990). Each contour line encloses one more species than the one immediately outside it.

on to a dot-distribution map, as Strid (1970) has done for the uppermost internode length of various populations of *Nigella arvensis* in the Aegean region. The position of a rectangle indicates the location of the population, while its height shows the relative length of the internode. Pie charts can be used to display the relative frequency of a character in different places. They are commonly used to show allelic frequencies in populations at different geographic locations, but have also been used to display morphological variation, e.g. stigma colour in *Crocus scepusiensis* (Rafinski, 1979). New (1958) used pie charts to demonstrate seed-coat variation in *Spergula arvensis*, the relative size of the circles indicating the size of the population sampled; different sized circles could also be used to indicate the number of database records from which the frequency has been estimated. Pie charts can also be used to compare the distribution of specimens with physical characteristics, e.g. altitude, temperature or soil type. Daday (quoted by Jones, 1973) used pie charts to show the relationship between the distribution of cyanogenic forms of *Lotus corniculatus* in Europe and January isotherms. Again, there is more on this in Chapter 15.

Mapping software can considerably simplify the production of distribution maps. Large scale maps of the world are available in digitized form, but if more detailed maps are required they may need to be customized or even digitized from scratch. Mapping programs will allow the user to import latitude and longitude coordinates from the ecogeographic database and plot them on to customized maps. The import facility allows the locality details to be transferred from the ecogeographic database directly, without re-entering data, while the ability to customize the base map allows a suitable scale to be used to display the distribution. More sophisticated mapping programs have built-in databases. These are called geographic information systems (GIS), computer hardware and software packages designed to store, analyse and display spatially referenced data (Haslett, 1990). The development of GIS will increasingly prove useful in the analysis of ecogeographic data. Chapter 16 deals with GIS in more detail.

Phase III – Product generation

Data synthesis

The final production phase of the project commences with the synthesis of all the disparate data collected during the study. The researcher should be aware of the degree of completeness of the database, or the collections on which it was based, in terms of how fully the target region has been effectively covered. If a particular habitat is under-represented in the database, it may be because the taxon is really not found there or because that habitat has not been sampled, or even because the target taxon has not been recognized in such a habitat. This problem must be considered if the results of the analysis and the inferences drawn from them are not to be misleading.

The database contents are summarized, together with the other data abstracted from the literature, into an ecogeographic conspectus. The pattern of the data included in both the database and conspectus can then be interpreted, the results of the analysis displayed and suggestions for appropriate target taxa and regions which warrant conservation discussed in the ecogeographic report.

The ecogeographic database, conspectus and report

The ecogeographic database, conspectus and report are the essential products of an ecogeographic study. The ecogeographic database contains the raw data. The conspectus summarizes the available taxonomic, geographic and ecological information for the target taxon through part or the whole of its range. The report interprets the data held in the other products and will help the ecogeographer identify conservation priorities.

The conspectus is arranged by plant names, which can be listed either alphabetically or systematically. In both cases it is helpful to provide an index to the taxa included in the study. The conspectus should summarize information from both the database and the literature survey. An abbreviated version of the ecogeographic conspectus, basically an annotated checklist containing, in coded form, the information shown in bold below, will be useful in the field when it comes to the actual germplasm collecting (Hammer, 1991). If possible, the following information should be included in the full conspectus:

- **accepted taxon name**, author(s), date of publication, place of publication;
- reference to published descriptions and iconography;
- short morphological descriptions or keys for important taxa or those that may be difficult to identify;
- **phenology, flowering season**;
- ethnobotanical notes, especially **vernacular name** and **local uses**;
- **geographic distribution**, i.e. countries, provinces or districts from which the taxon is recorded, including reliable records from the literature;
- distribution maps (preferably dot-distribution) produced directly from the latitude and longitude data held in the database;
- geographic notes, including an interpretation of the taxon's geographic distribution;
- ecological notes, including: altitude (minimum and maximum); habitat, topographic, soil, geological, climate and microclimatic preference; biotic interactions;
- taxonomic notes, including notes on any distinct genotypic and phenotypic variation within the taxon;
- conservation notes, containing an assessment of the variation currently conserved *ex situ*, the potential genetic erosion faced and the conservation status of the taxon in the field.

If the scope of the investigation is broad in the geographic or ecological sense, it may be necessary to provide a summary of the ecogeographic data for each geographic or ecological subunit. This can be illustrated with reference to Edmonds' (1990) survey of African *Corchorus*. The survey covered a vast geographic region and to increase the predictive value of the survey she lists the flowering time for each species in each country rather than providing one time range for the whole of Africa.

A listing of the specimens used during the study and a summary of the synonyms which have recently and frequently been used in the target region may also prove useful. Both these listings would significantly expand the size of the conspectus and so should perhaps be included as appendices.

The ecogeographic report discusses the contents of the database and conspectus and must draw general conclusions concerning the group's ecogeography, presenting a concise list of conservation priorities. If possible, the following points should be covered:

- the delimitation of the target taxon;
- the classification of the target taxon that has been used, and why;
- the mode of selection of representative specimens;
- the choice of hardware and software;
- the ecogeographic database file structures and inter-relationships;
- database contents;
- target taxon ecology;
- target taxon phytogeography and distribution patterns, with a summary of the distribution in tabular form;
- any interesting taxonomic variants encountered during the study;
- current and potential uses of the target taxon;
- the relationship between the cultivated species and their wild relatives;
- any particular identification problems associated with the group (identification aids to vegetative, floral and fruiting specimens should be provided);
- *in situ* and *ex situ* conservation activities associated with the target taxon, including the extent of diversity already conserved;
- genetic erosion threat faced by the group;
- priorities and suggested strategy for future conservation of the target taxon.

As discussed earlier, it is less easy to obtain ecological data from herbarium specimen passport data than it is to obtain geographic or taxonomic data. This may hamper drawing firm ecological conclusions from the study products. However, whatever information is available is a valuable asset and will aid in selecting conservation priorities. An example of the level of detail that might be included in an ecogeographic conspectus is given in Appendix 14.1.

The ecogeographic survey may yield information that is not of direct use in the identification of conservation priorities for the target taxon but which may, at a later date, be incorporated into purely ecological, taxonomic or other products. Maxted found that, during the ecogeographic survey of the Vicieae and Cicereae of the southern republics of the former Soviet Union, he gathered a large quantity of data on localities (place names, latitude, longitude and altitude) within the region. This will form the basis of a gazetteer for the region, which will prove useful to subsequent ecogeographic studies.

Identification of conservation priorities

The principal aim of the ecogeographic survey must be to provide a sound basis for the identification of conservation priorities and strategies, which includes collecting priorities and strategies. During the survey process, data from the literature, herbarium specimens and germplasm accessions are collated, summarized and synthesized into the three ecogeographic products. The pattern of variation within the target region and the target taxon is investigated and an estimate of potential genetic erosion and current conservation status made. On the basis of the various products of the ecogeographic survey or study, the ecogeographer can formulate future conservation priorities and strategies for the target taxon.

Within the target region, areas may be identified which are of particular interest either because of the plants found there or because of local conditions, e.g. areas with high concentrations of diverse or endemic taxa, low rainfall, high frequency of saline soils or extremes of exposure. If a taxon is found throughout a particular region, then the researcher can use the ecogeographic data to actively select a series of diverse habitats to sample. If a taxon has been found at one locality, but not at others with similar ecogeographic conditions, then a possible suggestion is that these localities be searched. Within the target group, specific taxa, populations and variants can be identified which warrant specific consideration, e.g. poorly known taxa, species whose potential as crops has not previously been noted, populations with particular adaptations or rare and endangered taxa.

Having ascertained the level of variation within the target taxa and the potential target region in the process of ecogeographic data compilation and analysis, this information must be assessed in the context of current conservation activities. Is sufficient genetic material of a particular taxon from a particular, interesting ecogeographic niche already safely conserved either *in situ* or *ex situ*? If not, should effort be expended to collect this material? Analysis of herbarium material may indicate that there is a rare relative of the winged bean (*Psophocarpus tetragonolobus*) growing on a small edaphic enclave in western Kenya, but, if a review of conservation activities indicates that large collections of this species are conserved in the Kenyan National Gene Bank and the material is also duplicated elsewhere, re-collecting would probably not be justified.

The ecogeographic survey or study should conclude with a clear, concise statement of the proposed conservation strategy for the target group and proposed conservation priorities. Questions should be considered such as whether population levels should be monitored to assess the threat of genetic erosion, whether a national or international collecting team should be directed to collect the priority target taxa, whether it is possible to conserve the taxa *in situ* and so on. Edmonds (1990), for example, proposes five specific missions to collect rare and endangered *Corchorus* species. If the ecogeographic data have been gathered solely from herbarium specimens, the ecological data obtained may be insufficient to draw detailed conclusions on the target taxon's habitat preferences. A survey mission to the target region may therefore be advisable, to obtain a clearer idea of the pattern of genetic variation and of the appropriate habitat types and clarify the conservation strategy to be proposed.

Once specific areas have been selected for collecting, a route that covers the maximum number of such areas in the minimum time can be suggested. With many species there is a narrow 'collecting window' during which collectors must find the target population. If they are too early the seeds will not be ripe, too late and the material may have shattered. The phenological data (in combination with both climate averages and weather data from the year of collecting) will indicate approximately when a collecting team should visit the target region, an estimate that will need to be refined on the basis of up-to-the-minute information on local weather conditions (Chapter 16).

Conclusion

Herbaria, gene banks, botanical gardens (and their associated libraries) are storehouses of botanical data as much as of plants, pressed and live. These data can be used to facilitate plant conservation. Analysis of a taxon's ecology, geographic distribution and taxonomy is a necessary prerequisite for assessing its conservation status and permits the prediction of which areas and habitats the taxon is likely to be found in. Once located, populations of the taxon can be monitored, sampled if necessary and effectively conserved.

Ecogeographic studies will always be limited by time and resources and it will be impossible to collate every piece of information available. However, if the study is planned carefully and undertaken efficiently, the data that are collated can be predictive. The results of an ecogeographic study will not always lead the collector to the exact localities of desired plant populations, although this is by no means impossible, but they can certainly identify the likely areas of current occurrence. The ecogeographic database and conspectus will also provide sufficient information to permit the conservationist to assess collecting and conservation priorities.

Acknowledgements

We are particularly grateful to T. Hodgkin, D.G. Debouck and L. Guarino for their comments on this manuscript and to R. Reid for discussion of the concept of ecogeographic studies. M. Perry and E. Bettencourt contributed substantially to the section on database design. K. Painting wrote the section on the Internet. D.M. Spooner added some important references.

References

Bisby, F.A. (1984) The Vicieae Database Project: products and services. *Webbia* 38:639–644.

Brummitt, R.K. and C.E. Powell (1992) *Authors of Plant Names*. Royal Botanic Gardens, Kew.

Cocks, P.S. and Ehrman, T.A.M. (1987) The geographic origin of frost tolerance in Syrian pasture legumes. *Journal of Applied Ecology* 24:678–683.

Date, C.J. (1981) *An Introduction to Database Systems*. Addison-Wesley, Reading.

Davis, P.H. and Heywood, V.H. (1973) *Principles of Angiosperm Taxonomy*. Krieger, New York.

Davis, S.D., Droop, S.J.M., Gregerson, P., Henson, L., Leon, C.J., Lamlein Villa-Lobos, J., Synge, H. and Zantovska, J. (1986) *Plants in Danger: What Do We Know?* IUCN, Gland.

Edmonds, J.M. (1990) *Herbarium Survey of African* Corchorus L. *Species*. Systematic and Ecogeographic Studies on Crop Genepools 4. IBPGR, Rome.

Ehrman, T. and Cocks, P.S. (1990) Ecogeography of annual legumes in Syria: distribution patterns. *Journal of Applied Ecology* 27:578–591.

Fhrman, T.A.M. and Maxted, N. (1990) Ecogeographic survey and collection of Syrian Vicieae and Cicereae (Leguminosae). *FAO/IBPGR Plant Genetic Resources Newsletter* 77:1–8.

Fitter, R. and M. Fitter (eds.) (1987) *The Road to Extinction. Problems of Categorizing the Status of Taxa Threatened with Extinction*. IUCN and UNEP, Gland.

Forman, L. and D. Bridson (1992) *The Herbarium Handbook*. Revised edition. Royal Botanic Gardens, Kew.

Frodin, D.G. (1984) *Guide to the Standard Floras of the World*. Cambridge University Press, Cambridge.

Funk, V.A (1993) Uses and misuses of floras. *Taxon* 42:761–772.

Hammer, K. (1991) Checklists and germplasm collecting. *FAO/IBPGR Plant Genetic Resources Newsletter* 85:15–17.

Haslett, J.R. (1990) Geographical Information Systems: a new approach to habitat definition and the study of distributions. *Tree* 5:214–218.

Hintum, Th.J.L. van (1993) A computer compatible system for scoring heterogeneous populations. *Genetic Resources and Crop Evolution* 40:133–136.

Hollis, S. and Brummitt, R.K. (1992) *World Geographical Scheme for Recording Plant*

Distributions. Hunt Institute for Botanical Documentation, Pittsburgh, for the International Working Group on Taxonomic Databases for Plant Sciences.

Holmgren, P.K. and N.H. Holmgren (1992) Plant specialists index. *Regnum Vegetabile* 124:1–394.

Holmgren, P.K., Holmgren, N.H. and Barnett, L.C. (1990) *Index Herbariorum I: The Herbaria of the World*. 8th edition. International Association for Plant Taxonomy and New York Botanical Garden, New York.

IBPGR (1985) *Ecogeographic Surveying and* In Situ *Conservation of Crop Relatives*. Report of an IBPGR task force meeting held at Washington DC. IBPGR, Rome.

International Standards Organization (1981) *Listing of ISO Codes*. ISO 3166. British Standards Institute, London.

IUCN (1994) *Draft IUCN Red List Categories*. Version 2.2. IUCN, Gland.

Jones, D.A. (1973) Co-evolution and cyanogenesis. In: Heywood, V.H. (ed.) *Taxonomy and Ecology*. pp. 213–242. Systematics Association Special Volume No. 5. Academic Press, London.

Krol, E. (1992) *The Whole Internet User's Guide and Catalog*. O'Reilly & Associates Inc., Sebastopol CA.

Mace, G., N. Collar, J. Cook *et al.* (1992) The development of the new criteria for listing species on the IUCN red list. *Species* 19:16–22.

Maxted, N. (1990) An herbarium based ecogeographic study of *Vicia* subgenus *Vicia*. Internal report. IBPGR, Rome.

Maxted, N. (1991) A revision of *Vicia* subgenus *Vicia* using database techniques. PhD thesis. University of Southampton, Southampton.

Maxted, N. (1993) An ecogeographic study of *Vicia* subgenus *Vicia*. Internal report. IBPGR, Rome.

Miller, R. (ed.) (1994) *Mapping the Diversity of Nature*. Chapman and Hall, London.

New, J.K. (1958) A population study of *Spergula arvensis* L. *Annals of Botany, New Series* 22:457–477.

Painting, K.A., M.C. Perry, R.A. Denning and W.G. Ayad (1993) *Guidebook for Genetic Resources Documentation*. IBPGR, Rome.

Pankhurst, R.J. (1991) *Practical Taxonomic Computing*. Cambridge University Press, Cambridge.

Peters, C.R., E.M. O'Brien and R.B. Drummond (1992) *Edible Wild Plants of Sub-Saharan Africa*. Royal Botanic Gardens, Kew.

Rafinski, J.N. (1979) Geographic variability of flower colour in *Crocus scepusiensis* (Iridaceae). *Plant Systematics and Evolution* 131:107–125.

Rhoades, A.F. and L. Thompson (1992) Integrating herbarium data into a geographic information system: requirements for spatial analysis. *Taxon* 41:43–49.

Rihan, J.R. (1988) An herbarium based ecogeographic survey of forages *Medicago* species in the Mediterranean and adjacent arid/semi-arid areas. Internal report. IBPGR, Rome.

Room, A. (1979) *Place Name Changes Since 1900*. Scarecrow Press, London.

Sánchez, J.J.G. and Ordaz, S. (1987) *El Teocintle en Mexico*. Systematic and Ecogeographic Studies on Crop Genepools 2. IBPGR, Rome.

Slageren, M.W. van (1990) *Genetic Resources Unit. 1989 Annual Report*. ICARDA, Aleppo.

Stace, C.A. (1989) *Plant Taxonomy and Biosystematics*. Edward Arnold, London.

Strid, A., (1970) Studies in the Aegean flora. XVI. Biosystematics of the *Nigella arvensis* complex with special reference to the problem of non-adaptive radiation. *Opera Botanica* 28:1–169.

Stressey, T.F. (1993) The role of creative monography in the biodiversity crisis. *Taxon* 42:313–321.

Tatian, P.A. (1993) The use of computers in field research. *Journal for Farming Systems Research–Extension* 3:97–118.

Times Books (1988) *Atlas of the World.* 7th edition. Times Books, London.

Westman, W.E. (1991) Measuring realized niche spaces: climatic response of chaparral and coastal sage scrub. *Ecology* 72:1678–1684.

Appendix 14.1
An example of ecogeographic conspectus construction (Maxted, 1993)

Vicia L. sect. *Atossa* (Alef.) Asch. & Graebner, Syn. Mitteleur. *Fl.* 6(2):949 (1909).

Ref. Pub. Description: Kupicha, *Notes Roy. Bot. Gard. Edinburgh* 34:320 (1976).

Perennial; erect or climbing; stem slender or stout. Stipules entire or semi-hastate; 2.5–9 x 1–5 mm; edge entire or with 1–6 teeth. Leaf 25–154 mm; apex tendrilous or mucronate; 2–28 leaflets per leaf; leaflet 10–85 x 3–38 mm; symmetric; margins entire. Peduncle 7–32 mm; with 1–8 flowers. Calyx mouth oblique; lower tooth longer than upper; base gibbous; pedicel 1–3 mm. Flowers 12–22 mm; all petals approximately equal length; standard yellow, blue or purple; shape platonychioid (limb and claw same width); claw bowing absent; upper standard surface glabrous. Wing marking absent; wing limb with slight basal folding. Legume 16–43 x 6–9 mm; oblong; laterally flattened; sutures straight or curved; valves glabrous (without hairs); septa absent; 2–7 seeds per legume. Seeds 2.5–4 x 3–5 mm; round or oblong; not laterally flattened; hilum over half seed circumference; lens positioned near hilum; testa surface smooth.

Number of taxa: 4

Chromosome numbers: 12, 14, 16, 18

Geographic distribution: Europe, Near East and Asia eastward to the Pacific.

Geographic notes: This section is relatively widely distributed from Iceland to Japan, but the extent of the distribution is largely due to a single species, *V. sepium.* The other three species of the section are much more restricted. There are two centres of diversity, one concentrated in northern Yugoslavia and the other in the Caucasus.

Ecological notes: All four species are encountered in similar habitats, stable semi-shaded woodland (coniferous, mixed or deciduous), woodland edges or hedgerows. They show no preference for a particular soil type, but are more commonly encountered at altitudes over 500 m (except *V. sepium*). The four species can be found in open or dense vegetation, under dry or moist conditions.

Taxonomic notes: The four species easily form three series. *V. balansae* and *V. abbreviata* are closely related.

Series *Truncatulae* (B. Fedtsch. ex Radzhi) Maxted, *Kew Bull.* 47(1):130 (1991).

Number of taxa: 2

Chromosome number: 14

Geographic distribution: southeast Europe and west Asia.

Geographic notes: The two species of this series are commonly found in northeast Turkey and the Caucasus, though *V. abbreviata* is also found further west in southeast Europe.

Taxonomic notes: Stankevich (1988) considers the two taxa of this series to be subspecies of *V. abbreviata*. After studying natural populations in the Caucasus she concluded that the two taxa intergrade from one to the other. This, she considered, was especially apparent in the subalpine zone between Karmadon and Chmi in North Ossetia, Russia. She argues that the two taxa have been able to remain morphologically distinct due to their preference for different ecological niches. While collecting in the Caucasus (spring–summer 1989), I located six populations of *V. balansae*. At five of these localities, *V. abbreviata* was equally abundant. Within the five sites, where both species were found, neither species showed a clear niche distinction and no putative hybrid forms were encountered. Therefore Maxted (1993) concluded the specific distinction should be retained.

V. balansae Boiss., *Fl. Orient.* 2:569 (1872)

Ref. Pub. Description: *Fl. Tur.* 3:304; *Fl. USSR* 13:457; *Illust. Fl. Iran* 32, Fig. 3.

Phenology: May–August

Chromosome number: 12, 14

Geographic distribution: SUN, TUR.[2]

Ecology: Alt. 550–2700 m; Hab. moist alpine pastures and forests.

V. abbreviata Fischer ex Sprengel, *Pl. Min. Cog. Pug. Prim.* 1(86):50 (1813).

Common synonym: *V. truncatula* Fischer ex Bieb., *Fl. Taur.-Cauc.* 3:473 (1819).

Ref. Pub. Description: *Fl. Iran.* 43–44; *Fl. Tur.* 3:303–304; *Fl. USSR* 13:456–457; *Illust. Fl. Iran.* 32, Fig. 3.

Phenology: May–August

Chromosome number: unknown

Geographic distribution: AUT, BGR, DEU, IRN, ROM, SUN, TUR, YUG.

Ecology: Alt. 100–2400 m; Hab. mountain forest and forest margins.

---

[2] These are ISO country codes.

Useful addresses

Harmonization of
  Environmental
  Measurement
  Programme
  (UNEP–HEM)
c/o GSF – Research
  Centre for
  Environment and
  Health
Ingolstädter
  Landstrasse 1
W-8042 Neuherberg b.
  München
Germany
Tel: +49 89 31875488
Fax: +49 89 31873325
E-mail: keune@gsf.de

International Union of
  Biological Sciences
  Commission on
  Taxonomic Databases
  (TDWG) Secretariat
Dr James L. Zarucchi
Missouri Botanical
  Garden
PO Box 299
St Louis
Missouri 63166
USA

Oxford Forestry
  Institute
University of Oxford
South Parks Road
Oxford
UK

Mapping the ecogeographic distribution of biodiversity

15

L. Guarino[1]
Worked example: G. Maggs[2] and L. Guarino
[1]*IPGRI, c/o International Laboratory for Research on Animal Diseases, PO Box 30709, Nairobi, Kenya:* [2]*National Botanical Research Institute, Private Bag 13184, Windhoek, Namibia.*

Diversity is not evenly distributed within the geographic range of plant gene pools. This has been recognized since Vavilov described centres of origin and of diversity for a number of crops and has been amply demonstrated in numerous recent studies of variation in both crops and wild species in morphological, physiological, biochemical and molecular traits, though patterns at these different levels do not always coincide (Chapter 6). Some areas are richer in taxa within a given gene pool than others, a given species may be more genetically diverse in some areas than in others, ecotypes and clines have developed within species, variation in a species may be concentrated within or among populations, some alleles and characters are found in some populations and not others. Some of this heterogeneity in the distribution of diversity is explained by selection (natural and artificial) and gene flow and reflects ecogeographic adaptation and the preferences of farmers. Some is due to chance factors such as founder effect and genetic drift. Not only is diversity not uniformly distributed in ecogeographic space, it is also not uniformly distributed taxonomically, for example among the primary, secondary and tertiary gene pools. These contain widespread as well as narrowly endemic, and variable as well as relatively homogeneous, taxa.

It is up to the collector to use knowledge of the actual or presumed distribution of species and of genetic diversity within gene pools, and of how effectively this has been covered by previous collecting, to plan expeditions which will be efficient in the amount of new and useful diversity that they obtain. In particular, to plan a cost-effective mission, the collector must identify areas:

- where particular material of interest is known or suspected to occur (e.g. taxa missing from germplasm collections, material with particular adaptations or characters);

- which have high species or genetic diversity;
- which are significantly different from each other agriculturally, ecogeographically, floristically (i.e. in terms of representation of different taxa) or genetically;
- which have been undercollected;
- which are under threat of genetic erosion.

Other chapters are concerned with obtaining and organizing ecogeographic data. This chapter deals with some map-based techniques that may be used to explore the available data to identify the kinds of areas listed above. Chapter 5 discusses how such an analysis can be used to modify a basic collecting strategy, in particular how much extra weighting should be given to different areas and what the dangers of such bias are likely to be. Analysis of ecogeographic data will clearly be much simpler if a computer is used; some analyses, indeed, would be almost impossible without one. However, an attempt is made here to document manual, relatively low-technology solutions whenever possible. A geographic information system (GIS) is by no means essential to carry out many of the analyses described. The use of GIS technology is described in Chapter 16.

Preliminary data handling

Relevant data on germplasm accessions, herbarium specimens and other records of plant populations will include passport, characterization and evaluation descriptors. Passport data include both population descriptors and collecting site descriptors. There will also be data from sites located in the collecting region but where no collecting took place, e.g. climatic data from meteorological stations. Preliminary manipulation and analysis of some of these data prior to mapping may be useful. Each 'object' (plant population, collecting site or other kind of site) will often have data for several descriptors associated with it. Passport data will include information on a number of different aspects of the plant population and collecting site ecogeography, including climate, soil and vegetation. Germplasm accessions may have been characterized for various morphological descriptors or genetic loci. It may be possible to measure a number of morphological characters on herbarium specimens. Meteorological stations will have data on several temperature and rainfall parameters. So-called multivariate statistical techniques will therefore be particularly relevant for data analysis. These aim to simplify such multidimensional data in various ways.

Objects may be classified using the multivariate technique of *hierarchical clustering*. This will define groups of objects, called clusters, such that the overall similarity between two objects within the same cluster (taking all descriptors into account simultaneously) will be greater than that between two objects in different clusters. Clustering may be agglomerative (in which objects are joined into progressively larger clusters) or divisive (in which the group of all the objects is progressively divided). In general, the agglomerative method will bring out the best groupings and the divisive method will highlight discontinuities. The former has been most commonly used in numerical taxonomy, the latter in phytosociology and will not be discussed further here (but see below for references).

The first step in agglomerative clustering is measuring the similarity or dissimilarity (distance) between two objects. The method depends on the type of data (Dunn and Everitt, 1982):

- If all descriptors are recorded on a nominal scale (i.e. with discrete states the order of which is arbitrary, as may be the case with leaf shape, for example), the simple matching coefficient is usually used. This is defined as the number of characters that match, divided by the total number of characters.

- In the case of binary descriptors, negative matches may not be considered meaningful, and the Jaccard coefficient may be more useful. This is defined as the number of positive matches, divided by the number of mismatches.

- If all descriptors are continuous (or discrete and multistate, but with the order of states meaningful), Euclidean distance is commonly used as a measure of dissimilarity. Based on Pythagoras' theorem, it is calculated by adding up the squares of the differences between the two objects for each descriptor, and taking the square root of the sum.

- If some descriptors are nominal and some continuous, Gower's general similarity coefficient may be used (Gower, 1971).

There are other similarity and distance measures (Clifford and Stephenson, 1975; Dunn and Everitt, 1982), and software packages that perform multivariate analysis often allow a wide choice. Nei (1975; 1987) gives definitions of genetic identity (I) and distance (D) among populations based on allele frequency data, as might be produced by biochemical or molecular studies (see also Hoelzel and Dover, 1991; Hoelzel and Bancroft, 1992).

Having calculated similarities among all pairs of objects, the most similar objects are united into a cluster and similarities recalculated between this cluster, taken as a new object, and all other objects. The process is then repeated until all objects are in one cluster. To produce geographically compact clusters, a so-called contiguity constraint can be imposed on the clustering, such that two objects which would have been grouped into a cluster on the basis of their overall similarity are only actually allowed to do so if they are also geographically closer than a predetermined distance (Charmet, 1993). There are several different ways of choosing from which point within two clusters to measure the similarity between them (the sorting strategy), for example nearest neighbour, furthest link and centroid. Different combinations of similarity measure and sorting strategy will give slightly different results and the final choice among them is usually subjective. Pankhurst (1991) discusses different ways of objectively comparing and/or combining different clustering methods. Again, clustering programmes normally allow a choice of methods.

A second type of multivariate statistical technique that is often useful in the analysis of ecogeographic data is *ordination* or *multidimensional scaling*, of which the best-known example is principal components analysis (PCA). It will not explicitly produce groupings of objects, as clustering does, but will bring out the overall similarity among them by defining new variables, independent among each other, which are linear combinations of the original variables, but which differ in that the correlations among them, and therefore the redundancies, are taken into account. In this way, a long list of descriptors may be reduced to just a few new 'synoptic' variables (the first few so-called principal component axes), which despite being fewer nevertheless conserve most of the variability present in the original descriptors. The analysis will reveal how much of the original overall variability is represented by each principal component; the first will account for the largest single amount, the second for the second largest, and so on.

PCA is suitable only for quantitative data (but see Hill and Smith, 1976), as it is based on a matrix of the correlations among variables. However, a similarity matrix of the same kind as is used in clustering, which may be calculated whether descriptors are qualitative or quantitative, can also be used to plot the position of the objects in two- or three-dimensional space, in this case by using principal coordinates analysis (PCO). A minimum spanning tree of the similarity matrix superimposed on such a plot will reveal how well the original similarities are preserved (Dunn and Everitt, 1982).

Another widely used ordination technique is canonical variate analysis (CVA). This is similar to PCA but differs in that objects must be assigned to groups a priori and the analysis produces new axes (canonical variates, the equivalent of principal components in PCA) such that maximum discrimination is obtained not among objects, as in PCA, but among groups when tested against the variation within groups. Each a priori group could be a different taxon, a set of landrace accessions from different countries or a set of accessions of a wild species collected on different soil types, for example. Each group could also be a population, with observations on a number of different individuals per population being recorded.

Other relevant ordination techniques are non-metric multidimensional scaling (Kruskal, 1964; Clifford and Stephenson, 1975) and detrended correspondence analysis (Hill and Gauch, 1980; Gauch, 1982). The latter has been much used in community ecology but also has application in ecogeography (Myklestad and Birks, 1993) and, indeed, numerical taxonomy.

A brief introduction to multivariate statistical techniques as they have been used to describe and analyse plant variation is provided by Stuessy (1990). A more detailed but still relatively brief account is provided by Dunn and Everitt (1982). Pankhurst (1991) is also useful. Sneath and Sokal (1973; updated by Sokal, 1986), Blackith and Reyment (1971) and Clifford and Stephenson (1975) are standard works. The manuals of software packages are also good sources of information on the relative merits, capabilities and competencies of different methods. There are programs available which carry out just one kind of analysis, for example DECORANA (Hill, 1979a), TWINSPAN (Hill, 1979b) and CANOCO (Braak, 1988). Other program packages carry out a variety of multivariate analyses; examples include CLUSTAN IV, SYN-TAX III and NTSYS (Pankhurst, 1991). More general statistical packages with multivariate options include GENSTAT, SAS, UNISTAT and STAT-GRAPHICS. It is not possible to carry out some multivariate statistical procedures if there are gaps in the data set. Though some packages do provide routines for estimating missing data values, this can be a major problem in the use of multivariate techniques.

Data on the variation in morphological characters within populations, already mentioned in connection with CVA, are not often available in herbarium-based studies. Similarly, variation within germplasm accessions is only occasionally measured for morphological characterization descriptors (frequently the character is simply noted as being variable), though more often for evaluation descriptors (but see Hintum, 1993). However, as mentioned earlier, various biochemical and molecular techniques can be used to calculate genetic similarities among populations. Different measures of genetic diversity can also be calculated from such data both for populations and for groups of populations (Nei, 1975, 1987; Brown and Weir, 1983; Hoelzel and Dover, 1991; Hoelzel and Bancroft, 1992). Some examples are given in Box 15.1. One-locus gene diversity (h) may be averaged over all the loci studied to obtain a measure of mean gene diversity in the population (H_T). This is perhaps the most widely used genetic diversity parameter.

Mapping ecogeographic data

Point maps

Having carried out any appropriate preliminary analysis of the variation among plant populations and/or collecting sites, the next step will usually be the plotting of distribution data and the visual interpretation of the resulting maps. Chapter 14 discusses the different kinds of distribution maps that can be prepared and suggests that point maps showing exact localities are usually the most appropriate in the context of planning germplasm collecting missions. Though they are certainly the simplest starting-point, and thus the most logical, they are capable of considerable elaboration, enabling the display of more information

---

Box 15.1
Some parameters used to measure genetic diversity

Percentage of polymorphic loci: P

Mean number of alleles per locus: n_e

Within-population gene diversity at a locus: $h = 1 - \Sigma p_i^2$

where p_i is the frequency of the ith allele or haplotype

---

than just geographic location. This additional information on each plant population could include:

1. the type(s) of plant record(s) associated with it;
2. collecting site and population data;
3. the taxonomic category to which the population belongs;
4. the state or value of characterization or evaluation descriptors, taken one at the time;
5. membership of classes defined by combinations of characterization and evaluation descriptors;
6. the value of 'synoptic' characterization and evaluation descriptors;
7. within-population variation in characterization and evaluation descriptors.

Type of plant record

A first and basic elaboration of simple point distribution maps is to show the location of germplasm samples and of herbarium specimens (or literature records) in different ways, usually using two different symbols. It is most effective to represent the former by a filled symbol and the latter by an empty one, perhaps plotting them on separate transparencies, which can then be overlaid on the base map. Areas where the target taxon is known to occur but from which no germplasm is available often stand out very clearly in such an exercise.

Collecting site and population data

Plant records may also be distinguished according to various characteristics of the site and population to which they refer, i.e. passport data. Site data such as soil type or altitude, for example, could be shown by using symbols of different shapes, colours, shades or sizes (Fig. 15.1a, b). Such parameters as the size of the population and the threat of genetic erosion it faces could be shown in similar ways. In some cases, it may be useful to distinguish plant records on the basis of date, for example to show changes in distribution over years or differences in flowering time in different areas.

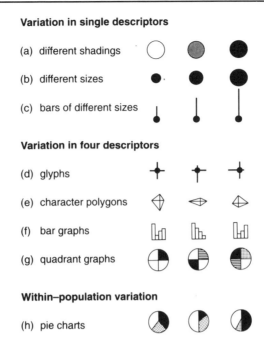

Fig. 15.1. Some symbols which may be used on dot maps to show variation in populations. See text for details.

Membership of different taxonomic categories

Different 'taxa', taken in a wide sense to include not just formal categories like species in a genus but also cultivar groups or landraces in a crop (Chapter 7), may also be displayed on the same map by using different symbols. Such a technique is useful in showing areas of high taxonomic diversity and areas where rare or endemic taxa may be found. Symbols must be chosen which can be superimposed and still remain identifiable, so that co-occurrence can be shown, and which can be filled to represent germplasm accessions or left unfilled to represent other kinds of records. If different symbols are being used to show site or population data, different taxonomic categories could be shown by lines emerging from the symbol at different angles (glyphs, see below).

Characterization or evaluation descriptors taken one at the time

Data on physiological, morphological or genetic descriptors can be added to a point map to investigate the spatial distribution of diversity. If a single descriptor measured on a discrete scale is being considered, its different states may be shown by different symbols or by the same symbol filled in with different shadings, patterns or colours (Fig. 15.1). The symbol may reflect the descriptor very closely. Variation in leaf shape among populations, for example, may be shown on a map by reproducing small

representations of the actual leaf shape at each site of observation, and
flower colour by differently coloured dots. If the descriptor is measured
on a continuous scale, variation may be shown by the same symbol at
different sizes or with different-sized bars emerging from it (Fig. 1b, c).

Membership of classes defined by combinations of descriptors

Combinations of plant descriptors may be displayed by lines (glyphs)
emerging at different angles to the central dot (or different symbols, if
site or population data are also being displayed), so that the length of
a line at 45° to the vertical might represent mean petiole length in the
population and that of a line at 90° mean leaf length, for example, and
so on (up to a maximum of eight characters in this case). These are called
radial bar graphs or rose diagrams (Richardson, 1985). A recent example
of their use is Francisco-Ortega et al. (1993). An example involving four
characters is given in Fig. 15.1d. The ends of adjacent glyphs may be
joined up and the enclosed 'character polygon' filled in for ease of comparison (Fig. 15.1e). An example of their use is given by Jones (1972).
Variation in a number of characters may also be shown in side-by-side
bar graphs, as in Fig. 15.1f and Morley (1971).

Another method that has been adopted to display the distribution
of more than one character on maps is to use a suitably large circle to
mark localities and divide this up into two or more equal slices, each
representing a descriptor (Fig. 15.1g). The descriptor state can be shown
by filling in the appropriate quadrant or leaving it unfilled (in the case
of binary descriptors), or by the use of different shadings, patterns or
colours in the case of multistate descriptors (e.g. Seyani, 1988). Both
these methods can also be used to show the co-occurrence of taxa. A
plant population's membership of different clusters may be shown on
point maps, again by using different symbols, as if it were a descriptor
with discrete states.

'Synoptic' descriptors

The score of each plant population along each of the first few principal
components (usually the first two to four) can be mapped as any other
continuous descriptor, though each principal component in fact represents a combination of several of the original descriptors, each differently weighted (the exact weights will be specified in the analysis
output). Plant populations with similar principal component scores will
tend to be similar for the original descriptors contributing most to that
principal component.

Within-population variation

The pattern of distribution of within-population variation for continuous
variables may be displayed by mapping a measure of dispersion such as
standard deviation in the same way as any other continuous variable,
for example with differently sized dots. One could thus have a point map
of mean leaf length of populations, say, along with a point map of the

standard deviation of leaf length within each population. Values of the genetic parameters in Box 15.1 may also be mapped in the same way as any other continuous descriptor. For descriptors with discrete states, including such biochemical data as the occurrence of different alleles at a locus, pie charts giving the frequency of each state in the population may be used (Fig. 15.1h).

Contour maps

The distribution of classes of plant populations (be the classes defined taxonomically, by a single character, by the joint occurrence of characters, by ordination or by clustering) can be related to environmental factors by searching the map visually for associations. For example, mean leaf length in a species, as shown by the length of a particular glyph, may seem to be associated with higher altitudes or higher rainfall, as shown by the size of the central dot. Contour, or isoline, maps may facilitate the identification of such patterns.

To produce contour maps, a line is drawn joining all the points where the variable of interest takes a particular value and the process is repeated for a set of (usually) equally spaced values (Richardson, 1985; Burrough, 1986). When the variable is number of taxa, the contour lines are sometimes called isoflor lines (Chapter 14). One could also draw lines of equal genetic diversity; a suitable name might be isodive lines. Lines linking populations with similar values for morphological or other kinds of plant characters are called isophenes. For climate data, lines of equal temperature are called isotherms, lines of equal rainfall isohyets. The particular advantage of contour maps 'lies in the fact that they may be easily correlated with environmental and/or geological variables, and intercorrelations between these surfaces (whether they be morphological characters, chemical characters, abstract factors, etc.) are readily visualized by the reader' (Adams, 1970). What contour maps do is render a continuous variable discrete. Associations between discrete variables are often much easier to notice than relationships between continuous variables.

It is not impossible to draw contour lines by eye. Indeed, up to relatively recently that was the only way to draw them. The development of computer mapping has meant that the subjectivity that was the main drawback of such an approach has been overcome (Adams, 1970). Whether done by hand or by computer, the first step in the generation of contours is the formation of a regular mesh of points for which values must be estimated on the basis of interpolation from the points where measurements were actually made. Mesh point values are then ranked consecutively so that appropriate isoline values can be chosen. Finally, the isoline values are located by interpolation between mesh point values and the resulting jagged lines smoothed (Richardson, 1985). The areas between contour lines are often filled in using shadings or patterns of increasing intensity. There are different methods of interpolation (e.g. Adams, 1970; Burrough, 1986), and software mapping packages will specify which one is being used and may offer a choice.

Using different base maps

If point distribution maps (however elaborate) or contour maps are drawn on a clear medium such as very thin paper or acetate transparencies, they can be used in conjunction with different base maps, for example not just topographic maps, but also ones showing agroclimatic zones, soils and vegetation types. This allows one to look for associations of distribution, characterization and evaluation data with different environmental features without having to plot distributions anew each time. This overlaying facility is of course one of the most useful features of a GIS. The trouble with the low-technology solution is that maps come in a variety of scales and projections. A photocopier with a reducing facility is very useful in overcoming the former problem. A pantograph or an overhead projector may also be used to change the scale of maps. The problem of projections is much more complicated and the only really viable methods are computer-based. Most GIS packages have the capability of making maps comparable in scale and projection.

Further analysis

Visual analysis of point and contour maps can be taken quite far. The human brain is reasonably good at distinguishing patterns, certainly better than most computers. However, various analyses can be carried out on mapped ecogeographic data to test any hypotheses on the factors controlling the pattern of distribution of diversity generated by visual inspection. Some of these have already been alluded to.

The association between a continuous characterization and/or evaluation descriptor (which could be scored along a principal component) on the one side (the dependent variable) and an environmental site descriptor (e.g. altitude) on the other (the independent variable) may be tested using regression analysis. Multiple regression may be used when it is wanted to explain variation in a dependent variable with reference to more than one independent variable. The ordination techniques of canonical correlation analysis and canonical correspondence analysis (Braak, 1986) may be used to explore the structure of a data set with multiple dependent variables as well as multiple independent variables. When the data are classified (as in contour maps), association may be tested using χ^2 or other non-parametric methods.

Mantel's test may be used to compare similarity matrices (Sokal, 1979; Heywood, 1991). Thus, a classification of plant population localities based on collecting site data may be compared with one of the populations themselves based on characterization and/or evaluation data. Autocorrelation statistics such as Moran's *I* may be used to quantify the morphological or genetic similarity between pairs of plant populations (or groups of plant populations) as a function of the geographic distance between them (Sokal and Oden, 1978; Heywood, 1991).

To combine more than one characterization or evaluation descriptor into a single contour map, different approaches are possible. One is to map contours of principal component scores. Another is to use *differential systematics*. Proposed by Womble (1951) and also described by Adams (1970) and Kirkpatrick (1974), this allows the simultaneous consideration of multiple contour maps. What is involved is the calculation of a slope (differential) perpendicular to the contour at each mesh point for each descriptor. This is a facility available in many GIS packages. The absolute values of the slopes for all descriptors are then summed and divided by the number of descriptors. Contours can then be drawn of this so-called 'systematic function'. The ridges on the contour map reveal areas where maximum change over distance is occurring and thus separate plant populations which are markedly different in the expression of the characters included in the analysis. Gentle clines do not show up as forcefully as more abrupt patterns, but areas where the systematic function is high are the ones where most variation is concentrated in the least space. In a similar way, genetic or morphological distances may be divided by geographic distances and the resulting matrix used to detect 'genetic boundaries' where several gene frequencies or characters vary abruptly (Monmonier, 1973; Pigliucci and Barbujani, 1991).

Areographic methods

How is one to pick out on a map the kinds of potentially high-priority areas listed at the beginning of the chapter? Some ways of dealing with the problem have already been mentioned, for example the use of isoflor maps and differential systematics for areas of high diversity. Nabhan (1990) provides a useful introduction to another type of approach, one based on so-called areographic methods.

Defining subregions

The first and fundamental step in areographic analysis is the definition of subregions within the general target or study region. It is by making comparisons among these with regard to parameters such as diversity and extent of previous collecting that it becomes possible to set priorities within a target region. Subregions could be demarcated by administrative boundaries, and numerous examples exist of the analysis of the distribution of variation in a crop among countries, for example, usually using CVA (e.g. Spagnoletti Zeuli and Qualset, 1987; Erskine *et al.*, 1989; Morden *et al.*, 1989; Porceddu and Damania, 1992). For full comparability, however, subregions should be 'equivalent', i.e. similar in size and shape (Nabhan, 1990). In practice, this often means some sort of grid array of squares, for example defined by lines of latitude and longitude. The number and size of subregions defined within the target area will depend crucially on the density of plant records and on the heterogeneity of the environment, if this is known or can be estimated. Subregions should be large enough for the number of plant records in each to allow meaningful comparisons of diversity among subregions, but not so large – and hence their number so small – that resolution suffers.

Choosing subregions for sampling: preliminary surveys

In an initial survey, when little is known about either target region and taxon, a grid can simply be superimposed on a map of the target region, and subregions (i.e. grid squares) chosen to be visited according to various methods, subject of course to accessibility and other logistical constraints. Subregions could be chosen at *random*, by assigning each a number and picking a proportion out of a hat or by flipping a coin or throwing a die for each. In *systematic* sampling, in contrast, subregions would be visited in a regular pattern. This would be equivalent to collecting every so many kilometres along a chosen route. Since there is often strong correlation between neighbouring measurements in the natural environment, systematic sampling may be a better method to capture variation than random sampling, though care must be taken that the sampling pattern does not coincide with pattern in the environment. Also, a regular grid of measurements carries more spatial information than an equal number of random points, which makes the drawing of contours easier, as already noted. Certain kinds of statistical analysis, however, assume that the data have been collected from a random arrangement of sampling units.

When more is known of the environment in the target region or of the target taxon, a different strategy may be more appropriate. Thus, subregions may be picked out at random (or systematically) separately and independently in areas of the target region that are known a priori to be different in some relevant feature. They could have different kinds of soils or show marked differences in annual rainfall or be inhabited by different subspecies. An agroclimatic classification may also be used to define homogeneous areas, either one of the standard ones (Chapter 9) or a more specific one, for example the cluster analysis or PCA results of data from meteorological stations in the target region. In Chapter 5, the suggestion is made that the weighting that each environmental subunit receives in the sampling should be in proportion to the commonness of the target taxon. Such a *stratified random* approach reduces the possibility of missing out an obviously important aspect of variation purely by chance and allows more flexible allocation of resources. *Gradset* or gradient-directed sampling is another example, involving the sampling of grid squares (chosen randomly or systematically) on replicate transects along some perceived environmental gradient or cline. The disadvantage is that any variation perpendicular to the gradient may be undersampled, depending on the relative allocation of resources between and within parallel transects.

Austin and Adomeit (1991) have carried out a cost–benefit analysis of different vegetation sampling methods by simulation. The cost was measured in terms of time and effort in the field, the benefit in terms of the number of species recorded. It could equally well have been genotypes within a species. Their conclusions are worth quoting:

- Simple statistical sampling designs (i.e. methods like random and systematic sampling) are unlikely to be cost-effective at detecting the range of community variation in a regional landscape.
- Environmentally stratified surveys can be much more effective, particularly if the stratification is based on the most important variables for species distribution. Maximal use of existing information for stratification will reduce costs and improve effectiveness.
- Logistical constraints indicate that gradset sampling stratified by topographic units is the most practical and effective strategy.

In crop collecting, the stratification may involve a combination of environmental and cultural or socioeconomic parameters. Thus, if an area of broadly homogeneous mean annual rainfall, or falling in a single agroclimatic zone, is inhabited by two or more distinct ethnic groups, for example, this should be taken into account to define areas that are not only ecogeographically homogeneous, but also relatively homogeneous in terms of their human population.

These considerations are applied here in choosing where to have collecting sites within a region, but they are no less relevant within collecting sites. Chapter 5 discusses sampling of individuals within a population, and also recommends a stratified random approach at this level in cases where a subpopulation structure exists.

Choosing subregions for sampling: more advanced surveys

If sufficient information is available on a study region, it is useful to build up a matrix giving the number of plant records belonging to each class in each subregion. As already pointed out, the term class could mean here a formal taxonomic unit such as species, but could also refer to categories derived from other kinds of classification: groups of plant populations distinguished by having particular states for a given descriptor, groups of plant populations distinguished by the joint occurrence of particular states of more than one descriptor, clusters derived from multivariate characterization and/or evaluation data or, indeed, traditionally recognized landraces. (It must be remembered, however, that the analysis for classes within the target gene pool defined in a particular way will not necessarily be valid for other classes.) If coverage is patchy or generally low, rather than the actual number of plant records, simple presence/absence of the class in the subregion could be recorded. A third possibility is to divide up each subregion, again in a regular grid, and record the number of sub-subregions occupied by the class within each subregion. Separate matrices could be built up for each level in the taxonomic hierarchy. For example, diversity within a family could be investigated through a matrix of the number of records for each species in each subregion together with further matrices giving the number of species or records in each genus and subregion.

Picking plant records at random one after the other from within a subregion and plotting cumulative number of classes against cumulative number of plant records can give an idea of whether a maximum has been reached, that is, whether collecting a further population is likely to result in an additional class being represented in the sample (Peeters, 1988). The total number of classes found in a subregion can be regressed against the total number of plant records recorded from that subregion to decide which subregions are relatively undercollected. According to Nabhan (1990), in any subregion falling above the upper 95% confidence limit of the regression line of classes on plant records, the number of classes recorded is higher than would be expected on the basis of the total number of collections made in the past, that is, the thoroughness of sampling. Further collecting in such a subregion is therefore more likely to turn up classes not previously collected there than further collecting in subregions falling within the 95% confidence limits or, even more so, falling below the lower confidence limit.

Rare, widespread variants can also be useful in assessing the completeness of collecting in a given area. Thus, for example, if a rare allele (occurring in less that 5% of accessions overall, say) occurs in only one accession in one subregion but in several accessions in others, the former may be relatively undercollected (Weeden *et al.*, 1988).

In addition to undercollectedness, the data in the class vs. subregion matrices can be used to calculate the class diversity of subregions. This may be measured in a number of different ways that bear different and not always obvious interpretations. Magurran (1988) provides a comprehensive survey of the calculation, interpretation and relative merits of several diversity indices. The definitions of some of the more widely used ones are given in Box 15.2.

Number of classes (richness) is the simplest one to use and understand. Another simple richness measure is Margalef's diversity index D_{Mg}. As has been mentioned, areas with equal numbers of taxa may be shown using isoflor maps. This is a useful method for picking out areas of high species richness, but three points should be kept in mind.

First, two areas may have equal numbers of species, but the species may be closely related in one case and taxonomically widely separated in the other. Isoflor maps of both species and sections within the genus might need to be compared to get an accurate impression of the distribution of diversity. Some ways of taking into account the similarities among classes in calculating class diversity are reviewed by Cousins (1991) and Krajewski (1994). Vane-Wright *et al.* (1991) and Williams *et al.* (1991, 1993) have produced a computer program (WORLDMAP) which can be used to identify areas of high biodiversity defined according to four different weightings of simple species richness derived from different measures of taxonomic relatedness. High species richness does not necessarily coincide with high genetic diversity within individual species and different measures of genetic diversity (based on morphological, biochemical or molecular characters) may not give corresponding

Box 15.2
Definitions of class diversity indices

Margalef diversity index: $D_{Mg} = (S - 1)/\ln N$

Shannon–Weaver diversity index: $H = - \Sigma \, p_i (\ln p_i)$

Simpson's index: $D = \Sigma \left(\dfrac{n_i(n_i - 1)}{N(N - 1)} \right)$

Log series diversity α: $S = \alpha \ln (1 + N/\alpha)$

where p_i is the proportion of objects in the ith class, n_i is the number of objects in the ith class, N is the total number of objects and S is the total number of classes

results (Chapter 6). It is important to remember what level of diversity is being mapped and analysed. The results may not be applicable at other levels.

Secondly, areas with equal numbers of species (or, indeed, landraces) are not necessarily floristically similar, i.e. the species involved may be completely different. This is discussed further below.

Thirdly, the number of species present in an area says nothing of the relative abundance of each. The distribution of plant records within classes, or of classes within higher-level classes, is an important aspect of diversity. Ten plant records may be distributed evenly among five species, or one species may be represented by six plant records and the other four by one each. Intuitively, the former case should score the higher diversity.

The two most commonly used measures of diversity which take into account the proportional abundances of different classes are the Shannon–Weaver diversity index H and Simpson's index D. The latter gives relatively more weight to common species; otherwise there is little to choose between them on theoretical grounds (Goldsmith et al., 1986). The gene diversity parameter calculated from allele frequency data (Box 15.1) is the genetic equivalent of Simpson's index; the Shannon–Weaver index can also be applied to such data (Brown and Weir, 1983). Hutchenson (1970) has provided a test for comparing diversity scores computed using the Shannon–Weaver method.

These diversity measures are also called heterogeneity indices, as they take both evenness of distribution and richness into account, and non-parametric indices, to distinguish them from measures which assume an underlying abundance distribution (Magurran, 1988). The most widely used of these is α, which assumes that the number of classes of different sizes will form a log series, though it can still be used if such a distribution is not in fact the best description of the underlying class abundance pattern. Southwood (1978) has suggested that α is the best candidate for a universal diversity statistic in ecology. It and Simpson's index are the least sensitive to sample size (Magurran, 1988).

Diversity within a subregion can be measured for individual continuous descriptors as the standard deviation of the individual plant population values in a subregion, which may then be compared using statistical tests of the homogeneity of variances. Overall diversity for several descriptors together may be measured as:

- the standard deviation of the principal component scores of the plant populations in a subregion;
- the mean of the standard deviations for individual continuous characters;
- the mean of the diversity index scores for individual categorical characters.

Biochemical or molecular data can be used to calculate genetic diversity parameters for subregions, factoring out within-population variation; thus, total gene diversity among populations (H_T) within a subregion may be partitioned as within-population diversity (H_S) and among-populations diversity (G_{ST}) (Nei, 1975, 1987; see also Hoelzel and Dover, 1991; Hoelzel and Bancroft, 1992).

It is useful to produce maps in which subregions are marked in different ways according to their environmental or biological diversity. Maps that show the distribution of qualitative or quantitative information in a geographic region by segregating that information as it is found within subregions are called choropleth maps (Richardson, 1985; Burrough, 1986). Grid squares can be shaded or coloured in different ways or filled circles of different sizes can be placed at their centres. Chapter 5 discusses on what basis collecting effort should be divided up among subregions of different diversities.

Having identified subregions that are undercollected or particularly diverse (on whatever definition), the next step is to investigate the similarities among subregions. Clearly, it may only be necessary to visit both of two equally diverse subregions if different classes are contributing to the high diversity scores of the two areas. The WORLDMAP software has a facility for the exclusion of particular areas from the analysis. If the area with highest diversity is so excluded, and the diversity measurement recalculated for the remaining areas, and so on, a set of areas can be chosen which will optimally cover overall diversity (see also Rebello and Sigfried, 1992). Another useful approach is calculating a similarity matrix for subregions and applying a clustering method. Membership of different clusters of subregions can be shown by using different shadings, patterns or colours.

Similarities among subregions may be calculated not only on the basis of characterization and/or evaluation data but also for various environmental factors. If the presence/absence (or extent) of particular soils or vegetation types in each subregion is recorded, for example, clustering can be carried out to reveal subregions that are environmentally similar. Such data could come from passport data or from soil and vegetation maps, meteorological stations, etc. In the absence of data on the similarity among subregions in terms of the classes of plant populations found there, it will be most efficient to target for priority collecting subregions that are maximally dissimilar on environmental grounds. Indeed, as already mentioned, these clusters are a good starting-point for the definition of the distinct but internally homogeneous areas within which stratified sampling could take place.

A final element that will have to be considered in deciding where to collect – besides completeness of collections, diversity within subregions and similarity among subregions – is risk of genetic erosion in a subregion. Population characteristics such as size and age structure will be important in determining this, but also a variety of external factors which it may be possible to quantify for a particular area. One way of doing this is the genetic erosion prediction system of Goodrich (1987) (Chapter 4).

How can one bring all these data together to compile a list of subregions ranked in order of priority for collecting? An approach that may be useful is to prepare a table listing subregions and record in separate columns the score of each subregion for several variables: number of classes, number of endemic classes, diversity (however measured), undercollectedness and presence/absence of classes missing from germplasm collections or otherwise of particular significance. The scores for each of these variables should be transformed to a discrete scale (e.g. 0–4) or simply given a rank. The system is flexible in that, if particular variables are thought to be more or less important than others, they can be given more or less weight by appropriate changes in the scale. See Chapter 5 and Nabhan (1990) for an assessment of the benefits and dangers of prioritizing on these different criteria. Adding up the transformed variable scores for each subregion would give an idea of rank in overall collecting priority.

Worked example

The Cucurbitaceae are a predominantly tropical family of about 118 extant genera and 825 species, the overwhelming majority of which are distributed in three main areas: Africa and Madagascar, Central and South America, and Southeast Asia and Malesia (Bates *et al.*, 1990). There are nine cultivated cucurbit species of major importance in four genera and six minor crops (Whitaker and Bemis, 1976). In addition, some wild species are exploited as food, fodder or water sources by local peoples, and there is interest in their possible domestication. An example of the latter is *Acanthosicyos horridus*, a shrub of the Namib Desert of the southwestern African coast. Southern Africa is also important as the

native home of wild forms of the watermelon, *Citrullus lanatus*, which
are extremely diverse in the Kalahari region. Both these areas are
represented in Namibia, which makes this country an important one for
cucurbit diversity. Some 34 cucurbit species in 11 genera are so far
recorded from Namibia, which has been fairly well collected from the
point of view of herbarium material but where germplasm collecting of
cucurbits has only very recently started in earnest.

This section presents an analysis of the distribution of cucurbit
diversity at the species level in Namibia which could be used as the basis
for the planning of germplasm collecting in the country. It is based on
the areographic principles outlined in the preceding parts of this chapter
and in particular by Nabhan (1990).

Materials and methods

Definition of subregions

The study region, i.e. the territory of Namibia, was divided into a
number of arbitrarily defined subregions of equal size. In this case, 17
subregions of 2° latitude by 2° longitude were defined, plus two further
subregions of approximately the same size but irregular outline
(Fig. 15.2). These subregions will be referred to as SR1–19.

Data collection

The raw data for the analysis were the locality information recorded on
herbarium sheets held at the National Botanical Institute, Pretoria
(PRE), and the National Herbarium, Windhoek (WIND). As collecting
continues, the distribution patterns of some species will need to be
amended, and therefore also some of the results. The herbarium data
presented here were collated in 1991. All the $\frac{1}{4}$° squares where each
species was collected were recorded. These data were worked up into a
matrix of 34 species by 19 subregions (Matrix A), giving in each cell the
number of $\frac{1}{4}$° squares within each subregion where each species was
present (Table 15.1). A further matrix was derived from Matrix A of 11
genera by 19 subregions (Matrix B), giving in each cell the number of
species within each genus recorded from each subregion.

Data analysis

Several analyses were carried out on the data in Matrices A and B.

1. From Matrix A, diversity values were calculated for each subregion
using the Shannon–Weaver index (H_{spp}). From Matrix B were calculated the numbers of species and genera per subregion and a second
measure of diversity, again using the Shannon–Weaver index (H_{gen}).
Zeros were taken as 0.001, which is standard practice in these cases.
2. Number of species per subregion was plotted against number of

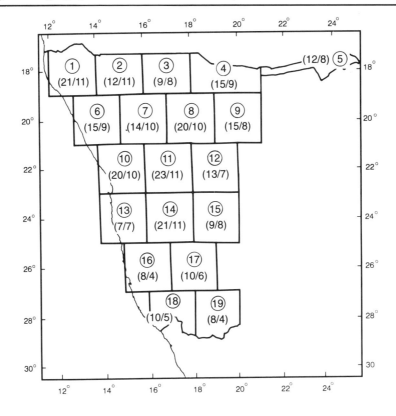

Fig. 15.2. Map of Namibia showing the 19 subregions used in the analysis (numbers enclosed by circles). In brackets are given the number of species/genera recorded from each subregion.

records per subregion (the column totals of Matrix B and Matrix A respectively). A regression line and 95% confidence limits were calculated.

3. Single linkage (nearest-neighbour) cluster analysis was used on a data matrix of the presence/absence of each species in each subregion.

Results

Distribution of diversity

Figure 15.2 gives numbers of species and genera for each subregion. The number of genera per subregion ranged from four (out of 11, or 36%) in SR16 and SR19 in the south to 11 (i.e. all Namibian genera) in SR1–2 in the northwest and SR11 and SR14 in the centre of the country. Number of species per subregion ranged from seven (out of 34, or 21%)

Table 15.1. Number of $\frac{1}{4}$° squares within each subregion (SR1–SR19) where each cucurbit species is present.

	1	2	3	4	5	6	7	8	9	10	11	12	13	14	15	16	17	18	19	Tot.
Acanthosicyos horridus	2	–	–	–	–	–	–	–	–	5	1	–	5	1	–	3	–	1	–	18
Acanthosicyos naudinianus	–	2	3	3	3	–	2	3	3	–	5	1	–	–	2	–	1	–	–	28
Citrullus lanatus	1	2	–	1	–	1	2	1	2	5	7	1	1	1	3	1	1	2	1	33
Citrullus ecirrhosus	2	–	–	–	–	1	–	–	–	6	–	–	–	1	–	5	2	–	–	17
Coccinia sessilifolia	–	–	–	–	–	–	3	1	2	2	8	1	–	2	1	–	–	–	–	20
Coccinia rehmannii	3	1	–	1	–	1	2	2	2	4	6	3	1	6	2	–	1	–	–	35
Coccinia adoensis	–	–	–	1	3	–	–	1	2	–	–	–	–	–	–	–	–	–	–	7
Corallocarpus bainesii	–	–	1	1	3	–	2	–	4	–	3	1	–	1	–	–	–	–	–	16
Corallocarpus triangularis	–	–	–	–	–	–	–	–	–	–	–	1	–	–	1	–	–	·–	–	2
Corallocarpus dissectus	–	–	–	–	–	–	–	–	–	–	–	–	–	–	–	–	1	2	–	3
Corallocarpus welwitschii	6	3	2	–	1	2	–	3	2	5	5	1	1	2	–	2	2	1	1	39
Corallocarpus schinzii	–	–	–	–	–	1	–	–	–	1	1	–	–	1	–	–	–	–	1	5
Cucumella cinerea	2	1	–	–	–	1	2	1	–	5	3	–	–	–	–	–	–	–	–	15
Cucumella aspera	6	–	–	–	–	3	–	–	–	4	–	–	1	2	–	–	–	–	–	16
Cucumis metuliferus	–	–	–	1	2	–	–	–	–	–	–	–	–	–	–	–	–	–	–	3
Cucumis mecusei	3	–	–	1	–	4	1	2	–	3	9	–	–	5	–	4	2	2	1	37

Species																			Total
Cucumis sagittatus	6	–	–	–	–	1	–	1	–	3	4	6	–	8	1	3	3	3	8	46
Cucumis rigidus	–	–	–	–	–	1	–	1	–	–	1	–	–	–	1	2	–	4	2	9
Cucumis africanus	2	–	–	–	3	7	1	3	–	2	7	–	13	10	2	2	2	2	2	58
Cucumis kalahariensis	–	–	–	1	1	1	–	2	1	2	–	4	–	–	–	3	–	1	–	12
Cucumis humifructus	1	–	3	1	–	–	–	2	–	2	–	–	–	–	–	–	–	–	–	6
Cucumis anguria	1	2	–	–	1	2	3	2	–	2	5	6	6	1	5	–	1	1	–	26
Dactyliandra welwitschii	10	1	–	–	1	8	–	–	–	–	1	–	1	4	–	–	–	–	–	36
Kedrostis foetidissima	1	–	3	3	–	–	2	1	1	1	1	1	4	–	–	–	–	–	–	13
Kedrostis hirtella	1	1	1	–	–	–	1	–	4	1	2	–	2	1	–	–	–	–	–	13
Kedrostis africana	–	–	–	–	–	–	–	–	1	–	1	–	1	1	–	–	–	6	1	3
Kedrostis capensis	–	–	–	–	–	–	–	–	–	–	1	–	–	–	–	–	–	–	–	9
Momordica welwitschii	7	–	–	–	–	–	–	2	–	–	1	–	1	–	–	–	–	–	–	7
Momordica boivinii	1	3	–	2	–	–	6	1	1	1	5	–	5	3	1	–	1	–	–	3
Momordica balsamina	1	1	1	–	7	3	1	1	1	7	3	–	3	2	–	–	–	–	–	29
Momordica humilis	4	1	1	1	1	–	1	3	–	2	4	–	4	1	–	–	–	–	–	31
Trochomeria macrocarpa	1	4	–	–	–	1	–	2	1	–	1	–	3	2	–	–	–	–	–	22
Trochomeria debilis	2	2	–	–	–	–	1	1	1	1	–	–	3	2	–	–	–	–	–	12
Zehneria marlothii	3	3	–	2	1	–	2	5	3	1	–	–	3	1	–	–	–	–	–	21

in SR13 to 23 (68%) in the nearby SR11. The strongest concentration of subregions with a large number of species recorded is in the centre of the country (SR8, SR10, SR11 and SR14). SR1 and SR8 had both ≥3 *Momordica* species and ≥5 *Cucumis* species. SR4 and SR10 also had ≥5 *Cucumis* species. The range of number of genera and number of species per subregion was divided into three sections with approximately equal numbers of subregions. A value of 1 was assigned to subregions at the bottom of the range and a value of 3 to subregions in the top category. Table 15.2 shows these data. Subregions with ≥5 *Cucumis* species are given a score of 1 in a separate column in the table, other regions scoring 0.

The ranges of H_{spp} and of H_{gen} values were each divided into four ranges in such a way that each range contained approximately equal numbers of subregions and values of 1 to 4 were assigned to subregions accordingly (Table 15.2). For H_{spp}, the subregions at the top of the range were SR1, SR8–11 and SR14. This corresponds to the central portion of the country and the northwest. In these subregions, a given number of records will represent more species than in other subregions. Subregions in the southern portion of the country generally have low H_{spp} values. An essentially similar pattern was evident for H_{gen}, with the highest values occurring in SR1–2 in the northwest and SR11 and SR14 in the centre and the lowest in the south. A higher value for this diversity index means that a given number of species will represent more genera.

Undercollected areas

Subregions that fall above the 95% confidence limits of the regression line of species per subregion against records per subregion may be considered to be relatively undercollected. SR4, SR6, SR8, SR9, SR12 and SR14 may be thus described, subregions mainly concentrated in the northeast and centre. They are given a score of 1 in a separate column of Table 15.2, all other regions scoring 0.

Subregional similarities

Clustering of the presence/absence data revealed the existence of two main floristic zones. One comprises SR2–5 and SR7–9 in the central and eastern parts of the north, the other includes SR13 and SR16–19 in the south. Membership of each subregion for one of six clusters (A–F) is shown in Fig. 15.3 and Table 15.2.

Distribution of important species

In addition to the distribution of diversity, the distribution of particular taxa is often of importance. An area of high diversity for the family as a whole might not contain the most interesting species, for example

Table 15.2. Various cucurbit diversity parameters for each subregion (SR1–SR19). See text for details.

SR	Number of genera	Number of species	Number of *Cucumis* species	H_{spp}	H_{gen}	'Collectedness'	Presence of *Citrullus lanatus*	Cluster membership	Total
1	3	3	1	4	4	0	1	A	16
2	3	2	0	3	4	0	2	B	14
3	2	1	0	1	3	0	0	B	7
4	2	2	1	3	2	1	1	B	12
5	2	2	0	2	3	0	0	B	9
6	2	2	0	3	3	0	1	C	12
7	3	2	1	3	3	1	2	B	13
8	3	3	0	4	3	0	1	B	16
9	2	2	0	4	2	1	2	B	13
10	2	3	0	4	3	0	2	C	15
11	2	3	1	4	4	0	2	D	15
12	1	2	0	3	3	1	1	E	11
13	1	1	0	1	2	0	1	F	6
14	3	3	0	4	4	1	1	C	16
15	2	1	0	2	3	0	2	E	10
16	1	1	0	2	1	0	1	F	7
17	1	1	0	2	1	0	1	F	6
18	1	1	1	2	1	0	2	F	8
19	1	1	0	1	1	0	1	F	5

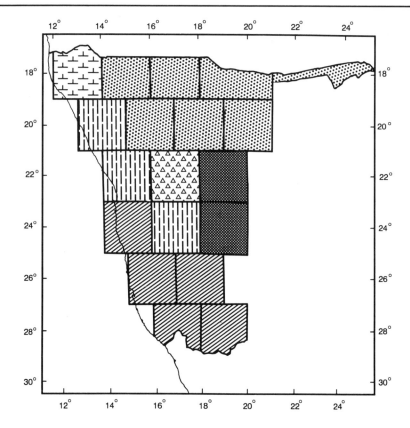

Fig. 15.3. Map of Namibia showing to which of six floristic clusters each subregion belongs.

those in the primary or secondary gene pool of a crop. Table 15.2 therefore also records the occurrence of wild *Citrullus lanatus* in each subregion. A single record in the subregion is given a score of 1, two or more records a score of 2.

Discussion

To rank subregions in terms of the importance of their contribution to overall cucurbit species diversity in Namibia, the scores of each subregion for the seven parameters shown in Table 15.2 were added together. The nature of the categorization of the data gives weights in the ratios of $4:4:3:3:2:1:1$ to, respectively, the scores for the two diversity measures, genera and species numbers, occurrence of *Citrullus lanatus*, number of *Cucumis* species and 'undercollectedness'.

Clearly, these weightings can easily be changed to reflect different collecting aims. The five subregions scoring highest are SR1, SR8, SR10–11 and SR14 (Table 15.2). Bringing in the cluster analysis results reveals that SR10 and SR14 are in the same cluster. SR14 would be the higher collecting priority if *C. lanatus* occurrence were to be ignored. With the weight attached to this particular variable here, however, SR10 has the higher total score. This makes the point that different collecting aims and objectives will result in different weights being attached to a given criterion, and possibly therefore different priorities being accorded to a given subregion.

The clustering results also reveal that clusters E and F are not represented in this initial selection of priority subregions. Based on their total scores, subregions SR12 and SR18 can be identified as the higher priorities within clusters E and F respectively.

If it was only possible to visit a limited number of subregions, a possible final selection for first-priority collecting (in numerical order) would thus be as follows: SR1, SR8, SR11, SR12, SR14 and SR18. Germplasm collecting within this limited selection of subregions would, according to the analysis presented here, result in the capture of species diversity within the family adequately representative of the study area as a whole, with something of a bias towards subregions with good representation of the genus *Cucumis* and the species *C. lanatus*. Alternatively, the total scores in Table 15.2 could be used to weight the amount of collecting time or effort in each subregion or, indeed, the number of samples collected in each subregion.

Conclusion

The worked example in this section should be seen as presenting a generalized model capable of being applied to a wide variety of different situations. Thus, though species were considered, it could equally well have been formal or informal infraspecific taxonomic categories within a crop (Chapter 7), traditionally recognized landraces or even genotypes. The aim of the analysis was to devise a strategy for the maximal sampling of species diversity within a botanical family. A separate analysis would be needed if the target were diversity within *C. lanatus*, say, in which case a measure of genetic diversity could be used.

To complement the information on the plants themselves, some measure of environmental diversity within the different subsections could have been calculated, for example by noting the extent of different vegetation types or agroecological zones in each. In the same way, though similarities among subregions were calculated here floristically, on the basis of species representation, they could be calculated as genetic or environmental similarities in other cases. Groups of environmentally similar subregions will form the basis of a stratified sampling strategy in situations when little is known of genetic or taxonomic variation within the target group.

Acknowledgements

Mark Perry and Tom Hazekamp (IPGRI) made valuable comments on early drafts of this paper.

References

Adams, R.P. (1970) Contour mapping and differential systematics of geographic variation. *Systematic Zoology* 19:385–390.

Austin, M.P. and E.M. Adomeit (1991) Sampling strategies costed by simulation. In: Margules, C.R. and M.P. Austin (eds) *Nature Conservation: Cost Effective Biological Surveys and Data Analysis.* CSIRO, Australia.

Bates, D.M., R.W. Robinson and C. Jeffrey (eds) (1990) *Biology and Utilization of the Cucurbitaceae.* Cornell University Press, Ithaca.

Blackith, R.E. and R.A. Reyment (1971) *Multivariate Morphometrics.* Academic Press, London.

Braak, C.J.F. ter (1986) Canonical correspondence analysis: a new eigenvector technique for multivariate direct gradient analysis. *Ecology* 67:1167–1179.

Braak, C.J.F. ter (1988) CANOCO – an extension of DECORANA to analyze species–environment relationships. *Vegetatio* 75:159–160.

Brown, A.H.D. and B.S. Weir (1983) Measuring genetic variability in plant populations. In: Tansley, S.D. and T.J. Orton (eds) *Isozymes in Plant Genetics and Breeding.* Part A. pp. 219–239. Elsevier Scientific Publishing, Amsterdam.

Burrough, P.A. (1986) *Principles of Geographic Information Systems for Land Resources Assessment.* Clarendon Press, Oxford.

Charmet, G. (1993) Geographically constrained clustering: a tool for establishing a core from a large collection of wild populations in forage grasses. In: IBPGR *Report of the Fourth Meeting of the ECP/GR Forages Working Group.* IBPGR, Rome.

Clifford, H.T. and W. Stephenson (1975) *An Introduction to Numerical Classification.* Academic Press, New York.

Cousins, S.H. (1991) Species diversity measurement: choosing the right index. *Trends in Ecology and Evolution* 6:190–192.

Dunn, G. and B.S. Everitt (1982) *An Introduction to Mathematical Taxonomy.* Cambridge University Press, Cambridge.

Erskine, W., Adham, Y. and Holly, L. (1989) Geographic distribution of variation in quantitative traits in a world lentil collection. *Euphytica* 43:97–104.

Francisco-Ortega, J., M.T. Jackson, A. Santos-Guerra and B.V. Forde-Lloyd (1993) Morphological variation in the *Chamaecytisus proliferus* (L.f.) Link complex (Fabaceae: Genisteae) in the Canary Islands. *Botanical Journal of the Linnean Society* 112:187–202.

Gauch, H.G. (1982) *Multivariate Analysis in Community Ecology.* Cambridge University Press, Cambridge.

Goldsmith, F.B., C.M. Harrison and A.J. Morton (1986) Description and analysis of vegetation. In: Moore, P.D. and S.B. Chapman (eds) *Methods in Plant Ecology.* pp. 437–524. Blackwell Scientific Publications, Oxford.

Goodrich, W.J. (1987) Monitoring genetic erosion: detection and assessment. Internal report. IBPGR, Rome.

Gower, J.C. (1971) A general coefficient of similarity and some of its properties. *Biometrics* 27:857–872.

Heywood, J.S. (1991) Spatial analysis of genetic variation in plant populations. *Annual Review of Ecology and Systematics* 22:335–355.

Hill, M.O. (1979a) *DECORANA – A FORTRAN Program for Detrended Correspondence Analysis and Reciprocal Averaging*. Cornell University, Ithaca.

Hill, M.O. (1979b) *TWINSPAN – A FORTRAN Program for Arranging Multivariate Data in an Ordered Two-way Table by Classification of Individuals and Attributes*. Cornell University, Ithaca.

Hill, M.O. and H.G. Gauch (1980) Detrended Correspondence Analysis: an improved ordination technique. *Vegetatio* 42:47–58.

Hill, M.O. and A.J.E. Smith (1976) Principal components analysis of taxonomic data with multi-state discrete characters. *Taxon* 25:249–255.

Hintum, Th.J.L. van (1993) A computer compatible system for scoring heterogeneous populations. *Genetic Resources and Crop Evolution* 40:133–136.

Hoelzel, A.R. and D.R. Bancroft (1992) Statistical analysis of variation. In: Hoelzel, A.R. (ed.) *Molecular Genetic Analysis of Populations*. pp. 297–305. IRL Press, Oxford.

Hoelzel, A.R. and G.A. Dover (1991) *Molecular Genetic Ecology*. Oxford University Press, Oxford.

Hutchenson, K. (1970) A test for comparing diversities based on the Shannon formula. *Journal of Theoretical Biology* 29:151–154.

Jones, S.B. (1972) A systematic study of the fasciculate group of *Vernonia. Brittonia* 24:28–45.

Kirkpatrick, J.B. (1974) The use of differential systematics in geographic research. *Area* 6:52–53.

Krajewski, C. (1994) Phylogenetic measures of biodiversity: a comparison and critique. *Biological Conservation* 69:33–39.

Kruskal, J.B. (1964) Nonmetric multidimensional scaling: a numerical method. *Psychometrika* 29:28–42.

Magurran, A.E. (1988) *Ecological Diversity and its Measurement*. Chapman and Hall, London.

Monmonier, M. (1973) Maximum-difference barriers: an alternative numerical regionalization method. *Geographical Analysis* 3:245–261.

Morden, C.W., J.F. Doebley and K.F. Schertz (1989) Allozyme variation in Old World races of *Sorghum bicolor* (Poaceae). *American Journal of Botany* 76:247–255.

Morley, T. (1971) Geographic variation in a widespread neotropical species, *Mouriri myrtilloides* (Melastomataceae). *Brittonia* 23:413–424.

Myklestad, Å. and H.J.B. Birks (1993) A numerical analysis of the distribution of patterns of *Salix* L. species in Europe. *Journal of Biogeography* 20:1–32.

Nabhan, G.P. (1990) *Wild* Phaseolus *Ecogeography in the Sierra Madre Occidental, Mexico*. Systematic and Ecogeographic Studies on Crop Genepools 5. IBPGR, Rome.

Nei, M. (1975) *Molecular Population Genetics and Evolution*. North-Holland Publishing Co., Amsterdam.

Nei, M. (1987) *Molecular Evolutionary Genetics*. Columbia University Press, New York.

Pankhurst (1991) *Practical Taxonomic Computing*. Cambridge University Press, Cambridge.

Peeters, J.P. (1988) The emergence of new centres of diversity: evidence from barley. *Theoretical and Applied Genetics* 76:737–752.

Pigliucci, M. and G. Barbujani (1991) Geographical pattern of gene frequencies in Italian populations of *Ornithogalum montanum* (Liliaceae). *Genetical Research* 58:95–104.

Porceddu, E. and A.B. Damania (1992) *Sampling Strategies for Conserving Variability of Genetic Resources in Seed Crops*. Technical Manual No. 17. ICARDA, Aleppo.

Rebello, A.G. and W.R. Sigfried (1992) Where should nature reserves be located in the Cape Floristic Region, South Africa? Models for the spatial configuration of a reserve network aimed at maximizing the protection of diversity. *Conservation Biology* 6:243–252.

Richardson, G.T. (1985) *Illustrations*. Humana Press, Clifton.

Seyani, J.H. (1988) The taxonomy of *Dombeÿa burgessiae* complex (Sterculiaceae) in Africa. Monograph. *Systematic Botany, Missouri Botanical Garden* 25.

Sneath, P.H.A. and R.R. Sokal (1973) *Numerical Taxonomy: The Principles and Practice of Numerical Classification*. W.H. Freeman, San Francisco.

Sokal, R.R. (1979) Testing statistical significance of geographical variation patterns. *Systematic Zoology* 28:227–232.

Sokal, R.R. (1986) Phenetic taxonomy: theory and methods. *Annual Review of Ecology and Systematics* 17:423–442.

Sokal, R.R. and N.L. Oden (1978) Spatial autocorrelation in biology. 1. Methodology. *Biological Journal of the Linnean Society* 10:199–228.

Southwood, T.R.E. (1978) *Ecological Methods*. Chapman and Hall, London.

Spagnoletti Zeuli, P.L. and C.O. Qualset (1987) Geographical diversity for quantitative spike characters in a world collection of durum wheat. *Crop Science* 27:235–241.

Stuessy, T.F. (1990) *Plant Taxonomy*. Columbia University Press, New York.

Vane-Wright, R.I., C.J. Humphries and P.H. Williams (1991) What to protect? Systematics and the agony of choice. *Biological Conservation* 55:235–254.

Weeden, N.F., B. Wolko, A.C. Emmo and J. Burr (1988) Measurement of genetic diversity in pea accessions collected near the center of origin of domesticated pea. Internal report. IBPGR, Rome.

Whitaker, T.W. and W.P. Bemis (1976) Cucurbits. In: Simmonds, N.W. (ed.) *Evolution of Crop Plants*. pp. 64–69. Longman, London.

Williams, P.H., C.J. Humphries and R.I. Vane-Wright (1991) Measuring biodiversity: taxonomic relatedness for conservation priorities. *Australian Systematic Botany* 4:665–679.

Williams, P.H., R.I. Vane-Wright and C.J. Humphries (1993) Measuring biodiversity for choosing conservation areas. In: LaSalle, J. (ed.) *Hymenoptera and Biodiversity*. CAB International, Wallingford.

Womble, W.H. (1951) Differential systematics. *Science* 114:315–322.

Geographic information systems and remote sensing for plant germplasm collectors

16

L. Guarino

IPGRI, c/o International Laboratory for Research on Animal Diseases, PO Box 30709, Nairobi, Kenya.

Introduction to GIS

A geographic information system (GIS) is a 'computerized information storage, processing and retrieval system that has hardware and software specifically designed to cope with geographically referenced spatial data and the corresponding attribute information' (FAO, 1988). In other words, a GIS is a database management system dedicated to the simultaneous handling of spatial data in graphics form and of related, logically attached, non-spatial data. For example, if the spatial data are the location of cities or districts in a country, the associated attributes could be the name, current population, past population and population growth rate of each. If any modification is made in one kind of data, an appropriate modification is automatically made in the other.

GIS technology has obvious applications in such fields as urban planning and natural resources management, but until fairly recently relatively expensive mainframes or minicomputers were necessary to run the software. Since the 1980s and the advent of the 80386 microchip, however, both hardware and software have become more cost-effective and easily available, increasing the potential of GIS as a problem-solving tool. Many national and regional institutions in developing countries have set up GIS facilities and are building up spatial databases relevant to their mandates. Such multilateral agencies as the Food and Agriculture Organization (FAO), the United Nations Environment Programme (UNEP)'s Global Resources Information Database (GRID), the United Nations Training and Research Institute (UNITAR) and the United Nations Sudano-Sahelian Office (UNSO), as well as bilateral donors, have supported these efforts through the provision of hardware, software, data and training. The planned national database of China is an instructive example, as it has been conceived specifically as a tool to aid biodiversity conservation. As proposed by the World Wide Fund for Nature (WWF), there will be seven regional databases, one in each of the biogeographic divisions of the country, plus a central repository in Beijing, at the Commission for Integrated Survey. Information on species richness, endemism, habitat threat, protected areas, watersheds, human pressure and the physical environment will be mapped and combined to derive maps of 'gene-pool sensitivity' and 'environmental sensitivity' for each of the provinces and autonomous regions of the country (McNeely *et al.*, 1990).

Some examples of GIS software packages are listed in Box 16.1. GRID (1992) gives the results of a survey of the characteristics and capabilities of a wide range of GIS software and hardware packages, and also lists other sources of such technical information. The main components of GIS software are (Burrough, 1986):

- data input, verification and editing;
- data storage and database management;
- data manipulation and analysis;
- data output;
- user interface.

These fit together as shown in Fig. 16.1.

Data input and data sources

'Data input is the procedure of encoding data into a computer-readable form and writing the data to the GIS database' (Aronoff, 1989). It includes the linking of the spatial and attribute data and the verification of data quality standards. Data entry is the major constraint on GIS implementation, as it is usually labour-intensive, time-consuming and

---

Box 16.1
Commonly available GIS software

Name	Publisher
Atlas*GIS	Strategic Mapping
GisPlus	Caliper
MapInfo for Windows	MapInfo Corps
PC ARC/INFO	Environmental Systems Research Institute
SPANS	Tac Systems
MARS	Montage Information Systems
IDRISI	Clark University
GRASS	Public domain
ERDAS	ERDAS Inc.
Hyperdyne's Mapix	Montage Information Systems

Data input	Data storage	Data manipulation	Data output

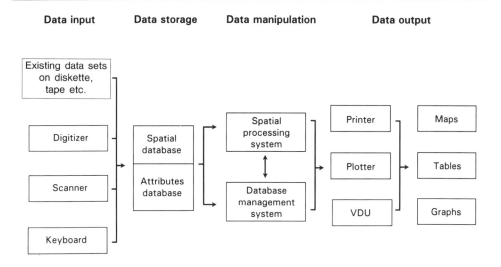

Fig. 16.1. The components of a GIS and their relationship to each other.

thus expensive. There are four ways of entering data into a GIS (e.g. Burrough, 1986):

- Attribute data are usually entered from a computer *keyboard*. This can be done as part of the same operation during which the geographic data are entered (e.g. during digitizing of maps) or as a separate operation. Thus, collectors could enter passport and other data into a separate database and import this into a GIS at a later date.
- In *digitizing*, the most common method of entering geographic data, a map is mounted on a digitizing table or tablet and the features to be entered are traced using a cursor or pointer. The coordinate data produced are either fed directly into the GIS or stored in a computer file for later use.
- *Scanning*, a kind of automated digitizing, involves generating a digital image of the map by moving an electronic sensor over its surface.
- *Remotely sensed* data (i.e. data recorded at a distance from the object of interest) can come from aerial photographs, electro-optical scanners or microwave receptors (see Table 16.1 for a summary of the more widely used systems).

For any given application, the logical first step is to ascertain whether suitable digital data sets already exist. Various data sets are held by organizations such as FAO, UNEP/GRID, the International Soil Reference and Information Centre (ISRIC), the World Conservation Monitoring Centre (WCMC) and the international agricultural research centres (IARCs), though these are mostly (though not

---

Box 16.2
Some FAO digital databases

Africa

   1. Integrated terrain units (FAO soils, US Department of Agriculture (USDA) soils, United Nations Educational, Scientific and Cultural Organization (Unesco) geology, physiography, geomorphology, landform, surface forms, potential vegetation, land use).
   2. Desertification study (soil hazard, wind hazard, water hazard, salinity hazard, population pressure, animal pressure).
   3. Template (boundaries, etc.).
   4. Mean annual rainfall.
   5. Number of wet days per year.
   6. Agroecological zones.
   7. Countries and provinces.
   8. Rivers.
   9. Watersheds.
   10. Roads.
   11. Cities.
   12. Ecofloristic zones.
   13. Vegetation (actual).
   14. Irrigation and water potential study (potential irrigable soils, aquifer rank, water availability).

Other developing countries

   1. Desertification study (soil texture, slope, pedogenic factors).
   2. FAO soil map.
   3. Agroecological zones.
   4. Countries.
   5. Template.
   6. Vegetation (S and SE Asia).
   7. Ecofloristic zones (S and SE Asia).

---

exclusively) at a regional or global scale (scales 1 : 1 million down to 1 : 100 million). As an example, some of the digitized data sets held at FAO and at GRID that could be of interest to collectors are listed in Boxes 16.2 and 16.3. The WCMC databases are introduced in Chapter 10. There are also relevant regional sources of data sets, such as the Regional Centre for Services in Surveying, Mapping and Remote Sensing in Nairobi, the Regional Remote Sensing Programme in Bangkok and the Regional Remote Sensing Centre in Burkina Faso. Local-level data sets (up to scales of about 1 : 20,000) may be available from such centres and from national agriculture, environment, planning and cartography services, as well as from private companies. Conditions of availability will vary.

   GRID's Meta-Database is an interactive electronic catalogue of spatial environmental data sets archived at GRID centres throughout

Box 16.3
Some GRID digital data sets

Global

1. Political and natural boundaries.
2. Elevation.
3. Soils and soil degradation.
4. Vegetation.
5. Human population.
6. Cultivation intensity.
7. Ecosystems.
8. Life zones.
9. Wetlands.
10. Temperature and moisture availability surfaces.

Africa

1. Political and natural boundaries.
2. Elevation.
3. Slope and aspect.
4. Soils.
5. Soil degradation and desertification.
6. Human population.
7. Roads and railways.
8. Hydrology and watersheds.
9. Protected areas.
10. Cattle and buffalo distribution.
11. Ecoclimatic suitability indices.
12. Vegetation and land cover.
13. Cultivation intensity.
14. Ecosystems.
15. Temperature.
16. Rainfall.
17. Evaporation.

National data sets are also available for a number of countries in Africa, Latin America and Asia. However, prior approval from the respective country is needed, prior to distribution.

the world. It is planned that it will in future also hold details of data (both electronic and otherwise) held in other institutions. The National Aeronautics and Space Administration (NASA) Master Directory provides a similar service, and GRID also has access to this. The GRID Meta-Database may be consulted at GRID centres and via the Internet, but there is also a personal computer (PC) version. Dangermond (1988) provides a review of the practical problems involved in entering existing data sets into a GIS.

The most common data requirement will be for base-layer data sets, showing country boundaries, internal administrative boundaries, rivers and lakes, altitude (contour and spot heights) and so on. The most commonly used medium-resolution data set of this type has been *World Data Bank II*, but the *Digital Chart of the World* (DCW), based on the 1 : 1 million Operational Navigational Charts and recently released on four CD-ROMs, is of potentially greater use. However, the extraction of data subsets can be difficult and data quality is variable (Anon., 1992). The EROS Data Center is using the DCW to develop a global digital terrain model (a quantitative model of landform in digital form) on a 1 km grid basis. The Centre for Resource and Environmental Studies (CRES) at the Australian National University has or is developing high-resolution digital terrain models and climate surfaces for Africa, Australia, New Guinea, China and southeast Asia (H.A. Nix, pers. comm.). On soils, FAO's revised 1 : 5 million *Soil Map of the World* is the best available strategic-level digital data set. ISRIC is developing a worldwide digital database for soils and terrain on a scale of 1 : 1 million, but the time frame for this work is 10–20 years. 'Skeleton' continental coverages are being prepared.

As for satellite data, GRID can give advice on how to obtain data sets from commercial and other sources, though it cannot procure these for users. FAO's Remote Sensing Centre has a comprehensive database of reference maps and imagery which is available to member nations and FAO programmes. Resolution varies from kilometres to tens of metres, depending on the system (Table 16.1). The cost of obtaining these data sets can be very high. Also, analysis (which will involve image restoration or correction, image enhancement and information extraction) is complex, requiring specialized software, hardware and skills. Careful ground truthing is necessary for many applications. However, for inaccessible areas for which there are no detailed maps, for example, remote sensing may be the only source of some data. It is also often the only source of data on changes in vegetation and land use, whether from year to year or from week to week in a given growing season. If the raw remote sensing data itself are beyond the reach of the collector, for either financial or logistical reasons, publications may be available analysing such data and presenting the results in maps and other potentially useful hard-copy forms. Hilwig (1987) lists numerous published examples of the application of remote sensing to agroecological characterization, classification and mapping (refer to the list of specialized journals at the end of this chapter).

Data storage

There are two main types of GIS software, which differ in how they store data. Vector-based systems store geographic data as points. Series of connected points make up lines, and lines enclosing an area make up polygons. In contrast, raster-based systems store data as grid cells, each

Table 16.1. Summary of characteristics of five satellite-based remote sensing systems, compared with aerial photography (adapted from Hilwig, 1987). Note that the smallest mappable area on a map (3 × 3 mm) corresponds to 900 ha at a scale of 1 : 1 million and 2.25 ha at a scale of 1 : 50,000.

System	Scale range	Spatial resolution	Temporal resolution	Property detected
Aerial photography	1 : 70,000– 1 : 120,000	1–7 m	10–20 years	Reflectance of visible and infrared solar radiation
Landsat TM (Thematic Mapper)	1 : 100,000–1 : 1 million	30 m	16 days	Reflectance of visible, photographic infrared and thermal radiation
Landsat MSS (Multi-Spectral Scanner)	1 : 250,000–1 : 1 million	80 m	18 days	Reflectance of visible, photographic infrared
SPOT HRV (High Resolution Visible) panchromatic and MSS	1 : 50,000–1 : 1 million	10–20 m	26 days	Reflectance of visible, photographic infrared
NOAA AVHRR (Advanced Very High Resolution Radiometer)	1 : 3.5 million– 1 : 10 million	1 or 4 km	12 hours	Reflectance of visible, photographic infrared and thermal radiation
Meteosat	Not applicable	2.4–5.0 km	30 min	Reflectance of visible, photographic infrared and thermal radiation

representing a memory location in the computer. Lines are rows of grid cells and polygons are groups of adjacent grid cells. Remote-sensing data are in raster form, each picture element (pixel) being a grid cell. Vector systems require more computing power but less storage memory than raster systems. They represent map data better, because lines on the map remain lines, rather than becoming rows of grid cells.

It is possible to convert vector files to raster files, but incorporating raster data into a vector system is not usually as efficient. Vector systems will be best for some applications, raster systems for others. Burrough (1986) recommends the use of vector data structures for archiving phenomenologically structured data (e.g. topographic units, soil types) and for the highest-quality output, and raster methods for the rapid overlay and combination of maps and for spatial analysis.

Data manipulation

The spatial processing system and database management system of a GIS allow one to bring together diverse data sets, make them compatible among themselves, analyse and combine them in different ways and display the results as a map or statistics on a computer screen or hard copy. Some standard GIS capabilities include (Burrough, 1991):

- *Geometric correction.* The scale, projection, etc. of different maps may be changed to make them comparable.
- *Digital terrain model analysis.* The altitude contours on a topographical map may be used to produce maps of slope, aspect, intervisibility, shaded relief, etc.
- *Interpolation.* Point data may be used to create isopleth (equal-value contour) maps.
- *Overlay analysis* (Fig. 16.2). Different maps of the same area may be combined to produce a new map, e.g. maps of slope, soil, wind speed and vegetation cover may be overlaid to synthesize a map of potential soil erosion.
- *Proximity analysis.* Buffers may be generated around features.
- *Computation of statistics.* Means, counts, lengths and areas may be calculated for different features.
- *Location.* Entities having defined sets of attributes may be located.

How can such features be of use to the germplasm collector? There are clearly applications beyond simplifying the production of high-quality maps of sample distributions and the like. All of the analyses

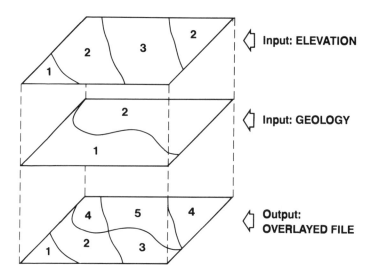

Fig. 16.2. Schematic representation of the GIS overlay facility.

described in Chapter 15, in fact, will be facilitated by the use of GIS. Some particular applications are summarized below under the separate headings of mission planning and the documentation of collections.

Planning

A GIS can delimit areas where particular conditions or combinations of conditions are found

This will be important when material with particular characteristics is being targeted for collecting. Thus, for example, areas can be identified where the occurrence of a particular soil type is combined with that of a particular rainfall regime. The raw rainfall data from meteorological stations would be transformed by the GIS itself into isohyets as part of this analysis. Data on the distribution of different states of a characterization descriptor could be superimposed on the initial output to investigate possible associations.

A GIS can combine data sets to generate derived data sets of predictive or enhanced descriptive value

This is not simply identifying the co-occurrence of conditions, but actually combining different kinds of data in some sort of mathematical model. Thus, data on human population, accessibility (road networks, etc.), economic development, potential for irrigation and agroclimatic suitability for cash crops or modern varieties could be used to develop a model for the risk of genetic erosion among landraces. Potential for irrigation and agroclimatic suitability are themselves predictive models that will be of interest to the collector. In particular, agroclimatic or agroecological classification of the target region into areas which are environmentally homogeneous and different from each other can be an efficient way of organizing a stratified collecting strategy. Data for different characterization descriptors can also be combined using a multivariate statistical procedure, for example to derive maps of principal component scores or systematic function scores.

A GIS can calculate climatic 'envelopes' for species, landraces or genotypes

In conjunction with digital terrain models and the climatic surfaces that can be derived from these, species distribution data can be used to calculate ranges for different climatic parameters, defining the adaptation of the taxa under consideration. Though still at the research stage, the BIOCLIM program of Nix, Busby and Hutchinson (Busby, 1986), for example, uses climatic interpolation surfaces to estimate conditions at each of a set of sites given latitude, longitude and altitude; it then derives a climatic envelope for the set of localities. Once the ranges of up to six important rainfall and temperature variables represented by the input sites have been calculated, programs such as AFRMAP (Booth *et al.*, 1989) and WORLD (Booth, 1990) can be used to display all the regions (if interpolation surfaces are available) or localities (if only point data are available) where such conditions obtain. This is not only useful in species introduction (i.e. the identification of homoclimes), the application for which the programs were initially developed, but also, for example, in identifying areas where the taxon has perhaps not been collected but where it might still be expected to be found on the basis of climate.

Remote sensing can be used to locate areas of interest

Different vegetation types can often be recognized in aerial photographs and satellite imagery. Isolated areas of cultivation can also be identified, and in some cases the types of crops grown recognized.

Satellite imagery can provide data on vegetation development with very short lag times

Use of such data would allow collectors to be very precise in timing their visit to areas showing vegetation flushes. This is particularly important in the arid and semiarid tropics, where rainfall is unpredictable in both space and time. For example, Meteosat and NOAA Advanced Very High Resolution Radiometer (AVHRR) data on rainfall and the state of vegetation (as measured by the Normalized Difference Vegetation Index, or NDVI) are analysed by FAO's Agricultural Real Time Environmental Monitoring Information System (ARTEMIS) to allow up-to-the-minute surveillance of the state of crops and vegetation in Africa and Asia. The information is available not only at FAO headquarters in Rome but, via Intelsat and the Data and Information Available Now in Africa (DIANA) system developed in collaboration with the European Space Agency (ESA), also in three regional centres in Africa (the Regional Centre for Services in Surveying, Mapping and Remote Sensing in Nairobi, the National Meteorological Service in Harare and Centre AGRYHMET in Niamey). See, for example, Justice *et al.* (1987) for descriptions of the annual course of NDVI in a variety of East African vegetation types, and how this measure relates to the phenology of rainfall and plant growth (see also Justice *et al.*, 1985; Davenport and Nicholson, 1993). One could conceivably read off latitudes and longitudes for a set of potential target collecting sites from satellite imagery and use a Global Positioning System (GPS) receiver to locate them in the field a matter of days later.

Satellite imagery and other remote-sensing systems can provide information on long-term vegetation change in an area

Actual trends and developments in deforestation and desertification can be monitored using remote-sensing data stretching back over many years. Examples include the work described by Skole and Tucker (1993) for the Amazon and Gastellu-Etchegorry *et al.* (1993) for Sumatra. This will help identify areas threatened with, or actually experiencing, habitat modification, and therefore perhaps genetic erosion.

Documentation

Fuller and more accurate passport data can be obtained
> Using GIS, the collector would be able to record latitude, longitude, altitude, soil type, vegetation and other attributes for collecting sites automatically by overlaying their locations on different digitized base maps. The capability to generate isopleth maps, and in particular digital terrain models, allows the values of particular attributes (e.g. climatic factors) at collecting sites to be accurately estimated on the basis of data collected at other points, in this case nearby meteorological stations.

A GIS can assist in data verification
> For example, a GIS can spot outliers due to miskeying of latitude, longitude or other attribute data.

A GIS can produce listings of samples satisfying different criteria
> All collecting sites with particular attributes (of the environment and/or germplasm) or combinations of attributes can be picked out. For example, a map of collecting site distribution may be overlaid on soil and rainfall maps to identify those collecting sites where the material collected might be expected to be adapted to particular combinations of edaphic and climatic conditions.

Data output

The ability to produce high-quality hard copies of the results of analyses is an important feature of GIS software and hardware. The software usually allows such manipulations as selecting particular areas or layers of a map for output, scale change, colour change, etc. (Burrough, 1986). As for the hardware, there are various possible options. A 35-mm camera attachment allows photographs to be taken of the visual display unit (VDU) screen, for example. Black-and-white maps can be prepared by dot matrix printers, which are cheap but relatively low-quality. Colour plots can be generated by ink-jet printers and pen plotters. Laser printers produce the highest-quality output, but are expensive. The results of numerical analyses can usually be output as histograms and tables and can also be exported to other software for further analysis.

Acknowledgements

I am grateful for the discussions I have had with Johannes Akiwumi, Mick Wilson (both GRID), Russ Kruska and Jim Scott (both International Laboratory for Research on Animal Diseases). I wish also to thank the organizers and participants of the UNEP/Rockefeller Foundation/ILRAD Workshop on 'Increased Sustainable Agricultural

Productivity in Africa Through the Use of Intelligent Geographic Information Systems' (Nairobi, Kenya, 14–18 January 1991). Tom Hazekamp and Mark Perry (IPGRI) made valuable comments on an early draft.

References

Anon. (1992) Report of the CGIAR/NORAGRIC/UNEP Meeting on Digital Data Requirements for GIS Activities in the CGIAR. September 1992. Arendal, Norway.

Aronoff, S. (1989) *Geographic Information Systems: A Management Perspective*. WDL Publications, Ottawa.

Booth, T.H. (1990) Mapping regions climatically suitable for particular species at the global scale. *Forest Ecology and Management* 36:47–60.

Booth, T.H., J.A. Stein, H.A. Nix and M.F. Hutchinson (1989) Mapping regions climatically suitable for particular species: an example using Africa. *Forest Ecology and Management* 28:47–59.

Burrough, P.A. (1986) *Principles of Geographic Information Systems for Land Resources Assessment*. Clarendon Press, Oxford.

Burrough, P.A. (1991) Intelligent geographical information systems. Paper presented at the UNEP/Rockefeller Foundation/ILRAD Workshop on 'Increased Sustainable Agricultural Productivity in Africa Through the Use of Intelligent Geographical Information Systems'. 14–18 January 1991. UNEP, Nairobi.

Busby, J.R. (1986) A bioclimatic analysis of *Nothofagus cunninghamii* (Hook.) Oerst. in southeastern Australia. *Australian Journal of Ecology* 11:1–7.

Dangermond, J. (1988) A review of digital data commonly available and some of the practical problems of entering them into a GIS. In: *Proceedings of the 1988 ACMS-ASPRS Annual Convention*. Vol. 5, pp. 1–10. American Society of Photogrammetry and Remote Sensing, Falls Church.

Davenport, M.L. and S.E. Nicholson (1993) On the relation between rainfall and the Normalized Difference Vegetation Index for diverse vegetation types in East Africa. *International Journal of Remote Sensing* 14:2369–2389.

FAO (1988) *Geographic Information Systems in FAO*. FAO, Rome.

Gastellu-Etchegorry, J.P., C. Estreguil, E. Mougin and Y. Laumonier (1993) A GIS based methodology for small scale monitoring of tropical forests – a case study in Sumatra. *International Journal of Remote Sensing* 14:2349–2368.

GRID (1992) *A Survey of Geographic Information System and Image Processing Software 1991*. GRID Information Series No. 18. UNEP, Nairobi.

Hilwig, F.W. (1987) Methods for the use of remote sensing in agro-ecological characterization and environmental monitoring. In: Bunting, A.H. (ed.) *Agricultural Environments*. pp. 221–245. CAB International, Wallingford.

Justice, C.O., J.R.G. Townshend, B.N. Holben and C.J. Tucker (1985) Analysis of the phenology of global vegetation using meteorological satellite data. *International Journal of Remote Sensing* 6:1271–1318.

Justice, C.O., B.N. Holben and M.D. Gwynne (1987) Monitoring East African vegetation using AVHRR data. *International Journal of Remote Sensing* 7:1453–1474.

McNeely, J.A., K.R. Miller, W. Reid, R.A. Mittermeirer and T. Werner (1990) *Conserving the World's Biological Diversity*. WRI, IUCN, Conservation International and World Bank, Washington DC and Gland.

Skole, D. and C. Tucker (1993) Tropical deforestation and habitat fragmentation in the Amazon: satellite data from 1978 to 1988. *Science* 260:1905–1910.

Further reading

Articles on GIS and remote sensing commonly appear in the following publications (FAO, 1988):

Cartographica
GIS World
IEEE Transactions of Geoscience and Remote Sensing
International Journal of Remote Sensing
International Journal of Geographical Information Systems
Photogrammetric Engineering and Remote Sensing
Proceedings of the IGU International Symposium on Spatial Data Handling
Proceedings of AUTO-CARTO
Remote Sensing of Environment

There are GIS news groups on Usenet (e.g. comp.infosystems.gis) and also various relevant list server mailing lists (e.g. acdgis-1@awiimc12. imc.univie.ac.at). For a full list, see 'A Biologists's Guide to Internet Resources' by Dr Una Smith, Department of Biology, Yale University, New Haven, CT, USA (smith-una@yale.edu).

Useful addresses

FAO
Remote Sensing Centre
FAO
Via delle Terme di Caracalla
00100 Rome
Italy
Tel: +39 6 57975583
Fax: +39 6 57973152, 57975155 or 5782610
Telex: 625852 FAO I

GRID
There are a number of GRID centres spread throughout the world. Though each has particular regional responsibilities, data and information can be obtained from any centre. No charges apply, but users are requested to supply media. In addition to the centres listed below, there are plans for further centres in Russia (boreal forests), Fiji (South Pacific environmental data), Brazil (Amazonian data sets) and Canada (freshwater data). In the list below, both Dialcom and Internet electronic mail addresses are provided, in that order.

Africa, South-West Asia,
   Latin America and the
   Caribbean
GRID-Nairobi
UNEP
Box 30552
Nairobi
Kenya
Tel: +254 2 230800
   ext. 4187
Fax: +254 2 226491
E-mail: 141:UNE008;
   hcroze@
   nasamail.nasa.gov

Asia and the Pacific
GRID-Bangkok
GPO Box 2754
Bangkok 10501
Thailand
Tel: +66 2 5162124
Fax: +66 2 5162125
E-mail: 141:UNE096

Global and Europe
GRID-Geneva
6, rue la Gabelle
Carouge CH 1227
Geneva
Switzerland
Tel: +41 22 438660
Fax: +41 22 438862
E-mail: 141:UNE060;
   hebin@cgegrdll.bitnet

ISRIC
International Soils
   Reference and Infor-
   mation Centre (ISRIC)
Postbus 353
NL-6700 AJ
   Wageningen
The Netherlands
Tel: +31 8370 19063
Fax: +31 8370 24460
Telex: 45888 INTAS NL

Himalayan and
   Hindukush
ICIMOD
PO Box 3226
Kathmandu
Nepal
Tel: +977 1 526313
Fax: +977 1 524509

Japan
GRID-Tsukuba
Centre for Global
   Environmental
   Research
National Institute for
   Environmental Studies
16–2 Onogawa
Tsukuba, Ibakari 305
Japan
Tel: +81 298 516111
Fax: +81 298 582645

North America
GRID-Sioux Falls
EROS Data Centre
US Geological Survey
Sioux Falls SD 57198
USA
Tel: +1 605 5946107
Fax: +1 605 5946589
E-mail: O:OMNET;
   SN:EROS.DATA.
   CENT; FN:OMNET;
   SITE:TELENET

Polar zones
GRID-Arendal
TK-Senteret, Longum
   Park
PO Box 1602
Myrene N-4801
Arendal
Norway
Tel: +47 41 35500
Fax: +47 41 35050
E-mail: 141:UNE061;
   hesjedal@grida.no

Poland (Baltic basin)
GRID-Warsaw
ul. Jasna 2/4, 00–950
Warsaw
Poland
Tel: +48 22 264231
Fax: +48 22 270328

Plant health and germplasm collectors

17

E.A. Frison[1] and G.V.H. Jackson[2]

[1]*IPGRI, Via delle Sette Chiese 142, 00145 Rome, Italy:*
[2]*87 Market Street, Randwich, Sydney, NSW 2031, Australia.*

Introduction: the need for healthy germplasm

The movement of plant germplasm can potentially spread pests, defined in the *International Plant Protection Convention* (IPPC) as encompassing all harmful or potentially harmful biotic agents from viroids to weeds (in some cases, the plant species collected is itself actually or potentially a pest). In recognition of this danger, most countries have legislation to regulate the entry (and sometimes the internal movement) of plants, plant parts and their products. Consignments of germplasm infested with pests or of plant species considered pests and material without proper documentation may be refused entry and destroyed or reconsigned. Collectors of plant germplasm need to be aware of these facts, or they may fail to accomplish their goals. Germplasm should always be collected, processed and shipped with the necessary phytosanitary precautions to avoid accidental transfer of pests.

There are other, perhaps less obvious, reasons why pests should be given attention when germplasm is collected. Pests may affect the quality, and therefore the usefulness, of germplasm samples. Infection by pathogens can reduce the viability of seeds during storage. When material is multiplied, growth may be distorted, colours altered and disease susceptibility increased. These changes may make it difficult, if not impossible, to collect characterization and preliminary evaluation data, and some important characteristics, crucial for plant improvement schemes, may go undetected. In addition, infested samples are unlikely to be distributed. They cannot be grown out and regenerated and if stored they will remain unused and deteriorate.

It is therefore important to know what pests are likely to be associated with the target gene pool. This will allow an assessment of the risks associated with moving the germplasm and for appropriate measures to be devised to reduce the risk to a minimum. It is also important to document the pests present on the target species at the time of collecting. This information, part of the passport data of the sample, will improve the usefulness of the germplasm and will also help during quarantine examination.

For all these reasons, it will often be useful to include a plant protection specialist in collecting teams if funds and logistical considerations allow. Preferably, this should be a plant pathologist experienced in the species to be collected, as pathogens are more difficult than insects and mites to detect during collecting and to eradicate from plant samples. If a plant pathologist cannot participate in the collecting mission, collectors should become familiar with the major pests of the target species. In all cases, collectors will have to ensure that the proper documentation has been assembled so that plant samples reach their intended destination unhindered.

This chapter gives guidelines on how these issues may be addressed. It considers what must be done at the planning stage, while collecting in the field and, finally, just before samples are dispatched.

Planning the collecting mission

At the planning stage, attention must be given to the pests that might be encountered on the target species and to the regulations governing plant movement. The following questions need to be considered when assembling information on plant pests:

- What pests have been recorded on the target species in the country of collecting, especially in the target area?
- What plant parts are they found on?
- How are the pests transmitted?

The following questions need to be answered to ensure compliance with phytosanitary regulations:

- What is the final destination(s) of all subsamples?
- What are the phytosanitary import requirements of the country(ies) of destination?
- What are the procedures for obtaining a phytosanitary certificate in the country of collecting?

Assembling information on pests

Pest surveys should be consulted to determine which pests have been recorded on the target species in the collecting region. However, in many countries such surveys are far from complete. Sometimes, they have not been done at all, are outdated or do not cover the entire country, concentrating on the more easily accessible areas. Another problem is that surveys mostly record pests of crop plants, neglecting wild relatives.

This lack of information is a major barrier to the formulation of quarantine regulations appropriate to the exchange of germplasm of some crops.

Texts that may be consulted for information on the pests of specific crops are listed under References. Holliday (1989) includes a list of major texts on plant pathology, including crop-specific disease compendia and references to these under crop names. On pest distributions, the following are important sources:

Distribution Maps of Pests	IIE (1968 *et seqq.*)
Distribution Maps of Plant Diseases	IMI (1942 *et seqq.*)

Collectors should confirm with the relevant institutes that these maps contain the most up-to-date information. Detailed descriptions, including notes on the transmission of many of the pests figured in the maps, can be sought from the following publications:

Descriptions of Fungi and Bacteria	IMI (1964 *et seqq.*)
Descriptions of Plant-parasitic Nematodes	IIP (1972 *et seqq.*)
Descriptions of Plant Viruses	CMI/AAB (1970–1984), AAB (1985 *et seqq.*)

On viruses, CAB International (CABI) and the Australian National University have collaborated on a major database, Virus Identification Data Exchange. *Viruses of Tropical Plants* (Brunt *et al.*, 1990) is an output of the database, and the publication of a further comprehensive book (*Viruses of Plants*) is planned by CABI in 1995. The commodity international agricultural research centres (IARCs) also publish useful illustrated guides to the pests of their mandate crops. These are particularly useful in the field.

In addition, a series of booklets of crop-specific *Technical Guidelines for the Safe Movement of Germplasm* is published jointly by the Food and Agriculture Organization (FAO) and the International Plant Genetic Resources Institute (IPGRI) (formerly IBPGR)). Each booklet is divided into two parts. The first part makes recommendations on how best to move germplasm of the crop concerned and lists institutions recovering and/or maintaining healthy germplasm. The second part covers the important pests and diseases of quarantine concern, giving a description of therapy and indexing methodologies. So far, guidelines have been produced for the following crops:

aroids (edible)	Zettler *et al.* (1989)
Citrus spp.	Frison and Taher (1991)
Cocos nucifera	Frison *et al.* (1993)
Dioscorea spp.	Brunt *et al.* (1989)
Ipomoea batatas	Moyer *et al.* (1989)
legumes	Frison *et al.* (1990)
Manihot esculenta	Frison and Feliu (1991)
Musa spp.	Frison and Putter (1989)

Saccharum officinarum	Frison and Putter (1993)
Theobroma cacao	Frison and Feliu (1989)
Vanilla spp.	Pearson *et al.* (1991)
Vitis vinifera	Frison and Ikin (1991)

Guidelines for small fruits (*Fragaria, Ribes, Rubus* and *Vaccinium*) were published in 1994, and booklets on *Allium, Brassica, Oryza, Solanum* potatoes, *Zea* and forestry trees are planned. Hewitt and Chiarappa (1977) is an earlier, crop-by-crop analysis of the problems and risks attendant on the transfer of plant germplasm.

The Plant Protection Service of FAO, in collaboration with regional plant protection organizations, has developed a database of plant pests worldwide, which is updated and distributed regularly to member countries. This Global Plant Quarantine Information System describes the geographic distribution of each pest, specifies hosts and commodity type affected, comments on quarantine status and describes preferred treatments. Lists of pests affecting particular hosts may be generated. For each country, the database provides lists of pest records, digests of plant quarantine regulations and information on the national plant quarantine organization. Based on the information in the database, the software can assess the quarantine risk associated with particular consignments. Updating of the database is coordinated with the regional plant protection organizations, CABI and other research institutes involved in verification of pest distribution and damage reports.

Assembling the required plant health documents

It is essential to begin making phytosanitary arrangements as soon as possible. Delays in obtaining the appropriate documents are common, but without these documents the mission may be postponed or, worse, the samples destroyed. It is the responsibility of collectors to obtain the necessary documents in order to transfer plant germplasm. Two documents are commonly required for international transfer: an import permit and a phytosanitary certificate.

The import permit

The import permit is obtained from the country or countries of destination of the germplasm before the mission sets out. At this stage, information is also gathered on how to obtain a phytosanitary certificate in the country of collecting and whether other authorizations are required to export germplasm. The FAO/IPPC Secretariat has a list of plant protection services worldwide with contact addresses of the authorities responsible for issuing these documents. This information is also available in the Global Plant Quarantine Information System and as a hard-copy directory (FAO, 1993).

An import permit, issued by the quarantine authorities of the importing country, stipulates the conditions governing plant introductions. If samples are to be collected and sent to more than one country for safety duplication, study or use, each country has to be approached to supply an import permit.

No general rules apply. Regulations differ among countries according to the perceived risks involved in making the importation. There are a number of different possibilities and the conditions of entry will be detailed on the import permit. Even when no conditions apply and germplasm is allowed unconditional entry, a document from the plant protection service of the importing country to that effect should be obtained.

Collectors should ensure that the number of samples and the approximate size of each are known to the importing country well ahead of arrival, so that the quarantine inspection service can properly plan to process the samples. This is particularly important if samples are to be grown in post-entry quarantine in the importing country or in a third country.

Usually, two copies of the import permit are obtained. The top copy should always accompany the consignment. A photocopy is usually allowed for multiple consignments. A copy of the import permit must be retained by the collector.

The phytosanitary certificate

The phytosanitary certificate is issued by the quarantine authority of the exporting country, certifying that the product meets the phytosanitary regulations of the importing country. Consignments are inspected and the certificate issued if they are 'free from quarantine pests and practically free of injurious pests' (see the IPPC model phytosanitary certificate, Appendix 17.1 at the end of this chapter). A 'quarantine pest' is different from a merely 'injurious pest' in this jargon in that it is of potential national economic importance to the country and not yet present there, or present but not widely distributed and being actively controlled.

In some instances, in order to reduce the overall pest risk, germplasm consignments will need to be given phytosanitary treatments in the country of origin (but see Chapter 20 on the potential risks for seed viability of such treatments). Fumigation may be requested or the samples may be dipped or dusted in an insecticide or fungicide, given a hot-water treatment, or whatever is considered appropriate by the importing country. The treatments should be applied exactly as requested. The permit may seek additional endorsements. These, as well as the treatments, should be detailed on the phytosanitary certificate. Finally, the certificate should be signed by the duly authorized government representative.

Under no circumstances should treatments be applied as alternatives to those of the import permit without first requesting the authority of the importing country. Likewise, if no treatments are requested, none should be given, since importing countries may wish to inspect or test germplasm consignments, and treatments already applied to seeds may mask symptoms of seed-borne pathogens and interfere with laboratory tests. Alternatively, treatments already applied may be ignored, in which case a second treatment could reduce viability. If seeds are pretreated prior to entry, against the conditions of the permit, this could seriously jeopardize their importation.

Where germplasm samples are to be sent to more than one country, it is necessary to obtain a phytosanitary certificate for each destination. It is important that the certificate(s) should be issued without amendment or erasure. Many countries refuse to accept altered certificates. A fee may be charged for fumigation or disinfection treatments and, occasionally, for inspection.

Two copies of the phytosanitary certificate should be obtained. The original should accompany the consignment.

Documentation and intermediate quarantine

Collectors are also responsible for arranging the documentation for germplasm samples that have to be grown in intermediate (third-country) quarantine. Such arrangements are necessary when it is unsafe to make transfers directly to the importing country, but quarantine multiplication may be necessary even if the material is not to leave the country of collecting. Procedures are essentially similar to those outlined above. An import permit must be obtained from the quarantine authority of the intermediate country. A copy of this must accompany the consignment, together with a phytosanitary certificate showing any treatments or endorsements requested on the permit. After the samples have been grown in intermediate quarantine and declared safe for further transfer, an import permit must be obtained from the country of final destination and a new phytosanitary certificate issued by the intermediate country.

Planning the identification of pests

Misidentifications of pests can seriously jeopardize the usefulness of consignments. Identification services for fungi, bacteria, nematodes, insects and mites are provided by CABI. Costs vary depending on whether or not a country is a member of CABI. CABI also publishes useful directories of organizations, for example Hall and Hawksworth (1990). It may be possible to arrange for the identification of important seed-borne diseases of tropical countries by the Danish Government Institute of Seed Pathology for Developing Countries, Hellerup, Denmark. Identification of virus and virus-like infections is more problematical. Specimens will need to be sent to institutes specializing in particular crop plants. Lists of institutes providing this service can be found in the appropriate booklet in the FAO/IPGRI series of safe transfer guidelines. The Tropical Virus Unit at the Institute of Arable Crops Research, Rothamstead Experimental Station, UK, is an example.

In all cases, arrangements must be made well ahead of dispatch to allow the orderly processing of specimens. Import permits may be needed. If so, these must be obtained from the appropriate authorities in the country where specimens are to be examined. Collectors should ensure that the institutes making the identifications know where to send the results.

In the field

Minimizing the pest risk

Familiarity with the symptoms caused by pests and with which plant parts are most likely to be contaminated by the different pests of concern is essential. In general, the risk of spreading pests with germplasm is greatest if rooted plants are moved. This is because of the likelihood that nematodes and other soil-borne pathogens will be present: these are difficult to treat without destroying the plant tissues. Other types of vegetative propagating material (e.g. stems, bulbs, corms, etc.) also present a risk, mainly because of infection from systemic pathogens. The international movement of seeds and pollen is considered safer, as fewer pests are harboured by these plant organs. Phytosanitary considerations may therefore contribute to the decision as to what plant part(s) to collect.

It may be possible to apply curative treatments to lessen or eradicate the pest risk. For surface-borne pathogens and insects, pesticide treatments and fumigation may be tried. Where virus, virus-like organisms and internally borne fungi and bacteria are a threat, thermotherapy and shoot-tip culture are most appropriate.

For vegetatively propagated species, transfer of germplasm as *in vitro* cultures will greatly reduce the pest risk. Nevertheless, it should be stressed that *in vitro* culture *per se* does not eliminate the risk entirely. It should be complemented by indexing (testing) for viruses and virus-like organisms that are suspected to be present in the area where the germplasm was collected.

The safe transfer guidelines give general advice to collectors on the type of germplasm considered safe to move internationally, and detailed technical recommendations on how the germplasm may be treated to ensure that it is free of pests. In some instances, because of the severity of the pest and the difficulty of collecting healthy material from the field, the guidelines advise transfer of material through a third country, where therapy and indexing procedures can be carried out to ensure freedom from internally borne pathogens. The general recommendations of the guidelines are useful even for crops not specifically covered in the series to date.

Recording data on pests

It is important for collectors to record the pests present on their target species and whether other pests are present in the target region. Noting that plants are free of pests in an area where pests are common is equally important. Collectors should attempt to describe the symptoms caused by pests. It is, however, often difficult for someone untrained in plant pathology or entomology to do this. Symptoms may be caused by a combination of several pests, or the causal agent may be obscured by the presence of a minor one or by an opportunistic saprophyte. Symptoms due to root attack or internal pathogens are often particularly difficult to interpret. Where there is doubt as to the identification of pests, plant specimens showing typical symptoms should be collected and dried or preserved by other means, as appropriate (see below).

A description of symptoms should include information on the following (Sonoda, 1979):

- the general condition of the plant;
- the plant part(s) affected;
- the type of damage;
- the stage of growth affected.

Rating the severity of attack, both in terms of its effect on the individual plant(s) affected and in terms of the percentage of the population affected, will increase the value of the information. Descriptor lists are published by IPGRI for many crops, and these catalogue the important pests and give scales of severity. Colour photographs showing the full range of symptoms, including close-ups of damaged areas and of the pests themselves, are often useful diagnostic tools (Sonoda, 1979).

Farmers' knowledge of pests can be extensive and detailed. Some examples are given by Altieri (1993). Collectors can often complement the kinds of observations described above with discussions with knowledgeable local people.

Preservation of pests associated with germplasm samples

Correct identification of pests depends on the quality of the specimens prepared in the field. Collectors should be equipped at least with specimen bottles, alcohol (75% isopropyl alcohol) and formalin for preserving insects, mites and nematodes, and with newspapers and plant presses for making dried herbarium specimens of plants with fungal and bacterial diseases (Chapter 27). Specimens may need to be shared among several institutes, and sufficient material should be collected to allow this.

Sonoda (1979) gives guidelines on capturing, killing and storing insects and other pests in the context of germplasm collecting. For insect pests, representative specimens of all life stages may be necessary for taxonomic identification. Insects can be captured using nets, by beating plants over a cloth or by using an aspirator. They can be killed using potassium cyanide or ethyl acetate, both of which are dangerous and should be clearly labelled and properly stored. Some must be pinned (e.g. Diptera and Hymenoptera), others can be stored in alcohol (e.g. beetles) and others can be stored in small envelopes (e.g. Lepidoptera).

Dried specimens of diseased plants should include as much of the plant as possible, showing both old and new lesions if possible. Fresh specimens can also be collected and stored in plastic bags. They will remain useful longer if refrigerated. Fungal and bacterial pathogens may be isolated from diseased plants in the field, but this requires sterile technique and will not often be feasible in the context of plant germplasm collecting.

Plants infected with viruses or virus-like organisms present the collector with the greatest challenge, as the material needs to be processed in different ways according to the type of pathogen. Where tissues are thought to contain non-cultivable mollicutes (formerly referred to as mycoplasma-like organisms), they need to be fixed in glutaraldehyde, whereas tissues for virus examination may be sent fresh, dried as thin sections over calcium chloride or as sap stained on electron-microscope grids. Because of the complexity of the subject, it is essential that prior to departure collectors seek advice on the preservation of specimens from the institutes where the specimens are to be sent for examination.

Details of methods of preserving various kinds of diseased material can be found in The *Plant Pathologist's Pocketbook* (Johnston and Booth, 1983). Methods for collecting and preserving different insect groups can be found in Bland and Jacques (1978), British Museum (Natural History) (1974) and Borror *et al.* (1976).

Back at base: treatment and dispatch of germplasm samples

This section gives a summary of the phytosanitary procedures involved in handling plant germplasm after it has been collected and brief notes on the dispatch of specimens for pest identification. For other aspects of the tasks that will need to be undertaken once back at base, see Chapter 28.

Inspection

Missions should carefully prepare germplasm samples before they are presented to quarantine authorities for inspection, treatment and certification.

- Germplasm samples should be carefully inspected for pests, insects and mites as well as for lesions or colour patterns which may denote fungal, bacterial or viral pathogens. Where such pests, or symptoms of pests are present, the pests and/or the symptom-bearing seeds should be removed.
- Bare-rooted plants should be thoroughly washed to ensure they are

free of soil, which might harbour nematodes and other soil-borne pathogens.
- Seeds and pollen should be free of debris. If this is present, it should be removed.

Phytosanitary treatments and certification

- If mandatory treatments are prescribed on the import permit or endorsements are required, these should be given by the relevant government authority exactly as requested.
- After treatments have been applied, they should be detailed on the phytosanitary certificate, together with any other endorsements, as requested by the importing country.
- Two copies of the phytosanitary certificate should be obtained, one of which should be the original. Each copy should bear the stamp of the organization issuing the certificate, and be signed by an authorized officer.
- Collectors should ensure that the phytosanitary certificate contains the following information:

    name and address of the exporter;
    name and address of the consignee;
    number of samples of each species in the consignment;
    botanical name of each species;
    phytosanitary treatments applied;
    additional endorsements required by the import permit.

Documents accompanying germplasm consignments

- The original copy of the phytosanitary certificate, plus a copy of the import permit, should accompany the consignment. Photocopies may be allowed if there are multiple shipments. This will have to be ascertained from the quarantine authorities of the importing country. The top copy of the permit should be placed on the outside of the package so it can be forwarded to the plant quarantine authorities without the need to open the package. A photocopy should be included inside the package in case of damage to the outside copy. However, this may vary from country to country. For example, US regulations specify that all documents should be inside the package.
- A copy of all documents sent with the consignments should be retained by the collector.

Preparation of samples for pest identification

Arrangements should be made in advance of the fieldwork with the institutes that are to receive samples for pest identification. Permits may have to be obtained to comply with the quarantine requirements of the country where samples are to be sent. Some additional points to note are:

- All material sent for identification purposes, whether preserved insects and mites, dried plant voucher specimens of diseased plants or living plant material for diagnosis of internal pathogens, should be labelled with:
    - a reference number;
    - the botanical name of the host plant;
    - the locality where collected;
    - the date of collecting;
    - the name of the collector(s).
- Collectors should keep a copy of the information accompanying each specimen.
- Samples of seeds and pollen may have to be sent for viability testing, as well as for inspection for internally borne pathogens and weeds. Samples should be properly dried before dispatch.
- Collectors should include the name of the person (and address) to whom the identification(s) should be sent.

References

General

AAB (1985 *et seqq.*) *Descriptions of Plant Viruses*. Association of Applied Biologists, Wellesbourne.

Altieri, M.A. (1993) Ethnoscience and biodiversity: key elements in the design of sustainable pest management systems for small farmers in developing countries. *Agriculture, Ecosystems and Environment* 46:257–272.

Bland, R.G. and H.E. Jacques (1978) *How To Know Insects*. W.C. Brown Company Publishers, Dubuque.

Borror, D.J., D.M. Delong and C.A. Triplehorn (1976) *An Introduction to the Study of Insects*. Holt, Rinehart and Winston, New York.

British Museum (Natural History) (1974) *Insects. Instructions for Collectors No. 4a*. British Museum (Natural History), London.

Brunt, A.A., K. Crabtree and A. Gibbs (1990) *Viruses of Tropical Plants*. CAB International, Slough.

CMI/AAB (1970–1984) *Descriptions of Plant Viruses* (Sets 1–18). CAB International and Association of Applied Biologists, Slough and Wellesbourne.

FAO (1993) *Directory of Regional Plant Protection Organizations and National Plant Quarantine Services*. AGPP/Misc/93/1. FAO, Rome.

Hall, G.S. and D.L. Hawksworth (1990) *International Mycological Directory*. CAB International, Wallingford.

Hewitt, W.B. and L. Chiarappa (1977) *Plant Health and Quarantine in International Transfer of Genetic Resources*. CRC, Cleveland.

Holliday, P. (1989) *A Dictionary of Plant Pathology*. Cambridge University Press, Cambridge.

IIE (1968 *et sqq.*) *Distribution Maps of Pests*. Series A (Agriculture). International Institute of Entomology (formerly Commonwealth Institute of Entomology) and CAB International, Wallingford.

IIP (1972 *et sqq.*) *IIP Descriptions of Plant-parasitic Nematodes*. International Institute of Parasitology (formerly Commonwealth Institute of Helminthology) and CAB International, Wallingford.

IMI (1942 *et sqq.*) *IMI Distribution Maps of Plant Diseases.* International Mycological Institute (formerly Commonwealth Mycological Institute) and CAB International, Wallingford.

IMI (1964 *et sqq.*) *IMI Descriptions of Fungi and Bacteria.* International Mycological Institute (formerly Commonwealth Mycological Institute) and CAB International, Wallingford.

Johnston, A. and C. Booth (1983) *Plant Pathologist's Pocketbook.* CAB International, Wallingford.

Sonoda, R.M. (1979) Collection and preservation of insects and pathogenic organisms. In: Mott, G.O. and A. Jiménez (eds) *Handbook for the Collection, Preservation and Characterization of Tropical Forage Germplasm Resources.* CIAT, Cali, Colombia.

FAO/IPGRI Technical Guidelines for the Safe Movement of Germplasm Series

Brunt, A.A., G.V.H. Jackson and E.A. Frison (1989) *FAO/IBPGR Technical Guidelines for the Safe Movement of Yam Germplasm.* FAO/IBPGR, Rome.

Frison, E.A. and E. Feliu (eds) (1989) *FAO/IBPGR Technical Guidelines for the Safe Movement of Cocoa Germplasm.* FAO/IBPGR, Rome.

Frison, E.A. and E. Feliu (eds) (1991) *FAO/IBPGR Technical Guidelines for the Safe Movement of Cassava Germplasm.* FAO/IBPGR, Rome.

Frison, E.A. and R. Ikin (eds) (1991) *FAO/IBPGR Technical Guidelines for the Safe Movement of Grapevine Germplasm.* FAO/IBPGR, Rome.

Frison, E.A. and C.A.J. Putter (eds) (1989) *FAO/IBPGR Technical Guidelines for the Safe Movement of* Musa *Germplasm.* FAO/IBPGR, Rome.

Frison, E.A. and C.A.J. Putter (eds) (1993) *FAO/IBPGR Technical Guidelines for the Safe Movement of Sugarcane Germplasm.* FAO/IBPGR, Rome.

Frison, E.A. and M. Taher (eds) (1991) *FAO/IBPGR Technical Guidelines for the Safe Movement of* Citrus *Germplasm.* FAO/IBPGR, Rome.

Frison, E.A., L. Bos, R.I. Hamilton, S.B. Mathur and J.D. Taylor (eds) (1990) *FAO/IBPGR Technical Guidelines for the Safe Movement of Legume Germplasm.* FAO/IBPGR, Rome.

Frison, E.A., C.A.J. Putter and M. Diekmann (eds) (1993) *FAO/IBPGR Technical Guidelines for the Safe Movement of Coconut Germplasm.* FAO/IBPGR, Rome.

Moyer, J.W., G.V.H. Jackson and E.A. Frison (1989) *FAO/IBPGR Technical Guidelines for the Safe Movement of Sweet Potato Germplasm.* FAO/IBPGR, Rome.

Pearson, M.N., G.V.H. Jackson, F.W. Zettler and E.A. Frison (1991) *FAO/IBPGR Technical Guidelines for the Safe Movement of Vanilla Germplasm.* FAO/IBPGR, Rome.

Zettler, F.W., G.V.H. Jackson and E.A. Frison (1989) *FAO/IBPGR Technical Guidelines for the Safe Movement of Edible Aroid Germplasm.* FAO/IBPGR, Rome.

Appendix 17.1
IPPC model phytosanitary certificate

No. _____

Plant Protection Organization of _____

To: Plant Protection Organization(s) of _____

Description of consignment

Name and address of exporter _____

Declared name and address of consignee _____

Number and description of packages _____

Distinguishing marks _____

Place of origin_____

Declared means of conveyance _____

Declared point of entry_____

Name of produce and quantity declared _____

Botanical name of plants _____

This is to certify that the plants or plant products described above have been inspected according to appropriate procedures and are considered to be free from quarantine pests, and practically free from other injurious pests; and that they are considered to conform with the current phytosanitary regulations of the importing country.

Disinfestation and/or Disinfection Treatment

Date _____ Treatment _____ Chemical (active ingredient) _____

Duration and temperature _____ Concentration _____

Additional information _____

_____

Place of issue _____

(Stamp of organization)   Name of authorized officer _____

Date _____ _____

(Signature)

_____

No financial liability with respect to this certificate shall attach to _____

(name of Plant Protection Organization) or to any of its officers or representatives.*

_____

*Optional clause

II

IN THE FIELD

Collecting plant genetic resources and documenting associated indigenous knowledge in the field: a participatory approach

18

L. Guarino[1] and E. Friis-Hansen[2]

[1]IPGRI, c/o International Laboratory for Research on Animal Diseases, PO Box 30709, Nairobi, Kenya: [2]Herluf Trolles Gade 26A.2tv, DK1052 Copenhagen K, Denmark.

Chapter 12 deals with the importance of indigenous knowledge (IK) of landraces, crops, wild plants and the environment in plant genetic resources collecting. It also describes possible secondary sources of the ethnographic data that germplasm collectors will find helpful, if not essential, when planning their missions. Once in the field, however, collectors will need to document IK for themselves, in Prain's (1993) words 'turning the passport [data] into a potted biography' of the germplasm they are gathering. As suggested in Chapter 12, this requires a participatory approach. Various suitable methodologies have been developed by social scientists, but these may be unfamiliar to many biologists. The aim of the present chapter is to introduce the basics of a participatory methodology for collecting plant genetic resources and the IK associated with them. Chapter 19 discusses in more detail the individual topics that it will be necessary to document in the field. Chapter 38 describes a case-study in which some of the principles and techniques discussed here were actually put into effect during a collecting mission.

Introduction to social science methods

Over the past decade, social science theories of development and the practical methodologies that underpin them have proliferated. Rather than a few comprehensive theories of development (and underdevelopment), we now have a whole variety of interlinked concepts, such as sustainability, environmental protection, gender relations, institutional capacity and empowerment. This has provided a justification for methodological pluralism and flexibility, resulting in the complementation (or replacement) of time-consuming conventional quantitative survey methods with novel, mainly qualitative analytical techniques. The significant expansion in development assistance during the same period and the subsequent need for rapid methods of impact assessment have also contributed to the development of new methods, and new ways of using old methods.

A list of some qualitative methods is given in Box 18.1 (Chambers, 1990, 1992, 1993; Theis and Grady, 1991). These largely visual methods stimulate discussion, allow the illiterate and the otherwise marginalized to contribute and facilitate communication between insiders and outsiders. They are increasingly seen as ideal not just for tapping IK, but for ensuring that communities benefit from the exercise. Practitioners select from such a repertoire the tools most relevant and appropriate to their particular situation and problem. Several will be of direct relevance to germplasm collectors, and these are described in more detail below and in Chapter 19. Secondary socioeconomic sources are discussed separately in Chapter 9 and secondary ethnographic sources in Chapter 12.

Two main strategies for using these qualitative, visual methods have emerged: rapid rural appraisal (RRA) and participatory rural appraisal (PRA). RRA was developed in the 1970s as a tool for development workers. Meant to replace expensive and time-consuming approaches such as formal large-scale questionnaire surveys, RRA involves relatively short but intensive visits by multidisciplinary teams, the members of which interact among themselves and with the community by means of a whole variety of qualitative techniques to answer specific, preset questions on a particular topic (McCracken, 1988). According to Chambers (1992), there were three reasons for the development of RRA: (i) dissatisfaction with the biases inherent in development work based on hasty visits to the countryside by mainly urban-based professionals; (ii) disillusion with the quantitative questionnaire methodology; and (iii) increasing interest in methods of tapping IK.

Chambers (1990, 1992) describes how RRA evolved into PRA in the 1980s, largely in the hands of activist non-governmental organizations (NGOs) seeking to integrate the whole of a local community (women as well as men, the poorest as well as the not-so-poor, etc.) in a development process generated from within rather than imposed from outside. The term PRA was probably first used in Kenya, to describe the village-level investigations undertaken by the National Environment Secretariat in collaboration with Clark University in the USA. It was introduced into India in a joint exercise of the Aga Khan Rural Support Programme and the International Institute for Environment and Development (IIED) in 1988. It has since spread widely there, mostly in the NGO sector but also in some training institutes. Box 18.2 lists the principles shared by RRA and PRA and the additional points stressed by PRA (after Chambers, 1992, 1993). The differences between the two approaches are summarized by Chambers (1992) as follows:

Box 18.1
Qualitative data collection, analysis and communication techniques

- Review of secondary sources:
  - historical documents;
  - official reports and statistics;
  - ethnographies.
- Direct observation.
- Do-it-yourself.
- Key local indicators (e.g. of wealth, health, etc.).
- Semi-structured interviews:
  - key individuals;
  - key probes;
  - focus groups, homogeneous or mixed groups;
  - chain of interviews;
  - simple questionnaires.
- Ranking and scoring exercises.
- Contrast comparisons and analysis of difference.
- Construction and analysis of maps and models:
  - interpretation of aerial photographs;
  - on-the-spot sketch mapping;
  - three-dimensional landscape models.
- Group treks and transects.
- Diagramming:
  - time lines, trend analysis;
  - seasonal diagrams (e.g. of crops, climate, etc.);
  - activity profiles and daily routines;
  - bar, flow and Venn diagrams;
  - decision trees.
- Case-studies and stories:
  - life histories;
  - oral or written stories by key people.
- Drama, games and role plays.
- Possible-future and scenario workshops.
- Brainstorming.

RRA is [still] mainly extractive. Outsider professionals go to rural areas, obtain information, and then bring it away to process and analyze. PRA, in contrast, is participatory. Outsider professionals still go to rural areas, but their role is more to facilitate the collection, presentation and analysis of information by rural people themselves. With RRA the data are owned by the outsiders, and often not shared with rural people; with PRA they are owned by rural people, but usually shared with outsiders.

Valuable comprehensive lists of sources of information on RRA/ PRA, of relevant organizations and of practitioners in several countries are provided by Chambers (1992). Some of the more important sources

---

Box 18.2
A comparison of RRA and PRA

Principles shared by RRA and PRA
- Learning directly from people, in the field and face to face.
- Learning rapidly and progressively, not following a rigid blueprint but adapting and improvising.
- Offsetting biases by being receptive and responsive, not imposing and dictating, and actively seeking out excluded groups such as poorer people and women.
- Optimizing trade-offs between the quantity, relevance, accuracy and timeliness of information, which includes being aware of what is not worth knowing (optimal ignorance) and not measuring more than necessary (appropriate imprecision).
- Triangulating, i.e. using a range of methods, investigators, types of information and/or disciplines to cross-check results.
- Seeking diversity rather than averages, i.e. deliberately looking for, noticing and investigating contradiction, anomaly and differences.

Additional principles stressed by PRA
- Encouraging investigation, analysis and presentation by rural people themselves, so that they have a stake in the results, the outsider adopting a low-profile catalytic role.
- Accepting and adopting the pace of the community.
- Self-critical awareness and responsibility, i.e. facilitators are continuously examining their behaviour, embracing error as an opportunity to learn to do better.
- Sharing of information and ideas among rural people, between them and the facilitators and between different facilitators.

---

are listed in Box 18.3. The Qualitative Research Methods Series (Sage Publications, Newbury Park) gives practical advice on specific techniques.

Towards a method of exploring IK of plant genetic resources

Memory banking

The principles of RRA and PRA have been used to develop a number of thematic methodologies, e.g. environmental impact assessment, institutional capacity analysis, human rights analysis and gender analysis. Warren and Rajasekaran (1994) have produced a general manual for the documentation of IK systems in development work. The first attempt to develop a methodology for participatory IK data collecting in the specific area of plant genetic resources conservation is memory banking (Sandoval, 1994).

Memory Banking entails three phases. During the first, which may

---

Box 18.3
Information sources on RRA/PRA methodologies

- *Agricultural Administration*'s 1980–81 special issue on RRA, Vol. 8(6).
- Charity Kabutha (National Environment Secretariat of Kenya), Richard Ford and Barbara Thomas-Slayter's (both of Clark University) *Participatory Rural Appraisal Handbook* (Kabutha *et al.*, n.d.).
- Proceedings of the 1985 International Conference on Rapid Rural Assessment. Rural Systems Research and Farming Systems Research Projects, Khon Kaen.
- *Forests, Trees and People Newsletter*, published by the International Rural Development Centre of the Swedish University of Agricultural Sciences at Uppsala and the Food and Agriculture Organization (FAO) Community Forestry Unit, in particular issue 15/16 (1992).
- FAO Community Forestry Unit's *Community Forestry Notes* (Nos. 2, 3 and 5) and *Community Forestry Field Manuals* (No. 2) and other FAO publications, such as *Participatory Monitoring and Evaluation. Handbook for Training Field Workers*.
- *RRA Notes* and other IIED publications, for example various PRA and RRA workshop reports, Theis and Grady (1991) and Gueye and Schoonmaker Freudenberger (1991); this last is one of the few publications on RRA in French; according to Chambers (1990), Ruano (1989) is a useful source in Spanish.
- The *PRA/PALM Series* of MYRADA, an NGO based in southern India.
- The PRA bibliography and various other publications of the Institute of Development Studies (IDS), University of Sussex, UK.
- Various publications of the Popular Participation Programme, Development Studies Unit, Department of Social Anthropology, Stockholm University.
- ILEIA's (1992) series of training handbooks, *Learning for Participatory Technology Development*. Also, the proceedings of the workshop 'Operational Approaches for Participatory Technology Development in Sustainable Agriculture' (ILEIA, 1989).

---

last a month or more, an effort is made to establish a rapport with the community and 'to arrive at a working knowledge of the agricultural system and the different players involved' (Nazarea-Sandoval, in prep.). Contextual information on the environment, on the human population and on farming systems is also gathered during this period. Specimens of landraces are collected and maintained as herbarium vouchers and perhaps also live in a demonstration garden. This will 'provide a physical record of types against which local names and evaluations can be checked' and an initial indication of the extent and character of genetic diversity within the target crop. This initial phase can be seen as one of 'participant observation', an approach that has long been used by anthropologists in which researchers interact closely with the community by immersing themselves in it for a considerable period, living with their respondents and helping them in their tasks (e.g. Jorgenson, 1989). A conventional structured benchmark socioeconomic survey may follow the initial familiarization phase of memory banking, perhaps administered by specially trained local assistants or enumerators.

The next step in the memory banking process is an attempt to reconstruct local history. The techniques involved in this process are:

- Interviews with so-called community 'gatekeepers'. These are people in authority who can provide an 'official' version of history (which can be presented in the form of time lines of important events, for example) as well as introductions to key informants.
- The elicitation of life histories from knowledgeable, usually elderly, users of the germplasm (i.e. local experts, or so-called key informants) in open-ended interviews, concentrating on changes in varietal composition and agricultural practices during their lifetimes (Crapanzano, 1984).
- Asking key informants to make drawings of different varieties from memory, in an effort to isolate those features most significant to different categories of local users. This is sometimes called cognitive mapping.

Finally, in the systematization phase of memory banking, the relationships among varieties are investigated with the key informants. Triads tests and sorting/ranking tests may be used to elucidate folk taxonomies. A triads test involves presenting sets of three stimuli (e.g. plant specimens) to informants and asking them which one does not belong in the group and why. Nazarea-Sandoval (1992), working with sweet potatoes in the Philippines, found that local criteria commonly used in discriminating among varieties were of five main types: morphological, gastronomic, life habit, familiarity and functional (for example, whether the main product is the roots or the leaves). Informants may also be asked to rank or sort a set of objects, and then to reveal the basis for their choice, or to rank objects according to a parameter suggested by the researcher. Friedberg (1968) describes other methods that have been used to discover the criteria by which classes of a folk taxonomy are distinguished; the informant can be asked to construct a dichotomous key or objects can be presented in pairs and similarities and differences listed. Nazarea-Sandoval (1990) describes other techniques. The result is documentation of the discrimination, characterization and evaluation criteria used by everyday users of the germplasm.

Limitations of memory banking

Memory banking perhaps comes most completely into its own in the context of on-farm, community-level conservation, for which it is ideally suited. The participatory philosophy which underlies it is also relevant to collecting for *ex situ* conservation, of course, but carrying out some of the exercises stipulated by the full protocol may not always be feasible in the context of a 'typical' germplasm collecting programme. The rhythm of memory banking is very much imposed by the cultural context. The problem is that cultures are complex, and collectors' time in the field is often short, for example in the case of emergency rescue missions or because of lack of resources. A month is a long time in botanical and genetic resources fieldwork, whereas a year is a short time in anthropology. A prolonged period of participant observation is a luxury most collecting programmes will not be able to justify, despite the fact that an extended sojourn within a community will have important practical advantages over the conventional short visit, for example allowing the collector to sample material throughout the growing season, thus capturing earlier- as well as later-maturing crops and landraces. A full-scale socioeconomic survey may be equally difficult to find time and resources for. Even the kind of sequence of long interviews and participatory exercises with several informants which is prescribed by memory banking may not be feasible in some cases.

It should, however, be possible to devise a procedure based on the ideal of memory banking but flexible enough to adapt to whatever limitations of time and/or resources may exist in any particular case. For example, an attempt can be made to make up for the lack of a prolonged participant observation stage (at least in part) by literature-based familiarization with the local culture and local history. Secondary sources may be sufficient to provide a working idea of the way a community is organized socially and economically and of relevant rules, customs and restrictions. The close involvement should also be sought of the local community not just in sanctioning the mission, but also in planning it. Indeed, the inclusion of farmers and other rural people in the collecting team itself should be considered.

There is scope for flexibility once in the field too, and a suitably adaptable field methodology is presented in the next section. It is a collecting system which perhaps tends more towards the RRA than the PRA pole, and is certainly not in the tradition of participant observation, while remaining basically participatory and informant-led. Such an approach has been adopted by the User's Perspective with Agricultural Research and Development (UPWARD) in collecting sweet potatoes in Irian Jaya, as described in Chapter 38. In such an exercise, the informant becomes a veritable 'germplasm consultant' (Prain, 1993). An important tool of RRA/PRA is multidisciplinary teams, and the inclusion of a social scientist in crop collecting teams, as in the pioneering UPWARD work, is clearly as desirable as that of local farmers as advisers, guides and go-betweens. The idea of agricultural and social scientists working together in 'sondeo' teams has also been used in the preparatory stages of on-farm research (Hildebrand, 1981).

A model for informant-led collecting

On arrival in a village or other population centre a meeting with the local authority figure(s) should be arranged, at which the purpose of the visit is explained in detail. Such gatekeepers may include religious leaders, administrators, village elders, extension workers, teachers and the leaders of community groups. Although in many cultures the community leaders are predominantly men, it is important to recognize that women may also be gatekeepers. Particular care must be taken to follow local etiquette and protocol during this initial encounter, and, indeed, at all times thereafter. In particular, it may be necessary to follow a definite procedure when entering homesteads or the village itself. If it appears from initial discussions that landraces of the target crop are being grown in the village, or that target wild species may be found in the land used by the villagers, the gatekeepers can then be asked for permission to undertake germplasm collecting. They can also be asked for introductions to those people most likely to be willing to participate in the work by: (i) locating and donating local germplasm; and (ii) sharing what they know about the material.

These two groups need not necessarily correspond, or even overlap. Within a community, tasks (including different aspects of agriculture and plant gathering and use), property and knowledge may be divided on the basis of social, economic, ethnic, age and/or gender grounds. Access to particular resources, including land and plant genetic resources, may be confined to one social category, and control of the resources to a different one. Whereas access usually implies intimate knowledge, control need not. Information must be asked of the appropriate people. For example, competitions have been organized among schoolchildren in Botswana to collect seeds: it is they who know the location of the trees with the tastiest fruits (F. Taylor, pers. comm.). Women are the farmers and the main custodians of farming knowledge in many cultures, and thus the best key informants on specific crops and agricultural practices, a fact that has been overlooked in the past. Richards (1985) points out that men and women within the same household may be responsible for different crops, landraces and fields. Cash crops, for example, are often a male preserve. Men and women may even have different tasks within the same field, and also often evaluate landraces according to very different criteria (Ashby et al., 1989). Thus, asking men for information about some crops, in particular subsistence food crops, could lead to misleading answers, though it may still be the men who have to be asked for permission to collect germplasm, whether from cultivated fields or the wild. This is not always the case, however, as the anecdote recounted by Berg et al. (1991) shows:

> Travelling with an ICRISAT germplasm collection team [in Eastern Equatoria region in southern Sudan] in the fall of 1983, we came to a [Lotuho] village and asked the people for one sorghum head. They said 'yes' and we picked a head from a ripe field. But immediately a woman came shouting ... at us. It took us a long time to calm her down and find out what had provoked her reaction. It turned out that, as the mother of the house, she was responsible for selection of seed for the next year. It was taboo for anybody else to start the harvest before she had walked through the whole field and done her selection.

Sometimes, particular individuals within the community control, or at least dominate, the production and exchange of planting material. Some farmers are more innovative than others: sometimes these are the poorest, in other cases those with better access to resources. Older people are likely to be the ones most in touch with traditions and will clearly be the best sources of information on the past. There may be specialist trade groups of resource users (guilds), such as craft workers or local healers, and some such societies are secret. For example, a special class of people, called 'explorers', are the repositories of expertise on pasture quality among the Fulbe, and have the task of seeking out flushes of particularly prized species (A.M. Bonfiglioli, pers. comm.).

Gatekeepers will be crucial in identifying key informants on different topics. However, care should be taken to avoid bias towards the richer, more influential members of the community, such as friends of the village head, or progressive or educated people. Key indicators may be useful in avoiding such bias. House type may give important clues as to socioeconomic status or ethnicity, for example (Chambers, 1987).

In some societies, it may be difficult for a collector to meet the 'best' consultants. This is the kind of information that would emerge at the participant observation stage of a full memory banking study, but could also come out of literature-based research during the mission planning stage. The problem can be overcome by careful selection of the collecting team, for example as regards to gender balance. Where women are known to play key roles with respect to the target species, a team member experienced in working with women farmers is essential. The help could be sought of female members of home economics or community development departments. *Tools for the Field* (Feldstein and Jiggins, 1993) is one of the few published sources for methods and applications which overcome the barriers (which male researchers in particular might face) to reaching women and engaging their active collaboration. The general problems of selection of informants are discussed by Johnson (1990).

It should be possible during the initial meeting with gatekeepers to use brainstorming techniques to compile checklists of the forages or medicinal plants used by the villagers, or of the crops and landraces that are currently grown in the village and surrounding areas, or that were grown at different times but can no longer be found. The reasons why some have disappeared should be discussed, if possible. Such lists can be amplified in the course of visits to local markets. The problems and opportunities presented by markets are discussed more fully in the next section. Gatekeepers should be asked when and where markets are held in the surrounding area.

Some data on 'collecting site context' (Chapter 19), in particular village-level information on land use and farming systems, should also be collected at this stage. In addition to simply answering questions, gatekeepers and others can be involved in drawing seasonal calendars, idealized resource maps and transects of the village and its surroundings. (However, this will be somewhat time-consuming, and may also be kept for a final group meeting of all collaborators.) Conway (1989) points out that these exercises can be used 'to identify the major problems and opportunities in the agroecosystem, and where they are located' in space and time, but they can similarly be used to identify

where wild species, crops and individual landraces fit in the environment, forming the basis of a user-defined characterization of collecting sites. Actually walking along transects with villagers, gathering
information on the different ecological zones encountered along the
way, is an important PRA tool which can be usefully adopted by germplasm collectors, particularly if combined with an assessment of the
changes in land use that have occurred over the years in the different
zones.

An idea of official village or community history can also be gathered
fairly rapidly at this stage. The timing of some important events may
be known from the analysis of secondary sources. What the gatekeepers
see as key events in the history of their village can now be recorded, for
example the advent of major changes and innovations in land use and
agriculture, emigrations and immigrations, natural disasters, years of
bad and bumper harvests, and so on. On such historical profiles or time
lines can be superimposed time trends (which can also be derived from
secondary sources) in important variables, such as population, prices of
agricultural commodities (in particular, of the target crop), rainfall, productivity and, crucially, the extent of cultivation of particular crops and
landraces. Did yields go up or down between two particular key events
on the time line? What did prices do during the same period? Was the
population of the village expanding at the time? What happened to the
extent of cultivation of landraces vs. improved varieties, and of the more
important landraces? These kinds of questions about trends will be
important in explaining changes in genetic diversity, cultural practices
and technologies, and thus in assessing the extent and threat of genetic
erosion.

Collectors should set aside at the very least a couple of hours for this
initial stage of the process. The potential key informants, or germplasm
consultants, should then be visited with a gatekeeper or a delegate in
attendance, possibly as part of a group participatory trek or transect,
and germplasm collected with their permission and assistance. Brief
consultations should be held with the key informants while collecting
and immediately afterwards, in the form of semistructured interviews.
Whereas structured interviews use a standard schedule where questions
may be open-ended (respondents give their own answer) or closed-ended
(respondents choose among a limited number of possible responses),
semistructured interviews allow new questions to be added in response
to particular answers if necessary. Detailed discussions of interviewing
technique are provided by Bernard (1988) and Foote Whyte (1982).
Useful practical tips may also be found in Kearl (1976). The consultations with farmers should aim to document:

- the household farming system and what the farmer sees as the
  important features of the exact site of collecting (i.e. the plot or field);
- the farmer's view of the main characteristics of the material
  collected;

- the methods whereby the farmer produced, selected, stored, processed and exchanged the germplasm.

Cognitive mapping of landraces (and useful wild species, such as forages) and triads/ranking tests may be carried out now and/or at a final group session. This information will find a place on collecting forms in the sections on 'collecting site description' and 'population information', along with pH of the soil, for example, and other scientific observations and measurements (Chapter 19).

Some general guidelines may be proposed for the process of consultation (or dialogue) with local users of germplasm, which is the core of this proposed participatory field methodology. In summary, collectors should make sure they know who to talk to, what exactly they want to know and how to ask. Choice of consultant has already been discussed. The timing of the consultation is also important (Rhoades, 1987). Farmers are busy people, especially at harvest time, when most collecting will be done, and their routine should be disrupted as little as possible. Women may not be available for consultation at the same time as men. A full explanation should be given for the collecting to each consultant, and what will happen to both the germplasm and the data collected described. There needs to be not just consent (i.e. permission) for the work, but informed consent. This will apply to collecting wild species as much as to collecting landraces. It should always be made clear that participation is voluntary. Donor/informants have the right to say no.

It may be useful to keep an alphabetized file of profiles of gatekeepers, key informants and other people met in the field as an aid to memory (Bernard, 1988), but some people might not want their names recorded, in which case an alias should be used. It may be a good idea not to start writing notes and filling in collecting forms until some way into the consultation, or just after it has finished. If there is any doubt, informants should be asked whether they object to notes being taken down during conversations.

Questions need to be framed in an appropriate way. The notion of human health and disease, for example, can be highly culture-specific. Balick and Mendelsohn (1992) point out that disease concepts in Belize, of Mayan origin, often have no equivalents in Western medicine. This underlines the holistic nature of IK. Documenting ethnobotanical data on medicinal plants will require an understanding of local disease concepts, not just local botanical concepts. The collector should be familiar with the local units of time, distance, area, weight and volume (Rhoades, 1990). For many cultures, history is cyclical: the Gabbra pastoralists, for example, think in seven-year cycles (Herr, 1992). A question about the crop varieties grown ten years ago may have to be couched in such terms as 'when your children were young' or with reference to some important happening. The year may be divided on the basis of the waxing and waning of the moon or the movements of the stars, or complex combinations. It is also often important that informants be shown plants (or pests, etc.)

as they are used to seeing them. Asking informants to name herbarium specimens or photographs may not always work. It should also be remembered that local people will not necessarily interpret maps and diagrams according to the collector's own conventions of scale and perspective.

Questions should be precise, but not restricting or leading. A name or a use may apply to only a part of a plant, or to the plant at a certain stage of growth. The mode of preparation of a drug may be very specific. A grass may be palatable only in a particular season, or only to a particular kind of livestock. The collector should be aware of the ways that an ethnobotanical statement may be qualified and 'serially narrow down or focus on the desired information' (V.D. Nazarea-Sandoval, pers. comm.) during interviews, using the so-called 'six helpers' – what, where, when, why, who and how. 'Probes' should be memorized that can be used during interviews to guide the conversation in particular directions and elicit specific information (Nazarea-Sandoval, 1990). Asking for a description of how something is done can be followed up by asking for a demonstration or by asking to actually do it oneself. Croom (1983) lists points that need to be recorded in collecting ethnobotanical information on medicinal plants (see also Waechter and Lejoly, 1990). A list of questions asked in the course of interviews on cassava is given by Boster (1985). Chapter 38 includes a topic list for sweet potato. It is best to see such questionnaires rather as a guide to the collecting of IK, however, than as a rigid schedule. Though it may be more difficult to code and analyse the data afterwards, interviews should be semistructured and questions open-ended.

Farm sketches are helpful in showing the way crops and landraces are arranged in space relative to each other and to important environmental features (e.g. streams, different soil types). It should be possible to produce an idealized sketch showing the typical arrangement after a number of visits. Examples may be found in Esquivel and Hammer (1992). If time allows, the individual consultations can also attempt to document relevant changes that have occurred during the informant's lifetime, in particular the flux of landraces, which can then be related to the official, communal version of history obtained from gatekeepers. The lists of crops and landraces begun at the initial meeting should also be updated continuously during this phase, as necessary.

Sometimes, the whole of the collecting in a particular village or locality can be organized as a group activity, rather than being articulated in separate interviews with individual farmers. This could take place in the form of participatory transects or group treks through the fields (Mathema and Galt, 1989). However, such a strategy will perhaps be especially apt if the visit is taking place at a time other than harvest, when collecting will be mostly from farmers' stores. In such cases, it might be a more efficient use of time to ask interested farmers to bring their material to a central place, rather than the collecting team moving from household to household. The meetings of established local institutions, which could be anything from a traditional village elders' council to a women's self-help group, pastoral association, or village development committee, may be suitable venues for group discussion of plant genetic resources.

Group discussion will not only permit some immediate cross-checking of information, but debate among the participants may also allow additional details to emerge. It will also save time, and it may be possible to identify people with special knowledge for follow-up individual interviews. Disadvantages include the possibility of domination by certain individuals (or groups, such as men at the expense of women) and the logistical problems involved in getting enough people together in one place at the same time. The locale should be as neutral as possible (Chapter 38). Group size and composition are also important variables. It is probably best if the participants see each other as of fairly equal social standing. Very large and varied groups can be broken up into smaller, more homogeneous working groups, e.g. separate men's and women's groups.

Even when the collecting takes place as a series of separate individual consultations, if possible a group discussion with farmers should always be organized at the end. The evening is often the best time for such a focus meeting. Chambers (1993) recommends that participants in PRA exercises should be available in the evening and early morning, and try to spend the night in the village. In any case, departure should not be abrupt and business-like. Time should be taken to bid farewell in the traditional way (Rhoades, 1987).

Farmers often have very well-defined, traditional systems for obtaining novel planting material. Germplasm collectors should know about and try to fit into such an existing system. In the highlands of East Africa, for example, when farmers want seeds of a particular bean variety they have seen in a neighbour's field, they offer seeds of their own in exchange. Germplasm collectors could do the same (J. Voss, pers. comm.). This would not absolve them of their obligations to the (national) authorities sanctioning the work, any more than simply paying farmers and market stallholders the going rate for planting material would. It would, however, help establish a relationship between collector and farmer based on reciprocity and mutual dependence rather than mere extraction.

IK and some specific practical problems of crop germplasm collecting

Crop collectors face particular practical problems that IK will be essential in helping them solve. The problem of mixed fields is one that will often come up. 'Sometimes, people prefer to maintain diversity in the form of one heterogeneous seed mixture instead of discrete varieties, but often with names for the different types which occur in the mixture' (Berg et al., 1991). For example, Asfaw (1990) points out that only 10% of the barley fields he surveyed in Ethiopia could be considered pure stands of a single locally recognized phenotypic entity. The key point is how the different entities have been maintained. If seed crops of the different entities were derived from separate, 'pure' fields and only mixed after harvesting to provide the planting material for the particular field being sampled, a case can be made for making a separate random sample of each entity (i.e. landrace) present, or for asking the farmer for stored seeds of each entity. If there has been no separate maintenance of the locally recognized entities for a considerable period, the field should be treated as a single population and a single random sample made, particularly if there is significant outcrossing in the crop. It is then the whole population that represents the landrace, though individual morphotypes within it may be given different names by farmers. Separate samples of these could be taken in addition to the population sample. In Rwanda, farmers grow *Phaseolus* beans (outcrossing frequency 2–5%) as carefully controlled mixtures of varieties. They recognize the individual varieties as different but only grow them in monoculture when testing them out (Voss, 1992). Clearly, a random sample of the mixture will be just as important as pure samples of each of its constituent varieties. In cases such as these, participation of the farmer in the collecting process will be necessary to ensure that the correct sampling method is chosen and that it takes place accurately.

Farmers can also help ensure that modern varieties and recently introduced material are avoided. They will be aware of the origin of the material they are growing. The issue of variation within a landrace among households or villages may be more complicated to resolve. Such variation may be considerable, and of great potential interest to breeders, but confined to characters that are non-obvious, or at least not involved in local recognition of the landrace. In work with a vegetatively propagated crop (sweet potato), G.D. Prain and co-workers relied on a local consultant to identify duplicates within a plot and also avoided similar material with the same name in nearby plots (Chapter 38). However, in more distant plots, and if discrepancies arose, the material and the associated knowledge were collected. Duplicate sampling of each named landrace will be necessary, if not in every village, at least in different agroecological zones and possibly in different districts or market areas, depending on information from farmers and others on the pattern of germplasm movement and exchange in the region.

Collecting from farmers' stores also needs the input of IK. Farmers often actively select the material that they will plant. This may take place at harvesting (the seeds from particular plants being stored separately from the grain that will be eaten) or after the harvest is in, perhaps just before sowing. Collectors need to know if the material they are sampling has been subjected to such selection, and if so the criteria for selection must be documented. Voss (1992) found significant differences between bean samples before and after farmer selection. Material on market stalls may also have been subjected to selection, or may consist of mechanical mixtures of varieties, which could be sampled separately. Markets can be very diverse and serve a large hinterland, making them ideal early stops for both information and germplasm in a target area (e.g. Berg, 1985). However, farmers often grow some land-races exclusively for home consumption, so that the material available in markets may be a biased sample of the total diversity of a crop in an area. Whether this is the case, and the actual geographic origin of the material, can only be ascertained in discussions with local people.

Corroboration of information

Clearly, 'there is more to plant–human relationships than drawing up a list of actual and potential uses' (Given and Harris, 1992). Documenting IK of wild plants and crops is much more complicated than that, and the accuracy and validity of the information obtained will to a great extent depend on collectors having an understanding of the community in which they are working. Understanding will in turn depend on involving the local user of germplasm fully in the collecting process; hence the importance of a participatory approach.

However, while some thematic methodologies using RRA/PRA methods are quite advanced, this is not yet the case for documenting IK of plant genetic resources. This chapter is merely an early attempt to introduce the theory and practice of RRA/PRA to plant germplasm collectors. Considerable work needs to be done to refine the methodology. Most biologists lack training in the skills of RRA/PRA, and it is not yet common to include a social scientist in collecting teams. Until more experience is available, there will be dangers – as well as opportunities – in using RRA/PRA methods in the context of plant collecting missions. For example, richer farmers and officials may take over and marginalize other groups, such as the illiterate, the very poor, the landless, women and ethnic minorities. Even if the process is properly guided and does not become biased in this way, there will be uncertainty in the qualitative data gathered. In so far as it is possible to do so, information should be corroborated.

A standard interview schedule or checklist (perhaps no more than a set of probes), to be followed for each informant but flexible enough to allow occasional departures, will enable the collector to keep track of which information has been confirmed, and how, and which has not. Corroboration can be done by triangulation. This may mean asking the same questions of more than one person (perhaps in a focus group discussion), or different members of the collecting team interviewing the same informant at different times, or using different techniques to investigate the same topic (e.g. triads tests and ranking to elucidate a folk taxonomy). There is also corroboration by actual observation, for example of the use of a plant or the preparation of a product. Photographs, audio recordings and video recordings can be used to document such observations, though in some cultures this may not be possible. Kirk and Miller (1990) discuss the problem of qualitative data reliability in detail.

It may also be possible to cross-check some information recorded from consultants in the field with material from secondary sources. This leads to the issue of validation of IK by scientific means. Examples of this might be the chemical analysis of folk herbal preparations (possibly followed by clinical trials), the comparison of folk nomenclatures with assessments of the differences among landraces using biochemical or molecular methods and the evaluation of locally favoured landraces for nutritive value. By means of such exercises, scientists can add value to IK, prior to feeding it back to the community. This is the basis on which a true partnership can be built.

Ethics and the collecting of plant genetic resources and associated IK

The implementation of a PRA strategy as part of plant germplasm collecting implies something of a departure from most current practice. Crucially, it means allowing farmers to influence the agenda and pace of collecting missions. It may be necessary to spend a whole morning – even a whole day – collecting in a village growing a dozen landraces of the target species. People more familiar with the cadences of conventional collecting may find this much too time-consuming. Many would say that a couple of hours should be more than enough.

In fact, the methodology proposed here is flexible enough to be abbreviated, if there is time pressure. It can also be confined to a preliminary, exploratory visit, with germplasm collecting taking place during a second visit sanctioned and organized during, and on the basis of, the first. However, given the determining role that local subsistence farmers have played in the generation and maintenance of the genetic diversity of crops, it will surely be generally recognized as necessary that crop collecting should be driven just as much – if not more – by the donor/informants and their requirements as by collectors and theirs. Indeed, some form of participatory collecting will be necessary for wild species too, plants known, used and perhaps actively conserved by local communities, occurring on land they may have communal rights to – not just necessary, but also fitting. Participation will not only ensure that the data gathered will be useful to plant breeders and other formal sector users. It will not only ensure that any statutory or customary rights of ownership or access to land and resources are respected. It will also help to ensure equity, by empowering the local community. A participatory approach that builds on local knowledge, initiatives and resources strengthens the self-help capabilities of farmers and other local people. It confirms or renews their faith in the value of their own knowledge and how it can be combined with outside knowledge or limited external inputs to solve their current problems.

Participation should mature into partnership. At its most basic, partnership in the collecting process means feeding back the information gathered and the results obtained (e.g. after some kind of validation exercise) to the people from whom it was collected. Numerous techniques may be used for this, including public meetings, workshops, exchange visits, demonstrations and such 'popular persuaders' as plays. Television, radio and other mass media have been used to feed back local knowledge. Such sharing is important, but partnership should reach beyond this. It should help ensure 'resource conservation, local economic development and distribution of the benefits from biodiversity to those who pay the direct or opportunity costs for its maintenance and development' (Reid *et al.*, 1993).

How is it possible to guarantee that, in Bell's (1979) words, there is indigenous exploitation of knowledge (and of genetic resources) rather than merely the exploitation of indigenous knowledge? The Food and Agriculture Organization (FAO) International Code of Conduct for Plant Germplasm Collecting and Transfer addresses the issue, but is mainly aimed at regulating the relationship between collector and national authorities (Chapter 2). Collectors of germplasm and IK are increasingly recognizing that they have responsibilities to local communities too and are engaging in formal or informal negotiations with them for access to the germplasm and information they hold. Various professional and other organizations have developed sets of rules to guide such negotiations and, indeed, the whole relationship between collector and local communities. The Society for Economic Botany has developed a code of ethics for ethnobotanists. One of the goals of the International Program on Traditional Ecological Knowledge is to promote the development and implementation of a code of ethics and practice on the acquisition and use of IK. Some relevant international agreements in the field of cultural property are discussed by Downes *et al.* (1993).

An attempt has been made here to show how at least some of the concepts generally recognized as being of importance in this context at the local community level no less than at the national – such as prior informed consent, equity, the voluntary nature of participation and the right of anonymity of informants – could actually be applied in practice. It has been stressed that the collector should attempt to fit into the indigenous crop germplasm exchange system, and should certainly at all times conform to customary rules regarding access to land and resources. The respect for the local communities who know and have shaped the germplasm, and in whose hands it has been brought down to us, which all this shows, though now enshrined in the concept of farmer's rights and, increasingly, in national and international laws, has perhaps been too long in coming.

Acknowledgements

Janice Jiggins, Virginia Nazarea-Sandoval and Gordon Prain made important comments and suggestions. L.G. would also like to thank Tony Cunningham, Joachim Voss, Monica Opole, Toby Hodgkin and all the participants of the Wageningen Agricultural University/ENDA-Zimbabwe/CGN/GRAIN Seminar on 'Local Knowledge and Agricultural Research' (Brodensbury Park Hotel, Nyanga, Zimbabwe, 28 September–2 October 1992) and of the Intermediate Technology Development Group/Plan International Workshop on 'Collecting and Feeding Back Existing Local Knowledge' (Embu, Kenya, 27–30 September 1993).

References

Ashby, J.A., C.A. Quiros and Y.M. Rivers (1989) Experience with group techniques in Colombia. In: Chambers, R., A. Pacey and L.A. Thrupp (eds) *Farmer First. Farmer Innovation and Agricultural Research*. pp. 127–132. Intermediate Technology Publications, London.

Asfaw, Z. (1990) An ethnobotanical study of barley in the Central Highlands of Ethiopia. *Biologisches Zentralblatt* 108:51–62.

Balick, M.J. and R. Mendelsohn (1992) Assessing the economic value of traditional medicines from tropical rain forests. *Conservation Biology* 6:128–130.

Bell, M. (1979) The exploitation of indigenous knowledge or the indigenous exploitation of knowledge: whose use of what for whom? *IDS Bulletin* 10:44–50.

Berg, M.E. van den (1985) Ver-o-Peso: the ethnobotany of an Amazonian market. In: Prance, G.F. and J.A. Kallunki (eds) *Ethnobotany in the Neotropics*. Advances in Economic Botany Vol. 1. pp. 140–149. New York Botanical Garden, New York.

Berg, T., A. Bjornstad, C. Fowler and T. Skroppa (1991) *Technology Options and the Gene Struggle*. NORAGRIC Occasional Papers Series C. Development and Environment No. 8. NORAGRIC, Aas.

Bernard, H.R. (1988) *Research Methods in Cultural Anthropology*. Sage Publications, Newbury Park.

Boster, J.S. (1985) Selection for perceptual distinctiveness; evidence from Aguaruna cultivars of *Manihot esculenta*. *Economic Botany* 39:310–325.

Chambers, R. (1987) Shortcut methods in social information gathering for rural development projects. In: *Proceedings of the 1985 International Conference on Rapid Rural Assessment*. Khon Kaen University, Thailand. pp. 33–46. Rural Systems Research and Farming Systems Research Projects, Khon Kaen.

Chambers, R. (1990). Participatory rural appraisals; past, present and future. *Forests, Trees and People Newsletter* 15/16.

Chambers, R. (1992) *Rural Appraisal: Rapid, Relaxed and Participatory*. IDS Discussion Paper 311. IDS, Brighton.

Chambers, R. (1993) Participatory rural appraisal. In: Hudson, N. and R. Cheatle (eds) *Working with Farmers for Better Land Husbandry*. pp. 87–95. Intermediate Technology Publications, London.

Conway, G.R. (1989) Diagrams for farmers In: Chambers, R., A. Pacey and L.A. Thrupp (eds) *Farmer First. Farmer Innovation and Agricultural Research*. pp. 77–86. Intermediate Technology Publications, London.

Crapanzano, V. (1984) Life histories. *American Anthropology* 86:953–960.

Croom, E.M. (1983) Documenting and evaluating herbal remedies. *Economic Botany* 37:13–27.

Downes, D., S.A. Laird, C. Klein and B. Kramer Carney (1993) Biodiversity prospecting contracts In: Reid, W.V., S.A. Laird, C.A. Meyer, R. Gámez, A. Sittenfeld, D.H. Janzen, M.A. Gollin and C. Juma (eds) *Biodiversity Prospecting: Using Genetic Resources for Sustainable Development.* pp. 255–287. WRI Publications, Baltimore.

Esquivel, M. and K. Hammer (1992) The Cuban homegarden 'conuco': a perspective environment for evolution and *in situ* conservation of plant genetic resources. *Genetic Resources and Crop Evolution* 39:9–22.

Feldstein, H. and J. Jiggins (eds) (1993) *Tools for the Field.* Kumarian Press, West Hartford.

Foote Whyte, W. (1982) Interviewing in field research. In: Burgess, R.G. (ed.) *Field Research: A Sourcebook and Field Manual.* pp. 111–122. George Allen and Unwin, London.

Friedberg, C. (1968) Les méthodes d'enquête en ethnobotanique. *Journal d'Agriculture Tropicale et de Botanique Appliqué* 15:297–324.

Given, D.R. and W. Harris (1992) *Techniques and Methods of Ethnobotany As an Aid to the Study, Use and Conservation of Biodiversity. A Training Manual.* Commonwealth Science Council, London.

Gueye, B. and K. Schoonmaker Freudenberger (1991) *Introduction à la Méthode Accelerée de Recherche Participative (MARP). Quelques Notes Pour Appuyer une Formation Pratique.* IIED, London.

Herr, R. (1992) *Pastoralism in Africa: Paths to the Future.* Mennonite Central Committee, Nairobi.

Hildebrand, P. (1981) Combining disciplines in rapid appraisal: the sondeo approach. *Agricultural Administration* 8:423–432.

ILEIA (1989) *Proceedings of the ILEIA Workshop on 'Operational Approaches for Participatory Technology Development in Sustainable Agriculture'.* ILEIA, Leusden.

ILEIA (1992) *Learning for Participatory Technology Development. A Training Guide.* ETC Foundation, Leusden.

Johnson, J. (1990) *Selecting Ethnographic Informants.* Qualitative Research Methods Series 22. Sage Publications, Newbury.

Jorgensen, D.L. (1989) *Participant Observation.* Sage Publications, Newbury Park.

Kabutha, C., B. Thomas-Slayter and R. Fox. (n.d.) *Participatory Rural Appraisal Handbook.* WRI in collaboration with Egerton University, National Environment Secretariat of Kenya and Clark University, Njoro, Kenya.

Kearl, B. (ed.) (1976) *Field Data Collection in the Social Sciences: Experiences in Africa and the Middle East.* Agricultural Development Council, Inc., New York.

Kirk, J. and M.L. Miller (1990) *Reliability and Validity in Qualitative Research.* Sage Publications, Newbury Park.

McCracken, J.A. (1988) A working framework for rapid rural appraisal: lessons from a Fiji experience. *Agricultural Administration and Extension* 29:163–184.

Mathema, S.B. and D.L. Galt (1989) Appraisal by group trek. In: Chambers, R., A. Pacey and L.A. Thrupp (eds) *Farmer First. Farmer Innovation and Agricultural Research.* pp. 68–73. Intermediate Technology Publications, London.

Nazarea-Sandoval, V.D. (1990) Potentials and limitations of ethnoscientific methods in agricultural research. In: Rhoades, R.E., V.N. Sandoval and C.P. Bagalanon (eds) *In-Country Training Workshop for Farm Household Diagnostic Skills.* pp. 90–101. CIP, Los Baños.

Nazarea-Sandoval, V.D. (1992) Of memories and varieties: complementarity between cultural and genetic diversity. Working paper. UPWARD, Los Baños.

Prain, G.D. (1993) Mobilizing local expertise in plant genetic resources research. In: de Boef, W.S., K. Amanor, K. Wellard with T. Bebbington (eds) *Cultivating Knowledge: Genetic Diversity, Farmer Experimentation and Crop Research.* Intermediate Technology Publications, London.

Reid, W.V., S.A. Laird, R. Gámez, A. Sittenfeld, D.H. Janzen, M.A. Gollin and C. Juma (1993) A new lease on life. In: Reid, W.V., S.A. Laird, C.A. Meyer, R. Gámez, A. Sittenfeld, D.H. Janzen, M.A. Gollin and C. Juma (eds) *Biodiversity Prospecting: Using Genetic Resources for Sustainable Development.* pp. 1–52. WRI Publications, Baltimore.

Rhoades, R.E. (1987) Basic field techniques for rapid rural appraisal. In: *Proceedings of the 1985 International Conference on Rapid Rural Assessment.* Khon Kaen University, Thailand. pp. 114–128. Rural Systems Research and Farming Systems Research Projects, Khon Kaen.

Rhoades, R.E. (1990) The coming revolution in methods for rural development research. In: *Proceedings of the Inaugural Planning Workshop on the User's Perspective With Agricultural Research and Development.* pp. 196–210. CIP, Los Baños.

Richards, P. (1985) *Indigenous Agricultural Revolution.* Unwin Hyman, London.

Ruano, S. (1989) *El Sondeo: Actualizacion de su Metodologia para Caracterizar Sistemas Agropecuarios de Produccion.* IICA, Costa Rica.

Sandoval, V.N. (1994) Memory Banking Protocol: A Guide for Documenting Indigenous Knowledge Associated with Traditional Crop Varieties. In: Prain, G.D. and C.P. Bagalanon (eds) *Local Knowledge, Global Science and Plant Genetic Resources: Towards a Partnership. Proceedings of the International Workshop on Genetic Resources.* pp. 102–122. UPWARD, Los Baños.

Theis, J. and H.M. Grady (1991) *Participatory Rapid Appraisal for Community Development. A Training Manual Based on Experiences in the Middle East and North Africa.* IIED and Save the Children, London.

Voss, J. (1992) Conserving and increasing on-farm genetic diversity: farmer management of varietal bean mixtures in central Africa. In: Mook, J.L. and Rhoades, R.E. (eds) *Diversity, Farmer Knowledge and Sustainability.* Cornell University Press, Ithaca.

Waechter, P. and J. Lejoly (1990) PHARMEL: banque de données de médecine traditionelle et de pharmacopée. In: Fleurentin, J., P. Cabalion, G. Mazars, J. Dos Santos and C. Younos (eds) *Ethnopharmacology: Sources, Methods, Objectives.* ORSTOM and Société Française d'Ethnopharmacologie, Paris.

Warren, D.M. and B. Rajasekaran (1994) *Utilizing Indigenous Knowledge Systems for Development: A Manual and Guide.* CIKARD, Ames.

Addresses of some relevant organizations

Clark University
Program for
  International
  Development and
  Social Change
950 Main Street
Worcester, MA 01610
USA
Tel: +1 508 7937201
Fax: +1 508 7938820

Egerton University
Office of Research and
  Extension
PO Box 536
Njoro
Kenya

Institute of
  Development Studies
  (IDS)
University of Sussex
Brighton BN1 9RE
UK

International Institute
for Environment and
Development (IIED)
Programme in
Sustainable
Agriculture
3 Endsleigh Street
London WC1H 0DD
UK

International Rural
Development Centre
Swedish University of
Agricultural Sciences
Uppsala
Sweden

MYRADA
2 Service Road
Domlur Layout
Bangalore 560 071
India

Stockholm University
Popular Participation
Programme
Development Studies
Unit
Department of Social
Anthropology
S-106 91 Stockholm
Sweden

World Resources
Institute (WRI)
From the Ground Up
Centre for International
Development and
Environment
1709 New York Avenue,
NW
Washington, DC 20006
USA

Gathering and recording data in the field

19

H. Moss[1] and L. Guarino[2]

[1]63 End Road, Linden Extension, Randberg 2194, Republic of South Africa: [2]IPGRI, c/o International Laboratory for Research on Animal Diseases, PO Box 30709, Nairobi, Kenya.

Introduction

The successful use of genetic resources conserved *ex situ* depends to a large extent on the availability and quality of data – both conventional 'scientific' data and indigenous knowledge – on the sample and on the physical, biotic and human environment at the collecting site recorded by the collector in the field. It is at least partly on the basis of such data that samples in different collections can be recognized as duplicates, that appropriate conditions for regeneration, characterization and evaluation can be identified, that material now extinct in the field can be reintroduced to the area where it was originally collected, and that users of conserved germplasm are able to make an initial decision regarding the suitability of the material for inclusion in breeding, introduction or screening programmes. These so-called passport data are also important in studying the phenology of material and the distribution of variation with respect to ecological and socioeconomic factors, which will help in the planning of future collecting. In addition, it will be useful in making an assessment of the threat of genetic erosion.

Germplasm lacking passport data is still usable, of course, but much less valuable to the user than material that is better documented. This is because resources will need to be directed to identifying its general adaptation, for example, which could be spent on detailed evaluation if something of the origin of the material were already known. The more data a germplasm sample has associated with it, the more valuable it is to users. However, there is often a trade-off between collecting germplasm and collecting data. Time in the field is usually limited, and collecting more data may therefore mean collecting less germplasm, either fewer or smaller samples. Typically, in most wild species collecting, putting together a germplasm sample of reasonable size is more

367

time-consuming than filling in a collecting form. Especially if more than one species is being collected at the site, time spent at the site will most certainly not be limited by form-filling. This will emphatically not be the case with crops: documenting indigenous knowledge (i.e. identifying, locating and consulting donor/informants), an essential part of crop collecting, may take considerable time (Chapter 18).

A balance will therefore need to be struck between exhaustively documenting a population and site and moving on to the next. Deciding on this is up to the collector, and will depend to some extent on the immediate purposes of the collecting. For example, an initial survey of a previously unexplored area may tend to concentrate more on data gathering. Emergency rescue collecting may leave little time for anything but the most basic documentation. Whatever the reasons for collecting may be, however, it should be remembered that the collector is also working for unknown potential end-users with unknown requirements in a possibly remote future, when the environments and cultures from which the material was collected and which helped to shape it may no longer exist, along with the germplasm. Though there will always be complaints of too few data (or the wrong data!) having been collected in the field, no user of conserved germplasm has ever accused a collector of recording too much data.

Passport descriptor lists

Many organizations which carry out germplasm collecting have developed and use their own individual, more or less specialized, collecting forms. This has led to difficulties in data exchange and to some mutual incomprehension. To promote standardization, the International Plant Genetic Resources Institute (IPGRI), in cooperation with crop experts, produces crop-specific (sometimes gene-pool-specific) descriptor lists covering passport data as well as characterization, preliminary evaluation and, in some cases, further evaluation data. A full catalogue of such lists is given in Chapter 8. IPGRI has also produced a basic generalized list of collecting data, arranged in a collecting form, and supporting database software, the Collecting Form Management System (Chapter 28). The descriptors and their codings in these lists should not be regarded as definitive, but they do provide a format for data exchange that is internationally understood. Data need not actually be recorded in the field according to these lists, but they may then have to be converted to fit the recommended format for exchange purposes.

Despite the multiplicity of collecting data lists and forms, there are a number of basic passport descriptors which are generally regarded as essential, the absolute bare minimum. The descriptors that must appear on any collecting form and that must be filled in are listed in Box 19.1.

Media for data recording

The descriptors for which data are to be recorded in the field are normally laid out on paper forms, which should be as easy as possible

---

Box 19.1
List of essential collecting data descriptors

Sample labelling
• Expedition identifier (or collecting organization).
• Name(s) of collector(s).
• Collecting number (or collector's number).
• Collecting date.
• Type of material.

Sample identification
• Genus, species, subspecies, botanical variety.
• Vernacular name (and language).
• Herbarium voucher number.
• Identification numbers of other associated specimens.
• Status of sample.

Sampling information
• Number of plants sampled.
• Sampling method.
• Collecting source.

Collecting site localization
• Country.
• Primary administrative unit.
• Precise locality.
• Latitude of collecting site.
• Longitude of collecting site.
• Altitude of collecting site.

---

to fill in, update and correct. They should be preprinted and designed to fit on a single sheet of paper (front and back, if necessary), with clear captions and enough space for text. Using more than one sheet for a single sample may result in data becoming separated and lost. *It is best to have one data sheet per germplasm sample.*

Some organizations bind their collecting sheets in a book, which means that sheets cannot go astray. Others use loose sheets. One way of handling these in the field is to keep them in a hard-backed double folder which has a clipboard on one side, where all the unused sheets can be kept, and a pocket on the other side for keeping the completed forms. Another way is to file them in a ring-binder.

Alternatives to filling in paper forms include using dictaphones or tape recorders, data loggers and small portable computers (laptops and notebooks). All have the problem that power is needed to run them. Dictaphones are very convenient in the field but a fully written-out list of descriptors will in any case be necessary to make sure that none is overlooked, and transcription of spoken notes can be difficult and time-consuming. Despite these technical problems, tape recorders may be extremely useful in some situations, for example when documenting indigenous knowledge in consultations with local users. As data will eventually have to be entered into some sort of computer database if any serious organization, analysis and searching is to be done, data loggers and computers would seem to be ideal options, especially when data are numerous. However, these also have their problems. Data loggers are entirely solid-state and thus sturdy, but many have no capability for the inputting of free text. Computers are more flexible in this respect, but also more vulnerable, although they are now becoming available with solid-state storage, increasing reliability in field conditions. Data loggers and computers are of course expensive relative to paper, and taking them across some international borders may be difficult. Whatever system is used for data recording, it is a good idea to keep a paper backup.

It is useful to keep a field notebook in addition to the collecting forms, in the form of a daily log or diary. Along with an account of the day's happenings, including general observations and sketch maps of the region through which one is travelling, some basic information can be recorded on each collecting site, in particular how it was reached. A list of the collecting numbers of all the samples collected at each site should be kept, along with taxonomic identification, local name and any unusual or interesting features of the sample. At the end of every day, this information should be checked against the date, site number, collecting number, etc. recorded on each collecting form filled in during the day. Numbering mistakes are best corrected as soon as possible. Further information which it may be useful to record in the field notebook is pointed out in the discussion of individual passport data descriptors. Bernard (1988) suggests that each day in the field can be represented by two facing pages in the log, with the page on the left showing what was planned for the day and the page on the right what actually happened.

A field notebook is a quick, easy reference to what was collected, when and where. It will be far easier to refer to in the field than a large pile of collecting sheets and is a useful backup in case collecting forms go astray. Field notebooks should be small (but not too small: A5 size is ideal) and hardbound, and should always be kept in the collecting bag, perhaps wrapped in clear plastic. The paper of the field notebook and the collecting forms should be acid-free, long-lasting and marked in permanent ink. Notes should be written in pencil or indelible pen. All writing should be clearly legible.

Collecting passport data

In order to ensure the continuing usefulness of a germplasm sample, more descriptors than just the minimal set listed in Box 19.1 will be required, covering a wide range of factors relating both to the collecting site and its surroundings and to the population from which the sample was taken. The essential descriptors and some of these additional ones are described in detail in this section, along with their importance to both user and collector. Recommendations on the most efficient ways of obtaining the required data are also made. More general guidelines on documenting indigenous knowledge are given separately in Chapter 18, though details of some relevant participatory methods are presented here. The aim in deciding which descriptors to include here has been to develop a list with general applicability. Some species may require specific descriptors in addition to the ones presented here, or very particular descriptor states. This may be ascertained by consulting more specialized descriptor lists.

Certain basic principles underlie the design of lists of descriptors to be recorded in the field. Such lists should be kept as simple as possible, but not to the point where flexibility is lost and the inclusion of important additional data made difficult by the rigidity of the format. Multiple choice and binary descriptors are preferable to ones requiring text to be written out, though there should always be room for comments, and in some cases free text is necessary, for example in the case of ethnobotanical observations. In multiple choice descriptors, it is better for states to be mutually exclusive, but, if they are not, it should be possible to record more than one choice. An option (perhaps labelled 'Other') which allows space for free text is often useful, included within the relevant data field rather than left until the end of the form. Some collecting forms list all the states of each descriptor, and the collector ticks or circles one or more as appropriate. To save space, other collecting forms simply have the descriptor name followed by a blank space, and the collector fills in the appropriate choice for each descriptor by consulting a separate master form on which all states allowed are fully spelled out.

The descriptors have been grouped below into convenient classes for ease of discussion, arranged in logical order, though the order in which descriptors are laid out on the collecting form within each class need not rigidly follow the sequence adopted here. Descriptors marked with a '▶' are on the list of essential descriptors in Box 19.1. The classes are:

1. Sample labelling.
2. Sample identification.
3. Sampling information.
4. Collecting site localization.
5. Collecting site context and description.
    (a) Physiography.
    (b) Soil.
    (c) Biotic factors.
    (d) Other.
6. Population information.

The definitions of sample, site and population are given in Chapter 3. The descriptors in classes 1–5 are common to both wild and cultivated material, though some biotic site descriptors, in 5c, are perhaps of more relevance in one case than in the other. Many of the descriptors in 6, however, are specific to either wild species or crops. *It is common practice to have separate collecting forms for wild and cultivated material.*

Note that the descriptors in classes 1–3 and 6 are sample-specific and need to be filled in for each sample, whereas descriptors in classes 4 and 5 are site-specific and will have the same value for all samples collected at a given site. *It is a good idea to arrange sample-specific and site-specific descriptors in separate sections on the collecting form, for example on the two sides of a sheet.*

Data should be recorded immediately, at the collecting site, pooling information from all the collectors involved. Leaving the filling of collecting forms until 'later', whenever that might be, usually results in forgotten detail and missing data. Collecting trips can be unpredictable, and there may not be a chance to catch up on missed notes for many days. A few descriptors, however, can be left until the end of the day or the end of the mission, for example latitude and longitude and other data which can be read off a map on which the location of the site was marked in the field. Descriptors such as these, which can and will eventually be filled in, must be distinguished from descriptors for which data were not collected in the field, for whatever reason, and which it will not be possible to fill in later. The best way is to tick off or otherwise flag descriptors as they are being completed in the field, to make sure that a blank means that data are genuinely not available and not that the descriptor was overlooked. When an interdisciplinary team is collecting, it will often be a good idea to hold meetings at the end of each day to summarize and discuss each other's data.

Recording information supplied by farmers and other local people needs to be approached with tact. In some situations, it may be best to start taking down notes or filling in forms immediately after a meeting, or some way into it, when confidence has been established, rather than right from the beginning (a suitable methodology is described in Chapter 18). Still and video cameras and audio recorders can greatly facilitate the collecting of some kinds of ethnobotanical data, but permission should always be asked before these are used. A demonstration may also be useful. There may be difficulties about taking photographs and making video recordings of people and their activities in some cultures. There may also be legal restrictions on the subsequent use of such material. Curtin (1968) provides some useful practical guidelines on the use of tape recorders for collecting oral data in the field, and then archiving and analysing them. Collier and Collier (1986) review the use of photography as a research method in anthropology. FAO (1991) gives technical guidelines on the use of video in the field.

A list of the basic equipment that will be necessary for data recording (using paper collecting forms) is given in Box 19.2. Some have already been mentioned; the rest are described and discussed more fully below under the appropriate headings, along with some additional pieces of equipment which may be desirable but not essential.

Box 19.2
Basic equipment for data gathering and recording

- Collecting forms.
- Field notebook.
- Pencils, pens, markers.
- Maps.
- Checklists and identification aids.
- Binoculars.
- Hand lens.
- Altimeter.
- Compass.
- Clinorule or clinometer.
- Tape-measure.
- pH kit.
- Colour chart.
- Single lens reflex (SLR) camera, tripod, flash and film.

Sample labelling
► *Expedition identifier (or collecting organization)*

Collecting expeditions normally have a name and/or code decided on in advance. This makes keeping track of samples much easier. The name or code can be stamped or printed on the collecting forms before departure, and will include a reference to the organization(s) involved in the mission, as well as the country in which the collecting is taking place.

► *Name(s) of collector(s)*

This is a very important field, which together with the collecting number gives a unique identity to each sample. The collector's name and the collecting number should stay with the sample and any subsamples wherever they go. Although new accession numbers may be issued on entering successive gene banks, they will always be cross-referenced to the original collector's name and collecting number, and it is this which will remain the unique identifier essential for tracing duplicates held in different collections around the world. There will frequently be more than one collector, in which case the surnames of each should be used, or a general team designation (e.g. the expedition identifier). Half a line of space is usually adequate for this field, or the collectors' names can be printed on the forms before departure if the composition of the team is unlikely to change during the course of the mission.

▶ *Collecting number (or collector's number)*

As detailed above, this descriptor, together with the collector's name, forms the sample's unique identifier. Collectors sometimes start a new number series for each different collecting mission, prefixed by different codes for new countries, areas, sites or even species. These codes may sometimes be dropped or lost at a later stage, leaving several identical number series. Since in practice it often happens that the collecting number is the only piece of information accompanying a sample as it moves between institutes (at least initially), this numbering practice may result in serious confusion and hence effective loss of samples.

A better, more reliable, numbering procedure is for the collector to give the number 1 to the first sample collected on his or her first collecting mission, and to continue numbering samples from mission to mission in a consecutive sequence which, although it may be broken, is never repeated. When a new trip is begun, the sequence commences with the number following the last collecting number of the last mission. If the last number has been forgotten, then a margin of safety should be left to avoid unintentional duplication, and the series commenced at the next hundred, for instance. Gaps of this nature in a collector's series are of no consequence, but duplications cause serious complications.

If it has been decided to collect both seeds and vegetative material of a vegetatively propagated species, numbering of samples should be according to the following guidelines (Hawkes, 1980): give the same collecting number if seeds are taken from the same plant from which a vegetative sample has also been taken; give a separate collecting number if seeds are taken from more than one morphotype and bulked. In the former case, an alternative is to give separate numbers, but cross-reference the two samples. Random and selective samples from the same population should be given separate collecting numbers. Again, it will be useful to cross-reference 'related' samples. This can be done in a separate descriptor, which could be labelled 'Other samples collected from the same population'. The different components of stratified random sampling procedures (i.e. samples from different microsites) should also be given separate collecting numbers.

Any other material that is collected as an adjunct to a germplasm sample, for instance pest specimens (Chapter 17), *Rhizobium* samples (Chapter 26) or herbarium specimens (Chapter 27), is best given the same number as the germplasm. Trying to reconcile several different numbering series for different types of specimens is time-consuming, tedious and frequently problematic.

If several collectors are involved in a single mission, each with their own interests and numbering system, it should be agreed in advance whose system will be followed. Separate numbering systems for the different members of a team should be avoided. An alternative is to use the expedition identifier or a similar general team designation instead of individually listing the names of all team members.

The collecting number should be written in indelible ink on whatever container is being used to hold each sample, e.g. cotton or paper bags, glass vial etc. It should also be written in pencil on a paper or plastic tag and placed inside the container. Cuttings, tubers and other vegetative samples can be labelled with tie-on plastic tags. Labelling should be done immediately.

► *Collecting date*

Recording the date on which a sample was collected is useful to the collector in keeping track of his or her activities. It allows the receiving gene bank to calculate the time that has elapsed between collecting and registration and hence estimate what deterioration there may have been in the quality of the material. Also, if the sample was collected very early or very late in the season, it may indicate unusual or biased genetic make-up. If it was collected in a year of unusual environmental conditions such as drought or flood, the population may have become biased in the direction of individuals possessing particular tolerances or traits of interest. Date of collecting can also help in deciding the timing of future collecting of the same species and in monitoring genetic changes in populations. It should be recorded in full using the format day/month/year (the format should always be specified).

► *Type of material*

This descriptor is important because the type of material collected will determine where a sample is to be sent and how it will be treated on arrival. Also, more than one type of material may be collected for a given population, and in some cases the same number may be given. An example of this might be collecting both seeds and tubers from the same plants in a wild population. It will be important in subsequent work to know what form the original sample took. Germplasm may be sampled as:

☐ seed
☐ vegetative material (e.g. herbaceous cuttings, budwood cuttings, storage roots, tubers, seedlings, etc.)
☐ *in vitro* material
☐ pollen

Sample identification

▶ *Genus, species, subspecies, botanical variety*

This information is usually best accommodated in separate data fields, each with a line of space, which allows for a provisional identification in the field and subsequent changes or confirmation. Before setting out, annotated checklists (of species and/or landraces, either already published or compiled by the collector during planning), relevant Flora accounts, botanical drawings and photographs of plants, botanical keys to the target taxa and other identification aids will have been gathered together to take on the expedition (Chapters 10 and 11). Hand lenses (×10 and ×20) are often necessary for the field determination of material. However, collectors may not have the time to fully verify all scientific names in the field, particularly on multi-species collecting missions. Indeed, in some cases a field determination may not be possible, for example if microscopic examination is necessary. A provisional name should then be entered on the collecting form (e.g. *Acacia* cf. *sarcophylla*), or different taxa may simply be given arbitrary labels (e.g. A, B and C, or *Vigna* X, Y and Z), so that the collector can at least keep track of how many samples of each taxon, whether actually named or not, have been collected. This can be done in the collecting notebook, where a checklist of species collected can be kept. Alternatively, a previously prepared checklist can be annotated with collecting numbers.

If there is any doubt about identifications, this should be noted, perhaps in a separate yes/no field indicating whether confirmation is required. It is the responsibility of the collector to ensure that all germplasm samples eventually have confirmed names, hence the need for herbarium voucher specimens. Seed samples suspected to be mixtures of species should be labelled as such on the collecting form. It may be possible to sort out the components back at base.

▶ *Vernacular name (with name of language and dialect)*

The importance of local names to collectors and users of germplasm is discussed in Chapter 12, along with some of the problems associated with collecting and using them. As with all ethnobotanical data, the source of a local name needs to be recorded, though this perhaps belongs more in the field notebook than on the collecting form. A translation of the name should be provided on the form, however, if the word or phrase has a meaning. A recognized standard system should be adopted for rendering local sounds, and words should also be written out in the local script, if one exists. A standard system of transliteration will usually be set out in dictionaries and, if such works exist for the languages likely to be encountered during collecting, it will be worthwhile taking them along. Much confusion has been caused in the past by different people rendering the same local word in different ways. If no recognized system exists, phonetic renderings and transliterations should at least be internally consistent. The language and dialect (and/or name of the ethnic group) should always be specified in a separate field. The degree of corroboration should be noted, perhaps in the field notebook. It should be made clear to which level of the taxonomic hierarchy the local name refers. In particular, when collecting cultivated material, both the local name for the crop and that of the landrace should be recorded. A line or so of free text will be necessary.

A running list of the local names collected can be kept in an alphabetically indexed notebook. This can be annotated with morphological information and observations on genetic erosion, for example. An initial checklist may be derived from research at the planning stage, and this can be added to as the collecting proceeds.

► *Herbarium voucher number*

Though it is recommended that herbarium vouchers be collected for all germplasm of wild species, this is essential if there is any doubt over determinations (Chapter 27). Accessions which cannot be named for lack of a herbarium voucher will tend to remain unused longer than material for which a confirmed name is available. The herbarium voucher for a particular germplasm sample should bear the same number as the germplasm, but whether this is the case or not the voucher number should be written on the collecting form, along with the destinations of all duplicates of each specimen. This last piece of information could be added at the end of the mission.

► *Identification numbers of other associated specimens*

Pest specimens and *Rhizobium* samples are also occasional adjuncts to germplasm samples. The number identifying the material, which again should be the same as the collecting number for the germplasm, and the destination of the material need to be recorded on the collecting form in the same way as for herbarium vouchers. Half a line of free text is usually adequate for each kind of associated material.

Photograph number(s) and subject(s)

Note should be taken in the field notebook of all photographs taken (whether of people, landscapes, collecting sites, plants, etc.), giving the roll number (all film cartridges should be labelled), frame number, subject and location (site number). The identifying numbers (roll and frame) of all photographs relevant to the site or the material collected should then be recorded on the collecting form.

▶ *Status of sample*

This descriptor refers to whether the plant population sampled is wild, weedy or cultivated. It is usually broken down into the following choices, though in most cases only the first three will be relevant:

☐ wild
☐ weedy
☐ landrace
☐ obsolete improved variety
☐ advanced improved variety
☐ breeding/research material
☐ interspecific derivative
☐ other

'Weedy' species in this context are those species related to crops but not actually cultivated themselves, which require the disturbance often caused by human activity for establishment and reproduction and which are therefore found on the edges of cultivation, often in close proximity to their cultivated relatives. A 'landrace' may be defined as a set of populations or clones of a crop species originally developed by farmers, maintained by them over a long period and recognized by them as belonging to a single entity. The term is used here as synonymous with 'traditional variety' (the terms 'primitive variety' and 'farmer's variety' are also occasionally used). 'Improved varieties', in contrast, are the products of scientific plant breeding, though of course this is not to imply that landraces have not also been 'improved' by farmers. Crop collectors will almost exclusively be interested in landraces and wild relatives, but will occasionally also collect older improved varieties ('obsolete') if these have been grown in the area for a considerable period and farmers have been maintaining seeds from year to year, rather than obtaining new stocks regularly from private or public seed suppliers.

'Other' could include recent farmers' selections and crosses, off-types, material descendent from the small seed samples brought back by travellers, grain saved from food aid etc. Details can be recorded under 'History of sample and landrace' (see below). Information on the status of crop samples will normally come from discussions with farmers.

Sampling information

▶ *Size estimates of population and of sample*

It is important for users to know the approximate size of the sample in relation to the size of the original source population (measured in numbers of individuals). If, for instance, the population was large but only a very small sample was taken, then the implication is that there may be more genetic diversity available in the population than is present in the sample. If the original population consisted of only a few plants, all of which were sampled, then the population will probably not be worth re-collecting in the hope of finding new genetic diversity, but the information will clearly be important in assessing the danger of the plant disappearing at the site. The size of the sample is also important to gene-bank curators. Some collectors, particularly foresters, record the total number of seeds collected and/or the weight of their seed samples in the field. Though this is not usually necessary, it can be useful to flag in some way small samples that will require multiplication prior to entering a gene bank.

In some cases, it may be useful to also indicate the proportion of the population sampled in area terms. Such estimates may be given in square metres or hectares. Though they are often necessarily very approximate, they can give the user an indication that there may be more diversity available, for example if only a small area of an extensive population was sampled. An effective method of calculating the area of an irregularly shaped field is to divide it notionally into squares and right triangles.

These descriptors could be set out as follows, and either completed with exact numbers or with a choice from among those set out below:

Number of plants in population is _____, covering ____ m^2
Number of plants sampled is _____, covering ____ m^2

Area	Number
☐ <1 m^2	☐ 1 plant
☐ 1–10 m^2	☐ 2–10 plants
☐ 10–100 m^2	☐ 11–100 plants
☐ 100 m^2–0.1 ha	☐ 100–1000 plants
☐ 0.1–1 ha	☐ 1000–10,000 plants
☐ 1–10 ha	☐ >10,000 plants
☐ 10–100 ha	
☐ >100 ha	

Estimating the number of plants in a population and the area it covers is relatively easy when the unit of collecting is a well-defined agricultural field, with plant density fairly uniform and individual plants clearly visible. It is straightforward in these situations to count plants in a number of small areas of known size within the field (located at random or along a transect, for example) and extrapolate up to the number of plants in the whole field.

For many wild species, this procedure will be neither so easy to apply nor so accurate, because populations may cover large areas that are not clearly delineated, density may be very low, individuals may be difficult to define and plants may not be particularly obvious. There are various plotless methods of density estimation that may be useful in these cases. The closest-individual method is probably the easiest to use. Sampling points are chosen at random and the distance to the nearest individual of the target species recorded. If the mean of these distances is d, $4d^2$ is the mean area occupied by each individual, and its reciprocal will be the density. Cottam and Curtis (1956) describe more complicated plot-less methods. Techniques particularly suitable for forestry species are discussed by Freese (1962).

If estimating total population is problematic, one option is to simply record the number of plants sampled as a proportion of the number encountered in searching a given area. In the case of plants that reproduce solely by vegetative means, there is no easily definable population structure to sample. Only the number of clones sampled, which could be just one, need be recorded. Population size estimates are not applicable if the source of the sample was a farmer's store or market. This is another reason why recording the source of a sample is important.

For perennial wild species (and some perennial crops) it is useful and instructive to note not only the size of the population, but also its age structure, for example in terms of the proportion of the population in different size classes. In forestry, diameter at breast height (DBH) is often used to define size classes. In particular, knowing whether recruit-ment to a population is taking place will affect any assessment of the threat of genetic erosion. This will be impossible to ascertain during a single visit, but the presence of seedlings may give a clue.

Frequency of populations in the area

The frequency of populations of the target taxon (wild species, crop and/or landrace) in the area around the collecting site, together with estimates of the size of these populations, can be used to assess the threat of genetic erosion. It is also useful to know what proportion of populations found have actually been sampled. It will obviously make a difference to the amount of genetic diversity recovered whether the population sampled was the only one found in an area or, in contrast, whether it was only one of several, chosen at random or for a particular reason.

The frequency of a taxon in an area can be recorded in a number of ways. The simplest is perhaps to use an arbitrary scale of 1–5 (sporadic to common, say) to record the relative number of sightings of the species: (i) during a given period of travelling through the area (for crops); or (ii) as a proportion of the number of stops made to look for it (for wild species). Local people can also be asked about the rarity or otherwise of plants, and about any changes in the frequency of occurrence of land-races, crops and wild species (see 'History of sample and landrace').

There are more complicated schemes. Rabinowitz (1981) discusses an eight-category system based on geographic distribution, habitat specificity and local population size which can be used to compare species. The World Conservation Union (IUCN) has a system of cat-egories of species conservation status (extinct, critical, endangered, vulnerable, susceptible, etc.), which, though aimed at comparing species globally, could also be used for a given species in different areas

(Mace *et al.*, 1992; IUCN, 1994). The Taxonomic Databases Working Group (TDWG, also known as the International Union of Biological Sciences Commission on Taxonomic Databases) is preparing a standard scheme for recording the state of a plant in an area.

Sampling method

Whether the plants from which a sample was collected were chosen at random, systematically or in a selective or biased manner from the population can have a profound effect on the amount and type of genetic diversity present in the sample. The number of seeds collected from each individual sampled will also be significant, and whether this was in fact the same for all individuals (Chapter 5). Such information will be necessary if comparisons are to be carried out, for example of the diversity at a site on two sampling occasions, or of the diversity at two separate sites. Some collecting forms simply have the random/selective choice, but this is unduly restricting. Note must be made of the details of the procedure used. For example, if sampling is along transects, the technique might be described as:

> 5 transects across the field, starting at random points along the S edge and proceeding in different, randomly chosen directions, sampling 10 seeds from a single plant every 5 m

In stratified random sampling at a site, note must be taken of the basis of the stratification, e.g. the nature of the different microsites sampled (this could be recorded under 'Specific habitat'). If selective sampling is carried out, then a note must be made of the characters on which selection was based, e.g. a few disease-resistant plants in an otherwise heavily infected field, locally recognized phenotypes in a mixture, etc. This data field is probably best left for free text, allowing one line of writing.

It is also worthwhile noting whether there was any bias in the collecting apart from that imposed by conscious sampling strategy at the site. Was the sample collected at the very end of the season? In the case of forages, this would indicate sampling biased in favour of genotypes which remain green after the rest of the population has died back. Was it collected in a period of drought? This might mean a bias in the sample in favour of drought tolerant genotypes.

Though this descriptor deals mainly with the sampling procedure employed at a given site to choose individuals for inclusion in the sample, it could also be used to record how the population itself was chosen. It could be that stops are being made at regular intervals of a given number of kilometres. On the other hand, particular micro-environments may be being targeted, for example areas of standing water to collect waterlogging-resistant genotypes of a particular species.

▶ *Collecting source*

Collecting wild species from habitats undisturbed by man and crops from fields and home gardens just at harvesting time are ideals that it may not always be possible to realize. Farm stores may have to be visited, if all the crops in a rotation are of interest or if the material has already been threshed or not yet sown. Local markets are good sources of information on the diversity available in the area they serve. Occasionally, they are the only sources of germplasm. However, markets often contain a biased sample of the agricultural diversity available in an area, farmers growing some crops and landraces exclusively for home consumption.

It is important to note the source of a sample because this conveys important information about the possible genetic structure of the material. For instance, the genetic structure of a disturbed population along a roadside, say, is likely to be different from that of a population of the same species growing in a relatively undisturbed natural habitat. A sample collected in a market may come from an area agroecologically quite different from the collecting site. When collecting from piles of threshed material, from farm stores and from markets, it will be impossible to record many morphological details, some postharvest selection by the farmer may already have occurred, and the material may consist of mechanical mixtures. Compared with freshly harvested material, material collected in farm stores and markets may have low viability.

This data field can have the following choices:

☐ undisturbed natural habitat
☐ disturbed natural habitat
☐ weedy habitat (roadside, field margin)
☐ farmer's field, plot or orchard
☐ backyard, kitchen or home garden (urban, periurban or rural)
☐ threshing floor
☐ farm store
☐ market or shop
☐ seed company (family or large-scale)
☐ institute, experimental station
☐ other

If the collecting site is not the place where the material was grown, as might be the case for material collected in markets, separate note should be taken of the actual source of the germplasm, if this can be ascertained. This may be recorded in a separate descriptor (see 'History of sample and landrace').

Collecting site localization

Detailed and thorough notes on the geographic origin of a sample are an essential part of germplasm collecting. Such information allows re-location of populations for future collecting. Researchers may wish to sample certain areas or even specific populations again more thoroughly on the basis of initial characterization and evaluation trials or screenings. Information on the geographic origin of material can also provide clues to the requirements and adaptation of the material, which is important at all stages from multiplication to use. Locality data are best dealt with in a number of separate descriptors.

► *Country*

The name of the country in which collecting is to take place can be stamped or printed on to the collecting forms before the trip. The official name should be used, or a standard abbreviation.

► *Primary administrative unit*

This field specifies an official first-level administrative division of the country. These can be obtained from cadastral and other maps, or from local people. Whether the administrative unit is called a province, a state, a 'wilayat', or whatever, should also be recorded, ideally in the local language. If secondary and even tertiary subdivisions exist and are known, they should be recorded after the primary, separated by a '/' or similar delimiter, or even in a separate descriptor. An example, taken from Kenya, might be as follows:

Province: Rift Valley / *District*: Kericho / *Division*: Belgut / *Location*: Soin

► *Precise locality*

Some collecting forms use the following format to record the exact locality of collecting:

_____ km from _____ (village name or other landmark) in a ____ direction

When collecting at a village, say, the first and third blanks would be filled in with a dash. They should not be left empty, as this could suggest that the field had been overlooked. In writing down the names of places given by local people, the observations made earlier with regard to local plant names also apply. Whenever possible, the same orthography used on the map being marked and annotated in the field should be adopted when entering names on the collecting form.

This format only allows for the broadest description of locality, however. In many instances, particularly in wild species collecting, there may be no village or precise landmark in the vicinity. An alternative is to allow free text for recording how the site was reached. A good general procedure for recording locality data might then be as follows:

1. Describe in the field notebook in detail how the site was reached and its setting in the surroundings, with sketch-maps of the area (showing major natural features and infrastucture) as necessary. The description

should be accurate enough for the site to be located easily by another collector on a later occasion. This will clearly be necessary if the mission is an exploratory one for wild species, carried out at flowering time, during which the site will be marked, to be revisited at the end of the season. However, if the material collected proves interesting, or if it is wanted to monitor genetic erosion at the site, repeat visits may well also be warranted. For distances, vehicle odometer readings can be used. The odometer should be reset at the beginning of each day at a fixed point clearly identifiable on a map (e.g. intersections).

The following is an actual example, from the Moroccan Atlas, of a detailed description of how a wild species collecting site previously targeted from herbarium labels was reached:

> On the main road RS501 from Marrakech to Tizi-n-Test, proceeded SW until 5 km SW of Talat-n-Yakoub. Turned right on the track to Tafreghoust and proceeded for 3 km to the village. Continued north for about 8 km (about 2 hrs by mule) up the valley to the third summer village (azib), at 2100 masl. Continued uphill on foot towards Djebel Gourza in a N direction; it is possible to reach 2800 m in about 3 hrs.

2.   Locate and mark the position on a large-scale map, labelling with the site number (see below). The position of sites should be marked with an 'X': dots tend to get lost as the map gets crumpled and dirty with use. If sites are close together, to avoid crowding of markings on the map, site numbers should be written at some distance from the 'X' and enclosed in a circle, with an arrow going from the circle to the location of the site.

3.   Record the precise locality on the collecting form as free text summarizing the information in the notebook or coded as suggested above. If the coding method is used, the landmark quoted should be easily identifiable on the map (i.e. its name should be printed on the map). In the Moroccan example given above, since the village was marked on the 1 : 250,000 map that was being used, the following was appropriate to record the location of the site on the collecting form:

> 10 km from Tafreghoust in a N direction

4.   Read off latitude and longitude from the map (this can be done later if necessary) and altitude from the map or an altimeter, and record them on the collecting form, as detailed below.

► *Latitude, longitude and altitude*

These descriptors greatly facilitate mapping distributions and will be essential if a geographic information system (GIS) is to be used in analysing the results of the collecting. They are also frequently used in computer searches for appropriate accessions, for example when looking for holdings likely to have a particular day-length sensitivity, which would come from particular latitudes. Latitude and longitude should be recorded in degrees, minutes, seconds and a hemisphere (N, S, E or W). Altitude should be in metres above sea level (masl).

Latitude, longitude and altitude can all be read off maps if locality is reasonably accurately known. Ideally, only maps of scale 1 : 250,000 or larger should be used for this, which would give accuracies of better than 1 km. It is good practice to take to the field one or two original copies of large-scale topographic maps of the specific target area or areas (rolled up and kept safe in a cardboard tube) and two or three reproductions of these on A4 sheets (perhaps in a ring-binder or in a plastic folder) for marking sites, plus one or two original copies of smaller-scale road maps of the whole region to be covered. Reproduction of published maps may require the publisher's permission.

Altitude can also be measured using an altimeter, which should have a range of 0–5000 m. The altimeter should be reset whenever possible, at sea level or when altitude is known from reliable sources.

As a result of recent advances in satellite technology, it is now possible to accurately locate one's position on the surface of the earth by means of small hand-held devices called Global Positioning System (GPS) receivers. These are rapidly becoming less expensive, and are already in use at several institutes involved in germplasm collecting. Prendergast (1993) reviews one of the models currently on the market. Latitude and longitude are commonly fixed by these instruments to a precision of about 100 m root mean square (RMS) (meaning that 63% of fixes will be within 100 m of the true position) in less than a minute. A reading of altitude can also be provided, though this will take somewhat longer and is of considerably poorer accuracy; in general, altimeters are preferable.

Transferring a GPS reading to a map requires knowledge of the 'map datum', essentially a mathematical description of the earth or part of the earth. The latitude and longitude coordinates of a given position differ from one datum to another. The datum that was used in making the map will usually be specified in the map legend, and should be entered into the GPS receiver if the locations of sites are to be accurately shown.

A limitation on the use of GPS receivers is their battery requirement, but some models come with a device which can be used to obtain power from a car cigarette lighter. GPS receivers require unimpeded sight of three of a constellation of 21 satellites, so their usefulness may be limited in heavily wooded and/or deeply dissected terrain.

Map reference

It is essential, in the case of map-derived positions, to fully identify the map used, in particular its scale. One line of free text is adequate for this. The information to be recorded includes map series number, sheet number, edition, scale and map grid reference (Lazier, 1985).

Site number

The site number is not a necessary piece of information for sample identification, collector's name and number being sufficient. However, it can assist in data handling and management. It is common to collect more than one sample per site. In such cases, all the site data will apply to all samples collected there. If each site is given a unique number, these site data need only be recorded once, and the site number cross-referenced with the appropriate collecting numbers. In compiling data-bases of collecting information, site number is a convenient linking field between a file containing sample-specific data and one containing site-specific data. It is also much easier to label collecting localities on a map with a single site number instead of enumerating the collecting number of every sample collected at that locality. Microsites are sometimes distinguished on collecting forms by appending different letters to the site number (as in 12a, 12b, 12c, etc.), though some collectors give entirely new site numbers.

This field is most conveniently located on the collecting form next to the collecting number and is a short numerical field. There are no strict rules about site numbering systems, but a common practice which works well is to begin at 1 on each new collecting mission. Unlike the collecting number, the site number is not used in sample identification and repetition is not a problem. It is useful to keep in the field notebook a listing of all the samples collected at each site, as a cross-check to the collecting forms should any of the latter go astray.

Some collectors write the site number on each sample container, circled or differentiated in some other way to prevent confusion with the collecting number.

Farmer's name (or owner's name)

In the case of crops, recording the name of the farmer or market stall-holder from whom the material was collected can help in finding material again, quite apart from acknowledging the part played by the farmer in the development, maintenance and conservation of the germplasm. However, any wish for anonymity should be respected. When collecting wild species, a note should be made of the person(s) and/or organization which gave permission to collect at the specific site (e.g. a national park administration, private landowner or village council). Cultural anthro-pologists and other social scientists often keep a separate, alphabetized file of short profiles of informants and other people met, noting eth-nicity, caste, gender, age, occupation etc. (Bernard, 1988).

Collecting site context and description

The unit of collecting (the population) may be defined pragmatically as those conspecifics inhabiting a restricted area under relatively homo-geneous ecological conditions (the collecting site). It is important to document these conditions because they will help to deduce the adaptation of the material, and the collecting form is the place for such site data. However, the surroundings of the site – its setting or context – will also need to be noted, in particular whether soil, vegetation, etc. are markedly different at the site compared with the surrounding area and with the region as a whole (Bunting and Kuckuck, 1970). Notes on the area and region may be made in the field notebook while travelling through it, and any differences between site and surroundings recorded on the collecting form. A pair of binoculars will be useful not only in locating potential collecting sites, but also in documenting their surroundings.

Describing collecting sites and their context, no less than their precise locality, may be facilitated by making reference to different kinds of thematic maps in the field. The kinds of maps that it may be useful to take to the field are discussed in Chapter 9.

Physiography

The natural form of the land surface around the collecting site – the physiography of the area – affects both soil and microclimate at the site itself. The general topography of the region should be given, followed by a more detailed description of the setting of the collecting site. Photographs are often used to supplement any written description; 35–50 mm lenses are the most appropriate for this application. Sketch maps and profiles are also useful. These should always include an indication of scale and orientation. Maps, diagrams and profiles drawn by farmers and other local people, or at any rate in participation with them, are increasingly being recognized as important ways of arriving at descriptions of the environment which are relevant to local needs (Conway, 1989).

Topography of region

Topography refers to the variation in elevation of the land surface on a broad scale. It commonly takes the following states (FAO, 1990):

- ☐ flat or almost flat   slopes <2%
- ☐ undulating   steepest slopes 2–10%
- ☐ rolling   steepest slopes 10–15%
- ☐ hilly   steepest slopes 15–30%, range of elevation moderate
- ☐ steeply dissected   steepest slopes >30%, range of elevation moderate
- ☐ mountainous   steepest slopes >30%, range of elevation >300 m

Landform at the site

It is necessary to detail the exact location of the site within the overall topography or landscape. FAO (1990) suggests that the position of the site within a land element of a landform be recorded. The major landforms are:

☐ mountain
☐ hill
☐ upland
☐ plain
☐ plateau
☐ basin
☐ valley

Land elements are subdivisions of landforms, though 'depending on magnitude, the same geomorphological feature may be described as a landform or as a land element' (FAO, 1990). It is not possible to give an exhaustive list of land elements here. Examples are flood plain, lagoon, interdunal depression, valley floor, etc. If a land system study of the target region has been carried out, it may be possible to use it to develop a more manageable list of land elements for inclusion on collecting forms.

Position of site in topography

Position within a landform or land element may be recorded as follows (FAO, 1990):

In undulating to mountainous terrain
☐ crest
☐ upper slope
☐ middle slope
☐ lower slope
☐ bottom (flat)

In flat or almost flat terrain
☐ higher part
☐ intermediate part
☐ lower part
☐ bottom (drainage line)

TDWG is developing sets of standards that can be used to record the landscape where a plant is growing and its habitat.

Description of site

Describing the topography of the region and locating the site in the landscape may not give a full picture of environmental conditions at the site. In an arid plain, one might be collecting wild species along a seasonal stream, for example. In a valley bottom, the cultivated field being sampled may be adjacent to the river or some distance from it, or in the type of restricted, specialized microenvironment described by Chambers (1990). It is clearly important to record the details of such collecting sites or misleading conclusions may be reached about the likely tolerances and requirements of the material collected there.

Though to some extent the problem of describing the site of collecting is dealt with in the land element description, it is nevertheless a good idea to also characterize the site in detail with free text. In the arid plain example, the kind of information recorded in this field might be 'seasonal watercourse in gravelly plain' (a more detailed description than this may be possible, and necessary, in the case of microsites: see 'Specific habitat'). Often, local words exist which describe a site quite precisely, for example 'wadi in hamada' in the Middle East in this case. Riverine strips and seasonally flooded areas, crucial to subsistence agriculture in Zambia and Zimbabwe, are called 'dambos' locally. Such terms can be used to save time in filling in forms in the field, but may not be understood by users, so translations or explanations should be substituted before the forms are sent for distribution. One line of text is normally enough for this field, though it could be supplemented by sketch-maps and profiles. In crop collecting, transect profiles through the village area and farm sketches, to which the farmers themselves should contribute, can be particularly useful in describing collecting sites.

The description of a site should include an indication of its size. This will be different from the area covered by the sampled population, another important descriptor which is discussed below, because more than one species may be sampled at a site and because a larger area may be searched than actually contained target species. In wild species collecting, it is useful to compare the size of a site with the 'minimal area' of the vegetation at the site. This is the smallest area that provides enough environmental space for a particular community type or stand to develop a full and characteristic species complement and structure. It can be measured empirically by counting species in progressively larger quadrants, but is generally 1–25 m^2 for herbaceous vegetation, 25–100 m^2 in short woody vegetation and 200–500 m^2 for the tree layer in forest (Goldsmith *et al.*, 1986).

Slope: magnitude, form and aspect

An estimate of the slope on which the site is located is often included as part of the physical description of a site, as it affects drainage, soil stability and microclimate. Slope is measured up from the horizontal by means of a clinorule or a clinometer. A clinorule is a simple piece of equipment that looks like a ruler with a joint in the middle. One segment of the ruler is held horizontal with the assistance of a built-in spirit level, and the second section is rotated to an angle parallel to that of the slope to be measured. The slope may then be read off. A clinometer is often built into a compass, and allows slope angles to be measured by the free movement of a suspended arm along a degree-graduated scale. Both instruments give measurements in degrees. Slope may be measured as a ratio, degrees or percentage; the units of measurement should always be specified. If clinometer readings are not possible, field visual estimates of slope gradient should be matched against calculated gradients from contour maps. Field estimates are perhaps most easily done by estimating the vertical distance between objects (two people, for example) located at a known distance apart along the slope. In addition to the extent of slope, some indication of its form may be given, as follows (FAO, 1990):

☐ straight
☐ concave
☐ convex
☐ terraced
☐ complex (irregular)

The aspect of a slope is the direction in which the slope faces. This is often a key factor determining absence or presence of a species. At high latitudes, the difference of insolation between south- and north-facing slopes may be considerable, for example. There may also be differences in the rainfall received by mountain slopes facing into and away from the prevailing winds. Aspect is usually measured with a compass but may also be estimated from maps.

The overall aspect and slope of the site should be recorded, disregarding irregularities. Thus, the aspect and slope of a site consisting of a terraced field on a mountainside are those of the mountainside, not the level field.

Soil

The soil is an important feature of a plant's environment and should ideally be described in detail. All of the soil descriptors listed below can be recorded directly in the field with a minimum of effort and without specialist knowledge. Some collectors gather more detailed soil data using specially designed soil testing kits and/or take samples for laboratory analysis. This is not often part of the collecting routine, but will be of value in some cases, for example if adaptation to specific edaphic conditions is being sought. See Ball (1986) and EUROCONSULT (1989) for useful introductions to the field description of soils.

Parent material

The composition of rocks directly affects soil type and soil chemistry, so the collector needs to consider the origin of the material from which the soil at the collecting site is derived. This field should include information both on the origin (aeolian, alluvial, colluvial, *in situ* rock) and nature of the material from which the soil at the site is derived. An example might be 'colluvial material derived from granite'. There are clearly a large number of possible states. If information on the general geology of the target area is available, this can be used to cut the list down to manageable size for inclusion on preprinted collecting forms. Otherwise, a line or so of free text will be needed. If parent material is simply read off a geological map, this should be noted, and the map name and scale specified.

Soil taxonomic class

This field refers to the name of the soil according to a recognized local or international system such as that of the Food and Agriculture Organization and the United Nations Educational, Scientific and Cultural Organization (FAO–Unesco) (FAO–Unesco–ISRIC, 1988), the US Department of Agriculture (USDA) (USDA, 1975) or the Commission de Pédologie et de Cartographie des Sols (CPCS, 1977) (Young, 1987). Soil classification can be complex and requires a knowledge of the entire soil profile. FAO has published guidelines on soil profile description (FAO, 1990). Collectors may not have the training to be able to apply these systems, and attempting to classify each soil would in any case be too time-consuming to warrant the effort in most cases. If classification is attempted, however, the system used should be specified. Young (1976) provides a useful simplified key to tropical soil classes in the FAO–Unesco system. Soil type may be read off soil maps, but, if this is done, it should be noted, and the map name, scale etc. specified. Note that TDWG is developing standards that can be used worldwide to characterize the soil type in which a plant occurs.

Local knowledge of soils is often systematized by farmers and others into indigenous classifications. These distinguish between soils on the basis of those characteristics which are most important to the everyday user of the land, in particular suitability for different crops, and can be an important complement to scientific taxonomies (Tabor *et al.*, 1990).

Soil texture

In the field, soil texture, the relative amounts of primary particles of different size classes in the fine earth fraction of the soil, is normally assessed by sight and feel. Texture is one of the more important characteristics of soil, usually giving a good indication of the edaphic preferences of a species. Particle size classes making up the fine earth fraction are: clay (diameter <0.002 mm), silt (0.002–0.05 mm) and sand (0.05–2.0 mm). Soils with an approximately equal contribution of all three classes are called loams. The percentage contribution of each fraction can be measured in the laboratory and plotted in a triangular diagram. EUROCONSULT (1989) describes a simple manual texture test for use in the field. A small heap of about 2–3 cm diameter is formed from about one tablespoon of soil. Water is then slowly dripped on to the soil until the material just starts to stick to the hand. The extent to which the moist soil may be shaped by hand is used to describe its texture, as follows (intermediate classes, such as 'sandy clay loam', are also possible):

☐ sand   soil remains loose and can only be heaped into a pyramid
☐ loamy sand   soil can be shaped into a ball that easily falls apart
☐ silty loam   soil can be shaped by rolling into a short, thick cylinder
☐ loam   soil can be rolled into a cylinder about 15 cm long that breaks when bent

☐ clay loam   soil cylinder can be bent into a U, but no further, without breaking

☐ clay   soil cylinder can be bent into a circle

Stoniness

The total extent and size (average or maximum) of coarse surface fragments in the soil and of rocky outcrops is sometimes recorded, usually separately. Extent can be recorded as none, low, medium or high; or as percentage cover. Size fractions for surface fragments are: fine gravel (2–6 mm), medium gravel (6–20 mm), coarse gravel (20–60 mm), stones (60–200 mm), boulders (200–600 mm) and large boulders (>60 cm). When describing crop collecting sites, stoniness is sometimes recorded in terms of the effects it has on cultivation: tillage may be unaffected, affected, difficult or impossible.

Soil colour

Soil colour can give indications as to leaching and fertility. It is, however, difficult to record accurately. Soil colour changes according to whether the soil is wet or dry, and it also appears different at different times of day, with the sun at different angles. The most accurate way to record this parameter is to compare moist unrubbed soil with the small squares of standard colours provided in colour charts. The colour chart used should always be specified. These are reviewed by Tucker *et al.* (1991). For many purposes, simply recording whether the soil is black, brown, grey, orange, yellow or white may be sufficient. The main matrix colour should be noted, plus the intensity and extent of mottling and of other secondary colours, if present. Bingham and Ciolkosz (1993) discuss the significance, causation and measurement of soil colour in detail.

Soil depth

This is a factor which can be difficult and time-consuming to determine, often requiring a soil pit to be excavated. Shallow soils or superficial bedrock are usually easy to detect, but for deep soils the only way to get data is often from fortuitous profiles, such as might be found along stream sides or road cuttings. When collecting crops, soil depth is sometimes given relative to plough depth.

Soil pH

Most soil pH values fall within the range 3.5 to 12, with 3 being very acidic, 7 neutral and 12 strongly alkaline. Tolerance of highly acid or alkaline soils can be an important agronomic trait, and plant introduction teams are continuously searching for genotypes with these tolerances.

Soil pH may be measured in the field by means of a dual glass-calomel electrode. This is usually the most accurate method, but the instrument is delicate and requires calibration with buffers. There are also methods which rely on the change of colour of reagents, though these tend to be less precise. It is important to choose a light, compact and durable soil testing kit for use in the field. Samples for pH measurements are usually taken from the top 20 cm of soil, but deeper samples are sometimes also taken.

Salinity

Salinity can be an important determinant of plant performance, and tolerance of high soil salt concentrations is being sought in many crops (often among the wild relatives) and forages. Salinity is measured in the field or laboratory by means of a conductivity meter. It can also be recognized visually, however, by a whitish crusting on the soil surface (the cover and/or thickness of which may be recorded) or by the degree of inhibition of salt-sensitive crops.

Drainage

Drainage covers a subjective estimate of the balance between water arriving at the soil surface and leaving it by run-off and infiltration. It depends on site relief, ground surface conditions and soil permeability. It may be divided into site drainage and profile drainage (Lazier, 1985). The former refers to the movement of water horizontally along the surface of the soil, the latter to vertical movement down the profile. Agronomists frequently seek plants with high tolerance to permanent or seasonal waterlogging in crops and forages, and poor drainage at the collecting site can give a good indication of this. Soils are most often simply described as poorly to well drained on an arbitrary scale. Such features as the frequency, duration and depth of flooding can also be recorded here, and the depth of the water table. If the site is on crop land, such information will be available from the farmer.

Identification number(s) of soil sample(s)

Isbell and Burt (1980), León *et al.* (1979) and Ball (1986) describe soil sampling methods. Soil samples are most efficiently taken with a screw auger, but a small digging implement such as a graduated planting trowel can also be used, and these are perhaps better in loose soils. Implements should be of stainless steel to avoid contamination. A screw auger is composed of a wood-boring large-diameter bit with a screw thread (2.5 × 20 cm long), welded to a T-shaped steel handle about 1 m long. The auger is screwed into the soil and then withdrawn by a strong upward pull. The top 20 cm or so of soil from 5–10 points should be sampled, bulked and mixed, taking care to discard loose plant litter and other debris. Samples are usually taken from the immediate vicinity of the plants being collected, but samples from adjacent areas may also be collected for comparison. About 500 g per sample is sufficient for most purposes. Soil samples should be kept in strong polythene bags, and are usually labelled with the site number both on the outside of the bag and inside. The number of the soil sample should be written on the collecting form. Soil samples should be air-dried, avoiding contamination from fertilizers and the like, and then the containers sealed.

Additional notes on soil
Other features of soils that may be important to germplasm collectors are (FAO, 1990):

- extent of bare ground;
- extent and type of soil erosion;
- size and frequency of surface cracking;
- thickness and consistency of surface sealing;
- soil fertility;
- organic matter content;
- groundwater quality;
- differences between the soil at the site and in the surrounding area.

If information on these is not recorded in separate predetermined descriptors, comments may be made in a general field for soil notes. Two lines of free text will usually be adequate.

Biotic factors

Vegetation type
Vegetation is the assemblage of plant species growing at a site. It may be described in terms of its component species i.e. floristically) or in terms of its appearance (i.e. physiognomically). Often, a combination of the two methods is used.

Species may have very precise associations with particular vegetation types. Recording vegetation type at the collecting site will thus assist future collectors in deciding where to look for more material. Being the result of the interaction of many climatic and soil factors, vegetation can also help in characterizing the general adaptation of material. Though it is usually only considered in wild species collecting, it is useful to record the dominant natural vegetation in the region even when collecting crops, as it can provide useful indications as to climate and soil in the absence of more direct information.

Terms such as 'forest' and 'grassland' are physiognomic descriptions of the size and spacing of the main components of the vegetation, which may be qualified and subdivided at various levels, for example by a statement of periodicity or phenology, as in 'evergreen forest', and/or climatic requirement, as in 'tropical rain forest'. Unfortunately, terms such as these, which are in general use, may be interpreted in various different ways, and precise definitions are therefore necessary. One way around the problem is to record the percentage cover of various life-forms, plus bare ground. Life-form categories could just be trees, shrubs, broad-leaved herbs and grasses, possibly subdivided into size classes, or Raunkiaer's categories based on the position of perennating buds relative to ground level could be used (e.g. Goldsmith et al., 1986).

However, various widely used and widely applicable descriptive systems based on physiognomy do exist which actually name and define different vegetation types. These include Fosberg's (1961) classification using spacing and vertical stratification, Unesco's (1973) authoritative attempt to produce a generally applicable system (see also Ellenberg and Mueller-Dombois, 1967), the not wholly standardized classification system used by the Institut de la Carte Internationale du Tapis Végétal (ICITV) and White's (1983) system used for the AETFAT/Unesco/ United Nations Sudano-Sahelian Office (UNSO) map of the vegetation of Africa. There is a standard classification system for wetlands, adopted as part of the *Convention on Wetlands of International Importance Especially as Waterfowl Habitat* (also known as the *Ramsar Convention*). A synopsis of the Unesco system is provided in Box 19.3, mainly giving only the first level of the classification, the formation class: each level is considerably subdivided in the full classification. A project is under way to develop a global classification scheme for vegetation (UNEP/GEMS, 1993).

Such a broad physiognomic classification at the formation class level is a useful first step in describing vegetation on collecting forms. There could be an additional vegetation type to the ones listed above, labelled 'arable land' or similar, to represent man-made landscapes. A broad physiognomic description may be augmented or qualified by floristic information, usually on the dominant or most abundant species, as in '*Acacia–Commiphora* woodland' or '*Themeda* grassland with scattered *Acacia* shrubs'. Alternatively, the more detailed levels of the full classification system may be used. However, as with the physiographic description of the site, listing all possible choices on a collecting form at this level of detail will only be feasible when dealing with a relatively restricted target area. It will then be possible to compile a full but still manageable listing of the vegetation types in the area with reference to published vegetation surveys and maps. Otherwise, a line or so of free text will be needed. In the case of Africa, for example, vegetation could be recorded at a first level according to the main vegetation types of White (1983), and it would be possible to produce lists of his mapping units for a given target area, among which a choice could be made for a more detailed vegetation description.

Vegetation type may be read off maps, but, as in the case of soil and geology, the name and scale of the map used will need to be specified. The vegetation classification system that is used should always be noted by reference to a publication. It is useful to record whether the vegetation at the collecting site is markedly different from that in the surrounding landscape, as observed during travelling or recorded on maps.

Vegetation types may have local vernacular names, some of which have actually entered into general botanical usage, such as 'fynbos', 'kefkalla', 'chaparral' or 'caatinga'. Such words or phrases may be used to save time in the field, but will need to be defined or substituted for more widely recognized terms when the forms are circulated to

---

Box 19.3

Unesco vegetation classification formation classes (Unesco, 1973)

I. Closed forest	stand of trees >5 m tall, with interlocking crowns
II. Woodland	stand of trees >5 m tall, with crowns not interlocking but tree cover >40%
III. Scrub (fourrés)	stand of caespitose woody perennials $\frac{1}{2}$–5 m tall
• shrubland	scrub with crowns not interlocking
• thicket	scrub with crowns interlocking
IV. Dwarf scrub	stand of caespitose woody perennials $\sim\frac{1}{2}$ m tall
• dwarf shrubland	dwarf scrub with woody perennials isolated or in clumps
• dwarf shrub thicket	dwarf scrub with woody perennials with interlocking crowns
V. Herbaceous communities	
• tall graminoid	stand of grasses or graminoids >2 m tall
• medium tall grassland	stand of grasses $\frac{1}{2}$–2 m tall
• short grassland	stand of grasses <$\frac{1}{2}$ m tall
• forb vegetation	stand of broad-leaved herbaceous species

VI. Deserts and other sparsely vegetated areas
VII. Aquatic plant formations

Each vegetation type in I–IV may be further qualified as
  • evergreen
  • semideciduous
  • deciduous
  • xeromorphic
Each vegetation type in III–V may be further qualified as having
  • trees (>5 m tall) contributing to 10–40% of cover
  • trees contributing to <10% of cover
  • shrubs
  • tuft plants
  • no woody plants

---

collaborating organizations. Indigenous vegetation classifications are a useful complement to scientific systems, often highlighting features that, though highly significant to local people who know and use the vegetation daily, have not been considered in more formal treatments.

Land use and farming system

If the climax vegetation actually or potentially present at a site is an integrated, synoptic expression of the natural environment, land use in an area is an integrated expression not just of the climatic and edaphic situation, but also of cultural and socioeconomic conditions (Oram, 1987). FAO (1990) lists the following main categories of land use:

☐ settlement, industry
☐ crop agriculture
☐ animal husbandry

☐ forestry
☐ mixed farming
☐ extraction and collection
☐ nature protection
☐ not used and not managed

A general impression of the extent of each category in the area surrounding the collecting site, and land use at the collecting site itself, should be recorded.

To further characterize protected areas, the system recently adopted by the IUCN Commission on National Parks and Protected Areas, a modification of that of IUCN (1978), is useful:

☐ strict nature reserve/wilderness area
☐ national park
☐ natural monument/natural landmark
☐ habitat and species management area
☐ protected landscape/seascape
☐ managed resource protected area

In crop collecting, the 'crop agriculture' and 'mixed farming' categories will also have to be further characterized, to the level of farming system, in the same way that vegetation type will further characterize some of the other categories in the list. A farming system may be defined as 'a reasonably stable arrangement of farming enterprises that the farm household manages according to well-defined practices in response to the physical, biological and socioeconomic environments and in accordance with the household's goals, preferences and resources' (Shaner *et al.*, 1982).

The FAO scheme provides for further description of farming systems (by noting whether crops are annual or perennial, rainfed or irrigated etc.), but more detailed classification frameworks are possible. That of Boserup (1965), for example, essentially ranks farming systems on a scale of increasing intensity of land use, from shifting cultivation to irrigated multicropping by way of fallow systems of decreasing duration. A similar scheme is followed by Okigbo and Greenland (1976). Altieri (1987) recognizes seven main types of agricultural systems in tropical environments:

☐ shifting cultivation
☐ semipermanent rain-fed farming
☐ permanent rain-fed farming
☐ arable irrigation
☐ perennial crop farming
☐ grazing systems
☐ systems with regulated ley farming

Going into a little more detail, Beets (1990) defines and describes seven major crop-based smallholder farming systems in the tropics:

☐ shifting cultivation
☐ lowland rice-based farming
☐ cereal-based farming
☐ smallholder mixed farming
☐ irrigated smallholder farming
☐ smallholder farming with plantation (perennial) crops
☐ agroforestry

Each of these can be detailed further and subdivided. One way is in terms of the dominant crop. Middle Eastern cereal farming may be based predominantly on barley, bread wheat or durum wheat, for example. A further way is in terms of crop growing environment. For example, IRRI (1984) provides a comprehensive review of classification systems for rice growing environments and suggests a generally applicable terminology based on a combination of factors, including water regime, drainage, temperature, soils and topography. Carter (1987) is a similar study on cassava. Nair (1985) describes the International Centre for Research in Agroforestry (ICRAF)'s classification of agroforestry (i.e. home-garden) systems (the descriptions are in a database; see Oduol et al., 1988), though his definition of the term is not exactly the same as Beets' (1990). More information on farming systems worldwide is available in Grigg (1974) and Ruthenberg (1980).

Of course, the situation is more complex than such classifications may imply. Farming system classifications often overlook Chambers' (1990) microenvironments, the sort of restricted, specialized, hard-to-find sites where much subsistence agriculture takes place. A single household (let alone a single village) may practise more than one of the farming systems in a classification, for example a permanent, home-garden-type 'infield' and a shifting 'outfield', perhaps with different people responsible for each. The character of the individual subsystems and the linkages among them need to be documented. In Francophone studies, the village level and the farm (or household) level are often distinguished as 'système agraire' and 'système de production', respectively (Beets, 1990). When the unit of analysis is the field or plot, one may speak of the cropping system used on that piece of land, or the 'système de culture'. This will be documented under 'Cultural methods'.

Many countries use their own standard national nomenclature and classification system for land use and farming systems, for example in producing land use maps. These systems may be very specific to the country. For example, the Land-use Map of China has a 'cultivated land' category divided into 'paddy', 'irrigated field' and 'non-irrigated field' (all are further subdivided on the basis of whether they are terraced or not) and a 'garden' category divided into 'orchard', 'tea garden', 'mulberry field', 'tropical crops' and 'diked pond' (Editorial Committee of Land-use Map of China, 1990). The Atlas de la Nouvelle Calédonie et Dépendances has such land use categories as 'coconut grove' and 'small-scale food crop garden', but also 'alluvial meadow with traces of old yam ridges' and 'former irrigated taro garden' (ORSTOM, 1981). Such maps, agricultural censuses, household surveys and ethnographic studies may be used to develop manageable lists of the land use types and farming systems occurring in a particular target area for inclusion on the collecting form. The source(s) used should always be specified.

Otherwise (or in addition, if secondary sources are deemed not sufficiently precise and time permits) the farming system at the household level will need to be described on the basis of observation and consultations with the members of the household. This can be done in many different ways. Conway (1985) describes a framework for farming systems description based on the analysis of spatial and temporal patterns, flows (e.g. of energy, materials, information, etc.) and decision-making. The annotated checklist of survey questions provided in an appendix by Richards (1985) 'to assist agricultural extension workers assess local skills and R&D priorities, and open up the possibility of participatory approaches to agricultural development' is a good model for farming system description. Another is the Worksheets for Land Use System Description used by ICRAF (Raintree, 1987). See Fernandes and Nair (1986) for the checklist used to describe and characterize home-gardens by ICRAF. Shaner *et al.* (1982) present a system for documenting farming systems based on the description of: (i) household structure and decision-making; (ii) household resources (land, labour, capital and management); and (iii) farming enterprises. An exhaustive farming system description would need to include information on at least the following:

- farm access, the size and fragmentation of the holding and the size and composition of the household which works it;
- the character of land tenure for the home compound, off-compound gardens, crop land, grazing land, woodlands, etc. (freehold, tenancy, communal control, state ownership); see Raintree (1987) for a full list of land tenure types;
- the main cultivated and semicultivated species, their relative importance and their spatial (horizontal and vertical) and temporal relationship to each other (intercropping, relay cropping, rotation, etc.);
- the main wild species used;
- the associated livestock species, their management (free-ranging, herding, paddocking, stall-feeding), source of feed and contribution to the system (milk, meat, manure, draught);
- the disposal of plant and animal produce and residues (home consumption, sale, barter, social uses);
- the seasonal calendar of temperature and rainfall, cropping activities for each species, collecting of different materials from the wild, agricultural labour demand, crop and livestock pest and diseases, diet, livestock feed availability, etc. (Box 19.4);
- the sharing out in space and time of farming responsibilities within

---

Box 19.4

Participatory preparation of seasonal calendars (after Theis and Grady, 1991)

- Draw an 18-month calendar either on a large piece of squared paper or on the ground, labelled with local names for months and seasons (how these are defined, i.e. by the moon, stars, etc., should be specified).
- Ask community members to use seeds, stones, twigs of different sizes or other small counters to indicate the relative magnitudes of different variables (e.g. rainfall, labour demand) at different times of the year.
- Ask community members to indicate planting, harvesting, etc. dates for different crops, e.g. using seeds of the different species. The ranges of dates should be shown, and reasons for differences between years investigated.
- Combine all seasonal patterns into one diagram to bring out correlations and discuss these with the group.

---

the household, i.e. the particular roles of men, women, children, paid hired labour;
- the method of land management (microcatchments, mounding, ridging, terracing) and soil fertility management (burning, manuring, mulching, fallowing); see Raintree (1987);
- the method of water management (rain-fed cropping, residual soil moisture cropping, flood-recession cropping, tidal irrigation, groundwater pumping, etc.); see Underhill (1984) and Adams and Carter (1987) for useful lists and classifications of water management practices;
- the character and extent of modern inputs (pesticides, fertilizers, mechanization, improved varieties, extension advice);
- the most significant constraints and bottlenecks in the system and changes in this and any of the above over the years.

To what extent the collector will be able to – and need to – collect information on all these aspects will vary. Ideally, in a preliminary or reconnaissance survey of an area, a number of households, reflecting the range of socioeconomic variation, should be documented fairly fully, whether or not germplasm is actually collected at each household. One could then specify whether many, some or only a few households within the village fall into a given class. Such data may be useful in formulating a sampling strategy and in identifying key informants for future collecting. On other occasions, the main crops and rotations and the most common land management practice and water management method in the village may be recorded (in a sentence or two, or using a broad classification category), in addition to more details on the particular household from which material is being collected.

For the non-cropping land use categories, any regular, artificial treatment of the environment or plants at the collecting site should be described. This could be grazing, burning, thinning, mowing or some other management practice. The frequency, intensity, extent and history of management are all important factors to note. In pastoral areas, a description of the system of land tenure obtaining at the collecting site will largely define the management the vegetation has received. Useful (though confined to Africa) reviews of traditional vegetation management systems – and frameworks for their description – are provided by Niamir (1990) and Shepherd (1992). Bruce (1990) describes rapid rural appraisal (RRA) methodologies for the assessment of tenure systems in a forestry context. Different species at a site in natural or seminatural vegetation may be managed in specific ways. This could be documented in this section, but, if some of these species are being collected, data pertaining to them should be included in a separate 'Population management' descriptor in the 'Population information' section.

Local people often have their own indigenous nomenclature for land use and farming systems (and perhaps a taxonomy). For example, systems of shifting cultivation are referred to as 'jhum' in the northeastern hill region of India and as 'podu' in Andhra Pradesh. Similar systems are called 'rosa' by some native communities in Mexico. However, as in the case of local names for landscape, soil and vegetation types, such appellations will need to be translated into generally understood phrases, carefully defined, or at least a reference given to an explanatory source, before wider distribution of the data.

Three-dimensional models, farm and village sketches and diagrammatic profiles through the village and surrounding areas drawn with farmers' participation can be used to relate particular ways of using the land – and, indeed, the distribution of particular crops or landraces – to specific features of the environment. Constraints and opportunities can be quite effectively pin-pointed in space in this way. Any changes that may have occurred in the pattern of land use over time can also be documented by inviting local people to draw a series of profiles through the village and surrounding areas showing the situation as it was at different times in the past (historical transects). Again, this is information for the notebook rather than the collecting form.

If there has been agroecological characterization of the target area, note could also be taken here of where the collecting site fits in the system, though this will not always be entirely possible in the field. Johnson (1974) gives an example of an indigenous system of ecological characterization. Describing a collecting site according to such multi-dimensional classifications (scientific or indigenous) can be an efficient way of defining the overall adaptation of the germplasm.

Dominant species/crop/landrace

Recording the dominant wild species at a site is only necessary if a non-floristic vegetation description is used. In crop collecting, the most important crop (and variety, whether local or modern) or rotation used by the household can be specified in this descriptor. Half a line of free text is probably adequate, especially if a code is used to record species in the field, such as the first two or three letters of genus and species. If such a code is used, it is essential to keep a key, so that collecting forms can be decoded prior to distribution.

Associated species/crops/landraces

A knowledge of associated species is often important in wild species collecting. For example, forage grasses and legumes which grow well together may be sought. Species may also be listed here that are not dominant or even common but nevertheless significant in characterizing the site. Examples would be forest emergents, endemics and ecological indicators. Target species which it was not possible to collect, for example because of incorrect timing or overgrazing, should be mentioned. If a representative inventory of the flora at a site is needed, an area at least equal to the minimal area of the vegetation should be surveyed.

Comprehensive lists of crops and varieties (both landraces and modern material) for the village (and, indeed, the surrounding area) may be compiled from visits to markets as well as in consultations with several farmers and other local users, who will also be able to say if any crops and landraces are no longer being grown or are being grown less than in the past, to what extent, and why. Not being tied to a particular germplasm sample, this information is perhaps better recorded in the collecting notebook and report. A list for the household under consideration can be recorded here. Information on the actual changes in the crops and landraces being grown by the household, and on probable future changes, can find a place under this descriptor and/or under a separate 'Genetic erosion' descriptor.

Additional notes on biotic factors

Notes on the degree of shading at the site can be made in a separate descriptor or in a comments field. Information on the extent and character of differences in vegetation or land use between the site and the surrounding area can also be recorded here.

Other

Site disturbance

A record of the degree (intensity and frequency) to which a site has been disturbed, and the type of disturbance, can give very valuable information on the genetic structure of the population. If a wild population is growing in a disturbed environment it is likely that the surviving individuals will represent only a proportion of the original genetic diversity of the population. This does not imply that the population should not be sampled, but could serve to suggest that there may be further diversity available elsewhere in the area. Information on disturbance could also indicate possible threats to the site or the population, and impending genetic erosion, although many species are adapted to regular disturbance and may actually require it for regeneration.

Fires, floods, landslides, high winds and drought may be termed 'natural' disturbance factors in that they can occur without human help. So are grazing and trampling by wild herbivores. Purely 'artificial' disturbance ranges from management to complete destruction of the habitat, for example for mineral exploitation or construction or by pollution. Artificial disturbance is perhaps best noted under 'Land use'. Some assessment of the extent of this threat may be recorded in a separate 'Genetic erosion' descriptor, added to this section (Chapter 4). Information on both natural and artificial disturbance can come from direct observation and secondary sources, but local people will usually be the main source.

Population information

Attributes of the population from which the sample is taken need to be described just as much as the ecological conditions at the site which the population inhabits. Attributes such as phenology, pest resistance and morphology will have some kind of genetic basis, and can be seen as preliminary characterization. As such, they will provide an important guide to users.

Information on some of the subjects in this section can only really be obtained from consultations with local people, especially in the case of crops. Some general guidelines on the methodology for documenting indigenous knowledge are given in Chapter 18. Checklists of descriptors such as collecting forms may be ideal for recording sample archival data and site ecological data in the field, but they can be somewhat restricting in the recording of indigenous knowledge. Nevertheless, some structure will still need to be imposed at some stage on the mass of data coming out of such procedures as semistructured interviews, life-history elicitation, ranking tests, cognitive mapping (of landraces and of the environment) and audio/visual recordings if useful information is to emerge. What are enumerated here are perhaps better thought of as general topics around which consultations with farmers and other local people and observation can take place, rather than as a set of rigid questions to be asked and answered or blanks to be filled in.

Inevitably, it will not be possible to find a place on the collecting form for all the information gathered in the course of an ethnobotanical investigation. Indeed, some information will not be directly referable to a particular germplasm sample or collecting site. Such general information should be recorded in the field notebook and included in the mission report.

Phenology

A knowledge of the proportions of the population at different stages in the phenological cycle when collecting took place will be important to breeders looking for genes for early maturation date in a crop, for instance, or for forage material suitable for seed production which produces seeds in a short period rather than throughout the season. It will also help in timing future collecting.

Estimates of this information are normally presented as the percentage of the population which is in the following states:

- vegetative
- flowering
- fruiting
- finished fruiting
- with sterile seed

Note that 'vegetative' refers to plants which did not flower or seed at all, while 'with sterile seed' refers to plants which flowered but produced no seed, as can be seen in empty heads or hollow seeds. The latter is a frequent and very deceptive occurrence in certain grasses and care must be taken not to collect such material. Godron and Poissonet (1970) provide much more detailed subdivisions of the phenological cycle for annuals, biennials and perennials. There are also more specific systems, for example for cereals.

Pests and diseases

Chapter 17 deals with collecting data on the plant damage caused by pests. It also discusses collecting specimens of pests and of the damage they have caused. Local people are often extremely important sources of information on the susceptibility of landraces to different pests and on what pests pose problems in an area, and to what extent. They can also describe the plant protection measures that they adopt.

Uses

The importance of information on the particular uses made by local people of landraces or wild species is discussed in Chapter 12. In addition to data from consultations with local people, this descriptor will also include information from other sources, for example field observations of grazing of forage species or of visits by bees for honey species. It is usually possible to decide from observation whether a species is being grazed or browsed, and a note should be made of what the herbivore is likely to be. It is sometimes possible to tell if a species is particularly palatable, for example if it is only found surviving within thorny bushes. One or two lines of free text should be adequate for this field. Alternatively, a list of non-mutually exclusive options may be provided on the collecting form, for example as follows, based on the main categories developed by the Survey of Economic Plants of the Arid and Semi-arid Lands (SEPASAL) (e.g. see Aronson, 1989):

- ☐ food and drink
- ☐ domestic products

- [ ] timber (including fuelwood)
- [ ] forage
- [ ] land use (e.g. shade, soil improvement and stabilization, ornamentals)
- [ ] fibres
- [ ] toxins
- [ ] medicinal
- [ ] chemicals (e.g. gums, resins, dyes)
- [ ] ritual and religious uses
- [ ] other

If such a list of broad groupings is used, it is important to allow space for details, in particular specifying what part of the plant is used for each purpose (including the use of crop residues), the method of preparation, etc. Alternatively, the full classification may be used, in which each of the above categories is subdivided at two or three levels, though the full SEPASAL list runs to several pages. Medicinal plants are often particularly difficult to categorize, as traditional views of health and sickness may be radically different from those of modern medicine. TDWG is developing standards for recording the economic uses of plants.

A plant or plant part may be used in different ways by different sections of a community, e.g. men and women. Gender-disaggregated benefits analysis is a useful tool for documenting this (Thomas-Slayter *et al.*, 1993).

The documentation of methods of preparation can be supplemented with photographs and audio/video recordings. Collier and Collier (1986) provide a scheme for the documentation of technology that may be useful in this context (Box 19.5); 35–50 mm lenses will again be most suitable.

Morphological description

A brief morphological description of the material collected can be very valuable to future users of conserved material. Much time and effort can be saved by being able to make an initial selection of what material to evaluate in field trials on the basis of a search through brief descriptions of the morphology of a number of holdings, if the characters are highly heritable.

There are different, complementary approaches to recording morphological data in the field. One is to provide several lines for free text on the collecting form, where unusual features and those of particular agronomic significance can be noted. Another approach is to use characterization descriptors specific to a taxon or gene pool, such as are provided in IPGRI descriptor lists (Chapter 8). Such lists are often very long, and some selection will be required. Characters identified in previous characterization work as best differentiating among landraces

Box 19.5
Documenting technology by photography

- Environmental location of the technology.
- Raw materials.
- Tools of the trade.
- How tools are used.
- How the craft proceeds.
- End result.
- The function of the technology.
- Social context of the technology.

or that have been used in infraspecific classifications (Chapter 7) will be favoured. The characters most frequently cited by farmers themselves in their description of landraces or in discriminating among landraces might also be a good starting point. It should be remembered, however, that farmers do not necessarily use only morphological features in differentiating among landraces: gastronomic, life habit, familiarity and functional criteria may be just as important (Nazarea-Sandoval, 1992).

When collecting wild plants, each species should be described as to life-form, size, life span and habit, either in a brief phrase, or by going through a list such as the one below, an abbreviated form of that quoted in Chapter 27 in the context of the notes required to document herbarium specimens.

- plant type: tree, shrub, herb or vine (or Raunkiaer categories)
- free-living, epiphytic or parasitic
- plant height
- life span: annual, biennial or perennial
- dry-season deciduous, wet season deciduous or evergreen
- direction of stem growth: climbing, erect, geniculate, decumbent, prostrate, creeping, etc.
- stem structural type: pachycaulous, succulent, bulb, corm, stolon, rhizome, etc.
- perennating organs
- thorns and spines

The models of Hallé et al. (1978) can be used to describe the growth form of trees. Forestry workers often record DBH and bole and/or total height of the trees they collect. Height can be measured directly or estimated visually, but the former method is laborious and the latter can be inaccurate. Using a clinometer and trigonometric conversion is probably the most convenient compromise for tall trees. TDWG is in the process of developing standard life-form descriptors.

Taking photographs of the material can be a useful additional method of data collecting in the field. Close-ups of flowers and fruits taken with a macrolens can complement herbarium specimens by recording details that will be lost or distorted on drying. The general habit of plants can also be shown, which will be particularly important for trees. For crops, Marchenay (1987) recommends three views:

- a general view of the field;
- a general view of the entire plant;
- a close-up of the part consumed.

For fruits and tubers, the following views will be needed:

- the entire organ from the side;
- the entire organ from the top;
- an equatorial section;
- a vertical section.

These can of course be arranged in a single frame, against a neutral background (a piece of grey card or cloth), taking care not to include shadows. It is important to include a scale at all times (a 10 cm ruler or a pencil is suitable). Colour transparency film is recommended, of speeds ASA 64, 100, 200 or 400. A flash and a tripod may be necessary in some situations, for example when collecting in forests.

An important reason for recording easily visible morphological information on collecting forms is that it can help keep track of phenotypic duplicates in the field. However, there are other ways of monitoring the material that is being collected that may be easier than comparing piles of paper forms. One possibility is to take Polaroid photographs, for example of the wheat ears, *Phaseolus* beans, apple fruits or sweet potato tubers (in section and whole) found in a field or village. This will be quite expensive, however, and a cheaper alternative is to make drawings on pieces of card. A farmer's drawing or description of a landrace, emphasizing the salient features of the material from the everyday user's point of view, may make it easier for the collector to remember it. If such cognitive mapping of landraces has been carried out, the results should be recorded here (Chapter 18). Another approach is to retain representative subsamples of each seed sample collected. These can be stored glued to pieces of card, in small transparent plastic bags or the pockets of slide holders (Debouck, 1988), labelled with the collecting number and the site number. In this way, they can be quickly referred to and compared with newly encountered material.

It is useful to keep a running count in the field notebook of the different kinds of material collected. Alphabetically arranged running checklists of local and scientific names (and some essential distinguishing features) have already been mentioned. In addition, two- or three-way tables can be constructed using the most important field characterization descriptors and collecting numbers added to the appropriate cell in the table as each sample is collected. For example, wheat landraces could be described in the field in terms of spike density, awnedness and glume colour, as in the example in Table 19.1. Local names for each sample could also be recorded in the table. Instead of the characters themselves, the categories used in an infraspecific classification of the crop could be used (Chapter 7).

Table 19.1. Example of description of a wheat landrace. Numbers refer to collecting numbers of different samples.

	Awnless	Awnletted	Awned
Spike lax	1024 – glume white	1025 – black 1032 – white	
Intermediate			1011 – white
Dense		1037 – brown 1056 – brown	

Morphological variation

Obvious variation in highly heritable, especially qualitative, characters within a population should be recorded. On the strength of this information, it may sometimes be decided to selectively collect separate samples of 'unusual' or otherwise interesting individuals. Variation may be recorded for individual characterization descriptors or simply an overall indication or impression given, in which case a line of free text should be adequate. Photographs are sometimes taken showing the range of variation in a sample. For example, it is common to choose an example of each of the various different ear types in a cereal field and photograph them together against a neutral background. Again, a scale should always be included in such photographs.

Proximity of close relatives

When collecting either crops or wild species, the presence of wild, weedy or cultivated relatives in the vicinity could point to the possibility of gene exchange and introgression. This is often encouraged by traditional farming practices, which should be documented. Note should be made of whether there are related forms nearby, and if so what they are and whether it is considered likely that gene exchange is taking place. The presence of intermediate forms would be evidence of this. Specialized collecting forms may ask for the distance from the collecting site to the nearest field of the cultigen or the nearest population of a wild relative. One line of free text is usually adequate for this data field.

Wild species

Specific habitat

It will sometimes be necessary to be very specific about the microsite occupied by a wild plant or crop. Using the seasonal watercourse example quoted above under 'Description of site', for instance, one species (or ecotype, in the case of stratified sampling) might be found 'along the sides of the wadi, among rocks' and another 'in the middle of the wadi bed, on sandy alluvium'. Site context, slope, aspect and vegetation will probably be recorded as identical for the two samples, but strictly speaking the collector is dealing here with two separate collecting microsites, as ecological conditions will in fact be somewhat different. In this case there might be a difference in soil texture, perhaps, which it will be possible to document on the collecting form in the appropriate descriptor, but it is by no means always the case that it will be possible to find a descriptor among the ones included on the form for which a difference can be recorded. Another relevant example might be collecting both in clearings or paths and in the surrounding woodland. Again, these should be treated as separate collecting sites, or at least microsites, though the nature of the ecological difference may not be specifically revealed by any of the site descriptors on the form, unless there is one for trampling or shading. If only one species is being collected, then the full habitat description can be accommodated under 'Description of site'. When more than one species or potential ecotype are being collected, however, it may be easiest to simply describe in this separate descriptor the microenvironmental differences involved. Free text is the most satisfactory way of recording this information, and two lines will usually be adequate.

Abundance

The abundance of a species or phenotype within the plant community may give an indication of such factors as its competitive ability or degree of adaptation in a particular environment. This information can be of interest but is not usually of prime importance, unless the population is small and threatened.

There are a number of ways of measuring the abundance of a species at a site, e.g. biomass, density, frequency, basal cover or crown cover. The simplest and most convenient method for germplasm collecting purposes is to give a rough estimate of crown cover. This entails imagining that the site is being viewed from above. The amount of ground occupied by the vertical projection of the aerial parts of a species is its crown cover. This is normally estimated as a percentage, ranges being grouped together into units. Several cover scales have been developed but the main two are the Domin and the Braun–Blanquet scales (Shimwell, 1971). A slight variant of the latter is given below.

☐ a single individual
☐ scarce, ⩽5%
☐ common, 6–25%
☐ moderately common, 26–50%
☐ abundant, 51–75%
☐ dominant, >75%

General frequency in the region surrounding the collecting site (y) may be combined with abundance at the site itself (x) in some composite index $x|y$ (Isbell and Burt, 1980).

Spatial pattern of population

This descriptor refers to how crown cover of the target species is distributed within the habitat, whether at random, regularly or contagiously (as clumps). Randomness is normally taken as the null hypothesis and departure from randomness (i.e. the existence of pattern) tested. The causes of departure from randomness may be environmental or intrinsic to the plant. Both are interesting to the plant collector. If the cause is environmental, this will help define the adaptation of the material. If intrinsic, it may give information on the distance of seed dispersal in the target species, for example, or the extent of vegetative propagation, which may be relevant to the future user and will in any case affect sampling strategy at the site.

Pattern may be detected by counting the number of individuals in a set of quadrants and comparing the results with the expectation from a Poisson distribution. This will rarely be possible or necessary in the context of germplasm collecting, when a simple visual assessment will usually be sufficient. The extent or scale of any clumping may be described according to a sociability scale such as the following, which also includes an element of abundance (Goldsmith et al., 1986):

☐ growing once in a place, singly
☐ grouped
☐ in troops or small patches
☐ in small colonies, extensive patches or forming carpets
☐ in great crowds or pure populations

Separation from other populations

The spatial relationship between the population being sampled and other populations of the same species is important because, in the absence of precise information on the distance that pollen and seeds can move, it can assist in inferring the degree of genetic isolation of a particular population. If populations are distinctly separated, there may sometimes be obvious physical reasons for this. An example would be a species that only occurred around the edges of widely separated freshwater pools in an otherwise dry environment.

This field is adequately covered by a straightforward question such as 'Is the population well separated from others of the same species?' If the answer is yes, details of any obvious barriers may be added as free text. The distance to the nearest conspecific population may also be given.

Population management

Individual wild and semicultivated species may be managed by local people quite specifically, independently of the general vegetation at the collecting site. Particular species within the vegetation, for example medicinal plants and fruit trees, may be managed and protected in specific ways. In an area to be cleared for shifting cultivation, for example, selected trees may be spared, some self-propagated trees actively protected and other species actually planted. In some societies, people may have private tenure over individual trees. A particular grass may be selectively harvested as fodder. If it is such species that are being collected, the specific methods of management need to be documented here. A useful framework for describing tree management is provided by Mathias-Mundy *et al.* (1992). Such information will mostly come from consultations with local people.

Crops

Cultural methods

The specific way in which a landrace, crop or semicultivated species is grown and managed can be an important determinant of its success: it will probably have become very closely adapted to particular cultural methods (cropping system, or 'système de culture') within the farming system. It is thus important that detailed information on the cultural methods associated with the population sampled accompanies the sample. The information will come both from consultations with farmers and actual observation (which could be photographed or video-taped). In general, free text will be needed to describe cultural practices, though the field could be broken up into separate topics. Examples include Byerlee *et al.*'s (1980) checklist of crop management practices and Mathias-Mundy *et al.*'s (1992) framework. The techniques, materials and tools used in carrying out each of the activities listed below, and the people involved (men, women, adults, children, etc.), should be described and the terms for them in the local language recorded. Timings and frequencies can be recorded here or in a separate 'Growing season' descriptor.

- site selection and seedbed preparation;
- planting and transplanting (type of planting material, density and spacing, associated crops and landraces, rotations);

- thinning (to achieve desired plant density, but also including roguing, removal of off-types, etc.), pruning, etc.;
- nutrient management;
- plant protection measures, including weed management;
- water management;
- harvesting;
- farmers' selection methods (including special seed production methods);
- threshing, cleaning, drying and other postharvest management;
- storage;
- disposal of major product and of crop residue.

Farmer's selection methods will be a particularly important factor to document. Selection is usually done after harvest, but some farmers mark those plants that are to be used as sources of next year's planting material while they are still in the field. Selection criteria may be for uniformity in particular character(s), or aimed at maintaining a degree of variation. Planting material may be produced in special plots, isolated from the main fields and treated in different ways (Linnemann and Siemonsma, 1989).

Growing season

Information on the timing of agricultural activities relating to the sample, in particular when the field was planted, transplanted and harvested, is important to breeders and other end-users. Characters such as early maturity or short growing season are often keenly sought by breeding programmes. These data also provide information on temperature tolerances, frost sensitivity and day-length sensitivity.

Timings will eventually have to be recorded in terms of months of the year on the collecting form, but will probably be collected in the first instance according to some local system, perhaps involving stars or the moon. Seasons will have specific local names, which may be incorporated into the local names of landraces. A seasonal calendar in which the activities listed under 'Cultural methods' are related to each other and to environmental factors in time is a useful way of presenting such data. Such calendars can be made gender- and age-disaggregated to show how activities are divided up within the community (Thomas-Slayter *et al.*, 1993). Box 19.4 describes a participatory method for obtaining information on the timing of activities through the year. The local system should be used for recording seasonal calendars, and this may mean that the yearly cycle starts not in January but with the rains, say, or the appearance of a particular star. It is often easier to detect seasonal patterns visually in 18-month than conventional 12-month calendars. Even longer calendars have been used.

User evaluation

The reason why a particular landrace is being grown is not always apparent to outsiders. Conventional agronomic factors such as yield can be of secondary importance to the subsistence farmer. A rice variety may be liked despite its relatively low yield because the endosperm does not break when pounded, for example, or for its taste. It is clearly important to know what different categories of local users (e.g. men and women) like and do not like about each landrace collected. Farmers' evaluation tends to be relativistic: this is why triads and sorting/ranking tests are used in this context. In contrast to conventional evaluation, where an absolute value for, say, yield, could be arrived at, in farmer evaluation the properties of a particular landrace will be expressed relative to those of other landraces, and will need to be recorded in such terms. Box 19.6 describes two participatory methods of germplasm evaluation by local users (they are also applicable to evaluation of land, soil, etc.).

History of sample and of landrace to which it belongs

People move crop germplasm around. It is not uncommon to find that in fact the population sampled can be traced back to planting material which was recently introduced from some other, perhaps very distant, place. Perhaps the original introduction was of only a few seeds or a single cutting, which would have consequences for the genetic base of the material. It is important to document this because collectors often work in a particular area in the expectation of finding material adapted to the environmental conditions prevalent there. If material was brought in from outside the area, its actual origin should be noted, as it may not have the required adaptation (though of course it might nevertheless be worth collecting). If produced locally, the method of selection and seed production needs to be documented. Origin is often alluded to in the vernacular name of a landrace. However, farmers will usually know the history of their planting material in detail, often going back many years.

Whether cultivation by the household of the crop or landrace represented by the sample is likely to decline, and why, may also be recorded here (or under 'Genetic erosion'). More general information on trends in the village or surrounding areas may also emerge in the course of consultations with local users, particularly life-history elicitation, and may be recorded (probably in the collecting notebook) as free text or in the form of notes on trends appended to time lines of important events in village history (Chapter 18).

Box 19.6
Participatory ranking (after Theis and Grady, 1991, and Kabutha et al., n.d.)

Direct matrix ranking

- List the landraces or species under consideration (three to eight items), and display examples of each to the interviewee or group.
- Elicit criteria by asking 'What is good about this item? What else?' and 'What is bad about this item? What else?' until there are no more replies.
- List all the criteria, turning negative criteria to positive ones, e.g. 'attacked by pests' into 'resists pests'.
- Draw up a matrix of criteria by items.
- For each criterion, ask which item is best, next best, worst and next worst. Of the remaining, ask which is better. Assign scores, and add up the score for each item. If a group exercise, people could be asked to vote for their preference, and the number of votes added up.
- Ask which criterion is most important.
- Ask which item is best overall: 'If you could have only one, which would you choose?'

Pairwise matrix ranking
- Draw up a matrix with the items in the same order along the side and the bottom.
- Each square represents a paired comparison. Ask the informant(s) which of the items is the better, and why.
- When the matrix is complete, add up the number of items where the item was identified as more important, and arrange them in the appropriate order.

References

Adams, W.M. and R.C. Carter (1987) Small-scale irrigation in sub-Saharan Africa. *Progress in Physical Geography* 11:1–27.

Altieri, M.A. (1987) *Agroecology: The Scientific Basis of Alternative Agriculture.* Westview Press, Boulder.

Aronson, J.A. (1989) *HALOPH. A Data Base of Salt Tolerant Plants of the World.* Office of Arid Lands Studies, University of Arizona, Tucson.

Ball, D.F. (1986) Site and soil. In: Moore, P.D. and S.B. Chapman (eds) *Methods in Plant Ecology.* pp. 215–284. Blackwell Scientific Publications, Oxford.

Beets, W.C. (1990) *Raising and Sustaining Productivity of Smallholder Farming Systems in the Tropics.* AgBé Publishing, Alkmaar.

Bernard, H.R. (1988) *Research Methods in Cultural Anthropology.* Sage Publications, Newbury Park.

Bingham, J.M. and E.J. Ciolkosz (eds) (1993) *Soil Color.* Crop Science Society of America, Madison.

Boserup, E. (1965) *The Conditions of Agricultural Growth.* Allen and Unwin, London.

Bruce, J. (1990) *Rapid Appraisal of Land and Tree Tenure.* Community Forestry Note 5. FAO, Rome.

Bunting, A.G. and H. Kuckuck (1970). Ecological and economic studies related to plant exploration. In: Frankel, O.H. and E. Bennet (eds) *Genetic Resources in Plants –*

Their Exploration and Conservation. IBP Handbook 11. pp. 181–188. Blackwell Scientific Publications, Oxford.

Byerlee, D., M. Collinson *et al.* (1980) *Planning Technologies Appropriate to Farmers. Concepts and Procedures.* CIMMYT, Mexico.

Carter, S.E. (1987) Collecting and organizing data on the agro-socio-economic environment of the cassava crop: case study of a method. In: Bunting, A.H. (ed.) *Agricultural Environments.* pp. 11–29. CAB International, Wallingford.

Chambers, R. (1990) *Microenvironments Unobserved.* Gatekeeper Series No. 22. IIED, London.

Collier, J. and M. Collier (1986) *Visual Anthropology.* University of New Mexico Press, Albuquerque.

Conway, G.R. (1985) Agroecosystem analysis. *Agricultural Administration* 20:31–55.

Conway, G.R. (1989) Diagrams for farmers In: Chambers, R., A. Pacey and L.A. Thrupp (eds) *Farmer First. Farmer Innovation and Agricultural Research.* pp. 77–86. Intermediate Technology Publications, London.

Cottam, G. and J.T. Curtis (1956) The use of distance measures in phytosociological sampling. *Ecology* 37:451–460.

CPCS (1977) *Classification de sols.* Documents des Laboratoires de Geologie et Pedologie. École Nationale Supérieur Agronomique, Grignon.

Curtin, P.D. (1968) Field techniques for collecting and processing oral data. *Journal of African History* 9:367–385.

Debouck, D.G. (1988) *Phaseolus* germplasm exploration. In: Gepts, P. (ed.) *Genetic Resources of* Phaseolus *Beans.* Kluwer Academic Publishers, Dordrecht.

Editorial Committee of Land-use Map of China (1990) *1 : 1,000,000 Land-use Map of China.* Beijing.

Ellenberg, H. and D. Mueller-Dombois (1967) Tentative physiognomic–ecological classification of plant formations of the Earth. *Berichte des Geobotanischen Institutes der Eidgenoessischen Technischen Hochschule Stiftung Rubel Zuerich* 37:21–46.

EUROCONSULT (1989) *Agricultural Compendium.* Elsevier Scientific Publishing, Amsterdam.

FAO (1990) *Guidelines for Soil Description.* FAO, Rome.

FAO (1991) *Using Video in the Field.* Development Communication Guidelines. FAO, Rome.

FAO–Unesco–ISRIC (1988) *Revised legend of the FAO–Unesco Soil Map of the World.* World Soil Resources Reports No. 60. FAO, Rome.

Fernandes, E.C.M. and P.K.R. Nair (1986) An evaluation of the structure and function of tropical homegardens. *Agricultural Systems* 21:279–310.

Fosberg, F.R. (1961) A classification of vegetation for general purposes. *Tropical Ecology* 2:1–28.

Freese, F. (1962) *Elementary Forest Sampling.* Agriculture Handbook No. 232. USDA Forest Service, Washington DC.

Godron, M. and Y. Poissonet (1970) Standardization and treatment of ecological observations. In: Frankel, O.H. and E. Bennet (eds) *Genetic Resources in Plants – Their Exploration and Conservation.* IBP Handbook 11. pp. 189–204. Blackwell Scientific Publications, Oxford.

Goldsmith, F.B., C.M. Harrison and A.J. Morton (1986) Description and analysis of vegetation. In: Moore, P.D. and S.B. Chapman (eds) *Methods in Plant Ecology.* pp. 437–524. Blackwell Scientific Publications, Oxford.

Grigg, D. (1974) *The Agricultural Systems of the World: An Evolutionary Approach.* Cambridge University Press, Cambridge.

Hallé, F., R.A.A. Oldeman and P.B. Tomlinson (1978) *Tropical Trees and Forests: An Architectural Analysis.* Springer-Verlag, Berlin.

Hawkes, J.G. (1980) *Crop Genetic Resources Field Collection Manual.* IBPGR and EUCARPIA, Rome.

IRRI (1984) *Terminology for Rice Growing Environments.* IRRI, Los Baños.

Isbell, R.F. and R.L. Burt (1980) Record taking at the collection site. In: Clements, R.J. and D.G. Cameron (eds) *Collecting and testing tropical forage plants.* pp. 18–25. CSIRO, Melbourne.

IUCN (1978) *Categories, Objectives and Criteria for Protected Areas.* IUCN, Gland.

IUCN (1994) *Draft IUCN Red List Categories.* Version 2.2. IUCN, Gland.

Johnson, A. (1974) Ethnoecology and planting practices in swidden agricultural system. *American Ethnologist* 1:87–101.

Kabutha, C., B. Thomas-Slayter and R. Fox. (n.d.) *Participatory Rural Appraisal Handbook.* WRI in collaboration with Egerton University, National Environment Secretariat of Kenya and Clark University, Njoro.

Lazier, J.R. (1985) Theory and practice in forage germplasm collection. In: Kategile, J.A. (ed.) *Pasture Improvement Research in Eastern and Southern Africa.* Workshop Proceedings. 17–21 September 1984. Harare, Zimbabwe. IDRC, Ottawa.

León, L.A., W.E. Fenster and P.A. Sánchez (1979) Soil sample collection and procedures. In: Mott, G.O. and A. Jimenez (eds) *Handbook for the Collection, Preservation and Characterization of Tropical Forage Germplasm Resources.* CIAT, Cali.

Linnemann, A.R. and J.S. Siemonsma (1989) Variety choice and seed supply by smallholders. *ILEIA Newsletter* 4/89:22–23.

Mace, G., N. Collar, J. Cook *et al.* (1992) The development of the new criteria for listing species on the IUCN red list. *Species* 19:16–22.

Marchenay, P. (1987) *A la Recherche des Variétés Locales de Plantes Cultivées.* PAGE-PACA, Hyères.

Mathias-Mundy, E., O. Muchena, G. McKierkan and P. Mundy (1992) *Indigenous Technical Knowledge of Private Tree Management: A Bibliographic Report.* Technology and Social Change Program, Iowa State University, Ames.

Nair, P.K.R. (1985) Classification of agroforestry systems. *Agroforestry Systems* 3:97–128.

Nazarea-Sandoval, V.D. (1992) *Of Memories and Varieties: Complementarity Between Cultural and Genetic Diversity.* Working Paper. UPWARD, Manila.

Niamir, M. (1990) *Herders' Decision-making in Natural Resources Management in Arid and Semi-arid Africa.* Community Forestry Note 4. FAO, Rome.

Oduol, P.A., P. Muraya, E.C.M. Fernandes and P.K.R. Nair (1988) The agroforestry systems database at ICRAF. *Agroforestry Systems* 6:253–70.

Okigbo, B. and D.J. Greenland (1976) Intercropping systems in tropical Africa. In: Papendick, R.I., P.A. Sanchez and G.B. Triplett (eds) *Multiple Cropping.* American Society of Agronomy, Madison.

Oram, P. (1987) Combining socio-economic data with biophysical environmental data. In: Bunting, A.H. (ed.) *Agricultural Environments.* pp. 261–270. CAB International, Wallingford.

ORSTOM (1981) *Atlas de la Nouvelle Calédonie et Dépendances.* ORSTOM, Paris.

Prendergast, H.D.V. (1993) Product review: the Magellan GPS NAV 5000 PRO. *FAO/IBPGR Plant Genetic Resources Newsletter* 91/92:60–61.

Rabinowitz, D. (1981) Seven forms of rarity. In: Synge, H. (ed.) *The Biological Aspects of Rare Plant Conservation.* pp. 205–218. John Wiley & Sons, New York.

Raintree, J.B. (1987) *D&D User's Manual. An Introduction to Agroforestry Diagnosis and Design.* ICRAF, Nairobi.

Richards, P. (1985) *Indigenous Agricultural Revolution.* Unwin Hyman, London.

Ruthenberg, H. (1980) *Farming Systems in the Tropics.* Oxford University Press, London.

Shaner, W.W., P.F. Philipp and W.R. Schmell (1982) *Farming Systems Research and Development. Guidelines for Developing Countries.* Westview Press, Boulder.

Shepherd, G. (1992) *Managing Africa's Tropical Dry Forests. A Review of Indigenous Methods.* Overseas Development Institute, London.

Shimwell, D.W. (1971) *The Description and Classification of Vegetation.* Sidgwick and Jackson, London.

Tabor, J.A., D.W. Kilambya and J.M. Kibe (1990) *Reconnaissance Survey of the Ethnopedology in the Embu, Meru, Machakos and Kitui Districts of Kenya's Eastern Province.* University of Missouri and USAID, Nairobi.

Theis, J. and H.M. Grady (1991) *Participatory Rapid Appraisal for Community Development. A Training Manual Based on Experiences in the Middle East and North Africa.* IIED and Save the Children, London.

Thomas-Slayter, B., A.L. Esser and M.D. Shields (1993) *Tools of Gender Analysis.* Clark University, Worcester, MA, USA.

Tucker, A.O., M.J. Maciarello and S.S. Tucker (1991) A survey of colour charts for biological descriptions. *Taxon* 40:201–214.

Underhill, H.W. (1984) *Small-scale Irrigation in Africa in the Context of Rural Development.* FAO, Rome.

UNEP/GEMS (1993) *Vegetation Classification.* Report of the UNEP–HEM/WCMC/GCTE Preparatory Meeting, Charlottsville, Virginia, USA. GEMS Report Series No. 19. UNEP, Nairobi.

Unesco (1973) *International Classification and Mapping of Vegetation.* Ecology and Conservation 6. Unesco, Paris.

USDA (1975) *Soil Taxonomy.* US Agricultural Handbook 436. USDA, Washington DC.

White, F. (1983) *The Vegetation of Africa. A Descriptive Memoir to Accompany the Unesco/AETFAT/UNSO Vegetation Map of Africa.* Unesco, Paris.

Young, A. (1976) *Tropical Soils and Soils Survey.* Cambridge University Press, Cambridge.

Young, A. (1987) Methods developed outside the international agricultural research system. In: Bunting, A.H. (ed.) *Agricultural Environments.* pp. 43–63. CAB International, Wallingford.

Collecting and handling seeds in the field

20

R.D. Smith

Royal Botanic Gardens, Kew, Wakehurst Place, Ardingly, Nr. Haywards Heath, West Sussex RI176TN.

Introduction

If gene banks are to be successful, the seed samples placed into storage need to be of the highest initial quality. It has been known for a long time that initial seed quality determines the absolute longevity of seed collections, but few studies have been made of the effect of different seed collecting methods on quality. The seed collector is thus left to make important judgements about how to handle seeds in the field with little well-researched advice, when also facing a daunting array of decisions from plant identification and sampling strategies to navigation, team morale and menu planning. This chapter updates earlier attempts (Smith, 1984, 1985) to synthesize laboratory studies of the factors that influence the quality of seed samples and to relate such studies to the practical options available under field conditions.

Angiosperm seeds are borne in a wide variety of types of fruits (Heywood, 1978, Plate VI, pp. 18–19), each of which presents its own particular problems for the collector. Seeds borne in dry dehiscent fruits will only be available for collecting before they are dispersed. They can easily be detached to reduce their bulk for transport. The availability of seeds in dry indehiscent or schizocarpic fruits will depend on how and when the fruits themselves are dispersed from the parent plant. Here the separation of fruit parts and seeds for ease of transport during the collecting mission will be more difficult. However, the weight of fruit parts relative to seeds will still be small and in practice the fruit can be considered as part of the seeds. Major cereal and pulse crops are unique among angiosperms in having been selected against natural seed dispersal. The seeds of these species will remain within the fruit and the fruit will remain attached to the mother plant until harvested by farmers or collectors. The most problematic seeds for collectors are those borne in fleshy fruits, whether they are derived from single flowers (as in berries, drupes, hesperidia or pseudocarps) or from an inflorescence (soroses, syconia and cocnocarpia). Here the ratio of the weight of flesh to seed can be in excess of 10:1, which can place the collector in the predicament of either cleaning the seeds in the field or overcoming the logistical problems of transporting large quantities of material.

Before answers to the practical problems of seed handling in the field can be attempted, we need to know more about seed storage behaviour. To date, three types of seeds have been identified, in terms of their storage characteristics:

- desiccation-intolerant (recalcitrant);
- partially desiccation-tolerant (intermediate);
- fully desiccation-tolerant (orthodox).

As the advice is different for each seed storage type, the first problem for collectors is to recognize which type they are dealing with.

Seed storage types

For well-known crop species, the problem of determining seed storage type can be solved by consulting published work. *Seed Abstracts* is a useful entry to the literature. This should quickly produce an answer, sensible though not necessarily correct. Unfortunately, the literature on the storage behaviour of seeds of wild plants such as forages or forest species is much less extensive. Collectors may then need to make some educated guesses. What information can they base such guesses on?

The use of evolutionary relationships to extrapolate from the known behaviour of species is of little help (Fig. 20.1). For example, all storage physiologies can be found in the genus *Araucaria*, with sympatric species exhibiting different behaviour (Tompsett, 1984). Aldridge and Probert (1993) have shown differences in storage physiologies between the closely related grass genera *Porteresia* and *Oryza*. Fruit type is an equally poor indicator, with both desiccation-tolerant and intolerant seed being known from schizocarpic samaras, caryopses, legumes and drupes.

Improved prediction can be achieved using a combination of seed weight, provenance and morphology. Figure 20.2 shows that, as seed weight rises from 0.001 g to 100 g, the probability of the seed being desiccation-intolerant rises from less than 1% to 100%. Similar correlations between storage behaviour and seed weight have been suggested for the families Dipterocarpaceae and Araucariaceae (Tompsett, 1984, 1988). While this correlation should be treated with caution due to the relatively small size of the sample (546 out of well over 250,000 plant species), a lack of caution can perhaps be justified by invoking the spirit of the *Convention on Biological Diversity*, which states that a 'lack of full scientific certainty cannot be used as a reason for postponing measures to avoid or minimise a threat of significant reduction or loss of biological diversity'. Insufficient species with intermediate seed storage behaviour are known for any meaningful general conclusions to be drawn, other than perhaps to note that their seed weights are also intermediate.

The ability of seed weights to predict seed storage behaviour can be enhanced by also considering the habitat of species (Jurado *et al.*, 1991). The implication of this is straightforward: in wetter habitats seed is likely to be larger. Combining the two observations, seed from wetter habitats are also more likely to be desiccation intolerant. However, even in the driest habitat there are species that produce seeds of sufficient weight for desiccation intolerance to be as likely as tolerance. Similarly, there are species in the wettest habitats whose seeds fall in the (low) weight classes for which only desiccation tolerance is so far known.

Plant habit can also be added to the equation. In both mesic British (Waller, 1988) and arid Australian (Jurado *et al.*, 1991) vegetation, seed weights of herbaceous species are lower than those of trees, suggesting a higher frequency of desiccation-tolerant seeds in the former group. Among the trees of the rain forest reserve of Los Tuxtlas in Mexico, the pioneer species have lighter seeds, while the dominant species of the mature forest have heavier seeds (Ibarra-Manriquez and Oyama, 1992). Storing seeds of dominant species would appear to be more problematic than storing seeds of pioneer species. Another expected 'hot spot' of desiccation intolerance is among aquatic species, based on the early work of Muenscher (1936) and the subsequent confirmation of many of his findings by Probert and Longley (1989).

To determine the practical implications of seed storage behaviour for germplasm collecting, a brief résumé is necessary, for each seed storage category, of the relationship between seed viability and the relevant environmental variables, i.e. temperature, relative humidity, atmospheric composition and time (both physical and physiological).

Desiccation-intolerant (recalcitrant) seeds

As the name implies, desiccation-intolerant seeds cannot withstand drying. Species as varied as aquatic tropical grasses and dominant temperate trees have seeds that lose viability when dried to water potentials in the range from -5 to -3 MPa (Probert and Longley, 1989; Pritchard, 1991). Over the range of normal biological temperatures, these water potentials would be in equilibrium with a relative humidity of $>95\%$. Such high ambient relative humidities are rarely found in nature. Consequently, recalcitrant seeds will dry to lethal water potentials even if they are held under the ambient conditions in which the species occurs in the wild. Not surprisingly, the seed-coats of some of these species are adapted to restrict dangerous moisture loss, as in *Acer pseudoplatanus* (Dickie *et al.*, 1991) and *Araucaria hunsteinii* (Tompsett, 1982). However, other desiccation-intolerant seeds, such as those of *Porteresia*, have no such adaptation, and drying

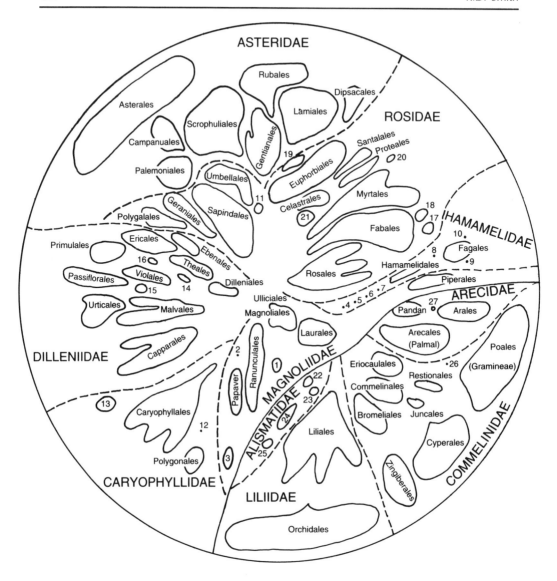

Fig. 20.1.(a) Evolutionary diagram showing the relative degree of specialization of the orders of Angiosperms (after Heywood, 1978). Smaller orders are indicated by numbers as follows: (Dicotyledons) **(1)** Nymphaeales; **(2)** Sarraceniales; **(3)** Aristolochiales; **(4)** Trochodendrales; **(5)** Cercidiphyllales; **(6)** Didymeleales; **(7)** Eupteleales; **(8)** Eucommiales; **(9)** Casuarinales; **(10)** Leitneriales; **(11)** Juglandales; **(12)** Batales; **(13)** Plumbaginales; **(14)** Lecythidales; **(15)** Salicales; **(16)** Diapensiales; **(17)** Podostemales; **(18)** Haloragales; **(19)** Cornales; **(20)** Rafflesiales; **(21)** Rhamnales; (Monocotyledons) **(22)** Alismatales; **(23)** Triuridales; **(24)** Najadales; **(25)** Hydrocharitales; **(26)** Typhales; **(27)** Cyclanthales.

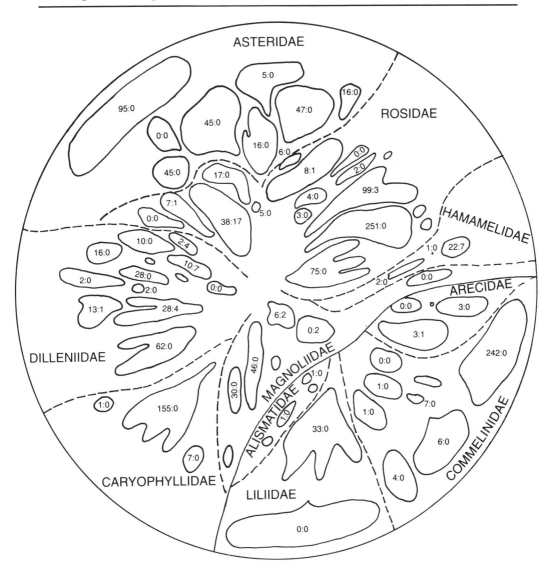

Fig. 20.1.(b) The same diagram as Fig. 20.1(a), showing the numbers of fully desiccation tolerant (first number) and desiccation intolerant (second number) species known to occur in each order from the literature. Where no numbers are shown for the smaller orders, the data are 0:0.

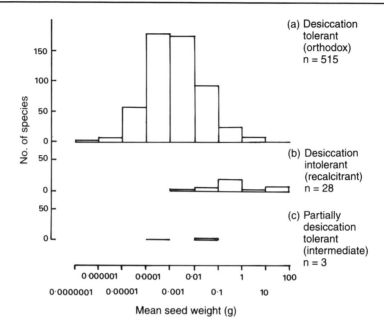

Fig. 20.2. The distribution of individual seed weight among species, showing the three kinds of seed storage behaviour. Data from Cromarty *et al.* (1985).

to lethal water potentials on the plant has been observed (R.J. Probert, pers. comm.).

If the seed is in a fleshy fruit, such as in mango or mangosteen, the flesh will act as a reservoir of moisture, effectively 'buffering' the seed against drying. The problem of maintaining high seed moisture can be overcome most easily for recalcitrant seeds in fleshy fruits by keeping them in their fruits and paying the logistical, physical and financial price of moving the seeds together with their highly adapted buffers.

One way collectors can minimize loss of moisture from seeds is by holding them in impermeable containers such as plastic bags. However, this raises the problem of controlling the gaseous atmosphere inside the bags. At the water potentials at which the seeds retain their viability, they are metabolically active. They are therefore consuming oxygen and releasing carbon dioxide and water. In desiccation-intolerant species, the rate of loss of seed viability increases as the oxygen concentration falls in the surrounding atmosphere. In practice, this depletion of oxygen can be avoided by keeping the seeds in relatively light-gauge polythene bags and regularly reoxygenating the internal atmosphere by deflating and reinflating the bags on a weekly basis (Tompsett, 1983). Even so, under such a regime the seed moisture content gradually increases, presumably due to uptake of the water respired into the atmosphere by the seeds themselves. This reduces a restraint on germination.

Little is known of the effect of temperature on the longevity of desiccation-intolerant seed, as experimental studies have often confounded this with the effects of temperature on the rates of drying and of germination. For temperate species of oak, reducing the temperature as low as $-1.5\,°C$ increases seed storage life, but further reduction to $-5\,°C$ is lethal (Suszka and Tylkowski, 1980). However, even modest reductions in temperature risk rapidly killing the seeds of tropical species through so-called chilling injury. Despite its name, this can occur in *Inga*, *Shorea* and other tropical genera between $21\,°C$ and $16\,°C$. While chilling injury is unlikely at the ambient conditions where these species generally occur, injurious conditions could well be encountered in excessively air-conditioned rooms and during air freight shipment, for example in light aeroplanes and during transfers from one plane to another at intermediate airports during indirect intercontinental flights. Labelling seed consignments as 'perishable' will risk their refrigeration by customs officials while they await import clearances.

Holding desiccation-intolerant seeds at higher temperatures, well above those at which chilling injury can occur, can cause problems of its own. Most such tropical seeds are non-dormant and germination will be hastened under these conditions. The more the radicle has protruded in *Avicennia marina* seeds, the less tolerant of desiccation they become and so the more demanding in their storage requirements (Farrant *et al.*, 1986). Efforts to discover a phase of greater desiccation tolerance have proved unsuccessful in aquatic grasses (Probert and Brierley, 1989). Work on seeds of *Aesculus hippocastanum* (Tompsett and Pritchard, 1993) and of oak (Finch-Savage, 1992) showed that their desiccation tolerance increased as they developed and was at its maximum during peak seed fall. So harvesting desiccation-intolerant seeds just before natural dispersal is better than collecting fallen fruits, because in the latter case the seed lot contains both individual seeds close to maximum desiccation tolerance and also older seeds already advanced in their germination and so losing desiccation tolerance. Where tree climbing and hand picking is impractical, and limb felling unacceptable, this collecting problem can be overcome by laying sheeting beneath the tree and collecting the seeds/fruits that fall over a short time period.

In practice, when dealing with desiccation-intolerant seeds the most easily manipulated variable is time. Working quickly between harvest and dispatch will not only reduce the practical difficulties of keeping the seeds ventilated and alive but will also ensure that the material is received still ungerminated, which will maximize the options for conservation. The evidence suggests that 30 days is the longest period which should be allowed between harvesting and receipt by the seed bank.

Partially desiccation-intolerant (intermediate) seeds

Variants of both major seed storage types exist. *Zizania palustris* seeds, for example, share many of the characteristics of desiccation-intolerant seeds and yet under highly specific conditions can be dried to low moisture content and subsequently germinated (Kovach and Bradford, 1992). However, under the less controlled conditions likely to be available to collectors, these seeds may behave as recalcitrant. They should always be treated as such. On the other hand, there are species that can be partially dried with ease but where full desiccation is lethal. They include coffee (Bacchi, 1956; Vossen, 1979), *Araucaria columnaris* (Tompsett, 1984) and oil-palm (Grout *et al.*, 1983). Recently, they have been categorized as of 'intermediate' storage behaviour (Ellis *et al.*, 1990). Under field conditions their behaviour is expected to be that of desiccation-tolerant species. Only active seed drying (i.e. in the sun or using silica gel) needs to be treated with caution.

Few such variants have been identified. Their existence demands that their storage characteristics be carefully specified and raises the possibility of a continuum of storage types. However, the imperative to act despite our uncertainties means that the advice offered here will assume that deviation in seed storage behaviour from the two major types is rare and that these two will describe most species.

Desiccation-tolerant (orthodox) seeds

Seed development

Considerably more is known about orthodox seeds than about the recalcitrant or intermediate types. The accepted general model for development is shown in Fig. 20.3a. It should be treated with due caution. The literature is dominated by studies of annual crops, but the duration of seed development can range from a few days in ephemeral species to over two years in conifers. Also, the model is derived from studies of seed populations, with mean values used to model the behaviour of individual seeds.

Soon after fertilization, which can be delayed by up to 14 months after pollination in *Pinus*, the weight of water in the seed reaches a constant, maximum value. This is followed by a period in which embryo development and the deposition of storage products result in seed dry weight increasing sigmoidally, until it too reaches a maximum. As dry matter increases while the amount of water in the seed remains constant, so percentage seed moisture content necessarily declines. At maximum dry weight, the moisture content of seeds from dry fruits falls rapidly, due to the net loss of water that follows the severing of the vascular connection between seed and mother plant. The seed is now hygroscopic, its moisture content coming into equilibrium with the ambient atmosphere. Inside fleshy fruits, seed moisture content remains close to its

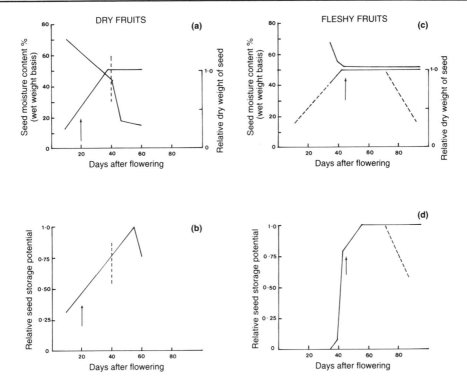

Fig. 20.3. Diagrammatic representation of seed development and seed storage potential in fully desiccation-tolerant species, with seeds borne in dry fruits (**a** and **b**) and fleshy fruits (**c** and **d**). The arrows indicate the acquisition of desiccation tolerance, the vertical dashed line the start of water loss from the seed. The dashed lines in (**c**) and (**d**) indicate the likely course of events if fruits begin to dry.

value at maximum dry matter and seeds only behave hygroscopically if taken from the fruit.

There are deviations from this general pattern. In the drupe fruits, the development pattern has a quiescent period of dry matter accumulation interpolated in the sigmoidal growth pattern. It is during this period that the embryo grows (Tukey, 1934). In the wetland herb *Ranunculus scleratus*, even though maximum seed dry weight is achieved, no phase of rapid drying takes place before the seeds are shed (Wechsberg *et al.*, 1993).

Where known, seed moisture content at the time of maximum dry weight ranges from about 30% on a wet weight (WW) basis for dry-fruited cereals up to 65% for the fleshy-fruited coffee berry. Across all species and fruit types, the mean value is close to 45%. However, for dry dehiscent, indehiscent and schizocarpic fruits, these moisture contents are unlikely to be those at which collecting will be done. The moisture content of seeds of domesticated cereals and pulses will oscillate in equilibrium with ambient conditions until harvested and go up as the seeds take up moisture from any rain falling on them (Yaklich and Cregan, 1987). Among naturally dispersed British species, seeds of *Lathyrus sphaericus* have moisture contents in the pod close to 15% at the point of explosive dispersal, while the grasses *Melica uniflora* and *Bromus sterilis* shed their seeds at 38% and 54% moisture contents respectively. In the wetland herb *R. sceleratus*, seed dispersal takes place at 56% moisture content. Even higher values (67%) have been recorded for the seeds of the tropical timber tree *Agathis macrophylla* as the cone shatters to release them (Smith, 1985). The range of seed moisture contents at dispersal is even wider than at maximum dry weight. This variation provides collectors with a considerable challenge to ensure the appropriate handling of each species following collecting.

Furthermore, the seed's desiccation tolerance and its storage potential vary through its development. Seeds that are desiccation-tolerant when mature are intolerant of drying at earlier stages of development. This is shown in a generalized way in Fig. 20.3b. Initially, developing seeds are desiccation-intolerant. Then, at least in the case of *Ricinus communis*, about halfway through seed development they become tolerant of slow desiccation. Finally, close to maximum dry weight, they become tolerant of rapid desiccation, such as that imposed by silica gel drying (Kermode and Bewley, 1985). This increasing tolerance of seeds to desiccation with maturation also manifests itself in *Acer* and coffee as a progressive lowering of the moisture content to which the seed population can be dried without loss of viability (Ellis *et al.*, 1991; Hong and Ellis, 1992). While this may not directly affect the survival of seeds under collecting conditions, it will certainly affect their subsequent prospects for long-term storage.

In other studies, not using different drying rates, tolerance to drying in dry-fruited cereals, pulses and crucifers is achieved about halfway through seed development. In cereals, seed storage potential continues to rise until well after maximum seed weight is reached and seed moisture has fallen to equilibrium with the surrounding air (Pieta Filho and Ellis, 1991). Seed storage potential is subsequently lost as the seed ages in the field. The dry-fruited achene of the wild wetland herb *Ranunculus sceleratus* differs from this pattern in two respects. First, the rapid drying phase does not occur. Secondly, the storage potential of seeds is still rising on the mother plant until they are shed (Wechsberg *et al.*, 1993). In contrast, in the fleshy-fruited tomato, tolerance to rapid desiccation does not occur until just before or just after maximum seed weight has been reached. Seed storage potential then continues to increase for the next ten days and remains constant for a further 40 days (Demir and Ellis, 1992).

There has been too little research to take these patterns as settled. Different authors working on the same species have produced different results and even the same author working on the same species has on different occasions come to different conclusions. None the less, collectors would do well to bear in mind that in dry-fruited species the early period of desiccation intolerance is followed by a period of drying sensitivity before maximum storage potential is achieved. After the seed has dried to equilibrium with ambient conditions, it then begins to age. This initial period of drying sensitivity may last longer in seeds borne in fleshy fruits, but, as long as the seeds remain in the fruits, the storage potential is maintained.

Morphological markers of seed maturity

The colour changes that are associated with fruit and seed ripening do not appear to correlate sufficiently well with the achievement of maximum storage potential to be of much use to collectors in the field. For example, the skin colour of tomato (*Lycopersicon esculentum*) fruits varies from between 20 and 25% red to completely green at the same level of seed maturity, namely attainment of maximum seed weight. In barley, the colour of the palea and lemma is not correlated with subsequent seed longevity (Ellis and Roberts, 1981). Developmental studies of grass species suggest that endosperm characteristics would also be of little value as markers of seed maturity (Komatsu *et al.*, 1979). In the drupe fruits, only final fruit size correlates with the completion of seed development. More obvious markers, such as pit (pericarp) hardening, take place just as the embryo begins its development. Dehiscence or dispersal remains uninvestigated as a natural diagnostic of maximum seed storage potential.

Rapid gravimetric techniques for measuring seed moisture content have been developed for field use, for example using hot exhaust gases from motor vehicles (Klein and Harmond, 1971). However, the relatively constant moisture content of tomato seeds over the period during which they achieved first desiccation tolerance and then maximum seed storage potential suggests that single moisture content measurements would be of little use when dealing with fleshy fruits.

Where seeds are being collected from regeneration plots, say, seed development can be monitored and harvest timed to coincide with the attainment of maximum storage potential. However, collectors working in the field cannot afford to remain for sufficient periods at any one site in this way to monitor the rate of change in the various indicators. Consequently, the knowledge that seed storage potential can rise or fall dramatically over a few days (so that bank storage lives will first double and then halve over the same period) will remain of little value to collectors until a simple, reliable and rapid field diagnostic is developed.

Seed longevity at high moisture content

Earlier discussions of the problems faced by seed collectors in keeping their material viable in the field (Smith, 1984) were hampered by the lack of comprehensive data on the rate of loss of seed viability over the range of moisture contents, temperatures and gaseous environments which seed samples could expect to experience during a collecting trip. Such data sets are now available for commercial seed lots of lettuce (Roberts and Ellis, 1989). Figure 20.4 summarizes these relationships schematically.

The whole response can be divided into two phases, determined by seed moisture content. In both phases, the effect of temperature appears to be very similar, higher rates·of loss occurring at higher temperatures. The effect of the gaseous environment is dramatic in the first phase and insignificant in the second. Under aerobic conditions, as seed moisture content falls over the first phase from full imbibition to about 15% moisture content, the rate of loss of viability increases. Further drying below this value decreases the rate of loss of viability until it is considerably lower than that at full imbibition. The rates of loss of viability are approximately equal at 40% and 8% moisture content, or in equilibrium with 99% and 65% relative humidity respectively.

In the first phase, the rate of loss of viability is 50 times higher under anaerobic conditions than under aerobic conditions at the same moisture content of 40% (close to full imbibition). At 25°C, seed viability could be expected to fall from 84% to 16% in seven days under anaerobic conditions, compared with 300 days under aerobic conditions. As the seeds dry further under anaerobic conditions during the first phase, the rate of viability loss decreases only slightly, until the seeds have dried to around 22% moisture content. The rate of viability loss then decreases sharply until at 15% moisture content it is the same as in seeds held aerobically.

This comprehensive study on the dry indehiscent fruit of lettuce is

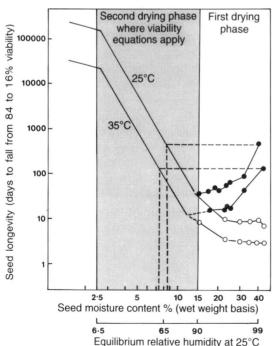

Fig. 20.4. Diagrammatic representation of the viability behaviour of lettuce seed at 25°C and 35°C over the range from 40 to 2.5% moisture content, under aerobic (filled circles) and anaerobic (open circles) conditions. The equivalent equilibrium relative humidity values are shown for four moisture content levels, i.e. where maximum longevity under hydrated condition (99%) occurs, where minimum longevity occurs (90%), where longevity under air dry storage is the same as maximum longevity under hydrated conditions (65%), and where the benefit of further drying decreases (6.5%).

matched very well by the much less extensive data for coffee presented by Smith (1984). The equilibrium relative humidity value for the switch between the two phases is acceptably close in both species to 90% relative humidity. A slightly higher value of close to 93% relative humidity has been reported for durum wheat (Petruzelli, 1986), while *Phleum* exhibits an intermediate value of about 91% (Smith, 1984).

Three important practical conclusions can be drawn from these limited data on viability loss in desiccation-tolerant seeds at high moisture contents:

- Great care must be taken to keep seeds aerated when they are held at high moisture contents, for example those at which many non-domesticated species are shed. Holding the seeds in sealed conditions, such as in a plastic bag, or packing the seeds tightly into limited space will result in an anaerobic environment rapidly developing around the seeds, due to their respiration. This will dramatically increase the rate of loss of viability.
- The switch to air-dry storage behaviour appears to occur at relative humidities close to 90%.
- The benefits of drying will only be realized if ambient relative humidities are <65%.

Relative humidity values as high as 90% are rare in nature. Consequently, if samples of seeds or dry indehiscent fruits are kept aerated, their hygroscopic nature will soon result in the seeds shifting quickly to the second, air-dry storage phase. However, seeds from fleshy-fruited species will either shift much more slowly or not at all if the fruit itself does not dry.

The behaviour of air-dry seeds has been much studied, with the viability constants determined for tens of species. The responses of interest to seed collectors are summarized below.

The behaviour of air-dry seeds

The effect of temperature on the storage of air-dry seeds

First, the effects of temperature and seed moisture content on seed longevity are independent. The effect of a given change in temperature is the same at any seed moisture content. Over the range −13°C to 80°C the effect of temperature is the same for seeds from eight widely different species, including monocotyledons and dicotyledons, herbs and trees, annuals and perennials, and temperate and tropical species, as well as grains, leafy vegetables and timber crops (Dickie *et al.*, 1990). The relationship between temperature and the rate of loss of viability is quadratic. At the ambient temperatures likely to be encountered while collecting, warming will cause greater increases in the rate of seed viability loss than cooling by the same amount will reduce it. In practical terms, this means that if cooling is possible (e.g. in air-conditioned rooms), it should be taken advantage of. However, the bigger concern should be to prevent seeds being heated by the sun, either directly or

indirectly, as in a parked vehicle. Vehicles fitted with reflective false
sun roofs are less prone to such heating. A roof rack full of luggage will
serve much the same purpose.

The effect of moisture content on the storage of air-dry seeds

In contrast, the effect of moisture content on the rate of viability loss
in air-dry seeds is much more variable among species. However, where
a wide range of air-dry moisture content has been studied, the relationship between seed moisture content and the rate of loss of viability has
been shown to be logarithmic (Ellis *et al.*, 1986; Tompsett, 1986; Kraak
and Vos, 1987; Dickie *et al.*, 1991). This means that for each progressive
similar reduction of seed moisture content, say 1%, an ever-increasing
reduction in the rate of loss of viability is achieved. In practice, this
makes lowering moisture content a potent tool in managing seed
viability loss during collecting.

What controls seed moisture content is thus important to collectors.
Desiccation-tolerant seeds are hygroscopic and so give out or take up
moisture until they are in equilibrium with the surrounding air. Consequently, seed moisture content is, in part, determined by the temperature and relative humidity of that air. Species from some plant families,
such as the Leguminosae and Malvaceae, produce 'hard' seeds, which are
adapted only to desorb, i.e. give out moisture when the ambient air is
drier than the equilibrium relative humidity of the seed and prevent
moisture uptake when the air is wetter. This characteristic accounts for
the extreme longevity of hard seeds and makes collecting such species
relatively straightforward.

For fully hygroscopic seeds, oil content is the other factor that determines the final equilibrium moisture content. The role of oil is passive,
acting as a constant but hydrophobic element of the seed weight. Consequently, seeds with high oil contents equilibrate to lower moisture
contents under the same conditions than seeds with low oil contents.

The relationship between seed moisture content, oil content, temperature and relative humidity has been described in an empirical equation
(Cromarty *et al.*, 1985), which can be combined with the viability equation
of Ellis and Roberts (1981) to predict the rate of seed viability loss under
ambient atmospheric conditions. For many species, this reveals that, at
a constant temperature, similar changes in the equilibrium relative
humidity produce the same relative changes in the rate of seed viability
loss. While this allows the construction of a simple tool to predict relative
seed viability losses for a majority of species under ambient collecting
conditions (Fig. 20.5), it is not absolutely applicable for all species. *Ulmus
carpinifolia* is one species for which it is inappropriate and one that also
differs in the relative humidity at which the minimum rate of loss of
viability is achieved (Smith, 1992). Figure 20.5 presents a standard
psychrometric chart, on which the combinations of temperature and
relative humidity calculated to bring about the same monthly viability
losses have been identified. (See also Box 20.1.) When planning collecting

Fig. 20.5. Psychrometric chart onto which the temperature and relative humidity conditions have been plotted, which under equilibrium conditions would be expected to bring about viability losses from 0.01 to 1.00 probit units in 31 days in chickpea seeds. The polygon delimited by the numbers 1 to 12 represents the losses expected in each month of the year (i.e. January = 1 etc.) at Tahoua, Niger (annual precipitation 444 mm). Note that highest rates of loss of viability are expected during the September harvest period. In moist tropical locations monthly differences are much reduced, but values are close to the stippled area, which indicates the humidities at which equilibrium moisture content would allow aerobic respiration to begin. This would make the gaseous environment experienced by the seed of importance.

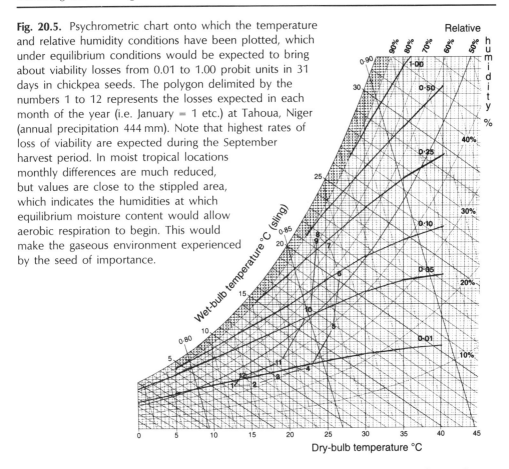

trips, plotting mean monthly temperature and relative humidity values, taken from meteorological tables, on to such a chart allows collectors to quickly identify the relative rates of loss of seed viability likely to be encountered in the field.

This method overcomes many of the drawbacks associated with the earlier approach of Smith (1984). However, its purpose is much the same, i.e. to help collectors calculate whether the expected monthly rates of loss of viability are sufficiently large (i.e. in excess of 0.1 probit/month[1]) for either: (i) active field drying of seeds to moisture levels below the likely equilibrium value to be necessary; or (ii) the time spent in the field before a base is reached where seed drying facilities exist to have to be reduced. Indeed, once the timing of the trip has been set, maps of the collecting area can be produced showing contours of equal levels of

---

[1] At this stage, the 0.1 probit/month is a somewhat arbitrary value which is discussed further below. However, note that this rate of loss of seed viability is 20 times greater than would be expected under recommended seed drying-room conditions. The meaning of 'probit' is discussed in Box 20.1.

Box 20.1

Percentages, probits and seed viability

As long as the germination test conditions adopted overcome any dormancy, seeds in viability monitoring tests can be classified as either viable (if they germinate) or dead (if they do not germinate). When the results of a time series of monitoring tests are plotted as percentage viability values, the graph takes the form of a reversed S-shaped or negative sigmoidal curve (opposite).

Given that the classes used were 'all' (germinating and therefore viable) or 'nothing' (non-germinating and therefore dead), the negative sigmoidal graph approximates a negative cumulative normal distribution of individual life spans within the seed lot (Smith, 1984).

A negative cumulative normal distribution is cumbersome to manipulate as a curve, but it can be converted into a straight line by transforming the raw percentage values to probit or normal equivalent deviate values, in the same way as transformation to logarithms is used in some applications. Variation in individual attributes (in this case life spans of seeds) about the population mean may be expressed in terms of standard deviations. The proportion (% age) of the population which lies within each standard deviation unit is fixed (see figure). Using values related to standard deviations rather than raw percentages thus gives a linear scale of viability loss with time. Tables, not unlike log tables, can be produced to convert percentages to probits (see table).

The value of this transformation is that it makes clear that, though the same damage is caused in any single seed lot by similar exposure to identical adverse storage conditions, at high viabilities this damage will be much more difficult to detect than at lower viabilities. For example, if we assume that the damage is sufficient to reduce viability by 1 probit, the % viabilities will be from 99.9 to >97.7, >84 and >50 as the probit values fall from 8 to >7, >6 and >5. The fall from 99.9% will be statistically undetectable in a standard germination test using 400 seeds, but the fall from 84% will be very significant.

Table 20.A The probit transformation.

Response rate	0.00	0.01	0.02	0.03	0.04	0.05	0.06	0.07	0.08	0.09
0.00	–	2.67	2.95	3.12	3.25	3.36	3.45	3.52	3.59	3.66
0.10	3.72	3.77	3.82	3.87	3.92	3.96	4.01	4.05	4.08	4.12
0.20	4.16	4.19	4.23	4.26	4.29	4.33	4.36	4.39	4.42	4.45
0.30	4.48	4.50	4.53	4.56	4.59	4.61	4.64	4.67	4.69	4.72
0.40	4.75	4.77	4.80	4.82	4.85	4.87	4.90	4.92	4.95	4.97
0.50	5.00	5.03	5.05	5.08	5.10	5.13	5.15	5.18	5.20	5.23
0.60	5.25	5.28	5.31	5.33	5.36	5.39	5.41	5.44	5.47	5.50
0.70	5.52	5.55	5.58	5.61	5.64	5.67	5.71	5.74	5.77	5.81
0.80	5.84	5.88	5.92	5.95	5.99	6.04	6.08	6.13	6.18	6.23
0.90	6.28	6.34	6.41	6.48	6.55	6.64	6.75	6.88	7.05	7.33

Response rate	0.000	0.001	0.002	0.003	0.004	0.005	0.006	0.007	0.008	0.009
0.97	6.88	6.90	6.91	6.93	6.94	6.96	6.98	7.00	7.01	7.03
0.98	7.05	7.07	7.10	7.12	7.14	7.17	7.20	7.23	7.26	7.29
0.99	7.33	7.37	7.41	7.46	7.51	7.58	7.65	7.75	7.88	8.09

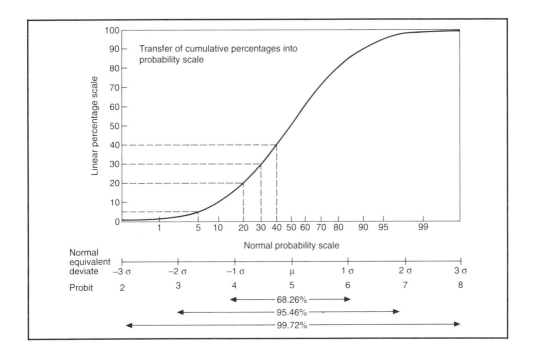

viability loss. This will be important in itinerary planning. For example, Fig. 20.6, from Prendergast *et al.* (1992), shows that in Mali in October, the normal harvest month, there is a ten- to 40-fold difference in the expected rate of viability loss from the west to the east. Thus, all other things (e.g. seed availability) being equal, collecting from west to east will be a good idea, as the seed will dry during collecting. If collecting is from east to west, the seed will absorb moisture during the mission, and the quality of samples therefore deteriorate.

Drying seeds in the field

Where the ambient conditions are such that the expected viability losses are acceptable, collectors will only need to make sure that seeds from dry dehiscent or indehiscent fruits are stored in permeable containers such as cotton or paper bags. These should be held in such a way that air circulates freely between and through them. It could be argued that seeds are adapted to such conditions and damage should be minimal. However, extrapolation of the research findings presented earlier suggests a more cautious approach. The risk of damage caused by rapid drying of immature seed must be balanced against that caused by mature seed drying slowly between high moisture contents, where viability loss is slow, and much lower moisture contents, where viability loss is again slow. In lettuce, this involves drying seed from 40% to 8% moisture content. This problem remains little researched, and advice given is therefore uncertain, though less so for dry-fruited than

Fig. 20.6. Days for pearl millet
(*Pennisetum americanum*) seed to fall
in viability from 84% to 50% under
prevailing conditions in Mali in
October. Higher ambient mean relative
humidity is shown by denser stipple.

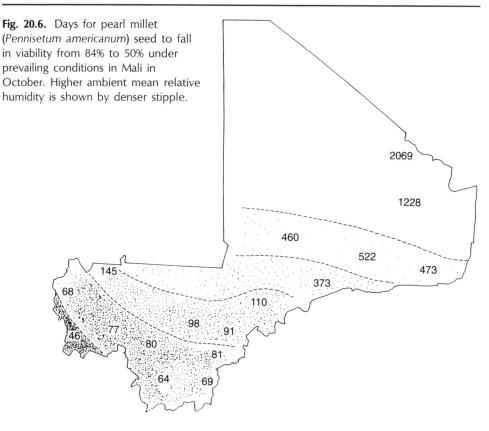

for fleshy-fruited species. First, tolerance to rapid desiccation appears
to occur sooner in seed development, so seed maturity at the time of
collecting may exert a smaller influence, and secondly, provided they
have achieved their maximum dry weight, the seeds of dry-fruited
species may already have undergone drying to ambient equilibrium
moisture content and so will be in the air-dry phase. Further drying will
probably be less stressful. Indeed, where rapid drying was found to
induce desiccation stress, this took place in seeds of high initial moisture
content and so was likely to be in equilibrium with relative humidities
above the transition zone.

   Therefore, dry-fruited seed lots can probably be dried to below
ambient moisture content safely, the only risks being to immature seed.
To overcome this final risk, holding the fruits loosely packed, and thus
aerated, under ambient conditions for a further three to seven days after
collecting would cause only tolerable damage in the mature seed while
providing an opportunity for the immature seed to mature. Following
this period of air-drying, the seed could then be dried further.

   For small, many-seeded fleshy fruits, if the logistics of the trip do
not prevent holding the seeds inside the fruits, this is the most practical
and most successful option. Apple seeds held in the fruits survived much

better than those extracted from the fruit and held at the same moderate temperature in a collecting bag (Dickie, 1988). Seed longevity in the fruit was also reduced if the fruits were held at higher temperatures or under semisealed conditions where the oxygen supply was limited. The advice appears to be: keep the fruit 'sound' and the seed will also keep sound. What is not clear, and would be helpful for collectors to know, is whether seed development can continue within detached fruits and if so under what conditions this process is most successfully completed.

Successfully dealing with large, many-seeded fruits or relatively small, single-seeded fruits, when the logistics prevent keeping the seeds in their fruits, provides the most difficult challenge. In fleshy fruits, desiccation tolerance occurs late in seed development, useful morphological markers of seed development do not appear to be displayed by the fruit and the seeds themselves will be at high moisture contents on extraction from the fruits. All that can be recommended is to extract the seeds carefully by hand, spread them into a thin layer to maximize aeration and allow them to dry under ambient conditions in the shade until they have achieved equilibrium. However, as seed drying theory predicts that over the initial drying phase the rate of moisture loss under most ambient conditions will be close to that expected over silica gel, the risks are high. Yet transporting constant humidity solutions into the field to slow seed drying rates to lower levels of damage in the same way as in laboratory studies is probably more logistically demanding than holding the seeds in the fruits.

When drying seed to below equilibrium moisture content, whether from fleshy or dry fruits, one of the least demanding methods is direct sun-drying. However, despite its common use in traditional agriculture, the advantages and disadvantages of the technique remain unquantified. Therefore, sun-drying cannot be recommended, despite the expectation that evaporative cooling should probably counteract the effects of increased temperature.

Another, more logistically complicated, possibility involves the use of desiccants such as silica gel. This raises the questions of how much to dry each seed lot and how much silica gel is required in total on the trip. Figure 20.7 allows the estimation of the weight of silica gel needed to reduce the moisture content of a sample to 5% from its fresh weight, provided seed oil and moisture contents are known. Fresh weight can be measured in the field using a spring balance, but the latter are more difficult. However, Earle and Jones (1962) analysed the seed oil content of 900 species in 501 genera in 113 families representing 35 angiosperm orders and two gymnosperm orders. The average seed oil content for this wide sample is about 25%. If 90% relative humidity is taken as the upper limit for the air-dry storage phase and 20°C as a reasonable global average temperature, an approximation of the worst likely equilibrium moisture content can be calculated. The value is 15% moisture content (WW basis). From Fig. 20.7, the weight of silica gel required to dry 1 kg of wet seeds to 5% moisture content is about 0.7 kg. This value rises to

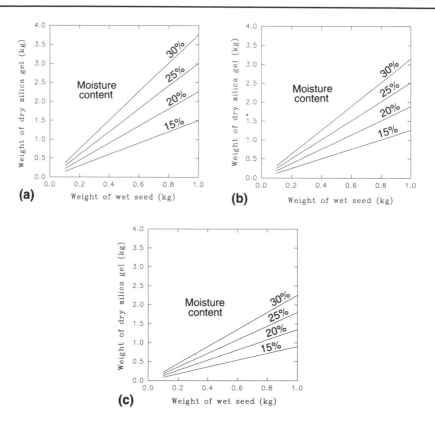

Fig. 20.7. Calibration charts to allow rapid estimation of the weight of silica gel required to dry seed lots of known weight and moisture content to 5% moisture content at 20°C. **(a)** 1.0% oil content; **(b)** 10.0% oil content; **(c)** 25.0% oil content. Adapted from Cromarty *et al.*, 1985.

1.4 kg as the seed oil content falls to 1%, as in the major grains. In practice, three parts seeds to two parts silica gel seems to be an acceptable compromise.

Seed moisture content will be reduced most quickly if the seeds and silica gel are sealed in a container that minimizes the air volumes enclosed and maximizes contact between the two. Alternate layers of packeted silica gel and packeted seeds, all enclosed in a sealed plastic container, may be an option. Rapid drying will result in the container becoming warm to the touch as the absorption of the water by the silica gel releases latent heat. This heating will be offset to some extent by evaporative cooling of the seeds as they dry.

If the transport of the necessary weights of desiccant is a problem and the logistics of collecting and transporting fruit overwhelming, then a more complex drying protocol can be used. This involves the regular reactivation of smaller weights of silica gel so that the seed moisture content is 'stepped down' to its final low value. This requires blue silica gel, which can be reactivated when it has turned pink by heating to 175°C for 6 hours or to 125°C for 16 hours. When reactivated, the silica gel returns to its original colour and can be used again. Only when the silica gel stays blue for several days will the seed have been dried to sufficiently low levels. The slower drying rate of this technique will mean that greater levels of damage are accumulated. When the seeds are dry, they can be packed tightly into a sealed heavy-duty polythene bag. If condensation is seen inside the bag, the seeds must be re-dried.

Seed cleaning in the field

If seed lots must be cleaned during a collecting mission for logistical or quarantine reasons, only manual methods that do not reduce subsequent longevity should be employed. Mechanical methods, even traditional ones such as threshing with flails (Saini *et al.*, 1982), reduce subsequent longevity by twice as much as is proposed as acceptable under ambient collecting conditions. Some mechanical methods are described in Chapter 23, but they are not recommended for seeds destined for long-term conservation. Flotation methods are sometimes used to separate viable seeds from unwanted material. The effect of this on viability has not been explicitly studied. However, there is evidence that most, if not all, orthodox seeds suffer imbibition damage when dried below 15% moisture content and then imbibing through contact with liquid water (rather than water vapour). There is also the need to redry the seeds back to their original or lower moisture contents for storage. The recommendation must again be to avoid such methods when collecting for long-term conservation.

Quarantine and transport effects

If seeds that have been collected through international collaboration are to be transferred from the country of origin, quarantine treatments and conditions during transport could reduce their viability further. While there are many studies on the immediate effects of fungicides and other quarantine seed treatments on germinability, most do not investigate their effects on subsequent storage. The literature is also ambivalent, some studies reporting that such treatments result in higher germination, others lower. Therefore, a strategy must be adopted that reduces the risk of damage to seed storage potential without increasing the risk of unwitting disease transfer.

An option is for the recipient organization to obtain permission from its own national authorities to carry out post-entry quarantine. Getting such agreement should be possible, as the seed lots will usually be small and not intended for general distribution and will enter the country in a sealed package destined for a single location, where they will remain banked and so isolated from the normal processes of disease transmission. Should such an arrangement be acceptable, collectors will need to have the necessary paperwork showing that post-entry quarantine has been agreed before the host country plant health officials can be expected to waive any treatment. The importance of quarantine inspection of germplasm should not be underestimated, as the conditions that prolong the longevity of desiccation-tolerant seeds also prolong the longevity of the desiccation-tolerant fungal and bacterial spores that are borne upon them (Lenné and Wood, 1991).

Seed samples should be transported to the seed bank for further drying as quickly as possible. Airmail is uncertain, with neither collector nor recipient able to trace the package from the moment of posting to delivery at its destination. It should only be considered as a last resort. Experience shows that air freight is not as certain or as quick a means of transporting seeds between countries as some agents might contend. Collectors need to take some defensive measures on behalf of their seeds if the vicissitudes of the air transport system are to be avoided. Ideally, the collector should accompany the package to its destination, so that delays between collecting and storage are kept to a minimum and loss of viability is minimized. On the surface, this may appear extreme. However, when agents' fees and air freight charges are added together, the comparison between excess baggage and air freight costs is acceptable. When the staff time of the recipient institute spent in monitoring the progress of the consignment towards its destination is added to the cost, the comparison is even more favourable. If air freight is the only practical way forward, the minimum defence is to pack the seed so that subsequent moisture uptake, either from being left standing in the rain or under conditions of high relative humidity, is unlikely. Collectors must inform the recipient institute by Telefax, telex or telephone of the airway bill number of their consignment(s), given to them by the freight agent. This will allow the progress of the consignment to be monitored through the international air freight computer system and the seed 'pulled' towards its destination and safety. Seed lots are forgotten and left to stand in airports or in agents' warehouses more often than might be hoped.

Why should the collector heed this advice?

Why is it important to minimize losses of seed viability between collecting and storage, as the advice given here attempts to do? The traditional argument is still valid and by itself is enough to justify the call for high viability seed lots. This points to declining seed lot viability as a reflection of the increased frequency of mutations in the surviving seeds. These mutations, though reduced by diplontic selection as the resulting seedlings develop into plants, do pass to the next generation (Dourado and Roberts, 1984). Thus, the surviving seed lot will gradually diverge from the population from which it was collected. Those responsible for the germination, establishment and subsequent cultivation of aged seed lots would also point to increased intolerance of stress, reduced growth rates and reduced seed yield of the surviving plants from such material. Those responsible for the funding of seed banks may also question why seed collected in the field was not directly storable and why an additional costly step of rejuvenation was required at all.

More disconcertingly for collectors, it is becoming apparent that the loss of seed viability can exert directed selection pressure. However, the selection pressures are not for 'seed bankability', as some have proposed. The arguments against this ill-considered view are straightforward. First, environmental conditions worldwide are such that desiccation-tolerant seed, when severed from the parental vascular supply and not in contact with soil water through capillarity, will soon air-dry to below the critical relative humidity and fall within the range of water potentials at which the viability equation applies. Here temperature (ambient or bank) and seed moisture content (ambient or bank) control the rate of loss of viability. There is no evidence of discontinuities in these relationships for temperatures over the range $-13°C$ to $80°C$ or for moisture contents in equilibrium over the range of 10% to 90% relative humidity. Therefore, the rate of seed viability loss can be deduced to be under the same control in all these conditions, being slowed and accelerated by changes in these parameters but not changed qualitatively. At a single pair of temperature and moisture content values throughout this wide range of conditions, seed survival curves can be described by a negative normal distribution of individual life spans. Some seeds will be more short-lived and others more long-lived than the average, but there is no evidence for systematic changes in the shape of the survival curve with storage conditions. This suggests that the relative longevity of individual seeds within these populations remains unchanged irrespective of the storage conditions under which they are kept. Hence the shortest-lived individuals under a single pair of storage conditions (ambient) will remain the shortest-lived individuals if moved to any other storage conditions (in the bank, say).

The selection pressure exerted during seed storage can be deduced from the seed viability equation to be for those genotypes which have the highest storage potential at the moment of collecting. The magnitude of this selection pressure is likely to be determined by: (i) whether the quantity of seed taken from each mother plant was similar or whether it reflected the seed available from each mother plant on the day of collecting; (ii) how great the differences were between the storage potentials of the seed lots taken from each mother plant; (iii) how high the storage potentials were of these seed lots; and (iv) how far seed viability is allowed to fall before rejuvenation is undertaken.

As has been mentioned earlier, barley and tomato seed storage potential varies with developmental age, and yet the rate of loss of viability between samples of different developmental age is the same when stored under similar conditions. A model can therefore be developed which reflects collecting practice by proposing that on the day of harvest each mother plant is different in development and so produces a seed lot which has different storage potential. The seed lot from each mother plant is then combined through normal seed collecting protocol into a single seed lot. Table 20.1 shows the expectation from theory for a seed lot drawn from individual plants producing seed of low storage potentials which were collected in equal amounts and then bulked together. Table 20.1 also shows the same model for seed lots taken at higher storage potentials, again in equal amounts. For all seed lots, the variation in storage potential of the sublots around the mean is the same, storage conditions are identical (so that the rates of loss of viability are the same) and the duration of storage is the same, as would be seed bank practice. In the low-storage potential lots the relative frequencies of surviving individuals from each mother plant are substantially different after storage from those which pertain at the time of collecting. However, this is not the case for the high potential lots.

Such a model seems well founded. Roos (1984) was able to show that observed experimental behaviour fitted the predictive model well when a synthetic population was constructed from beans of different initial viabilities (each with a different seed-coat colour). Through the use of the model, it is possible not only to predict the effects of higher and lower mean seed lot storage potentials on selection during ageing, but also to examine the effects of tolerating larger or smaller degrees of seed ageing following collecting on the selective deletion of genotypes, as well as the effects of collecting unequal seed amounts from each mother plant.

From Table 20.1, higher mean seed storage potentials will reduce to practically zero any deletion of genotypes with lower than average viability. However, at lower mean storage potentials, the same level of ageing will change the relative frequency of the surviving genotypes dramatically. At low mean levels of viability, this change in frequencies can be kept to acceptable limits by reducing the field ageing that is tolerated during expeditions. Clearly, the practical advice is to make every effort to collect seed lots at the highest possible initial mean storage potential.

The necessity of this advice in comparison with current practice would be easier to appreciate if the seed storage potentials of collections arriving at seed banks for storage were known. Regrettably, no values from actual collections are known. Experimental determinations have been made for very few species and vary greatly, from low values for seeds of *Malus domestica* taken directly from the fruits to moderate values for the dry dehiscent fruits of barley and wheat and high values for the desiccated berries of *Nicandra pysalodes*. Given so few data and such a large variation, there is no alternative but to base advice on the assumption that the relatively low storage potential (about 97.7%) found in a wide variety of apples by Dickie (1988) will be common and so only low levels of ageing acceptable.

Collecting unequal seed numbers from each mother plant can either increase or decrease the selective deletion of genotypes, depending on the initial viability of the seeds that are over-represented. If seeds with higher than average storage potential are over-represented, the relative deletion of the genotypes contributing the lower than average viable seed will be greater. If it is seed with lower than average viability which is overcollected, the effect will be less.

The genotypes that are vulnerable to deletion are clearly those that either directly or indirectly result in seed lots being contributed to the sample that are of lower than average viability. From the patterns of seed storage potential associated with maturation in the dry-fruited cereal barley, these would be those genotypes that matured much earlier or later than on the day of collecting. The more extreme their earliness or lateness, the more susceptible to deletion they would be. For the fleshy-fruited berry tomato, only the later-maturing genotypes would be at risk as within the fruit there is no subsequent fall in seed storage potential. This latter case is also more likely to be found in dry dehiscent and indehiscent-fruited species, where the seeds or fruits are dispersed as they mature and so are no longer available to be collected.

The preceding discussion suggests that genetic selection linked to seed maturity ought to be well known and commonplace in seed storage. Yet, when Roos and Rincker (1982) investigated selection among the four parental clones that make up the synthetic variety of *Dactylis glomerata* cv. Pennlate, no genetic shifts could be found following seed ageing and regeneration. This suggests that the four parental clones produce seed of substantially similar storage potentials. Two very different processes could produce such a result. The first involves all the clones sharing similar phenologies of seed maturation and fruit shedding. The second demands that consideration be given to the structure of inflorescences and the seed maturity patterns produced within them (Fig. 20.8). Clearly, any inflorescence reflects the temporal differences in the seeds' development in their spatial separation. For example, within a wheat spike seven weeks after anthesis, seed moisture contents differ from 30% in the apical grains to 60% in the basal grains, reflecting differences in stage of maturation (Wellington, 1956). Selective harvesting of individual seeds from different positions within inflorescences of different age could none the less result in collecting seeds of similar maturities (and hence storage potentials). Indeed, such selective harvesting will inadvertently result when mature seeds are shed from the plant and desiccation tolerance is achieved only late during seed maturation, as only seed with a more limited variation in storage potential will be available for collecting. In naturally occurring populations the variation in flowering time among the many inflorescences that can be borne on single plants may further 'indemnify' collectors who take random samples of the seed that is available on a single day from putting together samples where seed storage potential and maturation genotype are so linked that genetic selection occurs in subsequent storage.

Table 20.1. Genetic selection following seed collection, assuming differences in storage potential.

	Mother plant A	Mother plant B	Mother plant C[1]	Mother plant D	Mother plant E
Very high viability, high loss					
On collection					
Probit viability	4.0	4.5	5.0	5.5	6.0
% viability	99.99+	99.99+	99.99+	99.99+	99.99+
Relative frequency in survivors	1.0	1.0	1.0	1.0	1.0
After loss of viability (1 probit) in the field					
Probit viability	3.0	3.5	4.0	4.5	5.0
% viability	99.86	99.98	99.99+	99.99+	99.99+
Relative frequency in survivors	0.998	0.999	1.0	1.0	1.0
High viability, high loss					
On collection					
Probit viability	3.0	3.5	4.0	4.5	5.0
% viability	99.86	99.98	99.99+	99.99+	99.99+
Relative frequency in survivors	0.998	0.999	1.0	1.0	1.0
After loss of viability (1 probit) in the field					
Probit viability	2.0	2.5	3.0	3.5	4.0
% viability	97.7	99.38	99.86	99.98	99.99+
Relative frequency in survivors	0.98	0.99	1.0	1.0	1.0

Medium viability, high loss

On collection

Probit viability	2.0	2.5	3.0	3.5	4.0
% viability	97.7	99.38	99.86	99.98	99.99+
Relative frequency in survivors	0.98	0.99	1.0	1.0	1.0

After loss of viability (1 probit) in the field

Probit viability	1.0	1.5	2.0	2.5	3.0
% viability	84.1	93.3	97.7	99.38	99.86
Relative frequency in survivors	0.86	0.95	1.0	1.02	1.02

Low viability, high loss

On collection

Probit viability	1.0	1.5	2.0	2.5	3.0
% viability	84.1	93.3	97.7	99.38	99.86
Relative frequency in survivors	0.86	0.95	1.0	1.02	1.02

After loss of viability (1 probit) in the field

Probit viability	0.0	0.5	1.0	1.5	2.0
% viability	50.0	69.2	84.1	93.3	97.7
Relative frequency in survivors	0.59	0.82	1.0	1.11	1.16

Low viability, low loss

On collection

Probit viability	1.0	1.5	2.0	2.5	3.0
% viability	84.1	93.3	97.7	99.38	99.86
Relative frequency in survivors	0.86	0.95	1.0	1.02	1.02

After loss of viability (0.1 probit) in the field

Probit viability	0.9	1.4	1.9	2.4	2.9
% viability	81.6	91.9	97.1	99.18	99.81
Relative frequency in survivors	0.84	0.95	1.0	1.02	1.02

[1] Assuming equal seed numbers are taken from each plant, those values will also represent the values for the seed collection as a whole.

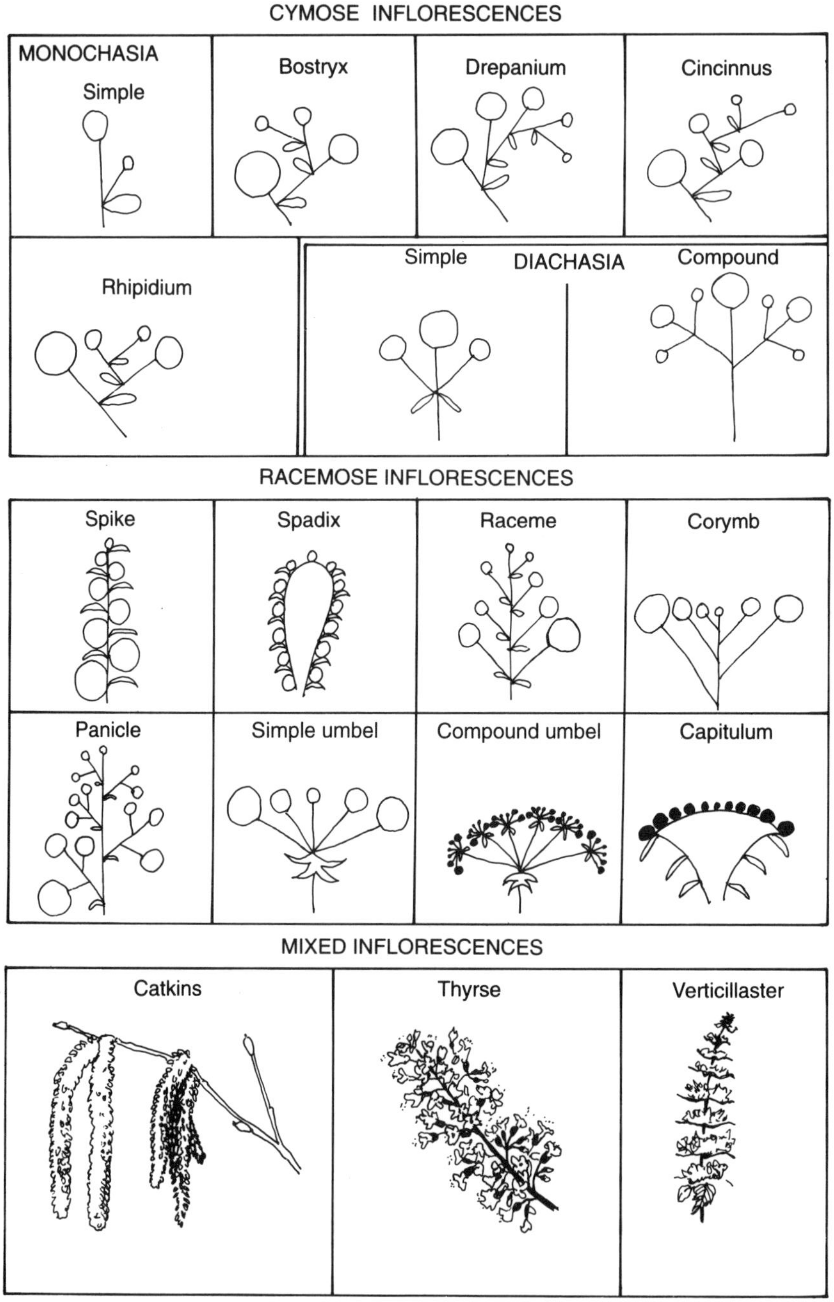

CYMOSE INFLORESCENCES

RACEMOSE INFLORESCENCES

MIXED INFLORESCENCES

Fig. 20.8. Types of inflorescence structures found among the Angiosperms and variation in seed maturity within each. The relative maturity is indicated by the size of the circle representing the flower (after Heywood, 1978).

In single plants of *Pastinaca sativa*, primary umbels are mature 10–14 days earlier than secondary umbels, which in turn mature 10–14 days earlier than tertiary umbels. For the whole population, seeds are dispersed naturally over a protracted period from August to October (Hendrix, 1984). For the weed *Echinochloa oryzicola*, the inflorescences, in this case panicles, borne on one plant varied by 11 days in their flowering dates and their glumous flower number varied from 27 to 493 per panicle. Within one of the earlier panicles to flower, seed ripened between 17 and 46 days after the first flowering of that panicle was observed. However, in the panicle that first flowered six days later, a similar delay was observed in the ripening of the earliest seeds compared with the earlier-flowering panicle. The latest seeds to ripen on the later panicle did so at much the same time as the latest seeds of the earlier panicle (Toshioka *et al.*, 1985).

The practical situations encountered by seed collectors in the field may therefore not be accurately reflected by the experimental situation reported in barley, where genetically homogeneous cultivars were used, and would be expected to flower and the seeds to mature over a very restricted period, while the plants themselves are grown at such densities that very few inflorescences are produced by each. In the same way, inflorescences possessing a more varied spread of flowering time than that of the barley spike may also reduce the peak in seed storage potential.

Overlying all these possibilities are the observations that pollination with pollen from different donors can result in the differential maturation of the fruits produced, while within individual plants ovules fertilized sooner are more likely to develop into seeds than those fertilized later, which could link male genotypes to seed maturation (Lee, 1988).

What is clear from this discussion is that the likelihood of genetic selection during storage will be unique for each seed lot, being an interaction between the phenology of the species in question, the maturity and hence storage potential of the seed populations on the day of collecting, the methods used by the collectors to assemble the sample and its subsequent storage. To minimize the risk of selective deletion of maturity types, collecting at least early and late during the period of seed dispersal may be necessary.

However, seed maturity need not be the only factor affecting seed storage potential. Different subspecies of rice have recently been reported to possess different seed storage characteristics (Ellis *et al.*, 1992), resulting in significantly different rates of loss of viability occurring under the same storage conditions. As all three subspecies were derived from a common ancestral stock, it is reasonable to assume that the differences in their seed storage behaviour also existed in the common ancestral stock and the selection for this character was passive. The trebling of seed oil content in *Zea mays* that can be achieved through 24 cycles of selection (Misevic and Alexander, 1989) is of similar concern. Here, either the differences in seed moisture contents which must result at the same equilibrium relative humidity will make a single moisture content term for the viability equation unacceptable or the relationship between equilibrium moisture content, oil content and relative humidity is not properly understood. Together, they draw into question the current assertion that, as all seed lots from the same species so far investigated experimentally respond in the same way to temperature and moisture content, the values of such viability constants apply to all genotypes within species. Clearly, this assertion now needs to be re-investigated over a wider range of the naturally occurring variation within species, both wild and cultivated.

The differences in storage behaviour of the rice subspecies under the same conditions have been modelled for seed lots of different storage potential as shown in Table 20.2. This model can also be used to illustrate the more general situation where seed storage characteristics vary between individuals in the population. As Table 20.2 shows, the practical solution of this problem is the same as for the variation in storage potential. The higher initial seed viability is at harvest, the less the separation will be of the different seed storage variants in the survivors for the same loss of viability. At lower initial viabilities the only technique that will reduce such a separation is to hold the loss of viability to relatively low levels. A similar model will apply to differing rates of viability loss resulting from the variation in individual seed moisture contents found within equilibrated rice (and so presumably all species) samples (Sieben-morgen et al., 1990). Whether this will exert a directed or random selection pressure remains unclear.

Accessions of genotypes of Zea mays have been ranked for their subsequent storability under constant conditions (Moreno Martinez et al., 1988). While different accessions of the same genotypes could be found in a variety of storage classes, suggesting environmental and harvesting effects, there were genotypes that regularly achieved higher and lower rankings. This work was followed up by Lozano et al. (1989), who showed these differences to be heritable and chose to seek direct differences in the seed's behaviour but found only slight supporting evidence. Reanaly-sis of their storage data shows that these differences can be just as validly explained by differences in storage potential. Genetically determined factors that affected subsequent seed drying rate, such as the tightness of the husks covering the cob and the thickness of the pericarp and testa could consistently affect the accumulation of damage within the seeds before they have dried to equilibrium. In turn, this would consistently affect seed storage potential. All such variants have been recorded among maize genotypes (Kang et al., 1986; Baker et al., 1991). Again, this would lead to selection pressures of the type described in Table 20.1.

Clearly, selection of some kind (though not for 'seed bankability') can occur as viability is lost. How frequent this is in practice remains unquantified. Consideration of the work of Scott (1981), who actively sought to select maize for improved response to accelerated ageing, suggests it may be a bigger problem in theory than in practice, though one that needs to be taken seriously none the less. Therefore, the earlier recommendation that viability losses should be restricted to 0.1 probit/ month in the field appears to be a reasonable practical balance. To allow more will undo much of the purpose of seed collecting.

Before leaving the problems of genetic selection, whether brought about by differences in seed maturity or genetically controlled differences in either seed storage behaviour or storage potential, one final point must be made. The advice offered here is of equal use to those who prefer to keep plant genetic resources *in situ* as plants, and collect seeds on demand as convenient disseminules, as it is to those who collect seed as convenient disseminules with which to build *ex situ* collections of germplasm resources, to both meet the demand and underwrite the continued existence of the population in the field. In both systems, the largest single annual loss of seed viability will occur between collecting the seeds and their receipt by either the user of an *in situ* resource or the *ex situ* collection. This will also be the period in which the greatest genetic selection is most likely. It is peculiar that genetic selection is thought of as an exclusive problem of *ex situ* conservation and that research is focused on the locations where annualized rates of selection should be lowest (i.e. in gene banks) and explanations sought based on inherent seed characteristics over which collectors can have no control.

Conclusion

In summary, what practical advice can be given to collectors? Not perhaps as much as one would have hoped. It is clear that the longevity of samples in the seed bank is determined by their quality at the moment of banking. However, there are very few direct investigations into the effects and interactions of seed development, seed maturation, subsequent storage in transit and cleaning under field conditions on subsequent long-term storage. Most of what is presented here would easily fit Albert von Szent-Györgi's conclusion that 'the basic texture of research consists of dreams into which the threads of measurement and reasoning are woven'. Only here seed collectors find themselves a little short of measurement and a little overdependent on dreams and reasoning. Yet again invoking the UN Conference on Environment and Development (UNCED) imperative to act rather than anguish, the advice is as follows.

All seeds

For all seeds:

- Attempt to collect equal numbers of seeds from each plant sampled, and at the same maturity, ideally when seed storage potential/ desiccation tolerance is highest. Avoid damaged seeds (mechanical damage, pest attack).

Table 20.2. Genetic selection in *Oryza sativa* subspecies following seed collection, assuming differences in storage characteristics.

		ssp. *indica*	ssp. *japonica*	ssp. *javanica*	Collection
Very high viability, high loss					
On collection	Probit viability	5.0	5.0	5.0	5.0
	% viability	99.99+	99.99+	99.99+	
	Relative frequency in survivors	1.0	1.0	1.0	
After loss of viability (1 probit) in the field	Probit viability	4.37	3.41	4.27	4.0
	% viability	99.99+	99.99+	99.99+	
	Relative frequency in survivors	1.0	1.0	1.0	
High viability, high loss					
On collection	Probit viability	4.0	4.0	4.0	4.0
	% viability	99.99+	99.99+	99.99+	
	Relative frequency in survivors	1.0	0.99	1.0	
After loss of viability (1 probit) in the field	Probit viability	3.37	2.41	3.27	3.0
	% viability	99.96	99.2	99.95	
	Relative frequency in survivors	1.0	0.99	1.0	
Medium viability, high loss					
On collection	Probit viability	4.0	4.0	4.0	4.0
	% viability	99.99+	99.99+	99.99+	
	Relative frequency in survivors	1.0	1.0	1.0	
After loss of viability (1 probit) in the field	Probit viability	3.37	2.41	3.27	3.0
	% viability	99.96	99.2	99.95	
	Relative frequency in survivors	1.0	0.99	1.0	

Medium viability, low loss				
On collection				
Probit viability	3.0	3.0	3.0	3.0
% viability	99.86	99.86	99.86	
Relative frequency in survivors	1.0	1.0	1.0	
After loss of viability (1 probit) in the field				
Probit viability	2.37	1.41	2.27	2.0
% viability	99.91	92.1	98.84	
Relative frequency in survivors	1.0	0.92	0.99	
Low viability, high loss				
On collection				
Probit viability	2.0	2.0	2.0	2.0
% viability	97.7	97.7	97.7	
Relative frequency in survivors	1.0	1.0	1.0	
After loss of viability (1 probit) in the field				
Probit viability	1.37	0.41	1.27	1.0
% viability	91.5	65.9	89.8	
Relative frequency in survivors	1.0	0.72	0.98	
Low viability, low loss				
On collection				
Probit viability	2.0	2.0	2.0	2.0
% viability	97.7	97.7	97.7	
Relative frequency in survivors	1.0	1.0	1.0	
After loss of viability (0.1 probit) in the field				
Probit viability	1.94	1.84	1.93	1.9
% viability	97.4	96.7	97.3	
Relative frequency in survivors	1.0	0.99	1.0	

- If seeds must be cleaned during the trip, do so by hand to minimize the chance of mechanical damage.
- If it is possible to avoid quarantine seed treatments without breaking quarantine regulations, for example through post-entry quarantine, do so.
- Personally ensure that seed arrives at the seed bank without undue delay.

Desiccation-intolerant seeds

For desiccation-intolerant seeds, which are more likely to be large seeds from dominant trees growing under wet conditions:

- Collect close to fruit fall. Do not collect from the ground unless you can be sure seeds are only recently dispersed.
- Keep seeds aerated and moist in inflated polythene bags, changing the air at least weekly by deflation and reinflation.
- Do not allow such seeds collected in the tropics to either cool below 20°C or heat up above ambient shade temperatures in the field or during transport.
- Plan your activities so that no more than one month elapses between collecting and reception by the seed bank.

Desiccation-tolerant seeds or their fruits

For desiccation-tolerant seeds or their fruits, which are more likely to be small seeds from herbs growing under dry conditions:

- If meteorological data suggest that more than 0.1 probit/month will be lost during the collecting trip, either modify your itinerary or prepare to dry actively with silica gel.
- For fleshy-fruited species, if logistically possible keep the seeds in the fruits and the fruits aerated and at ambient temperatures.
- If the above is not logistically possible, and for fruits that are dry dehiscent or indehiscent, air-dry the hand-cleaned seeds in a thin layer (to ensure aeration) under shade for three days or more (larger seeds need longer) to reduce the seed moisture content towards equilibrium with ambient relative humidities before packing, to use space more efficiently.

References

Aldridge, C.D. and R.J. Probert (1993) Seed development, the accumulation of abscisic acid and desiccation tolerance with aquatic grasses *Porteresia coarctata* (Roxb.) Tateoka and *Oryza sativa L. Seed Science Research* 3:97–103.

Bacchi, O. (1956) Novos ensaios sôbre a seca da semente de café ao sol. *Bragantia* 15:83–91.

Baker, K.D., M.R. Paulsen and J. van Zweden (1991) Hybrid and drying rate effects on seed corn viability. *Transactions of the American Society of Agricultural Engineers* 34:499–506.

Cromarty, A.S., R.H. Ellis and E.H. Roberts (1985) *Design of Seed Storage Facilities for Genetic Conservation.* IBPGR, Rome.

Demir, I. and R.H. Ellis (1992) Changes in seed quality during seed development and maturation in tomato. *Seed Science Research* 2:81–87.

Dickie, J.B. (1988) Prospects for the long-term storage of apple seeds. *Veröffentlichungen Landwirtschaftlich-chemische Bundesanstalt Linz* 19:47–63.

Dickie, J.B., R.H. Ellis, H.L. Kraak, K. Ryder and P.B. Tompsett (1990) Temperature and seed storage longevity. *Annals of Botany* 65:197–204.

Dickie, J.B., K. May, S.V.A. Morris and S.E. Titley (1991) The effects of desiccation on seed survival in *Acer platanoides* L. and *Acer pseudoplatanus* L. *Seed Science Research* 1:149–162.

Dourado, A.M. and E.H. Roberts (1984) Phenotypic mutations induced during storage in barley and pea seeds. *Annals of Botany* 54:781–790.

Earle, F.R. and Q. Jones (1962) Analyses of seed samples from 113 plant families. *Economic Botany* 16:221–250.

Ellis, R.H. and E.H. Roberts (1981) An investigation into the possible effects of ripeness and repeated threshing on barley seed longevity under six different storage environments. *Annals of Botany* 48:93–96.

Ellis, R.H., T.D. Hong and E.H. Roberts (1986) Logarithmic relationship between moisture content and longevity in sesame seeds. *Annals of Botany* 57:499–503.

Ellis, R.H., T.D. Hong and E.H. Roberts (1990). An intermediate category of seed storage behaviour? 1. Coffee. *Journal of Experimental Botany* 41:1167–1174.

Ellis, R.H., T.D. Hong and E.H. Roberts (1991) An intermediate category of seed storage behaviour? 2. Effects of provenance, immaturity and imbibition on desiccation tolerance in coffee. *Journal of Experimental Botany* 42:653–657.

Ellis, R.H., T.D. Hong and E.H. Roberts (1992) The low-moisture-content limit to the negative logarithmic relation between seed longevity and moisture content in three sub-species of rice. *Annals of Botany* 69:53–58.

Farrant, J.M., N.W. Pammenter and P. Berjak (1986) The increasing desiccation sensitivity of recalcitrant *Avicennia marina* seeds with storage time. *Physiologia Plantarum* 67:291–298.

Finch-Savage, W.E. (1992) Seed development in the recalcitrant species *Quercus robur* L.: germinability and desiccation tolerance. *Seed Science Research* 2:17–22.

Grout, B.W.W., K. Shelton and H.W. Pritchard (1983) Orthodox behaviour of oil palm seed and cryopreservation of the excised embryo for genetic conservation. *Annals of Botany* 52:381–384.

Hendrix, S.D. (1984) Variation in seed weight and its effects on germination in *Pastinaca sativa* L. (Umbelliferae). *American Journal of Botany* 71:795–802.

Heywood, V.H. (1978) *Flowering Plants of the World.* Oxford University Press, Oxford.

Hong, T.D. and R.H. Ellis (1992) Development of desiccation tolerance in Norway maple (*Acer platanoides* L.) seeds during maturation drying. *Seed Science Research* 2:169–172.

Ibarra-Manriquez, G. and K. Oyama (1992) Ecological correlates of reproductive traits of Mexican rainforest trees. *American Journal of Botany* 79:383–394.

Jurado, E., M. Westoby and D. Nelson (1991) Diaspore weight, dispersal, growth form and perenniality of central Australian plants. *Journal of Ecology* 79:811–830.

Kang, M.S., M.S. Zuber, T.R. Colbert and R.D. Horrocks (1986) Effects of certain agronomic traits on and relationship between rates of grain moisture reduction and grainfill during the filling period in maize. *Field Crops Research* 14:339–347.

Kermode, A.R. and J.D. Bewley (1985) The role of maturation drying in the transition from seed development to germination. 1. Acquisition of desiccation tolerance and germinability during development of *Ricinus communis* L. seeds. *Journal of Experimental Botany* 36:1906–1915.

Klein, L.A. and J.E. Harmond (1971) Seed moisture – a harvest timing index for maximum yields. *Transactions of the American Society of Agricultural Engineers* 14: 124–126.

Komatsu, T., N. Shimizu and S. Susuki (1979) [Studies on development and ripening of seeds in temperate grasses. 1. Seed development and ripening of Italian rye grass, and changes in germination and seeding vigour during seed development and ripening.] *Bulletin of the National Grassland Research Institute* 15:59–69.

Kovach, D.A. and K.J. Bradford (1992) Imbibitional damage and desiccation tolerance in wild rice (*Zizania palustris*) seeds. *Journal of Experimental Botany* 43:747–758.

Kraak, H.L. and J. Vos (1987) Seed viability constants for lettuce. *Annals of Botany* 59:353–359.

Lee, T.D. (1988) Patterns of fruit and seed production. In: Doust, J.L. and L.L. Doust (eds) *Plant Reproductive Ecology: Patterns and Strategies.* pp. 179–202. Oxford University Press, Oxford.

Lenné, J.M. and D. Wood (1991) Plant diseases and the use of wild germplasm. *Annual Review of Phytopathology* 29:35–63.

Lozano, J.L., S.H. Wettlaufer and A.C. Leopold (1989) Polyamine content related to seed storage performance in *Zea mays*. *Journal of Experimental Botany* 40:1337–1340.

Misevic, D. and D.E. Alexander (1989) Twenty-four cycles of phenotypic recurrent selection for percent oil in maize. 1. *Per se* and test cross performance. *Crop Science* 29:320–324.

Moreno Martinez, E., J. Ramirez Gonzalez, M. Mendoza Ramirez and G. Valencia Ramirez (1988) Comparison of Mexican maize races stored under adverse humidity and temperature. In: *Recent Advances in the Conservation and Utilization of Genetic Resources*. Proceedings of the Global Maize Germplasm Workshop. pp. 94–98. CIMMYT, Mexico City.

Muenscher, W.C. (1936) Storage and germination of seeds of aquatic plants. *Cornell University Agricultural Experimental Station Bulletin* 652:1–17.

Petruzelli, L. (1986) Wheat viability at high moisture content under hermetic and aerobic storage conditions. *Annals of Botany* 58:259–265.

Pieta Filho, C. and R.H. Ellis (1991) The development of seed quality in spring barley in four environments. I. Germination and longevity. *Seed Science Research* 1:163–177.

Prendergast, H.D.V., S.H. Linington and R.D. Smith (1992) The Kew Seed Bank and the collection, storage and utilisation of arid and semi arid zone grasses. In: Chapman, G.P. (ed.) *Desertified Grasslands: Their Biology and Management.* Academic Press, London.

Pritchard, H.W. (1991) Water potential and embryonic axis viability in recalcitrant seeds of *Quercus rubra*. *Annals of Botany* 67:43–49.

Probert, R.J. and E.R. Brierley (1989) Desiccation intolerance in seeds of *Zizania palustris* is not related to the developmental age or the duration of post harvest storage. *Annals of Botany* 64:669–674.

Probert, R.J. and P.L. Longley (1989) Recalcitrant seed storage physiology in three aquatic grasses (*Zizania palustris, Spartina anglica* and *Porteresia coarctata*). *Annals of Botany* 63:53–63.

Roberts, E.H. and R.H. Ellis (1989) Water and seed survival. *Annals of Botany* 63:39–52.

Roos, E.E. (1984) Genetic shifts in mixed bean populations. I. Storage effects. *Crop Science* 24:240–244.

Roos, E.E. and C.M. Rincker (1982) Genetic stability in 'Pennlate' orchard-grass seed following artificial aging. *Crop Science* 22:611–613.

Saini, S.K., J.N. Singh and P.C. Gupta (1982) Effect of threshing method on seed quality of soybean. *Seed Research* 10:133–138.

Scott, G.E. (1981) Improvement for accelerated aging response of seed in maize populations. *Crop Science* 21:41–43.

Siebenmorgen, T.J., M.M. Banaszek and M.F. Kocher (1990) Kernel moisture content variation in equilibrated rice samples. *Transactions of the American Society of Agricultural Engineers* 33:1979–1983.

Smith, R.D. (1984) The influence of collecting, harvesting and processing on the viability of seed. In: Dickie, J.B., S.H. Linington and J.T. Williams (eds) *Seed Management Techniques for Genebanks*. pp. 42–87. IBPGR, Rome.

Smith, R.D. (1985) Maintaining viability during collecting expeditions. In: IBPGR Advisory Committee on Seed Storage (ed.) *Report of Third Meeting*. pp. 13–20. IBPGR, Rome.

Smith, R.D. (1992) Seed storage temperature and relative humidity. *Seed Science Research* 2:113–116.

Suszka, B. and T. Tylkowski (1980) Storage of acorns of the English oak (*Quercus robur* L.) over 1–5 winters. *Arboretum Kornickie* 25:199–229.

Tompsett, P.B. (1982) The effect of desiccation on the longevity of seeds of *Araucaria hunsteinii* and *A. cunninghamii*. *Annals of Botany* 50:693–704.

Tompsett, P.B. (1983) The influence of gaseous environment of the storage life of *Araucaria hunsteinii* seed. *Annals of Botany* 52:229–237.

Tompsett, P.B. (1984) Desiccation studies in relation to the storage of *Araucaria*. *Annals of Applied Biology* 105:551–586.

Tompsett, P.B. (1986) The effect of temperature and moisture content on the longevity of seed of *Ulmus carpinifolia* and *Terminalia brassii*. *Annals of Botany* 57:875–883.

Tompsett, P.B. (1988) A review of the literature on storage of dipterocarp seeds. In: *Proceedings of the IUFRO Symposium on Forest Seed Problems in Africa*. pp. 348–365. Swedish University of Agricultural Sciences, Umea.

Tompsett, P.B. and H.W. Pritchard (1993) Water status changes during development in relation to the germination and desiccation tolerance of *Aesculus hippocastanum* L. seeds. *Annals of Botany* 71:107–116.

Toshioka, T., Y. Yamasue and K. Ueki (1985) [Seed ecological studies in relation to the asynchronous emergence of *Echinoctilea oryzicola* Vasing. 1. Variations in weight and ripening date of seeds among setting positions within one plant.] *Weed Research Japan* 30:58–64.

Tukey, H.B. (1934) Growth of the embryo, seed and pericarp of the sour cherry (*Prunus cerasus*) in relation to season of fruit ripening. *Proceedings of the American Society for Horticultural Science* 31:125–144.

Vossen, H.A.M. van der (1979) Methods of preserving the viability of coffee seed in storage. *Seed Science and Technology* 1:65–74.

Waller, D.M. (1988) Plant morphology and reproduction. In: Doust, J.L. and L.L. Doust (eds) *Plant Reproduction Ecology: Patterns and Strategies*. pp. 203–227. Oxford University Press, Oxford.

Wechsberg, G.E., C.M. Bray and R.J. Probert (1993) The development of longevity and response to priming of *Ranunculus sceleratus* seeds In: Côme, D. and F. Corbineau (eds) *4th International Workshop on Seeds: Basic and Applied Aspects of Seed Biology*. Vol. 3, pp. 845–850. ASFIS, Paris.

Wellington, P.S. (1956) Studies on the germination of wheat grains in the ear during development, ripening and after-ripening. *Annals of Botany* 20:105–120.

Yaklich, R.W. and P.B. Cregan (1987) A field study of moisture content of soybean pods and seeds after harvest maturity. *Journal of Seed Technology* 11:62–68.

Appendix 20.1
Collecting fern spores

L. Guarino

Roos and Verduyn (1989) discuss how to collect fern spores. The optimal time for collecting spores is when the sporangia are ripe or nearly ripe. The frond from which spores are to be collected should be placed in a paper pouch or envelope. If kept under dry conditions, most of the spores will have been shed after a couple of days, and the frond may be discarded. Most fern species have spores that are yellow or dark. These remain viable for considerable periods if kept dry and cool. In contrast, green spores (e.g. *Osmunda, Grammitis*) maintain their viability for only a few hours to a few days, and must therefore be sown as soon as possible after collecting.

References

Roos, M.C. and G.P. Verduyn (1989) Collecting and germinating fern spores. In: Campbell, D.G. and H.D. Hammond (eds) *Floristic Inventory of Tropical Countries.* pp. 467–468. New York Botanical Garden, New York.

Collecting vegetatively propagated crops (especially roots and tubers) 21

Z. Huaman[1], F. de la Puente[2] and C. Arbizu[2]

[1]Department of Genetic Resources, CIP, PO Box 5969, Lima, Peru:
[2]CIP, PO Box 1558, Lima, Peru.

Introduction

The most important root and tuber crops cultivated on a worldwide scale are potato (*Solanum tuberosum*), cassava (*Manihot esculenta*), sweet potato (*Ipomoea batatas*), yams (*Dioscorea* spp.) and taro (*Colocasia esculenta*), but there are many others that are of regional, national or local importance, in a total of over a dozen dicot and monocot families. Most of these originated in tropical or subtropical areas and are mainly used as sources of carbohydrates. Many minor root and tuber crops, such as turmeric (*Curcuma longa*) and arrowroot (*Maranta arundinacea, Tacca leontopedaloides*), are also used in folk medicine and as spices (Sastrapradja *et al.*, 1981). All these crops are vegetatively propagated. There are also vegetatively (or clonally) propagated crops that are not roots or tubers, e.g. bananas and sugarcane. A selection of the more important is given in Table 21.1.

Changes in the environment and in agricultural practices can cause particularly serious genetic losses in vegetatively propagated crops. This is because many set seeds only sparingly or not at all. Their capacity for adaptation is therefore lower than that of sexually propagated species, especially outcrossers, in which genetic recombination is much faster. Clonally propagated crops are also threatened by the accumulation of viruses in propagules (Lebot, 1992). It is therefore all the more important to conserve such species *ex situ* under controlled conditions. Among the Consultative Group on International Agricultural Research (CGIAR) commodity centres, the Centro Internacional de Agricultura Tropical (CIAT) maintains collections of cassava, the Centro Internacional de la Papa (CIP) potato and sweet potato, the International Institute of Tropical Agriculture (IITA) cassava and yams and the International Network for the Improvement of Banana and Plantain

Table 21.1. Selected vegetatively propagated herbaceous crops and the plant parts (●) suitable for germplasm collecting (Purseglove, 1968, 1972; Cardenas, 1969; Sastrapradja *et al.*, 1981; Tootill, 1984; Leon and Withers, 1986; National Research Council, 1989).

	Stools	Rhizomes	Bulbs	Offsets	Shoots	Suckers	Tubers	Stem cuttings	Bulbils	Stolons	Storage roots
Acorus calamus	●	●	–	–	–	–	–	–	–	–	–
Allium sativum	–	–	●	–	–	–	–	–	●	–	–
Ananas comosus	–	–	–	–	●	●	–	–	–	–	–
Agave spp.	–	–	–	–	–	●	–	–	●	–	–
Arracacia xanthorrhiza	–	–	–	●	●	–	–	–	–	–	–
Asparagus officinalis	–	●	–	–	–	–	–	–	–	–	–
Canna edulis	–	●	–	–	–	–	●	–	–	–	–
Colocasia esculenta	–	–	–	–	–	●	●	–	–	–	–
C. gigantea	–	–	–	–	–	●	–	–	–	–	–
Curcuma longa	–	●	–	–	–	–	–	–	–	–	–
Dioscorea spp.	–	–	–	–	–	–	●	–	–	–	–
D. bulbifera	–	–	–	–	–	–	–	–	●	–	–
D. esculenta	–	–	–	–	–	–	–	●	–	–	–
Eleocharis dulcis	–	–	–	–	–	–	–	–	–	●	–
Fragaria vesca	–	–	–	–	–	–	–	–	–	●	–
Hedychium coronarium	–	●	–	–	–	–	–	–	–	–	–
Ipomoea batatas	–	–	–	–	–	–	–	●	–	–	●
Lepidium meyenii	–	–	–	–	●	–	–	–	–	–	–
Manihot esculenta	–	–	–	–	–	–	–	●	–	–	–
Maranta arundinacea	–	–	–	–	–	●	–	–	–	–	–
Mirabilis expansa	–	–	–	●	–	–	–	●	–	–	●
Musa spp.	–	–	–	–	–	●	–	–	–	–	–
Oxalis tuberosa	–	–	–	–	–	–	●	●	–	–	–
Pachyrrhizus ahipa	–	–	–	–	–	–	●	–	–	–	–
Polymnia sonchifolia	–	–	–	●	–	–	–	●	–	–	–
Saccharum ssp.	–	–	–	–	–	–	–	●	–	–	–
Sansevieria ssp.	–	●	–	–	–	–	–	–	–	–	–
Solanum spp.	–	–	–	–	–	–	●	●	–	–	–
Tacca leontopedaloides	–	●	–	–	–	–	–	–	–	–	–
Tropaeolum tuberosum	–	–	–	–	–	–	●	–	–	–	–
Ullucus tuberosus	–	–	–	–	–	–	●	●	–	–	–
Xanthosoma nigrum	–	–	–	–	–	●	–	–	–	–	–
Vanilla planifolia	–	–	–	–	–	–	–	●	–	–	–
Zingiber officinale	–	●	–	–	–	–	–	–	–	–	–

(INIBAP) bananas and plantains. Collections of vegetatively propagated crops, whether in international centres or national programmes, are commonly stored in field gene banks or *in vitro*.

It is usually necessary to collect vegetative material of clonally propagated crops, normally the farmer's planting material. Even when possible, seed collecting is not necessarily desirable, as it may result in the loss of the specific gene combinations that make a clonally propagated variety distinctive. This presents several difficulties at different stages of collecting (Hawkes, 1975, 1980; Ford-Lloyd and Jackson, 1986), but all can be overcome by careful planning and appropriate techniques (Table 21.2).

Planning a collecting expedition

A particularly important aspect of the planning of expeditions to collect vegetatively propagated crops such as roots and tubers will be determining the most appropriate timing of the visit. Some of these crops, and especially their wild relatives, also produce seeds, though often only very few. An early decision that will therefore have to be made is thus whether the aim is to collect seed samples or vegetative material. The timing may be quite different. The decision will depend on:

*   *The breeding system of the species.* Autogamous species can be collected as seeds without breaking up gene combinations.
*   *Whether a population structure exists.* This will not be the case for traditional varieties clonally propagated for centuries but may still obtain for sexually fertile, outcrossing wild species.
*   *The purpose of the collecting.* Each seed in a collection is potentially a new variety (farmers routinely use such open-pollinated seeds to increase the diversity of their fields), whereas each vegetative sample of a crop will represent an established variety. The former will be useful if the germplasm is to flow directly into a breeding programme, for example, the latter if the purpose of collecting is the documentation of present varietal diversity or the testing of adaptation of existing landraces.

In most cases, the vegetative parts that are collected for genetic resources conservation will be those that are used by farmers to propagate the crop. If this is also the economically useful product, the harvest season will be the appropriate time for collecting. Collecting storage organs which are immature or which are already sprouting may result in significant loss of viability during transport. Because a certain amount of precision is therefore needed and because the harvest season may vary considerably both between areas and between years, local information is vital at the planning stage in deciding on the best time to collect, as indeed on which areas to visit. Of course, at the best time for tuber collecting in such crops as potatoes (i.e. during dormancy),

Table 21.2. Solutions to problems at different stages of collecting vegetative material.

Stage	Problem	Solution
Searching	As underground structures, roots and tubers can be difficult to find and laborious to collect. They must be collected mature, which usually means when above-ground parts are dead	Use local information to time and direct the collecting. Time collecting when above-ground parts are just dying down. If resources allow, mark sites during a preliminary visit and revisit
Sampling	Genetic variation is difficult to estimate: above-ground parts will have died off and the material to be examined will have to be dug up. Without careful examination of the available variation, there may be excessive sampling of some clones, and some may be missed altogether	For wild species, spend sometime investigating the average extent of clones, and then avoid taking samples too close to each other at any given site. For crops, consult farmers for a thorough account of the variation available locally and collect in selective fashion in consultation with them
Transport	Planting material is often bulky and would thus be awkward to transport even if no great care were necessary to keep it viable, which is often not the case. The risk of moving pests along with the germplasm is greater than with seed	Obtain information on quarantine risks and ways to minimize them. Collect sound, mature material and take simple precautions for storage during transport, keeping transit time to a minimum. Alternatively, collect *in vitro* or transit material through an intermediate quarantine centre, taking *in vitro* samples there. Follow *FAO/IPGRI Technical Guidelines for the Safe Movement of Germplasm*

there will be no above-ground parts visible. This is perhaps not much of a problem for crops, but becomes serious in the case of wild relatives, demanding either a two-stage collecting effort or, as suggested by Hawkes (1975), a single visit when the plants are only just dying down.

There are many different types of plant material that can be used to propagate crops vegetatively. Definitions and examples may therefore be helpful, and some are given in Box 21.1 (Purseglove, 1972; Tootill, 1984; Bell, 1991).

Table 21.1 shows what parts of a selection of vegetatively propagated herbaceous crops may be collected for germplasm conservation. Woody species are considered separately in Chapter 23. In some cases, the collector may have alternatives and a choice will have to be made. One important distinction is between perennating organs like bulbs and tubers and growing parts like stem cuttings and suckers, which are generally more difficult to keep alive. Perhaps the main choice to be made in this context, however, is that between collecting the entire organ or removing meristematic tissue and maintaining this *in vitro*. Though this alleviates the problem of transporting large quantities of bulky material,

> **Box 21.1**
> **Some definitions of vegetative propagating structures**
>
> - **Bulb** (e.g. *Allium*) – A short, fleshy, underground shoot, mostly made up of colourless, swollen leaf bases and scale leaves enclosing next year's bud. A **bulbel** or **bulbet** is a small bulb arising from a parent bulb.
> - **Bulbil** – A small bulb arising in an axil on the aerial part of a plant.
> - **Corm** (e.g. *Crocus, Gladiolus*) – A short, swollen, annual underground stem of several nodes and internodes, with that of the next year forming above it at the base of each flowering stem. A **cormel** is a corm arising vegetatively from a parent corm.
> - **Knob** (e.g. *Amorphophallus campanulatus*) – Sprout from the main tuber.
> - **Offset** (e.g. *Sempervivum tectorum*) – Small plantlet taken from the base of the above-ground part of the main stem.
> - **Rhizome** (e.g. *Acorus, Zingiber*) – An underground stem, usually horizontal and often elongated, that is distinguished from a root by the presence of nodes and internodes and/or the presence of buds in the axils of reduced, scale-like leaves.
> - **Shoot** (e.g. *Ananas*) – The apical or lateral meristematic dome together with the leaf primordia, emerging leaves and subadjacent stem tissue.
> - **Stem cutting** (e.g. *Ipomoea batatas*) – Section of a mature stem with several leaves and their axillary buds.
> - **Stolon** or **runner** (e.g. *Fragaria vesca*) – A trailing stem, capable of forming roots, shoots or both (and hence a new plant) from its axillary buds.
> - **Stool** (e.g. *Acorus*) – A clump of roots or rootstocks (short, erect underground stems, like vertical rhizomes).
> - **Storage root** or **root tuber** (e.g. *Mirabilis expansa*) – A fleshy, secondarily thickened root, without buds (except sometimes adventitious buds giving rise to suckers).
> - **Sucker** (e.g. *Agave*) – A shoot arising adventitiously from a root, often at some distance from the main plant.
> - **Tuber** (e.g. *Colocasia, Solanum*) – A swollen, subterranean stem, with numerous buds or 'eyes'.

and coupled with disease indexing may also overcome quarantine risks, it poses technical and logistical questions that need to be addressed at an early stage (Chapter 24).

Sampling procedures

Wild populations of root and tuber crop relatives are likely to be different from those of autogamous species not producing such structures only in so far as single clones may be quite extensive. Sampling vegetative material of the former will therefore not be any different from sampling seeds of the latter, except that care will have to be taken not to sample large clones excessively. It will also not be possible to collect as many propagules per plant. Hawkes (1980) suggests collecting as a bulk sample two to four vegetative propagules from each of 10–15 randomly chosen individual clones (more if time allows, up to 25; fewer if the propagules are very large). Individual clones may be identified by their appearance or duplication at any given site minimized by taking samples not too close together, based on an initial assessment of the likely average extent of clones. The latter approach may be taken to an extreme by collecting from only one or two plants per site but visiting many more sites. A bulk seed sample should also be taken, when possible, and given a separate collecting number.

In contrast to the situation in the wild, a field of a root or tuber crop will probably not display a population structure at all. There will be a mixture of many different genotypes in a traditional farmer's field (e.g. Jackson *et al.*, 1980), each being the result of intensive selection by farmers over many generations. Random sampling of such a field, the usual method for sexually reproducing species, will not be appropriate, as it will over-represent abundant clones at the expense of rare ones. The recommended procedure is selective sampling, i.e. to identify all 'morpho-types' in a field or market and collect two to four propagules of each one, repeating the process at each sampling site. Duplicates between sites can be identified later, when the material is grown out. Ford-Lloyd and Jackson (1986) define the morphotype as being composed of plants which are apparently phenotypically identical, without implying genetic iden-tity. If it is possible to collect seed, numbering of samples should be as follows (Hawkes, 1980): give the same collecting number if seeds are taken from the same plant(s) from which a vegetative sample has been taken; give a separate collecting number if the seeds are taken from more than one randomly chosen morphotype and bulked, the recommended sampling procedure for outcrossers.

In crops like potatoes, oca, ulloco and mashua, phenotypic variation in the tubers is high, which means that selecting potentially different cultivars growing in the same field is relatively easy. However, in many root and tuber crops, visible phenotypic differences among cultivars are few but one can find significant genetic variation when other, more cryptic, characters are considered. Thus, for example, morphologically similar cultivars may have different isozyme alleles at some loci, or display different reactions to biotic or abiotic stresses. In these cases, more samples need to be obtained, combining selective and random sampling techniques. Working in Irian Jaya, Prain *et al.* (Chapter 38, this volume) relied on the farmer to identify sweet potato duplicates within a plot and also avoided similar material with the same name in nearby plots; however, in more distant plots, and if there were any doubts, the material was collected.

Local knowledge is crucial to the sampling process, just as it is crucial in deciding when and where to sample in the first place. Farmers will be aware of the extent of variation in their field, village and district, i.e. the number of distinct cultivars available in a given area, their names, appearance and properties. Hawkes (1975, 1991) suggests that market areas should be the units within which sampling of morphotypes should be organized. Markets in towns serving a whole district will certainly be important sources of information on the varieties grown in the area, though not all varieties grown will be found on market stalls, some being kept solely for home consumption (Hawkes, 1991). On the basis of this kind of information, it may be feasible to attempt to collect all alleles occurring in the area, an ideal that is impossible to achieve in sexually reproducing crops. Chapter 38 describes how local knowledge has been brought to play during collecting of sweet potato in Irian Jaya.

Handling vegetative material in the field

Vegetative material can deteriorate rapidly after harvest. It is also easily damaged in transport. It must therefore be properly prepared and housed in suitable containers. Samples should also be free of insect damage and disease symptoms. If they consist of underground organs, they should be free of soil. Leon and Withers (1986) provide guidelines for the field handling of collected material of several vegetatively propagated crops (*Colocasia, Dioscorea, Hevea, Ipomoea, Manihot, Musa, Saccharum, Vanilla* and *Zingiber*).

Storage organs should be collected fully grown – dormant in crops like potatoes and yams. They should also be sound. Care should be taken to avoid mechanical damage when digging up underground parts with metal implements. Bruises can be cut out and treated with a mild alkali such as chalk, hydrated lime or wood ash, but in general damaged material should be avoided. If the tuber is too large, it may be possible to collect only a portion. In yams, this would be the head or proximal end, in taro and other aroids the crown of the tuber. Gentle drying in the sun can be followed by cleaning off excess soil and a fungicidal dusting or dip. Hawkes (1975) then advises wrapping tubers in a semipermeable material and keeping them under cool conditions, for example in insulated cool boxes. Expanded polystyrene containers covered with reflective mirror foil are suitable.

Storage organs that do not sprout soon after harvest generally require only to be placed individually in strong paper bags or newspaper and loosely accommodated in boxes with soft packing material. Material of some species is fragile and cannot survive if too many layers of samples are placed on top. In these cases, specially made crates are needed allowing the storage of single layers of samples in trays or shelves. There should be adequate ventilation and extremes of temperature should be avoided.

Growing tissues generally require more care. In general, stem cuttings should be taken from the middle portion of the stem and consist of several nodes. Leaf surface should be reduced to a minimum, but leaves should be snapped off at the junction with the stem rather than cut. Cut ends can be dipped in disinfectant to prevent rotting. Shoots, suckers and cuttings from fleshy stems keep reasonably well when the proximal end is wrapped in slightly wet paper towels or newspaper. They can then be stored in insulated cool boxes. Samples should be examined regularly to keep the wrapping paper moist and excise developing roots. Sweet potato stem cuttings are usually about 30 cm long and contain at least six nodes. The growing tip and large leaves are discarded. Cassava stem cuttings (called 'stakes', and usually 40–50 cm in length) are not so sappy and tender as those of species like sweet potato and can be simply wrapped in paper bags. In sugarcane, which is not a root or tuber crop but is clonally propagated from stem cuttings much like cassava, cut ends are sealed by dipping in hot paraffin after a short hot-water treatment (50°C for 30 minutes) (Coleman, 1986). Collecting vegetative material of *Musa* is described in detail in Chapter 33.

Fern perennating structures may be collected as live material following the same basic precautions as for those of flowering plants. However, aquatic ferns like *Azolla* spp., which are important nitrogen sources, may be collected as fronds. Watanabe *et al.* (1992) describe methods for collecting and transporting *Azolla*. They suggest gathering the fronds a handful at a time, squeezing out excess water and draining on toilet paper. The material may then be placed in Petri dishes or plastic bags, which should be sealed and maintained in a cool place until arrival at the laboratory. Under such conditions, samples may survive up to two weeks. *Azolla* material is conserved in the vegetative state in liquid medium. Watanabe *et al.* (1992) also give details of different ways of maintaining collections.

Chapter 24 describes the concept of *in vitro* collecting, gives general guidelines and provides some examples. Other examples involving root and tuber crops may be found in Chapter 32. Two further examples are worth mentioning here. The *in vitro* collecting method developed at CIAT for cassava (R. Chavez, pers. comm.) consists in taking actively growing vegetative buds or terminal stem cuttings from branches without flowering buds. Explants of 1.0–1.5 cm are immersed in 70% ethanol for 5–15 minutes and then surface-sterilized by immersion in a 0.5% solution of calcium hypochlorite for 5 minutes. Finally, they are rinsed with cool boiled water. Explants are inoculated into semisolid culture medium (MS or 4E) containing an antibiotic such as rifampicin in a small wick of filter paper. In contrast, the *in vitro* methods tested at CIP for collecting sweet potatoes have so far not produced high rates of survival of the cultures. A simple method that has been partially successful consists of taking cuttings containing one node with axillary buds; they are surface-sterilized and introduced into a test tube containing 1 ml of antibiotic solution (100 ml distilled water + 0.025 g streptomycin). Particularly high losses due to contamination have been noticed in sweet potatoes with thin or very pubescent stems.

After the mission

If there are transport and other delays, or if storage organs develop rots, it may be worthwhile planting out the material at some kind of way station in the field. Cuttings would be planted out as a holding operation, storage organs for conversion to stem cuttings.

FAO/IPGRI Technical Guidelines for the Safe Movement of Germplasm are available for several clonally propagated crops (Chapter 16). In general, they recommend shipment of collected material as sterile, pathogen-tested *in vitro* plantlets, if necessary via intermediate quarantine stations. Thus, cassava might be collected as stakes and transported in this way to an intermediate centre where the material would be planted out and meristem tips transferred to *in vitro* culture for eventual removal to the germplasm centre for indexing and storage in field and/or *in vitro* gene banks. Chapter 33 describes a similar procedure for *Musa*. If the material is not moved in this way, all samples collected should be inspected to eliminate organs or organ parts that are rotten or damaged, and further regular inspections will be necessary, with all infected material incinerated, as long as it is grown under quarantine conditions.

On arrival at its final destination samples of growing material will need to be planted out fairly quickly. Dormant tubers will need to be stored under appropriate conditions. *In vitro* plantlets will need to undergo propagation. All this requires that precise and timely preparations be made for receiving the material. This should be done at the planning stage if potentially damaging delays in dealing with the material are to be avoided.

References

Bell, A.D. (1991) *Plant Form.* Oxford University Press, Oxford.

Cardenas, M. (1969) *Manual de plantas economicas de Bolivia.* Imprenta ICTHUS, Cochabamba.

Coleman, R.E. (1986) Sugarcane. In: Leon, J. and L.A. Withers (eds) *Guidelines for Seed Exchange and Plant Introduction in Tropical Crops.* FAO Plant Production and Protection Paper 76. pp. 169–171. FAO, Rome.

Ford-Lloyd, B. and Jackson, M. (1986) *Plant Genetic Resources: An Introduction to Their Conservation and Use.* Edward Arnold, London.

Hawkes, J.G. (1975) Vegetatively propagated crops. In: Frankel, O.H. and J.G. Hawkes (eds) *Crop Genetic Resources for Today and Tomorrow.* pp. 117–121. Cambridge University Press, Cambridge.

Hawkes, J.G. (1980) *Crop Genetic Resources Field Collection Manual.* IBPGR and EUCARPIA, Rome.

Hawkes, J.G. (1991) Theory and practice of collecting germplasm in a centre of diversity. In: Engels, J.M.M., J.G. Hawkes and M. Worede. (eds) *Plant Genetic Resources of Ethiopia.* pp. 189–201. Cambridge University Press, Cambridge.

Jackson, M.T., J.G. Hawkes and P.R. Rowe (1980) An ethnobotanical field study of primitive potato varieties in Peru. *Euphytica* 29:107–113.

Lebot, V. (1992) Genetic vulnerability of Oceania's traditional crops. *Experimental Agriculture* 28:309–323.

Leon, J. and L.A. Withers (eds) (1986) *Guidelines for Seed Exchange and Plant Introduction in Tropical Crops*. FAO Plant Production and Protection Paper 76. FAO, Rome.

National Research Council (1989) *Lost Crops of the Incas*. National Academy Press, Washington DC.

Purseglove, J.W. (1968) *Tropical Crops. Dicotyledons*. Longman, London.

Purseglove, J.W. (1972) *Tropical Crops. Monocotyledons*. Longman, London.

Sastrapradja, S., N. Wilijarmi-Soetjipjo, S. Donimihardja and R. Soejono (1981) *Root and Tuber Crops*. IPBGR, Rome.

Tootill, E. (ed.) (1984) *Dictionary of Botany*. Penguin, London.

Watanabe, I., P.A. Roger, J.K. Ladha and C. van Hove (1992) *Biofertilizer Germplasm Collections at IRRI*. IRRI, Manila.

Collecting vegetative material of forage grasses and legumes

22

N.R. Sackville Hamilton and K.H. Chorlton
Plant Science Division,
IGER, Plas Gogerddan, Aberystwyth, Dyfed SY23 3EB, UK.

Introduction

In a number of situations it is preferable – and sometimes necessary – to collect germplasm as vegetative rather than seed samples. The decision whether to collect vegetative material or seeds will affect the genetic composition of the resulting collection in a way that depends on the population biology of the species. It will also affect both optimum sampling strategy and sampling procedure. Collectors must be aware of the consequences. They must match the choice of sample type both to the biology of the species and to the objectives of the collecting expedition, and must adapt sampling strategies and procedures accordingly.

This chapter outlines the factors that contribute to the decision to collect vegetative samples of forages and describes aspects of population structure that are likely to be associated with species chosen for vegetative sampling. It then considers the consequences – both the direct consequences of sampling vegetatively and the indirect consequences associated with population structure – and describes the procedures involved in planning and executing this kind of collecting. It also covers back-at-base issues in detail, because dealing with vegetative samples is very different from processing seeds. It is mainly outbreeding, clonal, perennial, natural and seminatural forage species that are dealt with. Indeed, the examples mostly involve *Trifolium repens* and other temperate species. However, the general concepts and conclusions are equally applicable to tropical forages, although much less is known about these, and exact guidelines are therefore more difficult to provide. In some cases, the concepts will also be applicable to other kinds of plants, for example vegetatively propagated crops and woody perennials. However, specific issues relevant to collecting vegetative material of these plants are discussed in Chapters 21 and 23 respectively.

Problems associated with collecting vegetative material

The collector of vegetative plant parts is faced with problems and constraints that are entirely different from, or at any rate more severe than, those faced by the seed collector. The main ones are as follows:

- *Taxonomic identification of material.* Most published keys and other identification aids rely almost exclusively on flower and fruit characters, though exceptions exist (e.g. Hubbard, 1968). If the target species are in the vegetative state, they may be particularly difficult to find and distinguish. It may be necessary for collectors to develop their own identification aids (Chapter 11).
- *Sample size.* The size of vegetative samples, especially if composed of rooted plants in soil divots, may restrict the total number of both individuals per population and samples that can be collected during an expedition. If it does, it will affect sampling strategy, as described below, and, to the extent that the amount of genetic variation sampled increases with sample size, may cause vegetative collecting to be less effective than seed collecting in acquiring the maximum diversity of alleles.
- *Sampling speed.* Collecting one vegetative unit usually requires more time and care than collecting one seed. This affects sampling strategy and, if time is limiting, may again cause vegetative collecting to be less effective than seed collecting.
- *Storage in transit.* In contrast to the seeds of most species, many types of vegetative material have a short life in storage. If arrangements cannot be made for each sample to be quickly returned to the gene bank to be processed while the expedition continues, an appropriate form of temporary storage must be arranged, such as a cool, moist atmosphere in a refrigerator in the collecting vehicle.
- *Quarantine.* It is usually necessary to place vegetative samples under stricter quarantine than seed samples, both because of the greater variety of pathogens and pests that are liable to be moved with vegetative material and in soil, and because cleaning procedures cannot be as effective as with seed.
- *Preparation of material for the gene bank.* In whatever form the samples are finally stored (as seeds, *in vitro*, as vegetative propagules or in field gene banks), accessions must be prepared from the collected material. The time taken to do this may actually be less than for seed samples, especially when seed dormancy has to be broken. In most cases in *Lolium* and *Trifolium*, seeds can be obtained from vegetative samples within one year.

Reasons for collecting vegetative material

Because of the extra difficulties associated with collecting vegetative samples, it is normally undertaken only when it is impossible to collect seeds or when collecting seeds introduces other problems. A range of situations in which collecting vegetative material will be an advantage are discussed below.

Vegetatively propagated crops

In some crops, the normal means of propagation is vegetative. The genetic variants produced by sexual reproduction may be of interest in a breeding programme, but the actual genotypes being grown by farmers will also need to be conserved. Collecting vegetatively propagated crops, with special reference to roots and tubers, is covered in Chapter 21.

Seed production unpredictable

In many perennial species, seed production is irregular or intermittent, varying in time and space and with genotype. Seed collecting may then be unsatisfactory, due to their poor quality and/or quantity. In many temperate pastures, for example, most plants produce no seeds in most years, largely as a result of agricultural practices. Hay and silage cuts are taken before the seeds ripen, removing immature inflorescences, because that is when feed quality is highest. In a well-grazed pasture, most inflorescences are eaten before maturity. Seeds mature in abundance only when the farmer manages a field to produce a seed crop. The absence of suitable insect pollinators can also depress seed set, as in the case of *T. repens* at the upper limits of its altitudinal range in the equatorial Andes (about 4600 m), where flowering is abundant but seed production virtually nil (pers. obs.).

Seeds mature at different times

In many populations of outbreeding species, seeds may mature over a long period of time. At any one time, only a small proportion of individuals may bear ripe seed. In temperate grasslands, for example, the seed-ripening season can extend throughout most of the growing season, with considerable differences among species. There may also be variation in seed-ripening among populations of a single species within a region.

Fast-shattering fruits

The fruit of some wild species shatter quickly after maturation, dispersing the seeds, which are therefore accessible to the collector for only a very brief period.

To avoid bias towards genotypes with ripe seeds

All of the above may introduce a genetic bias, in a seed sample, towards genotypes with ripe seeds at the time of collecting. For germplasm conservation this is unsatisfactory, and can be avoided by sampling vegetative material. For example, the two 'species' *Lolium multiflorum* and *L. perenne* are freely cross-compatible, producing fertile hybrids. In Europe, they form a hybrid swarm showing continuous variation between the two extreme parental types. The *L. multiflorum* extreme characteristically flowers throughout the year, whereas *L. perenne* has a more seasonal flowering period. A seed sample taken from a hybrid population between July and September is likely to contain more seeds of the *L. multiflorum* type and will therefore be biased towards that parent, whereas a vegetative sample will be unbiased.

To avoid bias towards large plants

Many perennial herbaceous plants are clonal. Some forms of vegetative propagation, such as stolons or rhizomes with an indeterminate growth habit and adventitious roots, can make a genetic individual, or clone, potentially immortal. A well-established population may contain some very large, old clones. For example, Harberd (1963) recorded a clone of *T. repens* 20 m across, with an estimated minimum age of 100 years. Oinonen (1967) recorded a single clone of *Pteridium aquilinum* covering 1500 m^2, and probably a thousand years old. A population of such species is likely to show a skewed size distribution, with few large and many small clones.

Sampling inflorescences or cuttings at random in such populations results in non-random sampling of individuals. Sampling is biased towards large clones, and a single large clone may unwittingly be sampled several times. For some purposes, the bias may be useful. Large clones are large because they have maintained a positive growth rate over a long period of time. Small clones are small either because they are young or because, though old, they have not grown well. Thus large clones have in a sense been 'tested' by natural selection and performed well, whereas small clones are either untested or have performed badly. If the sample is intended for use in breeding for the same environment, this preliminary on-the-spot selection may be an advantage. However, for other purposes the bias is undesirable. First, it might exclude genes of value in another location or in a future environment. Secondly, for a conservation collection, systematic exclusion of any genes is undesirable. Thirdly, the bias will invalidate studies of population genetics that require a random sample of clones.

Repeated sampling of the same clone is commonly minimized by spacing sampling points (for seeds or vegetative material) far apart in relation to the expected size of clones, but this does not eliminate the bias towards large clones. It is particularly unsatisfactory in species such as *T. repens*, in which individual clones may be large and yet closely intermingled. For example, Cahn and Harper (1976) found up to ten different clones of *T. repens* in a single dm^2 of pasture. Sackville Hamilton (1980) found up to three overlapping clones in 1 cm^2 of pasture in a population where individual clones were several metres in diameter.

The only complete solution is to take large numbers of cuttings and identify clones using whatever markers may be available (e.g. major gene loci for self-incompatibility, polymorphic enzyme systems or deoxyribonucleic acid (DNA) fingerprinting). However, such intensive sampling and post-sampling effort, essential for population genetics studies, cannot usually be contemplated as part of most large-scale germplasm collecting, where bias can be avoided only for species whose clones are readily distinguishable in the field. For example, the leaf markings of *T. repens* provide a useful, although not definitive, guide to the genetic identity of plants. They are expressed most strongly in the cool season, when the plants are not flowering, and so are most readily used when sampling vegetatively.

A sample may be biased both by seed presence and by plant size. For example, Table 22.1 compares seed and vegetative samples of two populations of *Festuca pratensis* in the hybrid zone between the diploid and the cross-compatible tetraploid (*F. pratensis* var. *appenina*). The two chromosome races are virtually indistinguishable morphologically in the field. Both vegetative samples contained a high proportion of triploid hybrids, which were absent from one seed sample and at low frequency in the other. The diploid was absent from the vegetative sample at site A but comprised 100% of the seed sample. At site B, the tetraploid was absent from the vegetative sample but comprised most of the seed sample. No single source of bias could explain all these results. Differences may be postulated in flowering times of diploid and tetraploid at the two sites to explain the biased compositions of the seed samples. However, to explain their absence from the vegetative samples, it is necessary to postulate size bias, i.e. that the triploid has high vegetative vigour but low fertility, producing large plants that are favoured in vegetative sampling.

Collecting outside normal ripe-seed season

It is essential that a seed collecting expedition be timed to coincide with the availability of ripe seed. This requires good local knowledge and may be particularly difficult if the targeted populations differ widely in their flowering times. The entire route of an expedition may need to be planned around flowering times, beginning with early-flowering and ending with late-flowering populations. The timing of vegetative collecting is likely to be less critical, depending on the species and the type of vegetative sample. For example, rooted cuttings of stolons of *T. repens* can be taken successfully at any time of year, even in midwinter (Sackville Hamilton, 1980). In such cases, collecting vegetative material guarantees success, and intensive collecting can be started whenever required and continued as long as necessary.

Within-population targeting

The genotypes and sizes of plants in a population reflect the outcome of natural selection acting on the gene pool of the population in the juvenile and mature vegetative phases. A seed sample represents the next generation's unselected recombination potential of a known female parent with a generally unknown male parent. The male parent may be

Table 22.1. Percentage of diploid, triploid and tetraploid plants in seed and vegetative samples taken from two populations of *Festuca pratensis* ($n = 7$) in the Carpathian mountains of Romania. The vegetative samples were taken in August 1980, and the seed samples in October of the same year.

Site	Sample type	Number of plants sampled	Percentage of samples in ploidy level		
			$2n = 14$	$3n = 21$	$4n = 28$
A	Vegetative	14	0	86	14
	Seed	42	100	0	0
B	Vegetative	16	0	100	0
	Seed	17	0	24	76

from a different microenvironment (see below) within the same site, if not from an alien site. In general, therefore, the genetic composition of a vegetative sample will more closely reflect population subdivision associated with microenvironmental variation, and may be used to target locally 'adapted' gene complexes.

Juveniles do not produce seed

Many perennials do not produce seeds in the first years of growth. If there is selective mortality before the age of first reproduction, the genotypic composition of the mother plants of a seed collection will not be the same as that of the total vegetative population including juveniles. If the population is in equilibrium, a seed sample would give rise to the same genetic composition of juveniles, and so this would not cause a genetic bias. However, if mortality before reproduction is causing a directional shift in genetic composition between successive generations, a seed sample will be biased against genotypes present only as juveniles.

Genetic structure of populations of clonal species

The genetic structure of populations will determine the amount of genetic variation recovered by a particular sampling strategy. Populations of clonal species are often very different from populations of species which do not spread appreciably by vegetative means in the number and spatial arrangement of the genotypes that comprise them.

Number of clones per population

Because the clones of some species are potentially immortal, being able to spread vegetatively indefinitely, it is theoretically possible for a population to comprise a single clone. For example, Sackville Hamilton (1980) found a number of populations of *T. repens* consisting of only one or two clones, apparently in locations near the limits of the ecological tolerance of the species for shade (i.e. in long grass or shrubland). In contrast, populations of the species in old pastures can contain many hundreds or thousands of clones (Sackville Hamilton, 1980; Gliddon and Trathan, 1985).

Population subdivision

It has long been established that in most plant species effective population size is much smaller than the apparent physical limits of any given population. For example, the reproductive genetic neighbourhood area of *T. repens* is $2\,m^2$ (Gliddon and Saleem, 1985), although a typical, apparently uniform pasture containing the species can be $10^4\,m^2$ to $10^6\,m^2$ or larger. This affords considerable opportunity for the development of genetically distinct subpopulations on a very small scale relative to the perceived extent of an apparently spatially continuous population.

Early work on genetic differentiation caused by strong selection pressures revealed sharp genetic boundaries between subpopulations, e.g. corresponding to differences in heavy metals levels in the soil. More recent studies have shown similar population subdivision in relation to a variety of different kinds of selection pressure. For example, Table 22.2 shows subdivision of a population of perennial ryegrass caused by a path traversing diagonally through a hay meadow. Progeny of samples from the path were more prostrate and more cold-tolerant and showed later spring growth and later flowering than those from the undisturbed meadow. A number of other examples in *Lolium* are given by Tyler and Chorlton (1976).

In another set of studies (reviewed by Sackville Hamilton, 1990), vegetative samples were taken from subpopulations of *T. repens* in four different microsites within a single old 0.75 ha pasture in the north of Wales (Turkington, 1975). The four microsites seemed to differ only in the dominant species of grass. The samples were multiplied in a glasshouse so that each could be transplanted back into all four original sites in the pasture. Each sample performed best when transplanted back into its native site, revealing a previously unexpected fineness of specialization within a single apparently uniform population. In the next stage of the study, the four samples of *T. repens* were grown in a glasshouse

Table 22.2. Characteristics of progeny of perennial ryegrass obtained from plants collected in 1967 from a hay meadow and a path traversing the same meadow in northern Italy.

Location within site	Seedling height (cm)	Date of flowering	Spring growth (kg ha^{-1} day^{-1})	Survival after freezing (%)
Hay meadow	11.4	1 May	20.5	57
Path	6.5	31 May	3.7	92

in all pairwise combinations with the four species of grass that dominated the four original collecting microsites. Again, three of the four samples gave the highest yield when grown with the grass species they were originally found with. Using other pastures of different ages, it was shown that this specialization takes three to ten years to develop. Effectively the same experiment, but at a still finer level of microsite distinction is reported by Gliddon and Trathan (1985). At each of four microsites within a single field dominated by *L. perenne*, a cutting was taken of *T. repens* and of the neighbouring clone of *L. perenne*. The four pairs of clones were multiplied and grown in all 16 grass–clover combinations, and again each tended to do best when grown in combination with its original 'partner'.

These studies show that *T. repens* is phenotypically responsive to remarkably fine microenvironmental variations. Moreover, one population is genetically variable for this responsiveness and divided into subpopulations specialized with respect to the fine-scale microenvironmental heterogeneity. This potential for substructuring is greatest in long-lived perennials with a stoloniferous or rhizomatous habit, i.e. in the very species most likely to be sampled vegetatively. Their growth habit enables them to 'forage' for and exploit favourable microsites. A stratified sampling strategy at each collecting site, recommended for general application in Chapter 5, will be especially appropriate for such species.

Sampling strategy

Marshall and Brown (1975) outline optimum sampling strategies for genetic resources conservation. This work is updated in Chapter 5. In this section those aspects specific to vegetative sampling are considered.

Number of sites and plants per site

The problem here is to choose, within the constraints of the resources available, how many sites should be sampled and how many plants should be taken at each site. In collecting for conservation, the aim is to collect maximum genetic variation. Oka (1969) estimated the proportion of the total genetic variation captured from a target area as

$$G = 1 - \{1 - P + P(1 - p)^n\}^N$$

where P = proportion of total variation represented by one population; p = proportion of one population's variation represented by one plant; N = number of populations sampled; and n = number of plants sampled per population.

Marshall and Brown (1975) present solutions of this equation maximizing G, the proportion of genetic variation captured, for a range of assumptions, but do not explicitly consider vegetative collecting of clonal plants. For outbreeding species, they suggest sampling 30–55 plants per site at six to nine sites per day (Chapter 5). This solution for outbreeding species is based on P between 0.50 and 0.75 and $p = 0.05$. Sackville Hamilton (1980) estimated $P = 0.37$ for variation in leaf area in a collection of *T. repens* from one area in the east of England, which would suggest sampling marginally fewer plants per site and more sites.

In a large population of a clonal species $p = 0.05$ may be accurate, but where the population contains only a few clones (see below) p will be much higher. In the extreme case of a population with one clone, $p = 1$. In practice, this simply means that the suggested sampling rates are accurate for large populations but impossible for small ones, and the number of plants sampled per population must be reduced.

The calculations also assume that size of the collection is limited by the time available for collecting, and that the time required to visit a site is 25 to 100 times greater than the time required to collect one plant. Both assumptions require modification for vegetative collecting. It usually takes longer to collect one vegetative sample than one seed, so we should expect smaller ratios and a smaller optimum sample size per population. In addition, time may not be the limiting factor in vegetative collecting. There may be a limit to the total number of plants that can be collected or subsequently processed, either because of the large size of each sample and limited temporary storage space in the collecting vehicle or the collector's rucksack, or because of limited labour for subsequent processing. If total number of samples is the limiting factor, then we have the constraint $Nn = k$ in Oka's formula, where k is a constant, the total number of plants that can be handled. Under this constraint, G is maximal at $n = 1$, and $N = k$ regardless of the values of p and P. That is, to maximize the diversity sampled, just one clone should be sampled from each of as many populations as possible. In practice, however, this is undesirable, because we would then have a poor estimate of the population mean and no estimate of diversity within the population. In obligate outbreeders it would also mean that no seed production would be possible except by crossing plants from different collecting sites. In any case, optimum sample size per population is considerably smaller than in seed collecting.

If the species is outbreeding and needs to be quarantined after international shipment, another constraint may be more limiting. Each population sample must be kept in its own isolation chamber in special quarantine facilities, isolated from cross-pollination with other samples and isolated from possible contamination of other plants. The total number of population samples that can be handled post-collecting may be the limiting factor for the number of sites visited, i.e. the number of sites visited is fixed. The number of plants sampled per population should then be increased until the next constraint becomes operative, whether that is the total number of plants that can be collected, the time available for collecting, or the number of clones in the population.

Site selection

Since the early days of genecology (the study of population genetics in relation to habitats) (Clausen *et al.*, 1940), a wealth of evidence has accumulated that genetic differences among populations of a species are associated with environmental differences in the sites they occupy. The genotypic composition of populations has been shown to be influenced by every environmental variable studied, whether biotic (e.g. competitors, mates, predators, pathogens) or physical (e.g. moisture, pH, soil structure, nutrients, temperature, quantity and spectrum of light). As emphasized by Marshall and Brown (1975) and others (Chapter 5), it is thus essential to seek the greatest diversity of environments in putting together a germplasm collection. A clustered distribution of collecting sites, besides decreasing the time wasted travelling between sites and improving the value of the collection for studies on ecogeographic variation, will: (i) increase the range of types of environmental diversity sampled, and (ii) 'force the explorer to search consciously for markedly different habitats within a region' (Marshall and Brown, 1975).

An illustration is given by Sackville Hamilton *et al.* (1979). In a small collection from all locatable natural populations of *T. repens* in one region in the east of England, variation for mean leaf length was greater than that observed in a large collection covering most of the global geographic range of the species, and much greater than that in the UK component of the global collection. The difference can be attributed to the fact that the global collection was targeted on a single specific habitat – agricultural grazed pastures – whereas the east of England collection, although geographically much more restricted, represented a much greater variety of habitats.

No type of site should be systematically omitted during collecting. For example, Sackville Hamilton (1980) made a two-level collection of *T. repens* from eastern England: an extensive collection of a few clones from every wild population located, followed by an intensive collection of many clones from six selected populations. The six populations were chosen partly for their distinctness, but partly also to enable the same stratified sampling procedure from a $10 \times 10\,\mathrm{m}$ plot to be applied at all six populations to facilitate sample comparisons. Although selected for distinctness, genetic variation among the six intensively sampled populations was far lower than among the extensive collection. Selecting populations with a structure sufficiently similar to allow the same sampling procedure to be applied appears to have had the negative effect of eliminating from consideration a great range of environments (e.g. paths, verges, gaps in shrubland, etc.) that support small but genetically distinct populations.

So, in collecting a species where population size can vary from one to several thousand, no attempt should be made to standardize the sampling strategy within populations, since that can be achieved only by excluding some sites. Moreover, since small populations are likely to occur in environments near the ecological limits of the species, they are likely to comprise extreme genotypes. As much, if not more, emphasis should be placed on sampling these small populations from unusual environments.

Population sampling

No single systematic procedure should be followed for sampling within sites. Rather, the sampling procedure must be matched to the population being sampled. In a population genetically subdivided by microenvironmental heterogeneity, as all larger populations will be (see above), targeting the various microenvironments within the site will extract more variation than random sampling. The sampling bias towards large plants (see above) is likely to improve the success of such targeting. The extent of population subdivision in relation to the apparently most trivial microenvironmental heterogeneity emphasizes the importance of noting and targeting even the tiniest variants. However, logistical constraints generally dictate that only samples from the more obvious microenvironmental variants (e.g. on and off a path, under and away from a hedge, fence, roadside verge, etc.) will be retained as separate accessions in the gene bank. Finer subdivisions will have to be bulked unless required for population genetics studies.

Collecting methodology

Equipment

The collecting vehicle must be capable of carrying the collecting team with their personal equipment and collecting equipment as well as up to 200 kg of vegetative samples. The vehicle used at the Institute of Grassland and Environmental Research (IGER), UK, is specially modified to keep vegetative samples alive. It has a fitted refrigeration system, consisting of two small refrigerators. These have been adapted to run off the vehicle alternator when travelling, a 240 V external alternating current (AC) supply when the vehicle is stationary and bottled gas in the absence of a suitable electrical supply point. The freezer compartments of the refrigerators are used to maintain a constant supply of frozen gel packs, which are then used to keep collected material cool in insulated expanded polystyrene boxes. For trips of relatively short duration, or in cool regions, fitted refrigeration may not be essential, though it will always be desirable.

Table 22.3 gives a list of the specialized equipment used by IGER teams in collecting vegetative samples.

Sampling technique

Different sampling techniques are used for different species. If the species produces vegetative propagules that are self-contained survival and dispersal structures (e.g. bulbils), sampling will be much the same

Table 22.3. Specialized equipment used by IGER teams in collecting vegetative samples.

Item	Quantity
Refrigerators	2
Gas bottles for refrigerators (15 kg)	2
Insulated boxes (capacity 240 l)	3
Gel packs	25
Thermometers	4
Extension cable (30 m)	1
Electrical adaptor	1
Polythene bags (450 × 300 mm)	1000
Plastic labels	250
Trowels	2

as for seeds. If, however, it is necessary to take cuttings of actively growing tillers, stolons, shoots, etc., more care is needed to prevent death of the cutting before it is processed at the gene bank.

For species whose cuttings do not root readily, the sample must be a rooted plantlet in a divot of soil. Where adventitious roots are readily produced, as from the stolons of *T. repens*, unrooted cuttings may be taken, although even in these cases survival and establishment can be increased to nearly 100% by including roots and soil in the sample. However, the inclusion of roots and soil increases sample bulk several-fold. Taking unrooted cuttings may increase the total number of samples that can be collected by so much that more samples may ultimately be established despite a lower survival rate. Which of the two methods is used depends on:

- to what extent survival is affected by including roots and soil;
- how severely the total number of samples collected is limited by their bulk;
- whether it is more important to establish many cuttings (allele conservation) or specific cuttings (population genetics or targeted conservation);
- whether the soil itself will be useful.

Including soil provides a plant-specific sample of rooting medium that can be analysed later. In addition, for legumes, host-specific strains of *Rhizobium* can be sampled from the nodules of a rooted sample and in some cases from the soil itself (Chapter 26).

To illustrate the collecting procedure, assume a team of three collectors sampling *L. perenne* and *T. repens* from a uniform seminatural grassland pasture defined by a stock-proof boundary, within which some type of agricultural management is applied. According to the owner or manager, the pasture has never been ploughed and reseeded with modern cultivars, at least not within living memory, and there has been a long history of agricultural management.

- The site is examined to confirm the presence or absence of the target species and to check for microenvironmental discontinuities. Detailed history and management details of the site, along with permission to collect, are obtained from the landowner. Once the presence of the target species is established and it has been confirmed that the site fits all the collecting criteria, sampling commences.

- A unique site number is allocated to the site. Each species to be collected from the site is given a different collecting number. These numbers are written on separate polythene bags (in waterproof ink) and also on plastic labels which go into the bags. These numbered bags are then carried around the site and filled with vegetative divots. Collector 1 collects stolons of *T. repens*, each with a small divot of soil. Collector 2 collects *L. perenne* divots as clumps of rooted tillers about 5 cm in diameter. Divots may contain both species, in which case both are extracted when the divot is processed on return to base. Collector 3 records general site details (Chapter 19) and then assists the other two collectors in sampling.

- Some 35–45 divots of *T. repens* and 25–35 divots of *L. perenne* are dug up using a sharpened knife or trowel. (There is a greater rate of loss of *T. repens* divots in transit.) Divots are collected over the whole area of the collecting site with due regard to spacing between points of collecting. Air is expelled from the polythene bags of divots and the bags are sealed. The filled bags are stored in a cooled, insulated box in the collecting vehicle.

- A collecting form is completed for each sample collected at the site. In this example, because there are no microenvironmental discontinuities within the site, two forms are filled in, one for each of the two species. The two forms will contain identical data for nearly all sections, except collecting number, species name, sampling details, associated samples (perhaps) and population data. The data are fully entered on site, with the help of the landowner to ensure accuracy. Information is pooled from the three collectors who have walked the site to sample it and the landowner who has managed it.

- The samples are packed into airtight polythene bags and placed into insulated boxes. Frozen gel packs from the refrigerator are packed around the bags to keep them cool and reduce plant respiration rates. Alternatives to plastic bags for storing samples include burlap sacks (kept moist) and styrofoam boxes (plant material wrapped in moist newspaper).

This process of collecting plants and data normally takes about 90 minutes for two separate vegetative samples from one site. The minimum time on site for one vegetative sample per site is about 1 hour and on a complex site with observable microenvironmental discontinuities this may increase to about 2 hours. Allowing for travelling time between sites, the average number of sites covered per day in the manner described is four or five.

At the end of the collecting day, the ideal situation is to return to a base, where the bags of divots can be opened and the divots (with identifying labels) laid out in separate trays for each sample to be watered if dry or allowed to drain if too wet. A secure area is required where the divots will not be disturbed and where separate samples can be kept apart. When the expedition is ready to return to its home station, the moist (not wet) divots are repacked in their respective bags and placed in cooled insulated boxes in the vehicle for transit.

Post-collecting

The post-collecting handling of vegetative material is very different from seed handling. It needs specialized facilities, in particular if quarantine is involved, and must be rapid.

Quarantine

The material collected may need to be sent abroad, for example for safety duplication or because adequate facilities for the conservation of vegetative material and/or the production of seeds for storage are lacking within the country. The transfer of live plant material from one country to another requires a phytosanitary certificate from the country of origin and an import licence from the country of final destination. These documents will need to be arranged at the early planning stage, in some cases at least six months in advance of the expedition departure date. Some countries do not allow the entry of soil. In such cases, soil will have to be washed off the roots and the plants wrapped in moist paper and stored in plastic bags before shipment. There may also be restrictions on the movement of plants and soil within a country, for example if a soil-borne disease is present in some districts and not others.

For a collection arriving at IGER on 1 September, the procedure, with specific reference to *L. perenne* and *T. repens*, is summarized in Table 22.4. For dealing with a collection originating from within the country, quarantine will usually not be necessary, but the rest of the procedure will be no different.

Seed production

All samples are maintained in quarantine while flowering to generate seeds for storage in the gene bank. The aim is to produce a maximum yield of high-quality seeds from each genotype, excluding foreign pollen. Both *L. perenne* and *T. repens* are self-incompatible, the former wind-pollinated and the latter insect-pollinated. A secondary quarantine facility is used at IGER for seed production, divided into isolation chambers to prevent cross-pollination from other populations (Table 22.5). As each

Table 22.4. IGER procedure after collection arrival.

Date	Stage	Action
1 Sept	Plant collection arrives	Copy of import licence given to customs officers at port of entry. Signed by customs and forwarded to Ministry of Agriculture as proof of entry of vegetative samples
2 Sept	Place divots in quarantine	Move sealed bags of divots to quarantine house. Each population placed in a separate plastic, compartmented tray, with soil of divots in contact with capillary matting
	Divots 'recover'	Spray with systemic insecticide and fungicide if necessary
16 Sept	Extract plants and soil from divots	Each divot is broken down and cleaned. One tiller and/or stolon per divot is retained and planted in population groups in sterile compost in plastic compartmented trays. Original soil collected from the divots, setting aside two soil samples per collecting site. One soil sample sent direct to microbiology lab for extraction of *Rhizobium*, the other dried and analysed. Waste soil and vegetation put in sealed polythene sacks for sterilization and safe disposal
	Promote active regrowth	In temperate latitudes, as natural day length and temperature decrease, supplement artificially to give an equivalent 16-hour day at 20°C. Cut back and fertilize as necessary
15 Dec	Promote flowering	Remove supplementary light and heat, to allow floral induction under normal short day length and low temperatures
	Control pests and diseases	Material monitored regularly, and sprayed if necessary
1 March	Material potted on	Transplant into 15 cm diameter plastic pots of sterile compost, maintaining genotype and population identity

population sample comes into flower, all genotypes from that population are transferred to a vacant isolation chamber. Cross-pollination within each population sample is promoted by the ventilation system for *L. perenne* and by bees for *T. repens*.

Clean, high-quality seeds are obtained from each genotype within the seed island and kept separately in labelled packets. These individual

Table 22.5. IGER procedure for seed production.

Date	Operation	Comment
End March	Monitor populations for onset of flowering	Emergence likely to be over several months even for a collection from a small geographic area
End April–early July	Immediately prior to anthesis, spray population against pests and diseases and transfer each population to a separate isolation chamber ('seed island')	Pots in random block design. Isolation chamber can be either in quarantine house or in a separate isolation block of similar design
4–6 weeks later	Seed island removed from isolation chamber 15 days after peak anthesis, the heads being covered with labelled pollen-proof bags	Allows increased throughput of seed islands
2 weeks later	Heads cut off individual genotypes (still in bags) and hung up to dry	Heads (in population groups) dried slowly in glasshouse, laboratory or ideally a drying room
4–6 weeks later	Dry heads initially pounded to remove most of the seed and then mechanically threshed to remove the remainder. Seeds hand-sieved and mechanically blown in a column blower	Dry seed heads essential for efficient threshing.

mother-plant seed packets, making up the seed island, are transferred to a drying cabinet where the seeds are dried passively over silica gel to 5–7% moisture content. After drying, each seed lot is weighed and divided up as follows:

1. Approximately 0.1 g of seeds from each mother plant is stored in separate packets at $-20\,^{\circ}$C in the long-term store. This is the basic seed. All seed samples are stored in sealed foil-laminate envelopes.
2. A balanced bulk for each population is made up of equal numbers of seeds from each mother plant. The number of seeds per mother plant is the lowest usable number produced by any one mother plant. The samples are stored at $0 \pm 2\,^{\circ}$C in medium-term store and used for research and exchange.
3. An unbalanced bulk is made up of the remainder of the seeds. These samples are stored at $0 \pm 2\,^{\circ}$C in medium-term store and are used for evaluation.

Conclusions

Collecting vegetative material has an important role in germplasm conservation. Although it raises some additional problems compared with seed collecting, it also has several specific advantages which make it preferable, or necessary, for some species and some purposes.

A decision to collect vegetative material has consequences for all stages of an expedition. The optimum sampling strategy changes, in terms of the optimum number of sites visited and plants sampled per site. The criteria for site selection and the method for sampling each population are also different. In general, more care must be taken, sampling fewer genotypes than in seed collecting. Different sampling techniques are required. Different equipment must be prepared to collect the samples and keep them alive. Different logistic arrangements for the temporary storage of samples must be made to keep the samples alive throughout the expedition, either by using whatever facilities are available *en route* or by frequent transfer back to the home institute during the collecting. More stringent phytosanitary regulations governing import of living material must be observed. Different post-collecting procedures for quarantine must be followed and seeds will have to be produced for storage and easier use and exchange of the germplasm.

In summary, optimum sample sizes per population are generally smaller than for seed samples. In some situations a sample size of one maximizes the diversity collected, although this extreme is not recommended because it eliminates information on diversity within populations. In many species, targeting the full diversity of microenvironments occupied by a population will extract greater diversity than a random sample. It is recommended that just two or three plants be taken from each microenvironment.

Acknowledgements

The editors would like to thank Drs Jean Hanson and Jean Ndikumana (both International Livestock Center for Africa (ILCA)) for valuable comments on an earlier draft of this paper.

References

Cahn, M.G. and J.L. Harper (1976) The biology of the leaf mark polymorphism in *Trifolium repens* L.: 1. Distribution of phenotypes at a local scale. *Heredity* 37:309–325.

Clausen, J., D.D. Keck and W.M. Hiesey (1940) *Experimental Studies on the Nature of Species. I. Effect of Varied Environments on Western North American Plants.* Carnegie Institute of Washington Publication No. 520. Carnegie Institute, Washington DC.

Gliddon, C. and M. Saleem (1985) Gene-flow in *Trifolium repens* – an expanding genetic neighbourhood. In: Jacquard, P., G. Heim and J. Antonovics (eds) *Genetic Differentiation and Dispersal in Plants*. Proceedings of the NATO Advanced Research Workshop on Population Biology of Plants. pp. 293–309. Springer-Verlag, Berlin.

Gliddon, C. and P. Trathan (1985) Interactions between white clover and perennial ryegrass in an old permanent pasture. In: Haeck, J. and J.W. Woldendorp (eds) *Structure and Functioning of Plant Populations*. pp. 161–169. North Holland Publishing Co., Amsterdam.

Harberd, D.J. (1963) Observations on natural clones of *Trifolium repens* L. *New Phytologist* 62:198–204.

Hubbard, C.E. (1968) *Grasses*. Penguin, London.

Marshall, D.R. and A.H.D. Brown (1975) Optimum sampling strategies in genetic conservation. In: Frankel, O.H. and J.G. Hawkes (eds) *Crop Genetic Resources for Today and Tomorrow*. pp. 53–80. Cambridge University Press, Cambridge.

Oinonen, E. (1967) The correlation between the size of Finnish bracken (*Pteridium aquilinum* (L.) Kuhn) clones and certain periods of site history. *Acta Forestalia Fennica* 83:1–51.

Oka, H.I. (1969) A note on the design of germplasm preservation work in grain crops. *SABRAO Newsletter* 1:126–134.

Sackville Hamilton, N.R. (1980) Variation and adaption in wild populations of white clover (*Trifolium repens* L.) in East Anglia. PhD thesis. University of Cambridge, Cambridge. Published (1982) by University Microfilms International, Ann Arbor.

Sackville Hamilton, N.R. (1990) Life history studies. In: Zarucchi, J.L. and C.H. Stirton (eds) *Advances in Legume Biology*. Monographs in Systematic Botany of the Missouri Botanic Garden. Missouri Botanic Garden, St Louis.

Sackville Hamilton, N.R., A.M. Evans and H.J. Harvey (1979) Genetic variation in wild white clover populations in eastern England. In: Zeven, A.C. and A.M. van Harten (eds) *Broadening the Genetic Base of Crops*. pp. 97–99. PUDOC, Wageningen.

Turkington, R. (1975) Relationships between neighbours among species of permanent grassland (especially *Trifolium repens* L.). PhD thesis. University College of North Wales, Bangor.

Tyler, B.F. and K.H. Chorlton (1976) Ecotypic differentiation in *Lolium perenne* populations. In: *1975 Annual Report of the Welsh Plant Breeding Station*. Welsh Plant Breeding Station, Aberystwyth.

Collecting woody perennials

23

Forest Resources Division, Forestry Department, FAO[1]
(based on the work of L. Thomson[2], Consultant)
[1]*Via delle Terme di Caracalla, 00100 Rome, Italy:* [2]*18 Etheridge,
Page, Canberra ACT, Australia.*

Introduction

Woody perennials have increasingly been targeted for germplasm collecting in recent years. This is associated with an upsurge in interest in a wide range of lesser-known species which have potential for agroforestry and small-scale, non-industrial plantings. At the same time, the importance of using the best available germplasm for industrial forestry plantations has also become widely appreciated. For many tropical species at an early stage of domestication, the best available germplasm remains that obtained through collecting seeds from wild stands of known superior provenances of proved adaptability to specific sites. 'Provenance' refers to the geographic origin of a particular seed source (Turnbull and Griffin, 1985; but see Jones and Burley (1983) for a discussion of different definitions and their limitations). In addition to seed collecting for evaluation trials and for large-scale use, a major challenge is to collect and conserve *ex situ* the genetic resources of many socioeconomically important woody species which are endangered in their native habitats.

This chapter will provide some specific guidelines on the planning and execution of field collecting of woody perennials. Germplasm collecting of wild populations of woody plants will differ from collecting other wild species and field crops in a number of important ways.

- Most tree species are preferentially outcrossing organisms with high levels of intraspecific variation. These characteristics will affect sampling strategy and frequency ('Sampling strategy').
- Most tree species do not flower and fruit gregariously every year. They flower at intervals typically of two to three years but sometimes of up to five to eight years ('Importance of collecting in most seed years' and 'Planning and reconnaissance').

- In the case of tall trees and those growing in denser formations the fruits are usually in inaccessible parts of the canopy. Therefore, various specialized harvesting techniques may need to be employed ('Seed collecting methods and equipment').
- The ratio of fruit to seed mass is often high for woody species. Therefore, when undertaking extended expeditions to remote localities, it is often necessary to partially process and clean the fruit to reduce bulk and weight, and also help prevent fungal attack ('Handling, processing and transport of fruits and seeds').
- Though germplasm collecting of woody plants will usually be accomplished most efficiently by collecting seeds, in some situations collecting other plant parts may be desirable or necessary ('Vegetative germplasm collecting procedures').
- The majority of woody species have coevolved with root microsymbionts, which may be essential for their healthy growth and long-term survival. Cosampling of symbiont germplasm during seed collecting expeditions may thus be necessary ('Cosampling of root microsymbiont germplasm').

Sampling strategy

Chapters 5 and 6 deal with collecting strategies in detail. This section discusses both the general principles of sampling and the practicalities with specific reference to woody species.

Sampling objectives

Germplasm may be collected for research, conservation, immediate use and/or evaluation for future use. For evaluation (e.g. provenance trials) and *ex situ* conservation, the aim in collecting is to obtain a genetically representative sample. Therefore, collectors must be careful not to bias sampling when selecting seed trees. For some specific purposes it may be advantageous to impose a level of phenotypic selection. However, this will only be the case when highly heritable traits are involved, e.g. wood grain orientation, leaf oil composition, flower colour. Usually, phenotypic selection will be either genetically undesirable or a waste of time and resources, especially if the material will be grown in areas remote from, and environmentally unlike, the collecting site. Turnbull (1978) and Willan *et al.* (1990) provide useful discussions of sampling strategies for woody plants.

Sampling approaches

Information is increasingly becoming available on the pattern of genetic variation for tree species with extensive, more or less continuous distributions, e.g. many *Eucalyptus* spp. (Green, 1971; Boland, 1978; Turnbull, 1980; Moran and Bell, 1983; Moran and Hopper, 1987), *Acacia mangium* (Moran *et al.*, 1989a), *Casuarina cunninghamiana* (Moran *et al.*,

1989b) and many northern hemisphere conifers (e.g. *Picea sitchensis*: Falkenhagen, 1977; *Pseudotsuga menziesii*: Merkle *et al.*, 1988). The proportion of genetic diversity among populations of insect- and bird-pollinated Australian trees has been found to be double that for northern hemisphere wind-pollinated conifers and angiosperms (Moran, 1992). For many Australian trees with widespread distributions, the bulk of genetic diversity resides within populations. National Research Council (1991) reviews what is known of the structure of genetic variation in forest trees.

The present practice of gene pool sampling of widespread species for provenance or progeny trials usually involves collecting 10–20 individuals per provenance, with individuals spaced at intervals of not less than the distance of normal seed dispersal. A common general rule has been to keep a minimum of 100–200 m between sampled individuals. When collecting seeds of preferentially outcrossing species, it is advisable to avoid sampling reproductively isolated individuals, because these may hold a disproportionately high proportion of selfed seeds. If the collecting is for *ex situ* conservation, the number of sampled individuals should be increased wherever possible to 25–50 per population. For improvement programmes and seed orchards the number of individuals sampled may be ⩾200 per population to avoid excessive narrowing of the genetic base at later cycles of selection. There have even been some suggestions of minimum numbers for *ex situ* and *in situ* conservation in excess of 2000 individuals (Krusche and Geburek, 1991).

The purpose of maintaining a reasonable distance between sampled trees is to avoid or reduce the likelihood of sampling related individuals. The degree of relatedness among neighbouring individuals will be influenced by several factors, including breeding system, stand size and history (recruitment events) and, in particular, the efficiency of pollen and seed dispersal. The sampling distance between individuals of species with poor seed dispersal (e.g. *Adansonia gregorii* and *Lodoicea maldivica*), those which form extensive clonal stands by root suckering (e.g. *Eucalyptus porrecta, E. ptychocarpa* and *E. jacobsiana*: Lacey, 1974; *Tamarix* spp.) and those which reproduce apomictically will need to be greater than that for species with efficient long distance seed dispersal by wind (e.g. *Alstonia* spp., *Casuarina* spp., *Cedrela toona, Securidaca longepedunculata*), water (e.g. *Acacia stenophylla, Eucalyptus camaldulensis* and most mangroves) or animal vectors (e.g. *Elaeocarpus* spp. and *Ficus* spp.).

In species occurring in small, disjunct populations, the major component of intraspecific variation will usually be among populations (e.g. Moran and Hopper, 1983, 1987). Accordingly, an optimal strategy to capture genetic variation will involve sampling relatively fewer individuals from as many populations as possible. This approach is also applicable to colonizing woody species (e.g. *Acacia cowleana*) which may occur as many small 'satellite' populations. Although within-population diversity may be highest in optimal areas central to the range of a species, marginal and isolated populations should also be sampled, as they will possess adaptations to extremes. Willan *et al.* (1990) suggest that one central and four peripheral populations is an absolute minimum number of provenances in first-stage trials of multipurpose trees (as, indeed, for *in situ* conservation), ranging up to $\geqslant 30$ populations for species with extensive and ecologically diverse ranges.

Optimizing the mix of collecting effort between and within populations is often problematic in the absence of information on the pattern and distribution of genetic variation for the target species. Growth and morphological characterization can be used to assess the extent of variation and help provide the basis of sampling strategies. Preliminary coarse-grid isozyme or even DNA surveys may be used to help define more efficient sampling strategies for comprehensive seed collecting and evaluation of lesser-known tree species (Moran, 1992). These are discussed in Chapter 6. A knowledge of the breeding system of the target species and its pollen and seed dispersal mechanisms will clearly be important in the development of a collecting strategy for first-stage sampling.

Individual tree versus bulk provenance collecting

The choice of whether to collect and maintain seeds of individual trees separately or bulk them in the field will largely depend on the purposes for which the seeds will be used. Seed collecting on an individual tree basis is required for studies of genetic variation and tree improvement programmes. Bulk provenance collecting will often suffice for first-stage sampling for provenance trials of less well-known species. However, bulking of seeds of different individuals should only be done when the collector is thoroughly familiar with the species and is certain of the taxonomic identity of each individual seed tree included in the bulk. Bulk collecting is often favoured for extended expeditions sampling a large number of species. In such cases, it may not be logistically possible to maintain and clean a large number of individual tree seed lots.

Approximately equal amounts of seeds should be collected from each tree included in the bulk (assuming approximately equal germination percentages). The bulked seed lot should be thoroughly mixed. When sampling a given individual, fruits from all parts of the crown should be collected, because these may have been pollinated by different pollen sources. Bulked seed lots must not be biased by the inclusion of a disproportionate number of seeds from a few heavier seeding or more easily collected individuals. In at least some *Acacia* spp., the seeds in each pod are full sibs. Other things being equal, this means that in such species each tree's contribution to a bulk sample should come from as many different pods as possible.

Importance of collecting in mast seed years

In wild stands the amount of fruit borne by a particular individual is not only under genetic influence, but is also strongly affected by age (e.g.

in forest communities the heaviest crops are borne by mature or over-mature dominants and emergents) and environmental factors (e.g. edge trees often bear heavily, while isolated individuals of outcrossing species may bear only very light crops). This may introduce a bias in sampling towards those individuals with precocious and/or heavier seeding habits. Such bias may to some extent be unavoidable, but it is often undesirable. An exception may be where flowers or fruits are the desired end-product, as in the case of nut or fruit species for human or stock food, honey species and ornamentals. Precocious and heavy-seeding genotypes are more likely to be less efficient wood producers. On the other hand, precocious genotypes can be used to shorten generation times in tree breeding programmes. Year-to-year variations in the factors that influence flowering and seed production, in particular climatic variables, also cause different trees to be represented as parents in different years (e.g. Houle and Filion, 1993). Undertaking seed collecting only in mast seed years, when many or most individuals are holding collectable numbers of seeds, will minimize these sampling biases.

Seed samples collected in years of sparse flowering and seeding will contain a higher proportion of empty or non-viable seeds and tend to be genetically less representative of the sampled population. For mainly outcrossing species, collecting in poor flowering and fruiting years and from isolated individuals may result in seed lots with abnormally high proportions of selfed seeds. These will usually give rise to genetically inferior plants, e.g. in *Eucalyptus regnans* (Eldridge and Griffin, 1983) and conifers (*Picea abies*: Eriksson *et al.*, 1973; *Pseudotsuga menziesii, Abies procera* and *Picea sitchensis*: Phillips, 1984). For such species there is a risk that provenance differences measured in field trials may be confused with or masked by collection year (i.e. percentage selfing) effects. Therefore, seed collecting in poor seed years is not only inefficient and wasteful of scarce resources, but may also be counterproductive.

Cultivated species

Zagaja (1983) has suggested that 'most of the fruit germplasm collections made in the past can be considered a biased sampling of the population', in that the aim of the collecting was to obtain material with particular characters, rather than to conserve genetic diversity. More recently, random population sampling of tree landraces has been recommended. Hawkes (1980, 1991) suggests that if the trees are being propagated by seeds, a bulk sample (of seeds or cuttings) of 10–15 individuals selected at random from throughout a village area or orchard is generally acceptable. When the trees are clonally propagated, they should be treated as a root or tuber crop, i.e. collected as vegetative material and each distinct morphotype (or traditionally recognized local variety) kept as a distinct sample. If seeds are collected from a clonally propagated cultivar at a given site, it can either be kept as single-tree samples (and given the same number as any vegetative sample) or bulked in a cultivar sample (and given a distinct number).

Collecting cultivated species will thus rely to a great extent on indigenous knowledge. For sampling to be appropriate to the situation, growers will not only need to be asked about the method of propagation of trees, the distinguishing features of different varieties and the total number of varieties in a village or district, but also about the genetic base of their material (i.e. whether seeds or other planting material came from a few or many mother trees) and their degree of adaptation to local conditions (i.e. how long they have been in the area) (Willan *et al.*, 1990). However, there is no less of a role for indigenous knowledge in collecting wild species. In particular, information on local uses can be critical in deciding on whether a given poorly known species might be worth collecting, a decision which would otherwise have to be made on the basis of probably inadequate field observations or simply taxonomic kinship to better-known species. Thus, the fact that the branches of a particular tree are said by local people to be lopped as fodder will be valuable information when collecting potential browse species. Without such information, one would have to actually observe animals browsing the foliage or guess that the foliage was palatable because that had proved to be the case for a closely related species.

Locating seed trees and seed stands

Planning and reconnaissance

A key factor in planning is to time seed collecting to coincide with peak maturation of abundant fruit crops. Accordingly, the flowering and fruiting pattern for the target species must be established. Information is needed on the time of the main flowering season and the time taken for fruits to mature. The period of maturation can vary considerably, even between closely related species. For example, in temperate-zone bipinnate acacias the maturation period varies from four to five months (*A. decurrens*) to 12–14 months (*A. mearnsii*). The time taken for *Eucalyptus* fruits to mature ranges from four to six weeks in *E. brachyandra* and *E. microtheca*, with fruits being shed shortly after maturation, up to 12–18 months for some species in the subgenus *Monocalyptus*. Some species may retain their fruits for several years after maturation.

It is also important to establish whether heavy flowering and seeding occurs annually, biennially or at irregular intervals related to climate and endogenous factors. Prolific flowering and heavy fruit set in many dry-zone species are dependent on particular rainfall conditions. For example, in *Acacia aneura* substantial quantities of fruit are only produced when flowering induced by summer rainfall is followed by good rains in winter (Davies, 1976). In subtropical dry-zone Juliflorae acacias (e.g. *A. colei* and *A. tumida*) substantial seed production has been observed to occur following heavy rainfall (totalling >350 mm) in the latter part of the summer rainy season. Examination of recent rainfall patterns using meteorological station data or remote sensing data can indicate those areas where good flowering and fruiting of dry-zone species are most likely to be occurring. Such information can greatly assist in planning seed collecting expeditions in arid zones and minimizing the time spent searching for seed crops.

Field reconnaissance of populations of the target species can help identify particular areas and individual trees from which seeds are most likely to be available. Reconnaissances are most warranted for species whose fruits take several years to mature and which bear heavily at irregular intervals (e.g. temperate *Pinus* spp. and *Picea* spp.). Reconnaissances are best undertaken either at the time of flowering or when the fruit crop is approaching full size and maturity. Newly formed, green fruit crops are easily overlooked. It should be borne in mind that reconnaissance missions, especially those to remote localities, are expensive and can only provide information on potential collecting sites. Developing fruit crops may abort or be destroyed by a variety of abiotic (strong winds, drought, fire) and biotic (disease and predation by mammals, birds and insects) agents.

Whether in planning or field reconnaissance, local people are often the major source of data on flowering and fruit development in target species, as well as on the distribution of species. These will include both forestry department staff and rural people making traditional use of the trees, either directly, for firewood, building, food or medicine, or indirectly, as fodder for livestock or nectar sources for bees.

In the field

Intraspecific variation in fruiting can occur at a regional level or on a finer scale. The collector will need to weigh up the relative merits of continuing to search in a particular area where seeding is generally sparse, as against moving to another region or aborting the mission. Quite often a population or a small group of neighbouring, and presumably closely related, individuals will be found holding heavy seed crops in a region where seed crops for that species are otherwise meagre or absent. Hence a seed collecting expedition should not be prematurely aborted if the first populations encountered are devoid of useful seed crops.

In more open plant communities, the ability to distinguish fruit-bearing trees of target species, especially from a distance, is a prerequisite for efficient seed collecting. This is a skill that should be developed by aspiring seed collectors. Fruiting trees can often be discerned from a distance through differences in the colour of their canopy. The canopy of trees holding a mature fruit crop will generally have a yellow, brown or reddish-brown tinge. Fruit crops are most easily spotted on sunny days, when the sun is at a low angle (i.e. early to mid-morning and late afternoon) and the light is behind the observer. Red wavelengths are more apparent in the late afternoon, making this the best time of day to locate fruiting trees of species with reddish brown or purplish fruits (e.g. *Acacia mangium* and *Argyrodendron actinophyllum*). A pair of binoculars or field glasses, with a moderate magnification of ×8 or ×10, is an essential piece of equipment for both locating and assessing crops on potential seed trees. Compact lightweight makes (i.e. 25 or 30 mm aperture) which can fit inside a pocket are especially recommended for work in hot climates. A critical safety aspect is that drivers should not also be involved in the task of locating seed trees while the vehicle is moving.

Once a potential seed tree is located it is important to quickly determine whether the seed crop is at the right stage of maturity and of sufficient quality and size to warrant collecting (Barner and Olesen, 1984c). For tall trees growing in forest formations, the first step usually involves scanning the canopy through binoculars to ascertain the extent and distribution of the fruit crop. This is followed by a search of the ground for fruiting material, which can be used to confirm species identity and to provide an indication of fruit maturity and the extent of insect attack or other damage.

During collecting it is important to continuously monitor the level of insect attack in a seed crop, as this can vary considerably between trees within a population (e.g. in tropical acacias such as *A. mangium* and *A. auriculiformis*). Seed collectors should be aware that some species of fig (e.g. *Ficus semicordata* from Nepal) may produce two types of tree, one bearing fruits with viable seeds and the other only wasp-infested gall figs (Amatya, 1990).

Seed collecting methods and equipment

The choice of seed collecting method(s) and equipment will depend on many factors, including:

- the number of seeds required;
- the relative size, number and distribution of fruits;
- fruit characteristics (including stage of ripeness);
- characteristics of the individual tree, stand and site.

Even within a population, a variety of collecting techniques may have to be applied to different individuals to reduce total collecting time. The use of high-technology machinery such as mobile hydraulic platforms and helicopters for collecting from wild populations of woody plants is only rarely warranted and these expensive options will not be discussed further.

The felling of individual trees for seed collecting purposes is not recommended. Such practices can erode the genetic resources of small populations and endanger their continued existence. If collecting is well planned, there is sometimes an opportunity to collect seeds from trees felled for other purposes. However, safety and such factors as rapid seed shed during warm weather often constrain the collecting that can be done in association with timber harvesting operations. In any case, for some species seeds or fruits collected from the canopy can be better conserved.

Removal of fruits from the canopy from the air is possible. Kitzmiller (1990), for example, describes collecting from large inaccessible stands by means of a cone rake attached to a helicopter. However, such a technique is clearly unlikely to be used very often by germplasm collectors. Most techniques for harvesting fruits and seeds in fact fall under the following three headings:

- collecting fallen fruits and seeds;
- removal of fruits from the canopy using ground access;
- climbing into the canopy.

The series of Technical Notes of the DANIDA Forest Seed Centre are particularly valuable in this context. The most relevant are listed separately in the References section. A checklist of specialized equipment is given in Box 23.1.

Box 23.1
Specialized equipment for collecting seeds of woody species

- Tree measuring instruments – diameter tape and height measuring instrument, e.g. clinometer.
- Calico collecting sheets (e.g. 2 × 2 m).
- Large, heavy-duty, canvas and nylon tarpaulins (for seed collecting, transport and camping).
- Secateurs.
- Rifles (plus cleaning equipment) and ammunition.
- Pruning saws (bow saw, commando saw with rope, long-handled pruner plus spare blades).
- Throwing ropes.
- Tree climbing equipment.
- Tree ladders.
- Chain-saw (chain-saw spanner, file, spare parts, fuel and oil).
- Safety gear (steel-capped boots, gloves, hard hat, safety belt).

Collecting fallen fruits and seeds

Collecting fruits and seeds from the ground is suitable for species with non-winged, heavy fruit growing where ground vegetation is sparse. Recently shed fruits and seeds may be raked or swept up and sieved clean. There is also the option of sucking material up with a portable car vacuum cleaner, light domestic vacuum cleaner (Bastide and Gama, 1980) or portable vacuum harvester.

For most research purposes, collecting fallen fruits and seeds is not generally recommended because of:

- uncertainties regarding their source;
- risks of contamination from morphologically similar seeds of nearby related species;
- their lower physiological quality, compared with those obtained by other methods, due to collecting a higher proportion of immature, empty and unsound seeds, and early onset of deterioration or germination;
- greater risks of contamination of the fruit or seed surface with soil-borne pathogenic fungi.

The main attractions of collecting from the ground are that large numbers of seeds can be gathered quickly and cheaply, without the need for highly skilled technicians or specialized equipment. Accordingly, this method is better suited to bulk provenance seed collecting for operational plantings. For example, the fruits of some northern temperate genera (e.g. *Castanea, Fagus* and *Quercus*) are commonly collected from the ground for large-scale plantings (FAO, 1985). The same is true of teak and *Gmelina arborea*. For bulk provenance collecting of hard-seeded species (e.g. *Robinia pseudoacacia, Acacia mearnsii* and *Prosopis tamarugo*) the seeds can be sieved from soil collected from under source trees and stands.

For research and *ex situ* conservation purposes, ground collecting of fallen fruits and seeds may be suitable for some dry-zone and rain-forest species.

Dry-zone species

There is often little ambiguity regarding the source tree of fallen fruits or seeds in plant communities in which individual shrubs and trees are widely spaced, as is the case in arid and semiarid zones. The large seeds and fruits of several dry-zone tree species are amenable to being collected from the ground. Examples are *Prosopis* spp. in Latin America (Ffolliott and Thames, 1983) and *Acacia sclerosperma, Owenia* spp. and *Terminalia arostrata* in Australia. Shed fruits may lodge in and around the base in bushy species. Large quantities of fruits of *Atriplex* spp. and *Acacia victoriae* are sometimes available in such situations.

Rain-forest species

There are several reasons why ground collecting of fallen fruits and seeds may be the preferred method for obtaining material of some rain-forest species.

- Many tree species in lowland tropical rain forests occur as scattered individuals within the forest, thereby allowing unambiguous identification of the source tree of fallen fruits.
- In thick rain-forest communities it may be difficult to sight fruit crops from the ground, reducing harvesting options.
- In some rain-forest species (e.g. *Flindersia*) the fruiting units are

sparsely and evenly distributed over the entire canopy, further restricting harvesting options.

- Fully mature, fallen fruits may be the easiest to handle and store. For example, de Muckadell and Malim (1983) found that fully mature fruits of dipterocarps were less susceptible to desiccation and rotting during handling and storage. The same authors reported that seedlings raised from such seeds demonstrated more vigorous initial growth.

A number of rain-forest leguminous species with large, hard-coated seeds (e.g. *Intsia* spp. and *Parkia javanica*) are well suited to ground collecting and this considerably extends the collecting period for such species. For most other tropical species the timing of collecting is critical. The first seeds to fall to the forest floor are often of poor quality. An indication of the soundness of fallen fruits can be obtained by cutting through the fruit with a pair of sharp secateurs and inspecting the seed embryo. Change in fruit colour can also be an important indicator of seed quality. For example, green and yellow fruits are preferred to brown fruits in *Gmelina arborea* (Woessner and McNabb, 1979), while greenish-purple fruits are preferred in *Terminalia platycarpa* (L. Thomson, unpublished data). Depulped fruits of *Azadirachta indica* (neem) collected at the yellow-green stage have been found to be suitable for long-term storage whereas seeds from more mature, yellow fruit rapidly lose viability (P. Tompsett, pers. comm.). Fruits containing sound seeds should be collected as soon as possible after shedding to minimize fungal, insect and animal attack and to reduce the incidence of mortality and germination. Early germination can be a major problem for species with recalcitrant seeds such as *Shorea platyclados*, *Syzygium suborbiculare* and *Macadamia* spp.

Seed collecting from the canopy using ground access

The harvesting technique adopted will depend on the species, plant form and stage of fruit maturity. Ground-based harvesting techniques include:

- hand picking or stripping;
- beating and shaking;
- sawing of small branches with long-handled pruning saws;
- sawing of branches using flexible saws;
- shooting down branches with rifles.

Hand picking or stripping

Hand picking is used for harvesting the fruits of shrubs and low-branching tree species in which the fruits are firmly attached to the stem. Hand picking is well suited to collecting fleshy-fruited dry-zone species such as *Boscia senegalensis*, *Grewia* spp., *Maytenus senegalensis*, *Securinega virosa* and *Vitex diversifolia* in West Africa and *Capparis* spp., *Carissa lanceolata*, *Eremocitris glauca*, *Eremophila* spp.

and *Santalum* spp. in Australia. Plants in some shrub genera (e.g. *Olearia, Cassinia* and *Maireana*) have light, fluffy seeds and fruits which may be suitable for harvesting from the stem with the aid of a portable car vacuum cleaner. This technique not only gives a very clean product, but also ensures that only fully mature seeds are collected.

If the fruit crop is sparse, fruits may be picked or stripped directly into a bag or other receptacle. For heavier crops, or where larger samples are required, it is usually more efficient to lay out a calico sheet or nylon tarpaulin on the ground adjacent to the plant and either hand-strip, or pick and drop fruits directly on to the sheet.

Secateurs can be very useful for cutting small fruit-holding branchlets, especially for those species with woody fruits which release their seeds on drying, but extreme care is required in their use. The retention of some leaf and twig material results in an open packing arrangement, giving better air flow around the fruits and thereby facilitating drying and reducing the hazards of mould and decay. However, for finer-foliaged *Casuarina* species (e.g. *C. cunninghamiana, C. equisetifolia* and *C. junghuhniana*), the cladodes (fine green stems) should be removed from cone collections as upon drying they easily break into short segments that are very difficult to separate from the samaras (winged seeds).

Thick rubber gloves are recommended when collecting fruits from species with caustic or irritant surfaces (e.g. *Arenga pinnata*: Masano, 1990; *Grevillea pyramidalis*) or which have toxic pulps (e.g. *Melia azedarach*). Leather gloves may provide some protection when collecting seeds of species with very large spines or with recurving brambles (e.g. *Acacia* subgenus *Senegalia* and *Acacia cuspidifolia*). For many spiny species fruits may be stripped along the branch using a continuous hand movement in the same direction as the spines are orientated.

Beating and shaking

Many shrubs and trees, including those from leguminous genera such as *Acacia, Leucaena* and *Sesbania*, have non-woody fruits which, when fully mature, are only very weakly attached to the bush. Fruits of these species can be very easily and efficiently collected by beating. The dry, mature fruits are beaten, usually with sticks, onto a large sheet or tarpaulin laid out beneath the bush. Sometimes, finer stems can be shaken vigorously to dislodge the mature fruits.

Beating has been found to be a very useful technique for seed collecting of many dry-zone *Acacia* spp. (Thomson and Cole, 1987), especially spiny or thorny species. However, near perfect synchronization of collecting with fruit ripening is required for optimal application of this technique. Nearly mature green fruits are not easily dislodged and may be damaged by beating, whereas fully mature and dried fruits are quickly lost from the bush during warm windy weather. Hence the period when fruits can be collected by beating is often rather limited, typically about two weeks for *Acacia* spp.

Shaking the bole may also be an effective technique for dislodging weakly held, fully mature fruits or seeds onto large sheets spread underneath the canopy. This technique has proved effective for *Cordia alliodora* (Stead, 1979) and *Acacia citrinoviridis* and *A. neurocarpa*.

Long-handled pruning saws and pole pruners

A variety of long-handled tools may be used to increase the canopy volume from which fruits can be harvested. These are illustrated in FAO (1985) and Robbins *et al.* (1981). A well-designed, long-handled pruning saw is an extremely efficient tool for removing fruit-bearing branches. The pole should be strong, rather rigid but light. An appropriate grade of 6 cm aluminium pipe is the preferred pole material. The optimum pole length is about 6 m. The pruning saw blade should be made from high-quality steel alloy, 35–40 cm long. It should be downward curving, with aggressive teeth set. It should be rigidly attached to the pole with a minimum of two fixing points, but the mounting should enable ready interchanging of blades in the field. Some commercially available pruning saws come with a cutting hook on the upper side, an arrangement that can be useful and eliminates the temptation to use the saw blade as a hook. Very long (8–10 m) poles with a hook can be used either to remove individual large fruits or to shake branches and loosen dry fruits, as is done very effectively by nomadic herders in dry tropical Africa for many acacias (e.g. *A. albida* and *A. nilotica*).

In some instances, pole pruners (consisting of 'parrot-beak' snippers mounted on a long pole) can be efficient harvesting tools and have the advantage of minimizing damage to the canopy. Pole pruners are most useful for small trees (<10 m tall) in which the fruiting units are large, with individual fruits or bunches of fruits distributed rather sparsely and evenly over the canopy.

Flexible saws

Various types of flexible saw can be used to sever fruit-bearing branches. They are illustrated in FAO (1985), Robbins *et al.* (1981) and Robbins (1984b). A weighted nylon cord is thrown over the desired branch and the saw, with heavier polypropylene cords tied at each end, is then drawn up into position. Once the saw is in position a person at each end of the rope can work it across the branch and quickly sever branches up to about 15 cm in diameter. The operation of one particularly efficient type of flexible saw is illustrated in Boden (1972). A potentially hazardous situation can arise if a flexible saw breaks in mid-use, as the broken blade may come hurtling back in the direction of the operators. Accordingly, flexible saws should be well maintained and regularly inspected for any signs of weakness.

Flexible saws are well suited for collecting seeds of medium-sized trees with a clear bole for 8–18 m and horizontal or semihorizontal branch habit, growing in open habitats. Examples include *Acacia sieberiana*, *Brachystegia spiciformis*, *Eucalyptus camaldulensis*, *Khaya* spp., *Sterculia quinqueloba* and *Sterculia africana*.

Rifle with telescopic sight

Severing fruit-bearing branches with a high-powered rifle is the preferred technique for collecting research seed samples from taller tree species. This method has being widely used by the Commonwealth Scientific and Industrial Research Organization (CSIRO)'s Australian Tree Seed Centre for sampling taller eucalypts and acacias and undoubtedly could be applied to many other tall tree species. However, extensive training and extreme care are essential. The technique has been described by Green and Williams (1969) and Kleinig and Boland (1977). The following account is derived from Boland *et al.* (1980).

The selection of branches and of the position along the branch at which the shots are to be placed is critical. It is important to select branches that will fall unobstructed to the ground and to position the shots at a point along the branch so as to take advantage of branch weight and leverage. In particular, horizontal and semihorizontal branches are more easily severed than erect branches.

The marksman should be positioned at a point so that the line of sight is at right angles to the selected branch and the rifle should be steadied against a fixed object such as a tree or the side of a vehicle. The best results are achieved by shooting to sever the bark on the underside of the branch with the first shot. In this way, branch hang-ups are avoided. Subsequent shots are placed at intervals across the branch. The recommended equipment is a .308 calibre rifle fitted with a ×4 to ×8 magnification telescopic sight and soft-nosed bullets.

Limbs with strong, fibrous bark may be difficult to dislodge completely because the severed branch may remain attached or hanging by a ribbon of bark. This is especially a problem for certain eucalypts (e.g. *E. globoidea*) and tropical acacias (e.g. *A. aulacocarpa* and *A. mangium*), which have very strong, fibrous barks. It is essential to shoot down any hung-up branches before moving on as these constitute a major hazard. Shooting from a different angle or position will often bring down branches which are otherwise difficult to dislodge.

Collecting by climbing into the canopy

Climbing may be a suitable alternative method of harvesting fruit from tall trees when rifles are either not available, not permitted (legal restrictions, closely settled areas) or not especially suitable (e.g. conifers and many rain-forest species). Well-trained climbers, operating with appropriate ancillary safety equipment, can make climbing a reasonably efficient and safe operation (FAO, 1985). Nevertheless, Landsberg and Gillieson (1982) have noted that conventional tree climbing with the aid of climbing spikes and safety belt is usually slow and can be hazardous.

Single-rope techniques for obtaining access to the canopy of tall trees have been found to be especially useful in the tropics (Perry, 1978; Perry and Williams, 1981; Landsberg and Gillieson, 1982). There is a wealth of published information on various aspects of tree climbing techniques, equipment and safety, which should be studied carefully by prospective climbers (Seal *et al.*, 1965; Pitcher, 1966; Dobbs *et al.*, 1976; Yeatman and Nieman, 1978; Whitacre, 1981; Barner and Olesen, 1983a, 1984a,b; Robbins, 1983a,b; Ochsner, 1984; FAO, 1985; Perry and Williams, 1985; Stubsgaard, 1987).

Handling, processing and transport of fruits and seeds

Chapter 20 gives information on the theory and practice of seed handling and storage in the field. Some specific examples are presented here to illustrate the main general points as they regard woody perennials.

Handling of fruit and seed samples

Fruits must be handled carefully after harvest to ensure maintenance of seed-lot viability. They must be kept in an environment that minimizes deterioration prior to seed extraction or return to base, with regular inspections, every two to three days or so, to allow early detection of deterioration due to fungi and insects.

The fruits of species with orthodox seeds are best stored under ambient, well-ventilated conditions in coarse cloth sacks, calico bags or securely tied calico sheets. In contrast, the fruits of species with recalcitrant seeds may need to be stored under cool, humid conditions, though it should be remembered that temperatures of $<20\,°C$ can cause chilling damage to recalcitrant seeds of tropical species.

The difficulties in handling the seeds of recalcitrant species may be illustrated by *Syzygium suborbiculare*. Seeds in the fleshy fruits of this species may germinate soon after collecting (61% after ten days) while being stored in a calico bag. In the field it may be possible to slow germination by storing the fruits in a moist, sterile medium such as peat, with added fungicide, under some kind of refrigeration (Turnbull *et al.*, 1986). The collecting of recalcitrant seeds must be accompanied by detailed forward planning and preparations for their immediate use as they have a limited storage life, even under optimum conditions. For example, de Muckadell and Malim (1983) have recommended that the time between seed collecting and sowing of Malaysian dipterocarps with recalcitrant seeds should not exceed six to seven days.

Field processing of fruit and seed samples

For extended collecting expeditions, it is usually necessary to extract the seeds or at least partially process the collected material in the field in order to reduce bulk and weight.

Methods of seed extraction will depend largely on the characteristics of the fruit. The first step for species with orthodox seeds is to dry the fruits in a sunny, well-ventilated environment. Sun drying has been found suitable for fleshy-fruited *Anthocephalus chinensis*. The seeds may be extracted later by rubbing them against a roughened surface and sieving (Chacko, 1981). In dry, sunny weather fruits can be dried on sheets or tarpaulins (check for holes!) spread on the ground in open locations. All sides of the sheet should be properly weighed down to avoid it being lifted by gusts of wind. Seeds should not be dried in close proximity to ant nests and sometimes application of insecticide around the drying area (not to the seed) may be necessary to prevent ants from removing seeds.

For bulk cone collecting of tropical pines a heavy-duty canvas tarpaulin of 5×10 m has been found satisfactory (Robbins, 1984a). The cones need to be raked frequently and released seeds should be collected each day. A calico sheet (2×2 m) has proved ideal for individual tree collecting of *Eucalyptus* species. Prior to seed extraction the fruit-bearing branchlets from each tree sampled are held in a separate sheet, which can be opened up for drying in sunny weather and quickly tied into a secure bundle in the late afternoon or for transport. It takes one or two sunny days to extract the seeds from the capsules of most *Eucalyptus* spp.

In cold and/or rainy weather it may be necessary to dry fruits under cover. A polythene tent can be quickly set up in the field for this purpose. Care should be taken to avoid excessively high temperatures building up in such structures. In some locations it may be possible to have access to open-sided sheds (e.g. hay sheds) or glasshouses for drying fruit samples.

Many tree species have dehiscent fruits which readily release their seeds upon drying. For most of these species (e.g. *Eucalyptus* spp.), the fruits will dry more quickly if cut with some twig (with or without leaves) attached. Some fruits shatter energetically upon drying, scattering their seeds widely. Care needs to be taken when drying the fruits of these species (e.g. *Acacia difficilis*, *Goodia lotifolia* and *Petalostigma* spp.) to ensure that their seeds do not inadvertently contaminate other seed lots. At the other end of the spectrum are species whose fruits require mechanical force to release the seeds. Even when the pods have been fully dried, the seeds of many *Acacia* species and other woody legumes may remain firmly attached to the pod or enclosed in sections of pod. Two species in the latter category, *Pterocarpus indicus* and *Dendrolobium umbellatum*, can be stored and germinated in this state. Various mechanical methods have been used to fully extract such seeds. However, though some examples of these will be given, it must be stressed that the recommendation is that only manual methods be used for seed lots destined for long-term gene-bank storage (Chapter 20).

The large, hard-coated seeds of some phyllodinous *Acacia* species (e.g. *A. anaticeps*, *A. pachycarpa*, *A. platycarpa* and *A. wanyu*) can be separated from the pod by placing the pods between a heavy-duty tarpaulin and running the wheels of a vehicle over them several times. For a large number of hard-seeded leguminous species (e.g. many species of *Acacia, Adenanthera, Albizia, Cathormium* and *Dendrolobium*: Searle, 1989), mechanical threshing has proved a suitable technique for separating the seeds from the pod. The Australian Tree Seed Centre has modified the 15 cm flailing thresher described in Doran *et al.* (1983) to make it belt-driven by a small 5 horsepower (HP) generator or electric motor and lighter and more readily assembled and disassembled in the field. The portable thresher has been successfully used for field processing of bulky *Acacia* seed samples, in both arid zones and the humid tropics. Operators wear protective breathing apparatus to minimize inhalation of the irritating dust produced during threshing. For *Acacia mangium*, threshing early in the morning when the pods are slightly moist can reduce the dust hazard. In contrast, the pods of *Acacia crassicarpa* need to be brittle dry to be sheared apart in the thresher: they are best processed in mid-afternoon. In the field, clean seed lots can be obtained from the threshed mixture by flotation, with viable seed sinking (Doran *et al.*, 1983). Wet processing also alleviates the dust hazard. There has been no experimental study of the effects of flotation seed cleaning methods on seed viability, but it is not recommended for long-term conservation (Chapter 20).

The extraction of seeds of tropical pines may be considerably hastened by rotating the partly opened cones in a tumbler. In Thailand, a manually operated tumbler consisting of a rotating drum (50 cm diameter, 240 cm long) made of iron rods at 10 cm intervals was very effective for *Pinus kesiya* and *P. merkusii* (Granhof, 1984).

A number of woody perennials (e.g. *Banksia* spp.) have massive serotinous cones and their seeds are only naturally released after exposure to fire. In the field the seeds of many *Banksia* spp. can be extracted by briefly (5–10 minutes) barbecuing the fruits on a grill over a fire, or by setting the cones alight for those species with persistent perianths or styles.

Removal of the pulpy part of fleshy fruit while still wet is usually done with a knife, and is very laborious. Lauridsen (1986) achieved rapid (⩽24 hours after collecting) and complete depulping of mature *Gmelina arborea* fruits by threshing, mashing or tumbling them to loosen the skin, followed by thorough washing. Searle (1989) reported successful depulping of a wide range of fleshy-fruited tropical trees using a concrete mixer and varying combinations of sand, rocks and water.

Clean seed lots can be obtained by sieving or winnowing to remove unwanted material. A set of interlocking sieves of different mesh size, with a base tray, is essential equipment for seed collecting of most woody plant species, especially those with round seeds. In order to obtain reasonably clean seed lots in the field, the sieve set should include a sieve of a mesh size which will just allow all seeds of the target species to pass through and a sieve which will just retain more than 95–99% of those seeds. A funnel and scoop can facilitate the sieving operation. It should be borne in mind that for some species seed size can vary both between years and between areas. For example, *Acacia mangium* seeds from Maluku Province, Indonesia, are smaller than seeds from other areas. In some species there may also be marked variation in the size of seeds collected from a given tree or population. The round seeds of *A. citrinoviridis* may vary between 3 and 6 mm in diameter from one individual to another. If examination of cut seeds shows that the smallest do not have a fully formed embryo and are not likely to be viable, these seeds can be sieved off.

Storage and transport of samples

After extraction and cleaning, dry orthodox seeds should be stored in calico bags out of direct sunlight and away from other heat sources. If seeds have been dried to acceptable levels of moisture content, the bags should be placed in air-dry containers. The seeds of some species (e.g. *Eucalyptus deglupta*) lose viability rapidly at ambient temperatures and samples are best maintained under refrigeration (about 4°C) after extraction.

During extended field missions it may be necessary to periodically send samples back to base. This would apply to extensive provenance collecting of species which fruit over a brief period and which have fruit crops that are bulky or that can only be processed slowly. Periodic dispatch of material from the field may also be required for recalcitrant seeds or in very humid areas. Some seeds and fruits require special packaging for long distance transport. The large pear-shaped fruits of the bamboo *Melocanna baccifera* perish rapidly and require careful packing in dry sand or charcoal for transport (Zabala, 1990). The fruits of *Gmelina arborea* need to be handled carefully to avoid damage to the skin and subsequent rapid fermentation. Fruits should be transported in open weave cloth sacks out of direct sunlight (Lauridsen, 1986).

When sending material back to base from the field it is vital that all necessary documentation and instructions are securely attached to the outside of the container and clearly visible. Documentation to accompany consignments of plant germplasm being transported across national borders will include an export permit, import permit and phytosanitary certificate. It is also advisable to include explicit instructions for handling at the receiving end (Willan, 1988). Collectors should be aware that even brief fumigation with methylbromide can be fatal for seeds with high moisture content (e.g. recalcitrant seeds) and many conifers.

Vegetative germplasm collecting procedures

Collecting seeds may not be possible or feasible for some woody species, so that collecting vegetative material may constitute the only available means of germplasm acquisition. Also, for certain purposes it may be desired to clone interesting phenotypes of mature trees directly (e.g. 'plus trees', which are individuals displaying outstanding vigour, disease resistance, outstanding bole form or good fodder value). However, the movement of vegetative material poses a greater quarantine hazard than the movement of seeds. Woody species may be collected vegetatively as:

- root cuttings;
- shoot cuttings;
- *in vitro* meristems;
- seedlings.

Root cuttings

A small number of woody species and populations have lost the ability to reproduce by sexual means. Examples include the Kalamunda (Western Australia) population of *Acacia anomala* and western Victorian populations of *Casuarina obesa*. These now only have the capacity to reproduce clonally by root suckering. More importantly, many other tree species with considerable socioeconomic value and high levels of genetic diversity (e.g. *Acacia albida, A. ampliceps, A. melanoxylon, Casuarina glauca, Pterocarpus indicus*) possess the ability to reproduce clonally by means of root suckering in addition to producing seeds. Collecting germplasm of these species can be undertaken by sampling suitable woody root material if seeds are not available.

To collect root cuttings, roots lying near the soil surface are followed outwards from the tree and segments are taken from them of diameter 8–15 cm and length 25–30 cm. After shaking off excess soil and washing, these are stored in a moist, sterile medium (e.g. peat moss) in plastic bags away from direct sunlight and other heat sources. For longer field trips it may be desirable to keep the root cuttings under cool storage (5–10 °C). Root cuttings are normally cut into smaller sections (5 cm) for planting.

Root cuttings have several advantages over shoot cuttings. They are more robust and easier to handle, yield a higher percentage establishment and provide physiologically juvenile plants with better developed root systems.

Shoot cuttings and in vitro techniques

Shoots suitable for taking cuttings should be healthy and of sufficient size to allow later division and trimming. The last fully mature growth flushes from the exterior canopy of the tree or shrub are ideal. It may be possible to stimulate the formation of adequate shoots by scoring the trunk in some species; these shoots could then be collected during a second visit (D. Boland, pers. comm.). Leafy shoots should be wetted as soon as possible after excision and leaf area reduced (by 75% or more) to minimize transpiration stress. Soaking for 10 minutes in soapy water, a 0.5–1.0% bleach (sodium hypochlorite) solution and/or a fungicidal and insecticidal dip will help reduce contamination. Cut ends can then be coated with a sealant such as melted low-temperature paraffin wax. Cuttings should be stored in closed plastic bags in a cool location (<22–$25\,^{\circ}$C) out of direct sunlight. It is preferable not to tie cuttings up in bundles, as this may lead to bruising. Aronson *et al.* (1990) give details of a method for the long-distance transport of cuttings under tropical conditions which involves placing the proximal cut ends in an aerated biogel solution in a polythene sack, the whole covered by a moistened paper sack. Green budwood cuttings of rubber trees are packed in boxes or cartons in single layers separated by 1–2 cm of damp, aged sawdust. Shoot material will be difficult to maintain in good condition for more than several days and should be rapidly transited to base for best results.

For the majority of woody species there are difficulties in rooting leafy cuttings taken from mature individuals. One solution is to collect scion material from mature trees in the field for grafting on to seedling stock plants of the same species. This approach has been used for various pines and *Terminalia superba*. It has proved useful for *Triplochiton scleroxylon* plus trees in southern Nigeria (A. Ladipo, pers. comm.) and high cineole phenotypes of *Eucalyptus camaldulensis* from northern Queensland, Australia (J.C. Doran, pers. comm.). It is fair to say that, for most forestry species, the best technique for collecting and regenerating shoot cutting has yet to be identified.

For extended collecting expeditions to remote localities it is likely that simple *in vitro* collecting techniques for buds and shoot apices, especially when coupled with micrografting, will prove more useful in the future than collecting large shoot cuttings, which are more difficult to transport and prone to rapid deterioration during transit. Various *in vitro* collecting approaches have been shown to be successful or are being explored for a wide range of woody genera including *Acacia, Citrus, Cocos, Eucalyptus, Gossypium, Pistacia, Prunus* and *Olea*. Information on *in vitro* techniques is given in Chapter 24 (see also FAO, 1993; 1994).

Seedlings

Collecting seedlings (wildlings) may constitute a useful method for those woody species which are either viviparous, have recalcitrant seeds, are shy seeders or otherwise present difficulties in seed collecting. It is evident that germinated seeds and seedlings should be planted as soon as possible after collecting.

In a small number of woody plants, notably various mangrove species in the genera *Aegiceras, Avicennia, Bruguiera* and *Rhizophora*, the seeds germinate while still attached to the mother plant. Therefore, germplasm sampling of such viviparous plants usually entails collecting germinated seeds. Following collecting, germinated and germinating seeds should be kept in a moist but well-aerated environment, i.e. they should not be allowed to desiccate nor should the water be allowed to become stagnant. A reduced light environment is desirable, but live plants should not be stored in total darkness for more than two or three days.

For some tree species collecting live seedlings may be a more feasible method than collecting seeds. One such example is *Oreocallis brachycarpa*, a fast-growing, valuable timber species from Papua New Guinea. This tree occurs in remote localities, sets seed in the wet season when access is very restricted and produces winged seeds that are shed as soon as they mature. Collecting in remote areas is expensive and resources may not be readily available for follow-up collecting if seed crops are immature, already shed or absent. In such cases it can be prudent to collect live seedlings. Howcroft (1978) collected large numbers of seedlings (100–500 per population) over a broad area in stands of *Araucaria cunninghamii* and *A. hunsteinii*, for which seeds were not available. These seedlings were used to establish *ex situ* conservation stands in more accessible locations.

Cosampling of root microsymbiont germplasm

The majority of woody plants form root symbioses with ectomycorrhizal and/or vesicular–arbuscular mycorrhizal (VAM) fungi. In addition, some woody angiosperms (some 200 species in 24 genera and eight families) form root nodules in association with the actinomycete *Frankia*. Root nodules are also formed by the bacteria *Rhizobium* and *Bradyrhizobium* on species in the family Leguminoseae and the genus *Parasponia*. These symbioses, which are often close genetic associations between host and microorganism, can be essential for good plant growth. Cosampling of plant and root microsymbiont germplasm during field missions is thus sometimes necessary, especially when the target area is remote, because funding for separate root microsymbiont germplasm collecting is often not available. It should, however, be noted that approaches to sampling of root microsymbionts generally require a high level of experience and specialist skills. Chapter 26 provides an introduction to the subject.

References

Amatya, S.M. (1990) Collecting Khanyu (*Ficus semicordata*) figs for seed: a practical application. *Forest Research and Information Centre Occasional Paper* 2/90:27–29. Department of Forest and Plant Research, Kathmandu.

Aronson, J., G. Wickens and E. Birnbaum (1990) An experimental technique for long-distance transport of evergreen or deciduous cuttings under tropical conditions. *FAO/IBPGR Plant Genetic Resources Newsletter* 81/82:47–48.

Bastide, R. and A. Gama (1980) Des aspirateurs légers pour le remassage des graines forestières au sol. *Revue Forestière Française* 32:296–304.

Boden, R.W. (1972) Plant propagation. In: *The Use of Trees and Shrubs in the Dry Country of Australia*. pp. 420–434. Australian Government Publishing Service, Canberra.

Boland, D.J. (1978) Geographic variation in *Eucalyptus leucoxylon* F. Muell. *Australian Forest Research* 8:25–46.

Boland, D.J., M.I.H. Brooker and J.W. Turnbull (1980) Eucalyptus *Seed*. CSIRO, Canberra.

Chacko, K.C. (1981) A simple method for extraction of seeds of *Anthocephalus chinensis. Evergreen – Newsletter of the Kerala Forest Research Institute* 7, 3.

Davies, S.J.J.F. (1976) Studies of the flowering season and fruit production of some arid zone shrubs and trees in Western Australia. *Journal of Ecology* 64:665–87.

de Muckadell, J.S. and P. Malim (1983) *Preliminary Observations on Harvesting, Handling and Storage of Seeds from Some Dipterocarps.* FAO/UNDP-MAL/78/009 Working Paper No. 18. Forestry Research Centre, Sepilok, and FAO, Rome.

Dobbs, R.C., D.G.W. Edwards, J. Konishi and D. Wallinger (1976) *Guideline to Collecting Cones of B.C. Conifers.* British Columbia Forest Service/Canadian Forestry Service, Joint Report No. 3, Victoria, Canada.

Doran, J.C., J.W. Turnbull, D.J. Boland and B.V. Gunn (1983) *Handbook on Seeds of Dry-zone Acacias.* FAO, Rome.

Eldridge, K.G. and A.R. Griffin (1983) Selfing effects in *Eucalyptus regnans. Silvae Genetica* 32:216–221.

Eriksson, G., B. Schelander and V. Akebrand (1973) Inbreeding depression in an experimental plantation of *Picea abies. Hereditas* 73:185–194.

Falkenhagen, E.R. (1977) Genetic variation in 38 provenances of Sitka spruce. *Silvae Genetica* 26:67–75.

FAO (1985) *A Guide to Forest Seed Handling (with Special Reference to the Tropics).* FAO Forestry Paper 20/2, compiled by R.L. Willan. FAO, Rome.

FAO (1993) Ex Situ *Storage of Seeds, Pollen and* In Vitro *Cultures of Perennial Woody Plant Species.* FAO Forestry Paper 113, based on the work of B.S.P. Wang, P.J. Charest and B. Downie. FAO, Rome.

FAO (1994) *Biotechnology in Forest Tree Improvement.* FAO Forestry Paper 118, by R. Haines. FAO, Rome.

Ffolliott, P.F. and J.L. Thames (1983) *Collection, Handling, Storage and Pretreatment of Prosopis Seeds in Latin America.* FAO, Rome.

Green, J.W. (1971) Variation in *Eucalyptus obliqua* l'Herit. *New Phytologist* 70:897–910.

Green, J.W. and A.V. Williams (1969) Collection of *Eucalyptus* branch specimens with the aid of a rifle. *Australian Forest Research* 4:19–30.

Hawkes, J.G. (1980) *Crop Genetic Resources Field Collection Manual.* IBPGR/EUCARPIA, Rome.

Hawkes, J.G. (1991) Theory and practice of collecting germplasm in a centre of diversity. In: Engels, J.M.M., J.G. Hawkes and M. Worede (eds) *Plant Genetic Resources of Ethiopia.* pp. 189–201. Cambridge University Press, Cambridge.

Houle, G. and L. Filion (1993) Interannual variations in the seed production of *Pinus banksiana* at the limit of the species distribution in northern Quebec, Canada. *American Journal of Botany* 80:1242–1250.

Howcroft, N.H.S. (1978) Exploration and provenance seed collections in Papua New Guinea 1976/77: *Araucaria cunninghamii* Lamb. and *Araucaria hunsteinii* K. Schum. *Forest Genetic Resources Information* 8:5–9.

Jones, N. and J. Burley (1983) Seed certification, provenance nomenclature and genetic history in forestry. In: Burley, J. and P.G. von Carlowitz (eds) *Multipurpose Tree Germplasm.* Deutsche Gesellschaft für Technische Zusammenarbeit (GTZ), IBPGR and ICRAF, Nairobi.

Kitzmiller, J.H. (1990) Managing genetic diversity in a tree improvement programme. *Forest Ecology and Management* 35:131–149.

Kleinig, D.A. and D.J. Boland (1977) Use of .243 and .308 calibre rifles for eucalypt seed collections. *Institute of Foresters of Australia Newsletter* 18:22–23.

Krusche, D. and Th. Geburek (1991) Conservation of forest gene resources as related to sample size. *Forest Ecology and Management* 40:145–150.

Lacey, C.J. (1974) Rhizomes in tropical eucalypts and their role in recovery from fire damage. *Australian Journal of Botany* 22:29–38.

Landsberg, J. and D.S. Gillieson (1982) Repetitive sampling of the canopies of tall trees using a single rope technique. *Australian Forestry* 45:59–61.

Lauridsen, E.B. (1986) Gmelina arborea *Linn.* Seed Leaflet No. 6. DANIDA Forest Seed Centre, Humlebaek.

Masano (1990) Sugar palm (*Arenga pinnata*) and its seed problems in Indonesia. In: Turnbull, J.W. (ed.) *Tropical Tree Seed Research*. pp. 18–20. ACIAR, Canberra.

Merkle, S.A., W.T. Adams and R.K. Campbell (1988) Multivariate analysis of allozyme variation patterns in coastal Douglas-fir from southwest Oregon. *Canadian Journal of Forest Research* 18:181–187.

Moran, G.F. (1992) Patterns of genetic diversity in Australian tree species. *New Forests* 6:49–66.

Moran, G.F. and J.C. Bell (1983) *Eucalyptus*. In: Tanksley, S.D. and T.J. Orton (eds) *Isozymes in Plant Genetics and Breeding*. pp. 423–441. Elsevier Scientific Publishing, Amsterdam.

Moran, G.F. and S.D. Hopper (1983) Genetic diversity and the insular population structure of the rare granite rock species, *Eucalyptus caesia* Benth. *Australian Journal of Botany* 31:161–172.

Moran, G.F. and S.D. Hopper (1987) Conservation of the genetic resources of rare and widespread eucalypts in remnant vegetation. In: Saunders, D.A., G.W. Arnold, A.A. Burbidge and A.J.M. Hopkins (eds) *Nature Conservation: The Role of Remnants of Native Vegetation*. pp. 151–162. Surrey Beatty & Sons, Sydney.

Moran, G.F, O. Muona and J.C. Bell (1989a) *Acacia mangium*: a tropical forest tree of the coastal lowlands with low genetic diversity. *Evolution* 43:231–235.

Moran, G.F, J.C. Bell and J.W. Turnbull (1989b) A cline in genetic diversity in river she-oak *Casuarina cunninghamiana*. *Australian Journal of Botany* 37:169–180.

National Research Council (1991) *Managing Global Genetic Resources. Forest Trees.* National Academy Press, Washington DC.

Perry, D.R. (1978) A method of access into the crowns of emergent and canopy trees. *Biotropica* 10:155–157.

Perry, D.R. and J. Williams (1981) The tropical rain forest canopy: a method providing total access. *Biotropica* 13:283–285.

Phillips, M.T.T. (1984) Small-scale seed collections can cause problems. *Forestry and British Timber* 13:26–27.

Pitcher, J.A. (1966) *Tree Climbers and Forest Genetics*. US Forest Service, Milwaukee.

Robbins, A.M.J., M.I. Irimeicu and R. Calderon (1981) *Recolección de Semillas Forestales*. Pub. Mics. No. 2. Escuela Nacional de Ciencias Forestales, Siguatepeque, Honduras.

Seal, D.T., J.D. Matthews and R.T. Wheeler (1965) *Collection of Cones from Standing Trees*. Forestry Record No. 39. Forestry Commission, London.

Searle, S.D. (1989) Seed collections of lesser-known trees and shrubs in Queensland, Australia. In: Boland, D.J. (ed.) *Trees for the Tropics: Growing Australian Multipurpose Trees and Shrubs in Developing Countries*. pp. 27–34. ACIAR Monograph No 10. ACIAR, Canberra.

Stead, J.W. (1979) Exploration, collection and evaluation of *Cordia alliodora* (R. & P.) Oken. *Forest Genetic Resources Information* 9:24–31.

Thomson, L.A.J. and E.G. Cole (1987) Woody plant seed collections in tropical, arid and semi-arid Australia and recommendations for international species trials. *Forest Genetic Resources Information* 15:37–48.

Turnbull, J.W. (1978) Seed collection and certification. In: *International Training Course in Forest Tree Breeding.* pp. 81–94. Australian Development Assistance Bureau, Canberra.

Turnbull, J.W. (1980) Geographic variation in *Eucalyptus cloeziana.* PhD thesis. Australian National University, Canberra.

Turnbull, J.W. and A.R. Griffin (1985) The concept of provenance and its relationship to infraspecific classification in forest trees. In: Styles, B. (ed.) *Infraspecific Classification of Wild Plants.* pp. 157–189. Oxford University Press, Oxford.

Turnbull, J.W., P.N. Martensz and N. Hall (1986) Notes on lesser-known Australian trees and shrubs with potential for fuelwood and agroforestry. In: Turnbull, J.W. (ed.) *Multipurpose Australian Trees and Shrubs: Lesser-known Species for Fuelwood and Agroforestry.* pp. 81–313. ACIAR, Canberra.

Whitacre, D.F. (1981) Additional techniques and safety hints for climbing tall trees, and some equipment and information sources. *Biotropica* 13:286–291.

Willan, R.L., C.E. Hughes and E.B. Lauridsen (1990) Seed collections for tree improvement. In: Glover, N. and N. Adams (eds) *Tree Improvement of Multipurpose Species.* pp. 11–37. Winrock International Institute for Agricultural Development, Arlington, VA, USA.

Woessner, R.A. and K.L. McNabb (1979) Large scale production of *Gmelina arborea* Roxb. seed – a case study. *Commonwealth Forestry Review* 58(2), 117–121.

Yeatman, C.W. and T.C. Nieman (1978.) *Safe Tree Climbing in Forest Management.* Forestry Technical Report No. 24. Canadian Forestry Service, Ottawa.

Zabala, N.Q. (1990) *Silviculture of Species.* UNDP/FAO BGD/85 011 Field Document No. 14. Institute of Forestry, Chittagong, and FAO, Rome.

Zagaja, S.W. (1983) Germplasm resources and exploration. In: Moore, J.N. and J. Janick (eds) *Methods in Fruit Breeding.* pp. 3–10. Purdue University Press, West Lafayette.

Technical Notes. DANIDA Forest Seed Centre, Humlebaek.

Barner, H. and K. Olesen (1983a) *Climbing into the Crown by Way of the Bole. 1. Portable Ladders.* Technical Note No. 3.

Barner, H. and K. Olesen (1983b) *Climbing into the Crown by Way of the Bole. 2. Tree Bicycles and Ladders.* Technical Note No. 5.

Barner, H. and K. Olesen (1984a) *Climbing into the Crown Directly. 2. Ladders.* Technical Note No. 8.

Barner, H. and K. Olesen (1984b) *Climbing Within the Crown. 1. Equipment and Safety Rules.* Technical Note No. 10.

Barner, H. and K. Olesen (1984c) *Seed Crop Evaluation.* Technical Note No. 19.

Granhof, J. (1984) *Seed Extraction. 2. Extraction of Pine Seed by Means of Sun Drying on Elevated Trays Followed by Tumbling.* Technical Note No. 17.

Ochsner, P. (1984) *Climbing Within the Crown. 2. Climbing Using the Prussik Knot.* Technical Note No. 11.

Perry, D.R. and J. Williams (1985) *Climbing into the Crown Directly. 3. Methods of Access into the Crown of Canopy Trees.* Technical Note No. 23.

Robbins, A.M.J. (1983a) *Climbing into the Crown by Way of the Bole. 3. The Use of Spurs for Tree Climbing.* Technical Note No. 6.

Robbins, A.M.J. (1983b) *Climbing into the Crown Directly. 1. The Advanced Line*

Technique for Gaining Access to Tree Crowns. Technical Note No. 7.
Robbins, A.M.J. (1984a) *Seed Extraction. 1. The Tarpaulin Method of Sun Drying Pine Cones and Seed Extraction.* Technical Note No. 9.
Robbins, A.M.J. (1984b) *Tools for Harvesting Tree Fruits.* Technical Note No. 13.
Stubsgaard, F. (1987) *Climbing into the Crown Directly. 4. Light-weight Catapult for Use in Connection with the Advanced Line System.* Technical Note No. 31.
Willan, R.L. (1988) *International Tranfser of Forest Seed.* Technical Note No. 35.

Collecting *in vitro* for genetic resources conservation

24

L.A. Withers
IPGRI, Via delle Sette Chiese 142, 00145 Rome, Italy.

The need for new collecting techniques

Experienced germplasm collectors will base their plans on the best information that is available on the constraints that can limit the effectiveness of the proposed project. There may only be a limited 'window of opportunity' within which the mission can be carried out, for example due to the weather or the limited availability of personnel. When the target germplasm is threatened by changing land use or other factors affecting its future availability, the uniqueness of the opportunity is further emphasized. Despite experience and good planning, however, a collecting project can still meet difficulties which have to do with biological features of the plant parts that would normally be collected. There may simply be insufficient material because of a poor growing season, general scarcity of the plant in the target region, grazing of the plant by animals (likely in the case of forage plants), damage by pests or diseases or immaturity of the material.

These problems can apply to both seeds and vegetative propagules that have a seasonal pattern of development. Plant parts that are not strictly organs of propagation or perennation, for example budwood, are more flexible in that some collectable material is likely to be available at all times. However, being growing tissues, they are less likely to be able to survive a long journey back to the gene bank. This problem of deterioration by natural processes and attack by pests is shared by recalcitrant seeds, which may germinate or deteriorate to a non-viable condition in transit. A final problem, shared to some degree by all types of plant material but most serious in the case of large vegetative propagules and recalcitrant seeds and their fruits, is excessive weight and bulk. The cost and inconvenience of transporting large amounts of material can be a severe limitation on the scope of a collecting mission.

In the face of these potential problems, collectors need as many options as possible. It is in this context that the possibility of using an *in vitro* collecting method began to be considered. In 1982, the International Board for Plant Genetic Resources (IBPGR) constituted an Advisory Committee on *In Vitro* Storage to examine the status of *in vitro* conservation technology and to highlight opportunities for developing and applying new approaches to solve bottleneck problems in the conservation of plant genetic resources. The difficulties associated with collecting germplasm of the two major categories of problem species – i.e. those with recalcitrant seeds and those that are vegetatively propagated – were recognized as such bottlenecks. It was also noted, however, that all types of plant could benefit from improved collecting techniques.

In order to examine this topic in detail, a subcommittee of the *in vitro* advisory committee met in 1983. This included experts in germplasm exploration, in the genetic resources of problem crops and in the application of *in vitro* technology to plant propagation and conservation. It was evident that *in vitro* techniques could assist in the safe conservation of problem crops (Withers, 1980, 1982). The basic premise that the subcommittee examined was that some aspects of *in vitro* technology, namely inoculation and plant regeneration, could also be adapted to allow problem material to be collected. This would, however, be a 'holding operation', rather than a full-scale *in vitro* propagation exercise.

With the help of some preliminary experiments carried out in the institute that hosted the meeting (School of Agriculture, University of Nottingham, UK), the subcommittee came to the conclusion that *in vitro* collecting had great potential. A report and recommendations were produced (IBPGR, 1984). The further development of collecting technologies for two model systems (*Theobroma cacao* budwood and *Cocos nucifera* embryos) ensued in research projects over the following years. These are described below. First, it is appropriate to analyse the principle of using *in vitro* inoculation and plant regeneration.

Adapting basic *in vitro* procedures

Useful reviews of *in vitro* culture methods include Sharp *et al.* (1983, 1984), Beversdorf (1990), Pollard and Walker (1990), Lindsey (1991) and Bonga and von Anderkas (1992). *In vitro* culture inoculation as carried out in the laboratory involves the following steps:

- selecting an appropriate inoculum tissue;
- cutting it to a suitable size;
- removing superficial soil and visible pests by washing;
- sterilizing the surface;
- washing to remove the surface sterilant;
- trimming away non-essential tissue and tissue damaged by sterilization;

- inoculating into the culture vessel containing nutrient medium and closure of the vessel;
- transferring to incubation conditions.

Sterilization and inoculation are carried out in an environment that protects the tissue from reintroduction of contaminants, usually a laminar air-flow cabinet. Incubation is carried out in a controlled environment in which temperature, light quality, light intensity and day length are regulated to allow optimum growth and development. The culture medium contains nutrients to support growth and development.

Development of the inoculated tissue will depend largely on the composition of the culture medium. Under some conditions the inoculum will be directed towards multiplication, and under others towards regeneration of an independent plant. Thus, a shoot explant could be induced either to produce multiple lateral shoots, which could be separated and multiplied further, or to produce roots, allowing its transfer to soil.

This pattern applies to 'non-adventitiously' regenerating systems, i.e. those capable of producing a plant from pre-existing shoot primordia, such as in a stem nodal explant or a zygotic embryo. Alternatively, it is possible to induce buds *de novo* (i.e. 'adventitiously') from virtually any tissue, opening up the possibility of rapid mass clonal propagation. Generally, genetic conservation applications of *in vitro* technology favour the non-adventitious propagation system. This choice is made on the assumption that genetic instability as a result of somaclonal variation is likely to be lower in such systems (Scowcroft, 1984). Although non-adventitious systems are more suitable, tissue that is likely to regenerate adventitiously should not be rejected if it is the only material available. This is an opportunity for imaginative use of the great flexibility of *in vitro* culture. For example, a piece of leaf tissue could produce plants via somatic embryogenesis or an unfertilized ovary could be cultured and pollinated *in vitro* (Dunwell, 1985; Tisserat, 1985).

In examining the stages of *in vitro* inoculation and plant regeneration from the point of view of germplasm collecting, the first point to remember is that *in vitro* collecting is a holding operation rather than propagation *per se*. It should be aimed at maintaining the material as well as possible under field conditions for a relatively limited period. A second point is that working in the field imposes limitations on what is feasible. Therefore, a decision should be made at the outset about which stages it is absolutely necessary to carry out in the field. Other operations can be done at an earlier or later time, at locations where more appropriate conditions can be provided. It may be necessary to introduce alternative or additional steps to overcome particular limitations imposed by working in the field. To illustrate how this kind of analysis can be carried out, each of the steps in the inoculation procedure listed above will be examined in turn.

1. *Selecting an appropriate inoculum tissue.* If possible, a tissue robust enough to withstand sterilization should be selected. The main choices

are: herbaceous shoots, woody shoots, zygotic embryos/embryo axes and seeds.

2. *Cutting it to a suitable size.* Surface injury should be minimized but the opportunity should be taken to remove very dirty, infested or damaged outer tissues.

3. *Removing superficial soil and visible pests.* Copious supplies of water for this purpose may not be available in the field but the water does not need to be very pure or sterile.

4. *Surface sterilization.* Sterilants should be safe and easy to transport and of low toxicity to the plant tissue in situations where adequate sterile water for subsequent washing is not available (mercury salts and strong oxidizing agents may be unsuitable in this respect). Unconventional sterilants, such as drinking-water purifying tablets or agricultural fungicides, and combinations of more than one sterilant, each at low concentration, could be used. Sterilization and subsequent stages will probably have to be carried out without benefit of an aseptic environment (but see below). This has implications for the container used for sterilization, one with a securely fitting lid being more suitable than, say, an open beaker. Sterilization is a step which can be repeated after the collected tissue reaches the laboratory. Therefore, it is worth considering simple, short-term methods to maintain adequate cleanliness rather than risking harsher treatments that might damage the tissue. Another consideration here is that many meristematic plant tissues are free of microbial contamination, being protected by overlying leaves, bracts, seed coat, etc. Advantage can be taken of this by selecting an explant that can be surface-sterilized to remove gross contamination and then dissected to remove outer tissues. The introduction of contamination into inner tissues can be avoided by frequent changes of dissecting instruments and by working carefully and swiftly.

5. *Washing to remove the surface sterilant.* The amount of washing necessary will depend on the strength and toxicity of the sterilant used. If post-sterilization washing is included in the procedure, adequate sterile water will be needed and the technique used must avoid reintroducing microbial contaminants. Also, any residual effects of sterilants (e.g. continuing fungicidal effect) will be lost by washing.

6. *Trimming away non-essential tissue and tissue damaged by sterilization.* This is another step which might reintroduce contaminants and should be avoided if possible.

7. *Inoculation into the culture vessel containing nutrient medium.* Several factors must be considered here, including the type of vessel, the number of pieces of tissue inoculated into each vessel and the type of culture medium used. The vessel must be portable and therefore robust but not excessively heavy. Plastic materials are more suitable than glass and, in certain circumstances, something as simple as a plastic bag may suffice. Placing several inocula in a single vessel will be more efficient of space but will also increase the risk of cross-contamination. The efficiency of sterilization, the length of time the tissue is in transit, the

susceptibility of the tissue to destructive injury by contaminants and the possibility of effective resterilization will all influence this decision. The culture medium must be designed to suit its purpose. If it is intended to promote development (e.g. germination of an embryo or out-growth of axillary buds), suitable growth regulators should be included along with other nutrients. This might be the case, for example, in a collecting mission of relatively short duration or when the collecting environment and the nature of the explant allow inoculation with a low probability of contamination. If it is more logical to try to suspend development, a minimal medium or even one containing growth retardants would be appropriate for the collecting period, to be replaced by a standard medium once the material reaches the laboratory. A minimal medium, particularly one low in sucrose, is less likely to support the growth of residual microbial contaminants than is a complete medium. Antimicrobial additives may also be included in the culture medium to retard the growth and destructive effects of bacteria and fungi, but possible side-effects on the inoculum should be borne in mind. A choice will have to be made between liquid or solid medium. A liquid medium is more accessible to the tissue but is less effective in retarding growth of microbial contaminants. Also, vessels containing liquid medium must be closed more securely.

8. *Transfer to incubation conditions.* This stage will be far more lengthy and hazardous than the simple room-to-room transfer carried out in the laboratory. Every effort must be made to protect the inoculated material in transit. This will influence the choice of culture vessels used and of the container used to transport them. Attention should be given to the conditions likely to be experienced in transit (e.g. fluctuating temperatures, physical disturbance). There is the possibility of using refrigeration to retard deterioration in collected material that is adequately chilling-tolerant.

The stages of the *in vitro* collecting procedure have been analysed above from the point of view of the demands of inoculation under field conditions. A second set of considerations will relate to the constraints imposed by the nature, scale and duration of the collecting mission, the extent to which *in vitro* collecting is a central activity and the expertise of personnel. There are many variables to examine here and only illustrative examples will be offered. For a multi-species collecting mission, it will be necessary either to design different facilities and procedures to match the different needs of each of the materials to be collected, or to use a very general collecting approach to be followed up by different treatments in the laboratory. If *in vitro* collecting is a backup rather than the main approach used, the levels of replication and the extent of the resources and time given to it can be reduced accordingly. If the collecting mission personnel have only basic *in vitro* experience, the field operations should be designed to match their level of expertise, leaving as much as possible to be carried out later in the laboratory by experienced people. This is, in fact the most likely situation, since it is more logical and feasible to train collecting experts in the principles of *in vitro* inoculation than to train *in vitro* technologists to enable them to carry out specialized collecting.

As well as adapting the inoculation procedure, it is necessary to adapt the equipment used, taking into account essential requirements and the constraints imposed by having to carry all items to the field, where services will be limited. What is certain is that the equipment used should be robust, simple to operate, as maintenance-free as it can be made and, if possible, multipurpose. Thus, a packing-case to hold instruments and culture vessels can also serve as a work bench and/or inoculation hood. The examples of the successful application of *in vitro* collecting given below illustrate the great flexibility of *in vitro* collecting in terms of the equipment that can be used and the degree of sophistication of the procedures that have been tested.

As the field collecting stage is a holding operation, an *in vitro* laboratory will be needed both before and after the collecting mission. As much as possible should be done in advance (e.g. preparation of culture medium, labelling of vessels, cleaning and sterilization of instruments), and as much as possible deferred to the receiving laboratory (e.g. resterilization of inoculated tissue, transfer of tissue to complex media).

Back at the laboratory

At the receiving laboratory, each type of explant will need to be handled by a different procedure to maximize its chances of survival. The other crucial variable is the destination of the material. Disease eradication, indexing and quarantine stages may need to be integrated into the process. Decisions will thus have to be made at the planning stage about how the material collected will be handled at the receiving institute. The important questions that need to be asked – and answered – are as follows.

Can the culture simply be placed in a controlled environment room
or, alternatively, does it require resterilization and further dissection
to yield an explant free of contamination and suitable for culture?

Obvious examples of the latter situation would be coconut zygotic embryos enclosed in an endosperm plug that need to be resterilized and dissected. Seeds might require resterilization and probably, since they are likely to be immature if selected for *in vitro* collecting, dissection to extract the embryo or embryo axis for rescue *in vitro*. For shoots, particularly woody ones, resterilization and dissection for grafting *ex vitro* or *in vitro* may be relevant. In the case of the *in vitro* option, this might be a method similar to conventional bud grafting but using an *in vitro* germinated rootstock or it might involve micrografting, which can be used to eliminate viruses (Navarro, 1981; Kartha, 1986). V.M. Villa-lobos (pers. comm.) and L.A. Withers (unpublished) have successfully grafted *in vitro* cacao shoots from somatic embryos and axillary buds from woody shoots respectively on to seedlings.

What type of culture is relevant to the intended destination of the material?

If the priority is to get the germplasm into storage under conventional conditions in a field gene bank or to produce seeds for storage in a seed gene bank, it is logical to produce an independently growing plant by the most direct route possible. Clearly, use of a particular *in vitro* collecting method will depend on the existence of a successful method of regenerating plants from explants. Information on this will have to be sought on a species-by-species basis from the relevant literature. If the material is so limited in quantity that it would be a good idea to multiply it before transfer to conventional conservation in the field or seed gene bank, or if its destination is the *in vitro* gene bank, then propagation *in vitro* to produce several plants should be considered.

For most species with orthodox seed, the conventional seed gene bank holding seeds at low moisture content and low temperature will be the eventual destination. This storage method is technically uncomplicated and offers a high degree of security (Justice and Bass, 1978; Ellis *et al.*, 1985). For recalcitrant seed, conservation is more problematic (Chin and Pritchard, 1988). Most such species are currently conserved in field gene banks, but new *in vitro* conservation methods, particularly cryopreservation (storage in liquid nitrogen), may offer a viable alternative in the near future. These alternatives are very welcome as the field gene bank is a costly and risky approach to conservation, the germplasm being exposed to loss by weather damage, disease and neglect (Withers, 1989; Withers and Engels, 1990).

For material unable to produce seeds because it is sterile (e.g. *Musa*) or for clones for which seed storage would break up valuable gene combinations (e.g. root and tuber crops, many fruit trees), storage in field gene banks is also the most common conservation method. However, *in vitro* conservation by cryopreservation for the long term or slow growth for the short to medium term again promises to provide a safer alternative. More details of *in vitro* storage methods are given by Withers (1980, 1985a,b, 1987b, 1992), De Langhe (1984) and Dodds (1991).

Given that the decision has been made to multiply the material *in vitro* before regenerating plants, what is the most suitable propagation system?

Generally speaking, non-adventitious systems offer greater genetic stability. Therefore, preference should be given to simple nodal cutting methods such as are common in potato (Espinoza *et al.*, 1984), cassava (Roca, 1984) and some woody species, or to multiple shoot formation, as used for some ornamentals and temperate fruits (Conger, 1981; Tisserat, 1985). Where this approach is impossible or impractical due to the nature of the explant and/or the state of propagation technology for the species in question, the next most suitable options of direct adventitious shoot formation or embryogenesis should be taken. These would be the likely routes for *Musa* shoot explants (Vuylsteke, 1989) or immature zygotic embryos of cacao (Pence *et al.*, 1980) respectively. The least suitable system would be one involving callus formation and indirect, adventitious regeneration (Scowcroft, 1984). Callus should only be used if no other option is available. However, some of the most suitable culture systems from the point of view of genetic stability may not be very amenable to *in vitro* conservation. Thus it may be necessary to compromise and choose a less suitable culture system so that the security of *in vitro* conservation can be exploited.

A further point to consider relates to the maintenance of genetic diversity. The basic objective of germplasm collecting is to acquire representative genetic diversity of the target gene pool from the target region. This should not be forgotten in the handling of the germplasm after it has been collected. Just as the collecting strategy should be designed to sample the maximum level of genetic diversity, the plant regeneration and/or multiplication strategy should be designed to maintain that level of genetic diversity. Thus, clonal propagation *in vitro* should be used if it is necessary to bulk up a rare genotype, but not to multiply one genotype at the expense of others. The plant regeneration procedures used should be widely applicable across the range of genotypes involved, so that propagation does not act as a genetic 'filter'.

Examples of *in vitro* collecting

It cannot be emphasized too strongly that the great advantage of *in vitro* collecting is its adaptability and flexibility. For this reason, no hard-and-fast rules or recipes are offered, only general guidelines to help in the adaptation of the concept to new species and situations. The purpose of this section is to present a number of examples, some published and some only reported informally, that illustrate the scope of the approach and how it could assist germplasm collecting missions (see also Withers, 1987a).

Theobroma cacao

One of the earliest species to be explored was cacao, for which an attempt was made to find an alternative to transporting budwood sticks from the collecting site to the nursery. Experiments by Yidana and colleagues (Yidana *et al.*, 1987; Yidana, 1988) demonstrated that a sterilization procedure for nodal stem segments using drinking-water purifying tablets (containing the active ingredient 'Halozone' (*p*-carboxy-benzenesulphondichloroamide) at a concentration of $0.4 \, \mathrm{g \, l^{-1}}$) and an agricultural fungicide (e.g. FBC Protectant Fungicide at 0.05%) was effective without subsequent washing.

Inoculation on to semisolid medium containing fungicide (e.g. Tilt MBC at 0.1%), with or without antibiotics (e.g. rifamycin plus trimethoprim, each at $30 \, mg \, l^{-1}$), would maintain the tissues in a relatively clean, although not absolutely contaminant-free, condition for up to six weeks. (Optimization of sterilants and culture media was carried out using a leaf disc assay system.) Shoot outgrowth and occasional rooting were reported. The field equipment was minimal, consisting of racks of plastic tubes of culture medium prepared in advance in the laboratory, jars of boiled water, preweighed/precounted sterilants, plastic forceps, and scissors/secateurs. Inoculation was carried out in the open air. The major limitation of this protocol was the lack of a good *in vitro* propagation method to process the material collected. However, it does illustrate the extent to which the rules of *in vitro* culture can be stretched.

Gossypium

Collecting germplasm of cotton and its wild relatives can be hindered by the unpredictable availability of viable seeds. Accordingly, Altman *et al.* (1990) attempted to develop an *in vitro* method for use in the field in Mexico. Stem nodal cuttings were surface-sterilized with 20% commercial bleach in 30% ethanol for 45 seconds and then inoculated directly into a medium containing half-strength salts, 1% glucose, the antibiotics rifamycin and trimethoprim each at $15 \, mg \, l^{-1}$, $1 \, g \, l^{-1}$ of the fungicide Tilt MBC, $1 \, mg \, l^{-1}$ naphthalenacetic acid, $0.5 \, g \, l^{-1}$ casein hydrolysate and $9 \, g \, l^{-1}$ agar. No special work bench or other protective environment was used, transfers being made in the open air. After being in transit for up to three weeks, the cuttings were resterilized with 4% bleach solution, treated with rooting hormone and planted in a sterile soil/sand/vermiculite mix supplemented with lime and slow-release nutrients.

Although it was clear that the collecting stage itself was successful, difficulties were experienced in rooting and raising plants from the cuttings. This again emphasizes the importance of developing or adapting a method for processing the collected material.

Forage grasses

Forage grasses share many of the collecting problems of cacao and cotton, as well as the risk of grazing by animals, but they are structurally very different. Thus, in the case of the grasses *Digitaria* and *Cynodon*, the available explant was a herbaceous plantlet. Despite the less robust nature of this material and its different *in vitro* culture requirements, it could be treated similarly to woody material. Ruredzo (1989) used a simple method similar to that of Yidana for cacao. In this case, the collecting site was sufficiently close to the collector's base (a hotel) for the inoculation to be carried out there. In such circumstances, it would be unnecessarily demanding to carry out the inoculation in the field.

Cocos nucifera

The final example in this brief survey of model plant systems and technical approaches is coconut. Two main points are important here. Firstly, coconuts are bulky and heavy, making them costly to transport. Secondly, coconut seeds are recalcitrant. Accordingly, several workers have adapted *in vitro* embryo culture techniques to the field collecting of coconut germplasm.

In all cases, the basic sequence of operations is as follows:

• dehusking and cracking open the nut;
• extracting a plug of endosperm containing the embryo using a cork borer;
• dissecting the embryo;
• inoculating into culture.

The differences between the various approaches lie in the degree to which an attempt is made to reproduce laboratory conditions in the field, the amount of culture work actually carried out in the field as opposed to the recipient laboratory and, consequently, the point at which sterilization is carried out.

The simplest method is that employed by Rillo and colleagues in the Philippines (E.P. Rillo, pers. comm.). The endosperm plugs were transported from the field in refrigerated plastic bags containing coconut water collected from some of the opened nuts. Sterilization and inoculation were carried out in the recipient laboratory (Rillo and Paloma, 1990). A simple insulated container kept the endosperm plugs at an adequately stable temperature in transit. The field requirements were minimal: implements to dehusk and open the nuts, cork borers, plastic bags, carrying containers. Personnel training was also undemanding. All that was required in addition to the normal collecting skills was the ability to recognize the position of the live embryo and extract it in the endosperm plug without damage. Instruction in this technique was included in a training course cosponsored by IBPGR and the Philippines Coconut Authority (PCA) in 1990. Students very quickly and easily learned the embryo extraction and inoculation process.

More complex protocols were used by Assy Bah *et al.* (1987) and Luntungan and colleagues (H. Luntungan and J.S. Tahardi, pers. comm.) in Côte d'Ivoire and Indonesia, respectively. They carried out surface sterilization of the endosperm plugs in the field using calcium hypochlorite at $45\,\mathrm{g\,l^{-1}}$ and inoculation into a simple salt solution (KCl at $16.2\,\mathrm{g\,l^{-1}}$) for transport back to the laboratory, where a second sterilization could be carried out if necessary, followed by dissection and inoculation of the embryo onto standard culture medium (Assy Bah *et al.*, 1989).

The field equipment requirements were more complex than in the first example. It was necessary to provide sterilants, sterile water for washing, containers of salt solution, a burner, a spirit lamp to sterilize instruments and flame openings of containers, forceps and a simple workbench. More training in the appropriate manipulations was also required. Assay Bah and colleagues also carried out direct inoculation into culture in the field. This obviously required more manipulative skills, greater care in handling the explants and a more protective working environment, such as an upturned box to keep out airborne contaminants.

The techniques described above accept the limitations of working in the field and take a relatively low-technology approach to solving them. The method of Sossou *et al.* (1987) follows a very different logic. In this case, every effort was made to overcome the inadequacies of the field environment and provide an inoculation facility that was almost as sophisticated as the laboratory. An inflatable glove box which could be sterilized with alcohol was used and the inoculation procedure was exactly as would be carried out in the laboratory, requiring the same level of manipulative skills. The glove box, pump, instruments, lamp, solutions and culture vessels all had to be transported with care to and from the collecting site.

In coconut, collecting, embryo culture, establishment of *in vitro*-germinated embryos in the field and germplasm distribution *in vitro* are all well developed (Assy Bah *et al.*, 1987, 1989). However, reproducible clonal multiplication techniques are not yet available. Somatic embryogenesis would appear to be the most likely approach. Zygotic embryos can be maintained for up to a year in slow growth but shoot or somatic embryo cultures would be preferable for this method. Cryopreservation of zygotic embryos, although at an early stage of development, appears promising (Assy Bah and Engelmann, 1992a,b). The model of oil-palm (Engelmann *et al.*, 1985) would suggest that, once a suitable somatic embryo system has been developed, its cryopreservation should be attainable.

Other species

All experience points to the *in vitro* collecting approach being widely applicable and easily adapted. In an IBPGR-sponsored training course held at the Centro Agronémico Tropical de Investigaçión y Enseñanza (CATIE), Costa Rica, in 1990, students experimented with adapting the basic methodology to a wide range of species. These included banana, coffee and citrus, all of which were successfully established in culture via either vegetative tissues or embryos. Methods for *Prunus* spp. and *Vitis* spp. based on the approach used for cacao described above have been developed (Elias, 1988).

Examples involving various root and tuber crops are briefly described in Chapters 21 and 32. In cassava, there is a well-established methodology for all stages of conservation from collecting to slow-growth storage and distribution, including, for example, the clonal propagation stage that is a problem with coconut. IBPGR (and now the International Plant Genetic Resources Institute (IPGRI)) and the Centro Internacional de Agricultura Tropical (CIAT) have been collaborating since 1987 to develop and test management procedures for *in vitro* active gene banks using cassava as a model (Chavez *et al.*, 1987). One hundred genotypes selected from CIAT's field gene bank have been introduced into culture via disease eradication and indexing and transferred to slow-growth storage. Characterization using morphological descriptors has been supplemented by isozyme analysis and these data are being used as standards against which to assess the stability of germplasm recovered from the gene bank for periodic monitoring.

Conclusions

These examples illustrate the flexibility of *in vitro* collecting. They involve different levels of complexity at the field stage and a range of explants suitable for different species. The spectrum of examples is given as a stimulus to the imagination rather than for exact adoption. In fact, the work of Sossou *et al.* (1987) with coconut, for example, is probably unnecessarily complex. It is important to judge very carefully the need for such complicated procedures because simpler alternatives, in this case coconut embryos enclosed in endosperm plugs, may be perfectly adequate. It should be carefully considered whether ingenious but intricate procedures are really necessary, whether they are compatible with other tasks to be accomplished on the collecting mission and whether they would, by their nature, place too great a burden on transport, training and technical backup.

Thus, the important message is that the operations carried out in the field should be only those that absolutely must be carried out there. These operations will be defined by the condition of the plant material, the nature of the collecting environment and the duration of the journey back to the laboratory. Any other operations should be delayed until the collected material has reached the laboratory.

The objective of this chapter has been to set the scene by showing how some of the problems of germplasm collecting could be overcome by the imaginative application of *in vitro* procedures. It is emphasized that this new approach to collecting is technically unsophisticated but cannot be undertaken without adequate background knowledge, preparation and planning, because it must be seen as part of a comprehensive conservation scheme that flows from quarantine and disease indexing to storage, distribution and use. *In vitro* collecting is not a means of circumventing quarantine and disease indexing procedures: it may reduce the risk of introducing pests and diseases (IBPGR, 1988) but this does not lessen the need for vigilance and the need to comply with appropriate phytosanitary procedures and regulations. After processing through quarantine, disease indexing and disease eradication procedures, the collected germplasm should be multiplied in preparation for storage. There must be a plan to store the germplasm safely by slow growth (short- to medium-term, active conservation) or cryopreservation (long-term, base conservation), making it available for use when necessary and in adequate quantities. For material stored *in vitro* to be used, plants must be regenerated: it would be useless to collect *in vitro* if no method of recovering plants were available. The material must also be distributed. It is logical to distribute *in vitro* conserved germplasm in the form of a culture and this can now be carried out for many species. In fact, the international distribution of clonal crop germplasm in the form of *in vitro* cultures from several centres of the Consultative Group on International Agricultural Research (CGIAR) is now routine (Espinoza *et al.*, 1984; Roca *et al.*, 1984; Ng and Hahn, 1985; Withers and Williams, 1985; Schoofs, 1991). *In vitro* collecting will only achieve its full potential if integrated into such a comprehensive system, and adequate planning is therefore critical.

References

Altman, D.W., P.A. Fryxell, S.D. Koch and C.R. Howell (1990) *Gossypium* germplasm conservation augmented by tissue culture techniques for field collecting. *Economic Botany* 44:106–113.

Assy Bah, B. and F. Engelmann (1992a) Cryopreservation of immature embryos of coconut (*Cocos nucifera* L.). *Cryoletters* 13:67–71.

Assy Bah, B. and F. Engelmann (1992b) Cryopreservation of mature embryos of coconut (*Cocos nucifera* L.) and subsequent regeneration of plantlets. *Cryoletters* 13:117–126.

Assy Bah, B., T. Durand-Gasselin and C. Pannetier (1987) Use of zygotic embryo culture to collect germplasm of coconut (*Cocos nucifera* L.). *FAO/IBPGR Plant Genetic Resources Newsletter* 71:4–10.

Assy Bah, B., T. Durand-Gasselin, F. Engelmann and C. Pannetier (1989) Culture *in vitro* d'embryons zygotiques de cocotier (*Cocos nucifera* L.). Méthode, révisée et simplifiée, d'obtention de plants de cocotiers transférables au champ. *Oléagineux* 44:515–523.

Beversdorf, W.D. (1990) Micropropagation in crop species. In: Nijkamp, H.K., L.H.W. van der Plas and J. van Aartrijk (eds) *Progress in Plant Cellular and Molecular Biology.* pp. 3–12. Kluwer Academic Publishers, Dordrecht.

Bonga, J.M. and P. von Anderkas (1992) In Vitro *Culture of Trees.* Kluwer Academic Publishers, Dordrecht.

Chavez, R., W.M. Roca and J.T. Williams (1987) IBPGR–CIAT collaborative project on a pilot *in vitro* active genebank. *FAO/IBPGR Plant Genetic Resources Newsletter* 71:11–13.

Chin, H.F. and H.W. Pritchard (1988) *Recalcitrant Seeds, a Status Report.* IBPGR, Rome.

Conger, B.V. (ed.) (1981) *Cloning Agricultural Plants Via* In Vitro *Techniques.* CRC Press, Boca Raton.

De Langhe, E.A.L. (1984) The role of *in vitro* techniques in germplasm conservation. In: Holden, J.H.W. and J.T. Williams (eds) *Crop Genetic Resources: Conservation and Evaluation.* pp. 131–137. Allen and Unwin, London.

Dodds, J.H. (ed.) (1991) In Vitro *Methods for Conservation of Plant Genetic Resources.* Chapman and Hall, London.

Dunwell, J.M. (1985) Haploid cell culture. In: Dixon, R.A. (ed.) *Plant Cell Culture. A Practical Approach.* pp. 21–36. IRC Press, Oxford.

Elias, K. (1988) *In vitro* culture and plant genetic resources. A new approach: *in vitro* collecting. *Lettere d'Information. Istituto Agronomico Mediterraneo (Valenzano, Italy)* 3:33–34.

Ellis, R.H., T.D. Hong and E.H. Roberts (1985) *Handbook of Seed Technology for Genebanks.* Vol. 1. *Principles and Methodology.* IBPGR, Rome.

Engelmann, F., Y. Duval and J. Derueddre (1985) Survie et prolifération d'embryons somatiques de palmier à huile (*Elaeis guineensis* Jacq.) après congélation dans l'azote liquide. *Comptes Rendus de l'Académie des Sciences* 301 Série III:111–116.

Espinoza, N.O., R. Estrada, P. Tovar, D. Silva-Rodriguez, J.E. Bryan and J.H. Dodds (1984) *Tissue Culture Micropropagation, Conservation and Export of Potato Germplasm.* Specialized Technology Document. CIP, Lima.

IBPGR (1984) *The Potential for Using* In Vitro *Techniques for Germplasm Collection.* IBPGR, Rome.

IBPGR (1988) *Conservation and Movement of Vegetatively Propagated Germplasm:* In Vitro *Culture and Disease Aspects.* IBPGR, Rome.

Justice, O.L. and L.N. Bass (1978) *Principles and Practices of Seed Storage.* USDA Agriculture Handbook No. 56. USDA, Washington DC.

Kartha, K.K. (1986) Production and indexing of disease-free plants. In: Withers, L.A. and P.G. Alderson (eds) *Plant Tissue Culture and Its Agricultural Applications.* pp. 219–238. Butterworths, London.

Lindsey, K. (1991) *Plant Tissue Culture Manual. Fundamentals and Applications.* Kluwer Academic Publishers, Dordrecht.

Navarro, L. (1981) Citrus shoot-tip grafting *in vitro* (STG) and its applications: a review. *Proceedings of the International Society for Citriculture* 1:452–456.

Ng, S.Y. and K. Hahn (1985) Application of tissue culture to tuber crops at IITA. In: *Biotechnology in International Agricultural Research.* pp. 29–40. IRRI, Los Baños.

Pence, V.C., P.M. Hasegawa and J. Janick (1980) Initiation and development of asexual embryos of *Theobroma cacao in vitro. Zeitschrift für Pflanzenphysiologie* 98:1–14.

Pollard, J.W. and J.M. Walker (1990) Plant cell and tissue culture. In: Pollard, J.W. and J.M. Walker (eds) *Methods in Molecular Biology.* Vol. 6. Humana Press, Clifton.

Rillo, E.P. and M.B.F. Paloma (1990) Comparison of three media formulations for *in vitro* culture of coconut embryos. *Oléagineux* 45:319–323.

Roca, W.M. (1984) Cassava. In: Sharp, W.R., D.A. Evans, P.V. Ammirato and Y. Yamada (eds) *Handbook of Plant Cell Culture.* pp. 269–301. Macmillan, New York.

Roca, W.M., J.A. Rodriguez, G. Mafla and J. Roa (1984) *Procedures for Recovering Cassava Clones Distributed* In Vitro. CIAT, Cali.

Ruredzo, T.J. (1989) *Progress Report on IBPGR–ILCA Tissue Culture Project.* IBPGR Report Number 89/11. IBPGR, Rome.

Schoofs, J. (1991) The INIBAP *Musa* Germplasm Transit Center. In: Musa – *Conservation and Documentation.* Proceedings of INIBAP–IBPGR Workshop. 11–14 December 1989. Leuven, Belgium. pp. 25–30. INIBAP/IBPGR, Montpellier.

Scowcroft, W.R. (1984) *Genetic Variability in Tissue Culture: Impact on Germplasm Conservation and Utilization.* IBPGR, Rome.

Sharp, W.R., D.A. Evans, P.V. Ammirato and Y. Yamada (eds) (1983) *Handbook of Plant Cell Culture – Crop Species.* Vol. 1. Macmillan, New York.

Sharp, W.R., D.A. Evans, P.V. Ammirato and Y. Yamada (eds) (1984) *Handbook of Plant Cell Culture – Crop Species.* Vol. 2. Macmillan, New York.

Sossou, J., S. Karunaratne and A. Kovoor (1987) Collecting palm: *in vitro* explanting in the field. *FAO/IBPGR Plant Genetic Resources Newsletter* 69:7–18.

Tisserat, B. (1985) Embryogenesis, organogenesis and plant regeneration. In: Dixon, R.A. (ed.) *Plant Cell Culture. A Practical Approach.* pp. 79–105. IRC Press, Oxford.

Vuylsteke, D.R. (1989) *Shoot-tip Culture for the Propagation, Conservation and Exchange of* Musa *Germplasm. Practical Manuals for Handling Crop Germplasm* In Vitro, 2. IBPGR, Rome.

Withers, L.A. (1980) *Tissue Culture Storage for Genetic Conservation.* IBPGR, Rome.

Withers, L.A. (1982) Storage of plant tissues. In: Withers, L.A. and J.T. Williams (eds) *Crop Genetic Resources – The Conservation of Difficult Material.* pp. 49–82. IUBS/ IBPGR/IGF, Paris.

Withers, L.A. (1985a) Cryopreservation of cultured cells and meristems. In: Vasil, I.K. (ed.) *Cell Culture and Somatic Cell Genetics of Plants.* Vol. 2: *Cell Growth, Nutrition, Cytodifferentiation and Cryopreservation.* pp. 253–316. Academic Press, Orlando.

Withers, L.A. (1985b) Cryopreservation and storage of germplasm. In: Dixon, R.A. (ed.) *Plant Cell Culture: A Practical Approach.* pp. 169–191. IRL Press, Oxford.

Withers, L.A. (1987a) Long-term preservation of plant cells, tissues and organs. *Oxford Surveys of Plant Molecular and Cell Biology* 4:221–272.

Withers, L.A. (1987b) *In vitro* methods for germplasm collecting in the field. *FAO/ IBPGR Plant Genetic Resources Newsletter* 69:2–6.

Withers, L.A. (1989) *In vitro* conservation and germplasm utilization. In: Brown, A.H.D., D.R. Marshall, O.H. Frankel and J.T. Williams (eds) *The Use of Plant Genetic Resources.* pp. 309–334. Cambridge University Press, Cambridge.

Withers, L.A. (1992) *In vitro* conservation. In: Hammerschlag, F.A. and R.E. Litz (eds) *Biotechnology of Perennial Fruit Crops.* pp. 169–200. CAB International, Wallingford.

Withers, L.A. and J.M.M. Engels (1990) The test tube genebank – a safe alternative to field conservation. *IBPGR Newsletter for Asia and the Pacific* 3:1–2.

Withers, L.A. and J.T. Williams (1985) Research on long-term storage and exchange of *in vitro* plant germplasm. In: *Biotechnology in International Agricultural Research.* pp. 11–24. IRRI, Los Baños.

Yidana, J.A., L.A. Withers and J.D. Ivins (1987) Development of a simple method for collecting and propagating cocoa germplasm *in vitro. Acta Horticulturae* 212:95–98.

Yidana, J.A. (1988) The development of *in vitro* collecting and isozyme characterization of cocoa germplasm. PhD thesis. University of Nottingham, Nottingham.

Collecting pollen for genetic resources conservation

25

F.A. Hoekstra

Department of Plant Physiology, Wageningen Agricultural University, Arboretumlaan 4 6703BD, Wageningen, The Netherlands.

Introduction

Pollen grains are the mature male (haploid) gametophyte of higher plants. These tiny organs (10–100 µm in diameter) develop in the anthers (Stanley and Linskens, 1974). When the flower opens the anthers become exposed to the environment and dehydration triggers a mechanism for release of the pollen. In the process, the pollen also dehydrates. Though the pollen of most species can withstand drying (Stanley and Linskens, 1974; Towill, 1985; Hoekstra, 1986; Hoekstra *et al.*, 1989b), it may lose some vigour and viability before it lands on a compatible female receptive surface, depending on the mode of dispersal, availability of vectors and environmental factors such as humidity and temperature. There, it produces a tube after rehydration, and this grows into the female tissues to ultimately deliver its two sperm cells to the embryo sac.

Their small size and desiccation tolerance render pollen grains particularly suitable for storage. In fact, pollen storage is a common practice in breeding programmes to bridge the gap between male and female flowering time and to improve fruit setting in orchards (Franklin, 1981; Towill, 1985). However, relatively limited use has been made thus far of pollen for long-term germplasm conservation. One reason for this is the short longevity of pollen grains relative to seeds. Longevity varies widely among species. Also, methods for testing viability are generally cumbersome and time-consuming. Furthermore, there is limited experience in the survival and fertilizing capacity of cryopreserved pollen that is more than five years old (Towill, 1985). Another serious drawback of using pollen in germplasm conservation for the time being is that mature pollen grains cannot develop independently into whole plants.

The advantages and disadvantages of the use of pollen for germplasm storage are discussed further below. The different factors that influence the rate of loss of pollen viability are then reviewed, followed by techniques for collecting and handling pollen in the field. Sampling strategies for pollen (Namkoong, 1981) are not dealt with in great detail, because they are essentially similar to those for seeds (Chapter 5), except as regards numbers of individuals sampled per site. Finally, methods for viability and vigour assessment are described, and a brief review is provided of storage pollen techniques.

Advantages of the use of pollen

Difficulties can be encountered with the seeds of some species that do not arise with their pollen. Examples of problems associated with seeds are dormancy and special light and cold requirements for germination and recalcitrance (Harrington, 1970; Roberts, 1975). A more general advantage of pollen grains over seeds is their relatively small size. The average diameter of pollen grains is 30–40 µm, which means that very large numbers can be collected and conserved (Akihama and Omura, 1986). Quite apart from the obvious advantage of better genetic representation, access to vast numbers of haploid individuals enables the breeder to make use of pollen selection for polygenic traits from the very first pollination. This is done by applying a surplus of pollen on styles under the conditions to which it is hoped to obtain tolerance, e.g. extreme temperatures or drought (Mulcahy, 1984). Pollen selection is the phenomenon of genetically better adapted tubes growing faster than others through the style, thus preferentially occupying the ovules (Mulcahy, 1979). Pollen selection can occur because a large number of structural genes are expressed in both the sporophytic and the gametophytic phase of the life cycle (reviewed by Mascarenhas, 1989).

The pollen of most species is desiccation tolerant, i.e. it can be dried to a moisture content (MC) of less than 5% on a dry weight (DW) basis, which can be produced by exposure to a relative humidity (RH) of approximately 30% (for reviews, see Towill, 1985; Hoekstra, 1986). This makes it very easy to store the material below 0°C for long periods, without the usual prerequisites for cryogenic storage of fully hydrated plant tissues. However, there are also species that cannot be dried without loss of viability. Species with recalcitrant pollen can be found in the Gramineae (Knowlton, 1922; Goss, 1968), Cucurbitaceae (Gay et al., 1987) and Araceae (Henny, 1980). The expectation is that species with recalcitrant pollen will be particularly common in very humid climates and niches. Species with recalcitrant seeds do not necessarily have recalcitrant pollen, and vice versa.

The international transfer of germplasm in the form of dry pollen is not generally restricted. While literature on the transfer of fungal and bacterial diseases through pollen is scarce, the case of viruses is different (Neergaard, 1977). Nineteen viruses have been shown to occur on the pollen wall or in the cytoplasm and to be able to infect the developing seed (Mandahar, 1981; Mandahar and Gill, 1984). If the virus survives the development and maturation of the pollen, it may reach the embryo sac via the pollen tube. Although pollen with virus in its cytoplasm seems less vigorous than uninfected pollen, there is still the chance that seeds could become infected. While pollen viruses borne externally have no epidemiological importance, a few of the internally borne viruses have. Therefore, in working with pollen as a source of germplasm, one should be alert to the fact that viruses and other plant diseases might be transmitted during fertilization (Miller and Belcher, 1981).

Disadvantages of the use of pollen

In a considerable number of plant species, the amount of pollen per flower is very small, not really sufficient for effective collecting and processing. Examples are cleistogamous plants and pioneer plants that are self-pollinating (Cruden, 1977). In such cases, pollen storage is unlikely to be a suitable method for germplasm conservation.

Another disadvantage of the use of pollen is a consequence of cytoplasmic inheritance. The number of mitochondria in the sperm cells differs among species, and plastids are usually absent (Wagner *et al.*, 1989; Theunis *et al.*, 1991). The fate of the organelles after fusion of the sperm cells with the egg cell and central cells is unclear. If the organelles survive the fusion events, they are usually outnumbered by the organelles of the egg cell. This would mean that the transmission of organelle genomes is not effective via pollen. This could be a drawback if pollen storage is the only method employed for germplasm conservation in a species. Sex-linked genes will similarly be missed in dioecious species.

In contrast to seed samples, a pollen sample cannot easily be renewed after a time in low temperature storage because pollen cannot independently develop into a plant. Therefore, new collections have to be made. An alternative is for the stored pollen to be used in pollinations with a particular mother, after which the resulting seeds or pollen from flowers of the first cross can be collected and stored, but the original material is then lost. However, it is probably only a matter of time until techniques become available for regrowth of pollen into whole plants. For example, work on the isolation and longevity of sperm cells from the pollen of various species has progressed so far in recent years that it is conceivable that it will soon be possible to force these cells into an embryonic or vegetative type of development (for a review, see Theunis *et al.*, 1991).

The life span of pollen is considerably shorter than that of seed, ranging according to species from just a few hours to several months at room temperature (Pfundt, 1910; Stanley and Linskens, 1974; Shivanna and Johri, 1985). The longevity of the pollen of a number of important crop species is shown in Table 25.1. Pollen of most species is also very

Table 25.1. Longevities of dry pollen (4–10% MC, DW basis) at 20–25°C and normal atmospheric conditions. Longevity is expressed here as the number of days until half or less of the original viability is left. Species are catagorized according to their respective families and to the number of nuclei in pollen.

Family	Species	Longevity (d)	Author(s)
Pinaceae	Pinus montana	272	Pfundt, 1910
	Pinus˙pinaster	275	Pfundt, 1910
	Pinus silvestris	279	Pfundt, 1910
	Pinus strobus	413	Duffield and Snow, 1941
Families with bi- or tricellular pollen			
Araceae	Arum maculatum (3)	46	Pfundt, 1910
	Dieffenbachia maculata (3)	2	Henny, 1980
	Spathiphyllum floribundum (2)	14	Henny, 1978
Rubiaceae	Coffea spp. (2)	25	Ferwerda, 1937
Families with bicellular pollen			
Aceraceae	Acer pseudoplatanus	55	Lichte, 1957
Caricaceae	Carica papaya	7	Ganeshan, 1985
Liliaceae	Lilium bulbiferum	142	Pfundt, 1910
	Lilium longiflorum	130	Hoekstra, 1986
	Narcissus poeticus	83	Hoekstra, 1986
	Trillium sessile	102	Holman and Brubaker, 1926
Malvaceae	Abutilon darwini	14	Pfundt, 1910
	Gossypium hirsutum	2	Rodriguez-Garay and Barrow, 1986
Palmae	Elaeis guineensis	107	Hoekstra, 1986
	Phoenix spp.	275	Visser, 1955
Papaveraceae	Papaver rhoeas	36	Hoekstra, 1986
Papilionaceae	Arachis hypogea	8	Vasil, 1962
	Lupinus latifolius	118	Holman and Brubaker, 1926
	Vicia faba	21	Pfundt, 1910
Rosaceae	Prunus padus	181	Pfundt, 1910
	Pyrus communis	114	Visser, 1955
	Pyrus malus	95	Visser, 1955
	Rosa moyesii	57	F.A. Hoekstra, unpublished
Scrophulariaceae	Antirrhinum majus	188	Lichte, 1957
	Digitalis purpurea	143	Pfundt, 1910
Solanaceae	Nicotiana tabacum	100	F.A. Hoekstra, unpublished

Table 25.1. *continued*

Family	Species	Longevity (d)	Author(s)
	Solanum melongena	50	Vasil, 1962
	Solanum tuberosum	18	Vasil, 1962
Families with tricellular pollen			
Asteraceae	*Chrysanthemum cinerariaefolium*	8	Hoekstra and Bruinsma, 1975a
	Onopordon illyricum	8	Pfundt, 1910
Chenopodiaceae	*Beta vulgaris*	81	Lichte, 1957
	Chenopodium bonus henricus	8	Pfundt, 1910
Cruciferae	*Brassica campestris*	5	Holman and Brubaker, 1926
	Brassica nigra	34	Vasil, 1962
Gramineae	*Alopecurus pratensis*	2	Pfundt, 1910
	Lolium perenne	1	Pfundt, 1910
	Pennisetum typhoides	177	Vasil, 1962
	Secale cereale	0.25	Lichte, 1957
	Triticum aestivum	<1	Vasil, 1962
	Zea mays	1	Pfundt, 1910
Umbelliferae	*Apium graveolens*	6	D'Antonio and Quiros, 1987
	Sanicula bipinnatifida	9	Holman and Brubaker, 1926

sensitive to water (i.e. rain) and cannot be dried after even short exposure to moisture without a considerable decline in viability (Lidforss, 1896; Hoekstra, 1983). There is much disagreement as to the success of different storage protocols (reviewed by Towill, 1985), probably mostly due to the occasionally low vigour of the starting material and to an ignorance of rehydration requirements. This has certainly contributed to the general confusion in the field and reduced the enthusiasm for practical use of cryogenic pollen storage.

Given these advantages and disadvantages, pollen collecting and storage are likely to be most useful in the foreseeable future as a complementary conservation strategy for species that are normally conserved in field gene banks and have large and/or recalcitrant seeds.

Stress on pollen grains

From maturation until the start of tube growth on a compatible style, pollen grains are entirely at the mercy of the environment. Rain, excessive dryness, extreme temperatures and oxidation all contribute to loss of pollen viability. The focus in this section is mainly on stresses during this independent part of the pollen life cycle. However, climatic conditions can already exert an influence during pollen development in the mother plant. There are reports, for example, that implicate spells of drought, high temperature (Halterlein *et al.*, 1980; Schoper *et al.*, 1987) and low temperature (Maisonneuve, 1983) in the release of bad-quality pollen.

Rain

Continuous rain may prevent anthers from opening. Once pollen has passed physiological maturity, ageing may start within the anthers under such conditions. Ultimately, pollen with reduced viability and vigour is released (Hoekstra and Bruinsma, 1975a; Linskens and Cresti, 1988; Linskens *et al.*, 1989), or the pollen may germinate inside the anthers (Pacini and Franchi, 1982). Rain is also generally detrimental to shed pollen, causing bursting, precocious germination, etc. If pollen is dried after a short period of soaking, it generally loses viability, with a few exceptions among the conifers (Hoekstra, 1983). One might therefore expect plants to protect their pollen from rain, but in a considerable number of species, shed pollen becomes wet during rainy weather (Lindforss, 1896). However, in species where pollen viability is at risk because of an intrinsically short life span or insufficient protection, this is often compensated for by the continuous production of fresh flowers and pollen, for example in a raceme or other type of indeterminate inflorescence (Hoekstra and Bruinsma, 1975a).

Humidity and temperature

Pollen longevity as measured by *in vitro* germination tests has long been known to be strongly affected by RH and temperature in storage (Pfund, 1910; Knowlton, 1922). Desiccation-tolerant pollen of *Typha latifolia* remains viable for about 150 days at 22°C at an internal MC of 6% (on a DW basis), brought about by equilibration in an atmosphere at 40% RH. At 17% MC (equilibrated at 75% RH) this period is reduced to only 16 days. In *Papaver rhoeas*, which has an intrinsically shorter storage life, the survival periods are 45 and four days for the same two MC levels, respectively. Apparently, ageing is considerably faster when MC is high. At 17% MC, respiration in hardly noticeable (Hoekstra and Bruinsma, 1975b), so loss of respiratory substrates is not to blame for this decline in viability. However, lipid breakdown occurs during dry and semidry ageing, most probably associated with the activity of free radicals (McKersie *et al.*, 1988; Bilsen and Hoekstra, 1993). Although the extent of fatty acid unsaturation of membrane lipids hardly changes, some new breakdown products of lipids appear, such as free fatty acids and lysophospholipids. These compounds can cause phase separation in the membranes, which leads to loss of integrity and excessive leakage of endogenous solutes during imbibition (Bilsen and Hoekstra, 1993;

Bilsen *et al.*, 1994). As a general rule, as with seeds (Priestley, 1986), longevity is doubled with every 2% reduction of the endogenous MC (keeping temperature constant).

High temperature reduces longevity as measured in *in vitro* germination tests (Hoekstra and Bruinsma, 1975b). Whereas *Nicotiana alata* pollen at 18% MC (on a DW basis) reaches 50% of its original viability in approximately 400 hours at 20°C, this value is reached in about 80 hours at 30°C. At high humidity, pollen viability in many species is usually at the 50% value after 24 hours at 30°C (Hoekstra and Bruinsma, 1975b). However, before the decline in viability has commenced under conditions of high humidity and high temperature (38°C), a decline in vigour can be observed on account of delayed tube emergence *in vitro* (Shivanna *et al.*, 1991a) and reduced rates of tube growth in the style (Shivanna *et al.*, 1991b). Exposing humid pollen to >45°C will denature the proteins and kill the cells. However, dry pollen is surprisingly tolerant to exposure to temperatures as high as 80°C for 1 hour (Marcucci *et al.*, 1982). Cryogenic temperatures are relatively harmless to dry pollen, whereas they cause injury to humid pollen because free water is turned into ice crystals that pierce the membranes (Towill, 1985). As in seeds (Priestley, 1986), longevity in pollen is doubled for every 5–6°C decrease of temperature.

Desiccation

When pollen is shed from the anthers, it has no control over its MC. A prolonged stay on the flower (e.g. due to the absence of pollinating insects) in combination with dehydrating winds can easily desiccate pollen to <30% MC. Pollen can also dry out on the body hairs of insect pollinators, or during its stay in the air in the case of wind-pollinated species. Already at 30% MC somatic tissue has irreversibly lost its viability. Below 30% MC, membranes and many enzymes undergo conformational changes which lead to loss of integrity and inactivation (Crowe *et al.*, 1987). However, mature pollen of many species is able to withstand an MC of <30%. One of the factors that is responsible for this desiccation tolerance is an elevated disaccharide content (Hoekstra and van Roekel, 1988; Hoekstra *et al.*, 1989a). Experiments have shown that sucrose is effective at preventing or postponing the conformational changes in dehydrating membrane lipids and proteins by replacing the water with –OH groups (Crowe *et al.*, 1987; Hoekstra *et al.*, 1991). Reducing monosaccharides are much less effective in this respect (Crowe *et al.*, 1986). In addition to producing disaccharides, desiccation-tolerant pollen must have mechanisms to withstand the considerable physical forces exerted due to the shrinking of the cell and the extreme solute concentration. It must also effectively cope with the free radicals generated by repressed respiration. These mechanisms are poorly understood at present.

Desiccation tolerance develops at the end of pollen maturation in the anthers (Hoekstra and van Roekel, 1988). Premature collecting may thus result in material with reduced viability and vigour. Full maturity is assumed to be reached at anther dehiscence in the flower under optimal weather conditions.

One part of desiccation tolerance is obviously the ability of cells to withstand dehydration, but an equally important factor is the successful rehydration of the dry cells. In all desiccation-tolerant organs and organisms, leakage may occur during imbibition of liquid medium, particularly at low temperatures (Hoekstra and van der Wal, 1988). If this leakage exceeds a certain level, reduced vigour results. In particularly severe cases, it is catastrophic for the cell (Hoekstra and van der Wal, 1988; Hoekstra et al., 1992). The suggestion by Simon (1974) that leakage during imbibition of dry seeds is due to transition of the membrane phospholipids from a non-bilayer to a liquid crystalline phase is also believed to hold for dry pollen (Shivanna and Heslop-Harrison, 1981). However, it has been demonstrated recently that leakage coincides with a change from a gel phase to a liquid crystalline phase (Crowe et al., 1989). Methods that frustrate such a phase change during imbibition are effective in reducing the leakage and maintaining vigour. These treatments consist of prehumidification in moisture-saturated air and/or heating during imbibition. Reduction of the rate of water uptake, by using osmotically active substances such as polyethyleneglycol or elevated gelatine and agar concentrations in germination media, is also beneficial.

In seed research there is currently a dispute as to whether storage at ultra-low MC is more or less desirable than storage at a slightly higher level of approximately 5% MC. It has been noticed that in very dry pollen, imbibition at elevated temperatures failed to reduce leakage and restore viability. In fact, leakage became more serious at elevated temperatures of imbibition (Hoekstra et al., 1989b, 1992). This may be explained by the behaviour of the phosphatidylethanolamine. This phospholipid has the tendency to form the above-mentioned non-bilayer, inverted hexagonal (H_{II}) phase at elevated temperatures and reduced hydration levels. Further drying brings the H_{II} transition temperature down, possibly into the physiological temperature region. In the case of very dry pollen, in which phosphatidylethanolamine makes up 30% of the phospholipid content, this may explain why leakage becomes more serious at elevated temperatures of imbibition. While in many species the H_{II} phase may still turn into a liquid crystalline phase by prehumidifying before the germination test in vitro, it may be detrimental if the H_{II} phase occurs on too large a scale. All this suggests that storing pollen at far below 5% MC (i.e. in equilibrium with an atmosphere at less than 20% RH) should be avoided.

There are a number of species whose pollen is recalcitrant, i.e. it cannot withstand drying. They can be found in the agriculturally important families Gramineae and Cucurbitaceae and several others. In the case of maize pollen, the endogenous moisture content at which viability is completely lost has been thoroughly investigated. This lies close to 10%

(on a DW basis; Hoekstra *et al.*, 1989a), which is much lower than, for example, a desiccation-sensitive meristem. Below 20% MC there is no freezable water left in the cells, and only tightly bound water is present. This offers possibilities for cryogenic storage in the range 10–20% MC.

Intrinsic sensitivity to ageing

Longevity differs widely among species. Out of 2000 species in 265 plant families, Brewbaker (1959, 1967) found that about 30% shed pollen in the tricellular stage (formerly called trinucleate). These species were mainly in phylogenetically advanced plant families (Brewbaker, 1967; Sporne, 1969). Bicellular (binucleate) pollen grains have still to perform a second mitosis in the style during tube growth, as a result of which the two sperm cells are formed. Families with tricellular pollen are the Asteraceae, Caryophyllaceae, Chenopodiaceae, Cruciferae, Gramineae, Juncaceae and Umbelliferae. It appears that the entire metabolism is relatively advanced in tricellular pollen (Hoekstra and Bruinsma, 1975b; Hoekstra, 1979), which adds to the sensitivity to stresses and ageing conditions. Tricellular pollen tends to be shorter-lived than bicellular, but faster-growing. However, binucleate pollen has been found with short storage life and extremely rapid tube emergence and growth, e.g. in the Balsaminaceae, Cucurbitaceae and Commelinaceae.

Pollen whose lipids are rich in unsaturated linolenic acid (18 : 3) tend to be short-lived and have a more rapid start of tube emergence (Hoekstra, 1986). Pollen linolenic acid content is much higher than in seeds and can reach 80% of total fatty acids (e.g. in the Gramineae and Balsaminaceae). Unsaturated lipids are more prone to peroxidation and breakdown than the more saturated types. The loss of total phospholipids is dependent on both the extent of unsaturation and the amount of lipid-soluble antioxidants present (Hoekstra, 1992). Strangely enough, the short-lived species are particularly low in endogenous lipid-soluble antioxidants. Apparently, longevity of highly unsaturated pollen is not a critical factor in fertilizing efficacy; otherwise this pollen would have been rich in antioxidants. Hoekstra (1992) has suggested that a high level of unsaturation and low level of antioxidants may be prerequisites for rapid tube growth.

The accelerating effect of oxygen on the rate of ageing has been firmly established through longevity experiments at elevated oxygen concentrations (Knowlton, 1922) and after the oxygen had been removed. Jensen (1964) has shown that storage in vacuum considerably extended longevity for all ten of the species he studied, at both 5°C and room temperature. It is to be expected that pollen with a high proportion of double bonds in its lipids should be particularly sensitive to oxygen. A recent example of dry storage under nitrogen gas at room temperature of such a high linolenic acid type pollen (*Impatiens*) shows that the life span is increased fourfold (Hoekstra, 1992).

Collecting and handling pollen in the field

The main rule to be observed in collecting pollen is to gather material
with good viability and vigour. This is simply stated, but, as discussed
in the section on stresses, many factors will make it difficult to achieve.
Contact of pollen and anthers with rainwater must be avoided, for example. One should thus refrain from collecting on rainy days. Also, only
physiologically mature pollen should be collected, which means one
should avoid collecting closed anthers from young flower buds.
Immature pollen may lack the capacity to germinate, have reduced
vigour or still be desiccation-sensitive. Old pollen from previous shed-
dings should also be avoided. It may be necessary to study the rate of
decline of viability of pollen in the particular species and conditions of
interest before embarking on collecting. The measurement of viability is
discussed in detail in the next section. If, under field conditions, pollen
viability has to be estimated without previous knowledge of specific
requirements for germination *in vitro* and without sophisticated equipment, the following simple procedure is recommended. A quick wash of
the stigma and style with water or ethanol a few hours after application
of pollen on the stigma removes those grains that were unable to grow
tubes into the stigma. The percentage of grains that are left is indicative
of the functional capacity of the pollen sample. Inspection with a simple
stereoscopic microscope at low magnification ($\times 40$) will be adequate.

Pollen needs to be dried for conservation. Special care will need to
be exercised when collecting species with metabolically active, rapidly
germinating, tricellular pollen than with the more quiescent, slowly germinating bicellular types.

Chapter 5 argues that a sample of 59 random unrelated gametes
from a population is sufficient to attain the objective of including in the
sample at least one copy of 95% of the alleles that occurred there at frequencies greater than 0.05. When collecting seed, this means sampling
30 randomly chosen individuals in a fully outbreeding, randomly mating
species, or 59 random individuals in a self-fertilizing species. When collecting pollen, the benchmark criterion will be 59 individuals per population. Self-incompatibility is not likely to cause problems when using
collected pollen in crosses. A population of strictly outbreeding, self-
incompatible plants must automatically carry a variety of different
S-alleles, irrespective of whether the gametophytic or sporophytic
system of self-incompatibility is operating (de Nettancourt, 1977). Thus,
pollinating a mother plant with the collected material will not lead to
mass rejection.

Pollen may be collected directly from plants in the field or from plant
parts brought in from outside, say at a base camp with a sheltered table.
Outside, pollinators and wind may be serious competitors for the collector, necessitating the bagging of flowers, though temperature and
humidity can become very high inside such bags. The collector must also
be aware that time of flower opening and anther dehiscence may be

earlier than normal in bags because of the higher temperature (Beers *et al.*, 1981). Pollen should be collected as soon as it is shed. Collecting can be done by shaking flowers over folded sheets of paper or into vials or by using small aspirators.

In some cases, it will be easier to bring pollen-containing plant parts (branches, racemes or just flowers on a stalk) into a sheltered place for collecting. It is advisable not to use cut branches and flowers too long, as pollen quality will decrease as these plant parts age. Supplemental nutrients, sugar and antibiotic compounds may be provided to extend the period during which collecting from cut parts can take place. In some cases, catkins or male inflorescences can be collected and transported for later gathering of the pollen. It is important that this material should not get too hot during transport. Back at base, the material should be spread out and dried by gentle ventilation. This will release the pollen, which can then be separated from the debris by passing through brass or nylon screens before further drying. In a number of species, still-closed anthers can be removed from young flowers or mature flower buds and left to dry. Upon dehiscence, pollen is separated from the anther remainders and cleaned by sieving.

Insects can also be used to collect pollen. Pollen pellets, usually mainly from one species, can be obtained by forcing bees to pass through a screen before they enter the hive. The viability of pollen in the pellets directly after collecting is good, and is maintained after drying and freezing (reviewed by Verhoef and Hoekstra, 1986). However, ageing is accelerated and longevity is reduced by substances added by bees to the pollen, such as reducing sugars, waxes and amylases. An antibiotic occurring in glands in bee heads (10-hydroxy-2-decenoic acid) inhibits pollen germination, but was found not to occur in pollen pellets (Verhoef and Hoekstra, 1986). The main problem with the dried pellets is the difficulty of reusing these hard lumps for pollination purposes. By mixing the pellets with an osmotic solution directly after collecting, filtering the washed pollen and (partly) drying it, a powder can be obtained that is suitable for pollination. However, after these treatments the washed pollen has reduced viability, since it has lost its desiccation tolerance, and will not be suitable for conservation.

It is always important to bring down the MC of the pollen as quickly as possible. Rapid drying can usually be done at 50% RH with gentle ventilation. If ambient RH and temperature are high, as is common in the hot humid tropics, oven drying may be considered (37°C) or the immediate exposure to desiccants such as silica gel or $CaCl_2$ for several hours. The drying should preferably be done at room temperature. Although it would seem appealing to reduce metabolism of the still-humid pollen during drying by exposure to low temperature, there may be a possibility of the rising membrane transition temperature coming close to the dehydration temperature, which may lead to injury. Although no clear evidence exists for this type of injury in pollen, it has recently been found in seeds. Furthermore, during drying at room temperature, desiccation tolerance may be enhanced due to some synthetic processes that obviously do not occur at low temperature (Hoekstra et al., 1989a).

In general, a moisture content of 5–8% is sufficiently low for the material to be stored in the deep-freeze or at cryogenic temperatures. If pollen is very sticky and not easily removed from the anthers, whole anthers can be dried, which may preserve the viability of the pollen (Crisp and Grout, 1984; Akihama and Omura, 1986). Large pollen samples may require long periods for drying. This may cause a problem, in that pollen that is dried too slowly may start to age, with the result of reduced vigour and viability. In forest tree breeding, special equipment is used for rapid drying and cleaning of the pollen (Sprague and Snyder, 1981). After drying, pollen can be stored in sealed plastic bags containing as little air as possible.

The exception to the rule of rapid drying, of course, is recalcitrant pollen. In grasses, for example, inflorescences can be put in vases and covered with plastic bags. Regular removal of the bags allows collecting of pollen of a homogeneously high MC. Pollen can then be stored, well covered, at low temperatures above 0°C, or first dried in a controlled way for physiological experiments or cryogenic storage (see below).

Methods for viability and vigour assessment

Several tests of pollen viability and vigour exist. Generally, the more accurate the test, the more time it takes to carry it out, with *in situ* pollination and subsequent determination of the quality of the resulting seeds as the ultimate test. The tests described here are mainly laboratory-based. A simple washing technique to remove non-functional pollen, suitable for use in the field, has already been described (see the previous section).

Staining tests

Acetocarmine and cotton blue are among the oldest stains used to assess pollen quality. Positive staining indicates the presence of cytoplasm and nuclei. Whereas lack of staining shows that pollen is non-viable, a positive staining is only a vague indication of viability because there are many factors that leave the cellular content intact but are nevertheless detrimental to survival. If pollen contains high levels of proline, as in maize and rye pollen, the content can be monitored cytochemically by using isatin reagent (Palfi and Köves, 1984). Low viability is associated with reduced proline levels and lacks the intense black staining.

Other stains, such as tetrazolium salts (Hauser and Morrison, 1964; Norton, 1966) and fluoresceindiacetate (FDA) (Heslop-Harrison *et al.*, 1984), are more accurate, results correlating closely with viability as determined by *in vitro* germination tests. However, there is a risk of false negatives if the following rules are not properly observed. Lipid-phase transitions should be prevented from occurring in the pollen membranes during imbibition in the staining solution (Crowe *et al.*, 1989). This can be achieved by rehydrating the dry pollen in water vapour prior to the test, which considerably increases the percentage of stained grains (Visser *et al.*, 1977; Shivanna and Heslop-Harrison, 1981; Hoekstra and van der Wal, 1988). The duration of the prehydration treatment depends on the species (from a few minutes to several hours). Too long a prehumidification may lead to ageing-associated reduction of membrane integrity, the manifestation of which is leakage of endogenous solutes from the grains and reduction of percentage staining. In such cases, the water vapour treatment should be done at low temperature, for example at 1°C (Hoekstra, 1979; Akihama and Omura, 1986). Protecting the membranes by prehydration is necessary because small molecules are involved in most staining procedures, which may leak out during the assay. Tetrazolium salts, for example, are acceptors of electrons from active dehydrogenases, during which reaction the water-insoluble coloured formazans are produced. The activity of the dehydrogenases depends, of course, on their molecular integrity, but also on the availability of the endogenous substrate, succinate. Loss of this substrate by improper prehydration can thus give false negatives. Non-viable pollen can give excellent formazan production if succinate is added to the staining solution.

The tetrazolium test differs from older staining methods such as the acetocarmine test in that extra information on the integrity of the pollen plasma membrane is obtained. The same is true for the vital stain, FDA (Heslop-Harrison *et al.*, 1984). This apolar compound readily passes through membranes into the pollen. The non-fluorescent FDA is de-esterified in the cytosol by endogenous esterases, yielding a polar fluorescent dye (fluorescein), which accumulates inside the grain unless the plasma membrane is leaky. Optionally, FDA can be applied in combination with other dyes such as lissamine green or propidium iodide, which cannot normally pass through a membrane unless it is leaky. Thus, staining by the two dyes is mutually exclusive. Lissamine green and propidium iodide can be used to check the effectiveness of the FDA test.

Tetrazolium and FDA have to be administered to the pollen in a suitable germination medium, or at least in a medium with an appropriate osmoticum, in order to protect the pollen from osmotic shock, which could cause membrane leakage. Lack of correlation between extent of staining and *in vitro* germination, which is commonly reported in the older literature, is probably due to membrane injury during application of the stains resulting from improper prehydration. The FDA test can show that membranes are sound but is not proof of germinability. Hydrated immature pollen grains, for example, stain very well, but may not yet have the capacity to form pollen tubes. The good correlation in aged pollen between percentage staining and germination is due to the dominant role of lipid breakdown in the ageing processes (Bilsen and Hoekstra, 1993; Bilsen *et al.*, 1994).

Germination in vitro

Germination in an *in vitro* system is superior to staining as a viability test. During the 18th century, it was found that pollen could be cultured in relatively simple media to form pollen tubes. Since then, numerous media and culture techniques have been published (for reviews, see Stanley and Linskens, 1974; Shivanna and Johri, 1985). These media usually contain boric acid and salts (Brewbaker and Kwack, 1963) and, depending on the species, different amounts of sucrose and agar or other osmotically active compounds. Media are often adjusted to pH 5–8, again depending on the species. In a number of cases other additives enhance germination and tube growth, e.g. amino acids (Palfi and Köves, 1984; Leduc *et al.*, 1990), casein hydrolysate (Mulcahy and Mulcahy, 1988) and micronutrients and vitamins (Leduc *et al.*, 1990). A popular system is the hanging drop technique, in which pollen in approximately 50 μl of a liquid germination medium is left hanging over a small chamber of vapour-saturated air. The small closed system allows continuous microscopic inspection without the risk of evaporation. Alternatively, pollen may be seeded on the partly dried surface of agar medium in Petri dishes or in shaking or aerated liquid media. The method of viability testing through germination *in vitro* is rapid. In general, pollen tubes can be counted within a few hours.

Plants differ in the amount of stigmatic exudate they produce (Bar-Shalom and Mattsson, 1977; Heslop-Harrison and Shivanna, 1977). Pollen of species with stigmas with profuse exudate produces tubes easily in artificial media with low amounts of sugar and agar, most probably because these conditions are very similar to the situation in the exudate. In these species, pollen is usually shed in the bicellular stage of development. Both dry- and wet-stigma species occur in families with bicellular pollen, however. It is considerably more difficult to germinate the pollen of species with dry stigmas because the water relations are more delicate (Bar-Shalom and Mattsson, 1977; Heslop-Harrison and Shivanna, 1977). Tricellular pollen is almost exclusively found in families with dry stigmas. The reputation of tricellular pollen for being difficult to germinate *in vitro* (Brewbaker, 1959) is probably associated with this. The problem of the poor *in vitro* germination of tricellular pollen is further compounded by its rapid ageing and short storage life.

For *in vitro* germination of tricellular pollen, high sucrose concentrations have been applied, up to 40% (Hoekstra and Bruinsma, 1975a), sometimes in combination with polyethyleneglycol (MW = 4000) (Leduc *et al.*, 1990). Alternatively, water availability can be controlled by high (up to 60%) concentrations of gelatine (Kubo, 1955, 1956), by cellophane sheets (Alexander and Ganeshan, 1989) or simply by restricting the addition of moisture (Bar-Shalom and Mattsson, 1977). Germination has been obtained in this way in pollen from species in the Asteraceae,

Cruciferae, Ericaceae, Gramineae and Chenopodiaceae, although pollen tubes remain short in some cases, additional nutrients then being required.

Occasionally, rapid ejection of tube-like structures in water or germination medium is observed, for example in bicellular cotton pollen (Barrow, 1981) and tricellular *Dipsacus* pollen (F.A. Hoekstra, unpublished results). This is an indication of nothing more than an intact plasma membrane, not necessarily of successful *in vitro* germination.

As in the staining tests, dehydrated pollen has to be treated in humid air before seeding it on to the medium. Failure to do so may have been the reason for the common phenomenon of seemingly non-viable pollen recovering in a later test after storage (Holman and Brubaker, 1926). Visser (1955) and Lichte (1957) have demonstrated in detail the beneficial effects of prehydration in humid air on growth performance of pollen *in vitro*. Apart from reducing the gel-to-liquid crystalline transition temperature of membrane lipids, water vapour may stimulate respiration and ageing. This is particularly risky for tricellular pollen.

The ability to form pollen tubes *in vitro* does not give an absolute answer as to whether pollen is vigorous enough for tube growth in the style towards the ovules. There are a number of reports on pollen with good germination *in vitro* failing to fertilize successfully (Shivanna *et al.*, 1991b). Also, pollen from the Umbelliferae, from some species of the tricellular families mentioned above and from a few other families cannot be successfully germinated *in vitro* at all. In such cases, germination *in vivo* is the only method suitable for viability assessment.

Germination in vivo

The most reliable method of pollen vigour determination is the regular monitoring of the progress of the pollen tubes in the style. This can be done by microscopic analysis of styles that are stained with aniline blue or Wasser blue (Linskens and Esser, 1957; Shivanna *et al.*, 1991b). In fleshy styles an additional bleaching in KOH, followed by extensive washing and squashing, has to be done before the fluorescing pollen tubes can be detected. If a fluorescence microscope is not available, it is possible to stain the tubes with an aqueous solution of cotton blue in combination with extensive bleaching (Janson, 1993). Although it remains difficult to determine accurately the percentage of pollen grains showing tube growth, a rough estimate is possible. A promising method is a semi-*in vivo* technique, in which, after pollination, the style is cut and the cut end placed in germination medium (Shivanna *et al.*, 1991b). The number of penetrating pollen tubes and their speed of growth can then be determined.

It is still an unanswered question whether hydration is required prior to pollination with dry stored pollen. Usually, conditions on the stigma will allow gradual rehydration of the pollen. A few papers report that prehydration was not effective, but emphasize that it was not harmful either (Matthews and Bramlett, 1983; Hecker *et al.*, 1986). However, beneficial effects have also been reported. It can be anticipated that in profuse exudate dry pollen may suffer imbibition injury, which would have been prevented by prehydration in humid air. Prehydration may also help in rendering pollen more adhesive, facilitating attachment to the receptive surfaces.

   In vivo tests can be done in a few days. If time is not a critical factor, the maturation of seeds can be awaited, 'and their germination tested, followed by screening of seedlings for normal growth performance.

Pollen storage methods

Pollen storage methods will not be discussed in great detail here, as some excellent reviews are already available (Binder *et al.*, 1974; Stanley and Linskens, 1974; Franklin, 1981; Shivanna and Johri, 1985; Towill, 1985). For practical aspects, the reader is referred to the appropriate chapter in the *Pollen Management Handbook* (Matthews and Kraus, 1981; Schoenike and Bey, 1981). For information specifically on woody perennials, see FAO (1993). What emerges from these studies is that desiccation-tolerant pollen generally does not survive longer than about one year at room temperature in the desiccated condition under normal atmospheric oxygen pressure. For longer storage the sample must be cooled or the oxygen removed. Storage of dry samples at $-20°C$ or lower can only be done successfully at below 20% MC (on a DW basis). A higher initial MC will lead to the deleterious formation of ice crystals. Storage in organic solvents is a promising option (Iwanami and Nakamura, 1972; Iwanami, 1973; Yabuya, 1983; Jain and Shivanna, 1990). For long-term storage (i.e. several decades or more) cryogenic storage is required. Storage longevity will be essentially infinite in liquid nitrogen (Bredemann *et al.*, 1947). For such storage, pollen must always be desiccated in advance, perhaps by freeze- or vacuum-drying (Towill, 1985).

   The pollen of grasses is notorious for its poor storability. This is due to its sensitivity to dehydration, the genus *Pennisetum* being an exception. In practice, this means that such pollen is stored at high MC which limits survival to days rather than weeks due to high metabolic activity and the associated ageing (extensively reviewed by Shivanna and Johri, 1985). A slight extension of longevity can be achieved by storage at low, above-freezing temperature. Some grass pollen can nevertheless be dehydrated to some extent. Maize pollen, for example, can be dehydrated to such a low MC that freezable water is removed (Kerhoas *et al.*, 1987), while still remaining viable as determined by *in vivo* germination (Hoekstra *et al.*, 1989a). Exploiting this, Barnabas and Rajki (1976, 1981) partially dried maize and rye pollen and stored it in liquid nitrogen and at $-76°C$. At the time of cooling, the pollen was at approximately 20% MC (on a DW basis). After rapid thawing it gave tube growth on the silks and seed set. This was true for both temperature conditions of storage at which the pollen was kept for at least three years. It is essential that, after application on the silks, the pollen is not further dried out. In other recalcitrant pollen the ability to survive partial drying may equally offer the possibility for long-term storage. If partial drying is not tolerated, the usual cryogenic protocols with controlled slow cooling and the use of cryoprotectant solutions might be carried out. However, there is hardly any experience with these methods in pollen (Towill, 1985).

Examples of pollen collecting

Most examples of pollen collecting have taken place as part of a programme aimed at the introduction of new genes into established crop species. Three case-studies are examined below, with the delay between collecting and pollination of mother plants ranging from a few days to three years.

*Pyrethrum (*Chrysanthemum cinerariifolium, *Asteraceae)*

Pyrethrum is grown for its insecticidal pyrethrins, which are extracted from the dried flower heads. It was introduced into Kenya in 1928 from semiwild plants growing along the Dalmatian coast. To broaden the supposedly narrow genetic base of the crop, pollen and seeds were collected in 1971 from 18 sites along the entire coast (Parlevliet *et al.*, 1979). Pollen longevity under dry conditions at room temperature was investigated by *in vitro* testing prior to collecting (Hoekstra and Bruinsma, 1975a) and found satisfactory for bridging the few weeks between collecting and pollination.

The collector faced intense competition from pollinating flies at the collecting site. Just as pollen began to be shed in the morning, it was tipped from the flower-heads into small vials and immediately dried for 12 to 14 hours over NaOH pellets at ambient temperature. The pollen samples were then packed in small plastic bags and sealed with as little air as possible. The bags were sent to Kenya by airmail. The viability of the dried tricellular pollen was still such that successful pollinations could be made with local clones of good general combining ability. The pollinations were made by hand (using a brush) in a greenhouse at high temperature and high RH.

Pyrethrum is self-incompatible, so that the profuse seeds derived from these pollinations had certainly received the Dalmatian genes via the foreign pollen. Plants grown from introduced seeds gave serious problems because they remained vegetative under Kenyan climatic conditions. In this respect, pollen collecting proved superior, since the plants derived from pollination with the introduced pollen gave few such problems.

Oil-palm (Elaeis guineensis)

The collecting and storage of oil-palm pollen was carried out in the context of breeding programmes (Ekaratne and Senathirajah, 1983). Male inflorescences with flowers in which the anthers had just begun to dehisce were shaken to release pollen on to paper sheets. The pollen was cleaned from adhering debris by sieving it through a 0.5 mm screen. The MC content was above the critical value of 20% (on a DW basis), so additional drying was required before freezing. This was done by placing the pollen for 2–8 hours in an oven at 37°C and then transferring it to airtight bottles at −15°C.

After four months of storage, viability as measured by *in vitro* germination was still approximately 90%. The simple germination medium consisted of sucrose (10%), agar (0.5%) and 100–200 ppm boric acid. The pollen was applied to a drop of the medium kept at high RH. Receptive stigmas were pollinated with pollen stored for 12 months at −15°C. Pollen survives for about six days on the flower. The female inflorescences were bagged before and after pollination to prevent contamination. Total fruit weight did not differ from control pollinations with fresh pollen, from which it can be concluded that oil-palm pollen can be stored dry for at least one year in an ordinary deep freezer and retain its fertilizing capacity.

Pecan (Carya illinoensis)

In pecan, pollen was bottled for cryogenic storage (Yates and Sparks, 1989, 1990). Catkins with dehiscing anthers were brought into the laboratory and spread on paper. Pollen was collected after 24 hours, during which time sufficient dehydration had taken place. Debris was removed by filtering the pollen through a brass screen. Pollen was then placed in cryovials and stored at −80°C and −196°C in liquid nitrogen.

Pollen survived at both temperatures for three years, as confirmed by *in vitro* germination and *in vivo* pollination and subsequent fruit set. Prehydration in humid air for 4 hours is required for successful *in vitro* tests. Germination *in vitro* was performed in a simple medium containing 15% sucrose, 150 ppm boric acid and 500 ppm $Ca(NO_3)_2 \cdot 4H_2O$. In the pollination experiments, the female flowers were carefully bagged to prevent contamination. The equally high fruit set after pollination with stored and fresh pollen indicates that cryogenic storage can be used for long-term germplasm conservation in pecan.

References

Akihama, T. and M. Omura (1986) Preservation of fruit tree pollen. In: Bajaj, Y.P.S. (ed.) *Biotechnology in Agriculture and Forestry*. Vol. 1: *Trees*. pp. 101–112. Springer-Verlag, Berlin.

Alexander, M.P. and S. Ganeshan (1989) An improved cellophane method for *in vitro* germination of recalcitrant pollen. *Stain Technology* 64:225–227.

Barnabas, B. and E. Rajki (1976) Storage of maize (*Zea mays* L.) pollen at −196°C in liquid nitrogen. *Euphytica* 25:747–752.

Barnabas, B. and E. Rajki (1981) Fertility of deep-frozen maize (*Zea mays* L.) pollen. *Annals of Botany* 48:861–864.

Barrow, J.R. (1981) A new concept in assessing cotton pollen germinability. *Crop Science* 21:441–443.

Bar-Shalom, D. and O. Mattsson (1977) Mode of hydration, an important factor in the germination of trinucleate pollen grains. *Botanisk Tidsskrift* 71:245–251.

Beers, W.L. Jr., J. Bivens and J.E. Mocha (1981) Pollen collection. In: Franklin, E.C. (ed.) *Pollen Management Handbook.* USDA Agriculture Handbook No. 587. pp. 30–32. USDA, Washington.

Bilsen, D.G.J.L. van and F.A. Hoekstra (1993) Decreased membrane integrity in aging *Typha latifolia* L. pollen: accumulation of lysolipids and free fatty acids. *Plant Physiology* 101:675–682.

Bilsen, D.G.J.L. van, F.A. Hoekstra, L.M. Crowe and J.H. Crowe (1994) Altered phase-behavior in membranes of aging dry pollen may cause imbibitional leakage. *Plant Physiology,* 104(4): 1193–1199.

Binder, W.D., G.M. Mitchell and D.J. Ballantyne (1974) *Pollen Viability Testing, Storage and Related Physiology.* Canadian Forestry Service Report BC-X-105, Ottowa.

Bredemann, G., K. Garber, P. Harteck and K.A. Suhr (1947) Die Temperatur-abhängigkeit der Lebensdauer von Blütenpollen. *Die Naturwissenchaften* 34:279–280.

Brewbaker, J.L. (1959) Biology of the angiosperm pollen grain. *Indian Journal of Genetics and Plant Breeding* 19:121–133.

Brewbaker, J.L. (1967) The distribution and phylogenetic significance of binucleate and trinucleate pollen grains in the angiosperms. *American Journal of Botany* 54:1069–1083.

Brewbaker, J.L. and B.H. Kwack (1963) The essential role of calcium ion in pollen germination and pollen tube growth. *American Journal of Botany* 50:859–865.

Crisp, P. and B.W.W. Grout (1984) Storage of broccoli pollen in liquid nitrogen. *Euphytica* 33:819–823.

Crowe, J.H., L.M. Crowe, J.F. Carpenter and C. Aurell Wistrom (1987) Stabilization of dry phospholipid bilayers and proteins by sugars. *Biochemical Journal* 242:1–10.

Crowe, J.H., F.A. Hoekstra and L.M. Crowe (1989) Membrane phase transitions are responsible for imbibitional damage in dry organisms. *Proceedings of the National Academy of Sciences* 86:520–523.

Crowe, L.M., C. Womersley, J.H. Crowe, D. Reid, L. Appel and A. Rudolph (1986) Prevention of fusion and leakage in freeze-dried liposomes by carbohydrates. *Biochimica et Biophysica Acta* 861:131–140.

Cruden, R.W. (1977) Pollen–ovule ratios: a conservative indicator of breeding systems in flowering plants. *Evolution* 31:32–46.

D'Antonio, V. and C.F. Quiros (1987) Viability of celery pollen after collection and storage. *HortScience* 22:479–481.

de Nettancourt, (1977) *Incompatibility in Angiosperms.* Springer-Verlag, Berlin.

Duffield, J.W. and A.G. Snow (1941) Pollen longevity of *Pinus strobus* and *P. retinosa* as controlled by humidity and temperature. *American Journal of Botany* 28:175–177.

Ekaratne, S.N.R. and S. Senathirajah (1983) Viability and storage of pollen of the oil palm, *Elaeis guineensis* Jacq. *Annals of Botany* 51:661–668.

FAO (1993) *Ex situ Storage of Seeds, Pollen and In Vitro Cultures of Perennial Woody*

Plant Species. FAO Forestry Paper 113, based on the work of B.S.P. Wang, P.J. Charest and B. Downie. FAO, Rome.

Ferwerda, F.P. (1937) Kiemkracht en levensduur van koffiestuifmeel. *Archief van de Koffiecultuur* 11:135–150.

Franklin, E.C. (ed.) (1981) *Pollen Management Handbook*. USDA Agriculture Handbook No. 587. USDA, Washington.

Ganeshan, S. (1985) Storage and longevity of papaya (*Carica papaya* L. 'Washington') pollen. I. Effect of low temperature and humidity. *Gartenbauwissenschaft* 50:227–230.

Gay, G., C. Kerhoas and C. Dumas (1987) Quality of a stress-sensitive *Cucurbita pepo* L. pollen. *Planta* 171:82–87.

Goss, J.A. (1968) Development, physiology, and biochemistry of corn and wheat pollen. *Botanical Review* 34:333–358.

Halterlein, A.J., C.D. Clayberg and I.D. Teare (1980) Influence of high temperature on pollen grain viability and pollen tube growth in the styles of *Phaseolus vulgaris* L. *Journal of the American Society for Horticultural Science* 105:12–14.

Harrington, J.F. (1970) Seed and pollen storage for conservation of plant gene resources. In: Frankel, O.H. and E. Bennett (eds) *Genetic Resources in Plants – Their Exploration and Conservation*. pp. 501–521. Blackwell Scientific Publications, Oxford.

Hauser, E.J.P and J.H. Morrison (1964) The cytochemical reduction of nitro blue tetrazolium as an index of pollen viability. *American Journal of Botany* 51:748–752.

Hecker, R.J., P.C. Stanwood and C.A. Soulis (1986) Storage of sugarbeet pollen. *Euphytica* 35:777–783.

Henny, R.J. (1978) Germination of *Spathiphyllum* and *Vriesea* pollen after storage at different temperatures and relative humidities. *HortScience* 13:596–597.

Henny, R.J. (1980) Germination of *Dieffenbachia maculata* 'Perfection' pollen after storage at different temperature and relative humidity regimes. *HortScience* 15:191–192.

Heslop-Harrison, J., Y. Heslop-Harrison and K.R. Shivanna (1984) The evaluation of pollen quality, and a further appraisal of the fluorochromatic (FCR) test procedure. *Theoretical and Applied Genetics* 67:367–375.

Heslop-Harrison, Y. and K.R. Shivanna (1977) The receptive surface of the angiosperm stigma. *Annals of Botany* 41:1233–1258.

Hoekstra, F.A. (1979) Mitochondrial development and activity of binucleate and trinucleate pollen during germination *in vitro*. *Planta* 145:25–36.

Hoekstra, F.A. (1983) Physiological evolution in angiosperm pollen: possible role of pollen vigor. In: Mulcahy, D.L. and E. Ottaviano (eds) *Pollen: Biology and Implications for Plant Breeding*. pp. 35–41. Elsevier Scientific Publishing, Amsterdam.

Hoekstra, F.A. (1986) Water content in relation to stress in pollen. In: Leopold, A.C. (ed.) *Membrane Metabolism and Dry Organisms*. pp. 102–122. Comstock Publishing Associates, Ithaca.

Hoekstra, F.A. (1992) Stress effects on the male gametophyte. In: Cresti, M. and A. Tiezzi (eds) *Sexual Plant Reproduction*. pp. 193–201. Springer-Verlag, Heidelberg.

Hoekstra, F.A. and J. Bruinsma (1975a) Viability of Compositae pollen: germination *in vitro* and influences of climatic conditions during dehiscence. *Zeitschrift für Pflanzenphysiologie* 76:36–43.

Hoekstra, F.A. and J. Bruinsma (1975b) Respiration and vitality of binucleate and trinucleate pollen. *Physiologia Plantarum* 34:221–225.

Hoekstra, F.A. and E.G. van der Wal (1988) Initial moisture content and temperature of imbibition determine extent of imbibitional injury in pollen. *Journal of Plant Physiology* 133:257–262.

Hoekstra, F.A. and T. van Roekel (1988) Desiccation tolerance of *Papaver dubium* L. pollen during its development in the anther: possible role of phospholipid composition and sucrose content. *Plant Physiology* 88:626–632.

Hoekstra, F.A., L.M. Crowe and J.H. Crowe (1989a) Differential desiccation sensitivity of corn and *Pennisetum* pollen linked to their sucrose contents. *Plant Cell and Environment* 12:83–91.

Hoekstra, F.A., J.H. Crowe and L.M. Crowe (1989b) Membrane behavior in drought and its physiological significance. In: Taylorson, R.B. (ed.) *Recent Advances in the Development and Germination of Seeds.* pp. 71–88. Plenum Press, New York.

Hoekstra, F.A., J.H. Crowe and L.M. Crowe (1991) Effect of sucrose on phase behavior of membranes in intact pollen of *Typha latifolia* L., as measured with Fourier transform infrared spectroscopy. *Plant Physiology* 97:1073–1079.

Hoekstra, F.A., J.H. Crowe and L.M. Crowe (1992) Germination and ion leakage are linked with phase transitions of membrane lipids during imbibition of *Typha latifolia* L. pollen. *Physiologia Plantarum* 84:29–34.

Holman, R.M. and F. Brubaker (1926) On the longevity of pollen. *University of California Publications in Botany* 13:179–204.

Iwanami, Y. (1973) Seed set of *Petunia hybrida* pollinated by stored pollen grains in organic solvent. *Botanique* 4:53–56.

Iwanami, Y. and N. Nakamura (1972) Storage in an organic solvent as a means for preserving viability of pollen grains. *Stain Technology* 47:137–139.

Jain, A. and K.R. Shivanna (1990) Storage of pollen grains of *Crotalaria retusa* in oils. *Sexual Plant Reproduction* 3:225–227.

Janson, J. (1993) Placental pollination in *Lilium longiflorum* Thunb. *Plant Science* 90:105–115.

Jensen, C.J. (1964) Pollen storage under vacuum. *Årsskrift den Kongeliges Veterinær og Landbohøgskole* 1964:133–146.

Kerhoas, C., G. Gay and C. Dumas (1987) A multidisciplinary approach to the study of the plasma membrane of *Zea mays* pollen during controlled dehydration. *Planta* 171:1–10.

Knowlton, H.E. (1922) *Studies in Pollen with Special Reference to Longevity.* Cornell University Agricultural Experiment Station Memoir 52. pp. 751–793. Cornell University Press, Ithaca.

Kubo, A. (1955) Successful artificial method of germination of Compositae pollen. *Journal of Science of Hiroshima University, Series B, Div. 2* 7:23–43.

Kubo, A. (1956) On the artificial pollen grain germination of Gramineae. I. *Triticum vulgare* Vill. *Journal of Science of Hiroshima University, Series B, Div. 2* 7:103–118.

Leduc, N., M. Monnier and G.C. Douglas (1990) Germination of trinucleate pollen: formulation of a new medium for *Capsella bursa-pastoris. Sexual Plant Reproduction* 3:228–235.

Lichte, H.F. (1957) Über die Physiologie von Angiospermenpollen und irhe Bedeutung für die Pflanzenzüchtung. *Angewandte Botanik* 31:1–28.

Lidforss, B. (1896) Zur Biologie des Pollens. *Jahrbuch für Wissenschaftliche Botanik* 29:1–38.

Linskens, H.F. and M. Cresti (1988) The effect of temperature, humidity and light on the dehiscence of tobacco anthers. *Proceedings of the Koninklijke Nederlandse Akademie van Wetenschappen* 91:369–375.

Linskens, H.F. and K. Esser (1957) Uber eine spezifische Anfärbung der Pollenschläuche in Griffel und die Zahl der Kallose-Pfopfen nach Selbstung und Fremdung. *Naturwissenschaften* 44:16.

Linskens, H.F., F. Ciampolini and M. Cresti (1989) Restrained dehiscence results in stressed pollen. *Proceedings of the Koninklijke Nederlandse Akademie van Wetenschappen* 92:465–475.

McKersie, B.D., T. Senaratna, M.A. Walker, E.J. Kendall and P.R. Hetherington (1988) Deterioration of membranes during aging in plants: evidence for free radical mediation. In: Noodén, L.D. and A.C. Leopold (eds) *Senescence and Aging in Plants*. pp. 441–464. Academic Press, New York.

Maisonneuve, B. (1983) Qualité du pollen formé au froid chez la tomate (*Lycopersicon esculentum* Mill.). *Agronomie* 3:873–878.

Mandahar, C.L. (1981) Virus transmission through seed and pollen. In: Maramorosch, K. and K.F. Harris (eds) *Plant Diseases and Vectors: Ecology and Epidemiology*. pp. 241–292. Academic Press, New York.

Mandahar, C.L. and P.S. Gill (1984) The epidemiological role of pollen transmission of viruses. *Zeitschrift für Planzenkrankheiten und Pflanzenschutz* 91:246–249.

Marcucci, M.C., T. Visser and J.M. van Tuyl (1982) Pollen and pollination experiments. VI. Heat resistance of pollen. *Euphytica* 31:287–290.

Mascarenhas, J.P. (1989) The male gametophyte of flowering plants. *Plant Cell* 1:657–664.

Matthews, F.R. and D.L. Bramlett (1983) Pollen storage methods influence filled seed yields in controlled pollinations of loblolly pine. *Southeastern Forest Experiment Station General Technical Report* 24:441–445.

Matthews, F.R. and J.F. Kraus (1981) Pollen storage. In: Franklin, E.C. (ed.) *Pollen Management Handbook*. USDA Agriculture Handbook No. 587. pp. 37–39. USDA, Washington.

Miller, T. and E.W. Belcher (1981) Pollen contamination and quarantine restrictions. In: Franklin, E.C. (ed.) *Pollen Management Handbook*. USDA Agriculture Handbook No. 587. pp. 70–71. USDA, Washington.

Mulcahy, D.L. (1979) The rise of the angiosperms: a genecological factor. *Science* 206:20–23.

Mulcahy, D.L. (1984) Manipulation of gametophytic populations. In: Lange, W., A.C. Zeven and N.G. Hogenboom (eds) *Efficiency in Plant Breeding*. pp. 167–175. PUDOC, Wageningen.

Mulcahy, G.B. and D.L. Mulcahy (1988) The effect of supplemented media on the growth *in vitro* of bi- and trinucleate pollen. *Plant Science* 55:213–216.

Namkoong, G. (1981) Methods of pollen sampling for gene conservation. In: Franklin, E.C. (ed.) *Pollen Management Handbook*. USDA Agriculture Handbook No. 587. pp. 74–76. USDA, Washington.

Neergaard, P. (1977) *Seed Pathology*. Vol. I. Macmillan, London.

Norton, J.D. (1966) Testing of plum pollen viability with tetrazolium salts. *Proceedings of the American Society for Horticultural Science* 89:132–134.

Pacini, E. and G.G. Franchi (1982) Germination of pollen inside anthers in some non-cleistogamous species. *Caryologia* 35:205–215.

Palfi, G. and E. Köves (1984) Determination of vitality of pollen on the basis of its amino acid content. *Biochemie und Physiologie der Pflanzen* 179:237–240.

Parlevliet, J.E., J.G. Brewer and W.G.M. Ottaro (1979) Collecting pyrethrum, *Chrysanthemum cinerariaefolium* Vis., in Yugoslavia for Kenya. In: Zeven, A.C. and

A.M. van Harten (eds) *Broadening the Genetic Base of Crops.* pp. 91–96. PUDOC, Wageningen.

Pfundt, M. (1910) Der Einfluss der Luftfeuchtigkeit auf die Lebensdauer des Blüten-staubes. *Jahrbuch für Wissenschaftliche Botanik* 47:1–40.

Priestley, D.A. (1986) *Seed Aging: Implications for Storage and Persistence in the Soil.* Comstock Associates, Ithaca.

Roberts, E.H. (1975) Problems of long-term storage of seed and pollen for genetic resources conservation. In: Frankel, O.H. and E. Bennett (eds) *Genetic Resources in Plants – Their Exploration and Conservation.* pp. 269–295. Blackwell Scientific Publications, Oxford.

Rodriguez-Garay, B. and J.R. Barrow (1986) Short-term storage of cotton pollen. *Plant Cell Reports* 5:332–333.

Schoenike, R.E. and C.F. Bey (1981) Conserving genes through pollen storage. In: Franklin, E.C. (ed.) *Pollen Management Handbook.* USDA Agriculture Handbook No. 587. pp. 72–73. USDA, Washington.

Schoper, J.B., R.J. Lambert and B.L. Vasilas (1987) Pollen viability, pollen shedding, and combining ability for tassel heat tolerance in maize. *Crop Science* 27:27–31.

Shivanna, K.R. and J. Heslop-Harrison (1981) Membrane state and pollen viability. *Annals of Botany* 47:759–770.

Shivanna, K.R. and B.M. Johri (1985) *The Angiosperm Pollen. Structure and Function.* Wiley Eastern Ltd., New Delhi.

Shivanna, K.R., H.F. Linskens and M. Cresti (1991a) Pollen viability and pollen vigor. *Theoretical and Applied Genetics* 81:38–42.

Shivanna, K.R., H.F. Linskens and M. Cresti (1991b) Responses of tobacco pollen to high humidity and heat stress: viability and germinability *in vitro* and *in vivo. Sexual Plant Reproduction* 4: 104–109.

Simon, E.W. (1974) Phospholipids and plant membrane permeability. *New Phytologist* 73:377–420.

Sporne, K.R. (1969) The ovule as an indicator of evolutionary status in angiosperms. *New Phytologist* 68:555–566.

Sprague, J.R. and E.B. Snyder (1981) Extracting and drying pine pollen. In: Franklin, E.C. (ed.) *Pollen Management Handbook.* USDA Agriculture Handbook No. 587. pp. 33–36. USDA, Washington.

Stanley, R.G. and H.F. Linskens (1974) *Pollen: Biology, Biochemistry, and Management.* Springer-Verlag, New York.

Theunis, C.H., E.S. Pierson and M. Cresti (1991) Isolation of male and female gametes in higher plants. *Sexual Plant Reproduction* 4:145–154.

Towill, L.E. (1985) Low temperature and freeze-/vacuum-drying preservation of pollen. In: Kartha, K.K. (ed.) *Cryopreservation of Plant Cells and Organs.* pp. 171–198. CRC Press, Boca Raton.

Vasil, I.K. (1962) Studies on pollen storage of some crop plants. *Journal of the Indian Botanical Society* 41:178–196.

Verhoef, H.C.M. and F.A. Hoekstra (1986) Absence of 10-hydroxy-2-decenoic acid (10-HDA) in bee-collected pollen. In: Mulcahy, D.L., G. Bergamini Mulcahy and E. Ottaviano (eds) *Biotechnology and Ecology of Pollen.* pp. 391–396. Springer-Verlag, New York.

Visser, T. (1955) Germination and storage of pollen. *Mededelingen Landbouwhogeschool* 55(1):1–68.

Visser, T., D.P. de Vries, G.W.H. Welles and J.A.M. Scheurink (1977) Hybrid tea-rose pollen. I. Germination and storage. *Euphytica* 26:721–728.

Wagner, V.T., C. Dumas and H.L. Mogensen (1989) Morphometric analysis of isolated
    Zea mays sperm. *Journal of Cell Science* 93:179–184.

Yabuya, T. (1983) Pollen storage of *Iris eusata* Thunb. in organic solvents and dry air
    under freezing. *Japanese Journal of Breeding* 33:269–274.

Yates, I.E. and D. Sparks (1989) Hydration and temperature influence *in vitro* germina-
    tion of pecan pollen. *Journal of the American Society for Horticultural Science*
    114:599–605.

Yates, I.E. and D. Sparks (1990) Three-year-old pecan pollen retains fertility. *Journal of
    the American Society for Horticultural Science* 115:359–363.

Collecting *Rhizobium*, *Frankia* and mycorrhizal fungi

26

R.A. Date

Division of Tropical Crops and Pastures, CSIRO, Cunningham Laboratory, 306 Carmody Road, St Lucia, Queensland, 4067 Australia.

Introduction

Symbioses between higher plants and bacteria or fungi are known to be often important, and perhaps essential in some cases, for good plant growth. This is generally recognized to be due to improved nutrition of the host plant. Satisfactory exploitation of new plant germplasm in new environments, therefore, may depend on the natural presence of a suitable microsymbiont or its simultaneous introduction. Experience in recent years suggests that there is a close genetic association between host plant and microsymbiont (Date *et al.*, 1979). Thus, when germplasm of some plant species is being collected for use in new environments, it is recommended that symbiotic microorganisms also be collected.

The most important microsymbionts include:

- the root-nodule bacteria, *Rhizobium* and *Bradyrhizobium*, for nitrogen fixation in legumes (Dobereiner and Campelo, 1977; Mulder *et al.*, 1977; Peoples and Herridge, 1990);
- vesicular–arbuscular mycorrhiza (VAM) for phosphorus supply in many plants (Hayman, 1982);
- actinorhizal associations (*Frankia*) for nitrogen supply in about 200 species, including forages and forestry species such as *Alnus*, *Allocasuarina*, *Elaeagnus*, *Hippophae*, *Purshia* and *Shepherdia* (Becking, 1977; Akkermans and van Dijk, 1981);
- ectomycorrhiza (over 1000 species of basidiomycetes and ascomycetes, mainly the former) for water and nutrient uptake in many forestry species in the families Pinaceae, Betulaceae, Salicaceae, Myrtaceae, Casuarinaceae and some Caesalpiniaceae and Dipterocarpaceae (Molina and Trappe, 1982).

Many of the early studies of inoculation with a microsymbiont involved the transfer of soil from established crops, plantations and forests. Today, however, most microsymbionts can either be cultivated on artificial media or maintained as 'enriched' cultures in association with an appropriate host nurse-plant in controlled conditions. These provide a good source of inoculum that can be introduced into the soil with the seeds at sowing, applied either separately or directly to the seed.

Some of the practical problems of collecting symbiotic microorganisms are outlined in this chapter as a guide to the plant germplasm collector undertaking this work for the first time. The approach and methodologies described define principles and can be altered to suit individual requirements and facilities.

Pre-sampling considerations

Suitable isolates of the target microsymbiont may already be available. To ascertain whether this is the case, the first step is to examine the catalogues of existing bacterial and fungal collections. Some idea of the holdings of collections may be obtained from the *World Directory of Collections of Cultures of Microorganisms* (Staines *et al.*, 1986). This information is also available both on-line and on disk. Bower (1989) reviews sources of information on microorganisms (see also Box 26.1). A Compact Disk Read-Only Memory (CD-ROM) called CD-STRAINS holds data on the holdings of the American Type Culture Collection, the Japanese Collection of Microorganisms and the Institute of Fermentation, Osaka. For information on root-nodule bacteria of legumes, perhaps a better starting-point is McGowan and Skerman (1986). Carlowitz (1991) lists suppliers of microsymbiont inoculants for multipurpose trees and shrubs. Hall and Minter's (1994) *International Mycological*

Box 26.1
Information sources on microbial collections

The World Data Centre of the World Federation of Culture Collections, located in Japan, has information on over 320 culture collections, forming the basis of Staines *et al.* (1986). It is a cosponsor of the Microbial Strain Data Network (MSDN).

Many collections worldwide, including informal research collections, are linked through MSDN. Its core service is access to databases, electronic bulletin boards and directories, via Dialcom.

The United Nations Environment Program (UNEP) and the United Nations Educational, Scientific and Cultural Organization (Unesco) have supported the development of a 20-country network of 24 microbiological resource centres (MIRCEN). These preserve, identify and distribute microbial germplasm, as well as providing training and resources in some cases. The MIRCEN in Stockholm has lately taken on an informatics role.

Directory brings together information on over 280 mycological organizations and institutions worldwide.

If potentially useful strains are not already available, one must decide where to collect them. Prospecting for microsymbionts should concentrate on the same geographic and ecological areas from which the plant germplasm to be inoculated was collected. Edaphic factors are particularly important in identifying areas where suitable strains would be most likely to be found (Date *et al.*, 1979; Molina and Trappe, 1982).

A perhaps more common situation, however, is for plant germplasm and microsymbiont collecting to occur in tandem. For example, a plant germplasm collector may also gather nodule samples and return them to a colleague in a microbiology department or institute for isolation of the microorganism. Joint collecting trips may be required when it is necessary to isolate from fresh nodule or root material.

Ideally, the microsymbiont and seeds should be collected from the same plant. However, this is often difficult, since seed collecting is often best done at times when the host plant has a reduced complement of active nodules or mycorrhiza. For example, nodules from annual legumes are best obtained from seedlings, though the nodule population in young plants may be composed of a different suite of strains from that of older plants as seasonal and soil conditions change (Caldwell and Weber, 1970; Weber and Miller, 1972; R.A. Date, unpublished data for strains of *Bradyrhizobium* forming nodules on *Stylosanthes*), whereas seeds are not available until plant maturity. In long-lived perennial plants, maximum seed set is often in the dry season, when nodules and fungi are not easily found. The presence of nodules on the root system in nitrogen-fixing symbioses is often seasonally dependent, especially if soil nitrogen is at levels adequate for good plant growth. For example, *Leucaena* growing as a closed-canopy stand was observed to have nodules at 98 days, but not at 205 or 274 days, with nodules reappearing on the root system at 423 days (Wong *et al.*, 1989). Sampling of soil and roots for ectomycorrhizal organisms is best done in the spring, when plants are actively growing. Young sporocarps are preferred for isolation but it is important to collect fully mature material for reliable identification of the species. Ectomycorrhizal numbers and sporocarp production also vary seasonally (Grand and Harvey, 1982).

Arrangements for the isolation and storage of the microsymbiont material should be made before setting out. When collecting abroad, it is essential to clear and coordinate microsymbiont collecting activities with the relevant local authorities, in the same way as is done for plant germplasm (Chapters 2 and 3). Microsymbionts are just as much part of biodiversity as the plants on which they live. Material collected by a national programme in its own country may need to be isolated or stored abroad, if adequate facilities are not available locally, and the isolates reimported as and when necessary. Quarantine clearance for the import and export of microsymbiont material may be necessary

and the question should be investigated before setting out to collect
(Chapter 17).

Types of sample

Microsymbionts can be isolated from nodules, roots or soil. For the root-
nodule bacteria of legumes and for many actinorhizal organisms, the
best source is obviously nodules obtained directly from the plant from
which seeds are also being collected. When nodules are not available,
collectors should sample a small piece of root and/or soil from near the
root. These can be used to inoculate seedlings of the host plant growing
aseptically to obtain fresh nodules, from which isolates can be made.
However, this is a method of last resort, because it will not necessarily
yield nodules of those strains that would have formed in the field. This
may be significant, depending on the exact purpose for which nodules
are being collected. If, for example, specific collecting is being carried
out for a strain with the ability to nodulate a host at low soil pH or for
a strain that is highly competitive for nodule formation in the presence
of large populations of ineffective strains, then reproducing these conditions for aseptically growing seedlings may not be possible.

   Collecting soil and/or root pieces is the only way of obtaining
material of those microsymbionts which do not form nodules or that cannot be grown easily in axenic culture, e.g. *Glomus* (VAM) and some
ectomycorrhizal organisms. In these cases, pure or enriched cultures
may be obtained by collecting spores by a wet sieving method (e.g.
Gerdemann and Nicholson, 1963; Hayman, 1982; Molina and Palmer,
1982; Beaton *et al.*, 1985).

Sampling guidelines

Plants typical of the area should be selected for nodule sampling. Collecting along roadsides or disturbed areas must be avoided, unless, of
course, such habitats are typical of the species. Ideally, each sample of
microsymbiont should be unique and originate from a single plant. It
should also be associated with a single-plant seed sample, as symbiotic
microorganism and host are often closely affiliated genetically (Date
et al., 1979). Recent experience indicates that 10–20 legume nodules per
plant should be collected to adequately sample the variation in strain
types in the nodule population. It is a good idea to collect two to five
such samples to represent a single site, as not all nodules will yield a
viable culture and only a small proportion (as low as 5%) of the nodules
may contain an effective nitrogen-fixing strain. Similar guidelines apply
to collecting *Frankia*, VAM and ectomycorrhizal fungi. Usually, it is
only necessary to process about half the sample, keeping the other half

as reserve against accidental loss during surface sterilization and isolation.

It is important in collecting microsymbionts to know where to look on or about the root system for nodules, rhizomorphs and sporocarps. The location of nodules on the root system is very species-specific and is further influenced by local conditions. In small, strongly tap-rooted species like *Trifolium semipilosum*, nodules can be found within 4–5 cm of the root crown. In *Stylosanthes*, nodules are more or less equally distributed along the length of the root system to a depth of 15 cm, but may be deeper in species like *S. capitata* growing in acid sandy soils (Venezuela) or oxisols (Brazil). *Acacia* and *Casuarina* species growing in free-draining deep sands have nodules at depths of 1–2 m, which is probably related to the level of the water-table.

Small herbaceous legumes (e.g. *Trifolium semipilosum, Lotus* spp.) can be sampled by digging a small core (diameter 10–15 cm, depth 10 cm) around the tap root. For shrubs and trees, it is rarely necessary to excavate the entire root system. Careful removal of soil from around the root crown with a knife or small trowel usually provides good results. Most nodules are located on adventitious roots in the top 5 cm of soil. For seedlings, a small core (as for herbaceous legumes) of soil around the root is usually adequate, but on mature plants nodules may be found further out and much deeper, depending on conditions. Usually, it is possible to collect nodules from the cores by sequentially fragmenting the soil by hand, but it may be necessary for dry or clay soils to soak the core of soil in water and allow the soil to fall away. Alternatively, and especially in species where nodules become detached readily from fine adventitious roots (e.g. *Macroptilium, Leucaena*), soil and roots with nodules can be washed over a sieve (1–2 mm openings). In either case, nodules can be recovered with forceps. When the sample is to be stored in vials with desiccant, it is advisable to dry the nodules with a paper towel before placing in the vial. (For equipment needed, see Box 26.2.)

Nodule size and shape are also very species-specific. Those of *Arachis* and *Stylosanthes* are 1–2 mm in diameter and firmly attached, whereas those of *Alnus, Casuarina, Macroptilium* and *Medicago* are typically larger and become detached very readily. The shape of legume nodules is governed by the extent and location of the meristem. Hemispherical peripheral meristems produce spherical nodules, as typified by *Arachis, Glycine, Lotus, Macroptilium, Stylosanthes* and *Vigna*. Elongated cylindrical forms result from growth of apical meristems, which may divide, giving digitate or even coralloid nodules, as observed in *Leucaena, Trifolium* and *Vicia*. When meristems divide laterally, subsequent growth of the nodule tends to surround the roots, as in *Lupinus* (Corby, 1971; Allen and Allen, 1981). Nodules with peripheral meristems are sometimes referred to as determinate and the other types as indeterminate. In cross-section, determinate nodules show a single circular active bacteroid (red) zone, whereas in indeterminate nodules there are one or more irregularly shaped active bacteroid

---

Box 26.2
Equipment required for collecting microsymbionts

- Digging implement (trowel, strong knife).
- Secateurs or strong scissors.
- Forceps, scalpel or small sharp knife.
- Hand lens.
- Collecting vials with desiccant.
- Waxed paper bags.
- Paper towels.
- 5 l plastic container.
- Permanent marker pen.
- Adhesive tape.
- Plastic bags.
- Containers for storage of samples.
- Cooler box (optional).
- See Date and Halliday (1987) for details of portable kits if isolation of microsymbiont is to be attempted in the field.

---

zones. Nodules on most non-leguminous plants are modified lateral roots with slow-growing meristems, usually branching dichotomously to give coralloid-type nodules up to 5–6 cm long, as typified by *Alnus*, *Casuarina* and *Ceanothus* (Becking, 1977).

Endomycorrhizal plants have no obvious macroscopic structures, but with experience hyphal growth on the roots may be recognized with a hand lens. There is wide variation in the morphology of ectomycorrhiza. They appear as a continuation of the root in *Pinus taeda*, but coralloid forms on *P. strobus* and *P. resinosa* are more like the root-nodules of *Lupinus* (Grand and Harvey, 1982). In other host species the mycorrhiza varies from a thin mantle of fungal hyphae to thickened zones of terminal roots (Chilvers and Pryor, 1965). In *Acacia* and *Nothofagus* species gasterocarps up to 3 cm in diameter may be found in the surface 10 cm of soil, usually on the side exposed to the prevailing winds, within 1.5 m of the trunk. They are usually most plentiful in late autumn, winter and early spring (Beaton *et al.*, 1985).

Storage guidelines

The viability of *Rhizobium*, *Bradyrhizobium* and *Frankia* in the nodule sample is variable and it is best to isolate on the day of collecting or soon thereafter. When this is not possible, nodules must be dried. The microsymbiont will usually remain viable in the dried state for several weeks, with 75–90% of nodules yielding viable isolates after reimbibition and surface sterilization.

Drying is best achieved by placing nodules in an airtight vial or screw-capped bottle with a desiccant such as anhydrous $CaCl_2$ or silica gel (Date and Halliday, 1987). Washing samples should be avoided, but, if it is necessary to wash, then nodules should be dried with absorbent paper before being placed in the vial. The desiccant can be kept in place by a plug of cotton or absorbent paper. Its volume should exceed that of the sample. A 10–20 ml capacity vial with 25% of its volume filled with desiccant and 10% made up of a cotton or paper plug is a good container for preserving and transporting samples. Vials should be kept cool.

There are some species, however, where the proportion of recovery is less than 5% of nodules and it is therefore necessary to isolate from fresh nodules. For example, no viable isolates of *Bradyrhizobium* were obtained from more than 150 dried nodule samples of *Stylosanthes capitata* collected in Brazil and processed in Australia six to ten weeks later, but isolation was 65–80% successful for fresh nodules processed the day of collecting (R.A. Date, unpublished data). When isolations are to be made on the day of collecting, the simplest method is to gather the whole or part of the root system into sealable polythene bags. On return to a laboratory, roots and nodules must be washed free of soil and the nodules excised with a short piece of root attached. This is important, especially for small nodules, as it reduces the entry of sterilant during surface sterilization. Portable isolating kits are described by Date and Halliday (1987).

Unlike nitrogen-fixing root-nodules, sporocarps should be kept dry, but not airtight. Waxed paper containers are usually used (Molina and Palmer, 1982).

Data collecting

Because of the close association between environmental, especially soil, features and various traits and properties of the microsymbiont (Date *et al.*, 1979; Molina and Trappe, 1982), it is important to record some descriptive information on the site. Experience suggests that the kinds of site data normally recorded by plant germplasm collectors are also relevant to collecting microsymbiont germplasm. These have been described in several publications (Mott and Jimenez, 1979; Date and Halliday, 1987; Stowers, 1987) and are also considered in detail in Chapter 19.

Isolation of microsymbionts

Successful isolation of microsymbionts depends on the quality of the sample. Damaged *Rhizobium*, *Bradyrhizobium* or *Frankia* nodules are not satisfactory and should be used only to inoculate aseptically growing seedlings of the host plant to obtain fresh nodules. Dried nodules need to be reimbibed for 30–60 minutes before surface sterilization. Methods of sterilization and isolation vary with individual requirements and with the microsymbiont. A number of laboratory procedures have been described for the following organisms:

- *Rhizobium* and *Bradyrhizobium*: Brockwell (1980), FAO (1983), Somasegaran and Hoben (1985), Date and Halliday (1987);
- VAM: Daniels and Skipper (1982), Hayman (1982);
- *Frankia*: Stowers (1987);
- ectomycorrhiza: Molina and Palmer (1982).

It is essential to confirm that isolates are representative, pure cultures and able to form nodules (or infect a host root system) on aseptically growing host plants and thus satisfy Koch's postulates regarding causal organisms. Isolates must also be adequately preserved, evaluated and documented. Methods and criteria are described by Vincent (1970), Brockwell (1980), FAO (1983) and Date and Halliday (1987).

Acknowledgements

I am indebted to Dr P. Reddell for advice on collecting mycorrhizal fungi.

References

Akkermans, A.D.L. and C. van Dijk (1981) Non-leguminous root nodule symbioses with actinomycetes and *Rhizobium*. In: Broughton, W.J. (ed.) *Nitrogen Fixation*. Vol. 1. pp. 57–103. Clarendon Press, Oxford.

Allen, O.N. and E.K. Allen (1981). *The Leguminosae. A Source Book of Characteristics, Uses and Nodulation*. University of Wisconsin Press, Wisconsin.

Beaton, G., D.N. Pegler and T.W.K. Young (1985) Gasteroid basidiomycota of Victoria State, Australia: 8–9. *Kew Bulletin* 40:827–842.

Becking, J.H. (1977) Dinitrogen fixing associations in higher plants other than legumes. In: Hardy, R.W.F. and W. Silver (eds) *A Treatise on Dinitrogen Fixation*. pp. 185–275. John Wiley & Sons, New York.

Bower, D.J. (1989) Genetic resources worldwide. *Tibtech* 7:111–116.

Brockwell, J. (1980) Experiments with crop and pasture legumes – principles and practice. In: Bergersen, F.J. (ed.) *Methods for Evaluating Biological Nitrogen Fixation*. pp. 417–490. John Wiley & Sons, New York.

Caldwell, B.E. and D.F. Weber (1970) Distribution of *Rhizobium japonicum* serogroups in soybean nodules as affected by planting dates. *Agronomy Journal* 62:12–14.

Carlowitz, P.G. von (1991) *Multipurpose Trees and Shrubs – Sources of Seeds and Inoculants*. ICRAF, Nairobi.

Chilvers, G.A. and L.D. Pryor (1965) The structure of eucalypt mycorrhizas. *Australian Journal of Botany* 13:245–259.

Corby, H.D.L. (1971) The shape of leguminous nodules and the colour of leguminous roots. *Plant and Soil* (Special Volume) 305–314.

Daniels, B.A. and H.D. Skipper (1982) Methods for the recovery and quantitative estimation of propagules from soil. In: Schenck, N.C. (ed.) *Methods and Principles of Mycorrhizal Research.* pp. 29–35. American Phytopathological Society, Minnesota.

Date, R.A. and J. Halliday (1987) Collecting, isolation, cultivation and maintenance of rhizobia. In: Elkan, G.H. (ed.) *Symbiotic Nitrogen Fixation Technology.* pp. 1–27. Marcel Dekker, New York.

Date, R.A., R.L. Burt and W.T. Williams (1979) Affinities between various *Stylosanthes* species as shown by rhizobial, soil pH and geographic relationships. *Agro-Ecosystems* 5:57–67.

Dobereiner, J. and A.B. Campelo (1977) Importance of legumes and their contribution to tropical agriculture. In: Hardy, R.W.F. and A.H. Gibson (eds) *A Treatise on Dinitrogen Fixation.* pp. 191–220. John Wiley & Sons, New York.

FAO (1983) *Technical Handbook on Symbiotic Nitrogen Fixation.* FAO, Rome.

Gerdemann, J.W. and T.H. Nicholson (1963) Spores of mycorrhizal *Endogene* extracted from soil by wet sieving and decanting. *Transactions of British Mycological Society* 46:235–244.

Grand, L.F. and A.E. Harvey (1982) Quantitative measurement of ectomycorrhizae on plant roots. In: Schenck, N.C. (ed.) *Methods and Principles of Mycorrhizal Research.* pp. 157–164. American Phytopathological Society, Minnesota.

Hall, G.S. and D.W. Minter (1994) *International Mycological Directory* 3rd edition. CAB International, Wallingford.

Hayman, D.S. (1982) Practical aspects of vesicular–arbuscular mycorrhiza. In: Subba Rao, N.S. (ed.) *Advances in Agricultural Microbiology.* pp. 325–373. Butterworths, London.

McGowan, V.F. and V.B.D Skerman (1986) *World Catalogue of* Rhizobium *Collections.* World Data Center, University of Queensland, Brisbane.

Molina, R. and J.G. Palmer (1982) Isolation, maintenance and pure culture manipulation of ectomycorrhizal fungi. In: Schenck, N.C. (ed.) *Methods and Principles of Mycorrhizal Research.* pp. 115–129. American Phytopathological Society, Minnesota.

Molina, R. and J.M. Trappe (1982) Applied aspects of ectomycorrhizae. In: Subba Rao, N.S. (ed.) *Advances in Agricultural Microbiology.* pp. 305–424. Butterworths, London.

Mott, G.O. and A. Jimenez (1979) *Handbook for the Collection, Preservation and Characterization of Tropical Forage Germplasm Resources.* CIAT, Cali.

Mulder, E.G., T.A. Lie and A. Houwers (1977) The importance of legumes under temperate conditions. In: Hardy, R.W.F. and A.H. Gibson (eds) *A Treatise on Dinitrogen Fixation.* pp. 221–242. John Wiley & Sons, New York.

Peoples, M.B. and D.H. Herridge (1990) Nitrogen fixation by legumes in tropical and sub-tropical agriculture. *Advances in Agronomy* 44:155–223.

Somasegaran, P. and H.J. Hoben (1985) *Methods in Legume–*Rhizobium *Technology.* Nif-TAL, University of Hawaii, Honolulu.

Staines, J.E., V.F. McGowan and V.B.D. Skerman (1986) *World Directory of Collections of Cultures of Microorganisms.* World Data Center, University of Queensland, Brisbane.

Stowers, M.D. (1987) Collection, isolation, cultivation and maintenance of *Frankia.* In: Elkan, G.H. (ed.) *Symbiotic Nitrogen Fixation Technology.* pp. 29–53. Marcel Dekker, New York.

Vincent, J.M. (1970) *A Manual for the Practical Study of Root-Nodule Bacteria.* Blackwell Scientific Publications, Oxford.

Weber, D.F. and V.L. Miller (1972) Effect of soil temperature on *Rhizobium japonicum* serogroup distribution in soybeans. *Agronomy Journal* 64:796–798.

Wong, C.C., J. Sundram, R.A. Date and R.J. Roughley (1989) Nodulation of *Leucaena leucocephala* in acid soils of Peninsular Malaysia. *Tropical Grasslands* 23:171–178.

Useful addresses

MSDN Secretariat
307 Huntingdon Road
Cambridge CB3 0JX
UK
Tel: +44 1223 276622
Fax: +44 1223 277605
E-mail: msdn@bdt.ftpt.br

Unesco Microbial Resources Center
Karolinska Institute
S 104 10 Stockholm
Sweden
Tel: +46 8 7287147
Fax: +46 8 331547
E-mail: eng-leong_foo_ba@kicom.ki.se

Collecting herbarium vouchers

27

A.G. Miller and J.A. Nyberg
Royal Botanic Garden, Edinburgh EH3 5LR, UK.

Introduction

A herbarium is a collection of plant specimens (vouchers) which is arranged systematically and stored in purpose-built cabinets. It may be housed in a special building or form part of a larger scientific institution or museum. A herbarium voucher is usually a dried and pressed plant or portion of a plant mounted on to a sheet of stiff card usually measuring 42 × 26 cm (frequently somewhat larger in the US). It can also be material stored in spirit or more bulky dried items such as samples of wood or large fruits. Vouchers are a permanent record of a plant. A specialist is able to name the plant or check any name previously applied to it. When named, vouchers act as a key reference point to further information on the plant. Of themselves, they are also sources of information on morphological (and possibly molecular) variation, phenology, distribution and autecology.

Germplasm collectors gather material for herbarium vouchers for various reasons:

- To 'determine' particular germplasm accessions. Such vouchers may be collected either at the same time as the germplasm or before, in the course of a preliminary reconnaissance survey carried out at flowering time. This may include marking sites or even individual plants for subsequent germplasm collecting. The vouchers will most often be used in the first instance simply to identify (or 'determine') the plants accurately, though they will of course remain as a source of other information. It may, for example, be possible to extract DNA from herbarium vouchers for genetic diversity studies (e.g. Liston *et al.*, 1990). Accessions undergoing numerous regeneration or multiplication cycles, or being grown out under a different climatic

561

regime, may change substantially in phenotype from the original population, of which it is therefore useful to have a durable record. All germplasm collections of wild species (forestry species, forages, crop relatives, etc.) should be accompanied by voucher specimens. For crops, only germplasm collections of rare or unusual types need to have accompanying voucher ·specimens.

- To show the range of variation in selected characteristics of the population sampled. A so-called 'mass collection' of herbarium specimens is sometimes made. This is a population sample, composed of a number of individuals large enough to be statistically useful, of critically selected corresponding plant fragments collected at a particular time and place.

- To identify pests, etc. Material showing evidence of pest or disease damage or of some abiotic problem may be collected to help in the identification of the exact cause of the problem, rather than of the plant itself (Chapter 17).

- To document the flora of a site or region. Collectors will usually gather germplasm of only a very small proportion of the plants found in an area, but may decide to collect herbarium specimens of other species, particularly if there are rare, endemic or otherwise interesting plants in the area, or if the region is poorly known botanically.

Vouchers will last indefinitely if properly prepared and stored and remain for study by future generations. This chapter will deal with how to collect, process, store and dispatch vouchers to ensure that good, well-documented collections are made.

Collecting specimens for vouchers

A small number of well-collected and clearly annotated specimens are of much greater scientific value than a large number of badly pressed, poorly labelled scraps. Ideally, specimens should be processed immediately after gathering, using one of the methods described below. In practice, there is rarely enough time to process material in the field but specimens can be kept fresh for a reasonable length of time in large, strong polythene bags (see Box 27.1 for basic equipment). A separate bag, clearly labelled, should be used for each locality. Within these large bags, smaller bags can be used for each plant collected (certainly for samples of small or fragile plants). However, delicate plants and those that wither quickly (e.g. many Boraginaceae) or whose petals fall quickly (e.g. many Papaveraceae and Cistaceae) should be pressed immediately into a field press or portfolio. This consists of two sheets of stiff card or thin plywood, slightly larger than the card on which the specimen will be mounted, held together by webbing straps and containing folded sheets of absorbent paper.

Box 27.1
Basic equipment for herbarium collecting

The final selection of equipment will depend on the conditions under which collecting will take place and the plants of interest (Boxes 27.2 and 27.3). The following is a list of the equipment that will be necessary no matter what the method of preparation of specimens. Equipment necessary for data gathering and recording is listed in Chapter 19.

- Gardeners' gloves.
- Secateurs.
- Knife.
- Pruning saw.
- Digger or entrenching tool (for collecting underground organs).
- Polythene bags (various sizes, e.g. 100×75, 50×25 and 30×15 cm).
- Field press.
- Jeweller's tags.
- Paper packets for seeds, loose flowers, etc.
- Polythene bottles and spirit.

What to collect

Care should be taken to collect plants that are representative of the population (or subpopulation) from which the germplasm collection is being made, and not simply those that fit most easily into the press or are easiest to reach. In making a general collection in an area, any plant may be of interest. The very large, the very small, the spiny and the succulent are all too often undercollected for no better reason than that they are awkward to press.

In general, sterile specimens are worthless, so normally only flowering and/or fruiting material should be collected. One should try to get flowers and fruits from the same plant. If this is not possible, different collecting numbers should be given to separate flowering and fruiting specimens. For monoecious and dioecious plants, flowers of both sexes should be collected. It is important to know which specific features or organs of the plant are important for identification. For instance, in some families (e.g. Cyperaceae, Umbelliferae, Cruciferae, Dipterocarpaceae, Eleagnaceae) ripe fruit is essential, in others (e.g. Orchidaceae) flowers are necessary and fruits unimportant and in others (e.g. Gramineae) the whole plant, including underground parts, must be collected. Forman and Bridson (1992) list a selection of families with short notes not only on the features of the plants which need to be noted but also on special collecting and processing techniques. References where fuller instructions for problem families can be found are:

- Agavaceae (Sánchez Mejorada in Lot and Chiang, 1986:107);
- Araceae (Nicolson, 1965; Nicolson in Womersley, 1981:115–119; Croat, 1985);

- Balsaminaceae: *Impatiens* (Grey-Wilson, 1980);
- Begoniaceae: *Begonia* (Logan, 1986);
- Bromeliaceae (Smith, 1971:23–24; Jorgensen, 1972; Aguirre Léon in Lot and Chiang, 1986:118–119);
- Cactaceae (Sánchez Mejorada in Lot and Chiang, 1986:106–107);
- Crassulaceae (Sánchez Mejorada in Lot and Chiang, 1986:107);
- Cyclanthaceae (Hammel, 1987);
- Gramineae: bamboos (McClure, 1965; McClure in Womersley, 1981:110–114; Soderstrom and Young, 1983; Koch in Lot and Chiang, 1986);
- Lecythidaceae (Mori and Prance, 1987);
- Lentibulariaceae: *Utricularia* (Taylor, 1977);
- Musaceae: bananas (Fosberg and Sachet, 1965:109–110; Womersley, 1981:100–102);
- Orchidaceae (Aguirre Léon in Lot and Chiang, 1986:114–117);
- Palmae (Tomlinson, 1965; Tomlinson in Womersley, 1981:103–109; Balick, 1989; Dransfield, 1986; Quero in Lot and Chiang, 1986); rattans (Dransfield, 1979);
- Pandanaceae (Stone in Womersley, 1981:94–97; Stone, 1983);
- Passifloraceae: *Passiflora* (Jørgensen *et al.*, 1984);
- Zingiberaceae (Burtt and Smith, 1976).

How much to collect

It is important to always bear in mind how much material one will need to collect, particularly if specimens are not being pressed immediately. How many duplicates will be needed? The absolute minimum will probably be three: one to be kept by the collector's institute (or an associated or nearby herbarium) for reference, one to be deposited in the country's national herbarium and one to be sent for verification to an expert, who will often be attached to a major regional or world herbarium. For each duplicate enough material should be collected to fill a herbarium sheet. This could mean several plants of a small annual or a single shoot of a shrub or tree. Of course, one should be careful not to collect so much material that the survival of the population is endangered. In certain plant groups two or even three sheets are needed for a single sample. It is often useful to collect extra flowers, which can be placed in small envelopes (sometimes called 'capsules') on the herbarium sheet. These flowers are used for dissecting, which is often necessary for identification, without damaging the main part of the voucher, which should be attached firmly to the sheet.

Processing vouchers

There are two main methods of processing herbarium vouchers:

- by drying in the field;
- by chemical treatment (the Schweinfurth method) followed by drying at a later date.

Using the first method the plants are pressed and then either they are air-dried or gentle heat is applied. In the second method the material is pressed and then stored in sealed plastic tubing with some spirit until it can be properly dried, usually back at the home institute.

Drying in the field

Drying in the field produces very good specimens and is particularly useful in dry climates. It is, however, very time-consuming and requires a lot of bulky equipment (Box 27.2). Some specimens must first be 'killed' to prevent the leaves becoming detached during drying. There is also the danger of material becoming mouldy before it is dried. Once dry it must be kept so, or mould may develop.

Plants for processing are placed directly into a flimsy. Small plants should be pressed whole and representative flowering and/or fruiting shoots with leaves should be selected from larger plants. Leaves that extend beyond the flimsy should be folded over. Pieces of stem that are too long for the flimsy can be broken and arranged in a 'V' shape; vines should be curved. Succulent plants and bulbs will dry more quickly if they are split lengthwise; spiny or intricately woody specimens can be initially flattened between boards or corrugates before being put into the plant press. Delicate flowers or flowers on thick woody stems are best pressed separately in small torn-off bits of flimsy.

The flimsy must be numbered (the collecting number) on the outside bottom right-hand (opening) corner. A jeweller's tag can also be attached to the specimen. The specimen will stay within this flimsy throughout the drying process. Notes about the plant are made in the collector's field book at this time.

The press is built up by alternating one or two drying papers with each flimsy. Corrugates added between the drying papers allow circulation of air and will speed the drying process. If enough corrugates are added the press can be as thick as the length of the straps will allow. However, for air-drying in the sun, the smaller the press the faster the plants will dry and the better the results. Finally, two straps are placed

---

Box 27.2
Equipment for drying specimens in the field

- Flimsies (folded sheets of thin, strong paper the same size as the press – e.g. airmail newspaper or ordinary newspaper).
- Drying paper (newspaper or felt).
- Presses and straps (presses can be slatted wood or metal lattice; straps can be of webbing, leather or nylon – the last is not suitable if heat is being used).
- Corrugates (aluminium or corrugated cardboard).
- Drying frame.
- Source of heat (paraffin stove, electric heater, etc.).

around the press (breadth wise) and tightened as firmly as possible. It is important not to allow the plants to rot by using a thick press and too few corrugates.

In dry climates the presses can be placed in the sun with as much air circulation as possible–a sunny, windy spot (e.g. a car roof-rack) is ideal.

Specimens can be rearranged after a few hours but should not be removed from the flimsy until they are ready to be mounted. Drying papers must be changed daily; if this is not done, plants soon become mouldy and blackened. Damp drying papers can be quickly sun-dried by spreading them about on the ground (Davis, 1961).

Alternatively, the press can be placed over a source of artificial heat on some sort of frame. This can be a purpose-built collapsible frame or can be made from two lengths of wood supported on the backs of chairs. The heat source can be paraffin stoves (Womersley, 1981), incandescent bulbs (Lawrence, 1951), coals (Forman and Bridson, 1992) or electricity, if available. An electric heater is very effective. A skirt of suitable material (e.g. canvas) should be placed around the bottom of the frame. Care must be taken that naked flames do not ignite the specimens and that the plants are not over dried. Drying using this technique can take anything from 4 to 48 hours.

Different plants take different lengths of time to dry. Presses should be checked at least twice a day or plants can become overdried and brittle. When dried, specimens should be removed from the press as soon as possible. It can be difficult to tell when a plant is dry. A good indication is when the leaves or stems are not bendable or marked when bruised with a fingernail.

Chemical treatment (the Schweinfurth method)

Chemical treatment using alcohol is particularly useful in areas of high humidity such as tropical rain forests. The spirit prevents fungal attack and not having to dry the plants immediately saves time in the field and generally requires less equipment (Box 27.3). However, specimens often dry black and alcohol-soluble substances are lost. Also, alcohol can be difficult to obtain.

Specimens are collected into the press as described above and left overnight. Still in their flimsies, they are then removed from the press and piled into conveniently sized bundles (about 12–15 cm thick). A length of polythene tube about 1.5 m long is cut and the bundle inserted into it. One of the opened ends of the tube is then folded over and sealed with non-alcohol-soluble tape. The tube is now stood upright with the opened end uppermost and about 0.5 to 1 litre of 60–70% (higher concentrations can be used on particularly succulent material but tend to make the specimens brittle) industrial alcohol or methylated spirits or 4–6% solution of formaldehyde added. Womersley (1957, 1981) describes how to prepare formaldehyde using paraformaldehyde powder. The use of formalin is not recommended, as it is carcinogenic and is not always

---

Box 27.3
Equipment for the Schweinfurth method

- Flimsies.
- Drying paper.
- Presses and straps.
- Polythene tubing (heavy gauge, 0.1 mm thick and about 25 cm wide).
- Tape (insoluble in alcohol).
- Spirit and spirit containers.

---

successful at preventing mould developing. The opened end of the plastic tube can now be folded over and sealed.

The spirit fills the bag with vapour and if air is excluded preserves the contents for weeks. If specimens are going to be dispatched by air, excess spirit must be poured off (airlines do not allow the carriage of alcohol). The most important thing to remember is that the process depends on the exclusion of air, so great care must be taken to ensure that the polythene tube is not punctured (Fosberg and Sachet, 1965; Womersley, 1981; Forman and Bridson, 1992).

Special techniques

- *Fragile plants or plants with delicate flowers* should be pressed immediately or placed in a field press until they can be transfered to the main press. Particularly delicate flowers (e.g. Orchidaceae, Zingiberaceae or Bromeliaceae) should be placed in spirit (see below). The corollas of certain flowers (e.g. petaloid monocots, *Impatiens* spp. and *Hibiscus* spp.) should be taken to pieces and their parts pressed separately in tissue paper or non-absorbent toilet paper.
- *Succulent or fleshy plants* (e.g. species of Asclepiadaceae, Crassulaceae, Cactaceae and Euphorbiaceae) need particular attention. In general, they can be cut in half and then 'killed' in boiling water or spirit to prevent the specimen continuing to grow in the press. They can then be air-dried, although this can take several weeks, particularly if the tissue has not been killed. If a microwave oven is available, this can be used to kill the tissue (Fuller and Barber, 1981; Leuenberger, 1982). The specimen is placed in the microwave for 30 seconds to 2 minutes, depending on the toughness of the plant; it can then be air-dried in the normal way. The leaves of *Aloe* spp. require special treatment. They should be carefully cut around the margin and the inner tissue scraped out; the upper surface with the margin intact is pressed. With Cactaceae, as much of the fleshy inner tissue as possible is scraped away and then salt applied to draw out the water.

- *Large or fleshy plants or those with complex flowers* (i.e. Orchidaceae, Zingiberaceae or Bromeliaceae) can be preserved in spirit (Fosberg and Sachet, 1965:117–119).
- *Bulky or cushion plants* can be sectioned before they are pressed and wads of paper added to even out the pressure on leaves and flowers and thus prevent shrivelling. Some flowers should in addition be pressed separately.
- *Very large leaves.* Measurements and drawings should be made of the whole leaf, which should then be sectioned and labelled in such a way that it can be reassembled later. The different parts (including usually the apex, middle and basal sections) can then be pressed separately. For notes on collecting large-leaved species from selected families, see the references for Araceae, Musaceae, Palmae and Pandanaceae listed earlier.
- *Aquatics with fine leaves* should be floated over drying paper which is submerged under the plant and then drawn out. For detailed instructions see Fosberg and Sachet (1965:107–109), Womersley (1981:98–99), Haynes (1984), Ceska and Ceska (1986), Lot (1986) and, for *Utricularia* spp., Taylor (1977).
- Collecting *epiphytes and flowering and fruiting specimens from the canopies of tall trees* presents particular problems. A light sectional rod with interchangeable knife and grapnel at the tip is handy for reaching high branches (Fosberg and Sachet, 1965). Alternatives are climbing into the canopy and using rifles. For detailed instructions see also Hyland (1972), Mitchell (1982) and Wendt in Lot and Chiang (1986). Chapter 23 describes techniques for collecting germplasm samples from tall trees, which clearly can also be applied to the collecting of botanical voucher specimens.
- *Ferns* are collected in basically the same way as higher plants. Part of the rhizome attached to the stipe together with basal, middle and apical parts of the lamina should be collected. It is also important to make notes on the habit (i.e. creeping, rosette-forming, etc.). See the section on collecting large leaves above and also Holttum (1957), Stolze (1973) and Forman and Bridson (1992).
- *Bryophytes* can be collected into paper packets (not polythene bags), made by folding sheets of paper, and dried by laying out in the sun or by application of gentle heat (Forman and Bridson, 1992).
- For instructions on collecting *fungi and lichenized fungi*, see Duncan (1970), Hawksworth (1974) and Forman and Bridson (1992).

Recording data

Vouchers without accompanying notes are of limited value, as indeed are germplasm samples without passport information. Such notes are included on a label, which is mounted with the specimen when this is prepared for inclusion in a herbarium collection. Many of the kinds of data that need to be recorded to document germplasm samples (Chapter 19) should also be recorded on the labels of herbarium vouchers, in particular: name(s) of collector(s) (and institute), collecting number (and number of associated germplasm sample), collecting date, (preliminary) taxonomic identification, locality, specific habitat and species abundance. In addition, notes and drawings should be made of any character of the plant which would be lost or not evident in a small dried specimen. The list provided by Forman and Bridson (1992), somewhat adapted, is as follows:

- Life-form, life span and habit:
    plant type: tree, shrub, herb or vine;
    free-living, epiphyte or parasite;
    life span: annual, biennial or perennial;
    direction of stem growth: climbing, erect, geniculate, decumbent, prostrate, creeping, etc.;
    stem structural type: pachycaulous, succulent, bulb, corm, stolon, rhizome, etc.
- Underground organs:
    roots: type (tap, fibrous, with tubers, etc.) and extent;
    rhizome: depth in soil, length, spacing of shoots;
    bulb, corm or tuber: size and shape.
- Stems and trunks (for instructions on collecting bark and wood samples, see Welle, 1989):
    size: total height, diameter at breast height (DBH), height of trunk or stem before branches;
    shape in cross-section (circular, fluted, with buttressing, etc.);
    bark: colour, texture, thickness, lenticel colour;
    wood: hardness, colour, grain type;
    sap or latex: colour, smell, consistency;
    thorns and spines.
- Leaves:
    deciduous or evergreen;
    texture, colours, smell, glossiness, hairiness;
    exudate or glands;
    orientation in relation to petiole or stem (pendulous, horizontal, etc.);
    outline (if large or complex);
    heterophylly.
- Inflorescence:
    exudate or glands;
    position or form (cauliflorous, ramiflorous, etc.);
    colour of axis.
- Flowers (for instructions on collecting pollen and karyological samples, see Le Thomas, 1989, and Morawetz, 1989, respectively):
    monoecious, dioecious or hermaphrodite;
    heterostyly;
    scent;

exudate or glands;
calyx colour and texture;
corolla colour and texture;
opening behaviour;
pollinators.
* Fruit and seeds:
smell;
fruit, seed-coat and aril: colour and texture;
fruit and seed: size and shape;
dispersal (animal, wind or water).

For tropical trees, Hallé and Oldeman (1970) and Hallé *et al.* (1978) have described 24 architectural 'blueprints' or 'models' for the pattern of branching and growth. Similar models have also been applied to herbs and lianes (Bell, 1991).

Photographs are often taken to complement herbarium specimens. These should show the habitat of the plant, its general habit and flowers and/or fruits in close-up.

Storing and dispatching vouchers

Field-dried specimens should be removed from the press in their flimsies, and placed in piles about 10 cm thick. These piles should be protected top and bottom by extra flimsies or cardboard corrugates and then firmly tied into bundles using thin string. Such bundles can withstand a fair degree of handling. It is essential that specimens are completely dry before they are packed, as mould can quickly spread throughout the bundle from a single damp specimen. In damp weather or in humid conditions the bundles should be placed in plastic bags. For storage or dispatch the bundles should be tightly packed into cardboard cartons or similar containers and liberally sprinkled with naphthalene or paradichlorobenzene. Screwed-up paper should be used to pad any spaces between the bundles. Dried specimens can be easily damaged if they are not packed tightly. Bundles of wet specimens packed in polythene tubing containing spirit can be stored in this form for several weeks. For dispatching they can be packed into strong hessian or plastic sacks. It is a good idea to pack specimen packages so that they can be relatively easily opened in case they need to be checked at customs. Suitable material to reseal the packages quickly in case they are opened should be taken along when dispatching vouchers. It can take a great deal of time to package specimens properly, but it is worth doing it carefully so that weeks of work are not destroyed for want of a little extra effort at the end of a long trip. For dispatch abroad, air freight is probably the best bet.

Care must be taken to ensure that the correct paper work for the export and import of duplicates has been done – delays can be very damaging to specimens. The procedures will usually be similar to those necessary for the movement of germplasm (Chapter 17). The agreement of herbaria and other institutes to accept herbarium specimens should be obtained before the start of the mission. Recipient institutes should be warned in good time that material has been dispatched.

References

Balick, M.J. (1989) Collection and preservation of palm specimens. In: Campbell, D.G. and H.D. Hammond (eds) *Floristic Inventory of Tropical Countries*. pp. 482–483. New York Botanical Garden, New York.

Bell, A.D. (1991) *Plant Form*. Oxford University Press, Oxford.

Burtt, B.L. and R.M. Smith (1976) *Zingiberaceae Flora*, Description, Sri Lanka. A.A. Balkema, Rotterdam.

Ceska, A. and O. Ceska (1986) More on the techniques for collecting aquatic and marsh plants. *Annals of the Missouri Botanical Garden* 73:825–827.

Croat, T.B. (1985) Collecting and preparing specimens of Araceae. *Annals of the Missouri Botanical Garden* 72:252–258.

Davis, P.H. (1961) Hints for hard-pressed collectors. *Watsonia* 4:6.

Dransfield, J. (1979) *A Manual of the Rattans of the Malay Peninsula*. Forest Department, Kuala Lumpur, West Malaysia.

Dransfield, J. (1986) A guide to collecting palms. *Annals of the Missouri Botanical Garden* 73:166–176.

Duncan, U.K. (1970) *Introduction to British Lichens*. T. Buncle & Co., Arbroath.

Forman, L. and D. Bridson (eds) (1992) *The Herbarium Handbook*. Revised edition. Royal Botanic Garden, Kew.

Fosberg, F.R. and M.-H. Sachet (eds) (1965) *Manual for Tropical Herbaria*. Regnum Vegetabile 39. International Bureau for Plant Taxonomy and Nomenclature, Utrecht.

Fuller, T.C. and G.D. Barber (1981) A microwave-oven method for drying succulent plant specimens. *Taxon* 30:867.

Grey-Wilson, C. (1980) Notes on collecting *Impatiens*. *Flora Malesiana Bulletin* 33:3435–3436.

Hallé, F. and R.A.A. Oldeman (1970) *Essai sur l'Architecture et la Dynamique de Croissance des Arbres Tropicaux*. Collection de Monographies de Botanique et de Biologie Végétale 6. Masson et Cie, Paris.

Hallé, F., R.A.A. Oldeman and P.B. Tomlinson (1978) *Tropical Trees and Forests: An Architectural Analysis*. Springer-Verlag, Berlin.

Hammel, B.E. (1987) The origami of botany: a guide to collecting and mounting specimens of Cyclanthaceae. *Annals of the Missouri Botanical Garden* 74:897–902.

Hawksworth, D.L. (1974) *Mycologists Handbook*. Commonwealth Mycological Institute, Kew.

Haynes, R.R. (1984) Techniques for collecting aquatic and marsh plants. *Annals of the Missouri Botanical Garden* 71:229–231.

Holttum, R.E. (1957) Instructions for collecting tree ferns. *Flora Malesiana Bulletin* 13:567.

Hyland, B.P.M. (1972) A technique for collecting botanical specimens in rainforest. *Flora Malesiana Bulletin* 26:2038–2040.

Jørgensen, P.M., J.E. Lawesson and L.B. Holm-Nielsen (1984) A guide to collecting pas-
    sion flowers. *Annals of the Missouri Botanical Garden* 71:1172–1174.
Jorgensen, V. (1972) The preparing, pressing and mounting of bromeliads. *Journal of the
    Bromeliad Society* 23:211–214.
Lawrence, G.H.M. (1951) *Taxonomy of Vascular Plants*. Macmillan, New York.
Le Thomas, A. (1989) Collection and preparation of pollen samples. In: Campbell, D.G.
    and H.D. Hammond (eds) *Floristic Inventory of Tropical Countries*. pp. 474–475.
    New York Botanical Garden, New York.
Liston, A., L.H. Rieseberg, R.P. Adams, Nhan-Do, G.L. Zhu and N. Do (1990) A method
    for collecting dried plant specimens for DNA and isozyme analyses and the
    results of a field test in Xinjiang, China. *Annals of the Missouri Botanical
    Garden* 77:859–863.
Logan, J. (1986) A pre-pressing treatment for *Begonia* species and succulents. *Taxon*
    35:671.
Lot, A. (1986) Acuáticas vasculares. In: Lot, A. and F. Chiang (eds) *Manual de Herbario:
    Administractión y Manejo de Collectiones Técnias de Recolección y Preparación
    de Ejemplares Botanicos*. pp. 87–92. Mexico.
Lot, A. and F. Chiang (eds) (1986) *Manual de Herbario: Administractión y Manejo de Col-
    lectiones Técnias de Recolección y Preparación de Ejemplares Botanicos*.
    Mexico.
Leuenberger, B.E. (1982) Microwaves: a modern aid in preparing herbarium specimens
    of succulents. *Cactus and Succulent Journal of Great Britain* 44:42–43.
McClure, F.A. (1965) Suggestions on how to collect bamboos. In: Fosberg, F.R. and M.-H.
    Sachet (eds) *Manual for Tropical Herbaria*. Regnum Vegetabile 39. pp. 121–122.
    International Bureau for Plant Taxonomy and Nomenclature, Utrecht.
Mitchell, A.W. (1982) *Reaching the Rainforest Roof – A Handbook on Techniques of
    Access and Study in the Canopy*. Leeds Philosophical and Literary Society and
    UNEP, Leeds.
Morawetz, W. (1989) Collection and preparation of karyological samples. In: Campbell,
    D.G. and H.D. Hammond (eds) *Floristic Inventory of Tropical Countries*.
    pp. 474–475. New York Botanical Garden, New York.
Mori, S.A. and G.T. Prance (1987) A guide to collecting Lecynthidaceae. *Annals of the
    Missouri Botanical Garden* 74:321–330.
Nicolson, D.H. (1965) Collecting Araceae. In: Fosberg, F.R. and M.-H. Sachet (eds)
    Manual for Tropical Herbaria. Regnum Vegetabile 39. pp. 123–126. International
    Bureau for Plant Taxonomy and Nomenclature, Utrecht.
Smith, C.E. (1971) *Preparing Herbarium Specimens of Vascular Plants*. USDA Agri-
    cultural Information Bulletin.
Soderstrom, T.R. and S.M. Young (1983) A guide to collecting bamboos. *Annals of the
    Missouri Botanical Garden* 70:12–136.
Stolze, R.G. (1973) Inadequacies in herbarium specimens of large ferns. *American Fern
    Journal* 63: 25–27.
Stone, B.C. (1983) A guide to collecting Pandanaceae (*Pandanus, Freycinetia* and
    Sararanga). *Annals of the Missouri Botanical Garden* 70:137–145.
Taylor, P. (1977) On the collection and preparation of *Utricularia* specimens. *Flora Male-
    siana Bulletin* 30:2831–2832.
Tomlinson, P.B. (1965) Special techniques for collecting palms for taxonomic study. In:
    Fosberg, F.R. and M.-H. Sachet (eds) *Manual for Tropical Herbaria*. Regnum
    Vegetabile 39. pp. 112–116. International Bureau for Plant Taxonomy and
    Nomenclature, Utrecht.
Welle, B.J.H. ter (1989) Collection and preparation of bark and wood specimens. In:

Campbell, D.G. and H.D. Hammond (eds) *Floristic Inventory of Tropical Countries*. pp. 467–468. New York Botanical Garden, New York.

Womersley, J.S. (1957) Paraformaldehyde as a source of formaldehyde for use in botanical collecting. *Rhodora* 59:299–303.

Womersley, J.S. (ed.) (1981) *Plant Collecting and Herbarium Development*. FAO Plant Production and Protection Paper 33. FAO, Rome.

Further reading

Anon. (1957) *Instructions for Collectors No. 10. Plants*. British Museum (Natural History), London.

Archer, W.A. (1945) *Collecting Data and Specimens for Study of Economic Plants*. USDA Miscellaneous Publications 568. USDA, Washington DC.

Hicks, A.J. and P.M. Hicks (1978) A selected bibliography of plant collecting and herbarium curation. *Taxon* 27:63–99.

Hollis, D., A.C. Jermy and R.J. Lincoln (1977) Biological collecting for the small expedition. *Geographical Journal* 143:251–265.

Savile, D.B.O. (1962) *Collection and Care of Botanical Specimens*. Canadian Department of Agriculture Research Branch 1113, Ottawa.

Steenis, C.G.G.J. van (ed.) (1976) *Flora Malesiana Series 1*. Noordhof International Publishing, Leyden, The Netherlands.

Steenis, C.G.G.J. van (1977) Three pleas to collectors – improve your field data. *Flora Malesiana Bulletin* 30:2843–2844.

III

BACK AT BASE

Processing of germplasm, associated material and data 28

J.A. Toll

IPGRI, Via delle Sette Chiese 142, 00145 Rome, Italy.

Introduction

The fieldwork may be over, but no collecting project is over until the germplasm samples and their associated reference materials and data are processed and deposited with the institutions undertaking their conservation and use and all reporting obligations have been fulfilled. Poor handling of the material and data on return to base can undermine the effort put into the difficult and costly task of collecting germplasm. Vegetative material and *in vitro* cultures may rot and seeds will lose viability if they are not properly handled and do not quickly arrive at the recipient gene bank. If information on the samples is incomplete or unreadable because the data forms have not been collated and edited, maintenance and future use of the material are jeopardized.

Time, facilities and funds must be set aside for back-at-base tasks when planning the mission. Arrangements must be made for the eventual distribution of the samples for duplicate conservation, research and use. Once back at base, samples must be split and distributed according to these plans. There are occasions when there may be no arrangements in place for the reception of the collected material or when the agreements made cannot be met. This may be the case for an emergency rescue mission or when the samples collected are too small for splitting. In these cases, the collector must make the necessary arrangements immediately on arrival back at base for follow-up research and conservation of the germplasm or arrange for its multiplication prior to distribution to the agreed recipients.

This chapter will deal in detail with the three major tasks of collectors on their immediate return from the field:

- sorting and preparing germplasm samples and any reference and ancillary specimens;
- collating, completing and editing the collecting data;
- distribution of the germplasm for conservation, study and use.

Reporting, which is a less immediate priority, is treated separately in Chapter 29.

Processing germplasm samples

On returning to base, immediate attention must be given to the germplasm samples since the survival of the material may be at risk. A germplasm sample may consist of seeds, whole plants, vegetative plant parts such as tubers, cuttings or *in vitro* explants or perhaps pollen. Sometimes a collection will include samples of more than one type. In previous chapters guidance has been given as to which parts of a plant to collect in different situations, the sampling strategy to use and how to handle the material in the field. This section describes the procedures to follow and techniques to use in processing and preparing for dispatch the various types of germplasm samples.

The processing required to prepare the germplasm for distribution after a collecting mission will depend on a number of factors:

- *The sample.* The type and condition of the sample at the time of collecting and its subsequent handling during the mission will determine what sort of processing is required and its urgency. The quantity of material collected will govern whether splitting can take place immediately or whether multiplication will be needed first.
- *The mission base.* The location of the base, the climate, and the facilities available will determine which kinds of treatments can be carried out.
- *The receiving institutes.* The whereabouts of the receiving institutes will determine how the samples will be sent on.

Germplasm samples always have data associated with them and may also have different kinds of reference material attached. Dealing with these is covered in separate sections, but it should be seen as an integral and necessary part of the processing of the germplasm.

Orthodox seeds

Seeds begin to age and deteriorate soon after they reach maturity. The rate of deterioration and loss in viability is influenced by a number of factors, most importantly temperature and seed moisture content. As described in Chapter 20, for the majority of species lowering temperature and reducing seed moisture content will increase the life span of the seeds. Species whose seed survival can be controlled in this manner are termed orthodox. In contrast, many tropical plantation and fruit crops, such as mango and coconut, as well as timber species, are categorized as recalcitrant because their seeds are killed by drying and are short-lived even when moist. These different kinds of seeds need to be handled in quite different ways.

Chapter 20 discusses the effects on the longevity of orthodox and recalcitrant seeds of their maturity when collected, of the climate in the collecting region and of the handling of the samples during the mission. Guidance is also given on ways to minimize loss of viability in the field. Back at base, precautions must be continued to safeguard against a decrease in the viability of the seed samples before receipt and storage at a gene bank. A drop in initial viability, though small when expressed as percentage germination, will result in an appreciable reduction of potential longevity in storage. The precautions to take involve:

- ensuring early drying of seed samples under conditions which optimize the preservation of viability;
- avoiding physical damage to seeds during the cleaning and processing of the samples;
- ensuring quick dispatch of the samples and their protection from high temperature and humidity during transport.

Drying and cleaning seeds

All the samples should be inspected immediately to determine their condition. Priority attention must be given to samples that are moist and those that are infested with pests or show symptoms of diseases.

Drying

Careful consideration should be given at the mission planning stage of the possible effects on the viability of the samples of the temperatures and humidities to which the seeds will be exposed during the mission, at base and in transit to the gene bank. Chapter 20 discusses in detail how ambient temperature and relative humidity (RH) can be used to predict the rate of loss of seed viability. If the climate in the collecting region is unfavourable, then arrangements must be made in advance for the rapid dispatch of samples back to base (or elsewhere) at intervals during the mission for drying.

The more hot and humid the climate, the more unfavourable the conditions are for drying seeds. At a RH of more than 75%, the equilibrium moisture content of non-oily seeds will be >13% and that of oily seeds >9%. Since viability loss in orthodox seeds is greatest at intermediate moisture contents of 15–30%, it is critical that the seeds stay at these dangerous moisture levels for the shortest possible period. Therefore, the samples must immediately be put under an artificial system of drying to quickly lower moisture content to the safer level of 15% or less. The alternative option of keeping seeds fully hydrated, aerobic and dormant is not usually practical, except possibly in the case of dormant seeds inside fleshy fruits.

In drier climates, where the RH is <50%, samples may be dried under ambient conditions to the relatively safe moisture content levels of <15%. Drying is most efficient if the seeds are arranged in thin layers with good ventilation (Cromarty et al., 1982; Cromarty, 1984). The cloth or paper collecting bags containing pods, seeds in heads and fruits awaiting cleaning should be laid out on racks in a breezy place in the shade. The samples must be kept away from extreme heat. Serious reduction in longevity can result from the seeds overheating in direct sun, particularly when seed moisture content is high. At the end of each day, the bags should be gathered up and packed in closed containers to prevent them reabsorbing moisture as humidity increases at night. Once seed moisture content has dropped below 15% and the seeds detach easily from the inflorescence and fruit structures, the samples can be cleaned, as described below.

After cleaning, the seeds can be exposed to further natural drying laid out in trays or hung out on racks in cloth or net bags. Eventually, they will attain a moisture content in equilibrium with the ambient RH, but this can take time, particularly if the seeds are large and aeration poor. Once a safe seed moisture content has been reached, it is preferable to pack and dispatch the samples for further drying under optimal conditions at a gene bank.

Artificial drying will be necessary if conditions at base are too humid for the samples to dry naturally to a safe moisture content or if the seeds are to be processed at base for immediate conservation and hence must be dried down to the very low moisture levels preferred for storage.

Hot air is a common method of drying seeds, but it is not recommended for genetic resources conservation. High drying temperature has a detrimental effect on viability, especially when the seeds are moist. In tropical environments hot-air driers are anyway unable to dry seeds sufficiently because the moisture content of ambient air is too high. The recommended alternative is to dry seeds by dehumidification at low temperature. Cromarty et al. (1982) describe the design and operation of a special cabinet or room able to maintain conditions of low temperature (15–25 °C) and RH (10–15%).

Probably the most feasible approach for most base locations, however, is to dry the seeds by using a desiccant like silica gel. Zhang and Tao (1989) describe the use of silica gel for drying seeds for genetic conservation (see also Chapter 20). General instructions are to use a ratio of seeds to silica gel of 3 : 2 and place the seeds in a thin layer just above the dehydrated gel, inside a desiccator or any suitable container with a moisture-proof lid and large area : volume ratio. Zhang and Tao (1989) report periods of two days to two weeks (depending on seed size) to dry seeds from >15% moisture content to <7%.

In order to monitor and manage the seed drying process, it is necessary to measure seed moisture content. Cromarty *et al.* (1982) and Hanson (1985) describe the method in detail.

Cleaning

Samples are often brought back to base with the seeds still in the heads (inflorescences), pods or fruits, and require threshing or extraction and cleaning in preparation for sample splitting, packaging and distribution. To minimize the risk of transmitting pests and disease, seed samples must be cleaned to eliminate soil, plant debris and seeds of noxious weeds and parasites. Legume seeds should be extracted from pods for international distribution (Frison *et al.*, 1990). Chapter 17 deals with these phytosanitary issues more fully.

Seeds detach more easily from dry inflorescences and the risk of mechanical damage is lessened if the seeds are neither too wet nor too dry. It is thus best to thresh and clean seeds when their moisture content is in the range 12–14%. Threshing and cleaning by hand is recommended as the safest and most practical way of processing genetic resources samples, which are typically small and need special care to minimize risks of damage, viability loss and mixing or contamination among samples (Chapter 20). Manual threshing normally involves rubbing, beating or breaking up the dried inflorescence and fruits. It is sometimes best done with the samples still inside the cloth collecting bags. More robust methods are needed in some cases: the pods of *Acacia* species may need gentle pounding, for example (Doran *et al.*, 1983; see also Chapter 23). In some species, it is difficult to free the seeds entirely from the fruit, for example the spikelets of wild grasses such as *Pennisetum* and the tight-fitting pods of some legumes species. In these cases, if there are no plant health risks or restrictions, it may be best to send the samples for seed extraction at the recipient institute, which will have the necessary experience and specialized equipment. Chaff and debris are removed by winnowing and sieving. Again, manual methods are usually the best. Flotation techniques are not recommended for seeds destined for long-term conservation. Though their use is also not recommended for genetic resources conservation, if mechanical threshers and seed cleaners are employed, great care must be taken not to lose seeds and to clean the equipment between samples to safeguard against contamination. Air-screen cleaners, air separators and specific-gravity separators are described and discussed in Young and Young (1986).

As a general rule, seeds should be extracted from fleshy fruits as soon as possible after harvest, and this may in fact have been undertaken in the field during the mission. However, inside ripe, sound, fleshy fruits, seeds are in a fully imbibed state and will maintain their viability for long periods if prevented from germinating by dormancy. If, on arrival at base, the fruits are undamaged, they can be left moist and whole while awaiting cleaning, or even dispatched in this state for seed extraction at the recipient gene bank. For transport, the whole fruits must be packed to keep them cool, moist and well aerated, since oxygen is required for hydrated seeds to maintain their viability.

In some species, some seeds may germinate, either because they have no dormancy or because dormancy has been broken. If this occurs or is likely, the seeds must be extracted from the fruits as soon as possible. If fruits are dried up, decomposed or damaged, they must also be cleaned quickly. Great care must be taken not to damage seeds when extracting them from fruits. Once seeds are removed from fruits they must be dried quickly to safe levels below 15% and as close as possible to the recommended moisture content for storage (3–7%). If delays in drying the seeds are likely, the samples should be maintained temporarily as whole fruits to avoid seeds remaining at damaging moisture contents. Notes on the cleaning of different types of fleshy fruits are given in Ellis *et al.* (1985) and in Chapter 23. Generally applicable instructions are (after Hawkes, 1980):

1. open the fruits with a knife or by hand and carefully tease or squeeze out the seeds;
2. remove pulp and gelatinous coatings by washing in water, if possible, and then draining, or otherwise by blotting;
3. put the seeds out to dry in the shade
4. remove any remaining gelatinous material when the seeds are dry.

Some fruits contain poisonous substances, which can irritate the skin and worse. Ellis *et al.* (1985) suggest the use of light plastic gloves when handling such material.

Seed treatment and phytosanitary inspection

Chemical treatments may adversely affect seed viability and should be avoided if possible and where quarantine regulations allow. However, treatment of samples that have become infested with insects may be unavoidable in order to save at least some material. If samples are treated, the packets should be clearly labelled with the name of the chemical used, as some products are hazardous to health and special precautions must be taken when handling seeds treated with them.

The dried, cleaned and treated (where required) seed samples should then be inspected and certified by the plant protection service in the country of origin. Arrangements for this should be made at the mission planning stage (Chapter 17).

Packaging and dispatch of seed samples

Once the samples have been dried to a safe seed moisture content level and cleaned and phytosanitary controls have been undertaken, their sharing out, packaging and dispatch should proceed as quickly as possible.

Sharing out

    Seed samples may need to be split and shared out among various institutions. This will follow the distribution protocols laid down at the mission planning stage. Care is needed to avoid contamination among samples and ensure accurate transcription of collecting numbers on to the different packets. Each sample should be thoroughly mixed before splitting.

    If the samples are very small or their viability is suspected to be very low (because the seeds are immature or in poor condition), it will not be possible to meet the sample splitting requirements. Interim multiplication or regeneration of these samples will then have to be arranged (see below). To guard against loss in transit, each sample to be dispatched may be split into two and each half sent in a separate shipment.

Packaging

    The samples must be protected from high humidity, high temperature and mechanical damage during transport. Sample packaging should be moisture-proof to prevent dry seeds taking up moisture at high ambient RH. This is particularly important if the samples are likely to transit through a humid tropical area. Aluminium foil plastic pouches are practical, especially if the samples are being sent directly to medium- or long-term storage and will not need further processing. However, they need to be heat- and pressure-sealed with a special machine. The laminate must be sufficiently robust to guard against moisture ingress and puncture by the seeds inside. The following specification is recommended: $17 \, g \, m^{-2}$ polyester; $33 \, g \, m^{-2}$ aluminium; $63 \, g \, m^{-2}$ polyethylene. For bulky samples and very large seed quantities, containers should be made of rigid plastic or metal and have tight-fitting lids that seal hermetically. If there is a risk of moisture ingress through the packaging, then a small quantity of silica gel in a cloth packet can be put inside with the seeds. Seeds have also been moved for short periods in simple plastic zip-locked bags, and this may be a more cost-effective option if further seed processing will be needed at the receiving institute.

    During transport, the seeds must be prevented from moving about and being damaged or breaking out of their bags. Legume seeds are particularly vulnerable to damage of the seed-coat and wounding of the embryo. Room inside the sample packages should be kept to a minimum by sealing the bags tightly or plugging excess space in containers with cotton wool or paper. The sample bags or containers should be packed firmly into crates or boxes so that they do not shift about.

Dispatch

    If seed samples are to be duplicated in several collections for safety, as generally recommended, they will have to leave the country of collecting. The phytosanitary and other regulations governing the export of seeds from the country of origin and their import into the countries of the recipient gene banks should be investigated at the mission planning stage. This is discussed in Chapter 17.

Samples should be sent to recipient organizations by a fast and secure means. This is discussed in Chapter 20. For international destinations, shipment should be by air. Air freight, recorded express airmail or air couriers are the alternatives: cost, time in transit and reliability will need to be taken into account in deciding. Precautions must be taken to minimize the time the samples spend in customs or awaiting delivery to the recipient. The documentation for the export, import and quarantine clearance of the shipment must be prepared in advance. Notice by Telefax or telex to the recipient institute of the date of dispatch, flight(s), airway bill number and number of boxes in the consignment will facilitate its speedy reclaim from the airport and early tracing in the event of a problem.

Each box or crate in the consignment must be clearly labelled with the address of the institute of destination, including telephone number. The boxes should also be marked as follows:

> SEED SAMPLES OF NO COMMERCIAL VALUE FOR SCIENTIFIC PURPOSES ONLY. FRAGILE.

They should be marked as perishable only if the seeds have been dried, because such a label may result in refrigeration of the consignment at customs. A label stating that material does not contravene the *Convention on International Trade in Endangered Species of Wild Fauna and Flora* (CITES) may also be necessary. Each shipment should be accompanied by:

- a letter describing the contents: i.e. listing species and number of samples of each; describing the condition of the samples (stage of drying and cleaning); and notifying of any chemical treatments (which samples treated and what chemicals used);
- a note of any special handling instructions;
- the passport data;
- the phytosanitary certificates.

Recalcitrant seeds

When recalcitrant seeds are to be collected, arrangements must be made at the mission planning stage to ensure the rapid receipt and early planting of the seeds by the recipient institutes. For very short-lived species, special procedures will have been followed during the mission to keep the seeds viable. These involve keeping the seeds moist, as near the fully imbibed state as possible, but in aerobic conditions and able to respire. The seeds may have been left within the fruits in the fully imbibed but dormant state, in which case they will probably need to be extracted back at base prior to dispatch (see above). If they have already been extracted and kept moist in plastic bags or charcoal, back at base these samples must be packed in suitable containers for transport and dispatched to the recipient organization as quickly as possible, allowing no more than one month between collecting and receipt. Temperatures below 20°C or much above ambient should be avoided. Marking consignments as perishable may risk their refrigeration by customs authorities.

Vegetative plant parts

Crops that are propagated vegetatively, woody perennials and species that produce few or no seeds may need to be collected as vegetative plant parts. The collecting of vegetative material and its handling in the field are dealt with in Chapters 21, 22 and 23. Back at base, these plant parts must be prepared for transport and distribution. For transport, vegetative organs should be wrapped in semiabsorbent material and packed firmly, but not too tightly, into a box or carton. A filling such as hay or crumpled paper can be added to protect against shocks.

The movement of vegetative material from one country to another entails a great risk of the transmission of diseases and pests. Whole plants and storage organs constitute the highest risk, stems and budwood a lesser risk. For *in vitro* culture, the risk is less still, but it is only through treatment therapy and indexing that there can be any assurance that the material is pathogen-free (IBPGR, 1988). Where treatment of the material or its *in vitro* culture is not sufficient or feasible, it may be necessary to organize quarantine in a third country (Chapter 17).

In vitro *material*

The techniques for *in vitro* collecting and the processing of *in vitro* material at the receiving laboratory are described in Chapter 24. At the mission planning stage arrangements should be made with a specialized laboratory for receipt and processing of the cultures prior to the eventual deposit of the accessions in gene banks. If there is no *in vitro* laboratory at the mission base or close by and the cultures have to be sent long distances, special care is needed in packing and transporting the culture vessels in order to avoid damage or deterioration of the material. Many of the guidelines for the packing and dispatch of seed samples are also relevant to the handling of *in vitro* material. When the laboratory is outside the country of collecting, the cultures must first pass through appropriate plant health testing and quarantine procedures.

As with all samples, the labelling of the culture vessels should be checked for legibility and the identifier (collecting number) verified against the corresponding collecting data form. The data forms should bear the name and address of the laboratory processing the cultures.

Processing reference samples

To obtain more information about the germplasm, the germplasm collector may also have gathered different kinds of reference material, including:

- herbarium voucher specimens;
- *Rhizobium* samples;
- soil samples;
- specimens of pests and pathogens;
- photographs of the plants or collection site.

Herbarium voucher specimens

The reasons for taking voucher specimens and the techniques for collecting them are covered in Chapter 27. Back at base, these voucher specimens and their associated documentation must be sorted and prepared for dispatch.

As with the germplasm samples, the amount and type of processing to do back at base will depend on the condition the specimens are in on leaving the field and on what is necessary to make sure that they will reach their destination in an acceptable state. As a rule, voucher specimens are sent to a herbarium with which a prior agreement has been made for their conservation. The essential tasks back at base are to:

- dry and treat the specimens;
- sort, label and pack the replicate specimens for dispatch;
- assemble information on the specimens and label them;
- record that voucher specimens have been taken and where they have been sent.

The first task is to *dry and treat* the specimens. Specimens of whole plants usually come in from the field either still in presses or in bundles, having been dried. Specimens that still contain moisture will have to be dried and bundled prior to dispatch. Specimens dried in the field or back at base are often subject to insect attack while they are being processed and in transit to the herbarium of destination. If the equipment is available at base, the specimens can be fumigated or given a low-temperature treatment to rid them of insects, larvae and eggs. Often, a simple treatment using formalin or a spray insecticide will suffice until the specimens can be properly treated on arrival at a herbarium. See Chapter 27 for further details on treatment and packing.

There may well be large numbers of replicate specimens to *sort, label, pack and distribute*. If the populations sampled were highly diverse, several voucher specimens, representing the range of morphotypes visible, may have been taken from each population (Hawkes, 1980). There will be voucher specimens to distribute with each germplasm subsample. In addition, it is standard practice to deposit a full set of specimens at the national herbarium in the country of collection and at any other herbarium that specializes in the flora of that part of the world. Usually it is to one of these international herbaria that specimens requiring taxonomic verification will be sent.

Replicate specimens must be examined to make sure that they are correctly labelled with the same number and are as complete as possible, with leaves, flowers and fruits. If there are insufficient flowers or fruits for replication, then the location of other, fuller replicates must be recorded on the collecting data forms or herbarium labels so that other botanists can request them on loan if need be.

The numbering of specimens needs careful checking back at base. In the field, a voucher specimen may be labelled with the same collecting number that was assigned to its corresponding germplasm sample. However, a separate numbering series will sometimes be used for herbarium specimens if, for example, they are part of a wider botanical survey that has set standard identification descriptors. Back at base it is important to ensure that, whatever numbering system is used, voucher specimens can be correlated with germplasm samples and the correct numbers are recorded on the corresponding collecting data forms.

Voucher specimens must be accompanied by adequate *documentation* for them to be accepted for identification or registration by a herbarium. The documentation of voucher specimens is reviewed in Chapter 27. Many of the relevant data descriptors will have been filled in on the collecting data form, but additional information on the specimen's appearance, such as flower structure and colour, may have been noted in a separate field book. Back at base all the information on each specimen must be collated. Copies of the collecting data forms, or the relevant data transcribed from them, must be compiled with any supplementary information recorded in the field. The International Plant Genetic Resources Institute (IPGRI) collecting form software (Box 28.1 below) provides for the production of herbarium labels from the computerized germplasm collecting forms.

Drawings may have been made or photographs taken to illustrate certain features of the plants, particularly those that are not observable in dried material, or attributes of their habitats. This must be noted on the collecting forms and herbarium labels, and copies should be distributed with each replicate specimen or, in the case of photographs, arrangements made to send on copies after the film has been developed.

Rhizobium *samples*

Collecting root nodules for the isolation of *Rhizobium* is generally considered an integral part of the collecting of many leguminous species. Having the strains that are genetically and geographically associated with the samples will ensure effective nodulation when the plants are evaluated in a new environment. *Rhizobium* specimens may also be important for the successful maintenance of wild legume species. Details of collecting methods are given in Chapter 26.

Rhizobium samples may be collected in the field as excised root nodules preserved in tubes containing a desiccant or as specimens of soil and root material. On arrival back at base, these samples must be sent on as soon as possible to a laboratory for the isolation, testing and storage of the *Rhizobium*. Arrangements to handle the *Rhizobium* samples should be made at the planning stage with an appropriate laboratory.

Each container of nodules or soil must be checked to make sure that it is clearly and correctly labelled with a collecting number. This may or may not be the same as the collecting number of the associated germplasm sample, though the former procedure is preferable. It should be verified that the corresponding collecting data forms record that a *Rhizobium* sample has been gathered, what its collecting number is and where it has been sent.

Prior to transport, the desiccant should be renewed in the nodule tubes and these or the containers of soil carefully packed to prevent breakage and exposure to humidity during transit. The consignment should include copies of the corresponding collecting data forms or a list of the essential descriptors that identify the species and collecting project.

Each subsample of germplasm distributed should be accompanied by a specimen of its corresponding *Rhizobium* strain, assuming that the receiving institute has the facilities for maintenance of the microbial material. More usually, perhaps, samples are sent to a central laboratory, which then supplies subsamples on demand to institutes maintaining the associated germplasm.

Soil samples

The majority of the soil descriptors on the collecting data form will have been filled in by visual observations or from field measurements using soil colour charts, portable pH meters, etc., as described in Chapter 19. However, more comprehensive information about the soil in which a plant is growing may be needed in order to help establish relationships between the occurrence of a species or ecotype and particular soil conditions. In these cases, specimens of the soil will have been taken at key collection sites for detailed determination of soil physical and chemical properties. Soil samples should be about 500 g in weight, packed in cotton or plastic bags and labelled with the germplasm collecting number or collecting site number.

During the planning stage, contact should be established with a local (government or private) laboratory for the subsequent analysis of soil samples. Sending the samples out of the country for analysis may be necessary in some cases, but this will incur quarantine problems and heavy transport costs.

Back at base, the soil samples must be checked before delivery to the laboratory for any errors in numbering, illegibility of labels and damage to bags. Included in the delivery will be a listing of the analyses to be performed. The types of tests to be done will have been decided in advance in discussion with a soil scientist but will also depend on what the laboratory is equipped to perform. The analyses performed may include pH, N, P, trace elements, organic matter, anion exchange capacity, conductivity and texture.

The results of soil analyses must be transcribed from the analysis sheets provided by the laboratory to the individual collecting data forms of all the germplasm samples taken from each site tested. The result sheets of the soil sample analyses can be included in the mission report.

Photographs

Photographs may have been taken in the field to illustrate the plants, the places they grow and the people that use them. As mentioned above, photographs can be useful in recording plant characters that will not be visible in dried voucher specimens, such as flower colour. They can also be used to portray the specific habitat where a population was found growing and panoramic views of the collecting site can aid recognition on subsequent visits. Such reference photographs should be labelled with the collecting number of the germplasm sample or the site number to which they correspond and should be copied for distribution with the germplasm and voucher specimens. Accessible and secure storage of the print negatives and transparency originals, usually in the photographic library of the collecting institute, should be arranged. Photographs are often used to illustrate the mission report and any ensuing articles about the expedition findings. For these purposes, black-and-white or colour prints are preferable since they can be reproduced in publications more successfully than transparencies.

Specimens of pests and pathogens

Dealing with specimens of pests and pathogens is described in Chapter 17. Much the same points need to be considered with regard to such specimens as with voucher botanical specimens and other reference material. Specifically, care should be taken to ascertain that each reference sample has been given a number which clearly associates it with a germplasm sample and that this has been noted on the collecting form along with the method by which the reference sample will be named (e.g. the name of the institute where it will be sent for identification). Clearly, when the mission is international, the usual conventions regarding the deposition of reference material in the country of collecting also apply to pest and pathogen samples.

Processing collecting data

Chapter 19 discusses the type of collecting information to be gathered in the field and shows how to fill in the collecting form. Back at base, the collecting data must be processed and prepared for distribution with the germplasm samples. The collecting data forms must be completed, checked for errors, made legible and duplicated as necessary. Once the samples are split and dispersed among different institutions, it becomes much more difficult to match data to samples and register additions and corrections. All this is best undertaken by the collector (at the very least, it should be checked by the collector), to minimize transcription errors.

Editing the collecting data forms

The procedures to follow in processing the collecting information and preparing the data forms for distribution with the germplasm samples are:

- sorting and checking the forms;
- completing the forms;
- adding information from reference sources;
- checking the botanical names and local words.

Sorting and checking the collecting forms

The unique collecting number assigned to a germplasm sample at the time it is collected will be recorded on its accompanying collecting form. The data forms and samples must be sorted and compared to ensure that their numbers correspond and that there are no errors or discrepancies in the numbering sequence.

For whatever reason, a number in the sequence may not have been assigned. Also, samples may have been lost. This is not a problem in itself, but to avoid possible confusion any missing numbers should be clearly pointed out to recipients of the germplasm samples. On the other hand, any differences in numbering between the data forms and the samples to which they relate, or the allocation twice of the same number, must be traced and corrected before material and data are distributed. These types of errors can lead to confusion, which may be difficult to rectify once the samples are dispersed among different institutes. If a diary or separate field book has been kept, this can help in resolving discrepancies.

Completing the collecting forms

Some collecting data descriptors may not change for the whole duration of the mission, for example country of collecting and collector's name. Others remain constant for a day (date of collecting) or several days (province, ethnic group). To save time in the field, these descriptors may have been entered on just the first form of the mission, of the day or of the start of work in a new region. Back at base these data descriptors must now be copied across all relevant collecting forms to complete each data form.

Often, more than one sample is collected at a site. In the field, the site descriptors may have only been completed on the data form of the first sample collected at that site, with just a site number recorded for the other samples to provide a way of referring back to the first form. Back at base, the site data must be entered on the collecting forms of all samples taken at the same place and bearing the same site number. It is very important that each germplasm sample is accompanied by a complete set of collecting data because samples of different species are often distributed to different institutes and each recipient requires all the information.

Adding information from reference sources

The location of a collecting site is described both by reference to the nearest village or some other feature and by its coordinates (longitude, latitude, map grid reference) taken from a map. Sometimes, to save time in the field, only the former is recorded, or site location is simply marked on a map. It will then be necessary to read off the coordinates back at base and enter them on the collecting forms.

Some types of information about the collecting sites can only be recorded on return to base because they require consulting reference sources. For example, climate descriptors such as annual rainfall and mean annual temperature have to be completed with data taken from reference records and maps which it is not always feasible to take to the field. Reference to the sources used must be made on the collecting forms. Chapter 9 describes sources of environmental data.

Checking taxonomic designations and local words

The data forms must be checked for consistency and accuracy in the use of the scientific names of plants. Floras, checklists, etc. should be used to check the taxonomic designations of wild species (Chapter 10). For cultivated species, Schultze-Motel (1986) is a good source. Where the taxonomic identification of a sample is uncertain and confirmation is to be sought, this should be noted on the collecting data form under the relevant descriptor or as a remark, together with the address of the institute to which the voucher specimen is being sent. Species identification takes time and confirmation is likely to be received only after the germplasm samples have been received by the institutes charged with their conservation, study and use. Therefore, it is important that it be clearly indicated on the relevant collecting forms that verification of sample taxonomic identifications has been sought, so that the recipients of the samples are able to follow up on any eventual alterations in the data initially entered on the forms.

The spellings of words in local languages or dialects for places, people, vegetation types, plants, crop varieties, soils, etc. should also be checked for confusing spelling variations of the same word. In general, a standard (or at least consistent) transliteration of repeatedly occurring foreign words should be decided on, and adherence to the standard should be checked at this stage. Lexicons, gazetteers and dictionaries may be useful, and any reference works used should be noted.

Computerization of the data

Computerizing the data has many advantages. It facilitates all the procedures described above for editing the collecting data forms: sorting, completing, adding information, checking and copying the forms are made easier and quicker. Clean, readable data forms can be produced quickly. In computerized form the data can be added to the documentation systems of recipient institutes and central crop databases more readily and with less risk of transcription errors, assuming that software and hardware are compatible, a point that should be verified early on. Furthermore, later amendments and additions to the information, for example verification of voucher specimens and the identification of pests, can be more easily registered and communicated. Computerization also facilitates analysis of the data and report generation in general.

If computerization of the forms is not possible, they should be typed or care should be taken to write them out very clearly. To aid in the computerization of collecting data, IPGRI has developed database software that mimics the format of the IPGRI collecting form. This software is freely available to interested germplasm collectors and details of its specifications and how it can be obtained are in Box 28.1.

Data duplication and dissemination

Its associated collecting data must always accompany each germplasm sample wherever it is sent. Each germplasm sample is generally split and the subsamples are dispersed among different national and international institutes for study, use and safety duplicate conservation. In whatever way a sample is split and distributed, each of the subsamples must be accompanied by a copy of the complete collecting data form. The data can be dispatched as hard copy (computer-generated, typed or handwritten) or on diskette, both if possible.

If conditions during the mission are unfavourable and facilities at base limited, or if the species or type of plant material collected is short-lived, it will be necessary to dispatch the samples to specialized institutes for their processing and preservation at intervals during the mission or as quickly as possible on arrival back at base. Under these circumstances, there may be insufficient time to complete the processing of the collecting data and prepare the forms for distribution with the samples. However, a minimum number of essential data descriptors must accompany the samples; otherwise there can be problems later in matching the germplasm samples to their corresponding collecting data. The collecting number plays a crucial role in this. The descriptors that are considered essential to accompany the samples are listed in Box 19.1 of Chapter 19.

Germplasm distribution

The germplasm samples, together with their collecting data and any reference specimens, will be distributed to gene banks, other institutes and scientists according to arrangements made at the planning stage. In international collecting, these agreements will include the standard

Box 28.1
IPGRI Collecting Form Management System

Introduction
The Collecting Form Management System (CF) is available from IPGRI free of charge to interested germplasm collectors. The system:

- provides a convenient and standard way to record and update information on collected germplasm and herbarium specimens in a manner that mimics the IPGRI collecting forms for both wild and cultivated material;
- allows additional site details to be recorded and updated in a file linked to the germplasm data;
- allows the printing of completed collecting forms for germplasm samples;
- allows printing of completed herbarium specimen labels;
- allows printing of completed site forms;

Hardware and software requirements
- An IBM[1] PC, XT, AT, PS/2, 386 or IBM-compatible computer.
- MS-DOS[2] version release 2.0 or higher.
- Minimum 512 kb RAM.
- One floppy disk drive.

Description
All database files are constructed using dBASE III Plus[3] and can be accessed using that software. The execution programme is menu driven (help fields give the allowed choices for each field) and has been compiled in CLIPPER (compiled version of dBASE III programmes). There are various time-of-entry checks for data-inputting errors.

Distribution of software
The software can be obtained from IPGRI headquarters upon request. It is distributed on a single 360 kb 5.25" diskette and accompanied by a user manual. Other disk sizes may also be obtained by clearly specifying the size and density that are needed.

---

[1] IBM is a trade mark of IBM Corp.
[2] MS-DOS is a trade mark of Microsoft Corp.
[3] dBASE III Plus is a trade mark of Ashton-Tate Corp.

procedure of depositing a part of each sample with an institute in the host country and sending subsamples of each accession to other gene banks, usually outside the country of origin, for security duplicate conservation. Third-country quarantine may be necessary in some cases. A number of national, regional and international gene banks have accepted responsibility for global or regional base collections of particular crops. At the mission planning stage, collectors should find out which gene banks are holding base collections of the species that are to be collected. This information is available from the Food and Agriculture Organization (FAO) and IPGRI. The gene banks should be contacted in advance and arrangements made for the deposit of subsamples at the end of the mission. Any quarantine restrictions or import regulations must be confirmed at this stage. It is recommended that each germplasm sample be duplicated in at least two base collections.

The final destinations of the subsamples must be decided well in advance so that the necessary import and other documents may be prepared. Sometimes this is not possible, for example if species are collected that were not part of the original objectives of the project. If the required arrangements cannot be made immediately on return to base, it may be necessary to send the samples to an intermediate destination for a period.

Seed samples too small to split or of poor initial quality will have to be multiplied prior to distribution and conservation. Usually, one of the gene banks that has agreed to conserve the germplasm will be able to carry out the multiplication and then undertake to forward subsamples to the other final destinations. The institutions which could undertake any interim multiplication should be identified at the mission planning stage. *In vitro* and vegetative material may also require interim processing. This may involve the preparation of new cultures or could entail seed production or the generation of fresh vegetative material (Chapters 21 and 24).

Follow-up study and use

Collecting may have been carried out as part of an active breeding, selection or research programme, with the material being used immediately. However, even in emergency rescue collecting, it will be important to make sure that the germplasm is rapidly studied. Follow-up research on the viability of the material and the diversity it represents is important for its management in conservation, for the targeting of subsequent collecting of the same gene pools and regions and for the assessment of the value and possible uses of the germplasm. Collectors must ensure that interested scientists and institutes both nationally and internationally are informed of the results of their work and the location of the germplasm, so that study and use of the material can begin as quickly as possible. Report writing and other methods of information dissemination are discussed in Chapter 29.

References

Cromarty, A.S. (1984) Techniques for drying seeds. In: Dickie, J.B., S. Linington and J.T. Williams (eds) *Seed Management Techniques for Genebanks*. IBPGR, Rome.

Cromatry, A.S., R.H. Ellis and E.H. Roberts (1982) *Design of Seed Storage Facilities for Genetic Conservation*. Handbooks for Genebanks: No. 1. Revised 1985 and 1990. IBPGR, Rome.

Doran, J.C., J.W. Turnbull, D.J. Boland and B.V. Gunn (1983) *Handbook on Seeds of Dry-zone Acacias*. FAO, Rome.

Ellis, R.H., T.D. Hong and E.H. Roberts (1985) *Handbook of Seed Technology for Genebanks*. Vol. I: *Principles and Methodology*. Handbook for Genebanks: No. 2. IBPGR, Rome.

Frison, E.A., L. Bos, R.I. Hamilton, S.B. Mathur and J.D. Taylor (eds) (1990) *FAO/IBPGR Technical Guidelines for the Safe Movement of Legume Germplasm*. FAO/IBPGR, Rome.

Hanson, J. (1985) *Procedures for Handling Seeds in Genebanks*. Practical Manuals for Genebanks: No. 1. IBPGR, Rome.

Hawkes, J.G. (1980) *Crop Genetic Resources Field Collection Manual*. IBPGR and EUCARPIA, Rome.

IBPGR (1988) *Conservation and Movement of Vegetatively Propagated Crops*: In Vitro Culture and Disease Aspects. IBPGR, Rome.

Schultze-Motel, J. (ed.) (1986) *Rudolf Mansfeld. Verzeichnis Landwirtschaftlicher und Gärtnerischer Kulturpflanzen*. Springer-Verlag, Berlin.

Young, J.A. and C.G. Young (1986) *Collecting, Processing and Germinating Seeds of Wildland Plants*. Timber Press, Portland, OR, USA.

Zhang, X.L. and K.L. Tao (1989) Silica gel seed drying for germplasm conservation – practical guidelines. *FAO/IBPGR Plant Genetic Resources Newsletter* 75/76:1–5.

Reporting on germplasm collecting missions

29

J.A. Toll[1] and H. Moss[2]

[1]*IPGRI, Via delle Sette Chiese 142, 00145 Rome, Italy:*
[2]*63 End Road, Linden Extension, Randberg 2194,*
Republic of South Africa.

The mission report

Having completed the tasks of processing and dispatching germplasm samples and data, the collector must then write a mission report. No collecting mission or programme can be considered over until a full report has been prepared. Its purpose is to record the mission's objectives, planning, execution and findings. It provides an important reference for follow-up collecting and also informs breeders and other users of the availability of germplasm that may be of interest to them.

It is very important, however, that the report is not seen as a substitute for properly filled-in collecting data forms. It is really only as sample-specific collecting forms that data can be readily related to the corresponding germplasm sample and copied, disseminated, entered into databases and accessed. As much information as possible about the individual germplasm samples and their collecting sites should be recorded on the individual collecting forms. The mission report is the appropriate place for more general observations and summaries of the collecting data.

The collecting report should include:

* a statement of the objectives of the collecting;
* a description of the environment of the target region;
* an account of the logistical and scientific planning;
* details of the execution of the mission (timing, itinerary, sampling strategy and collecting techniques);
* a summary of the results (germplasm collected and areas surveyed);
* details of the onward distribution of the germplasm and data;
* recommendations for follow-up;
* acknowledgements.

Objectives

The reasons for mounting the mission in the first place must be explained. The aim may have been to rescue germplasm at risk from changes in land use or cultural practices; the information sources suggesting a threat of genetic erosion should then be reported. Or the objective may have been to fill 'gaps' in existing collections or to sample populations identified as being particularly diverse or of potential use. Data on earlier collecting activities and on research carried out on the material will then need to be presented and discussed. Sometimes, the mission is part of an ongoing study of particular plant gene pool(s) or of the resources of a larger region. If it is a follow-up mission, or part of a programme, reference must be made to the reports and findings of the previous missions. If a collecting proposal had previously been prepared, for example in order to obtain funding for the work, it will certainly have included a statement of why the collecting was thought to be necessary. This section of the report will clearly draw extensively on the earlier document.

Box 29.1 is an example of the introductory section of a report on collecting forages in Lesotho and Box 29.2 is another extract from the same report, itemizing specific considerations with respect to the target groups and target area.

Target area and target species

Information should be provided on the geography, climate, soils and land use of the collecting region as a whole. This will include maps, which are discussed further below. In wild species collecting reports, there should be information on floristics and vegetation, including references to any poorly known local Floras, checklists, etc. Some mention should also be made of protected areas such as national parks. In crop collecting reports, background information should be provided on the history and character of the human population of the area, and on the agriculture that is practised there. The mission's target areas (whether administrative units or agroecological zones) must be defined, and the collecting priority assigned to each explained. The target species should also be listed and prioritized. Both target areas and target species are then described in more detail in the section of the report dealing with the germplasm collected.

The example in Box 29.3 illustrates the kind of information required in this section. It is taken from an actual report on collecting forages in Namibia, where it was accompanied by vegetation maps of the whole country and of the Etosha National Park and by isohyet and isotherm maps of the whole country.

Logistics

There must be a clear identification of all the people and organizations involved in the mission, whether in planning, funding or execution, and of their respective roles. A full list of expedition members must also be

---

Box 29.1
Collecting forages in Lesotho: introduction and objectives

The mountainous Kingdom of Lesotho represents a unique environment in that it is a temperate 'island' surrounded by an otherwise subtropical area. The high-altitude pastures are adapted to an inhospitable and extreme environment, and as such are valuable sources of genetically isolated and exceptionally hardy, cold-tolerant forage legumes and grasses.

This unusual reservoir of genetic resources has for a considerable period been under severe pressure from overgrazing and overuse, and many of the more valuable genotypes and even species are known to be disappearing at a rapid rate. An ongoing and comprehensive genetic resources collecting programme is thus of the utmost urgency in Lesotho.

Two collecting trips have already been undertaken by Lesotho's National Plant Genetic Resources Committee in collaboration with IBPGR (International Board for Plant Genetic Resources) during 1989 and 1990. Efforts have been concentrated on the mountain zone proper. About 400 seed samples and 350 herbarium specimens (in sets of four) have been collected. Of particular interest are the high-altitude legumes, *Trifolium africanum* and *T. burchellianum*, growing at almost 3000 m, as well as what appears to be a new variant of *Medicago laciniata*. An indigenous wild barley species, *Hordeum capensis*, is of interest to barley breeders because of its geographic isolation from the rest of the gene pool. Several species of *Lotononis* occur in dense, prolific mats on the barren, windy summits of the highest mountains. By virtue of their prostrate growth habit they survive heavy browsing and trampling, and also have exceptional soil-binding capabilities. Although these species have been collected in the two previous missions, only a small part of the available genetic diversity has so far been captured. Although 28 species of *Lotononis* are recorded in Jacot Guillarmos's *Flora of Lesotho* written 20 years ago, less than a quarter were found in the previous two missions. It is likely that overgrazing of this palatable genus has reduced its distribution severely. Also, certain palatable and nutritious grasses recorded at the turn of the century were not seen at all in spite of extensive searching.

In an attempt to locate and collect germplasm of some of the more palatable forages not found in earlier missions, it was decided to undertake a third collecting expedition in 1991. The objective was to explore the most remote and inaccessible high-altitude areas of the country along the eastern escarpment. It is here that there is the least overgrazing, and thus the best chance of finding relict populations of forage species. The summit of the Drakensberg in the vicinity of the eastern escarpment edge is grazed annually, but is reputedly not as densely populated as the lower areas west of the escarpment. Consequently, grazing pressure on the vegetation is not excessive. The vegetation on the summit of the Drakensberg has been observed to be in good condition, having a relatively high basal leaf cover and favourable species composition. For these reasons, this was the area targeted for collecting.

---

provided, with contact addresses and dates of their involvement. This should include temporary and *ad hoc* members of the team who helped out in a particular area. This often includes local extension workers. The role of community 'gatekeepers' must be acknowledged (Chapter 18). The advice and experience of guides and porters may also be useful to future collectors and their names and addresses should be recorded.

Box 29.2
Collecting forages in Lesotho: justification

The high-altitude rangelands of Lesotho are an unusual environment in southern Africa and in fact in Africa as a whole. As such, this ecosystem represents a valuable and unique reservoir of forage genetic resources. The specific factors of importance are detailed below.

1. The Afroalpine vegetation of Lesotho is adapted to withstand extreme cold (snow has been recorded from every month of the year), intense frost-heaving, severe physiological drought, high insolation levels, strong winds, considerable grazing pressure and leached, acid soils. Productive and palatable forage species and genotypes from these problem environments could be used in the rehabilitation of degraded parts of Lesotho, as well as in similar areas elsewhere in the world.
2. Genotypes growing in the higher areas of Lesotho are subjected to high levels of ultraviolet radiation. Adaptation to high UV is an important character. If levels of UV penetrating the atmosphere increase due to the hole in the ozone layer as some models predict, this feature will be increasingly sought after as a character to be transferred into forages and crops. Even slightly raised UV levels are expected to adversely affect the reproductive capacities of plants.
3. From a phytogeographical point of view, the high mountains of Lesotho can be viewed as an anomalous temperate 'island' surrounded by the very different vegetation of the nearby subtropical lowlands and plateau of southern Africa. The Afroalpine species which have evolved there have been effectively isolated, both geographically and genetically, for considerable periods. The species and genotypes from such islands are substantially different genetically from their nearest geographic or taxonomic relatives. They thus represent a unique and important genetic resource. Taxa of particular interest include: *Hordeum capense*; the genus *Poa*, which has a centre of diversity in temperate Europe; the genus *Trifolium*, which has a centre of diversity in the mountains of east and northeast Africa.
4. The effects of global warming cannot yet be accurately predicted, but it is possible that in southern Africa summer and winter temperatures may rise by up to 4°C in the next 30–50 years. Soil moisture may also decrease, and drought periods could be longer, more frequent and more severe. Because many of the alpine species are only found at the very tops of the mountains, even a small increase in temperatures could cause these species to disappear. It would certainly induce a genetic shift in the gene pool.
5. A number of forage genera have representatives in Lesotho. These include *Andropogon, Anthoxanthum, Brachiaria, Bromus, Chloris, Cynodon, Digitaria, Dolichos, Echinochloa, Ehrharta, Eleusine, Eragrostis, Festuca, Indigofera, Lotononis, Lolium, Medicago, Panicum, Paspalum, Pennisetum, Poa, Rhynchosia, Setaria, Themeda, Trifolium* and *Urochloa*. A well-planned collecting, characterization, selection and use programme may reveal genetic variants more suitable for harsh temperate environments than those currently available.
6. Lesotho represents a major centre of diversity for the genus *Lotononis*, and a secondary, outlying centre of diversity for *Trifolium*. Both are important forage genera.
7. Probably the greatest problem facing Lesotho is land degradation and erosion. Despite strict erosion control measures being applied in the last 50 years, soil erosion is taking place rapidly, as population pressure increases. Land degradation has resulted in genetic erosion so severe that a number of forage species recorded at the turn of the century appear to be no longer present.

Box 29.3

Collecting in the Etosha Pans, Namibia: physical and biotic environment

Namibia is a large country in southwestern Africa. It is bordered to the north by Angola, to the east by Botswana, to the south by the Republic of South Africa, and to the west by 1800 km of Atlantic coastline. Most of the country is made up of the arid and semiarid Namib and Kalahari deserts. Collecting was concentrated in the halophytic Etosha Pans system, which is located between these deserts.

The Etosha Pans are a system of highly halophytic salt-pans in the northern part of central Namibia, once a large, shallow lake, which has been drying up as a result of long-term climatic changes. The topography of the area is uniform, being flat and forming a slight depression relative to the surrounding regions. There is one very extensive main pan (4590 km^2), surrounded by numerous much smaller ones. These become seasonally flooded, and hold water for varying lengths of time depending on the extent of the rains. Average annual rainfall is 400–500 mm and occurs mostly in the form of convectional thunderstorms in November–March. Summer temperatures can be as high as 40°C, but winter temperatures are moderate.

Soil pH in the pans themselves is 8.5–10. The main pan is not vegetated, but many of the smaller ones are, the communities being dominated by salt-tolerant grasses. Surrounding areas support shrub veld and tree veld of various kinds, but usually dominated by *Collophospermum mopane*.

The entire pan area and surroundings are a game reserve, the Etosha National Park, one of the largest of its kind in the world, protected since 1907. It has been subject to large build-ups of various game species no longer able to migrate, and this has had a deleterious effect on the vegetation in some areas.

In describing the logistical arrangements for the mission it is important to state what expedition clearances and permits were required for travel to the target region, photography, plant collecting and the export of the germplasm samples, with details of the procedures that were followed to obtain them and the time involved. Such clearances will often be as necessary for national programmes as for outside collectors. All sources of information that were found useful should be specified. Information on the availability of vehicles and other forms of transport, such as boats or pack-animals, should be recorded and any special transport needs or expedition equipment should be pointed out.

Logistical problems encountered during the collecting expedition, either in exploring a region or in handling the germplasm samples, need to be reported. Any special arrangements made for the handling of the collected material should be described.

The extract in Box 29.4, taken from the report of the collecting trip in Lesotho, which is also quoted in Boxes 29.1 and 29.2, provides an illustration of the kind of logistical problems that may be encountered.

Box 29.4
Collecting in Lesotho: logistical problems

Most of this trip was undertaken on foot and horseback, using pack-animals to transport the equipment and seed samples. The mission was successful in so far as a number of populations of the more important forage species were located, frequently at significantly higher altitudes than on the previous two expeditions. However, there had been a severe drought, which meant that many species had not grown during that season, and those which were available had for the most part been so severely overgrazed that few or no seeds could be found. The drought was breaking at the time of the visit, and these late rains made it rather difficult to keep the seed samples dry while travelling on horseback. To avoid mould and further deterioration of the seeds, the trip along the eastern escarpment had to be shortened. This area, however, was less overgrazed than the rest of the country and yielded much interesting germplasm. Further visits to this area are recommended.

Scientific planning

Details should be provided in the mission report of the scientific preparations made for the mission. In particular, there should be reference to the people and other information sources that were consulted. If the collecting followed an in-depth ecogeographic survey, the results of the study should be summarized, as described in Chapter 14. If annotated checklists were developed during planning, and further elaborated in the field, these should be provided. For international missions, the report should mention if a preliminary visit was made to the country to consult local scientists, herbaria or other information sources and to draw up the collecting strategy and itinerary, or whether such preparations were made by correspondence. Herbaria in other countries may also have been visited, or relevant material requested from them.

In some cases the collecting expedition will have been preceded by a field survey to confirm plant maturity or to better plan the collecting. In the case of vegetatively propagated crops, it may have been necessary first to survey the plants when the aerial parts were mature, in order to better distinguish the different clones. There may have been preliminary diversity studies, or preparatory socioeconomic surveys of farmers. The mission report must include the findings of such survey missions.

Itinerary

A summary timetable for the mission should be included, giving the dates during which different areas were explored and the site numbers and collecting numbers from each area. A detailed itinerary giving a daily log of the places visited can be provided as an annex to the report. A sketch-map of the whole country should always be provided, showing the location of the target region. There should also be a rough map of the target region itself showing mission routes and at least some town names that appear on the generally available maps of the country. This will assist future users of the report, who may not be familiar with the areas concerned. An example is given in Fig. 29.1 (see also Mota *et al.*, 1983).

An example is given in Box 29.5 of the kind of itinerary information that would be suitable for appending to the report. It comes from an actual report on wild species collecting in Zambia. The summary table in the report itself just gave the dates during which collecting took place in different areas, and site and collectors' numbers, as in Table 29.1.

Sampling strategy and collecting techniques

The report must describe the approach used to sample the genetic diversity and the techniques employed to collect the germplasm. As discussed in earlier chapters, these methodologies will have been influenced by a number of factors: the biology of the target species, the environment in the target area, cultural and socioeconomic considerations, the logistics of the expedition and perhaps the eventual use of the material.

Mention should be made of any stratification imposed on the

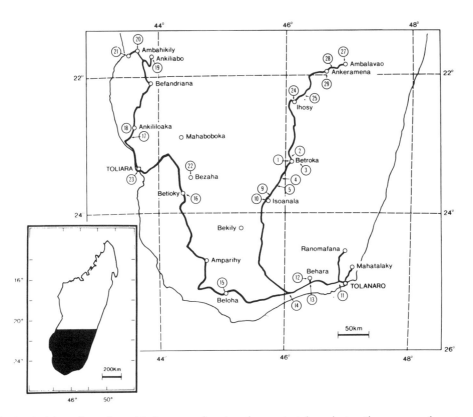

Fig. 29.1. Map of southern Madagascar showing the route taken during the course of a mission to collect rice landraces. The location of collecting sites is shown by the numbers.

Box 29.5
Collecting in Zambia: itinerary

5 June   Meeting with the Director of Agricultural Research to finalize logistical arrangements.
6 June   Collect in the Kafue Flats and flood plain.
7 June   Organize first leg of trip to Kasanka National Park. Buy supplies, etc.
8 June   Drive Lusaka–Kasanka.
9–13 June   Collect in Kasanka National Park. This area was selected because of its wide diversity of habitat types and its well-conserved and managed vegetation. The park rangers were very helpful collecting assistants.
13 June   Drive to the southern part of Bangweulu Swamps, and set up camp near Chikuni. Meeting with the local WWF [World Wide Fund for Nature] representative.
14–16 June Collect in the southern part of Bangweulu Swamps, both in the flood plain and in the main river channels. Much work was done from dugout canoes.
17 June   Drive through Lavushi Manda National Park to Mkushi, collecting along the way. Obtain accommodation on a farm.
18 June   Collect on Mkushi Kachana Farm. Return to Lusaka.
19 June   Prepare for second leg of expedition to Western Province. Buy supplies, etc., and have repairs undertaken to vehicle.

Table 29.1. Summary of itinerary.

Area	Dates	Sites	Samples
Kafue Flats and flood plain	6 June	1–4	1–12
Kasanka National Park	9–13 June	5–19	13–23

sampling. For example, in crop collecting an attempt may have been made to sample all agroecological zones in a region and all ethnic groups within each zone. Indigenous knowledge (IK) may have been used to locate particular material. The strategies adopted to sample the genetic diversity within populations must also be explained, since it is important for future users of the germplasm to know how a sample relates to the original population. This information will have been recorded in detail on the collecting data forms but in the report there should be a description of the basic strategy adopted and any particular exceptions to it. Thus, the basic strategy might be one of random population sampling, but there may have been additional selective sampling of particular phenotypes in some cases.

Any specialized methods used (e.g. the details of an *in vitro* collecting technique) should be described in the report and any problems in applying them should be pointed out, with any tips on their use or suggestions for improvement.

In the case of multi-species collecting missions, the above information may best be incorporated into the separate sections dealing with each of the different species collected (see below).

Results

In presenting the findings of the mission, the report should bring together and summarize the information gathered in the field and from reference sources about the species collected and the environments in which they were found. Much of this information will have been recorded on the individual sample collecting forms, but a major function of the mission report is to synthesize the data and present an overview of the ecogeographic distributions of the target species. This helps build up the base of general knowledge on the species and gene pools, and can guide users in the evaluation and use of the collection.

The report should include lists of the germplasm collected. Full lists of the individual samples in collecting number order and in species order, with such additional basic collecting data descriptors as collecting date, locality and site number (in crop collecting, also the local names of landraces), can be an annex to the report. A summary table listing collecting numbers by landrace, species or gene pool should be presented in the text. Another table could list collecting numbers arranged by collecting area (administrative unit, vegetation type, soil type, agroecological zone). A two-way table combining the above information, listing collecting numbers by taxon (or landrace) and collecting area, is often useful. There should always be a table summarizing the total number of samples of each species.

Also important are maps showing in detail the locations of all collecting sites. These could be photocopies of the maps annotated in the field. Collecting sites should be labelled with the site number or collecting number(s). The former option is usually more convenient, especially if several samples are collected at each site. Usually, the base map used to show collecting site locations will be a topographic map. However, climatic, soil and vegetation maps may be used as appropriate to provide additional, easily visualized, information on the likely adaptation of material. Collecting site locations may be marked on an acetate transparency or transparent paper, which could then be superimposed on different base maps as necessary. Maps showing the location of samples of individual species could also be included. Different species may be shown using different symbols on the same map, or separate maps may be prepared for each species. All maps should include a scale and compass directions.

General observations on the germplasm samples (IK, morphology, phenology, threat of genetic erosion, etc.) and where they were found can be presented in the text of the report by agroecological zone and/ or by landrace, species or gene pool. Thus, the report on a general crop collecting programme in a country or region could have both of the following:

- general accounts of the physical environment, farming systems and cultural and socioeconomic situation in each agroecological zone sampled;
- overviews of each crop (and/or landrace), including descriptions of specific cultural practices.

Box 29.6 gives an example of the former taken from a report on crop collecting in Zimbabwe; the extract discusses one of the several provinces in which collecting was carried out. As an example of species overviews, Box 29.7 is an extract from a report on collecting forages in Botswana, which shows how information gathered in the field can be combined with published information on individual species to provide guidance to prospective users of the germplasm. Box 29.8 is an extract from a report on collecting in Madagascar, which provides a similar service for a cultivated species; a more specialized report might have had different sections on each landrace collected.

Information from local people and other sources on the changes that have been occurring in the target area, particularly regarding landraces that are no longer grown in particular areas, should be recorded in this section. Two examples are given. Box 29.9 is taken from a report on a collecting programme in southern Arabia involving three countries. Box 29.10 reproduces a more detailed report relating to finger millet on the Yemeni island of Socotra, which illustrates how ethnobotanical information may usefully be incorporated into a collecting mission report.

Whether the timing and duration of the mission coincided with maturity of the plants and the availability of ripe seeds or mature roots and tubers should be commented upon and this should be related to the weather during that particular year. Detailed climatic data for the year of the mission (rainfall, temperature, river levels), if available, and averages, can be provided in an appendix.

Brief summaries of the descriptive information on each site (e.g. altitude, soil type, pH, slope, aspect, habitat, etc.), together with the collecting numbers of all samples collected at each site, provide a useful addition to collecting reports. These can assist users who do not have ready access to the full passport data set in assessing whether particular material may be of interest to them. Photographs of germplasm and collecting sites can be included here.

Processing and distribution of the samples, data and reference specimens

The report must describe what was done on return from the field to prepare and distribute the germplasm and data. It is important for the recipients of such material to know what procedures were used to clean, dry, treat, pack or culture (in the case of *in vitro* material or *Rhizobium* samples) the samples. Advice on any modification of standard techniques that were found to be necessary will greatly help future collectors. Mention should be made of the plant health and quarantine procedures

Box 29.6
Collecting in Zimbabwe: results

Victoria Province

Of the 17 Communal Areas in this province, 12 were explored well. The coverage of Matibi No. 2, Sengwe and Chikwanda was superficial and the small areas of Denhere and Serima were not visited.

The soils are mainly light granite sands. There is an area of red sandy loams in the northern parts of Bikita and Ndanga and a tract of black, basaltic clay in the southwest, in Sengwe, Matibi No. 2 and Sangwe. The province falls within Natural Regions IV and V, with a small area stretching from Bikita to Lake Kyle, east of Nyanda, which is classified in Natural Region III. Rainfall was less than normal this year throughout the province. Not surprisingly, it is the Communal Areas in Natural Region V that have been affected the worst by the drought.

Sangwe, Matibi No. 2 and Sengwe in the extreme southeast of the country were severely hit by the drought. The black, basalt clay soil had deep, wide cracks. Red Swazi was the most common sorghum variety found. It had performed moderately well in spite of the drought. There were still some local landraces of sorghum in these areas. They constitute an important source of genetic diversity, especially in terms of drought tolerance. The basalt soils are heavy and alkaline and generally unsuitable for pearl millet. A few landraces were found on the red sandy loams in the northern part of Sangwe.

Although still badly drought-affected, it was possible to collect various crops from the other Communal Areas (Chibi, Matibi No. 1, Maranda) of Natural Region V. Sorghum diversity in Matibi No. 1 was high. A number of different pearl and finger millet landraces were collected from all three areas.

In the Communal Areas of Region IV (namely Zimbub, Chikwanda, Bikita, Ndanga, Victoria, Nyajena Mtilikwe and Gutu), maize, groundnut and sunflower were being grown as cash crops but on a limited scale. Maize was of hybrid seed origin, but often not first-generation seed. Improved groundnut varieties, such as Makulu Red, were present, but it was still possible to collect some local types. Although sorghum was mostly the commercial varieties Red Swazi and Fremida, various local landraces were still being grown. Matsai was especially rich in sorghum diversity. Finger millet was variable and widespread throughout these areas. *Eleusine africana* was a common weed.

Throughout the province, cowpeas, watermelons, pumpkins, melons and cucumbers were typically intercropped with maize, sorghum and pearl millet. Most farmers also grew bambara nut. A wide range of variation was encountered in all these traditional legume and vegetable crops. Surprisingly, gourds were not as common or as variable as in the other provinces visited. Okra was also infrequent and the local leaf rapes were not found at all. However, it was only in Victoria Province that pigeon pea was encountered, but it was only found as a backyard plant at two sites in Ndanga Communal Area.

High priority should be given to the further exploration of Sengwe and Matibi No. 2 for sorghum.

Box 29.7
Collecting wild species in Namibia and Botswana: results

Citrullus ecirrhosus. A perennial creeper with a somewhat thickened rootstock. Stems to 2 m long, prostrate. Fruits ovoid, mottled greenish/yellow, about 20 × 10 cm.

This is a little-known but close relative of the cultivated watermelon. It is of particular interest in that it is perennial and grows in extremely arid areas. The two populations sampled in Namibia, # 1407 and # 1428, had produced prolific quantities of fruit in areas of the Namib Desert which had received around 60 mm of rainfall that year, but which are normally even drier. # 1428 was very different in appearance from most other populations and may represent an unusual variant. The large fruits are sought by game.

Cenchrus ciliaris. A broad-leaved, profusely branched, tufted perennial. It is about 1 m high and normally geniculate. This widespread climax grass occurs on most soil types, but is particularly characteristic of alkaline soil. Palatability seems to vary. As a pasture species it can give an exceptionally good yield but requires heavy fertilization both for high productivity and for palatability.

It has a remarkable ability to remain green and productive long after the surrounding grasses have withered and died back in the dry season. It often occurs in favourable microenvironments such as sandy washes or rock crevices in areas that would otherwise appear to be too dry. It is particularly deep rooted. Even under the most harsh conditions, this species appears to be highly productive. Seed set under natural conditions generally tends to be poor, and the spikelets shatter with varying ease. This is a very variable species, with a number of different cultivars available commercially.

The Botswana genotypes were unusual in showing tolerance to low P levels, stoloniferous habit, tolerance of exceptionally high and low pH values, and drought tolerance. According to Skerman and Riveros (1990), this species shows a preference for soils high in P. The Botswana genotypes may differ in this requirement since the soils of the Kalahari are notoriously low in P. A very wide pH tolerance is exhibited in the genotypes from Botswana, 3.2–9.5 being the range encountered. The population sampled under # 743 appeared to be slightly stoloniferous, a character not known in this species. It was growing on a slight elevation in a salt-pan at pH 8. # 745 was growing in a similar site at a pH of 9.5. # 753 was also a stoloniferous variant, and was growing at a pH of 5.3. This population appeared to be susceptible to smut. # 832 was thriving in exceptionally acid soil (pH 3.2). This did not appear to affect its palatability, since it had been heavily selectively grazed. Seed set, however, was poor. This population was growing at around the 250 mm isohyet, and # 862 was found at around 150 mm. These latter two populations may thus be particularly hardy variants, surviving in rainfall regimes very much lower than the 375–750 mm suggested by Skerman and Riveros (1990) as being the norm for the species.

followed. Any data preparation carried out should be described, noting if verification of some information has been requested (e.g. taxonomic determination of voucher herbarium specimens or pest specimens).

Recipients of the germplasm need to know if the samples have been divided and where duplicates are held. The report should describe the sample splitting and distribution protocols followed by the collector, indicating whether this was for duplicate conservation, interim

Box 29.8
Collecting in southern Madagascar: results

Some 1200 traditional rice landraces are thought to be grown in Madagascar. Highest diversity is in the High Plateau areas. In this collecting trip, about 45 distinct local names of traditional landraces were encountered. The most common landraces were 'Makalioka' and 'Tsipala Fotsy'. While largely the same suite of landraces was found throughout most of the drier areas of the southwest (Areas 1, 2–5), a different set seemed to be grown in the Ihosy and Ankaramena–Ambalavao areas (Areas 6 and 7). Some landrace names reflect grain shape and size, pericarp colour and foliage coloration. Others are more fanciful. Translations are given in the passport data.

Because most of the samples were collected from farm stores, it was not possible to record observations on plant characters. However, based on pericarp shape, coloration and hairiness, some 15 classes may be recognized. It is instructive to check local names against such an admittedly crude classification. Thus, six of the samples of 'Tsipala Fotsy' fall in one class and two in another. All four of the 'Kianga' samples fall in a single class. Three of the 'Makalioka' samples are in one class, two in another and one in a third. There therefore seems to be some variation among samples given the same local name in different areas. Looking next at a single class, the case of the distinctive striped pericarp type is interesting. The four samples falling in this class, collected at sites 10, 18 and 19, all have different local names ('Taimbalala', 'Ambatondrazaka', 'Manga Fototra' and 'Masokibo'). Thus the classification based on pericarp features is, not surprisingly, not sufficient in differentiating among local landraces, though it is a useful start. Incidentally, there is another 'Manga Fototra' in the collection (the name refers to its dark foliage), collected at site 12, but this falls in another pericarp class.

The following landraces or samples were highlighted by farmers for particular characteristics or traits: *long cycle* – Tsitelovola; *short cycle* – Kianga, Vary Malaky, Kianga Makalioka, Tsiroavolana, Maroroka, # 1178 (Kely Mena); *drought tolerance* – Tsy Mataho Paosa; *consistency* – Tsy Matahotra Osy, Tsy Mikoty, Soalava, Laniera; *can be grown in either season* – Kely Mena, Tsy Matahotra Osy.

Some other samples should be highlighted. The name 'Mamambary Tsipala' may be translated as 'Better than Tsipala', and farmers confirmed that it was indeed the case that this landrace performed better than the standard, reference landrace 'Tsipala' in favourable years. The sample # 1196, for which, uniquely, no local name seemed to be used, was said by farmers to be 'better' than the reference landrace in the area, 'Angika', though exactly in what respect was not clear. Finally, the name 'Korintsa' seems to refer to the rattling noise made by the ripe inflorescence in the wind, and this is said to help keep birds off the crop.

Collecting sites fell into two main groups with respect to temperature: *Group A* – Areas 2–5; *Group B* – Areas 1, 6 and 7. Sites in Group A have a mean maximum temperature of the hottest month (i.e. at the sowing/transplanting stage) of about 31°C, those in Group B of about 27–28°C. It has already been alluded to that the suite of landraces grown in areas 6 and 7 is different from that grown in the rest of the region covered. This may partly reflect the temperature difference, though water availability will also differ. However, this should not obscure the point that factors relating to the cultural make-up of different areas will also be extremely important. Areas 1, 6 and 7 fall mainly in the region of the Bara people, whereas the rice-growing areas around Tolanaro are dominated by the Antanosy people. Both groups grow rice around Toliara, and there are of course some ethnically mixed areas. Cultural reasons may be as important in determining the landraces grown in a particular area as climate, if not more so.

Box 29.9
Collecting wheat in southern Arabia: genetic erosion

A drastic decline in the area sown to wheat in the Dhahirah region of Oman from the mid-1970s has been documented. This has mainly been due to increasing competition from Australian grain imported for processing at the Oman Flour Mill. An additional pressure on the local landraces has come from the release of Mexipak in the early 1970s and Sannine in the early 1980s. Two ICARDA [International Center for Agricultural Research in the Dry Areas] varieties were released in 1985. No seeds of modern varieties were made available to Omani farmers through extension centres until 1980, when some 4 t were distributed; in 1983, the figure was 11 t and the trend is still certainly upwards. Local landraces, in Oman as elsewhere, are preferred for their taste and drought resistance, but they are low-yielding, susceptible to lodging and for the most part lack rust and smut resistance. Probably more than 50% of wheat grown in Oman (on an area basis) is landraces.

The situation is somewhat different in Saudi Arabia, which thanks to government policy is self-sufficient in wheat. In 1986–87, however, commercial wheat-growing projects of the sort which have helped bring this about amounted to less than 100 ha in the Aseer administrative region, compared with over 7000 ha of wheat on traditional farms, which nevertheless represents a marked decline from the 1982–83 figure of almost 20,000 ha. Unlike in other areas of Saudi Arabia, then, where the threat to local landraces comes from commercial agricultural projects and the influx of modern varieties, in the Aseer the problem is one of gradual abandonment of subsistence agriculture, in favour of cash crops or urban living. Modern varieties are grown by traditional farmers, but on a very limited scale. Landraces are preferred, though the state flour mills will accept grain of only a small number of modern varieties.

Genetic erosion of wheat in northern Yemen is also occurring mainly through the abandonment of land and the increased growing of cash crops such as *Catha edulis*. The spread of modern varieties (Sonalika was introduced in the 1970s and several other varieties since) is also significant, but perhaps of less importance. The main reasons given for the abandonment of traditional agriculture are much the same throughout the region: high cost of labour, difficulty of mechanization on steeply terraced land, greater profitability of other work, easier access due to expanding road system. A clear example of genetic erosion in northern Yemen is *T. dicoccon* ('alas' in the local Arabic). This species is highly prized by farmers for bread-making, but is said to need more preparation and more favourable growing conditions than other wheats and is no longer sown anything like as extensively as formerly. A similar situation also obtains in Oman.

In the Wadi Hadramawt area of southern Yemen, 65% of the wheat area is now sown to modern varieties, mainly Kalyansona, though Ahgaf has recently been released. However, these cannot be grown where groundwater is very saline. Another factor protecting the local varieties is that wheat straw is almost as important economically as the grain, as it is needed in brick-making. Landraces are thus relatively safe because of their salt tolerance and tallness. In Shabwa, Sonalika was released in 1973, and seems to have made bigger inroads than Kalyansona in Wadi Hadramawt. Wheat cultivation seems to have declined markedly in the high-altitude areas of southern Yemen, due to the ready availability and comparatively low price of imported grain. It was difficult in 1989 to find landraces; five were said by farmers to be common up to ten years ago but only one ('Hargadi') seems to be used now, though seed samples of another two were eventually obtained.

Overall, it seems safe to suggest that the area sown to landraces in the region is more than half of the total wheat area. It is inevitable that this will decrease. There will be loss of genetic diversity as rarer types fall out of cultivation and as fewer farmers sow the

continued

Box 29.9 *Continued*

commoner types. Some areas are more at risk than others, however, as the situation in Wadi Hadramawt shows. Even in the Riyadh region of Saudi Arabia, where commercial wheat farming is predominant, some farmers still grow small fields of local landraces for home consumption.

multiplication, quarantine, etc. The destination of associated material such as *Rhizobium* samples and voucher specimens must also be recorded. Detailed sample-by-sample distribution lists can be included as an appendix.

Further exploration and research

Suggestions can be made in the report for further collecting or particular studies to be carried out on the germplasm. Areas that it was not possible to explore or that were inadequately covered should be identified. If it was not possible to sample certain populations or species because the plants were immature or their fruits had already shattered, the report should identify these for future collecting and include recommendations on more appropriate timing.

Observations in the field or information from farmers may have drawn attention to the likely presence of important characteristics in the material, such as disease resistance, drought tolerance or preferred cooking quality or taste. This information will have been recorded on the relevant collecting data forms. The report should highlight particularly noteworthy samples for early study and evaluation, including those found in unusual situations, for example at higher altitude than normal.

If the mission was part of an ongoing study on a particular gene pool or area, then the report should include a description of the research that will be conducted on the germplasm and indicate where this work will be done. This could include characterization and evaluation of the material collected. It is important that the report describes any observations or findings made during the mission that could be of significance to the planned research.

Summary

Lengthy reports should include a summary or abstract. This should highlight samples needing immediate multiplication or taxonomic verification. It should also include the dates of the mission, the areas visited, the numbers of samples collected of each species or genus and the distribution of the material.

Disseminating the findings

Distribution of the mission report

The report of a collecting trip must be distributed to all scientists and organizations participating in the expedition and to the sponsoring

Box 29.10

Finger millet on Socotra: cultural practices and genetic erosion

Finger millet, called 'bombeh' in the Socotri language, is now grown only in some villages in wadis (seasonal watercourses) draining the Haggier mountains. However, according to local people, it used to be much more widely cultivated in the mountains and was also to be found elsewhere, for example on the eastern plateau. Its cultivation is traditionally the work of women. It takes place on terraces called 'mutereh' enclosed with date-palm frond fencing provided with a wicker gate. Many of the new date-palm plantations now seen are on the site of old finger millet terraces, giving the mistaken impression that bombeh was not all that common formerly. A site enclosed by a family belongs to that family, unless left unworked for a period of years. In the Haggier mountains, former bombeh terraces are kept enclosed and natural grassland allowed to develop within them to be used as fodder.

The soil is prepared for finger millet cultivation by setting fire to piles of wood and animal dung and then spreading out the ash and mixing it with the soil. The wood of *Boswellia ameero*, an endemic frankincense tree, is particularly valued for this. Sowing occurs when the cool winds of the NE monsoon begin to blow, normally in November. Irrigation is usually by hand, the site having been chosen on the basis of the ready availability of water. Often, the terrace is divided up into small contiguous areas with small rocks, each area receiving one container-full of water. The small seedlings ('saabi') are watered daily until they reach hand-span height, some 20 days after germination. At this time, the plants, now called 'chaabir', are dug up and most of the greenery is twisted off. They are then divided into tillers and these are planted out again. Three months after sowing, the main tiller is harvested; secondary tillers are not harvested, though they set grain. The stalks are cut and either fed green to livestock or dried. Livestock manure the fields as they graze the straw.

The threshing of the heads (by treading) and the winnowing (with date-palm fronds) is done by men. The grain is sun-dried and then stored in clay pots, baskets or sacks. It can keep for several months. It is ground to make a sticky porridge with boiled water (rarely milk) called 'muqdeyreh'. The best grain is saved for sowing the following year.

Since what used to be the People's Democratic Republic of Yemen (this merged with its northern neighbour the Yemen Arab Republic in late 1990) gained its independence, cheap substitutes for finger millet, like wheat flour and rice, have become increasingly easily available. This is the main reason for the decline in finger millet cultivation on Socotra, always an extremely labour-intensive enterprise, especially for the women. The situation is a common one. A similar process is at work in the highlands of the Yemen mainland, where wheat cultivation seems to be decreasing due to the availability of cheap imported flour. The case of finger millet on Socotra is, however, particularly dramatic.

The crop now seems to be grown only by some older women who have better access to water, for example due to the recent acquisition of a pump, making the main labour of watering easier. These women continue the practice perhaps mainly out of sheer habit and nostalgia. However, other reasons were also given during interviews. There is a belief in the health-giving properties of the crop, for example, and the taste is much preferred to that of recent substitutes by the older generation. Older people also doubt the reliability of the supply of cheap imports of wheat flour and rice, and grow finger millet as a precaution. There is no evidence that the crop is being abandoned due to a general shortage of water on the island compared with 20 years ago.

agencies. It may be most appropriate for each participant to write a separate report. Alternatively, the team leader may be given the responsibility of producing a first draft of the report, which would then be sent to all participants for revision and finally officially distributed. The report must also be sent to all the gene banks, institutions and researchers receiving germplasm samples or associated specimens. It then becomes the responsibility of recipient institutions and scientists to ensure that a copy of the report accompanies any subsequent distribution of the germplasm that they undertake.

Publishing the findings

Publishing an account of collecting activities will aid potential users of the germplasm and others planning to collect the same gene pools or in the same region. A useful venue for publication of collecting mission reports is the *Plant Genetic Resources Newsletter*, which is widely read in the plant genetic resources community. Chapter 8 lists other possibilities for the reporting of collecting activities. Published articles on germplasm collecting should make reference to where the original report and the complete passport data can be obtained.

References

Mota, M., L. Gusmão and E. Bettencourt (1983) Reporting on collecting missions. *FAO/IBPGR Plant Genetic Resources Newsletter* 55:32–39.

Skerman, P.J. and F. Riveros (1990) Tropical grasses. *FAO Plant Production and Protection Series No. 23*. FAO, Rome. 832 pp.

IV

CASE-STUDIES

Collecting tropical forages

30

R. Reid

*Pastures and Field Crops Branch, Department of Primary Industry,
Mt Pleasant Laboratories, PO Box 46, Kings Meadows,
Tasmania 7249, Australia.*

Systematic exploration for species of grasses, legumes and browse plants to be used in tropical pasture development is a relatively new phenomenon. It was not until the early 1950s that many tropical areas of the world began to be traversed in the search for new and useful pasture species, a search that has not stopped. This new germplasm and improved livestock management have led to a boom in tropical pasture development. There are even instances where formerly degraded and unproductive lands have been returned to pastoral production by adopting the new technologies. The relative newness of tropical pasture species development can be illustrated by two Australian cultivars. Until the mid-1960s, *Stylosanthes hamata* was known only as a somewhat weedy component of seminatural pastures in the Caribbean. It was certainly not considered to be a major pasture legume anywhere. However, an introduction from Maracaibo, Venezuela, into a *Stylosanthes* evaluation programme of the Commonwealth Scientific and Industrial Research Organization (CSIRO), Australia, was quickly recognized as having enormous potential in dry tropical pastures. It was released in 1973 as cv. Verano and has had an enormous impact in pasture development, first in northern Australia and now in the Sahelian zone of West Africa and parts of India. Its success has led to further detailed collecting of *Stylosanthes* in northwestern Venezuela in the hope of finding even more productive material.

Another example is the legume *Cassia rotundifolia*. Many early botanists in subtropical South America commented on its abundance in native pastures. In 1947–48 the first Australian mission to South America collected seed, but more importantly pointed out the potential of the species. Further introductions took place and some initial evaluation was done but it was not until 1984 that cv. Wynn was released. This example shows that, even if a plant is identified as being useful, it may

617

still take some time to make its way through the introduction, evaluation and selection system. There is no telling how many other species
are just waiting to make a similar impact to *C. rotundifolia*, for example
Aeschynomene villosa, *Demanthus virgatus* and *Digitaria milangiana*.

During the early years of collecting tropical forages, teams tended
to cover large distances quickly, sampling simply wherever an interesting plant or site was encountered. Strickland (1974) covered a wide range
of environments in eastern and southern Africa by following collecting
routes designed to traverse a range of soil types and altitudes in the
various vegetation types outlined by Rattray (1960). Altogether, almost
45,000 km were covered by four-wheel-drive vehicle, starting in
Mombasa, Kenya, and ending in Durban, South Africa. The resultant
grass collection amounted to 118 species in 36 genera, in particular
Anthephora, *Brachiaria*, *Cenchrus*, *Cynodon*, *Digitaria*, *Panicum* and
Urochloa. The legume collection totalled 138 species in 38 genera, the
more important of which were *Dolichos*, *Indigofera*, *Lotononis*,
Macrotyloma, *Rhynchosia*, *Trifolium*, *Vigna* and *Zornia*. A somewhat
more detailed strategy in exploration is to conduct a series of missions
covering essentially the same terrain and collecting sites. Reid (1983)
traversed Mexico in such a way that the maximum number of contrasting regions could be sampled on a recurrent basis over a period of
two years. Some sites were visited up to five times, not only to ensure
that ripe seeds of target species would be collected but also to guarantee
that all species of likely potential were in fact sampled.

The exploratory approach has been superseded as both geographic
regions and individual species have become better known, but it still has
value in regions that are relatively poorly known botanically and where
there is need to establish the general state of the environment, i.e. land
degradation, level of overgrazing, human encroachment, etc. There are
still significant regions that have not received any attention from forage
collectors and which may therefore warrant going through an initial
exploratory phase. Schultze-Kraft and Giacometti (1978) argue that:

> there are two important reasons for continuing tropical pasture
> germplasm collecting: (a) the need to obtain the maximum possible
> genetic variability, in order to proceed with the selection of the most
> promising materials; and (b) the need to guarantee the preservation of
> the available genetic resources, *while they still exist*. An important
> reason to stress the latter need is the very clear evidence that genes
> of tropical species are being lost. On more than one occasion, upon
> returning to a collection site of especially promising genetic material, it
> has been found that the material no longer exists.

Most experienced tropical forage collectors would agree as to which
regions remain to be explored. The order of priority may differ, and
indeed change as research develops, but institutions, both international
and national, with a long-term commitment to tropical pasture plant collecting (e.g. the Centro International Agriculture Tropical (CIAT),

CSIRO's Division of Tropical Crops and Pastures, etc.) have clear ideas as to which countries or regions remain to be explored. In Africa, for example, Angola, Mozambique and Sudan could be mentioned. Deciding what species to collect is more problematic. Some collecting missions deal with one or two genera which are required for particular purposes. Others will collect all variations in genera known to be used as pasture plants, and in addition species unknown in agriculture if they seem to be eaten by livestock and possess agronomically important environmental adaptations or morphological attributes. As demands on existing cropland increase and the more productive pasture lands are ploughed up, increasing emphasis will be placed on the upgrading of pastures by the introduction of such species. The overgrazing of traditional grazing lands and general land degradation are also creating the need for new species to fill 'niches' where at present no species can be recommended. While progress has been rapid, there are still many habitats, such as semiarid cracking clays, acid semiarid soils and salt-affected lands, where new species are urgently required. Further, as management improves (fencing, controlled herding, pasture spelling and fertilization), so the option of using improved species becomes ever more valid.

Broad-spectrum collecting puts a particular strain on the collector's skills in field identification of material. All too often material has been collected and stored in gene banks without adequate taxonomic characterization or verification (Marshall, 1989). Many accessions are also clearly mixtures of species when grown out. There may be various reasons for this, though few excuses. Identification aids may be lacking or of poor quality, especially for difficult and little-known groups. Many of the climbing legumes intermingle and it is difficult to physically separate them. Diagnostic features (e.g. flower colour and structure) are often absent at pod set and the material may be too dry to make a satisfactory herbarium voucher. Ideally, the collector would visit the collecting site on a number of occasions to overcome these problems, but in practice this is rarely possible. The collector should therefore be as familiar as possible with the species in the areas to be visited which are of known or potential forage value. This means surveying the literature, including Floras, ecogeographic studies and vegetation studies. Most importantly, however, it means visiting herbaria.

On visiting herbaria, many collectors are surprised to find how many of their target species have been collected extensively by botanists and are well known to them. Examination of herbarium specimens helps not only in becoming familiar with the material but in pin-pointing potential collecting localities and in establishing the optimum time for collecting. Unfortunately, it is often not possible for the collector to visit those herbaria that hold the major collections of the target species. For example, although the National Herbarium at Nairobi has excellent holdings from eastern Africa, any collector needing an in-depth picture would also need to spend some time at the Royal Botanic Gardens, Kew. The situation is somewhat similar with many Latin American floras, where the most extensive holdings are in the herbaria of either the Missouri or New York Botanical Gardens. An example of the value of preliminary herbarium work is given by Pengelly and Reid (1988). They examined approximately 800 herbarium specimens of legumes from Papua New Guinea considered to be of potential as forage plants and established a database containing such information as stage of maturity and time of collecting, geographic coordinates, altitude, soil type and rainfall. From this, a collecting route was planned and a mission undertaken by the senior author. The resultant collection contained samples that were agronomically interesting because of their climatic requirements, acceptability by livestock or occurrence on particular soils. This collection is likely to prove of great value to pasture workers in the African and American tropics.

Herbarium surveys are useful in determining the best time to collect. The timing of collecting missions is critical when seeds are to be collected but clearly more flexible if vegetative material is the target. In the former case, collectors have to time their activities in an area to coincide with the availability of ripe seeds of the target species on the plants, before shattering occurs. In general, this can be estimated from rainfall and day-length data but clearly more accurate information can be obtained from herbarium sheets and local informants. Recent developments in the use of *in vitro* collecting techniques not only offer alternative means of conserving and using plant germplasm but also solve many of the logistical difficulties associated with planning and executing collecting missions. In a recent collaborative project between the International Livestock Center for Africa (ILCA) and the International Board for Plant Genetic Resource (IBPGR), an *in vitro* technique was developed and applied to collecting germplasm of the forage grass genera *Cynodon* and *Digitaria* (IBPGR, 1990). By freeing the collector from the constraints imposed by the necessity to collect seeds it will be possible to be more flexible in timing the expeditions, also allowing more time in the field. Of probably even greater importance is the possibility of being able to collect, and thus evaluate, material from those grass genera that are routinely overgrazed or that set few viable seeds (e.g. *Digitaria*, *Echinochloa*, *Eriochloa*).

Acquiring pasture plants that will be used in specific environments usually involves exploring environmentally homologous regions. For example, on the broadest scale CIAT has targeted tropical, acid, infertile soils, initially in the Americas but lately in southeast Asia and Africa, as source areas of germplasm adapted to its reference areas in South America. Less extensive and more narrowly focused was the programme by Staples (1986) to collect tropical legumes in India adapted to clay soils (vertisols) and suitable for evaluation on similar soils in northern Australia. Prior knowledge of the occurrence of small areas of particular soils or vegetation types within larger soil or vegetation units makes it possible to include visits to such areas in the mission plan: discovering them at the time of collection often requires last-minute changes in the expedition schedule that it may not be possible to accommodate.

Some species are relatively common in all or part of the area sampled by a collecting mission or programme, while others require a great deal of effort to obtain even one sample. Schultze-Kraft *et al.* (1984) found that two of their target species (*Desmodium heterocarpon* and *Pueraria phaseoloides*) were among the most common native legumes of China's Hainan Island, but another (*Codariocalyx gyroides*) was quite rare. Reid (1983) attempted not only to acquire samples representing the broad range of variation in *Leucaena leucocephala* in Mexico, but also samples from all 14 *Leucaena* species. All were in fact acquired but two proved particularly problematic. *L. retura* is uncommon though very widely distributed through the arid regions of northeast Mexico; thus long distances had to be covered and many sites examined before samples could be obtained. In contrast, *L. cuspidata* was relatively easy to find, with the help of herbarium locality data, but proved to be a relic population of less than 40 plants.

A number of workers (Allard, 1970; Marshall and Brown, 1983) have suggested sampling strategies to achieve the conservation of the maximum amount of genetic variation without incurring the penalty of excessive sample numbers. The number of plants sampled per population depends on the breeding system of the target species; fewer plants need to be sampled for cross-pollinating species than for self-pollinators. However, the breeding system of many of the tropical forage species that are being collected is still not known, and it is self-evidently better to collect more seeds rather than fewer if circumstances permit (Reid and Strickland, 1983). This problem of how much to sample is without doubt the most vexing question faced by the collector. All collectors have problems putting sampling theory into practice, for a number of reasons. Few forage species are found in large, dense, evenly distributed populations. It is much more common, particularly in perennial legume species, to find individual plants thinly and patchily distributed across the habitat. It is not uncommon for collectors to report taking days or even weeks to find a single, much sought-after, ecotype.

Also, tropical forage plants are still wild and essentially weedy, and their seeding characteristics uneven. Most legumes have pods that shatter on ripening. This makes it very difficult to ensure that any given population is sampled adequately. Many experienced collectors have stories of combing over a population of legumes for hours to be rewarded with nothing more than the sight of shattered pods. Of course, it may be possible to keep returning to a site and to collect sufficient seeds at the time of maximum yield, as the author was able to do in Mexico (Reid, 1983). However, the reality is that, through lack of resources, most collectors are unable to return to a site; regrettably, the majority of collecting sites will only be visited once.

Finally, as many collectors are not only involved in the planning and execution of a collecting programme but also in the initial evaluation of the material they collect, their emphasis is likely to be on cultivar development rather than the conservation of genetic resources. Knowing that, even if only a small number of seeds are collected, the accessions will nevertheless be grown out, multiplied and evaluated, the collector will be satisfied with a smaller sample. Most tropical forage collectors adopt the attitude that it is far more important to gather diversity from as many different sites as possible than to spend valuable time gathering large seed samples from a few sites. From the point of view of pasture species development this approach has proved to be eminently successful, with many now well-advanced cultivars being produced from initially very small samples of less than 50 seeds. Examples include *Stylosanthes hamata* cv. Verano and *Aeschynomene americana* cv. Glenn (L.A. Edye, pers. comm.).

Information on infraspecific variation can also help in formulating an efficient collecting strategy, but very little is available on most wild forage species. Even where some data are available, the number of genotypes involved is usually very small. For example, in a collection of 121 accessions of *Alysicarpus* spp. classified by Gramshaw *et al.* (1987) into 19 morphological/agronomic groups, only four were represented by more than ten accessions. A number of the groups contained only one or two accessions, which would suggest that there is much diversity yet to be collected. Most of the tropical forage genera that have been studied are similar in this respect.

Each species and each area will present unique problems in sampling, but where sufficient genetic information is available it is possible to plan better sampling strategies. This is well illustrated by the research conducted on the genus *Stylosanthes*, which consists of about 44 species. Over the last 25 years much information has been accumulated on the agronomy and regional adaptation of many of these species. It has taken that long to adequately describe the available germplasm and its agronomic variation and to confirm that a number of additional species remain to be fully exploited by tropical pasture science. Starting in 1967, CSIRO began a detailed study of the genus *Stylosanthes* at Townsville based on the proved potential of *S. humilis* and *S. guianensis*. CIAT started collecting and evaluating in 1972, with the aim of developing cultivars adapted to acid soils in the humid tropics.

The first phase of this work emphasized collecting germplasm and consisted of a series of missions aimed at broadening the genetic base of the genus in cultivation. The material was characterized and evaluated through a series of national and international cooperative testing programmes, over a range of tropical environments (Schultze-Kraft *et al.*, 1984). The results of this work led to the second phase of the programme, when individual species and ecotypes were developed to the point of domestication. They were adapted to a wider range of environments than *S. humilis* and *S. guianensis* and greatly extended the boundaries of feasible pasture improvement. Such new species as *S. capitata* (adapted to the low-fertility acid soils of Brazil, Colombia and Venezuela), *S. hamata* (which grows well on a range of soils in the semiarid tropics) and *S. scabra* (which is particularly resistant to disease) became well known throughout the tropics. As researchers and farmers gained further experience with this material, and as limitations became apparent (e.g. disease susceptibility, lack of salt tolerance, low nutritive value), the third phase of the programme began. This entailed returning to collect from the original areas of succesful introduction, i.e. northeastern Brazil for *S. scabra* and northwestern Venezuela for *S. hamata*. In the latter case, detailed exploration has occurred in the upland areas of the Lara and Falcon searching (successfully) for ecotypes that are as drought-tolerant as the existing cultivar but more cold-tolerant. New variation continues to emerge in most species of agronomic importance (Edye, 1987). From an initial 167 accessions in the mid-1960s, over 8000 accessions have been collected by various organizations so far and are at various stages of characterization and evaluation.

For valuable germplasm to reach the user (i.e. the evaluator and ultimately the farmer) as quickly as possible, data on the collecting site are required. Some evaluators argue that only the most detailed site data should be recorded and that approximations may ultimately be misleading. This school of thought advocates the use by the collector of very detailed environmental descriptors, which are deemed to be indispensable for the accurate selection of ecotypes of forage plants adapted to particular conditions. Others disagree. While recognizing their responsibility for recording site information, many collectors claim that time and other logistical constraints will preclude the noting of all but the most basic site descriptors. Certainly, it is important that what ecological data are gathered be recorded in such a way that they can be used for comparative purposes. Most collectors make some observations on the general area of collection, even if only at the most basic level, e.g. 'forest edge', 'desert grassland' or 'swamp'. Others, perhaps with a greater knowledge of the land, may use such terms as '*Cenchrus–Chrysopogon* grassland', 'xerophilous open woodland' or '*Acacia* savannah'. Where very well-known communities are involved, it is possible to convey a great deal of information by using such local descriptive terms as 'caatinga' (Brazil), 'kunai' (Papua New Guinea) or 'miombo' (Zimbabwe).

It has rarely proved to be feasible in practice to describe in detail the overall community, or conduct a detailed analysis of the vegetation, during the course of collecting. However, associated species are important in determining the ability of the plant to compete successfully in a given environment and are useful guides to the possible companion species in an improved pasture. For example, *Pennisetum clandestinum* and *Trifolium semipilosum* occur together in the highlands of Kenya, and *Neonotonia wighti* and *Panicum maximum* in Zimbabwe. Both combinations have been used in improved pastures in other countries with broadly similar environments.

A description of the climate at the collection site is an important part of the environmental data required for evaluation. Data from the nearest metereological stations need to be interpolated or extrapolated to get an estimate of climatic variables at the site of collecting. This will prove difficult in areas such as steep mountainous terrain, where conditions can vary enormously over short distances. However, in extensive areas with little surface relief (e.g. the Sahelian zone or much of the Amazon Basin), annual rainfall and temperature regimes at any given site can be readily approximated. In such cases estimates with an error of ±25 mm in the annual rainfall total are quite acceptable and have proved invaluable to evaluators.

Of equal importance to basic climatic data is the description of the soil at the collecting site. Unfortunately, most tropical forage collectors have little training in soil science and even such simple descriptions as 'deep sands' or 'cracking clay' are rarely recorded. In an examination of passport data accompanying tropical forage species collected by IBPGR-funded missions, less than 20% of samples had any climatic data (usually annual rainfall) and less than 10% had edaphic data (usually drainage). Ideally, an indication of the soil type and its surface texture should be obtained at the same time as the collector is testing for soil pH. In addition, some indication of the depth to any clay layer is desirable and is readily obtained with a small auger. The nutritional status of the soil is important but relatively difficult to measure without taking samples for laboratory analysis. When this is possible, it should be encouraged as it is extremely useful in evaluation.

Until recently, most tropical pasture collectors were also involved in evaluation of the material they collected. If not engaged in the hands-on initial characterization stage, then at least they were fully aware of where, and for what purpose, the germplasm would be evaluated. As more organizations become involved in plant germplasm collecting, there is a danger of collecting becoming somewhat divorced from use. Species of little value are collected simply 'because they are there' and valuable samples languish in obscurity in gene banks. One possible solution is to link researchers interested in all aspects of tropical pasture germplasm in a network, through which they can communicate their interests and coordinate their activities. An example is the Dryland Pasture and Forage Legume Network, sponsored jointly by the International Center for Agricultural Research in the Dry Areas (ICARDA) and the International Plant Genetic Resources Institute (IPGRI). Various national and international organizations are currently investigating the possibility of a similar structure for tropical forages.

In summary, the successful collecting of tropical forage plants has been characterized by thorough planning, in particular the exploration of the herbarium prior to the landscape. Tropical forage collectors are usually covering new ground both figuratively and literally. They need skills in a wide variety of different fields, from taxonomy to climatology. They also need common sense, however. Wild forage plants are often difficult to find and identify, often few in number at a given site, often morphologically very variable and rarely well studied. The sampling procedures required will depend more on what is available than on sophisticated strategies based on the population structure of the target species. A large number of tropical forage accessions have been accumulated since planned exploration began. However, the fact remains that there are still numerous species and ecotypes that are known about but which remain inadequately collected or not collected at all. Some regions have yet to be explored even in a general way and most habitats need to be examined in further detail. Many useful plants are still to be discovered, if the evidence of the last 30 years is any indication, but collectors are all too often just ahead of the land clearers, and in many cases a long way behind.

References

Allard, R.W. (1970) Population structure and sampling methods. In: Frankel, O.H. and E. Bennett (eds) *Genetic Resources in Plants - Their Exploration and Conservation*. Blackwell Scientific Publications, Oxford.

Edye, L.A. (1987) Potential of *Stylosanthes* for Improving Tropical Grasslands. *Outlook on Agriculture* 16:124–130.

Gramshaw, D., B.C. Pengelly, F.W. Muller, W.A.T. Harding and R.J. Williams (1987) Classification of a collection of the legume *Alysicarpus* using morphological and preliminary agronomic attributes. *Australian Journal of Agricultural Research* 38:355–372.

IBPGR (1990) *1989 Annual Report*. IBPGR, Rome.

Marshall, D.R. (1989) Limitations to the use of collections. In: Brown, A.H.D., O.H. Frankel, D.R. Marshall and J.T. Williams (eds) *The Use of Plant Genetic Resources*. Cambridge University Press, Cambridge.

Marshall, D.R. and A.H.D. Brown (1983) Theory of forage plant collecting. In: McIvor, J.G. and R.A. Bray (eds) *Genetic Resources of Forage Plants*. CSIRO, Melbourne.

Pengelly, B.C. and R. Reid (1988) Collecting forage legumes in Papua New Guinea. *FAO/IBPGR Plant Genetic Resources Newsletter* 73/74, 43–46.

Rattray, J.M. (1960) *The Grass Cover of Africa*. FAO, Rome.

Reid, R. (1983) Pasture plant collecting in Mexico with emphasis on legumes for dry regions. *Australian Plant Introduction Review* 15, 2.

Reid, R. and R.W. Strickland (1983) Forage plant collection in practice. In: McIvor, J.G. and R.A. Bray (eds) *Genetic Resources of Forage Plants*. CSIRO, Melbourne.

Schultze-Kraft, R. and D.C. Giacometti (1978) Genetic resources of forage legumes for the acid, infertile savannas of tropical America. In: Sanchez, P.A. and L.E. Tergas (eds) *Pasture Production in Acid Soils of the Tropics*. CIAT, Cali.

Schultze-Kraft, R., R. Reid, R.J. Williams and L. Coradin (1984) The existing *Stylosanthes* collections. In: Stace, H.M. and L.A. Edye (eds) *The Biology and Agronomy of* Stylosanthes. Division of Tropical Crops and Pastures, CSIRO, Townsville, Queensland, Australia.

Staples, I.B. (1986) Pasture plant collection mission in India. *Australian Plant Introduction Review* 18(2), 1–4.

Strickland, R.W. (1974) Plant collecting mission to Africa 1971–72. *Australian Plant Introduction Review* 19.

Surveying *Mangifera* in the tropical rain forests of southeast Asia

31

J.M. Bompard

Laboratoire de Botanique, Institut Botanique, Université des Sciences et Techniques du Languedoc, 163 rue Auguste Broussonet, 34000 Montpellier, France.

Introduction

To date, genetic improvement of the Indian mango (*Mangifera indica*), a tropical fruit of major importance, has depended on the exploitation of intraspecific variation only. Yet there are some 60 species in the genus, displaying considerable diversity, especially in fruit characters, and occurring in a wide range of environmental conditions and over a large geographic area. Tapping this rich gene pool can be expected to lead to significant progress in mango breeding (Kostermans and Bompard, 1989). However, deforestation is occurring at alarming rates in many areas within the range of the genus, which extends from India to Melanesia and Micronesia. There is thus general agreement on the need for active measures to ensure conservation of the gene pool.

A sound taxonomic base is a prerequisite for any such conservation effort and for the informed use of germplasm. Accurate identification of vouchers and of living material in collections and in the field is the basis of genetic resources work, in *Mangifera* no less than in other groups. Unfortunately, existing taxonomic treatments of the genus are not completely satisfactory (Mukherjee, 1949, 1985; Hou, 1978; Kochummen, 1989). They are based on inadequate herbarium material, the flowers or fruits of several species being unknown. When not entirely lacking, descriptions of fruit characters are generally poor since they are often based on the study of dried herbarium material only. The determination of sterile material is very difficult due to extensive intraspecific variation in vegetative characters, showing intergrading between species in many cases.

Mangifera is not alone in this. The taxonomic treatments of many other fleshy tropical fruits (e.g. the genera *Baccaurea*, *Eugenia* and *Garcinia*) need to be updated or revised. Intensive collecting in poorly explored areas inevitably brings out new material and hence new data on the distribution and range of variation of these taxa.

The IBPGR/IUCN/WWF project

In view of the lack of adequate taxonomic knowledge and of probably rapid genetic erosion of the gene pool, the World Conservation Union (IUCN), World Wide Fund for Nature (WWF) and International Board for Plant Genetic Resources (IBPGR) agreed in 1984 to initiate field surveys of *Mangifera* with emphasis on Borneo and Peninsular Malaysia, the probable areas of maximum diversity. The main objectives of the project were as follows:

*   to draw up an accurate inventory of the *Mangifera* species in the region and their intraspecific variation;
*   to compile ecological and agronomic data on each species (distribution, habitat, morphological and fruit characteristics, actual and potential economic value);
*   to assess the conservation status of each species, and any threats of genetic erosion;
*   to explore the possibilities for *in situ* conservation, identify gaps in the present system of protected areas and recommend measures to ensure the long-term survival of this germplasm.

The emphasis of the project was on collecting herbarium specimens of the flowering and fruiting material necessary to solve taxonomic problems and clarify the taxonomic treatment of the genus. When possible, living material was also to be collected and established in field collections. Mango seeds are recalcitrant, which means that drying and long-term storage at low temperatures are not yet a viable conservation option. At best, the seeds of *M. indica* can be stored for about 100 days (Chin and Roberts, 1980). Conservation thus needs to focus on the establishment of field collections and *in situ* reserves.

Following a preliminary exploratory survey conducted in 1985 in Kalimantan, intensive surveys were carried out in 1986–88 in Kalimantan in cooperation with the Indonesian Institute of Science and the Indonesian Commission on Germplasm and in West Malaysia in cooperation with the Forest Research Institute of Malaysia (Saw, 1987; Bompard, 1988). Surveys were also carried out in Sabah by the Sandakan Forest Research Centre (Lee Ying Fah, 1987) and in cooperation with the Sabah Agriculture Department.

Constraints on collecting

The constraints imposed on collecting by difficult field conditions in inaccessible areas are well known and do not need to be reviewed here, but those inherent to the target species themselves do repay consideration. One such problem is the fact that target trees occur at very low densities in dense forest. Most wild *Mangifera* species found in Borneo and the Malay Peninsula are canopy or emergent trees of the tropical lowland rain forest. They are as a rule large trees, up to 50 m in height. Several species are exploited for their timber. A few species – e.g. *M. gedebe*, *M. griffithii* and *M. parvifolia* (syn. *M. havilandii*) – are gregarious in certain types of swamp forest (with densities of 20 trees per hectare). A couple of species occur in mountain forests between 1000 and 1800 m above sea level. The majority, however, occur as scattered individuals at very low densities in dry lowland forest, where the genus is represented by an average of one to three trees (>40 cm in diameter) per 10 ha. This kind of dispersion clearly means that finding trees is difficult and time-consuming. It also means that defining the population from which one is sampling is almost impossible.

Mangifera species (like those of many other genera in the West Malesian floristic region) flower and fruit very irregularly. Even if a tree is located, therefore, the chances are overwhelming that it will not be in flower or fruit. Mast fruiting at intervals of two to eight years is the dominant pattern. In mast years the ground beneath the trees can be covered with fruits, whose strong smell attracts many animals. This mast flowering can be widespread or restricted to certain areas. The rate of flowering of a few species (e.g. *M. lagenifera* and perhaps also *M. subsessilifolia*) is only once in five to eight years. So far, it has not been possible to collect fertile material of *M. subsessilifolia* despite monitoring of marked trees for four years. Isolated flowering may occur at shorter intervals and is generally followed by poor fruiting. Most of the *Mangifera* trees of wild origin growing in village areas (i.e. in a more open environment) tend to flower more regularly, though they have a flowering habit basically similar to that of trees growing wild in the surrounding forests. Two species (*M. rufocostata* and *M. swintonioides*) have the peculiarity of flowering and fruiting outside the main season.

These constraints highlight the need for intensive preparation, neglecting none of the possible sources of information, and for explorative surveys before undertaking collecting missions.

The preparation and planning of collecting missions

Herbarium data

Mangifera collections were studied in the major world herbaria and in several local herbaria. Herbarium data are valuable first-hand information not generally available in published works, continuously augmented by the addition of newly collected material. Data on distribution, habitat, reproductive phenology and vernacular names were systematically entered into a computer database from more than 2500 herbarium labels.

Herbarium data can be very useful in determining the optimum time for collecting. Unfortunately, the herbarium study was able to provide only limited information on the reproductive phenology of *Mangifera* species. Generally, fertile specimens from a given region from which data on flowering or fruiting time could be gathered were not numerous enough to convey an adequate idea of the optimal date for collecting in that region. In the best cases, they narrowed down the choice to a period of several months.

A thorough review of herbarium collections can also help in becoming familiar with the range of morphological variation present in the target taxa and it is thus a good idea that such a study be carried out by the prospective collectors themselves. Such knowledge is vital if one is to be able to check determinations and eventually assess deficiencies in the current state of taxonomic knowledge, as set out in Floras and monographs. Visiting collections of living material in botanic gardens is also helpful in this connection. A few cases of ecotypic variation could be identified during the course of the herbarium survey. For instance, *M. griffithii* growing in markedly different habitats (dry land or freshwater swamp forests) showed differences in leaf shape and texture.

Data from the literature

A broad knowledge of the proposed survey areas was gained by gathering together ecogeographic information on the target regions from the relevant literature and maps on geology, soils, climate and vegetation. These were used to identify ecological units and areas of particularly high environmental diversity. Information on human diversity, which can be as important a determinant of *Mangifera* diversity as the physical environment, was gathered from the specialized ethnographic literature. These works often contain such data as the vernacular names and local uses of plants.

Data on the occurrence of target species were collated from forest inventories and accounts of previous collecting. The inventories compiled by forestry services are usually of limited use in this context as *Mangifera* species are generally merged together under their generic local timber trade name. More information was found in inventories of forest research plots. Though these are scarce, the records are generally substantiated by herbarium specimens, allowing the verification of determinations and the collecting of additional data from herbarium labels. Permanent forest research plots make it easy to localize particular trees, which are mapped and numbered. They also contribute to a better knowledge of the reproductive phenology of species, as flowering or fruiting times are regularly recorded.

Identification of targets

Despite the gaps in the herbarium and literature data, it was possible to recognize specific regions, areas and taxa as being high priorities for *Mangifera* collecting. Two priority regions were identified. One was

Borneo, especially its Indonesian part (Kalimantan), which was insufficiently explored, and the other was West Malaysia (Peninsular Malaysia), clearly a major centre of diversity. Within these target regions, areas fulfilling the following conditions were selected for detailed study:

- areas likely to have particularly high species diversity or including distinctive ecological conditions (e.g. freshwater swamp forests);
- existing or proposed protected areas (national parks, nature reserves);
- areas insufficiently known from previous collections;
- sites threatened with imminent habitat destruction (e.g. limestone outcrops being exploited as quarries).

The collecting priority accorded to each taxon (whether groups of species, species or infraspecific taxa) must also be clearly defined. Not all the *Mangifera* species required the same intensity of collecting. Moreover, due to practical constraints, it was not always possible to devote the same effort to collecting every representative of the gene pool. The collector may have to decide, for instance, between devoting a few days to assessing the intraspecific variation in a couple of species seen fruiting in local forest gardens along a certain river and going further up-river to explore a forest where a rare wild mango, also fruiting at that time, is reported by local people. A clear understanding of priorities helped in making the right decision when a choice between several alternatives presented itself in the field.

High priority was given to finding insufficiently known taxa, notably little-collected species and those known only from poor material. For instance, the project succeeded in relocating *M. longipetiolata*, a species described more than a century ago from Larut Hill, in Perak, West Malaysia. Fortunately, the area is still forested, as it belongs to the Forest Reserve network of Malaysia. *M. whitmorei*, however, only collected once (in 1971), could not be found again. The original collecting site in the Upper Perak, West Malaysia, has been flooded by the Temenggor Dam and the species was not found in the forest remaining in the vicinity, which is being logged.

In such cases, the objectives are clear in terms both of taxa and of areas, but more generally an area was explored simply on the suspicion of the possible presence of certain species, for example based on information on habitat preferences. Very often during the survey work sites were visited based on no more than a guess as to what species a local informant might be referring to. Sometimes it happened that the guess was correct. It was often more exciting, however, when the guess was wrong. The 'odd wild mango' described by local people might well turn out to be a new species or a new record for the region, or, indeed, not a *Mangifera* at all. To finally arrive at the locality in question, after a few hours' walk, only to learn that the rare tree had been chopped down a few weeks earlier was, sadly, a not uncommon occurrence.

Two-phase collecting

Planning was of necessity usually a two-phase process, in which the determination of collecting areas, itinerary, routes and timing, based on the herbarium and literature work described above, was modified on the spot to take account of the actual situation in the field. The most appropriate collecting itinerary, for example, could often be definitively selected only on the spot on the basis of information that was only available locally and had to be up to date, such as the condition of roads or rivers at that time.

As for determining the optimum time for collection, it has already been mentioned that a period of several months could usually be approximately defined. Peak flowering over most of Kalimantan, for example, occurs from September to December, before the onset of the rainy season, or after a dry spell during the wet season (north of the Equator, in January–February). However, the exact period of flowering and fruiting and its intensity vary greatly from year to year and could not be predicted for a given area unless precise information was provided by local informants.

In some cases, both flowering and fruiting material of the same tree are needed to solve taxonomic problems. A preliminary survey mission during the flowering season can help to predict the ideal time for a second visit during fruiting. Following this procedure, it was possible to collect the fruits of 16 *Mangifera* species during a one-month collecting mission, from trees marked during a preliminary survey. By chance, this collecting trip coincided with a period of mast fruiting. It was even occasionally possible to ask people in the field to monitor certain mango trees and send back information about their phenology or even collect material.

When conventional sources of information proved to be inadequate, a roundabout approach sometimes paid off. For instance, it would have been useless to ask desk officers in forestry or agriculture departments in towns whether wild mango species were likely to be available on the local markets in the hinterland at that time. The same people, however, would be able to say whether the durian season had started and whether that year it was a poor or a good one, the durian (*Durio zibethinus*) being a very popular fruit in southeast Asia. This is a useful clue to the intensity of the fruiting season in a particular area for other forest fruits, including several mango species. People who have just arrived from up-river areas and middlemen trading local fruits were found to be important strategic informants, able to provide information which often proved to be very useful in determining which one of the preselected collecting routes it was best to concentrate on.

The role of indigenous knowledge

In view of the constraints, collecting adequate specimens of trees like *Mangifera* may sound like mostly a matter of luck. If it is actually not as bad as that, it is thanks to the knowledge local people possess of the forest and its products. Traditionally, shifting cultivators do not cut down useful species such as fruit trees or bee-trees when clearing the land, so that most wild mangoes can be found in the secondary forest surrounding settlements. Local people also plant seeds or seedlings collected in the forest in gardens near their homes. A high diversity of mango relatives in old forest gardens may thus partly reflect the diversity of edible wild mangoes occurring in the surrounding forest.

In view of this, surveys were first made in old village forest gardens and secondary forests, collecting as much information as possible from local people along the way. Valuable information was gained by exploring local markets in remote areas, for example. By questioning the stall-keepers, it was possible to trace trees that needed to be collected. In Borneo, no less than 16 species can be bought in local markets, though several are for sale only very occasionally or in limited quantities. The next step was to move on to primary forests, especially protected forest areas. In tall primary forest the crown of mango trees can rarely be seen and the trees must be detected by observing the forest floor closely for fallen leaves, seedlings or rotten seeds and watching for trunk and bark features. Here also, the task of hunting for mangoes can be made somewhat easier by the help of local people. The best informants were local people engaged in logging, hunting or collecting forest products (such as rattan, eaglewood and birds' nests) and such hunter–gatherer groups as the Orang Asli aborigines living in Taman Negara National Park in West Malaysia.

An important facet of indigenous knowledge is vernacular names. These can be extremely helpful. In the case of *Mangifera*, the degree of precision of a name (i.e. whether it refers to a group of species, a single species or a variety) can be a measure of the importance of the trees, as a source of food or in folklore or myth. A vernacular name can also suggest exotic origin (e.g. corrupted names borrowed from a different language or the designation 'Mango from the coast') or confirm local origin (e.g. 'Mango from the forest'). A checklist of the vernacular names collected in a particular area can provide some indication of the diversity of indigenous and introduced mango types known to occur there, though it should never be considered complete.

It is important, however, to be aware of the limitations of vernacular names. A given name will tend to be valid only within a specific area. Also, although local people possess a vast knowledge of the forest and its products, this is not equally shared by the members of a community. Names need to be cross-checked with several reliable informants, especially elderly people. Totally fanciful names (even insults) have been carefully noted down by collectors ignorant of the language and customs of the local people. It is notoriously difficult to agree on the transcription of local names and easy for errors and disagreements in pronunciation and spelling to accumulate in published lists. A basic knowledge of the languages and ethnography of the region is necessary to avoid these and other dangers.

Collecting and documenting germplasm

Having located a tree of a target taxon, herbarium material and, if possible, living material for *ex situ* conservation were collected. The latter consisted mainly of seeds, which can be kept for up to several weeks if cleaned. Whenever feasible, seed collections were sent by special delivery services from Kalimantan to Java. Otherwise, the collecting itinerary was so devised as to include visits to marked trees on the way back, to reduce the length of time living material spent in transit. In the very few cases when living material was especially desired but fruits were not available, budwood was collected. Cuttings were carefully cleaned and wax placed on the cut ends. They were wrapped in wet newspaper and stored in plastic bags kept open for ventilation. Varieties of *M. casturi* thus collected were successfully grafted on *M. indica* cv. Madu at the Kraton collection near Malang in East Java.

Following the requirements of the Indonesian National Commission on Germplasm, all living material was established at Kraton. Seeds were also distributed to the Bogor Botanical Garden, the Cipaku Horticulture Station near Bogor and the National Centre for Research, Science and Technology at Serpong near Jakarta, which keeps a garden of rare plants.

It is essential that collections, whether of herbarium material or germplasm, be adequately documented. The effort of locating rare trees in deep forest at the right moment in their irregular reproductive cycle will largely be wasted if precise field records are not kept. It is equally important for the information to be made widely available. A reference collection of the material and associated data collected during the course of the *Mangifera* surveys has been deposited in the national herbarium involved and duplicates sent to major world herbaria.

Among the information on each sample that was collected in the survey work were:

- geographic data on the locality precise enough to make it possible for future collectors to locate the same tree again;
- notes on the site, including vegetation type, habitat and target species frequency;
- morphological data, in particular detailed descriptions of flowers and fruits, especially features which disappear or change on drying;
- vernacular names and local uses, always with an indication of the language or dialect and the area of validity;

- information on the degree of genetic erosion and actual or potential threats to the habitat.

Floristic information was also gathered in each collecting area, in particular regarding useful hardwood species, other wild crop relatives and so on. This can be extremely useful in providing additional justification for the *in situ* conservation of a particular site.

Results and prospects

Herbarium specimens collected during the surveys have made a significant contribution to taxonomic knowledge of *Mangifera*, helping in the preparation of a monograph (Kostermans and Bompard, 1993). It is worth repeating that a sound taxonomic base, such as can only be provided by this kind of publication, is absolutely essential for the assessment of future use possibilities and for proposing appropriate conservation measures.

The Malay Peninsula, Borneo and Sumatra represent the areas of highest *Mangifera* diversity. The selection of Borneo as a target region proved to be fully justified as more than 20 species, including several new ones, were found there, as against the 11 recorded in the literature before the project. Out of the 30 or so species currently recorded in Borneo and the Malay Peninsula, 26 species (plus hybrids and varieties) were collected during the surveys. Herbarium specimens of 204 numbers were made, about one third from truly wild trees, the rest mainly from semicultivated trees of wild origin. Of the trees collected, 16% were found in flower, 25% with mature fruits, 5% with immature fruits and 54% with neither flowers nor fruits. As regards trees of primary forest, nearly 90% had neither flowers nor fruits at the time of collecting, highlighting again the special problems posed by the reproductive phenology of these species.

Wild species, semicultivated species on the path to domestication and primitive cultivars found in Borneo form an outstandingly rich gene pool which is unique to this region. An important part of such genetic diversity can be explained by the great ethnic diversity of the indigenous inhabitants. Several species, for instance, are 'semicultivated' in Borneo, but wild in West Malaysia. The cultivation of local races (and even species) is often very restricted. Some of these species and forms have direct potential economic value. Species with a peculiar taste might have potential for making juices and flavouring yoghurt, for example. Wild mangoes also offer great possibilities in mango breeding and improvement. The results of these surveys show that wild *Mangifera* species exhibit several desirable characters, such as: resistance to anthracnose, the ability to grow in inundated areas, the ability to grow at high altitude, the absence of fibre, high fruit setting rate and out-of-season fruiting. Wild mangoes also offer great possibilities in mango breeding and improvement (Bompard, 1993).

There is no doubt that this wealth of genetic resources is under threat. Besides the high rate of destruction of lowland forests (the major habitat of *Mangifera* species), there is a high degree of genetic erosion among mangoes occurring in man-made landscapes. Even in remote areas, old mango trees are being cut down and only a few species (*M. caesia*, *M. foetida*, *M. odorata* and *M. pajang*) are regularly replaced.

How is *Mangifera* germplasm in southeast Asia to be conserved? Long-term seed storage is not feasible at present, so *ex situ* conservation would have to involve living collections in field genebanks. There are very few such collections of native fruit trees in Borneo, the most important one that of the Sabah Department of Agriculture at Tenom (Lamb, 1987). There is a special need to establish representative collections of the variation in the species closely related to the Indian mango (e.g. *M. laurina* and *M. pentandra*) and of the primitive cultivars of species of potential economic importance (e.g. *M. casturi*). This material will then be more easily accessible to potential users. However, it might be that wild species established in field gene banks do not fruit for very long periods (much longer than in the wild) even when the edaphic and climatic conditions are more or less similar to those of their original homes. For instance, a mature tree of *M. macrocarpa* flowered only twice and fruited only once in 20 years at the Ulu Dusun Agricultural Station in Sabah and a similar situation is recorded for the same species in the Bogor Botanical Garden. It is also unrealistic to think that the entire range of existing diversity of semiwild and cultivated forms of mangoes found in Borneo can be adequately conserved in traditional living collections.

In situ conservation is thus perhaps a better option. The project has initiated inventories of *Mangifera* species occurring in several protected areas and the conservation status of each species has been assessed. The areas of maximum species diversity and high intraspecific diversity for different species have also been identified in a preliminary way. The bulk of Malesian *Mangifera* species, at least 21 out of 28 native species found in West Malaysia and/or Borneo, are present in protected areas located in these two regions. It should be noted, however, that not all the protected areas have similar legal status or comparable conservation importance (IUCN and UNEP, 1986). *In situ* conservation of wild mangoes requires effective protection of large areas of undisturbed forests. Selecting candidate genetic reserves is difficult due to a general lack of basic floristic data. Data on the occurrence of wild mangoes should ideally be considered together with those on other wild crop relatives (e.g. other important fruit trees, such as *Artocarpus*, *Dimocarpus*, *Durio*, *Garcinia* and *Nephelium*).

The results of the mango project demonstrate that a specialist group focusing on collecting a single-crop gene pool is fully justified, but *in situ* conservation measures will be easier to justify and better perceived by conservation planners once data for several groups of crop relatives are available. It is hoped that the results of the Inventory of Plant

Resources in Kalimantan, a project of the Arnold Arboretum, will contribute to assembling such a comprehensive picture. Considering the wealth of Bornean fruit tree diversity and the high degree of genetic erosion, an action plan for the conservation of Borneo fruit trees genetic resources bringing together all the concerned people – users, conservationists, botanists and foresters – is probably the best way forward.

Much collecting remains to be done. In-depth preparation and good planning are prerequisites for success, though chance and intuition, based on acquired field experience, surely play a part. But it is perhaps only strong motivation that in the end allows the collector to overcome the many difficulties, both practical and scientific, which inevitably arise in the field. It is in many cases a race against time. Plant collectors know that there is always a way to find what they are looking for. Provided that it is still there to be found.

References

Bompard, J.M. (1988) *Wild* Mangifera *Species in Kalimantan (Indonesia) and in Malaysia.* Final report on the 1986–1988 collecting missions. Report to IUCN/WWF/IBPGR.

Bompard, J.M. (1993) The genus *Mangifera* rediscovered: the potential contribution of wild species to mango cultivation. *Acta Horticulturae* 341:69–77.

Chin, H.F. and E.H. Roberts (eds) (1980) *Recalcitrant Crop Seeds.* Tropical Press Sdn. Bhd., Kuala Lumpur.

Hou, D.H. (1978) Anacardiaceae. 4. *Mangifera.* In: van Steenis, C.G.G.J. (ed.) *Flora Malesiana* 1(8):423–440.

IUCN and UNEP (1986) *Review of the Protected Areas System in the Indo-Malayan Realm.* IUCN, Gland.

Kochummen, K.M. (1989) Anacardiaceae. In: Ng, F.S.P. (ed.) *Tree Flora of Malaya. A Manual for Foresters*, 4:9–57. Longman, Petaling Jaya, Malaysia.

Kostermans, A.J.G.H. and J.M. Bompard (1989) Mangoes. In: Siemonsma, J.S. and N. Wulijarni-Soedjipto (eds) *Plant Resources of South-East Asia.* Proceedings of the First PROSEA International Symposium. 22–25 May 1989. Jakarta, Indonesia. PUDOC, Wageningen.

Kostermans, A.J.G.H. and J.M. Bompard (1993) *The Mangoes. Their Botany, Nomenclature, Horticulture and Utilization.* Academic Press, London.

Lamb, A. (1987) The potential of some wild and semi-wild fruit trees in Sabah and the progress made by the Department of Agriculture, Sabah in establishing a germplasm pool. Paper to a Forest Research Institute of Malaysia meeting, Kepong, Malaysia.

Lee Ying Fah (1987) *A Preliminary Survey of* Mangifera *Species in Sabah.* Report to WWF-Malaysia.

Mukherjee, S.K. (1949) A monograph of the genus *Mangifera* L. *Lloydia* 12:73–136.

Mukherjee, S.K. (1985) Mangifera *L.* Systematic and Ecogeographic Studies on Crop Genepools 1. IBPGR, Rome.

Saw, L.G. (1987) *Conservation of the Mango and its Wild Relatives in Peninsular Malaysia.* Report to WWF-Malaysia.

Collecting Andean root and tuber crops (excluding potatoes) in Ecuador

32

R. Castillo[1] and M. Hermann[2]

[1]*Estacion Experimental 'Santa Catalina', INIAP, 14 km, carretera Panamericana al Sur de Quito, Casilla Postal 340, Quito, Ecuador;* [2]*CIP, Apartado 5969, Lima, Peru.*

Introduction

The Andean region has long been recognized as a major centre of crop diversity. In pre-Columbian times, Andean civilizations domesticated some 70 crops, many of which are hardly known outside the region. Beside the potato, these include another eight species with edible underground parts, sometimes referred to as Andean root and tuber crops (ARTs) (Table 32.1). They are widely used by farmers throughout the Andes as subsistence crops, with the surplus production going to rural and urban markets. Ulluco (*Ullucus tuberosus*) and arracacha (*Arracacia xanthorrhiza*) in particular are sold in considerable quantities in markets in Andean countries.

Probably some 40 to 50 million people include ARTs in their diet. They are well adapted to the Andean environment, productive under low-input conditions and highly nutritious. Some ART species and varieties may have potential as novel sources of sugars, as raw material for the production of starch (e.g. achira, *Canna edulis*) and in the food-processing industry (e.g. arracacha). The underground parts and foliage of several species have been successfully tested as animal feeds. High yields can be achieved at comparatively low cost, giving ARTs potential in livestock production. Mashua (*Tropaeolum tuberosum*), for instance, produces on average $60 \, t \, ha^{-1}$ of tubers combining high levels of protein with high carbohydrate content.

Despite all this, the importance of ARTs has sharply declined since colonial times, a process which continues to the present day. When the Spaniards arrived in America, Old World crops such as wheat, barley, broad beans and a large number of vegetables and fruits were introduced and new dietary patterns were taken up by the indigenous population. Today, reasons for the underuse of ARTs include social prejudice against

Table 32.1. Andean root and tuber crops.

Botanical name	Common name[1]	Family	Edible part	Altitudinal range[2] (m)
Ullucus tuberosus	Melloco (ulluco, papalisa)	Basellaceae	Tuber	2060–(3050)–3700
Oxalis tuberosa	Oca (apilla, ibia)	Oxalidaceae	Tuber	2130–(3100)–3900
Tropaeolum tuberosum	Mashua (isañu)	Tropaeolaceae	Tuber	2560–(3240)–3900
Arracacia xanthorrhiza	Zanahoria blanca (arracacha, virraca)	Umbelliferae	Roots	1450–(2510)–3600
Canna edulis	Achira (chiri)	Cannaceae	Rhizome	2200–2800
Polymnia sonchifolia	Jícama (yacón, aricoma)	Compositae	Roots	2040–(2620)–2920
Mirabilis expansa	Miso (chago, mauca)	Nyctaginaceae	Roots	2470–(2610)–2800
Lepidium meyenii[3]	(Maca)	Cruciferae	Hypocotyl-root	3800–4500

[1] Common name in Ecuador. In brackets are given other names used in the Andes.
[2] Range in Ecuador. The average altitude is given in brackets. Data taken from INIAP's collection (Hermann, 1988).
[3] Only known in cultivation around Lake Junín, Central Peru.

their consumption, ignorance of their dietary value and the preference of urban populations for processed food, especially as incomes rise. Moreover, subsidies have eroded the competitiveness of ARTs. This has resulted in genetic erosion in almost all ART species. For example, oca (*Oxalis tuberosa*), once an important subsistence crop in Ecuador, can no longer be found in many areas. In other areas, varieties that were still available in farmers' fields and markets a generation ago have disappeared. Reasons for this include: market demands for a limited number of clones with characteristics desired by urban consumers, changes in diet and phytosanitary problems such as the oca weevil in Peru.

There is probably still time to conserve much of the variation historically available in these crops, but it may soon be too late. Unfortunately, national programmes in the Andean countries cannot give as much attention to ARTs as is necessary. With only modest resources, they struggle to cope with the basic tasks of sustaining work on staples such as potato, maize and beans.

The current status of collections

About 20 years ago concern over genetic erosion in ARTs led a few Andean scientists to start assembling germplasm. Much of this material has been collected with support from the Centro Internacional de la Papa (CIP), the International Board for Plant Genetic Resources (IBPGR) and the International Development Research Centre (IDRC), Canada. As a result, several local and regional collections have been established in Ecuador, Peru and Bolivia. Table 32.2 summarizes current germplasm holdings. This material is predominantly maintained in field collections. The Instituto Nacional de Investigaciones Agropecuarias (INIAP) of Ecuador maintains all accessions both in the field and *in vitro* (Castillo, 1989) and some of the Peruvian and all the Bolivian accessions are duplicated *in vitro* at CIP's field station in Quito for security reasons.

The holdings of the national programme of Peru account for 78% of the total. The diversity of ARTs which can be found in the country is certainly high, but the number of accessions held there is at least partly a consequence of the extensive exchange of material without proper documentation. Research undertaken jointly by the national programme, IBPGR and CIP has shown that duplication within this collection may be as high as 80% (Hermann and del Río, 1989; del Río and Hermann, 1991).

Bolivia, which has perhaps as much diversity as southern Peru, is currently holding only 98 ART accessions, the result of a single recent collecting missions by the Instituto Nacional de Tecnología Agropecuaria (IBTA) jointly undertaken with CIP. Bolivian institutions have lost a major part of their collections (well over 1000 accessions) over recent years. Fortunately, some 150 Bolivian ulluco accessions have been maintained in Finland at the University of Turku, from which duplicates will be repatriated. There are no reported collections from

Table 32.2. Numbers of clonally maintained accessions of Andean roots and tubers maintained by national programmes[1] (1990).

	Ecuador	Peru	Bolivia	Chile	Total
Ullucus tuberosus	210	515	14		739
Oxalis tuberosa	139	1184	47	4	1374
Tropaeolum tuberosum	54	259	18		331
Arracacia xanthorrhiza	78	123	6		207
Canna edulis			7		7
Polymnia sonchifolia	21	39	6		66
Mirabilis expansa	10				10
Lepidium meyenii		38			38
Total	512	2158	98	4	2772

[1]Ecuador: Instituto Nacional de Investigaciones Agropecuarias (INIAP). Peru: Centro de Investigaciones de Cultivos Andinos (CICA) of the Universidad Nacional San Antonio Abad del Cusco; Universidad Nacional de San Cristobal de Huamanga, Ayacucho; Instituto Nacional de Investigaciones Agrarias y Agroindustriales (INIAA), Baños del Inca, Cajamarca. Bolivia: Instituto Nacional de Tecnología Agropecuaria (IBTA), Cochabamba.

Colombia and Argentina and only very few from Chile. Though their diversity of ARTs is rather limited, these countries are nevertheless a high collecting priority in view of the unique traits that can be expected in such material. For example, in Chile oca has in the past been grown as far south as the island of Chiloé (latitude 43°S) and such germplasm is probably long-day adapted, in contrast to oca from the Andes (10°N to 24°S), which requires short days (<13 hours) to form tubers. Also, certain ullucos from Colombia can produce tubers after only four months, which is much less than the seven to eight months taken by material from the Central Andes (Peru and Bolivia).

Collecting missions in Ecuador

As in other parts of the Andes, ARTs in Ecuador are closely associated with native people, who still cultivate a wide variety of species, typically intercropped on small plots of land. For example, oca and mashua are generally intercropped with *Ullucus tuberosus*, the latter being economically the most important species and often grown in monoculture. Initially, collecting missions concentrated on areas with large native populations, such as the provinces of Cotopaxi and Chimborazo in central Ecuador. As evidence for the replacement of ARTs by barley, broad beans, forages and potatoes mounted, the emphasis was shifted to areas where the threat of genetic erosion was estimated to be highest (INIAP, 1985).

Figure 32.1 shows where INIAP has collected ARTs between 1982 and 1988. This material is maintained clonally at INIAP's experimental station south of Quito. Each dot in Fig. 32.1 represents a locality where ART samples have been collected; commonly, several samples are collected at each site. Ecuador comprises a minor part of the Andes and its geography does not pose the formidable barriers which are common in the central Andes, such as deep valleys, 'cordilleras' and sparsely populated plains. Consequently, most of the Ecuadorean highlands (above 2600 m) and valleys descending to the lowlands are comparatively easy to reach. The map shows that INIAP's germplasm is fairly representative of the Ecuadorean highlands in terms of geographic coverage. Some areas, however, still need to be visited, particularly the less accessible eastern highlands as well as some valleys to the west and south which can only be reached by foot or on horseback.

Eighty-one per cent of INIAP's holdings was collected in farmers' fields or stores. Only 9% was collected in markets. The rest is wild material, somaclonal variants and clones selected for breeding purposes. Basic passport data, such as locality, altitude, latitude and longitude, local name, etc., are available for all Ecuadorean ART germplasm in the INIAP collection (Castillo, 1991).

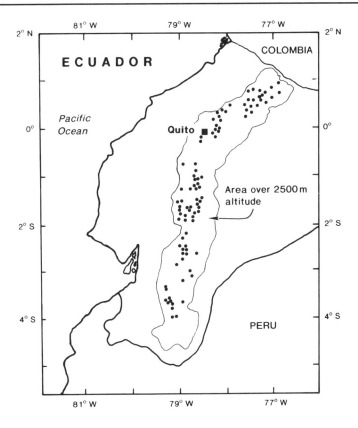

Fig. 32.1. Collection sites of Andean root and tuber crops in the Ecuadorean highlands (Hermann, 1988).

Collecting techniques

With the exception of maca (*Lepidium meyenii*), a biennial propagated by orthodox seeds, all other ARTs are vegetatively propagated perennials. Generally, the edible plant part serves as the propagule, but in the case of *Arracacia xanthorrhiza* and *Polymnia sonchifolia* other parts must be used as propagating material.

All vegetatively propagated ARTs have retained the ability to set functional flowers, but they rarely set viable seed. For example, most accessions of melloco flower abundantly, but seed set under field conditions is extremely rare. Germination in melloco is slow and the factors controlling it are as yet not well understood (Lempiäinen, 1989). In other cases, special procedures, such as scarification, particular germinating substrates and disinfection, are known to be required for germination (Muñoz and Castillo, 1991).

Consequently, only vegetative plant parts, if possible those the farmer uses for planting, are usually collected. In the case of oca, ulluco and mashua, the mature tubers are most appropriate. At ambient temperatures (10–12°C at 3000 m) they can be stored for up to six months. However, mature tubers are available only during a short period, i.e. between harvesting and planting (July–October), so shoot cuttings or immature tubers are sometimes collected. The former must be placed in a rooting substrate within a few days, and the latter tend to be easily bruised and are susceptible to bacterial and fungal decay.

Arracacha is propagated by cormel-like structures, called 'colinos' locally. These are composed of the basal parts of the leaf sheaths, which are inserted on a thick underground vertical stem, or rootstock. 'Colinos' are very similar to the cormels used in the propagation of taro (*Colocasia esculenta*) and can be stored in paper bags at ambient temperatures for up to two months. The roots cannot be used as planting material as they begin to decay very quickly after harvesting, like cassava roots.

Miso (*Mirabilis expansa*) is most easily propagated from stem cuttings, but sexual seeds can also be used. The collecting of achira and jícama (*Polymnia sonchifolia*) does not present major problems. Achira reproduces by rhizomes and jícama by a rootstock which can be taken from the plant at any time of the year.

Traditionally, farmers tend to grow a variety of clones of each species intercropped in the same field. In collecting, the tubers found in a field or store are sorted into morphologically distinct groups and an accession number is assigned to each group. Great care is taken to avoid clonal mixtures under one accession number. The participation of farmers is vital in this, though occasionally their indigenous classification systems do not take into consideration small but genetically determined differences. Occasionally gene-bank curators must split up accessions as such mixtures become apparent during subsequent grow-outs. However, preliminary isozyme studies have shown that material considered and vegetatively maintained as one accession in the germplasm collections of ARTs consists with very few exceptions of just one clone (Hermann, 1988).

The importance of passport data

A meticulous description of the locality may seem superfluous to the local collector, but germplasm may be very widely duplicated and exchanged and such data will be necessary to workers not familiar with the collecting region. The study of the ecogeographic distribution of variation is impossible without accurate passport data, including at least locality and altitude. Passport data will also be needed to re-collect a lost accession. There is some flexibility possible in recording data in the field, however. In one locality a more detailed description of the environment may be required, whereas in another place the mode of use or the agronomic features of germplasm may deserve more detailed comment. Unfortunately, even basic passport data are not available for a major part. of the collections of ARTs. Ideally, herbarium samples should accompany each accession, but again very little herbarium material of ARTs has been collected during germplasm collecting missions in the past. This is a source of particular regret when germplasm collections are lost.

Farmers can provide much information on clones which they may have been growing and observing for decades. An attempt is usually made when collecting ARTs to tap local knowledge on such topics as disease susceptibility, agronomic characteristics, mode of use and market acceptance. Unfortunately, farmers are normally busy people and do not always like being questioned. Certainly, technical jargon and leading questions must be avoided when talking with them. Native farmers in Peru and Bolivia apply folk classification systems to ARTs. These are not 'natural' systems but do provide meaningful descriptions which can complement technical accounts of variation. For example, 'huahuaquepe' is a popular class of melloco in Peru. It gives a strong yellow colour to meals and has rather large tubers much appreciated by consumers. The name is a Quechua word meaning 'carrying a child', an apparent reference to the tendency of this ulluco to form small tuberlets in the eyes of larger tubers, especially on light soils. Unfortunately, the knowledge that forms the basis of such classifications is in many places disappearing along with the crops themselves.

After the collecting mission: conservation and use

All samples are quickly planted in the field for multiplication. When enough planting material is available, the collection is characterized and evaluated under field conditions. Planting is repeated every year. For each accession, plots of about $12 \, m^2$ are required. Harvested material is stored at INIAP for three months in a traditional store at an ambient temperature of 10–12°C.

Such a system of periodic replantings is particularly vulnerable to phytosanitary and climatic problems (Hawkes, 1970). Today, most of INIAP's collections of *Ullucus*, *Tropaeolum* and *Oxalis* are being conserved *in vitro* (Castillo *et al.*, unpublished). The *in vitro* media for these species are based on the standard Murashige and Skoog (1962) medium supplemented with 10 ppm gibberellic acid and 3% sucrose. Good results have been obtained by adding 0.5% activated charcoal to the medium for *Oxalis*. Different concentrations of mannitol and sorbitol have been used in the culture medium to retard the growth of germplasm maintained *in vitro*. The best results have been obtained for *Ullucus* with 4% mannitol and for *Tropaeolum* and *Oxalis* with 6% and 4% sorbitol, respectively. With these additions, and at a temperature of 8°C, Andean tubers can be kept for up to two years before a new micropropagation is necessary (Muñoz, 1988; Tapia, 1991). Tissue culture work is under way with the other species and preliminary results with *Arracacia* are promising (Cevallos, 1991).

Evaluation has led to the identification of elite lines of ARTs with respect to yield, disease resistance and precocity. The Andean Crops Breeding Programme of INIAP is now using 20 elite lines of melloco, for example. Selection and further evaluation are carried out in order to produce new varieties suitable not only for the Andes but also for other tropical highlands.

References

Castillo, R. (1989) Andean crops in Ecuador: collecting, conservation and characterization. *FAO/IBPGR Plant Genetic Resources Newsletter* 77:35–36.

Castillo, R. (1991) Breve análisis sobre recursos fitogenéticos. In: Rios, M. and B. Pedersen (eds) *Las Plantas y el Hombre*. pp. 3–7. Editorial Aby Ayala, Quito.

Castillo, R., Munoz, L. and C. Tapia (unpublished). Conservación de germoplasma usando métodos *in vitro*. I Congreso de Cultivos de Tejidos. Ambato, Ecuador.

Cevallos, A. (1991) Respuesta a la introducción *in vitro* de zanahoria blanca (*Arracacia xanthorrhiza*). BSc thesis. Universidad Técnica de Ambato, Ambato.

del Rio, A. and M. Hermann (1991) Polimorfismo isoenzimático en oca (*Oxalis tuberosa* Molina). VII Congreso Internacional de Cultivos Andinos. La Paz, Bolivia.

Hawkes, J.G. (1970) The conservation of short lived asexually propagated plants. In: Frankel, O.H. and E. Bennet (eds) *Genetic Resources in Plants*. pp. 495–499. Blackwell Scientific Publications, London.

Hermann, M. (1988) *First Progress Report of the IBPGR Project on Andean Tuber Crops*. IBPGR, Rome.

Hermann, M. and A. del Río (1989) Polimorfismo isoenzimático en *Ullucus tuberosus*: su detección e importancia en la conservación de germoplasma. IX Congreso Latinoamericano de Genética. Lima, Peru.

INIAP (1985) *Recolección de Varios Cultivos Andinos en Ecuador*. Final report. INIAP-IBPGR, Quito.

Lempiäinen, T. (1989) Germination of the seeds of ulluco (*Ullucus tuberosus*, Basellaceae). *Economic Botany* 43:456–463.

Muñoz, L. (1988) Respuesta al establecimiento y conservación *in vitro* de melloco, oca y mashua. BSc thesis. Universidad Católica, Quito.

Muñoz, L. and R. Castillo (1991) Pruebas preliminares para la germinación de semilla botánica de oca (*Oxalis tuberosa*) y melloco (*Ullucus tuberosus*). In: Castillo, R., C. Tapia and J. Estrella (eds) *Memorias de la II Reunión Nacional Sobre Recursos Fitogenéticos*. pp. 97–101. Quito, Ecuador.

Murashige, T. and F. Skoog (1962) A revised medium for rapid growth and bio-assays with tobacco tissue cultures. *Physiologia Plantarum* 15:473–497.

Tapia, C. (1991) Conservación *in vitro* de oca (*Oxalis tuberosa*) y mashua (*Tropaeolum tuberosum*). BSc thesis. Universidad Central, Quito.

Collecting the *Musa* gene pool in Papua New Guinea

33

S. Sharrock

CARDI, University of West Indies, PO Box 64, Cave Hill Campus,
St Michael, Barbados.

Introduction

Bananas were one of the earliest crops to be cultivated and both cooking and dessert bananas are still extremely important to the economies of many tropical countries. Though the dessert banana export industry is based on only one cultivar, many hundreds of varieties of bananas and plantains are cultivated worldwide as a subsistence food crop. Over the last few years there has been renewed interest in the collecting, maintenance and use of *Musa* germplasm, particularly as a source of disease resistance. This is due to recent disease outbreaks, such as that of race 4 of Panama disease (*Fusarium oxysporum* f. sp. *cubense*), to which Cavendish clones are susceptible, and the spread of Black Sigatoka disease (*Mycospherella fijiensis* var. *difformis*) into Africa.

At an international workshop held in Cairns, Australia, in 1986 it was recommended that further collecting should take place of the primary gene pool of *Musa* in Papua New Guinea, concentrating on wild material. The workshop also emphasized the dangers inherent in the movement of germplasm, particularly in relation to the banana bunchy top virus (BBTV). Though this is no longer the case, there was at that time no reliable indexing method for BBTV and plants had to be observed for visual symptoms of the disease over several months. In cooperation with the International Network for the Improvement of Bananas and Plantains (INIBAP), the Queensland Department of Primary Industries (QDPI) in Australia agreed to act as the quarantine centre for all *Musa* germplasm collected in Papua New Guinea.

Following the Cairns workshop, a *Musa* germplasm collecting project was initiated in 1987 with the following aims:

- to collect indigenous banana cultivars and wild *Musa* species in Papua New Guinea;
- to gain a better knowledge of the genetic diversity within the wild species;
- to tissue-culture and disease-index the collected germplasm;
- to supply INIBAP with disease-free germplasm for long-term storage and for distribution to banana gene banks and to improvement programmes, including, of course, that of Papua New Guinea.

Planning the germplasm collecting mission

In order to obtain the maximum amount of information in the field, a multidisciplinary team was involved in each mission, typically including a banana taxonomist, a plant pathologist with particular experience of bananas and a person with local knowledge. The selection and coordination of the collecting team was an essential part of the planning process, which involved addressing the issues of:

- which species to collect;
- which areas to visit;
- how to collect and maintain the germplasm.

Target taxa

An understanding of the taxonomy of the target taxa and knowledge of their distribution are essential before undertaking any collecting work. The taxonomy of *Musa* is complicated, but there are numerous works both on the genus as a whole and on its representatives in different regions, including Papua New Guinea (Cheeseman, 1947–50; Simmonds, 1956, 1962; Argent, 1976; Stover and Simmonds, 1987; Shepherd, 1990; Simmonds and Weatherup, 1990).

The family Musaceae comprises two genera, *Ensete* and *Musa*. *Ensete* is considered to be an old genus which probably originated in Asia, spreading from there to Africa (Purseglove, 1985). It consists of six or seven species, which are equally divided between the two continents. One species (*E. glaucum*) is widely distributed through southeast Asia and is found in Papua New Guinea. The genus *Musa* contains approximately 40 species distributed throughout southeast Asia and the Pacific. The centre of diversity and probably of origin of the genus is the Assam–Myanmar–Thailand area (Simmonds, 1962). *Musa* is divided into four sections: *Eumusa*, *Australimusa*, *Callimusa* and *Rhodochlamys*.

Eumusa ($2n = 2x = 22$) is the largest and most diverse section of *Musa*, being distributed from southern India to Japan and Samoa. Almost all edible, cultivated bananas are in this group. Three species are found in Papua New Guinea (*M. acuminata* ssp. *banksii*, *M. schizocarpa* and *M. balbisana*). The section *Australimusa* ($2n = 2x = 20$) contains six or seven species distributed from Queensland in Australia to the Philippines. The Fe'i group of edible bananas (related to *M. maclayi*) belongs here. *M. textilis* (abaca or Manila hemp) is also included in this group. Five sect. *Australimusa* species are found in Papua New Guinea (*M. maclayi*, *M. boman*, *M. peekelii*, *M. bukensis* and *M. lolodensis*).

One other *Musa* species is present in Papua New Guinea. This is *M. ingens*, which has a chromosome number of $2n = 2x = 14$ and is reputed to be the largest known herb (Purseglove, 1985). It was placed in a new section (sect. *Ingentimusa*) by Argent (1976), the validity of which has been questioned by Simmonds and Weatherup (1990). No species from the sections *Callimusa* ($2n = 2x = 20$) and *Rhodochlamys* ($2n = 2x = 22$) have been found in Papua New Guinea. They are distributed from Indochina to Malaysia and from India to Indonesia respectively.

All edible bananas (with the exception of the Fe'i bananas) are derived from *M. acuminata* (A genome) and *M. balbisana* (B genome) in the section *Eumusa*. The centre of diversity of *M. acuminata* is the Malaysia–Thailand area, where four of the five currently recognized subspecies overlap. It is considered that wild-seeded forms of this species may have been used by early fishermen, who used the leaves and leaf-sheaths for fibre and as wrapping material and plates and who may have eaten the male buds and immature fruits. Edibility of the mature fruit of diploid *M. acuminata* came about as a result of two important changes, parthenocarpy and female sterility. As a result of the former, fruit pulp develops without the stimulus of pollination, but should pollination occur the latter ensures that seeds are not formed. These characters would have been deliberately selected for by humans and maintained by vegetative propagation.

Early cultivated bananas were diploids entirely derived from *M. acuminata* and are called AA diploids. Triploid *M. acuminata* (AAA; $2n = 33$) forms also appeared, apparently as a result of hybridizations in which partly sterile edible diploids crossed with male-fertile forms. Such triploids contain 22 chromosomes (AA) from the female gamete and 11 chromosomes (A) from the haploid pollen. Such triploid plants have larger fruits and are more vigorous, more productive and hardier than the diploids. They were selected in preference to the diploids, which they replaced in most areas. One exception to this is Papua New Guinea, where, due to its ancient isolation, AA diploids remain agriculturally important and AAA triploids are less common.

Diploid (AA) and triploid (AAA) cultivars were taken by humans to areas where the wild species *M. balbisiana* was native. In these areas natural hybridizations took place to produce progeny with the genomes AB, AAB and ABB. No edible diploid (BB) form of *M. balbisiana* is known but it is thought that triploid BBB forms may be present in the Philippines (Vakili, 1967; Valmayor *et al.*, 1981). *M. balbisiana* is indigenous to areas with a monsoon climate and a pronounced dry season. The B genome confers drought resistance and hardiness to the diploid and triploid hybrids as well as introducing variation in disease resistance and fruit quality.

Bananas are one of the most important food crops in Papua New Guinea. They are grown throughout the country, including the highlands, up to an altitude of 2000 m above sea level. As mentioned above, the unique feature of banana cultivation in Papua New Guinea is the importance of AA diploids. It is estimated that there may be as many different diploid clones growing there as in the whole of the rest of the world (Stover and Simmonds, 1987). The present situation in Papua New Guinea is considered to resemble that of Malaysia thousands of years ago, when poor-yielding, primitive diploid cultivars predominated. They have now been replaced by more vigorous and productive triploid (AAA) clones.

Banana and plantain improvement programmes are, on the whole, aimed at improving the production of either dessert bananas (AAA triploids) or plantain-type cooking bananas (AAB triploids). Most triploid AAA and AAB clones, however, are almost completely sterile, making banana breeding particularly difficult. Diploid cultivars, being frequently male-fertile and sometimes also female-fertile, are thus of particular importance. If economic pressures should cause changes in traditional farming practices in Papua New Guinea, the diploids may be lost; hence the urgent need for conservation.

The disease Black Sigatoka (*Mycospherella fijiensis* var. *difformis*) is also widespread in Papua New Guinea; indeed, the region including that country and the Solomon Islands is believed by some to be the centre of origin of the fungus (Stover, 1978). If this is the case, Black Sigatoka may have been exerting selection pressure as the predominant leaf spot for a considerable period, resulting in some level of resistance in indigenous cultivars. During the planning stages of the collecting work it was therefore decided to put emphasis on collecting indigenous diploid rather than triploid cultivars.

A national banana germplasm collection was established in Papua New Guinea in the early 1970s. Before carrying out any collecting work a visit was made to this field collection to ascertain how much diversity was already represented in the gene bank. The original collection contained 675 accessions but this had declined to 195 by 1988 (Sharrock, 1990). Most of the accessions that were lost during this period were AA diploids and wild species. Triploids AAA, AAB and ABB were well represented in the collection, which also contained several possible tetraploids. Only two wild species were present (*M. maclayi* and *M. balbisiana*). The visit to this germplasm collection confirmed that, to avoid duplication, the project should concentrate on collecting AA diploid cultivars and the wild *Musa* species.

Target areas

It was first necessary to identify those specific areas where AA diploids were known to be agriculturally important. Two areas, East New Britain and Madang, were initially selected on the basis of published accounts (Stover and Simmonds, 1987), information from the national banana germplasm collection (Shepherd and Ferreira, 1984) and consultations with local agricultural experts. These areas were covered during the first two collecting missions. A range of ecogeographical zones were then visited during subsequent visits. The identification of potential germ- plasm collecting sites was made in collaboration with local Department of Agriculture staff.

A comprehensive study of the wild bananas of Papua New Guinea was carried out by Argent (1976). This includes information on the distribution of each species. In order to ensure that germplasm of all the wild species was collected, areas where each species was known to be pre- sent were visited, as well as some areas not previously covered.

Germplasm collecting and movement

Banana cultivars are vegetatively propagated and have to be collected as suckers. The dangers of moving germplasm in this form are well recognized and guidelines for the safe movement of *Musa* germplasm have been published (Frison and Putter, 1989). For the purposes of this project, the QDPI agreed to provide the facilities for the tissue culturing and disease screening in quarantine of the collected germplasm. The following germplasm introduction procedure was developed:

1. Plants were imported as deleafed suckers and meristems were established in tissue culture to eliminate risks from fungal, bacterial and nematode pathogens.
2. Representative plantlets from each tissue-cultured meristem were grown in the quarantine glasshouse for visual screening for symptoms of BBTV and disease indexing for cucumber mosaic virus (CMV).
3. If screened plants were deemed free of disease, all material derived from that meristem was released from quarantine and *in vitro* cultures were sent to INIBAP.

Banana suckers can only be stored for a limited period of time after collecting, particularly in hot, humid conditions. To overcome this pro- blem, it was decided that the collecting work should be carried out in a series of short missions, each one lasting for no longer than one month, so that suckers could be returned to the tissue culture laboratory while still in good condition.

Wild species can be collected as seed rather than suckers and are for this reason easier to deal with from a quarantine point of view. Seed samples are also easier to carry and store during collecting missions and it was decided to collect wild species as seed whenever possible. No bacterial or fungal pathogens are known to be seed-borne in *Musa*, but the situation regarding virus diseases is less clear. It is possible that CMV may be seed-borne (Gold, 1972) and all plants grown from seeds were therefore indexed for this disease. There are no reports to date of BBTV or any other virus disease being transmitted via seed.

In the field

Wild bananas typically appear as early colonizers of cleared forests. Roadsides and the edges of cultivated land are thus ideal habitats, making these species relatively easy to find and collect. The only exception is *M. ingens*, which grows not only at a higher altitude than any other *Musa* species but also well within forested areas rather than on their edges.

Slightly different sampling strategies must be used depending on the breeding system of the species involved. Species that were known to be self-pollinating (i.e. those that have hermaphrodite basal flowers, such as *M. schizocarpa*) could be safely collected as seed as they breed true to type. Populations of such species consist of individuals which are more or less homozygous. Seeds from any one individual will give progeny that are very similar to each other and to the parent plant. To collect maximum diversity, small seed samples were taken from a number of individuals rather than large samples from any one plant. Many *Musa* species are cross-pollinated, however, with their basal 'hands' functionally female. Species such as *M. balbisiana* will produce variable progeny when grown from seed. Therefore, if a particular plant was seen with a desirable or uncommon characteristic, suckers were collected in addition to a population sample of seed, so as to ensure the preservation of that particular genotype.

Where the ranges of two species overlap, hybrids may be formed, as is the case with *M. acuminata* ssp. *banksii* and *M. schizocarpa* in Papua New Guinea. Such hybrids are generally sterile, producing poorly developed seed, and had to be collected as suckers. However, embryo culture techniques can be used to rescue embryos from poorly developed seeds which would not normally germinate, so both seeds and suckers were collected to enable the parent and its offspring to be compared during characterization. In areas where hybrids grew, care was taken when collecting each parent species to ensure that seeds were taken from plants well isolated from the other parent, so as to avoid collecting only seeds of hybrid origin.

While wild species can be collected as either seeds or suckers, cultivated types have to be collected as suckers. A good starting place was often the local market. Here it was possible to form an idea of the range of varieties being grown in the area, and discussions with local producers helped to identify places where these cultivars could be collected. The assistance of provincial Department of Agriculture extension officers was often essential, as they had extensive knowledge of the area and were able to arrange visits to farmers known to cultivate a range of varieties. Farmers were also more likely to cooperate in donating suckers if the collecting team included someone they knew.

One of the main difficulties in collecting cultivated *Musa* is distinguishing diploids and triploids. There are, of course, published identification aids, but including an experienced banana taxonomist in the collecting team mitigated this problem. The recognition of synonyms of vernacular names is more complicated. In a country such as Papua New Guinea, which has as many as 700 different languages, the local name for the same banana cultivar often changes from one village to the next. When this is compounded by phenotypic variation due to differences in growing environments and with the local traditions of covering bunches and removing male buds, it is easy for duplicates to be collected. In an attempt to minimize this, detailed descriptions were made and photographs taken of the plant as it was collected. These were then used in subsequent collecting missions. The knowledge of local banana varieties of farmers and local agricultural extension staff also proved useful in avoiding collecting duplicates.

Handling and care of collected material

Suckers were collected only from apparently healthy plants, with a minimum of damage being inflicted on the parent plant. Suckers had to be at least 10 cm in diameter at the base and ideally with little or no weevil borer (*Cosmopolites sordidus*) damage to the corm. Larger suckers were cut back before transportation but smaller suckers were avoided unless the period of storage was to be no more than a few days. Whenever possible, at least three suckers were collected per plant.

Once removed from the parent plant, the sucker was labelled and cleaned and any damaged outer layers of tissue were removed. The upper portion of the sucker was cut back to approximately 20 cm above the meristem and the corm trimmed to approximately 10 cm below the meristem. The cleaned and trimmed sucker was then allowed to dry before being wrapped in newspaper for storage. Suckers prepared in this way could be stored for up to three weeks, or longer if kept in cool, dry conditions.

One of the main problems encountered in the storage of suckers was rotting due to weevil borer damage. It was frequently difficult to obtain clean suckers and often what appeared to be relatively minor damage on the outside of the sucker was actually considerably worse towards the centre of the corm. Because of this, whenever possible during the collecting missions suckers were sent back to the tissue culture laboratory by air freight so as to minimize the period of storage.

Banana fruits contain from 30 to more than 150 seeds each, which must be mature (i.e. hard and black) when collected. The seeds mature before the fruit, so immature fruit may contain seeds suitable for collecting. Seeds are easier to store and carry than suckers and can be kept fresh inside the fruit after collecting if desired. Banana seeds exhibit dormancy and, although they germinate readily when freshly harvested, if dried they may take two to six months to germinate. If seeds had to be extracted from the fruit before returning to base, they were carefully cleaned, which was most easily done when the fruits were very ripe. The seeds were then kept moist if the period of storage was to be no more than one or two weeks. If a longer period of storage is required, seeds must be dried. The moisture content of banana seeds can be reduced to around 10% by drying at 20°C for three days in an air-conditioned room. At this moisture content they will remain viable for about three months at an ambient air temperature of around 22°C. For longer-term storage, the moisture content and temperature of storage must be reduced (see below).

It is possible in bananas to use the meristem contained within the male bud as an explant for initiating tissue cultures. In situations where neither seeds nor suckers were available, the male bud was therefore collected. It is not possible to store the male bud for more than a few days, but if it could be returned to the tissue culture laboratory within that time it was a viable alternative to suckers.

Data collecting

Data collected in the field were of crucial importance, particularly for the identification of duplicates, as already mentioned. Preprinted forms were taken to the field to record details of each sample. In addition to recording the usual passport data (e.g. location, altitude, collecting date, etc.), descriptions were made of the plant as a whole (size, colour, amount of suckering) and of the bunches (number of hands, angle of bunch, size of fruit; presence/absence, size, colour and shape of male bud; colour of male flower and presence of pollen). Distinctive features and the presence (or absence) of pests and diseases were also recorded. Local names and uses were documented from interviews with farmers.

This was not considered a definitive description of the material and absence of a disease or pest that was locally prevalent was not taken as necessarily indicative of resistance. Growing conditions can have a significant effect on the expression of many characters in bananas, and detailed descriptions would be carried out later, with the plants growing under experimental conditions.

Back at base

On arriving back at base the suckers were established in tissue culture as soon as possible. Full details of the methods used in the shoot-tip culture of *Musa* are given elsewhere (Cronauer and Krikorian, 1984; Vulysteke, 1989). Many accessions collected during the first mission were lost due to contamination of the cultures. As a result, a double surface-sterilization procedure was developed. The culture medium used was a basic Murashige and Skoog (1962) medium supplemented with $2.5 \text{ mg} \, l^{-1}$ 6-benzylaminopurine, $20 \text{ g} \, l^{-1}$ sucrose and $8 \text{ g} \, l^{-1}$ agar.

During the first few weeks in culture, blackening sometimes occurs

at the cut surfaces of the corm due to the oxidation of phenolic compounds in the wounded tissues. The extent of blackening varies with
species and cultivar but it is severe in cultures of *M. balbisiana* and ABB
cultivars. It was also found to be a problem with the sect. *Australimusa*
species and Fe'i cultivars. When blackening occurred the blackened
tissue had to be cut away and the explant transferred to fresh medium
regularly, possibly every few days, until the blackening was reduced. If
blackening was particularly severe explants were pretreated by immersion in a sterile solution of an antioxidant (e.g. cysteine, ascorbic acid
or citric acid) prior to inoculation in the medium, or an antioxidant was
added to the medium (Vuylesteke, 1989).

The method for initiating shoot-tip cultures recommends a double
surface-sterilization treatment and a relatively large final explant
(approximately 1×2 cm). The reason for this was that frequently only
one or two suckers per accession were available and losses due to contamination in culture had to be minimized. Having a large explant meant
that, should contamination occur, the explant could be subjected to a
further surface-sterilization treatment. Secondly, if blackening occurred
it was possible to remove the outer blackened layers of tissue without
damage to the meristem.

Seeds were dried in preparation for storage and a sample germinated
for characterization. If seeds were brought back to base in the fruit, they
were cleaned before drying. Because the base was outside the country
of collecting, the fruit was under quarantine restrictions due to the
possibility of introducing foreign types of fruit fly. To prevent this happening the fruit was immersed in an insecticide solution before the pulp
was removed from the seeds. After cleaning, seeds were dried, ideally to
6–8% moisture content, which can be done at 15°C and 15% relative
humidity or in an air-conditioned room at 20°C.

Seeds could then be stored at subzero temperatures in sealed,
vapour-proof containers (Kar Ling Tao, pers. comm.). The viability of
banana seeds in long-term storage is being investigated by Professor
Chin at the University Pertanian in Malaysia. It is known that they can
remain viable for at least one to two years under the above conditions
and cryogenic storage of seeds or embryos may be a possibility in the
future. Fresh seeds will germinate readily, but after drying the seeds
become dormant and may take two to six months to germinate at
ambient temperatures. However, seeds exposed to alternating temperatures of 35/20°C germinate more readily.

Embryo cultures were mainly used to 'rescue' embryos of hybrid
origin which would not germinate under normal conditions, but tissue
cultures of accessions with viable seeds were also required. Banana seeds
are small, generally 4–6 mm in diameter, although size does vary with
species, and the outer layer is very hard. The structure of seeds is similar
in all species. At one end of the seed is the micropyle, below which lies
the embryo, embedded in the endosperm. Below the embryo, at the other
end of the seed is the chalazal mass. For embryo culture, the seeds were

split open without damaging the embryos, which were then removed and placed on the culture medium using sterile techniques. The embryos, which are less than 1 mm in length, were removed with the help of a stereomicroscope. The medium used for embryo culture was the same as that used for shoot-tip culture. Germination of the embryos usually took place within one to two weeks. Cultures originating from different seeds were labelled separately, even though the seeds may have originated from the same plant, to take account of variation within the seed lot. More details of the embryo culture of bananas are given by Escalant and Teisson (1987).

Report writing

After each collecting mission a report was written giving background information on the collecting mission and details of what was collected and where. A map indicating the germplasm collecting sites and the distribution of the wild species collected was always included. A summary of the diversity in different areas was provided. The reports also highlighted the incidence and severity of pests and diseases encountered. Brief details of each sample were included as an appendix to the report. An example of a report written after a *Musa* collecting mission to Papua New Guinea is given by Sharrock (1990). Reports were circulated to the International Board for Plant Genetic Resources (IBPGR) and INIBAP as well as to other interested parties both within Papua New Guinea and elsewhere.

Results of the collecting work

A total of four collecting missions were made to Papua New Guinea during 1988–89. Two hundred and sixty accessions were collected, of which 116 have been tentatively classified as AA diploids, 51 accessions came from nine of the ten wild species present in Papua New Guinea and the remaining 93 accessions comprise a mixture of triploids, tetraploids and Fe'i cultivars. All these accessions have been established in tissue culture and representative plantlets screened for diseases in quarantine. Cultures of disease-free accessions have also been sent to the *in vitro* laboratory of INIBAP for storage and distribution, including return to Papua New Guinea for inclusion in the national banana germplasm collection.

Triploid clones, many introduced from other countries, are found growing close to houses and along roadsides and are not frequently replanted. In contrast, diploid bananas are planted in mixed food gardens, which may be located some distance from houses. These food gardens are replanted after each harvest and the diploid bananas are as a result treated more or less as an annual crop. This system does not allow diseases and pests to build up. The greatest diversity in diploid bananas is to be found in areas such as East New Britain, Madang and Kiunga in Western Province, where bananas form the staple food crop. In seasonally dry areas banana production is based on triploid ABB cultivars, which are more drought resistant than AA diploids. In the highland areas most of the banana cultivars are AAB triploids, unusual in their erect leaf habit.

The most widely distributed wild *Musa* species in Papua New Guinea are *M. acuminata* ssp. *banksii* and *M. schizocarpa*. These are probably the ancestral parents of many of the traditional Papua New Guinea cultivars. Variation within these species is limited, although some differences in bract colour and finger shape do occur. Natural hybridization between the species does occur but no signs of introgression with either parent have been observed. A few unknown hybrids were collected. These may have originated from crosses between wild species and cultivated diploids, the most likely parents being a male-fertile diploid and *M. acuminata* subsp. *banksii*, which is known to sometimes lack functional stamens in its basal hands and does hybridize with *M. schizocarpa*. Such hybrids were rare even in areas where wild species were growing in close proximity to cultivated diploids.

Five species from sect. *Australimusa* are found in Papua New Guinea (*M. maclayi*, *M. peekelii*, *M. lolodensis*, *M. boman* and *M. bukensis*). All of these were collected except *M. bukensis*, which is only recorded from the island of Bougainville, where, due to the political situation at that time, it was not possible to collect. A number of Fe'i cultivars were collected from various locations throughout Papua New Guinea. These are not a popular food source but are kept as a backup for when other food is scarce. They tend to be vigorous plants resistant to disease and require little attention. *M. ingens* and *Ensete glaucum* were also collected.

References

Argent, G.C.G. (1976) The wild bananas of Papua New Guinea. *Notes from the Royal Botanic Garden Edinburgh* 35:77–114.

Cheeseman, E.E. (1947–50) The classification of the bananas. *Kew Bulletin* 1947:97–117; 1948:11–28, 145–157, 323–328; 1949:23–28, 133–137, 265–272, 445–452; 1950:27–31, 151–155.

Cronauer, S.S. and A.D. Krikorian (1984) Multiplication of *Musa* from excised stem tips. *Annals of Botany* 53:321–328.

Escalant, J.V. and C. Teisson (1987) Comportements *in-vitro* de l'embryon isolé du bananier (*Musa* species). *Fruits* 42:33–342.

Frison, E.A. and C.A.J. Putter (eds) (1989) *FAO/IBPGR Technical Guidelines for the Safe Movement of* Musa *Germplasm*. FAO/IBPGR, Rome.

Gold, A.H. (1972) Seed transmission of banana viruses. *Phytopathology* 62:760.

Murashige, T. and F. Skoog (1962) A revised medium for rapid growth and bio-assays with tobacco tissue cultures. *Physiologia Plantarum* 15:473–497.

Purseglove, J.W. (1985) *Tropical Crops – Monocotyledons*. Longman, London.

Sharrock S.L. (1990) Collecting *Musa* in Papua New Guinea. In: Jarret, R.L. (ed.) *Identification of Genetic Diversity in the Genus* Musa. Workshop Proceedings. 5–10 September 1988. Los Baños, Philippines. pp. 140–157. IRRI, Los Baños.

Shepherd, K. (1990) Observations on *Musa* taxonomy. In: Jarret, R.L. (ed.) *Identification of Genetic Diversity in the Genus* Musa. Workshop Proceedings. 5–10 September 1988. Los Baños, Philippines. pp. 158–165. IRRI, Los Baños.

Shepherd, K. and F.R. Ferreira (1984) The Papua New Guinea Biological Foundation's banana collection at Laloki, Port Moresby, Papua New Guinea. *IBPGR/ Southeast Asia Newsletter* 8:28–34.

Simmonds, N.W. (1954) Notes on banana varieties in Hawaii. *Pacific Science* 8:226–229.

Simmonds, N.W. (1956) Botanical results of the banana collecting expedition, 1954–5. *Kew Bulletin* 1956:463–489.

Simmonds, N.W. (1962) *The Evolution of the Bananas*. Longman, London.

Simmonds N.W. and T.C. Weatherup (1990) Numerical taxonomy of the wild bananas (*Musa*). *New Phytologist* 115:567–571.

Stover, R.H. (1978) The distribution and probable origin of *Mycosphaerella fijiensis* in South East Asia. *Tropical Agriculture (Trinidad)* 55:65–68.

Stover, R.H. and N.W. Simmonds (1987) *Bananas*. Longman, London.

Vakili, N.G. (1967) The experimental formation of polyploidy and its effect in the genus *Musa. American Journal of Botany* 54:24–36.

Valmayor, R.V., F.N. Rivera and F.M. Lomuljo (1981) *Philippine Banana Cultivar Names and Synonyms*. IPB Bulletin No. 3. National Plant Genetic Resources Laboratory, Institute of Plant Breeding, University of the Philippines, Los Baños.

Vuylysteke, D.R. (1989) *Shoot Tip Culture for the Propagation, Conservation and Exchange of* Musa *Germplasm*. IITA, Ibadan, Nigeria.

Collecting the rice gene pool

34

D.A. Vaughan[1] and T.T. Chang[2]

[1]NIAR, 2-1-2 Kannondai, Tsukuba, Ibaraki, Japan-305: [2]IRRI,
PO Box 933, Manila, Philippines.

The rice gene pool

The grass genus *Oryza* consists of about 20 tropical and subtropical wild species and two cultigens, African (*O. glaberrima*) and Asian (*O. sativa*) rice (Chang, 1985b; Vaughan, 1989). The genus is in the subfamily Oryzoideae, which comprises 12 or 13 genera closely related to the bamboos (Duistermaat, 1987). There are no other major cereals that are taxonomically closely allied to the rices but, in the same subfamily, *Zizania aquatica* is harvested for grain in North America and *Z. latifolia* is used as a vegetable in east Asia (de Wet and Oelke, 1978).

The gene pool concept (Harlan and de Wet, 1971) is helpful in suggesting limits to the group of related species that should be of concern to rice breeders and therefore to rice gene banks (Chang and Vaughan, 1991). Because it is never possible to entirely predict what traits will be needed in the future, it is vital to conserve as broad a range of genetic diversity as possible. For example, ten years ago it could not have been foreseen that it would be necessary in the 1990s to find germplasm resistant to the golden snail (*Pomacea canaliculata*), hence the current interest in the conservation of at least some samples even of species in genera relatively distantly related to *Oryza*.

Despite numerous reports of remote hybridization involving rice (e.g. Yieh, 1964; Zhou *et al.*, 1981; Zhou, 1986), the main focus of concern for the rice breeder is the subfamily Oryzoideae. Characters found in other genera in the Oryzoideae but not in any *Oryza* species include seed survival in cold water in *Zizania*, tolerance to salt water in *Porteresia* and unisexual florets in a number of genera. However, most evaluation efforts involving wild species have centred on the genus *Oryza* itself. Hybridization between cultivated *O. sativa* and most other species in the genus has been successful. Table 34.1 lists some of the potentially useful

659

Table 34.1. Species in the genus *Oryza* and some useful attributes (expanded from Vaughan and Sitch, 1991).

Taxa	Genome group	Distribution	Number of distinct accessions in the base collection at IRRI	Examples of traits being used or of potential use in rice improvement
O. schlechteri	Tetraploid	Irian Jaya, Indonesia and Papua New Guinea	1	Stoloniferous
O. brachyantha	FF	Africa	11	Stem borer resistance
O. ridleyi complex				
O. longiglumis	Tetraploid	Irian Jaya, Indonesia and Papua New Guinea	5	Blast resistance
O. ridleyi	Tetraploid	Southeast Asia	11	Stem borer resistance
O. meyeriana complex				
O. granulata	Diploid	South and Southeast Asia	7	Shade tolerance
O. meyeriana	Diploid	Southeast Asia	7	Shade tolerance
O. officinalis complex				
O. officinalis	CC	Tropical Asia	155	Multiple pest resistance
O. rhizomatis	CC	Sri Lanka	18	Drought resistance
O. eichingeri	CC	Sri Lanka and Africa	18	Multiple pest resistance

O. *minuta*	BBCC	Philippines	47	Blast resistance
O. *malampuzhaensis*	BBCC	Southern India	5	Shade tolerance
O. *punctata*	BB	Africa	29	Multiple pest resistance
O. *schweinfurthiana*	BBCC	Africa	10	Mutiple pest resistance
O. *latifolia*	CCDD	Latin America	33	Tungro resistance
O. *alta*	CCDD	Latin America	10	High biomass production
O. *grandiglumis*	CCDD	South America	8	High biomass production
O. *australiensis*	EE	Australia	25	Drought resistance
O. sativa complex (equivalent to primary gene pool)				
O. *glaberrima*	A^gA^g	West Africa (mainly)	2397	Cultigen
O. *barthii*	A^gA^g	Africa	147	Drought avoidance
O. *longistaminata*	A^lA^l	Africa	108	High pollen production
O. *sativa*	AA	Asia originally, now worldwide	74,357	Cultigen
O. *nivara*	AA	Tropical Asia	318	Grassy stunt virus resistance
O. *rufipogon*	AA	Tropical Asia/Australia	494	Bacterial leaf blight resistance
O. *meridionalis*	AA	Tropical Australia	39	Drought avoidance
O. *glumaepatula*	AA	Central and South America	32	Deep-water rice

characters of each *Oryza* species, along with the present number of samples at the International Rice Research Institute (IRRI) gene bank, which serves as a repository of a base collection for the cultivated rices and their wild relatives.

Past collecting and the present status of collections

Cultivated rices

The collecting, study and exchange of rice germplasm has a long history. Perhaps the first large-scale introduction was that of Champa varieties from Vietnam to China's Yangtze Delta in AD 1012 by Emperor Chen-Tsung. This stimulated double-cropping of rice (Ho, 1956) and has been called the first rice 'green revolution' (Bray, 1984). One of the earliest systematic collecting efforts was undertaken in India. Watt (1891) reports examining 4000 samples of rice from Bengal on the occasion of an international exhibition in Calcutta. This and other early collecting efforts were geared towards either rice improvement or varietal classification and not conservation. Early rice breeding involved pure-line selection of the 'best' traditional varieties (Parthasarathy, 1972).

Collecting for conservation and use started in the 1950s and is well documented elsewhere (Chang, 1970, 1975; Oka, 1977; Ng *et al.*, 1983; Bezançon and Second, 1984). Collaborative efforts to collect traditional rices on an unprecedented scale in Asia and Africa were stimulated by the 1971 Rice Breeding Symposium held at IRRI (IRRI, 1972; Chang, 1985b) and lasted through the 1970s. The 1980s saw a decline in collecting, however, though extensive plans were made at the 1983 Rice Germplasm Conservation Workshop (IRRI and IBPGR, 1983). This was because of lack of funding, which is surprising given that a high return on investment from this kind of work has been demonstrated (Evenson, 1989; IRRI and IBPGR, 1991).

The present status of the main rice germplasm collections held at national and regional centres is summarized in IRRI and IBPGR (1991). There are about 300,000 rice accessions stored in gene banks worldwide. This may seem more than sufficient, but such a raw number can be a misleading indication of the diversity conserved because of duplication within and among collections as well as repeated collecting from easily accessible locations. Probably about 75% of the approximately 100,000 cultivars of *O. sativa* once grown worldwide (Chang, 1984) are in the base collection at IRRI. Gaps remain with regard to minor cultivars and remote areas. Some re-collecting is also necessary in cases where unduplicated stocks of earlier collections have been lost due to inadequate facilities. New gene banks in rice-growing countries such as Bangladesh, Myanmar, the Philippines and Sri Lanka have stimulated an interest in re-collecting traditional rices, where it is still possible, in these countries.

Wild relatives

Early collecting of wild relatives of rice was carried out by: the National Institute of Genetics, Mishima, Japan; the Central Rice Research Institute, Cuttack, India; and France's Institut de Recherches Agronomiques Tropicales et des Cultures Vivrières (IRAT) and Office de la Recherche Scientifique et Technique Outre-Mer (ORSTOM, now the Institut Français de Recherche Scientifique pour le Développement en Coopération) (Chang, 1975). Wild rices were surveyed and then intensively collected in China in the 1970s and early 1980s (Kwangtung Agricultural and Forestry College, 1975; Cooperative Team of Wild Rice Resources Survey and Exploration of China, 1984). In the 1980s, it became apparent that a broad and coordinated plan was necessary, particularly in view of increasing requests for germplasm for wide hybridization, evolutionary and other studies. Not only were there very few accessions of wild rices in gene banks, but many samples were unknowingly duplicated within the same gene bank since they had been received separately from different sources without clear passport information (IBPGR–IRRI Rice Advisory Committee, 1982; Oka, 1983; Sharma, 1983; Sharma *et al.*, 1988). Consequently, from 1987 to 1990, 13 countries participated in a collecting programme for wild relatives of rice in Asia and the Pacific coordinated by IRRI (Vaughan, 1991b).

The status of wild rice collections at the beginning of the 1990s has been reviewed by Ng *et al.* (1991) and Vaughan *et al.* (1991). There is a need for further duplication of wild rice germplasm. Most of the wild rice accessions conserved at the National Institute of Genetics in Japan, at the International Institute of Tropical Agriculture (IITA) in Nigeria, at the Rice Research Institute in Thailand, at the Central Rice Research Institute in India and at IRAT and ORSTOM in France are also in the IRRI gene bank, but some other smaller national collections are not well duplicated.

Further collecting is also needed. Based on a survey of *Oryza* specimens in the world's major herbaria, there are 24 countries in Africa and Latin America where wild relatives of rice have been recorded but from which no material is present in the rice germplasm base collection at IRRI (Table 34.2). Clearly, collecting this germplasm is a priority, especially in view of the speed with which habitats are being destroyed in many tropical areas.

Collecting rice germplasm

Target regions and taxa

A necessary first step in planning any systematic collecting programme is to set explicit priorities for regions and taxa. This will be even more crucial if the programme is to be an extensive international effort. For rice as for other species, broad consultation at international conferences and workshops has proved useful in achieving consensus on priorities

Table 34.2. Countries where wild relatives of rice grow[1] and accessions are not present in the base collection at IRRI.

Angola	*O. longistaminata, O. punctata*
Belize	*O. alta, O. latifolia*
Bolivia	*O. latifolia, O. glumaepatula, O. grandiglumis*
Burundi	*O. longistaminata*
Central African Republic	*O. brachyantha, O. longistaminata, O. barthii*
Dominican Republic	*O. glumaepatula, O. latifolia*
Ecuador	*O. latifolia, O. grandiglumis*
El Salvador	*O. latifolia*
Gabon	*O. longistaminata*
Guinea-Bissau	*O. barthii, O. brachyantha*
Haiti	*O. latifolia*
Honduras	*O. latifolia, O. glumaepatula*
Martinique	*O. longistaminata*[2]
Mauritania	*O. barthii*
Mozambique	*O. longistaminata, O. punctata*
Namibia	*O. longistaminata, O. barthii*
Peru	*O. latifolia, O. grandiglumis*
Puerto Rico	*O. latifolia*
Rwanda	*O. longistaminata*
Somalia	*O. longistaminata*
South Africa	*O. longistaminata*
Swaziland	*O. punctata*
Trinidad	*O. latifolia*
USA	*O. nivara*

[1] Based on herbarium records.
[2] Apparently a recent introduction; identification based on specimen in US National Herbarium.

prior to implementing a collaborative collecting programme. For example, specific recommendations were made during the 1990 Rice Germplasm Workshop for the collecting of both cultivated and wild material (IRRI and IBPGR, 1991). It is then possible to develop appropriate collaborative linkages and to seek the necessary funds. The training of collectors in national programmes has been an integral component of this approach. Further details are given by Chang *et al.* (1977) and Chang (1985a).

Given limited financial resources, the relative merits of single- versus multiple-objective collecting have had to be carefully considered at the planning stage. As an important, widespread and diverse crop, it has not been difficult to justify collecting only cultivated rice during a mission. Looking for a number of wild species at the same time as the cultigen can result in incomplete coverage of both. Considerable time may be spent searching for small, isolated populations of wild rices in difficult terrain. Collecting wild rice germplasm has often involved covering great distances between sampling sites, whereas collecting the cultigen has been carried out by methodical village-to-village surveying, including interviews with farmers in each village. Wild and weedy rices of disturbed habitats can be collected in the same places and at the same time as the cultigen, but even then it is populations growing away from cultivated material that may be genetically most useful, rather than ones highly introgressed with the cultigen. Finally, local contacts tend to be forestry or wildlife department personnel when collecting wild rice but agricultural extension officers when collecting cultivated material. In summary, it is only really possible to comprehensively collect both wild and cultivated rices in any given area by greatly extending the length of the collecting trip.

The problem does not arise if wild species and cultigen are not found together. A number of species of wild rice occur in areas where rice is not cultivated, for example the lowlands of Papua New Guinea, northern Australia and the Amazon basin. In rice-producing areas, some wild relatives occur in habitats where rice is not cultivated, such as the well-developed secondary forests where the shade-loving species of the *O. ridleyi* and the *O. meyeriana* complexes occur (Fig. 34.1). Protected areas such as forest reserves or national parks, where no farming is allowed, are also important sources of wild germplasm, sometimes the only areas where a particular species may be found (Vaughan, 1990; Vaughan and Chang, 1993). Collecting specifically targeted on wild relatives has been necessary in such cases.

The number of samples obtained during a mission has sometimes been used as a measure of the efficiency or success of a trip. However, the quality and the genetic potential of the samples are also factors to consider when proposing to embark on collecting, and when summarizing its results. When writing a project proposal, it has been found worthwhile to indicate the expected number of samples to be collected during the trip and the relative importance of the material being sought. For example, there are many reports of wild rice in southern Vietnam, Laos and south and southeast China, but only very few from northern Vietnam, where only one herbarium specimen of a wild rice of the AA genome has been collected (Dao The Tuan, 1985). Consequently, a mission sent to northwest Vietnam was not expected to obtain much wild material. As it turned out, wild rice of the AA genome (*O. rufipogon*) was abundant in the Dien Bien Phu Valley, geographically isolated from other populations. This collection became an important addition to conserved germplasm of the species and a significant contribution to our knowledge of wild rice distribution. Such isolated populations can be expected to have unique combinations of genes. The two-day walk solely devoted to finding one population of the previously uncollected *O. schlechteri* was worth the effort, since more 'new' genes would be expected from a population of this species than from a large number of populations of a species already in germplasm collections (Fig. 34.2).

A careful search for a specific trait may also occasionally be worthwhile. Thus, the successful search in 1970 for sterile plants of *O. rufipogon* in Hainan Island, China, led to a much-needed breakthrough in

Fig. 34.1. Habitat of *O. granulata* in the Parambikulum forest reserve, India.

Fig. 34.2. Population of *O. schlechteri* at the type locality, beside the Jamu River, Papua New Guinea.

hybrid rice breeding. The consequence of finding this excellent source of male sterility has been the expansion of the area of hybrid rice in China to 8.5 million hectares in 1985 (Lin and Yuan, 1980; Yuan and Virmani, 1988). Of course, new problems are arising in rice production all the time and old problems gaining or losing prominence. Analysis of data accumulated by past collecting missions (among other sources) can indicate 'hot spots', areas where a problem is particularly important or where a pest or pathogen is particularly diverse. These areas have been priorities for intensive sampling of germplasm for rapid evaluation. For example, brown planthopper is a particularly serious problem in southern India and Sri Lanka. The pest is diverse there and the region is also rich in resistant germplasm (Chang *et al.*, 1975; Khush, 1979). Similarly, the nature of the monsoon winds and the hilly topography result in a concentration of pests and diseases in the northeast part of south Asia, and this is an area rich in sources of pest and disease resistance (Sharma *et al.*, 1971; IRTP, 1980; Glaszmann *et al.*, 1989). Evaluation data have also been important in the planning of rice collecting missions. They can reveal areas where resistance to a pest is high in spite of the absence of the pest. In rice, good examples are resistance to hoja blanca virus (Buddenhagen, 1983) and green leafhopper (Chang *et al.*, 1977; Pathak and Saxena, 1980; Vaughan, 1991a).

In the field

During a collecting trip to the Kinabatangan River region of Sabah, Malaysia, the greatest diversity was found along the middle reaches of the river. However, the distinctive agroecosystems around isolated villages at the source and at the mouth of the river were found to be home to a number of varieties available nowhere else. Careful analysis during the first few days in the field of the available diversity and of farmers' knowledge of varieties is the best guide to the intensity with which sampling should be undertaken.

Time is often a precious commodity during a collecting trip, but it is important to be patient. Visiting a teashop at the time of day when a group of villagers are relaxing has often proved worthwhile in rice collecting. The more knowledgeable farmers can then be identified, the names of local varieties from a wide area noted, variation in cultural practices determined and even the location of populations of particular wild species ascertained. In regions where ethnic diversity is great, visiting as many different groups as possible has been important. The range of varieties grown in a village in northwest Vietnam depends on the ethnic group: the Humong hill tribe prefer non-glutinous rices, whereas Thai people in the region eat glutinous rice as their staple food.

Conditions can differ greatly from one region to another. Also, unpredictable factors, such as a plane's late arrival or impassable roads, require flexibility. An ability to adapt to unforeseen events and situations has been essential, as regards both the details of daily life and the scientific content of the mission. For example, guidelines for the number of seeds to collect at a site (Oka, 1975, 1988; Chang, 1985b) may not always be helpful in rice collecting. The collector may be limited by many unexpected or extraneous factors, quite apart from the problem of place-to-place and year-to-year variation in heading time. Poor seed set, small population size or religious taboos on collecting mature seeds from the field can all limit the size of rice germplasm samples. The guiding principle has been to try to collect enough seeds to allow the rapid placement of new samples into long- and medium-term storage, though not at the cost of sampling too few populations (Marshall and Brown, 1975; Oka, 1975; Yonezawa, 1985; Marshall, 1990).

Vegetative samples of wild rice have been collected if seeds were unavailable. Rootstocks and young tillers can be maintained in water (aquatic species) or damp sphagnum moss (mesophytic species), but need to be got back rapidly to base for transplanting. When vegetative material is to be sent to another country, the transfer has been made safer by using a quarantine station in a third country where rice does not grow to check for virus infection or other diseases.

Passport data

The accurate documentation of each sample is a crucial part of all germplasm collecting. The passport data that should be recorded for wild relatives of rice are substantially different from those of the cultigen. Although a common collecting form would perhaps be ideal, separate forms with maximum compatibility have been used by IRRI collectors.

One of the most useful items to record for a rice variety is the type of agricultural system to which it is adapted. This requires discussion with the farmer to determine conditions in the field during the growing season. A comprehensive review of the terminology used to describe rice-growing environments has been developed (IRRI, 1984). This has allowed comparative studies of germplasm adapted to different agroeco-systems (IRRI, 1989). Knowledge of the environment from which a rice sample was collected and of the details of cultural practices has been helpful in deciding on the best evaluation strategy and on which breeder or breeding programme the germplasm may be of most direct use to. Data on the rice plants themselves are also collected. The decimal coding system, initially proposed to describe the growth stages of cereals and later adopted by rice workers and others to characterize their germplasm (IRRI, 1988), has been found suitable for a broad range of plant characters. However, some of the information the germplasm collector gathers cannot readily be condensed into such a system, and it has often been found necessary to record free text comments on collecting forms.

A specialist is often necessary if data on such features as disease resistance are necessary. For example, the important rice collection made by Professor H. Conklin, an anthropologist from Yale University, in the Central Cordillera region of northern Luzon, Philippines (Conklin, 1967, 1980) was particularly well documented with respect to Ifugao vernacular names. These can provide information on harvest season and eating quality, but apparently not on pest and disease susceptibility, information that an agronomist or pathologist may have been able to supply.

Rice varieties are sometimes grown in deliberate mixtures. Such information is important to gene-bank staff handling incoming samples. Describing the grain characteristics of a variety or keeping back a few grains for the gene-bank seed file has been found helpful in sorting mixtures. Indicating the extent of variation in a population also helps future users.

New agronomic problems in isolated areas have sometimes been discovered and at the same time germplasm has been collected that might be useful in addressing the problem. In Vietnam's Dien Bien Phu Valley, for example, a new virus/mycoplasma disease of cultivated rice was reported to a collecting team. In the same valley, abundant wild rice was found growing vigorously. Evaluation of this wild material for resistance to the new disease would clearly be a high priority. Surprising uses of the germplasm have also been recorded. For example, *O. granulata* is used as a contraceptive by tribal people in Madhya Pradesh and Orissa, India. In China, *O. officinalis* is used in traditional medicine (CAAS, 1986).

The post-collecting phase

Rice germplasm collected by IRRI collaborative missions has normally been stored at the national centre of the country where it originated, at the international base collection at IRRI and, if possible, at a third centre with storage facilities meeting minimum international standards. The collector has the responsibility of ensuring that material reaches the gene bank and that the accession number eventually assigned there is linked to the correct collecting number and passport data. Copies of passport data closely accompany the germplasm to which they refer in all its movements. Field notes and trip reports are of historic value and are often very helpful in planning future collecting missions. In the case of rice collecting trips, ideally this documentation should also be duplicated at IRRI, whose library and other resources are accessible internationally.

In international collaborative collecting, germplasm will have to cross international boundaries during or at the end of a collecting mission. It then becomes the responsibility, for a variable period of time, of the quarantine officers of the exporting and importing countries. It is submitted to the quarantine officers of the exporting country to check for pests and pathogens and to undergo any treatment necessary to obtain a phytosanitary certificate. The germplasm, together with export and import permits, is then submitted to the quarantine officers of the importing country. This phase of the operation is required by law. When material from a collecting mission is brought to IRRI in the Philippines, it and the relevant permits are handed over to government quarantine inspectors. Guidelines have been developed by the Philippines Bureau of Plant Industry for post-entry growing of cultivated and wild rice. Seeds are tested for the presence of pests and diseases. In the case of vegetative material, the germplasm is placed in a phytotron and portions of the roots and leaves are tested for infection. Seeds of incoming cultivated rices are grown in an isolated and protected area in the dry season. Wild rices are always grown in a protected nursery area screened with fine wire mesh. Wild rices are grown in the wet season because they are generally sensitive to photoperiod. After harvesting, all vegetative parts of wild rices are burned. Further details of these procedures are given by Mew (1991).

The relative importance of different samples and the comprehensiveness of the collection as a whole are very clear to collectors and should be recorded. Collectors' trip reports can be a great help to the evaluator if samples are highlighted which were observed or locally reported to have interesting traits or were found in unusual ecological situations, such as at high altitude or on adverse soils (Chang, 1980). It is important that such material should quickly pass quarantine, undergo multiplication and receive an accession number, thus becoming available for evaluation. Periodic checking by the collector of the stage a set of samples has reached in this process has been important in ensuring that important rice germplasm reaches the user as quickly as possible.

Future directions

Evenson (1989) has pointed out that:

> The costs of adding to the collections of rice have probably not been more than 2 or 3 million dollars per year over recent years. With expenditures of, say, 4 million dollars per year, most rice landraces not now in collections could be collected within the next 10 to 15 years. Would it pay to do so? Almost certainly yes. An expenditure of 50 million dollars for collection would have to generate a benefit stream of only 25 to 50 million dollars 30 years from now. The Indian experience [... of increased rice production resulting from conserved germplasm ...] over the past 20 years strongly indicates that this is very likely.

This critical analysis of the costs and benefits resulting from collecting and conserving germplasm provides a strong argument for greater support for conserving the rice gene pool. One international and ten national gene banks, which preserve large numbers of accessions of rice, have been built since 1984 (IRRI and IBPGR, 1991). These new gene banks have given impetus to national efforts to collect or re-collect germplasm. Several national and regional programmes have also retrieved 9000 accessions from duplicate storage at IRRI to restock their gene banks. The plan for a collaborative scheme of duplicate storage made during the

1977 Workshop on Rice Germplasm Conservation (IRRI and IBPGR, 1978) proved to be important. It will be even more important in the future, especially in areas where re-collecting lost material is no longer possible.

Seed storage is not the only approach to rice genetic resources conservation, of course. Germplasm collectors should take an increasingly active role in the formulation of *in situ* conservation programmes to complement their *ex situ* work. Key farmers or villages with a noteworthy collection of traditional varieties, managed and manipulated in interesting ways, can be identified, encouraged and supported. The input of ethnobotanists will be important in this context. Protected areas such as national parks and wildlife reserves where populations of wild rice grow can be identified and their importance brought to the attention of conservation officers (Vaughan and Chang, 1992). The state of protection of a site, the potential for genetic erosion in the area, the genetic diversity of local populations and unusual features of the habitat are all relevant in identifying sites for *in situ* conservation. *O. schlechteri*, the rarest species in the genus, is only known from three widely scattered localities in Papua New Guinea and Irian Jaya, according to herbarium specimens collected in 1907, 1912 and 1974. The type locality was visited in 1990 and the species found again, apparently in the same place where R. Schlechter found it 83 years ago (Vaughan and Sitch, 1991). However, the small population (about 3 m × 3 m on the banks of a river) is in a young mountain range where numerous natural landslides occur. The herbarium specimen (LAE 212991) of this species from another locality records 'creeping grass on landslip'. These populations may be vulnerable to natural destruction. The only area where *O. officinalis* and *O. minuta* are known to grow sympatrically is in Leyte, the Philippines. This was highlighted in a recent collecting report as a site worthy of protection because of both its uniqueness and the risk of destruction due to nearby urban development. Following discussion of an unusually large population of wild rice in a lake near Ajigara, Nepal, by a germplasm collector in a national newspaper article, the government has endorsed the creation of a reserve at the site (G.L. Shrestha, pers. comm.).

At the moment, the gene pool which the rice genetic resources community is dealing with can be taken to include the whole subfamily Oryzoideae, since crossing barriers between rice and species in related genera are likely to be overcome in the near future. If present developments in biotechnology continue, however, traits found outside the subfamily may become the focus for ever more ambitious hybridization efforts. The Oryzoideae is usually allied with bamboos, which have C_3 photosynthesis. However, some workers ally the subfamily with the arundinoid grasses (Tzvelev, 1989), which include some genera with C_4 photosynthesis, e.g. *Asthenatherum*, *Aristida*, *Stipagrostis*, *Eriachne* and *Pheidochloa* (Watson *et al.*, 1985). Taxonomic considerations suggest that the intricate switch from C_3 to C_4 photosynthesis has occurred during evolution on several occasions in the Poaceae (Watson *et al.*, 1985). It may be possible to introduce this very complex trait or an intermediate system into rice in the future (Chang and Vaughan, 1991).

With both the exploitable gene pool and the international rice genetic resources network widening, critical genetic diversity studies of conserved germplasm, including the identification of duplicates, will be increasingly important in future. Such work will help in identifying diverse and undercollected regions and thus in setting an agenda for future collecting. Molecular techniques will be important tools in this effort. Collectors will clearly need to work very closely with specialists in such diverse fields as ethnobotany and molecular biology in the future to ensure that the full spectrum of rice genetic resources will be readily available to rice scientists worldwide (Chang, 1985b).

Acknowledgements

Many of the perspectives presented in this chapter have resulted from cooperative work with scientists in many countries over many years. The stimulating work the authors have undertaken with rice scientists worldwide, but particularly in tropical Asia, is gratefully acknowledged. The international team approach to collecting and conserving rice genetic resources has laid the foundation for future cooperation to further improve the collection and conservation of rice germplasm worldwide.

References

Bezançon, G. and G. Second (1984) Les riz. In: Pernes, J. (ed.) *Gestion des Resources Génétiques des Plantes*. Vol. 1: *Monographies*. pp. 107–156. Agence de Cooperation Culturelle et Technique, Paris.

Bray, F. (1984) In: Needham, J. (ed.) *Science and Civilization in China*. Vol. 6. Part II. *Agriculture*. Cambridge University Press, Cambridge.

Buddenhagen, I.W. (1983) Disease resistance in rice. In: F. Lamberti, J.M. Waller and N.A. van der Graaff (eds) *Durable Resistance in Crops*. NATO Advanced Studies Institute Series 55. pp. 401–428. Plenum Press, New York.

CAAS (1986) *Chinese Rice Science*. Agricultural Publishing Society, Beijing (in Chinese).

Chang, T.T. (1970) Examples of crop exploration – rice. In: Frankel, O.H. and E. Bennett (eds) *Genetic Resources in Plants – Their Exploration and Conservation*. pp. 267–272. Blackwell Scientific Publications, London.

Chang, T.T. (1975) Exploration and survey in rice. In: Frankel, O.H. and J.G. Hawkes (eds) *Crop Genetic Resources for Today and Tomorrow*. pp. 159–165. Cambridge University Press, Cambridge.

Chang, T.T. (1980) The rice genetic resources program of IRRI and its impact on rice improvement. In: *Rice Improvement in China and Other Asian Countries*. pp. 85–105. IRRI, Los Baños.

Chang, T.T. (1984) Conservation of rice genetic resources: luxury or necessity? *Science* 224:251–256.

Chang, T.T. (1985a) Collection of crop germplasm. *Iowa State Journal of Research* 59:349–364.

Chang, T.T. (1985b) Crop history and genetic conservation: rice – a case study. *Iowa State Journal of Research* 59:425–455.

Chang, T.T. and D.A. Vaughan (1991) Conservation and potentials of rice genetic resources. In: Bajaj, Y.P.S. (ed.) *Biotechnology in Agriculture and Forestry*. pp. 531–552. Springer-Verlag, Berlin.

Chang, T.T., S.H. Ou, M.D. Pathak, K.C. Ling and H.E. Kauffman (1975) The search for diseases and insect resistance in rice germplasm. In: Frankel, O.H. and J.G. Hawkes (eds) *Crop Genetic Resources for Today and Tomorrow*. pp. 183–200. Cambridge University Press, Cambridge.

Chang, T.T., A.P. Marciano and G.C. Loresto (1977) Morpho-agronomic variousness and economic potentials of *Oryza glaberrima* and wild species in the genus *Oryza*. In: *Proceedings of the Meeting on African Rice Species*. pp. 67–76. IRAT–ORSTOM, Paris.

Conklin, H.C. (1967) Ifugao ethnobotany 1906–1965: the 1911 Beyer–Merrill report in perspective. *Economic Botany* 21:243–272.

Conklin, H.C. (1980) *Ethnographic Atlas of Ifugao*. Yale University Press, New Haven.

Cooperative Team of Wild Rice Resources Survey and Exploration of China (1984) A general survey and exploration of wild rice germplasm resources in China. *Scientia Sinica* 6:27–34.

Dao The Tuan (1985) Types of rice cultivation and its related civilizations in Vietnam. *East Asian Cultural Studies* 24:41–55.

de Wet, J.M.J. and E.A. Oelke (1978) Domestication of American wild rice (*Zizania aquatica* L., Gramineae). *Journal d'Agriculture Traditionelle et de Botanique Appliqué* 25:67–84.

Duistermaat, H. (1987) A revision of *Oryza* (Gramineae) in Malesia and Australia. *Blumea* 32:157–193.

Evenson, R.E. (1989) Rice genetic resources: economic evaluation. Yale University. Mimeographed.

Glaszmann, J.C., P. Benyayer and M. Arnaud (1989) Genetic divergence among rices from northeast India. *Rice Genetics Newsletter* 6:63–65.

Harlan, J.R. and J.M.J. de Wet (1971) Towards a rational classification of cultivated plants. *Taxon* 20:508–517.

Ho, P.T. (1956) Early-ripening rice in Chinese history. *Economic History Review* 9:200–218.

IBPGR–IRRI Rice Advisory Committee (1982) Conservation of the wild rices of tropical Asia. *FAO/IBPGR Plant Genetic Resources Newsletter* 49:13–18.

IRRI (1972) *Rice Breeding*. IRRI, Manila.

IRRI (1984) *Terminology for Rice Growing Environments*. IRRI, Manila.

IRRI (1988) *Standard Evaluation System for Rice*. IRRI, Manila.

IRRI (1989) *Program Report*. IRRI, Manila.

IRRI and IBPGR (1978) *Proceedings of the Workshop on the Genetic Conservation of Rice*. IRRI, Manila.

IRRI and IBPGR (1983) *Rice Germplasm Genetic Conservation Workshop*. IRRI, Manila.

IRRI and IBPGR (1991) *Proceedings of the 1990 Rice Germplasm Workshop*. IRRI, Los Baños.

IRTP (1980) *Global Rice Improvement Network: Five Years of IRTP*. IRRI, Manila.

Khush, G.S. (1979) Genetics of and breeding for resistance to the brown planthopper. In: *Brown Planthopper Threat to Rice Production in Asia.* pp. 321–332. IRRI, Manila.

Kwangtung Agricultural and Forestry College (1975) The species of wild rice and their geographical distribution in China. *Acta Genetica Sinica* 2:31–36 (in Chinese, English summary).

Lin, S.C. and L.P. Yuan (1980) Hybrid rice breeding in China. In: *Innovative Approaches to Rice Breeding.* pp. 35–51. IRRI, Manila.

Marshall, D.R. (1990) Crop genetic resources: current and emerging issues. In: Brown, A.H.D., M.T. Clegg, A.L. Kahler and B.S. Weir (eds) *Plant Population Genetics, Breeding and Genetic Resources.* pp. 367–388. Sinauer Associates Inc., Sunderland, USA.

Marshall, D.R. and A.H.D. Brown (1975) Optimal sampling strategies in genetic conservation. In: Frankel, O.H. and J.G. Hawkes (eds) *Crop Genetic Resources for Today and Tomorrow.* pp. 53–80. Cambridge University Press, Cambridge.

Mew, T.W. (1991) Rice seed health evaluation at the International Rice Research Institute. In: Frison, E.A. (ed.) *Proceedings of the Inter-centre Meeting on Germplasm Health and Movement.* IBPGR, Rome.

Ng, N.Q., M. Jacquot, A. Abifarin, K. Goli, A. Ghesquiere and K. Miezan (1983) Rice genetic resources collecting and conservation activities in Africa and Latin America. In: Denton, R. and T.T. Chang (eds) *Rice Germplasm Conservation Workshop.* pp. 45–52. IRRI and IBPGR, Manila.

Ng, Q., T.T. Chang, D.A. Vaughan and C.Z. Altoveros (1991) African rice diversity: conservation and prospect for crop improvement. In: Ng, N.Q., P. Perrino, F. Attere and H. Zedan (eds) *Crop Genetic Resources of Africa.* Vol. II. pp. 213–227. IITA, IBPGR, UNEP and CNR, Ibadan, Nigeria.

Oka, H.I. (1975) Consideration on the population size necessary for conservation of crop germplasm. In: Matsuo, T. (ed.) *Gene Conservation – Exploration, Collection and Preservation and Utilization of Genetic Resources.* pp. 57–63. JIBP Synthesis 5. University of Tokyo Press, Tokyo.

Oka, H.I. (1977) The ancestors of cultivated rice and their evolution. In: *Proceedings of the Meeting on African Rice Species.* IRAT–ORSTOM, Paris.

Oka, H.I. (1983) Conservation of heterogenous rice populations. In: Denton, R. and T.T. Chang (eds) *Rice Germplasm Conservation Workshop.* pp. 11–19. IRRI and IBPGR, Manila.

Oka, H.I. (1988) *Origin of Cultivated Rice.* Japan Scientific Societies Press, Tokyo.

Parthasarathy, N. (1972) Rice breeding in tropical Asia up to 1960. In: *Rice Breeding.* pp. 5–29. IRRI, Manila.

Pathak, M.D. and R.K. Saxena (1980) Breeding approaches in rice. In: Maxwell, F.G. and P.R. Jennings (eds) *Breeding Plants for Resistance to Insects.* pp. 421–455. John Wiley & Sons, New York.

Sharma, S.D. (1983) Conservation of wild species of *Oryza.* In: Denton, R. and T.T. Chang (eds) *Rice Germplasm Conservation Workshop.* pp. 21–25. IRRI and IBPGR, Manila.

Sharma, S.D., J.M.R. Vellanki, K.L. Hakim and R.K. Singh (1971) Primitive and current cultivars of rice in Assam – a rich source of valuable genes. *Current Science* 40:126–128.

Sharma, S.D., A. Krishnamurti and S.R. Dhua (1988) Genetic diversity of rice and its utilization in India. In: Paroda, R.S., R.K. Arora and K.P.S. Chandel (eds) *Plant Genetic Resources – Indian Perspective.* pp. 108–120. National Bureau of Plant Genetic Resources, New Delhi.

Tzvelev, N.N. (1989) The system of grasses (Poaceae) and their evolution. *Botanical Review* 55:141–204.

Vaughan, D.A. (1989) *The Genus* Oryza *L.: Current Status of Taxonomy*. IRRI Research Papers Series 138. IRRI, Los Baños.

Vaughan, D.A. (1990) *In-situ* conservation of wild rices in Asia. *Rice Genetics Newsletter* 7:90–91.

Vaughan, D.A. (1991a) Choosing rice germplasm for evaluation. *Euphytica* 54:147–154.

Vaughan, D.A. (1991b) The gene pools of rice: recent collecting activities and future directions. In: *Proceedings of the 1990 Rice Germplasm Workshop*. IRRI and IBPGR, Manila.

Vaughan, D.A. and T.T. Chang (1992) *In-situ* conservation of rice genetic resources. *Economic Botany* 46:368–383.

Vaughan, D.A. and L.A. Sitch (1991) Gene flow from the jungle to farmers: wild-rice genetic resources and their uses. *BioScience* 41:22–28.

Vaughan, D.A., S.R. Almazan and M.D. Oliva (1991) Wild relatives of rice conserved at International Rice Germplasm Center, IRRI. *Rice Genetics Newsletter*.

Watson, L., H.T. Clifford and M.J. Dallwitz (1985) The classification of Poaceae: subfamilies and supertribes. *Australian Journal of Botany* 33:433–484.

Watt, G. (1891) *A Dictionary of the Economic Products of India*. Vol. 5. Periodical Experts, New Delhi.

Yieh, M.F. (1964) The distant hybridization between broad sheath bamboo (*Arundinaria latifolia* Keng) and wheat, barley and rice. *Genetica Sinica* 5:145–149 (in Chinese).

Yonezawa, K. (1985) A definition of the optimal allocation of effort in conservation of plant genetic resources with application to sample size determination for field collection. *Euphytica* 34:345–354.

Yuan, L.P. and S.S. Virmani (1988) Status of hybrid rice research and development. In: *Proceedings of the International Symposium on Hybrid Rice*. pp. 7–24. IRRI, Manila.

Zhou, G.Y. (1986) Distantly related hybridization and genetic engineering of crops. In: *Rice Genetics*. pp. 867–876. IRRI, Manila.

Zhou, G.Y., Y. Zen and W. Yang (1981) The molecular basis of remote hybridization – an evidence for the possible integration of sorghum DNA into rice genome. *Scientia Sinica* 24:701–709.

Collecting wild species of *Arachis* 35

J.F.M. Valls[1], C.E. Simpson[2] and V. Ramanatha Rao[3]

[1]*EMBRAPA, Centro Nacional de Recursos Geneticos e Biotecnologia, SAIN – Parque Rural – CP 10.2372, CEP 70.770 Brasilia DF, Brasil:* [2]*Texas Agricultural Experimental Station, Texas A & M University, PO Box 292, Stephanville, Texas 76401, USA:* [3]*IPGRI, c/o International Development Research Centre, Tanglin PO Box 101, Singapore 9124.*

In most angiosperms, seeds, fruits and other propagules are generally quite evident at the appropriate time for collecting. *Arachis* is an exception. The cultivated groundnut (*Arachis hypogaea*) is well known for its underground fruits, which generally remain attached to the mother plants by 'pegs' as the plants are removed from the soil at harvest time. Pegs develop soon after fertilization at the base of the calyx tube of aerial flowers, which arise from axillary inflorescences hidden in the leaf axils. The peg is geotropic, reaching the ground by growth of an intercalary meristem located at the base of the ovary, proximal to the ovules (Gregory *et al.*, 1973). Maintenance of peg rigidity with tight fruit attachment after maturation is an advanced character in the genus *Arachis*, obviously selected for in the process of domestication of the cultigen. As mentioned by Gregory *et al.* (1973), cultivated groundnuts from the Guarani region, where the common method of harvest in the past was to pull up the plants by hand, have very tough pegs and the pods adhere tightly to them. In regions where pods are frequently taken individually from the plant, they may adhere much less tightly to the peg. In the wild species of *Arachis*, as a rule, a meristematic tissue also occurs between the ovules, so that an isthmus (ocasionally two or three) is formed, separating one-seeded fruit segments in a distinctly lomentiform pod (Conagin, 1959; Pattee *et al.*, 1991). The basal and intercalary portions of the peg tend to collapse at maturity. At seed maturity, the best time for germplasm collecting, seeds of wild species of *Arachis* are therefore no longer attached to the plants, or break away very easily if the soil is disturbed.

Underground fruit development in *Arachis* means that it is impossible to detect even the presence of seeds until the soil under the plants is dug up and sifted, a time-consuming procedure, let alone estimate the quantity available and their developmental stage and degree of maturity. This is bad enough for the germplasm collector, but there is another consequence. In the absence of drastic soil disturbance, for example by animals or flooding, most seedlings will grow at only a short distance from their mother plants. The result of such relatively ineffi-cient dispersal and limited vegetative spread by rhizomes or stolons is that populations may be dense, but are often small in extent, their perimeter remaining quite stable for many years, even decades. Appar-ently trivial obstacles, such as sparse groups of trees and stands of tall grass, may obscure such compact, isolated populations, perhaps the only ones for many kilometres.

The problems of wild *Arachis* collecting are not confined to those caused by the mode of fruit development, however. Locating populations at the appropriate time for seed collecting may be quite difficult because of the phenology of the species. In annual species maturation tends to be complete only after wilting of the mother plants. Experienced collec-tors can identify wilted stems or leaflets lying on the ground as belong-ing to *Arachis* species, but sometimes even such slight clues may be missing, and indirect, intuitive evidence must be used to decide whether it is worth searching for seeds in a particular area. In northern Brazil, seeds of a yet undescribed species were searched for under palm trees more because just such a location had first yielded material than because of any obvious clues to the presence of the plants. In perennial species full maturation may occur long after the peak of flowering. Therefore, showy yellow or orange flowers may help to locate plants in the field, but will not necessarily mean that mature fruits will be available. Also, some species of the Brazilian sections *Ambinervosae* and *Extranervosae* (Resslar, 1980) present at first a small number of normal, showy flowers, immediately followed by abundant cleistogamic flowers, much smaller in size and very difficult to see. The period when populations of such plants can be noticed from a distance by their flowers is extremely short.

There is a further problem. Consulting herbarium labels is common practice among plant germplasm collectors as a means of acquiring a good knowledge of target species, in particular their distributions, prior to going into the field. In the case of wild *Arachis*, this has not been a very fruitful approach. Specimens are very scarce in herbaria, since most botanists will tend not to collect plants of which they do not see fruits, and whose flowers remain turgid for only a few hours a day. Present *Arachis* collecting efforts, however, stress the need for a herbarium specimen of all accessions, even when the plants are wilted or without fruits. There is an important group of vegetative morphological characters that provide good hints for identification at the species level. Information from local people, who are usually able to associate the wild species of *Arachis* with the cultigen, mainly due to leaf and flower similarities, is extremely important when an expedition visits a new area. For example, locally used vernacular names have been found helpful in locating populations.

Underground fruit development, small populations, difficult pheno-

logy and poor representation in herbaria have led wild *Arachis* collectors
to develop a strategy which often involves repeated visits to collecting
sites. A good example of this and of some of the problems encountered
in fieldwork is provided by accession VSGr 6416 (BRA-012726) (collectors: V = J.F.M. Valls; S = C.E. Simpson; Gr = A. Gripp), collected in
Mato Grosso, Brazil. Thus far, this represents the only known population of an undescribed species in sect. *Arachis*. It has a very short life
cycle and has already been included in a breeding programme as a potential source of earliness (Simpson, 1990). The species was collected for the
first time in August 1981, when a very distinctive and novel fruit segment, acute and strongly nerved, was found in the soil being sifted while
collecting germplasm of a plant of sect. *Extranervosae* which was very
frequent at the site. After some additional searching, only two very
young seedlings of the new plant were found. These were collected as live
plants in pots for seed production. Some 50 seeds were later harvested
from one surviving seedling and from the plant obtained by germination
of the first original seed, which had about one third of each cotyledon
eaten by an insect. In May 1985, another expedition visited the site.
After 2 hours of intensive search by four experienced *Arachis* collectors,
two of whom had made the original collection, a few additional plants
were spotted and produced a small seed sample. At a better time, in
October of the next year, a third collecting team reached the area, to be
surprised by the sight of a continuous carpet of seedlings of the species,
extending for some 500 m along both sides of the road. There were abundant seeds in the ground, this time easy to find in the burned grassland
of the roadside, but they were all either germinated or rotten. Only very
few plants were seen in the adjacent undisturbed grassland, obviously
due to intensive grazing. By then, it was clear that the species had a very
short and synchronous life cycle and that timing was critical. A fourth
expedition visited the area in November 1987, hoping to collect large
amounts of seed. However, it was found that the top soil had been
scraped off along both sides of the road, piled up and compacted for hundreds of metres, as the base of a new paved road!

Precise notes in field books and herbarium labels are important if collecting sites are to be visited again in this way. Latitude and longitude
are generally taken from maps in a somewhat imprecise way, and yet
they are important references to establish the general area of search.
They must be complemented by precise distances, preferably from stable
geographic landmarks. For example, distances taken from city limits
may result in errors of hundreds of metres in a few years. Even if a site
can be accurately located again, the vagaries of the weather can also con-
spire against the collector. Floods can make roads impassable and
submerge entire populations, for example. Annual species may not germinate and grow every year in dry climates, and even perennials may
simply disappear from sight in bad years.

Once a wild *Arachis* population is located in the field, seeds are collected and herbarium specimens of the population prepared. *Rhizobium*

nodules are also collected. A description of site and plant conditions is then made in a field book. Sometimes, soil samples are taken for chemical analysis and herbarium specimens of associated species are prepared. Digging and sifting along transects across a population theoretically help to maximize variation in the samples. Digging out entire plants (which will also be useful for making herbarium specimens and collecting nodule) and sifting the soil underneath with large sieves are the most efficient way to harvest seeds (in fact, fruit segments) of wild species of *Arachis*. However, fruits recovered from the soil may be from previous years and thus have very low germination. Seed dormancy may also hinder the multiplication of poor samples. Also, as seeds of wild *Arachis* are generally loose in the soil at maturity, sampling techniques based on any 'x seeds per mother plant' design cannot be applied. Different *Arachis* species sometimes grow in a mosaic pattern at the same site. Identification of the seeds collected may only be possible at the multiplication stage, when germinated plants show their differences. Another restriction on the use of a standard sampling procedure for all wild *Arachis* species is that very little is known of their pollination mechanisms. Although the wild species are generally considered autogamous (by analogy with the cultivated *A. hypogaea*), differences in stigma morphology and in pollination behaviour have been pointed out between annual and perennial species (Banks, 1990; Lu *et al.*, 1990). A single sampling technique will probably not be suitable for all species.

Seeds are put into cloth bags, which are then hung up to dry. Alternative or complementary collecting of entire live plants in pots, or preparation of cuttings, may be required. Collecting live plants is frequently the only way to preserve germplasm of a population *ex situ* when seeds are not produced or are not available. All collecting of live plants requires careful prior arrangements for the subsequent work of maintenance and multiplication. Rhizomatous and stoloniferous species, such as *A. glabrata* and *A. repens*, are easy to transport as cuttings wrapped in newspaper and stored in plastic bags, without additional water. They will survive for many days, especially when kept cool. They have also been transported for up to 40 days in sphagnum moss, cleaned and washed periodically. Other perennial species, which do not have rhizomes or stolons, are transplanted directly into pots, but this requires large amounts of space in the mission vehicle. A few annual species of *Arachis* do not survive for long when transplanted at the stage of fruit maturation, but may complete the maturation of a few additional fruits. *In vitro* meristem culture may be used to increase the number of individuals producing seeds (Pittman *et al.*, 1983). Complementary *in situ* conservation has been suggested for some species and situations (Valls, 1985). This may be very important in future in buying time until special techniques for the conservation of important germplasm of difficult species are developed (Simpson, 1988).

An understanding of the genetic variation encompassed by each accession and by the collection as a whole is only attained with subsequent characterization. This will not only affect decisions about use in breeding, but also allows the identification of geographic areas of high diversity, where additional collecting work may need to be carried out. Intensive efforts to characterize and evaluate wild *Arachis* germplasm are under way in many countries and institutions, involving taxonomy, cytogenetics, breeding behaviour, genome analysis, crossing behaviour, pest and disease resistance, potential for forage use and so on (Singh and Moss, 1982; Pompeu, 1983; Grof, 1985; Simpson *et al.*, 1985; Subramaniam *et al.*, 1985; Kretschmer and Wilson, 1988; Moss *et al.*, 1989; Nelson *et al.*, 1989; Cook *et al.*, 1990; Lu *et al.*, 1990; Stalker, 1990, 1991; Kochert *et al.*, 1991; Singh *et al.*, 1991; Stalker *et al.*, 1991). Many scientists directly involved in *Arachis* characterization and evaluation research have had an opportunity to participate in collecting mission(s) (Table 35.1). A good link thus exists between the collectors and users of the germplasm. Decisions on priority areas and priority species for collecting have been made by specialists who have had access to up-to-date characterization and evaluation information (Valls *et al.*, 1985; Simpson, 1990).

Wild species of *Arachis* are native to five countries, some 60 species occurring in Brazil, 15 in Bolivia, 12 in Paraguay, seven in Argentina and two in Uruguay. Many more countries grow the cultigen and are interested in its improvement. It is thus not surprising that, though expeditions have always included local scientists, international cooperation has been essential in the continuing build-up of a comprehensive collection of wild *Arachis*. This has required much careful planning. Each country has different legal requirements for collecting work and the possibilities for local support also vary widely. The planning of expeditions has taken months and sometimes even years (Simpson, 1984). Field activities have been supported by the International Board for Plant Genetic Resources (IBPGR), the US Department of Agriculture (USDA), EMBRAPA and other institutions. Subsamples of each accession are always deposited in the host country. They are usually also shared with other national institutions which did not participate in the mission but which have a clear commitment to the conservation of *Arachis* germplasm (Simpson, 1980, 1984, 1991; Williams, 1989). An international cooperative approach was also used for the development of a list of descriptors for wild *Arachis* germplasm (IBPGR and ICRISAT, 1992).

Special care has been taken to involve promising young scientists and even undergraduate students in collecting (Simpson, 1990). For example, from 1981 to 1992, a period of intensive collecting, especially in Brazil, 28 scientists or highly trained technicians participated in field missions, representing 14 institutions in six countries. They collected accessions from all eight taxonomic sections of the genus (Krapovickas, 1990) (Table 35.1). Despite all this effort, the general feeling among *Arachis* workers is that much fieldwork remains to be done. A reservoir of expertise is being built up: the solutions to the unique problems of wild *Arachis* collecting which have been developed over scores of

Table 35.1 Expeditions for collecting of wild *Arachis* germplasm in South America, 1981–92. Number of accessions collected in each country and sections of species collected in each expedition.

Year	Collectors	ARG	BOL	BRA	PRY	URY	AM	AR	CA	ER	EX	PR	RH	TR
		Countries					Sections							
1981	VW			6					+		+		+	
	VVeSv			7			+				+			+
	VSGr		1	40				+			+	+	+	
1982	VKRSv			38			+	+			+			
	ScVn	10					+							
	VSW			17			+		+		+			+
1983	VKVeSv			27			+	+			+			+
	KSSc	5	2				+							
	VSMGeSv			18				+				+		
1984	VRGeSv			76				+		+		+	+	
	VSStGdW			25				+			+		+	
1985	Mt					6		+						
	VVeSv			45			+	+			+			
	VKSSv			26				+			+	+	+	
	VPoBi			27				+		+	+	+	+	
1986	VSW			14	1			+		+	+	+	+	
	VPoJSv			11				+		+		+		
1987	VRSv			35			+				+			
	VeSv			5						+			+	
1988	VFdSv			11						+			+	
	Wi		4					+						
1990	Wi	5	4					+						
	VGaRoSv			9				+			+			
1991	VPmSv			22			+	+			+			
	VFaPzSv			13			+		+					+
1992	VPzVaW			29			+		+		+			
	VSPmWiSvVePzRs			17			+	+	+		+		+	

Key: Collectors: Bi = Bianchetti; Fa = Faraco; Fd = França-Dantas; Ga = Galgaro; Gd = Godoy; Ge = Gerin; Gr = Gripp; J = Jank; K = Krapovickas; M = Moss; Mt = Millot; Pm = Pittman; Po = Pott; Pz = Pizarro; Q = Quarin; R = Rao; Ro = Rocha; Rs = Santos; S = Simpson; Sc = Schinini; St = Stalker; Sv = Silva; V = Valls; Va = Valente; Ve = Veiga; Vn = Vanni; W = Werneck; Wi = Williams. Countries: ARG = Argentina; BOL = Bolivia; BRA = Brazil; PRY = Paraguay; URY = Uruguay. Sections: AM = *Ambinervosae*; AR = *Arachis*; CA = *Caulorhizae*; ER = *Erectoides*; EX = *Extranervosae*; PR = *Procumbensae*; RH = *Rhizomatosae*; TR = *Triseminalae*.

missions will need to be further refined and applied for many years to come.

References

Banks, D.J. (1990) Handstripped flowers promote seed production in *Arachis lignosa*, a wild peanut. *Peanut Science* 17:22–24.

Conagin, C.H.T.M. (1959) Desenvolvimento dos frutos nas espécies selvagens de amendoim (*Arachis* spp.). *Bragantia* 18:51–70.

Cook, B.G., R.J. Williams and G.P.M. Wilson (1990) Register of Australian herbage plant cultivars. B. Legumes. 21. *Arachis*. (a) *Arachis pintoi* Krap. and Greg. *nom. nud.* (Pinto peanut) cv. Amarillo. *Australian Journal of Experimental Agriculture* 30:445–446.

Gregory, W.C., M.P. Gregory, A. Krapovickas, B.W. Smith and J.A. Yarbrough (1973) Structure and genetic resources of peanuts. In: *Peanut – Culture and Uses.* pp. 47–133. American Peanut Research and Education Society, Stillwater.

Grof, B. (1985) Forage attributes of the perennial groundnut *Arachis pintoi* in a tropical savanna environment in Colombia. In: *Proceedings of the 15th International Grassland Congress.* pp. 168–170. Science Council of Japan and Japanese Society of Grassland Science, Nishi-nasuno, Tochigi, Japan.

IBPGR and ICRISAT (1992) *Descriptors for Groundnut.* IBPGR, Rome; ICRISAT, Patancheru.

Kochert, G., T. Halward, W.D. Branch and C.E. Simpson (1991) RFLP variability in peanut (*Arachis hypogaea* L.) cultivars and wild species. *Theoretical and Applied Genetics* 81:565–570.

Krapovickas, A. (1990) A taxonomic summary of the genus *Arachis*. In: IBPGR *Report of a Workshop on the Genetic Resources of Wild* Arachis *Species*. International Crop Network Series 2. p. 9. IBPGR, Rome.

Kretschmer, A.E. Jr and T.C. Wilson (1988) A new seed producing *Arachis* sp. with potential as forage in Florida. *Proceedings of the Soil and Crop Science Society of Florida* 47:229–233.

Lu, J., A. Mayer and B. Pickersgill (1990) Stigma morphology and pollination in *Arachis* L. (Leguminosae). *Annals of Botany* 66:73–82.

Moss, J.P., V.R. Rao and R.W. Gibbons (1989) Evaluating the germplasm of groundnut (*Arachis hypogaea*) and wild *Arachis* species at ICRISAT. In: Brown, A.H.D., O.H. Frankel, D.R. Marshall and J.T. Williams (eds) *The Use of Plant Genetic Resources.* pp. 212–234. Cambridge University Press, Cambridge.

Nelson, S.C., C.E. Simpson and J.L. Starr (1989) Resistance to *Meloidogyne arenaria* in *Arachis* spp. germplasm. *Journal of Nematology* 21:654–660.

Pattee, H.E., H.T. Stalker and F.G. Giesbrecht (1991) Comparative peg, ovary, and ovule ontogeny of selected cultivated and wild-type *Arachis* species. *Botanical Gazette* 152:64–71.

Pittman, R.N., D.J. Banks, J.S. Kirby, E.D. Mitchell and P.E. Richardson (1983) *In vitro* culture of immature peanut (*Arachis* spp.) leaves. Morphogenesis and plantlet regeneration. *Peanut Science* 10:21–25.

Pompeu, A.S. (1983) Cruzamentos entre *Arachis hypogaea* e as espécies *A. diogoi* e *A.* spp. (30006, 30035). *Bragantia* 42:261–265.

Resslar, P.M. (1980) A review of the nomenclature of the genus *Arachis* L. *Euphytica* 29:813–817.

Simpson, C.E. (1980) Collecting wild *Arachis* in South America. Past and future. In: *Report of a Workshop on the Genetic Resources of Wild* Arachis *Species.* pp. 10–17. IBPGR, Rome.

Simpson, C.E. (1984) Plant exploration: planning, organization, and implementation with special emphasis on *Arachis.* In: *Conservation of Crop Germplasm – An International Perspective.* pp. 1–20. CSSA Special Publication No. 8, Crop Science Society of America, Madison.

Simpson, C.E. (1988) Techniques for maintaining precious *Arachis* germplasm. *Proceedings of the American Peanut Research and Education Society* 20.

Simpson, C.E. (1990) Introgression of early maturity into *Arachis hypogaea* L. *Proceedings of the American Peanut Research and Education Society* 22:49.

Simpson, C.E. (1991) Global collaborations find and conserve the irreplaceable genetic resources of wild peanut in South America. *Diversity* 7:59–61.

Simpson, C.E., D.L. Higgins and W.H. Higgins, Jr (1985) Crossability and cross-compatibility of five new species of section *Arachis* with *Arachis hypogaea. Proceedings of the American Peanut Research and Education Society* 17:24.

Singh, A.K. and J.P. Moss (1982) Utilization of wild relatives in genetic improvement of *Arachis hypogaea* L. Part 2: Chromosome complements of species in section *Arachis. Theoretical and Applied Genetics* 61:305–314.

Singh, A.K., S. Sivaramakrishnan, M.H. Mengesha and C.D. Ramaiah (1991) Phylogenetic relations in section *Arachis* based on seed protein profile. *Theoretical and Applied Genetics* 82:593–597.

Stalker, H.T. (1990) A morphological appraisal of wild species in section *Arachis* of peanuts. *Peanut Science* 17:117–122.

Stalker, H.T. (1991) A new species of *Arachis* with a D genome. *American Journal of Botany* 78: 630–657.

Stalker, H.T., J.S. Dhesi, D.C. Parry and J.H. Hahn (1991) Cytological and interfertility relationships of *Arachis* section *Arachis. American Journal of Botany* 78:238–246.

Subrahmanian, P., D.H. Smith and C.E. Simpson (1985) Resistance to *Didymella arachidicola* in wild *Arachis* species. *Oléagineux* 40:553–556.

Valls, J.F.M. (1985) Groundnut germplasm management in Brazil. In: *Proceedings of an International Workshop on Cytogenetics of* Arachis. pp. 43–45. ICRISAT, Patancheru.

Valls, J.F.M., V.R. Rao, C.E. Simpson and A. Krapovickas (1985) Current status of collection and conservation of South American groundnut germplasm with emphasis on wild species of *Arachis.* In: *Proceedings of an International Workshop on Cytogenetics of* Arachis. pp. 15–35. ICRISAT, Patancheru.

Williams, D.E. (1989) Exploration of Amazonian Bolivia yields rare peanut landraces. *Diversity* 5:12–13.

Collecting rare species in Florida

36

S.R. Wallace

Bok Tower Gardens, PO Box 3810, Lake Wales, FL 33859-3810, USA.

Introduction

Rare plant destruction in tropical rain forests and other developing areas of the world has been the focus of much conservation attention, but the same development pressures and destructive phenomena are also at work in the relatively wealthy and developed (some would say too developed) state of Florida in the USA. Because of its unique geological history, during which ridges of high ground became isolated islands during interglacial periods, Florida is home to one of the highest concentrations of endemic species anywhere in the world. Although the numbers vary with the taxonomist doing the counting, there are about 2500 native plant species in Florida, of which 275 are endemic. In the last 60 years, citrus and cattle production, road building, residential development and a booming tourist industry have destroyed much of the natural habitat, leaving many of the rarest endemics close to extinction. The Florida Natural Areas Inventory records 385 species as rare, threatened or endangered; 169 of these are endemics.

Today many rare plants remain as remnant populations along roadsides and on private land near developments, sites too expensive to buy and too small to protect. Bok Tower Gardens' Endangered Plant Program was begun in 1986 in response to this critical situation. Genetically representative collections of the most severely endangered species are being grown in the protected setting of the garden, where they can be studied and propagated.

Collecting strategies

Working closely with other private conservation organizations and government agencies, the programme began with a reasonably accurate (though by no means complete) assessment of which plants are at risk and where they grow. Species were ranked using several factors: the number and size of the populations, the vigour and longevity of the plants, the number of protected sites and the degree of threat to unprotected sites. Species with fewer than five populations and no protected sites were deemed the highest priority. The development of good field inventories is an essential first step in understanding and meeting each species' individual needs.

After a population has been located and the necessary collecting permits and landowner permission obtained, the site is monitored until the most propitious time to collect propagules. If several populations are available, the one most seriously at risk is selected first. Unless the conservation situation is desperate, for example if bulldozers are about to destroy the site, only what it is thought that the population can reasonably afford to give up is taken, in order to have as little impact as possible on its persistence. How much material to take without having an adverse impact on the individual plants (in the case of cuttings) or the population as a whole (especially in the case of annuals) is a difficult judgement to make. When working with a population about which very little is known, but for which time is very short, the risk of making the species' already precarious existence even more doubtful is great.

Fortunately, experience can help with this decision, but ultimately it is as much a matter of intuition as of science. The methods used at Bok Tower tend to be very conservative, looking first at the size and overall vigour of the population, natural seedling generation, if any, and the likelihood of success with the material taken. One dares take very little from small populations of plants that are slow-growing or produce few seeds and which are known to be difficult to propagate. For example, *Lupinus aridorum* is known from only 15 sites, consisting of from one to 200 plants, of which fewer than half produced seeds in 1990. This species has never been grown to maturity in cultivation, so the chances are that all or nearly all the seeds collected will die. However, none of the natural sites is protected; the species has no long-term future on any of them. There are also rare species that seem to be limited more by habitat destruction than by biology. Often, these species set abundant viable seeds which can be collected with little danger to the persistence of the population. *Chrysopsis floridana* is a small, weedy composite with a very restricted geographic distribution on the edge of an expanding metropolitan area. Each plant produces hundreds of seeds which are readily cultivated.

One factor that can mitigate the removal of seeds from a rare plant population is the enhancement of the on-site germination of the remaining seeds. In the hot, dry, sandy scrub areas where most of Florida's rare plants grow, many seeds dry out or are eaten before they can germinate. Their chances are considerably improved if the collector simply pushes some of the remaining seeds into the soil.

Knowing whether a plant is better propagated vegetatively or by seeds takes some experience and often some experimentation. At Bok Tower both methods are generally tried when beginning work on an unfamiliar species, before deciding which is best. Several mint species whose seeds are either short-lived (*Dicerandra* spp.) or very time-consuming to collect (*Conradina* spp.) are included in the programme. Fortunately, both root readily from tip cuttings, but only at certain times of the year. None of the endangered woody shrubs roots well from cuttings (three Annonaceae, one *Chionanthus* and one *Prunus* species), and yet all grow readily from seed. One of Florida's rarest plants, *Ziziphus celata*, known from only five clones, will not root from tip cuttings and does not set seed. Fortunately, it sprouts readily from root cuttings.

Having decided what to collect, there is the question of how to collect. A lot has been written about random sampling methods, but unfortunately much of the theory is irrelevant under the pressure of very practical considerations in the field. Often, the best that can be done is to document the known faults in the sampling method used, in the hope that they can be rectified later. Time constraints often preclude visiting a population more than once in a season even if individual plants ripen seeds over a long period of time. The best seeds ever collected of *Chionanthus pygmaeus* were from a late-blooming plant whose flowers had escaped damage by the weevils which frequently decimate seed crops produced in mid-season. By making this additional late-season collecting trip, genetic material was probably collected which was not represented in the earlier sample.

Rattlesnakes, cactus plants, wild hogs and boundary fences can also limit the sample, as can the collector's own aesthetic sense. The natural inclination is to take from the largest, healthiest, best-looking plants while small, weak plants (those least able to give up seeds or cuttings) may be under-represented. If scrawniness is a survival mechanism (e.g. against herbivores), then these plants may be genetically significant.

Germplasm conservation

There are limitations inherent in *ex situ* conservation in living collections. It is very expensive in labour, land, laboratory and greenhouse equipment and other institutional resources. Especially considering the long-term costs, over many generations, conservation in living collections is often a poor substitute for preserving a species in its own habitat. It works spectacularly well for some species, less well for others and not at all for a few. Long-lived woody species are the easiest to preserve, even if they take more space, since whole collections can live and prosper with little care. Long-lived perennials with predictable life cycles may require periodic propagation, but with good care there should be few losses over time. Annuals and short-lived perennials, whose survival strategies rely more on numbers than longevity, often have high attrition rates; whole collections may die suddenly. These problematic species require more comprehensive solutions. Serious conservation efforts over sustained periods of time will require far larger numbers than have been dealt with so far. To really preserve a species it is necessary to think on a much larger scale, perhaps in terms of thousands of individuals.

Seed storage appears to be a low-cost alternative to garden cultivation, but experience at Bok Tower suggests several caveats. Often, very little is known about natural regeneration in these wild plants or how long seeds remain viable in storage. Many species appear to have low germination rates and high attrition rates over the life of the individual plants. Only by growing the species for several generations in numbers large enough to reflect natural systems is it possible to determine how many seeds will be necessary to reproduce a genetically representative, self-sustaining population. Merely putting seeds in storage without a comprehensive research programme is inadequate. A packet of frozen seeds is useless if no one knows how to grow them out.

Certainly, increased knowledge of these rare species is the greatest benefit of the Endangered Plant Program. By observing the plants in all stages of growth it is possible to learn to identify them in the wild when they are not in flower, which is often difficult to do with annual species and perennials which sprout from a dormant stage. Drawings of the plants in their infant and juvenile stages make it much easier to find and inventory them during other seasons of the year. Daily observations over a period of years provide continuous phenological data, much better than one-time observations in the field. It is possible to document a range of dates for flowering, seed set, dormancy and sprouting, which will assist in preparing long-term management schedules.

The successes, and also the failures, of trying to grow the plants in cultivation have yielded new information not only about the individual species but also about the systems in which they grow. Having so many rare plants together in one easily accessible place has been a powerful tool in educating the land managers who must care for the plants and the public policy makers who will ultimately make their preservation possible.

Strategies for the future

Ideally, the inclusion of a rare species in the Endangered Plant Program collection will be only the first step towards its recovery as naturally reproducing wild populations on protected land. The emphasis now is less on the long-term maintenance of the species in cultivation and more on developing the technology to enhance wild populations and to introduce new populations on to protected sites. Four such introductions have already taken place with some success, although it may be years before it will be certain that the plants will prosper and reproduce on the new sites.

Ironically, it has been easier to develop the techniques for introducing a plant than it has been to find suitable land. There is enough information now to begin introduction projects for most of the species in the collection and Bok Tower Gardens is actively soliciting the cooperation of land managers of appropriate sites. Unfortunately, rapid habitat destruction means that there are fewer and fewer such sites, meeting both the biological and administrative criteria to make them suitable for the introduction of rare plants.

It has been argued that garden cultivation of rare species can too readily be used as an excuse for the destruction of their natural habitats. In fact, the Bok Tower Gardens Endangered Plant Program makes a compelling case for preserving natural habitats. The limitations inherent in *ex situ* conservation (both practical and philosophical) argue forcefully for habitat preservation. What good are our conservation collections if they have no future beyond being a botanical curiosity in a flowerpot on a greenhouse bench?

KENGO's Genetic Resources Conservation Project

37

G. Arum[1] and K. Kiambi[2]

[1]*KENGO, PO Box 48197, Nairobi, Kenya:* [2]*IPGRI, c/o International Laboratory for Research on Animal Diseases, PO Box 30709, Nairobi, Kenya.*

Local rural people are generally very knowledgeable about the wild plants around them, many of which have local names and are important to them economically or feature in folklore. This knowledge is the best starting-point for effective conservation, which requires accurate and up-to-date information on the status of plant populations, on the extent and nature of plant use by local communities, and on the capacity of the resource base to support different economic activities. Indigenous knowledge can be used in the evaluation of the cultural, biological and economic importance of biodiversity. It is also useful in creating awareness of the importance of biodiversity, as it is generally easier for the general public to relate to than the results of scientific trials. Such public awareness is essential. Conservation projects conceived in a hurry and imposed from outside, with minimal participation by the target communities, have often actually resulted in the erosion of resources. Collecting indigenous knowledge about plants (ethnobotanical information) involves communities in conservation efforts, rekindling their interest in the plant genetic resources available in their surroundings.

A strong ethnobotanical component will ensure that conservation goes beyond basic authoritarian protective measures. It will help in developing conservation methods which are egalitarian and in harmony with the environment and local peoples' material and cultural needs. The Genetic Resources Conservation Project of the Kenya Energy Non-Governmental Organization (KENGO) was established in response to this perceived need for baseline ethnobotanical information on genetic resources.

The work at KENGO combines resource production, research and extension and training strategies. A team of specialists provides technical expertise and coordination. The principal outreach points are KENGO's Regional Resources Centres on Environment and

Development, spread through the various major ecological regions of Kenya. These centres provide support services and information to community groups in land use management, wood-fuel conservation and use technologies and natural resources conservation.

Research activities on plant genetic resources were initiated in 1983, with a focus on the indigenous trees of the arid and semiarid lands, where the need was (and is) perceived to be greatest. The impetus for the initiation of this project was the realization that the indigenous plant resource base of Kenya was being neglected at both the research and policy levels, and that this would result in severe genetic erosion, as clearing of forests and natural habitats to give way to agricultural development and infrastructure accelerated. Cultural erosion, the passing away of the older generation who are the depositories of indigenous knowledge base, added to the problem of genetic erosion. The erosion of cultural knowledge has been worsened by the lapsing of mechanisms for passing on the information to the younger generation. The project's initial task was therefore to carry out an inventory of the indigenous trees of the arid and semiarid lands of Kenya and to document indigenous knowledge on their actual and potential uses.

To date the project has carried out ethnobotanical surveys in eight districts in the arid and semiarid areas of Kenya. Technical information on the characteristics and uses of indigenous trees has been compiled for publication in local dailies, pamphlets and resource books, with the aim of promoting their use. This information, which covers the physical characteristics of species, their uses and environmental requirements, planting and management methods and the socioeconomic and cultural value of the trees, is provided in a format that is accessible to field workers and rural people.

For the botanical studies, the method employed at KENGO involves taking photographs of the whole tree, plus close-ups of flowers, fruits, leaves and bark. These are supplemented by herbarium specimens and a description of the plant. Taxonomic identification is carried out by the National Herbarium, Nairobi. A set of comprehensive questionnaires is used in describing the plant. First, there is the provenance and botanical questionnaire. This seeks to gather information on the specific locality of the population and its natural habitat, including physiographic features, pedology and associated plants. Data on the frequency of the tree are also recorded. Phenotypic characteristics of the tree, including general morphology and size, rooting and branching systems, and shapes and sizes of floral parts, fruits, seeds and leaves are recorded.

The second questionnaire records socioeconomic information. It is normally used to enter answers given by a local respondent. Very often, this is an elderly person, or one who has extensive knowledge of plants, such as a herbalist. The informant accompanies the researchers to the field and identifies the trees by their local names and then proceeds to provide ethnobotanical information, in particular the uses to which different parts of the plant are put by the local community. This may include direct uses such as timber, food, fodder, medicine, fuelwood, gums, dyes, and so on. It may also include information on indirect uses such as soil erosion control, honey production or swamp drainage. Data on fruiting and seeding time and seed dispersal methods are also entered.

All known propagation methods of the plant and pregermination treatments of the seeds are recorded. This information is normally based on either common practice or long-term observations. For example, on how to propagate *Markhamia lutea*, a typical answer, based on common practice, would be to cut a flexible stave and introduce both ends into the ground. On the other hand, an entry such as 'seedlings and saplings of this species will only be seen after there has been a bush fire or some severe flooding' is an example of long-term observation. It is typical of *Acacia* spp.

The socioeconomic questionnaire also records information on the agroforestry potential of the species, for example how well it mixes with crops, and such limitations of the plant as weediness, susceptibility to pests and pathogens and any toxicity. The questionnaire also seeks to assess and document the significance of the plant in the local culture, for example the existence of any taboos against the plant's use, propagation and cultivation. An attempt is made to ascertain which plants or plant communities have disappeared from the area, and why.

Public awareness campaigns on the need to conserve indigenous trees have resulted in numerous requests for seeds and other propagation material. This has led to the incorporation of a seed procurement and distribution component into the project. To ensure procurement of good-quality seeds, the project organized a seed collecting and handling training course in which 35 people from various institutions carrying out forestry activities in the arid and semiarid areas of Kenya participated. These people are the core of the project's seed acquisition programme. In order to become acquainted with pertinent problems associated with seed collecting in the field, the project also organizes seed collecting expeditions. Over the years, numerous high-quality natural seed sources have been identified in various parts of the country, which are recommended to the seed suppliers. The seed stands are mapped and fenced and their conservation is entrusted to the local Forest Officers. If the stands are on private farms, farmers are given an incentive to conserve them by the project purchasing the seeds from them. Records on these seed stands are kept, including such information as species, population size, physical and phenological characteristics (including seed collecting period), precise geographic location and name and address of custodian.

Before the project team organizes a seed collecting expedition, information is gathered on the fruiting periods for different species in different parts of the country. Sources of this information are the literature, District Forest Officers and the custodians of the seed sources. In the field, 50–100 seeds are harvested from each individual tree in a population and then mixed to make one bulk sample. Seeds are normally collected from two to ten randomly chosen trees, depending on population size and availability of physiologically mature seeds. Seeds are harvested from the crowns of healthy mother trees by climbing up the tree or using ladders and tree pruners to reach seed-bearing branches. Collecting seeds that have fallen on the ground is discouraged as they may be immature, aged or diseased. The number of trees from which a sample has been taken and phenotypic characteristics of the trees are recorded in a pre-prepared seed collecting data sheet.

All the seed collected by the project team or sent to the project by the trained collectors is documented with information on the botanical name, local name, place and date of collection, name and address of collector, uses of the tree and any other indigenous knowledge of the tree, particularly seed storage and propagation methods. Upon delivery, seeds are cleaned, dried, tested for viability, packaged and put through a registration process. This includes being assigned an accession number and being weighed. Seed samples are normally meant for distribution purposes and they are supplied free of charge. Users include research institutions, schools, individual farmers, non-governmental organizations (NGOs), community groups and development agencies. Seeds are packed in 50–200 g packages and the following information attached: source (collecting site), collecting date, name(s) of collector(s), viability, uses, known seed presowing treatment and propagation methods. The project recently developed a collaboration with the Kenya Forestry Research Institute (KEFRI) for the medium-term conservation of excess seeds received from large consignments.

In order to address some practical constraints in indigenous tree seed handling and propagation, the project initiated a pilot seed testing and growth trials programme for some indigenous tree species. The site comprises 7.5 ha donated by the Jomo Kenyatta University College of Agriculture and Technology. Seed samples for research purposes are processed and viability determined after storage under different environmental conditions and in different containers. Through the pilot trials, seed viability and storage conditions, presowing treatments and propagation methods for 40 economically important indigenous trees have been determined and recommended to farmers through the Agroforestry and Field Extension Programme. The establishment of a tree nursery has ensured the reintroduction and distribution of these trees back to the communities.

The trials site also includes seed stands and a botanical garden in which 96 different indigenous tree species are now conserved. This includes threatened taxa such as *Dalbergia melanoxylon* (African ebony) and *Diospyros* spp. The site has recently become popular for public education and demonstrations, particularly to schools and environmental organizations. It has also become an important source of seeds and propagation materials, even to herbalists.

Interdisciplinary collecting of *Ipomoea batatas* germplasm and associated indigenous knowledge in Irian Jaya

38

G.D. Prain[1], Il Gin Mok[1], T. Sawor[2], P. Chadikun[2],
E. Atmodjo[2] and E. Relwaty Sitmorang[2]

[1]*CIP/UPWARD, c/o International Rice Research Institute,
PO Box 933, Manila, Philippines;* [2]*Root and Tuber Crops Research
Centre, Cenderawasih University, Manokwari, Irian Jaya, Indonesia.*

Introduction

Sweet potato (*Ipomoea batatas*) constitutes a major food source in a number of areas of Indonesia, particularly in Irian Jaya, the western half of the island of New Guinea. Whereas annual per capita production for the whole country is currently around 12 kg (total annual production is a little over 2 million metric tons), in Irian Jaya per capita production stands at around 100 kg (total annual production 180,000 metric tons). In Irian Jaya sweet potato is consumed as a staple mainly in the mountains, for example in the Regencies of Jayawijaya, Jayapura, Panai, Manokwari and Sorong. In other parts of the mountains, taro (*Colocasia esculenta*) is the primary staple and, in much of the coastal areas, especially the swampland to the south, sago (*Cycas circinalis*) is a more common staple. Rice is increasingly consumed in urban centres.

Whereas taro and sago are indigenous to Asia, sweet potato is almost certainly of Central and South American origin (O'Brien, 1972). Archaeological research discussed by Yen (1982) suggests that it may have arrived in New Guinea from South America via Polynesia well before it was introduced into other parts of Asia by the Spanish. Since its introduction into New Guinea it has displaced taro from many areas and has been responsible for major shifts in social and political organization through its association with pig-raising. An enormous number of cultivars have been selected and conserved by New Guinea peoples, attesting to its economic importance.

Collecting and evaluation of this rich New Guinea diversity for improved sweet potato performance there and elsewhere has been conducted almost exclusively in Papua New Guinea (Takagi, 1988), with relatively little collecting in Irian Jaya. Almost no effort has taken place in either region to systematically document local knowledge of New

Guinea varieties in order to accelerate characterization and use of cultivars. This chapter records the preliminary efforts of members of the Root and Tuber Crops Research Centre of Cenderawasih University, Manokwari, Irian Jaya, supported by the User's Perspective With Agricultural Research and Development (UPWARD) and by the Centro Internacional de la Papa (CIP), to systematically collect both the wide range of sweet potato germplasm and its associated indigenous knowledge. It attempts to document both the benefits and the dangers of the methods developed for this purpose. The approach and methods discussed here resulted from a fieldwork-based training workshop held in Manokwari and Anggi, Irian Jaya, Indonesia, on 3–12 February 1992. The results of the fieldwork are reported by Sawor *et al.* (1993).

Philosophy and approach

At least three key reasons for collecting are commonly cited: to enlarge scientific understanding of biological and evolutionary processes; to preserve the world's biodiversity for future generations in the face of severe threats; and to make available new cultivars to farmers and new germplasm resources to plant breeders for the development of improved varieties. Curiously, most of these activities have been undertaken as if collecting is discovering. Whether wild or cultivated, plants are too often automatically treated as unknown genetic packages awaiting only science to reveal their secrets. Often left out of the picture are the originators of cultivated plant germplasm and the specialists of wild species, in other words, gatherers, hunters and cultivator groups the world over. Although many scientists recognize the role cultivators have played in developing the diversity of landraces, far fewer recognize that this is part of a broader expertise which rural people have of their local environments, and hardly anyone has yet tried systematically to incorporate this local knowledge into germplasm conservation and evaluation.

The approach taken in the work in Irian Jaya is to recognize that the assiduous cultivation of sweet potato diversity evident there reflects a sophisticated knowledge of the crop which has evolved with the germplasm. This knowledge is as much a resource as the physical material it illuminates, contributing to more effective use of sweet potato genetic material in the present as well as being of potentially vital importance in an uncertain future. However, like the germplasm itself, this knowledge is in danger of disappearing as new varieties and other modern technologies are introduced. Both are in urgent need of conservation.

Methods

The methodology draws on a combination of conventional collecting practices, rapid participative survey techniques and ethnobotanical elicitation.

Team size and composition

Mobility in Irian Jaya is complicated by lack of roads in the interior (except in the central, Baliem Valley region) and the consequent need to take light planes often limited to four or five passengers. This limits the possible size of the collecting team. In this example, the local team consisted of four people and, since the activity was also part of a training and methodology development exercise, two additional resource persons took part.

From other studies of sweet potato agriculture in Asia, we know that women are often important repositories of expertise and the principal managers of the crop. It is therefore essential to involve local women in the germplasm documentation process. The presence of women in the team can help to ensure female input. One member of the collecting team in Irian Jaya was female.

Interdisciplinarity is a key ingredient of group composition. The spread of disciplines should if possible cover social sciences as well as genetic resources and taxonomy. It is rare, however, for provincial root crop research centres to have taxonomic specialists and in the present case two members of the team were agronomists with exposure to genetic resources issues through attendance at courses. On the social sciences side, the original plan had been to include an anthropologist from the Social Sciences Faculty in Jayapura, but this proved logistically too complicated to organize. Both social scientists were trained in a broad socioeconomics course at the Faculty of Agriculture in Manokwari.

The distinction between multi- and interdisciplinarity needs emphasizing here. Whereas 'multi' implies several areas of expertise and points of view, which is good, 'inter' implies the interpretation, interdependence and sharing of those points of view, which is better. In other words, the team was not expected to establish an inflexible division of labour between germplasm collectors and knowledge collectors. Rather, in mixed pairs or as a group, specialists were expected to take the lead in their own area of specialization, while participating in all types of cultivar and data collecting.

Sampling

The unit for sampling was taken as the individual sweet potato plot rather than the whole of a farmer's holding. Plots were chosen for sampling if a large number of sweet potato cultivars appeared to be growing there. Only plots where the owner was known and could be located were sampled. There was also an attempt to sample plots in all the different hamlets of the village to maximize cultivar range and possibly ecological diversity. These criteria were not always compatible. This led to a slower collecting process than would perhaps have been the case in a more 'conventional' collecting expedition.

For each cultivar found in the plot, five proximal stem cuttings were taken, each approximately 50 cm long. Where only one or two plants of a cultivar could be identified, fewer cuttings were taken to avoid the risk of the farmer losing the variety. Collecting the same cultivar twice in the same plot was avoided, but no attempt was made to avoid collecting what appeared to be the same cultivars from different plots, simply because duplication could not be conclusively determined in the circumstances.

The concept of 'plot' that was used by the team turned out to correspond to a rather complicated notion in Anggi. Rights of access to particular areas of land are determined by tribal affiliation, but ownership or use of particular gardens ('ro') is by household, with a number of kinship-related households often opening a new area together, but separately planting adjacent gardens within the area. Ideally, the sample plot should coincide with the 'ro' cultivated by a single household, in order to make more specific the local knowledge elicited from the farmer. In several cases, however, the enthusiasm of farmer consultants led them to dig up cultivars planted by different households in adjacent gardens. This also complicated the collecting of data on the precise number of cultivars each household managed.

Documentation of information

There are at least three pools of information associated with a plant germplasm collection:

- the genetic make-up of the plant;
- cultural (or indigenous) knowledge about the plant;
- cultural, socioeconomic and ecological characterization of the plant's 'context' or environment.

At the time of collecting, little can be directly learnt about the first information pool, apart from the gross observable characteristics of cultivars. Simplified passport data sheets were used to register the usual passport information (collecting number, location, local name, etc.) and also basic information about each cultivar (skin and flesh colour of roots, root shape, earliness, uses, etc.). Some of the simpler physical characteristics of the sample plot, such as slope, soil type and stoniness, were also recorded on the passport data sheets.

The second and third information pools were handled through a topic guide sheet divided into three sections (Appendix 38.1):

- the ethnobotany of *Ipomoea batatas* and individual cultivars;
- characteristics and management of the sample plot;
- broader 'contextual information' about local cropping, farming and livelihood systems.

Topic guides were worked out during the pre-fieldwork planning sessions based on the available secondary information, the field experiences of the local team members and a reconnaissance trip conducted in the area some months earlier. These were then amended and improved during the fieldwork itself.

Data were collected using a range of different methods.

Informal interviews

Informal interviews with individual farmers were conducted during the extraction of roots of individual cultivars, to gather preliminary cultural information on the cultivar and on the history of the plot. This type of interviewing proved more successful for gathering data on the plot than on individual cultivars, since it was found that discussion of a cultivar in isolation is often difficult and that there was little time to get into great detail while also collecting and labelling specimens. It was therefore decided to limit questions and observations on a few major characteristics of the cultivar (local name, plant type, flesh and skin colour, time to first harvest, etc.) and to concentrate on gathering information on the plot. The rhythm of collecting proved to be an occasional source of conflict between team members from different disciplines. The collecting of specimens and the filling out of basic information in the passport form was relatively easy and fast. Elicitation of more detailed information on the plot and the cultivars was much more time-consuming, limiting the number of plots that could be sampled in a day. A second type of interviewing with individual farmers was conducted in their house later in the day, to go over the evaluation of the cultivars in detail.

Individual interviews were not initially very satisfactory. This could have been partly due to the elicitation method used at first, which involved discussions of individual cultivars in turn, rather than comparing and contrasting cultivars. It was also partly due to the sort of people whose plots the team were first taken to and who were the first interviewees, local political leaders and perhaps not necessarily the real experts on sweet potato cultivars. Obligations to these local leaders, who grant permission for collecting to take place at a locality, must of course be honoured. However, care should be taken that they do not derail the search for local expertise. Wider testing of comparative evaluations with a range of local people, especially women, should be attempted. On several occasions the team subdivided, and the female team member led interviews with women farmers. These worked quite well when conducted in the home, but in the field the men who were present tended to answer for the women.

Key informant interviews

Key informant interviews were used mainly to obtain contextual data and to a lesser extent to discuss the classification and other aspects of cultivars. An employee of the Kecamatan (local District Administrative Office) served as a key informant for the village of Iray. A Rural Secretary acted as a key informant for the village of Sururey. Both key informants were respected elders in the villages where they hold these posts. To understand local cropping practices and the farming and livelihood structures, a number of participatory rural appraisal (PRA) techniques were used as part of these key informant interviews:

- A personal biography was elicited to help understand the dynamics of the local communities and significant local events.
- Transects were drawn of both villages to show different local uses of land and resources and to highlight potential problems associated with resources degradation (Fig. 38.1).
- Social and resource maps were used to characterize land distribution and important institutions at the sites (Fig. 38.2).
- Seasonal and yearly calendars show cropping systems and farmer management activities and work diagrams show the division of labour between men and women (Fig. 38.3).
- Planting material flow diagrams help to illuminate the management of continuity in plantings (Fig. 38.4).

In retrospect it would have been possible, and perhaps more beneficial, to have involved groups rather than key informants in these PRA activities.

Group interviews

Group interviews with farmers were found most successful and quickest for the elicitation of salient characteristics of cultivars. Two modalities were tested. The first was in the field plot, immediately following the sampling of the different cultivars. Both cuttings and roots were harvested, labelled and documented in the passport forms. They were then laid out in an open area of the field. Once all the specimens were laid out in this way, the farmer, his family and any other people present in the plot (usually many!) were invited to gather around the cultivars to discuss them. Using the topic list as guide, the discussion moved through different aspects of the cultivars, from ease of establishment, rate of formation of roots, plant habits, reaction to stresses, ease of harvesting and then the whole postharvest area of ease of cooking, different aspects of taste and storability. The field setting proved very dynamic, with interventions and side discussions, so that recording of opinions and evaluations proved to be quite difficult.

The second modality was to gather a group of farmers together in the evening to discuss the cultivars in a more leisurely way in a public or private location. This allowed more concentrated discussion in an environment where it was easier to record the observations. The site chosen by village elders in Sururey was the unoccupied house of the local forestry technician, which was also where the team stayed. This may not have been the most neutral of sites and probably contributed to the largely male elder composition of the meeting, excluding a potentially important part of local knowledge.

The aim of these group discussions was to elicit comments on those cultivars which were most positively or negatively noteworthy with

Type of Land	Upland	Midland	Marshland	Lake
Soil	Compact clay soil	Loam with stones	Organic	
Crops	None	Planting on the agricultural land is divided into 2 cycles: Cycle I: kidney bean, maize, sweet potato, pumpkin. Cycle II: kidney bean, maize, sweet potato, pumpkin, shallot, garlic, onion, carrot, potato Homeyards fertilized with household rubbish and planted with potato, maize, sugarcane, garlic, shallot, onion, carrot, celery, sweet potato	Drainage canals built on the marshland to enable the planting of kidney bean, maize, sweet potato, pumpkin, shallot, garlic, onion, carrot, potato	
Livestock	*Kaskus*, birds	Dogs, pigs, chickens, ducks and goats	Dogs, chickens, goats	Fish, eels
Other products	Rattan, firewood, building materials			

Fig. 38.1. Transect of the village of Iray showing the distribution of natural resources.

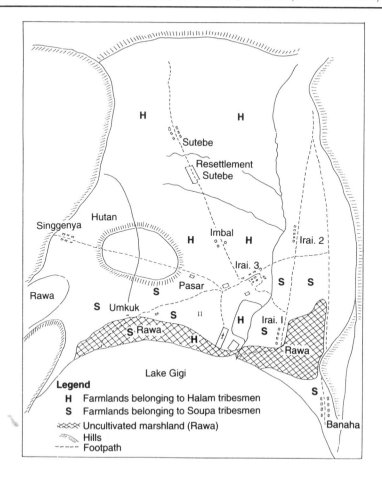

Fig. 38.2. Map of the village of Iray.

regard to a particular characteristic. Discussion of characteristics was iterative, the mention of one aspect (sweetness, say) provoking group members to raise and discuss related aspects (for instance fibrousness). This approach has the advantage of quite quickly establishing a consensus on the important characteristics possessed by certain cultivars (Appendix 38.2). Its disadvantage is the inability to characterize some of the more 'middle-of-the-road' cultivars, which are neither 'famous' nor 'notorious' for any particular characteristic. Ideally, this procedure should be more 'visually iterative'. Using a blackboard or flip chart, the assembled cultivars should be characterized by the group in the way described, so that the group can recheck and complete characteristics assigned to 'extreme' cultivars and can give more attention to 'mediocre' cultivars. This is an adaptation of a technique used to evaluate unfamiliar varieties in Latin America (Prain *et al.*, 1993). A similar argument is made in favour of the use of the triad test to identify salient evaluation characteristics (Sandoval, 1994).

	J	F	M	A	M	J	J	A	S	O	N	D

Year 1 (First cycle)

Maize

Kidney bean

Pumpkin

Sweet potato

Year 2

Sweet potato

Year 3 (Second cycle starts in May)

Sweet potato

Maize

Kidney bean

Shallot

Garlic

Onion

Carrot

Potato

Fig. 38.3. The cropping calendar at Iray and Sururey.

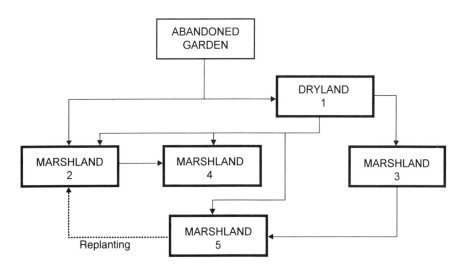

Fig. 38.4. Planting material flow diagram for the plots owned by Markus Ahoren at Sururey. Planting material is generally obtained from old gardens. The planting material for a particular variety normally originates from more than one plot. This is so that sufficient quantity and adequate quality of material can be obtained. Maintaining a constant supply of cuttings is complex and involves the sequential planting and maintenance of plots. This diagram gives a specific example of a household 'seed' supply system. Each box represents a garden plot. Numbers give the sequence in which plots are planted and the arrows show the flow of planting material. Plot 1 is the oldest existing garden. The material there came from a now abandoned plot.

Care needs to be taken that the zeal to document local characterization of cultivars does not result in forcing people to assign characteristics just to please collectors. In general, farmer knowledge of crops seems to be strongly focused at a 'generic' level (Berlin, 1992). That is, there is knowledge of sweet potato, of cassava or of taro, but knowledge of varieties within these crops is likely to be selective and tends to be comparative. This is why the group discussions of all varieties together produced most information.

An alternative approach to ascribing characteristics to cultivars is to start by eliciting what the locally important characteristics are. This can be done by open-ended interviewing, by systematic farmer evaluations in trials (Ashby et al., 1987; Prain, 1993) or by the triads test. In triads testing, farmers are asked to compare sets of three cultivars and to identify which pair is most closely related and which is the odd one out. The criterion for relating and discriminating these cultivars is left up to the farmer. In this way, the characteristics which are used for discrimination are identified (Sandoval, 1994). This procedure is time-consuming and therefore perhaps more suited to extended research in a single community or to 'multiple visit' collecting, rather than the 'single-visit' expedition of relatively short duration described here. This is even more the case with trials, which require a lengthy commitment by both farming families and researchers and could well be the next stage after collecting.

Data recording and synthesis

As already mentioned, a simple passport sheet was used to record basic identification data and some additional observational data on both the cultivars and the sample plots. Individual notebooks were used by all team members to record information coming from local people on particular cultivars, either during the collecting process itself or during the evaluation sessions. Additional local information on sample plots, contextual data on farming systems and local social and political structures as well as personal observations were also recorded in these notebooks.

Every night, members of the research team discussed, debated and brainstormed over the specimens collected and the individual notes and observations recorded during the day. From these discussions, a single daily record was made which was structured around the topic guide and linked to the specimens via a collecting number.

After the expedition, three simple relational databases were established for basic passport data, plot data and ethnobotanical data. Ideally, a fourth database including 'contextual' information amenable to matrix-type storage could also be added (Fig. 38.5). These databases were related via the accession code and a sample plot code.

The principal difficulty of data recording and data storage concerned what may be called textual or discursive information. In fact, it is rather difficult to pin down exactly how to describe this 'problem' information. It is sometimes referred to as 'qualitative', but this term is misleading

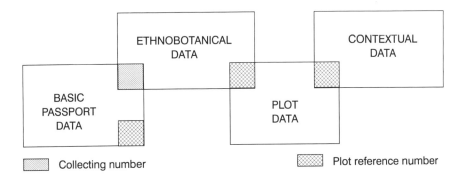

Fig. 38.5. Organization of sweet potato germplasm and indigenous knowledge data.

since some particular qualities, such as root colour or shape, are relatively easy to categorize and were recorded without too many problems by the team. This suggests that the issue is not whether information is qualitative or quantitative, but, rather, the ease with which data can be divided up into discrete 'bits' of information. Information already presented as numbers, such as dates, prices, yields, etc., is easily recorded and stored in databases. Non-numerical information, such as colour, taste or shape, which is associated with conventional classification schemes can also be digitalized in the process of recording and hence easily stored. There will often be information loss in this process, however, and sometimes distortion. A description such as

variety X has a blue-black colour

may become recorded as

variety X: colour = purple

to conform to a pre-existing classification scheme. This would defeat a major aim of the ethnobotanical elicitation work, which is to understand how local people structure and classify plants falling within the 'genus' sweet potato. The greatest problems, however, occurred where no guide existed for a 'translation' into a classification scheme. The practice of giving part cooked and part raw roots of certain cultivars to pigs and the reasons for doing this, for example, which were described by one farmer, were either not recorded, or were recorded as

variety X: use = pigfeed

This is a formulation which is readily included in a database but which loses most of the information in the original statement.

Part of the explanation for these difficulties lies in the training of technicians and researchers. Scientists expect to collect data that describes in the most precise way possible the reality under investigation, so that interpretations and analyses can be made 'from the outside'.

There is little preparation for collecting data that include local people's explanations of their practices and technologies. Explanations are least amenable to translation into classification schemes. They require the technique of précis or summarizing, which is unfamiliar to most technically trained researchers. A great deal more effort needs to be dedicated to expanding these skills if researchers are to succeed in fully documenting ethnobotanical knowledge of crop genetic diversity. At the same time, data storage techniques need to go beyond the use of 'memo' fields in conventional databases in order to really make use of textual information.

Handling of specimens

The cuttings of each specimen were wrapped in damp newspaper to preserve them during the period of the collecting mission. The perishability of cuttings means that sweet potato collectors have two options:

- plant out the specimens during their period in the field;
- stay in the field for short enough periods to allow the successful transfer of specimens to the *ex situ* gene bank.

A third possibility would be collecting roots rather than cuttings. The disadvantages of this are mainly the weight factor and the increased chance of spreading disease.

Conclusion

Though systematic plant collecting has a long history in Western science and the documentation of ethnobotanical knowledge has been for several decades a recognized subdiscipline of cultural and social anthropology, the integration of these practices is only just beginning. The work described here was a preliminary attempt to define a toolkit of methods which can contribute to that integration.

The dynamics of the 'interdisciplinary' relationship during collecting was very variable. Though the overall rhythm of the work was determined by collecting needs, documenting multiple spheres of knowledge clearly takes more field time than simply labelling a sample and giving it a number. The experience so far is that the biological team members find the discussions with farmers very rewarding and see the benefits of having this information at an early stage. Of course, the boundary dividing useful and superfluous information will depend on the perspectives of different disciplines and will always need to be negotiated. For example, biological scientists had difficulties justifying collecting data on ritual practices associated with crops and farming, even though anthropologists have clearly documented the seamless interconnectedness of 'economic' and 'ritual' knowledge and action in many societies (Sahlins, 1972). As it turned out, in this exploratory effort, ritual data were negotiated out of existence. In future, different intensities of data gathering will need to be established for different spheres of knowledge. We need a minimum core set of ethnobotanical and sociocultural data which will have to be defined, but there should also be opportunity for follow-up where important issues arise in other spheres lightly touched on.

Collecting germplasm is expensive and collecting indigenous knowledge as well considerably increases the costs. These extra costs can be justified by the increased utility of 'known' material for other farmers and scientists versus the relative uselessness of anonymous accessions. However, though the detail of documentation is constrained by the utility of what is documented, it should not be totally determined by it. Use value either for genotypes or for local knowledge cannot be fully known. Biological and cultural diversity need to be conserved for an uncertain future.

References

Acheta, A., H. Fano, H. Goyas, O. Chiang and M. Andrade (1990) *El Camote (Batata) en al Sistema Alimentario del Peru. El Caso del Valle de Canete*. Instituto Nacional de Investigaciones Agrarias y Agroindustriales (INIAA) and CIP, Lima.

Ashby, J.A., C.A. Quiros and Y.M. Rivera (1987) *Farmers Participation in On-farm Varietal Trials*. ODI Agricultural Administration (Research and Extension) Network Discussion Paper No. 22. Overseas Development Institute, London.

Berlin, B. (1992) *Ethnobiological Classification: Priciples of Categorization of Plants and Animals in Traditional Societies*. Princeton University Press, Princeton.

CIP/AVRDC/IBPGR (1991) *Descriptors for Sweet Potato*. IBPGR, Rome.

Mok, Il Gin and J. Schneider (1993) Collection and documentation of sweet potato germplasm: methodological experiences in Indonesia. In: Prain, G. and C. Bagalanon (eds) *Workshop Proceedings*. UPWARD, Los Baños.

O'Brien, P.J. (1972) The sweet potato: its origin and dispersal. *American Anthropologist* 74:342–364.

Prain, G. (1993) Involving farmers in crop variety evaluation and selection: an experience from the Peruvian highlands. In: Prain, G. and C. Bagalanon (eds) *Workshop Proceedings*. UPWARD, Los Baños.

Prain, G. and Mok, Il Gin (unpublished) Trip report on field visit to Irian Jaya, Indonesia, October 1991. UPWARD, Los Baños.

Prain, G., H. Fano, H. Goyes and M. Daza (1993) Using group assessment to determine farmer selection criteria for sweet potato varieties. In: Prain, G. and C. Bagalanon (eds) *Workshop Proceedings*. UPWARD, Los Baños.

Sahlins, M. (1972) *Stone Age Economics*. Tavistock Publications, London.

Sandoval, V.N. (in prep.) *Memory Banking Protocol: A Guide for Documenting Indigenous Knowledge Associated with Traditional Crop Varieties*. In: Prain, G. and C. Bagalanon (eds) *Local Knowledge, Global Science and Plant Genetic Resources: Towards a Partnership. Proceedings of the International Workshop in Genetic Resources*. pp. 102–122. UPWARD, Los Baños.

Sawor, T., P. Chadikun, E. Atmodjo, E. Relawaty Sitmorang, G. Prain and Il Gin Mok (1993) *Interdisciplinary Collection of* Ipomoea batatas *Germplasm Associated with Indigenous Knowledge in Anggi, Irian Jaya, Indonesia.* UPWARD, Los Baños.

Takagi, H. (1988) Sweet potato collections in Papua New Guinea. In: *Exploration, Maintenance, and Utilisation of Sweet Potato Genetic Resources.* Report of the First Sweet Potato Planning Conference. CIP, Lima.

Yen, D. (1982) Sweet potato in historical perspective. In: Villareal, R. and T.D. Griggs (eds) *Sweet Potato.* Proceedings of the First International Symposium. AVRDC Publication No. 82–172. AVRDC, Taipei.

APPENDIX 38.1
Topics and subtopics for collecting contextual and cultivar information

Contextual information

Contextual information is defined as information (obtained through observation and interviewing of different informants) on the functioning of the local agricultural system, with particular reference to common practices in the cultivation of sweet potato.

Location (district, village).
Altitude.
Source of information:

- group of farmers (random, local leaders, women, etc.);
- individual farmer (name);
- key informant (name and identity).

Population:

- approximate size, type and name of ethnic group;
- presence of migrants;
- history of the settlement;
- demographic features (migrant males, predominance of an age-group, etc.).

Transect of locality, with member of village if possible:

- main agroecological zones;
- associated soils;
- associated crops;
- associated livestock;
- associated problems;
- associated opportunities for improvements.

Calendar of activities (associated with rainfall and other climatic factors if possible):

- crops (especially sweet potato);
- livestock;
- male labour;
- female labour.

Sweet potato cultivation

Cultivars:

- approximate number of cultivars in the locality;
- why plant many cultivars?

- compare number with the past: more or less?
- why do cultivars disappear? is it important?
- interest in conservation of wide range of cultivars;
- who plants most cultivars these days?
- important outside sources of information.

Planting material:

- types used under what circumstances (be specific – tip cuttings, basal cuttings, root combinations);
- sources of material (individual maintenance system, links with neighbours, etc.);
- form of planting (number of cuttings, how placed in ground, etc.).

Land preparation, by zone.
Planting (use of mounds, on flat, 'pressing under' in garden, etc.).
Cultural practices:

- hilling up;
- weeding;
- use of organic/chemical fertilizer;
- presence of insect pests and diseases;
- use of synthetic and/or 'rustic' pesticides;
- presence of other stresses (water logging, drought, rats, etc.).

Ritual practices associated with sweet potato:

- at planting;
- use in rituals;
- use for curing;
- links to women;
- at harvest;
- with food preparation.

Uses of sweet potato tops and roots, by zone:

- estimation of percentages going to different uses;
- estimation of change in percentages during past 10, 20, 30 years;
- how does marketing system work?
- storage, if any;
- processing, if any;
- consumption.

Assessment of overall role, likely changes.

Sample plot for variety evaluation
Farmer details:

- name;
- number of family members;
- total area of farm;
- type of tenancy;
- most important crop;
- area of sweet potato.

Plot:

- agroecological zone and cropping system type;

- size;
- other crops.

General comments on sweet potato crop:
- production problems;
- diversity of cultivars: what is the advantage?
- grouping or classification of cultivars: what are the major categories?[1]
- has the farmer encouraged diversity by preserving new types?
- marketing issues, prices, etc. of sweet potato.

Comments on cultivars planted

Collecting number.
Local name(s) of variety:

- known by other names elsewhere?
- widely distributed over locality/other localities?

Physical characteristics (described by farmer):

- root shape and form;
- root skin and flesh colour;
- plant type (spreading, compact, where are roots deposited? is it a problem?);
- plant colour, texture.

Vegetative period (minimum and normal).
Productivity of roots:

- number of roots;
- size of roots: which important?
- performance in different soils;
- performance over the last few years.

Productivity of tops.
Quality of root: floury, watery, sweet, dry, fibrous, other, do family members seek it out on the plate?
Quality of leaves: do people/animals eat them? if so, what is important (succulence, non-hairiness, etc.)?
Effects of climate: special problems or advantages compared with other cultivars?
Effects of insect pests: special problems or advantages compared with other cultivars?
Effects of diseases: special problems or advantages compared with other cultivars?

APPENDIX 38.2
An example of indigenous knowledge of varieties recorded in Anggi, Irian Jaya

Classification and naming of varieties

Opportunities for collecting detailed information on the way varieties are classified and named were limited. Though important, this is complex and time-consuming and it was decided to give preference to the collecting of performance and quality data. When working in unfamiliar local languages, even just collecting variety names presents problems. In our completed list of varieties,

---

[1] This information will most effectively be elicited during the comparative evaluation of cultivars.

several names appear in more than one colour category, and others have very closely related orthographies, which could reflect mistakes in transcription. On the other hand, different varieties sometimes have the same name, and only careful characterization can eliminate duplicates and differentiate varieties.

Skin colour seems to be the principal criterion for classifying sweet potato varieties in Irian Jaya, though some informants use flesh colour. This ambiguity is also found in other areas (Acheta *et al.*, 1990). There are three main varietal classes based on skin colour: 'reds', 'whites' and 'yellows', plus a residual category. Information collected on an earlier visit to the area suggested that 'yellows' may be more recently introduced varieties (Prain and Mok, unpubl.). No information was collected on categorial characteristics of 'reds' or 'whites'.

Naming of individual varieties occurs through various ways. The simplest describes the person making the initial introduction or the place from where the variety was brought. For example, variety 'Mirer' was introduced by a missionary called Miller, whereas, 'Tiom' was introduced from the area of Tiom in the mountainous region of Jayawijaya. Other names derive from the morphology of the plant. Informants say 'Aug' mobatkej' (transcribed 'Aug' atkach' during the first visit), which means 'hard', referring to the flesh quality. Some names combine both external and quality characteristics, for example 'Bebau bob' (previously transcribed 'Aug' behop'), meaning 'white and hard'.

Although some names refer to the colour of either skin or flesh or other characteristics of the storage root, farmers seem to find it easier to identify the variety of a given sweet potato from its plant characteristics rather than its storage roots.

Indigenous characterization of varieties

Information on comparative agronomic performance was rather difficult to elicit. This may be related to the almost exclusively subsistence status of the crop. There are no major pests or diseases to which varieties may differentially respond. There was an interest in 'months to harvest', meaning the length of time it takes before large roots can be harvested. Five varieties were identified as maturing in four months, which is early for this altitude. No information is available at the present time on relative yield performance or on differential responses to abiotic stresses.

The identification of 13 varieties as 'spreading' is more an observation than cultural knowledge. In terms of formal descriptors (CIP/AVRDC/IBPGR, 1991), these varieties should probably be described as 'extremely spreading', since a preliminary *ex situ* evaluation of the material suggests that almost all the varieties from the area are spreading types (Mok and Schneider, 1993).

Local people consulted on the varieties showed much greater interest and willingness to differentiate varieties according to consumption characteristics. Characterizations were made according to ease of cooking and by cooked root texture and taste. As with most of these characterizations, only varieties most associated with particular qualities were mentioned, other, 'mediocre' varieties being simply left out. Only four varieties were identified as really notorious for their 'hardness' in cooking, while eleven varieties were noted as 'soft'. This latter characteristic needs further elucidation, since it has both positive and negative aspects. When associated with a high level of 'dryness' of texture (perhaps the case with 'Bekau ayosei' or 'Ayoseiya' and 'Bekau arpokmoi' or 'Arfokngoi', which are identified as 'dry'), it may have a negative connotation since the variety would easily disintegrate in the water. On the other hand, other varieties simply cook easily and quickly.

In terms of texture, local people evaluated the varieties by degree of 'fibrousness' and degree of 'wateriness' of roots. Fibrousness is also a quality which requires greater clarification to differentiate between a normally fibrous quality of the variety and the susceptibility of some varieties to become fibrous the longer they are left in the soil. These would be varieties which are not adapted to piecemeal harvesting.

Evaluation of taste focuses on the sweetness of the variety, with the majority regarded as 'sweet' and a few identified as especially sweet or as lacking sweetness.

Collecting by the Institute of Plant Genetics and Crop Plant Research (IPK) at Gatersleben

39

K. Hammer, R. Fritsch, P. Hanelt, H. Knüpffer and
K. Pistrick

IPK,
Corrensstr. 3, D-06466 Gatersleben, Germany.

Introduction

There is a long tradition of plant germplasm collecting at Gatersleben. H. Stubbe, the first director, collected in the Balkans in 1941–42, shortly before the foundation of the Institute of Cultivated Plant Research in 1943. Soon after the war, he encouraged further collecting work, for example in southern Italy (Maly *et al.*, 1987) and China. An intensive, systematic collecting programme was started in the 1970s (Table 39.1). The central parts of Europe were the first targets of this effort. In eastern Germany, where the Gatersleben institute is situated, genetic erosion was already so high at that time that it was not feasible to mount a conventional expedition. Instead, a series of articles was published in newspapers and popular gardeners' journals asking for samples of relic crops and rare cultivars (Hammer *et al.*, 1977). This approach was successful with garden plants (vegetables, herbs, fruit trees), but few accessions of field crops (cereals, pulses, oil crops) were obtained. Similar tactics were also later used in western Germany (Dambroth and Hondelmann, 1981).

The mountainous regions of Czechoslovakia (as it then was) and Poland were collected next. Here what was to become the standard Gatersleben collecting approach matured. This involves mounting multispecies missions for crops and their wild relatives, on the basis of ecogeographical and ethnobotanical data, in close cooperation with specialists in the host country, consulting local experts in the collecting areas. Each sample collected is shared among participating organizations. Other important aspects of the approach are:

- splitting of variable populations into phenotypically different lines when collecting;

Table 39.1 Collecting missions conducted by Gatersleben staff (1974–92) and accessions collected (after Hammer et al.,1994.

Collecting area (years)	Number of accessions
Czechoslovakia (1974, 1977, 1981)	1,153
Eastern Germany (1975–84)	694
Poland (1976, 1978, 1984)	442
Spain (1978)	344
Italy (1980–92)	2,414
Libya (1981–83)	468
Georgia, former USSR (1981–90)	2,873
Austria (1982, 1983, 1985, 1986)	265
Ethiopia (1983, 1984)	186
Korean DPR (1984–89)	530
Mongolia (1985, 1987)	67
China (1986, 1988)	67
Iraq (1986)	141
Cuba (1986–92)	661
Central Asia, former USSR (1987, 1988)	141
Colombia (1988)	112
Peru (1988)	37
Total	10,595

- the collecting of herbarium material, if possible during the missions, or at least from the first multiplication of the accessions in the Gatersleben experimental fields;
- preparation by the collectors of comprehensive collection lists, with preliminary botanical characterization, before the transfer of the material to the gene bank;
- preparation of the material for multiplication according to a specific standard (Lehmann and Mansfeld, 1957);
- botanical verification by the collectors in the multiplication fields;
- characterization and evaluation carried out jointly by the participating scientists and published in the form of catalogues, e.g. Kühn et al. (1976) on the collecting mission in Czechoslovakia in 1974;
- compilation of the information in checklists of cultivated plant species (Hammer, 1991; Hammer et al., 1992).

In the following years, collecting was extended to the classical Vavilovian gene centres, first the Mediterranean area and then the Near East, Ethiopia, East Asia and Latin America. The funding for some missions was provided by the International Board for Plant Genetic Resources (IBPGR). Most, however, were financed and organized by agreements between the former Academy of Sciences of the German Democratic Republic (GDR) and the equivalent academies or societies in other countries. From a total of nearly 100 missions (Hammer et al.,

1994), a few examples are described in the rest of this chapter to illustrate specific aspects of the Gatersleben collecting approach. The explorations of Italy, Georgia and Cuba are discussed first. The fourth example is the investigation of the genus *Allium*, carried out by the Department of Taxonomy, which illustrates some of the problems of collecting wild relatives of crop plants.

The Gatersleben gene bank is not only active in collecting, regeneration, maintenance, characterization and evaluation of plant genetic resources. It also encourages and stimulates use of the material in breeding programmes by other institutes (Hammer, 1993). In the collecting, determination and characterization work, the gene bank is strongly supported by the Department of Taxonomy and Evolution of the Institute of Plant Genetics and Crop Plant Research (IPK), which includes a herbarium housing the reference collection of the gene bank. The close connection between gene bank staff and taxonomists has been a strong factor in the development of the Gatersleben collecting approach.

Italy

Italy is part of Vavilov's Mediterranean gene centre (e.g. Vavilov, 1987). Agriculture was introduced into the peninsula from the eastern Mediterranean and the Near East about 5000–4000 BC (Zohary and Hopf, 1993). Extensive variation in the early crops has developed. Later, several new crops were domesticated in the area (Hammer *et al.*, 1992). Cultivated plants have also been introduced from most other centres of diversity (Hammer *et al.*, 1992). Crops from east Asia (after AD 1000) and from the New World (after AD 1500) play a particularly important role in Italian agriculture and horticulture, and in many cases have developed high levels of diversity. Italy has also been an important bridge between the Mediterranean and central and western Europe.

The exploration of plant genetic resources in Italy was much influenced by Vavilov. As his special interest was wheats, early Italian activities in the field mainly involved cereals (Strampelli, 1932). A first comprehensive exploration was carried out in parts of southern Italy and Sicily in 1950 (Maly *et al.*, 1987). The material collected is being maintained by the Gatersleben gene bank. The Laboratorio del Germoplasma (later Istituto del Germoplasma) was founded in 1969 within Italy's Consiglio Nazionale per la Ricerca (CNR). It started a comprehensive programme of collecting and evaluating indigenous plant genetic resources. In the first years, this concentrated on *Triticum*, but later the scope of activities was widened. In 1980, the Istituto del Germoplasma established a cooperative programme with the gene bank of the Gatersleben institute, within the framework of an agreement between the CNR and the Academy of Sciences of the GDR. Annual joint multi-species collecting missions have since been carried out in Italy. The material is duplicated at Bari and at Gatersleben, and is evaluated at both sites.

The exploration programmes are planned jointly and carried out using additional funds from the Istituto del Germoplasma. After the recent political change in Germany, the cooperation was continued with partial support from the European Community (now the European Union). The most recent joint exploration was carried out in Sardinia in the autumn of 1993. The exploration of Sardinia will be completed in the next two years. Two more years will be necessary to cover some smaller islands, so that within about four years the first general overview of Italian crop genetic resources will be possible. A summary of the collections made between 1980 and 1992 is shown in Table 39.2.

Much information has been gathered in the field in the course of this collecting programme, for example on genetic erosion. A comparison was made in selected areas of the current situation and the results of the Maly mission (Maly *et al.*, 1987). An increase in the extent of genetic erosion was observed from south to north, and field crops were found to be more vulnerable than garden crops. Assessments of the genetic erosion status of some 522 crops in southern Italy and Sicily have been made (Hammer *et al.*, 1992). It has also been possible to elaborate proposals for *in situ* and 'on-farm' conservation on the basis of the fieldwork. Crops growing sympatrically with wild relatives, and for which introgression could therefore be a factor, were given particular consideration. Examples include *Secale cereale/S. strictum*, *Beta vulgaris/B. maritima*, *Pyrus communis/P. amygdaloides* and *Brassica oleracea/B. rupestris* (Hammer *et al.*, 1992). The most striking findings have been introgressions involving the traditional wheat species einkorn and emmer in some areas of southern Italy (Hammer and Perrino, 1984). These wheats have long been neglected, with no literature reports since the end of the last century. The discovery of this material led to the development of a novel method of exploration, i.e. the indicator-crop method, in which the cultivation of crops like einkorn in certain areas of southern Italy is taken as an indicator of traditional agriculture, and therefore of the persistence in those areas of traditional landraces of other crops. It also resulted in increased public awareness of traditional crops. Ecologically minded, 'green' farmers throughout Italy are increasingly taking them up again (e.g. D'Antuono, 1989). This is a promising model for 'on-farm' conservation.

Table 39.2 Material collected in Italy (1980–92).

Crops	Number of accessions
Cereals	704
Leguminous crops	829
Vegetables, oil, medicinal and other plants	881
Total	2414

Georgia

The plant genetic resources of the Republic of Georgia were investigated by annual Georgian–German collecting missions between 1981 and 1990 (Beridze *et al.*, 1992). The region is characterized by an enormous diversity of natural conditions. The warm-temperate Colchis region of western Georgia, with an annual precipitation exceeding 2000 mm, contrasts sharply with the more continental east of Georgia, with its steppes and semidesert receiving less than 400 mm. Subtropical crops, such as *Citrus* spp. and tea, can be grown in large plantations in the Colchis, whereas cereal cultivation reaches its altitudinal and drought limits in eastern Georgia and in the mountains of the Great and Small Caucasus.

This variety offers optimal conditions for the differentiation of a broad spectrum of local landraces of cultivated plants (e.g. Beridze *et al.*, 1992; Menabde, 1958). Agricultural tradition of Georgia goes back to the sixth millennium BC (Schultze-Motel, 1989). Part of the classical Near Eastern gene centre described by Vavilov, where Old World agriculture probably began (Vavilov, 1987), the country has, during its long history, been settled and influenced by many different peoples. Recently introduced crops, especially those from the New World (*Zea mays*, *Phaseolus vulgaris*, *Cucurbita* spp.), have broadened the range of cultivated plants (Hanelt and Beridze, 1991).

Collecting in the Caucasus dates back to the time of Vavilov. Again, special emphasis was given to endemic wheat species, which have proved very important in breeding (Dekaprelevič, 1954; Vavilov, 1987). The new exploration programme covered nearly all parts of the country. A first visit to Tbilisi allowed initial contact with local partners and experts in cultivated plant research at the Institute of Botany of the Georgian Academy of Sciences. A cooperative programme was agreed in the framework of a general agreement between this institution and the Academy of Sciences of the GDR. It included joint multi-species collecting missions and evaluation of the material during regeneration in experimental fields at Gatersleben and Tbilisi, with joint publication of the results. Travel routes were determined each year on the basis of previous experience and local information. Local agricultural offices and the seed control laboratories of the 'rayons' (districts) were consulted before and during the missions to obtain detailed information about local cultivation practices and the location of farmers or cooperatives in remote settlements where material of special interest could be expected to be found in cultivation.

Farmers, both male and female, showed an intimate knowledge of the diversity of crop species. They have names for many cultivars and can distinguish among them on the basis of morphology, use and agronomic requirements. They can sometimes provide information on the history and origin of landraces. The older generation is aware that the number of crops and the variation within them has decreased in recent years, mainly due to agricultural development. They are critical witnesses of the process of genetic erosion and were able to explain, for example, the decline of the famous endemic cultivated Georgian wheat species (*Triticum timopheevii*, *T. zhukovskyi* and *T. macha*) after the Second World War. Various socioeconomic developments have contributed to crop genetic erosion: the organization of agricultural cooperatives after the war; the shift in focus towards animal husbandry in mountainous areas such as Svanetia, Tushetia and Khevsuretia during the last 15–20 years; and population movements caused by the abandonment of high mountain villages and resettling in the lowlands or by the politically motivated expulsion of national minorities during wartime. All these factors had to be considered during the collecting missions in order to understand the situation in the fields and gardens.

Collections were made directly from fields and gardens and in markets, but more often from recently harvested material and from farmers' seed stores. Variable seed samples, especially of grain legumes and cereals, were divided into phenotypic lines (e.g. Hanelt and Beridze, 1991; Table 39.3).

Cuba

The plant genetic resources of Cuba are described in detail in the various papers in Hammer *et al.* (1992–94), from which the material presented here is drawn. Cuba is well known for its mosaic of ethnic groups and cultures and also for its rich wild flora comprising about 6700 species of higher plants, about 50% of which are endemic. However, the flora of cultivated plants was largely neglected until the 1980s, with only a few exceptions (e.g. Roig, 1975). Even Vavilov, who spent a few days in Cuba, characterized the country as lacking any interesting plant genetic resources (Díaz Barreiro, 1977). A new perspective developed as a result of work with traditional root and tuber crops (Rodríguez Nodals, 1984) and from the cooperation between the Instituto de Investigaciones Fundamentales en Agricultura Tropical (INIFAT) in Santiago de las Vegas and the Gatersleben gene bank. A number of joint collecting missions were carried out (e.g. Esquivel *et al.*, 1987), resulting in a large collection, including very diverse material of beans, paprikas, pumpkins and tomatoes and some accessions of rare and neglected crops. This material is maintained in the gene bank of INIFAT. Selected samples are also duplicated at Gatersleben (Table 39.1). It has proved valuable in evolutionary and taxonomic studies (e.g. Castiñeiras *et al.*, 1991).

After seven years of exploration, it is clear that Cuba is extremely rich in crops, with more than 1000 species cultivated. (Ornamental plants and forest trees have not yet been considered at this stage of the study.) There are also many wild and weedy relatives of crop plants, wild medicinal plants, etc. (Hammer *et al.*, 1992–94). Exploration for plant genetic resources was combined with ethnographic, taxonomic and other

Table 39.3 Accessions collected during the joint missions in the Republic of Georgia (1981 to 1990). Further separations have been done during multiplication and evaluation at Gatersleben (Beridze *et al.*, 1987).

Crops	Number of accessions
Cereals	570
Zea mays	320
Triticum spp.	90
Hordeum vulgare	75
Secale cereale	31
Avena sativa	14
Legumes	1172
Phaseolus vulgaris	923
Pisum sativum	58
Vegetables	780
Allium spp.	264
Cucurbita spp.	138
Cucumis spp.	47
Beta vulgaris	41
Brassica spp.	63
Lycopersicon esculentum	37
Raphanus sativus	36
Spice plants and others	318
Anethum graveolens	25
Apium graveolens	28
Coriandrum sativum	43
Petroselinum crispum	34
Ocimum basilicum	33
Total	2873

studies. This integrated approach allowed a thorough characterization of the history of Cuban crops, using taxonomic, historical and ethnobotanical information from both the literature and the field (Hammer *et al.*, 1992–94):

1. Pre-Columbian period (e.g. *Manihot esculenta, Nicotiana tabacum, Capsicum frutescens, Phaseolus vulgaris, Ph. lunatus*).
2. Early introductions – the Spanish influence (e.g. *Allium cepa, A. sativum, Artemisia abrotanum, Brassica oleracea, Coriandrum sativum, Mentha spicata, Origanum majorana, Ruta graveolens*).
3. African influence (e.g. *Abelmoschus esculentus, Solanum melongena, Elaeis guineensis, Sesamum orientale, Coffea arabica*).
4. East Asiatic influence (e.g. *Benincasa hispida, Allium tuberosum, Diospyros kaki, Raphanus sativus, Vigna umbellata*).
5. Latecomers, mostly introduced from the USA (e.g. *Glycine max*, many temperate fruit trees and vegetables).

6. 'Mysterious immigrants', i.e. plants for which at the moment it is impossible to trace the provenance (e.g. *Allium* aff. *glandulosum*).

Several crops can be put in more than one group. An example is *Allium fistulosum*. Types introduced early from the Mediterranean (group 2) never flower in Cuba, while an east Asiatic type introduced later (group 4) flowers occasionally. Another example is *Brassica juncea*, which includes both material introduced from Africa with the slave trade (group 3), and now found mainly as a weed, and also an east Asiatic type introduced fairly late, possibly from the USA (group 5).

Another result of the integrated, multidisciplinary collecting approach has been the development of a method for the study of the typical Cuban home garden, the 'conuco' (Esquivel and Hammer, 1988). It has revealed them as places of active evolution, supporting high levels of variation within crops as well as related wild and weedy species. The 'conucos' can be considered as focal points for the *in situ* conservation of diversity (Esquivel and Hammer, 1992). A relatively small number of selected crops with high infraspecific variation can be conserved *ex situ* in the Cuban gene bank. The majority of the hundreds of crops grown in Cuba will have to be conserved 'on farm', using the traditional horticultural system of the 'conuco'. The monitoring of this system must be developed and will be a topic of further investigation.

Collecting wild *Allium* species in central Asia

Allium is a large genus of more than 700 species, including many useful plants. Onion (*A. cepa*) and garlic (*A. sativum*) are well-known crops throughout the world. Others are important in more restricted areas, such as leek (*A. porrum*) in Europe, or only locally, such as *A. hookeri* in Asia. The Department of Taxonomy of the Gatersleben institute has been investigating the taxonomy of *Allium* for more than ten years (Hanelt *et al.*, 1992). George Don, the 1827 monographer of *Allium*, wrote that 'the genus *Allium* can only be studied satisfactorily from living specimens'. An attempt has therefore been made in the last decade to take into cultivation at Gatersleben as many species as possible, representing all subgenera and sections, to give taxonomic research a sound basis. Some material has been donated by botanical gardens. However, only about 70 species could be obtained in this way, and there was much misnamed material. Collecting missions in areas rich in critical *Allium* species have been the main source for the Gatersleben *Allium* collection, which now comprises nearly 300 species (Fritsch, 1990; Fritsch *et al.*, 1994).

The genus is distributed in the wild almost exclusively in the northern hemisphere, with a centre of diversity in southwest and central Asia. Contacts were made with the botanical institutions of the former Soviet Academy of Sciences in the now independent central Asian republics of Tadzhikistan, Uzbekistan and Kazakhstan. These countries are in Vavilov's central Asian gene centre, where the wild ancestors of onion and garlic occur. Despite its rich flora, botanical exploration of this region began only in the second half of the last century. There have been advances, especially in the last 60 years, but much of the region is still insufficiently explored. It consists for the most part of rugged mountains and desert-like steppes. The lower alluvial regions are mostly under irrigation and planted with cotton. They are more densely populated. Onions and garlic form an essential ingredient of most local dishes and are extensively cultivated throughout the region. Several wild species are also regularly collected from the wild and used as vegetables or seasonings, or preserved (Fritsch, 1990). These are well known to local people, who can distinguish among them and have vernacular names for many.

Collecting missions were organized on the basis of continuing, long-term scientific cooperation. Each sample was divided into duplicates, one of which remained in the country of origin. Collections were established in Dushanbe, Tashkent and Alma-Ata. The support of local institutions was invaluable to the Gatersleben staff for various reasons. The available Floras of the Central Asian republics are partly out of date (Vvedensky, 1971). However, herbarium specimens, the results of botanical surveys of local areas and other unpublished data are available at local scientific institutions. Together with the knowledge of local botanists, these allow much sounder conclusions about where, when and how to collect. Also, though communication in the field with adult men is usually possible in Russian, young people and women mostly speak only local languages. Contact with local people is important not only in documenting germplasm samples but also to obtain information on roads, accommodation and sources of good water in an area where there are few detailed maps available.

Depending on the target area and species, collecting took place at different times of the year between April and August. Collecting at different times imposes different restrictions and offers different opportunities to collectors. These are summarized in Table 39.4.

Joint collecting missions and visits to research stations have taken place nearly every year since 1983. The routes are shown in Fig. 39.1, where important collecting sites are also marked. Though *Allium* was the main target, whenever possible other plants of interest to the collaborating gene banks were also collected (Table 39.5).

Table 39.4 Collecting restrictions and opportunities in central Asia.

Spring (April to May)	The leaves of all *Allium* species are still developing; taxa can be distinguished only by leaves, sometimes also by young scapes	Bulbs must be kept moist in plastic bags for several days, dry after the leaves wither. Rhizomatous species must be kept moderately moist. Many species do not flower in the following year	Plants can only be collected in markets if seeds, bulbs or fruits of last harvest are being sold
Early summer (May to early June)	Leaves of early-flowering *Allium* species begin to wither, but can still be used for differentiation. Flower and scape characters are observable. Leaves of late flowering species are fully developed, their scapes only partly	Bulbs should be held rather dry, rhizomatous species moderately moist. Most species will flower in the next year. Best time to collect	In June, young bulbs of onions and garlic start to be sold in markets
Late summer (July to August)	Early flowering *Allium* species difficult to recognize. Only scape and capsule fragments remain, often dislocated. Some late-flowering species still in flower or bear capsules. May be possible to reconstruct leaf shape. Best time to collect them	Bulbs should be kept dry, rhizomatous taxa slightly moist. Most will flower in the next year	Onions, garlic, etc. can be bought in the market. Best time to collect them there

Table 39.5 Accessions collected in central Asia (1983–93).

	Tadzhikistan	Uzbekistan	Kazakhstan
Allium spp.	318	100	124
Cultivated plants and wild relatives	134	14	10
Ornamental and other plants	153	45	41

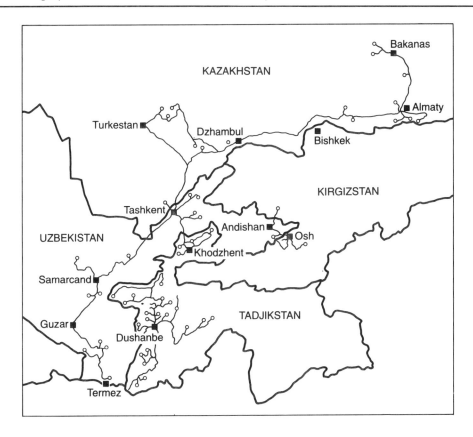

Fig. 39.1. Map of the collecting missions for *Allium* in central Asia (1983–93).

References

Beridze, R.K., P. Hanelt, D. Mandžgaladze and K. Pistrick (1987) Collection of plant genetic resources in the Georgian SSR 1986. *Kulturpflanze* 35:335–353.

Beridze, R.K., P. Hanelt, T.S. Girgvliani, V.N. Kandelaki and D. Mandžgaladze (1992) Collecting plant genetic resources in Georgia (South Ossetia, Dzhavakheti) 1990. *Feddes Repertorium* 103:523–533.

Castiñeiras, L., M. Esquivel, L. Lioi and K. Hammer (1991) Origin, diversity and utilisation of the Cuban germplasm of common bean (*Phaseolus vulgaris* L.). *Euphytica* 57:1–8.

Dambroth, M. and W. Hondelmann (1981) Some notes on the collection of landraces of cultivated plants indigenous to the territory of the Federal Republic of Germany. *Kulturpflanze* 29:41–45.

D'Antuono, L.F. (1989) Il farro: areali di coltivazione, caratteristiche agronomiche, utilizzazione e prospettive colturali. *Informatore Agricolo* 24:49–57.

Dekaprelevič, L.L. (1954) Vidy, raznovidnosti i sorta pšenic Gruzii [Species, varieties and cultivars of the wheats of Georgia]. *Trudy Instituta Polevodstva Akademii Nauk Gruzinskoi SSR* 8:3–58.

Díaz Barreiro, F. (1977) *Nicolás I. Vavilov. Primeras Relaciones Científicas de la URSS y Cuba*. La Habana.

Esquivel, M. and K. Hammer (1988) The 'conuco' – an important refuge of Cuban plant genetic resources. *Kulturpflanze* 36:451–463.

Esquivel, M. and K. Hammer (1992) The Cuban homegarden 'conuco': a perspective environment for evolution and *in situ* conservation of plant genetic resources. *Genetic Resources and Crop Evolution* 39:9–22.

Esquivel, M., L. Castiñeiras, B. Rodríguez and K. Hammer (1987) Collecting plant genetic resources in Cuba. *Kulturpflanze* 35:367–378.

Fritsch, R. (1990) Bericht über Sammelreisen in Tadzhikistan (1983–1988) zum Studium von mittelasiatischen Vertretern der Gattung *Allium* L. *Kulturpflanze* 38:363–385.

Fritsch, R.M., F.O. Khassanov and N.B. Zhaparova (1994) Collecting mission 1993 for wild *Allium* species in Central Asia (Kazakhstan and Uzbekistan Republics). Allium *Improvement Newsletter* 3:1–3.

Hammer, K. (1991) Checklists and germplasm collecting. *FAO/IBPGR Plant Genetic Resources Newsletter* 85:15–17.

Hammer, K. (1993) The 50th anniversary of the Gatersleben gene bank. *FAO/IBPGR Plant Genetic Resources Newsletter* 91/92:1–8.

Hammer, K. and P. Perrino (1984) Further information on farro (*Triticum monococcum* L. and *T. dicoccon* Schrank) in South Italy. *Kulturpflanze* 32:143–151.

Hammer, K., P. Hanelt and C. Tittel (1977) Sammlung autochthoner Kulturpflanzen auf dem Gebeit der DDR. *Kulturpflanze*. 25:89–99.

Hammer, K., H. Knüpffer, G. Laghetti and P. Perrino (1992) *Seeds from the Past. A Catalogue of Crop Germplasm in South Italy and Sicily*. Istituto del Germoplasmo, Bari.

Hammer, K., M. Esquivel and H. Knüpffer (eds) (1992–94) '. . . y tienen faxones y fabas muy diversos de los nuestros . . .' – *Origin, Evolution and Diversity of Cuban Plant Genetic Resources*. Vols 1, 2 (1992) and 3 (1994). Institut für Pflanzengenetik und Kulturpflanzenforschung (IPK), Gatersleben.

Hammer, K., H. Gäde and H. Knüpffer (1994) 50 Jahre Genbank Gatersleben – eine Übersicht. *Vorträge für Pflanzenzüchtung* 27:333–383.

Hanelt, P. and R.K. Beridze (1991) The flora of cultivated plants of the Georgian SSR and its genetic resources. *Flora et Vegetatio Mundi* 9:113–120.

Hanelt, P., K. Hammer and H. Knüpffer (1992) The genus *Allium* – taxonomic problems and genetic resources. In: Hanelt, P., K. Hammer and H. Knüpffer (eds) *Proceedings of an International Symposium*. 1991. Gatersleben, Germany. Institut für Pflanzengenetik und Kulturpflanzenforschung (IPK), Gatersleben.

Kühn, F., K. Hammer and P. Hanelt (1976) Botanische Ergebnisse einer Reise in die CSSR 1974 zur Sammlung autochthoner Landsorten von Kulturpflanzen. *Kulturpflanze* 24:283–347.

Lehmann, C.O. and R. Mansfeld (1957) Zur Technik der Sortimentserhaltung. *Kulturpflanze* 5:108–138.

Maly, R., K. Hammer and C.O. Lehmann (1987) Sammlung pflanzlicher genetischer Ressourcen in Süditalien – ein Reisebericht aus dem Jahre 1950 mit Bemerkungen zur Erhaltung der Landsorten *in situ* und in der Genbank. *Kulturpflanze* 35:109–134.

Menabde, V.L. (1958) Kul'turnaja flora Gruzii [The cultivated flora of Georgia]. *Botaniceskie Ekskursii po Gruzii* 1:81–89.

Rodríguez Nodals, A. (1984) Mejoramiento genético de los cultivos de raíces y tubérculos tropicales en la Republica de Cuba. PhD thesis. Academy of Sciences of Hungary, Budapest.

Roig, J.T. (1975) *Diccionario Botánico de Nombres Vulgares Cubanos.* Ed. Pueblo y Educación, La Habana.

Schultze-Motel, J. (1989) Archäologische Kulturpflanzenreste aus der Georgischen SSR (Teil 2). *Kulturpflanze* 37:415–426.

Strampelli, N. (1932) *Istituto Nazionale di Genetica per la Cerealicoltura in Roma – Origini, Sviluppo Lavori e Risultati.* Istituto Nazionale di Genetica per la Cerealicoltura, Rome.

Vavilov, N.I. (1987) *Proischoždenie i Geografija Kul'turnych Rastenij [Origin and Geography of Cultivated Plants].* Nauka, Leningrad.

Vvedensky, A.I. (1971) Luk [*Allium*]. In: Vvedensky, A.I. and S.S. Kovalevskaya (eds) *Opredelitel' Rastenij Srednej Azii. Kritičeskij Konspekt Flory [Key to Middle Asiatic Plants. Critical Conspectus of the Flora].* Vol. 2. pp. 39–89. Nauka, Leningrad.

Zohary, M. and M. Hopf (1993) *Domestication of Plants in the Old World.* Clarendon Press, Oxford.

Index

Note: page numbers in *italics* refer to figures and tables.